DIFFERENTIAL EQUATIONS
AN APPLIED APPROACH

DIFFERENTIAL EQUATIONS

J. M. CUSHING

Department of Mathematics
Interdisciplinary Program in Applied Mathematics
University of Arizona

PEARSON
Prentice
Hall

Upper Saddle River, New Jersey 07458

Library of Congress Cataloging-in-Publication Data

Cushing, J. M.

[Mathematical statistics]

    Differential equations: an applied approach / J. M. Cushing.—1st ed.

      p. cm.

    Includes index.

    ISBN 0-13-044930-X

    1. Differential equations. I. Title

    QA371.C83 2004

    515'.35–dc22   2003063207

Executive Acquisitions Editor: *George Lobell*
Editor-in-Chief: *Sally Yagan*
Production Editor: *Jeanne Audino*
Assistant Managing Editor: *Bayani Mendoza de Leon*
Senior Managing Editor: *Linda Mihatov Behrens*
Executive Managing Editor: *Kathleen Schiaparelli*
Vice President/Director of Production and Manufacturing: *David W. Riccardi*
Assistant Manufacturing Manager/Buyer: *Michael Bell*
Manufacturing Manager: *Trudy Pisciotti*
Marketing Manager: *Halee Dinsey*
Marketing Assistant: *Rachel Beckman*
Director of Creative Services: *Paul Belfanti*
Art Editor: *Tom Benfatti*
Creative Director: *Carole Anson*
Art Director: *Kenny Beck*
Interior Designer: *Kenny Beck/Dina Curro*
Cover Designer: *Geoffrey Cassar*
Editorial Assistant: *Jennifer Brady*
Cover Image Specialist: *Lisa Amato*
Cover Photo: *Glass Hall at Night of the Tokyo International Forum Akio Kawasumi*
Art Studio: *Laserwords Private Limited*
Composition: *Dennis Kletzing*

©2004 Pearson Education, Inc.
Pearson Prentice Hall
Pearson Education, Inc.
Upper Saddle River, NJ 07458

Printed in the United States of America

10   9   8   7   6   5   4   3   2   1

ISBN 0-13-044930-X

Pearson Education Ltd., London
Pearson Education Australia Pty, Limited, Sydney
Pearson Education Singapore, Pte. Ltd.
Pearson Education North Asia Ltd., Hong Kong
Pearson Education Canada, Ltd., Toronto
Pearson Educacion de Mexico, S.A. de C.V.
Pearson Education Japan, Tokyo
Pearson Education Malaysia, Pte. Ltd.

This book is dedicated to the memory of my father, Harry,
to my mother, Lorraine,
and to their granddaughters, Alina and Lara.

Contents

Preface

This book is an outgrowth of lecture notes I wrote for an introductory course on differential equations and modeling that I have taught at the University of Arizona for over twenty years. The book offers a blend of topics traditionally found in a first course on differential equations with a coherent selection of applied and contemporary topics that are of interest to a growing and diversifying audience in science and engineering. These topics are supplemented with a brief introduction to mathematical modeling and many applications and in-depth case studies (often involving real data). There is enough material and flexibility in the book that an instructor can design a course with any of several different emphases. For example, by appropriate choices of topics one can devise a reasonably traditional course that focuses on algebraic and calculus methods, solution formula techniques, etc.; or a course that has a dynamical systems emphasis centered on asymptotic dynamics, stability analysis, bifurcation theory, etc.; or a course with a significant component of mathematical modeling, applications, and case studies. In my own teaching I strive to strike a balance among these various themes. To do this is sometimes a difficult task, and the balance at which I arrive usually varies from semester to semester, primarily in response to the backgrounds, interests, and needs of my students. Students in my classes have, over the years, come from virtually every college on our campus: sciences, engineering, agriculture, education, business, and even the fine arts. My classes typically include students majoring in mathematics, mathematics education, engineering, physics, chemistry, and various fields of biology.

It is not intended, of course, that all material in the book be covered in a single course. At some schools, some topics in the book might be covered in prerequisite courses and some topics might be taught in other courses. For example, it is now often the case that slope fields, the Euler numerical algorithm, and the separation of variables method for single first-order equations are taught in a first course on calculus. This is the case at the University of Arizona, and therefore I treat these topics as review material. Laplace transforms are not taught in introductory differential courses at the University of Arizona (where they are taught in mathematical methods courses for scientists and engineers), but I do include Laplace transforms in the text because they are an important topic at many schools. All topics in the text are presented in a self contained way and the book can be successfully used with only a traditional first course in (single variable) calculus as a prerequisite.

At some schools linear algebra is prerequisite for a first course in differential equations (or students have had at least some exposure to matrix algebra), while at other schools this is not the case. At yet other schools first courses in linear algebra and differential equations are co-requisite and might even be taught in the same course. A first course in linear algebra is a prerequisite for my course at the University of Arizona. The book, however, develops all topics first without a linear or matrix algebra prerequisite and include follow-up sections that introduce the use of these subjects at appropriate times. I personally have found that my students who have previously studied linear and matrix algebra benefit nonetheless from an introductory presentation without use of these topics, succeeded by a follow-up that brings to play relevant matrix and linear algebraic topics. For a course without a matrix or linear algebra prerequisite, the follow-up sections can simply be omitted, or be used as a brief introduction of these topics (matrix notation and algebra, eigenvalues and eigenvectors, etc.).

A significant component of the course I teach at the University of Arizona consists of applications and case studies. Again, the issue of balance in topics is crucial. A semester or quarter course can reasonably cover only so much material. One option is to include no applications at all, or limit the course application-type examples whose purpose is to illustrate a mathematical point, but to do so in the context of a problem arising in another scientific discipline. Another option is to spend some time on modeling methodology and selected extended applications and case studies. One approach I have used to facilitate the latter goal utilizes semester projects. I typically have each student study a sequence of case studies over the course of the semester, chosen so as to synchronize with the mathematical techniques and procedures developed in the course. For pedagogical reasons I usually have an individual student focus on applications based on a single theme from the same scientific discipline (e.g., population dynamics in biology, objects in motion in physics, etc.). In this way students experience the full modeling procedure, as models are modified and extended in order to meet the challenges of a different set of circumstances and new questions. Other formats are of course possible. For example, one could devise instead a set of case studies that exposes a student to in-depth applications from several different disciplines in order to attain more interdisciplinary breadth.

Another way in which I utilize case studies is to set aside a small number of lectures throughout the course that are devoted to a detailed look at a few selected applications. However I treat applications in a particular course, I do so with three basic goals in mind. One, of course, is to illustrate the use of the mathematical techniques learned in the course. Another goal is to learn some interesting facts about a scientific topic that one obtains by the use of mathematics. For me, just as important as these two goals is the goal of illustrating the way mathematical modeling is done. For this reason when teaching an application I continually make reference to the "modeling cycle" discussed in the Introduction. Although somewhat simplistic, I have found this way of organizing one's thinking about a mathematical application is a useful compass for students when, in the thicket of the details in an extended application, questions arise such as "how do I start?", "what do I do next?", or "why am I doing this?".

For instructors who wish to incorporate some extended applications into their course, a set of case studies appears at the end of each chapter. I chose these studies to illustrate the kinds of equations, the solution and approximation techniques, and analytic methods covered in the chapter. Model derivations are discussed, assumptions are laid out, interpretations and punch lines are drawn from the analysis, and in some cases the results are compared to real data. Many of the applications are presented in a way that applied mathematicians typically work today: exploratory studies are carried out using a computer, conjectures are formulated, and then an attempt is made to corroborate the conjectures using some kind of mathematical analysis and proof.

I like to begin my course with a discussion of what it means to "solve" a mathematical problem and, in particular, a differential equation. Students usually view this question as the search for a solution formula. While solution formulas can be useful, I point out and stress that methods for calculating solution formulas are available for only specialized types of equations. An alternative to a solution formula is a solution approximation. There are many ways to approximate solutions: geometric approaches that approximate graphs of solutions, numerical methods that estimate solution values, formulas for solution approximations (e.g., Taylor polynomial approximations), and methods that approximate the equation by a simpler equation (e.g., the linearization procedure). All of these approximation procedures appear in the book (and are used in the applications). In an introductory course, it is of course also important to learn about some special types of equations for which methods are available for the calculation of solution formulas. There are several reasons for this. Equations of these specialized types do sometimes arise in applications and their solution formulas can be useful. Some special types of equations (e.g., linear equations) often serve as approximations to more complicated equations; however, an approximating equation is useful only if it is more tractable in some way than the original equation (e.g., a solution formula is available). Also, solvable types of equations serve as "targets" for transformations, that is, a change of variable might transform an equation to a special type for which a solution formula can be found. Finally, special kinds of equations serve useful pedagogical purposes as aids in learning about and understanding differential equations. For this reason the text covers several of the most important types of specialized differential equations and procedures for the calculation of their solution formulas.

Organization

The book begins with the study of a single first-order equations (in Chapters 1, 2, and 3) and then covers (in Chapter 4 through 9) systems of first order equations (which includes higher order equations). The treatment of systems deliberately parallels that of first-order equations. It is useful for the instructor to keep this in mind while covering the material in Chapters 1, 2 and 3 on first order equations and to take advantage of having introduced various concepts, definitions, theorems and methods in the first order case when it comes time to cover systems and higher order equations in later chapters. Examples include the fundamental existence and uniqueness theorem, slope and vector field analysis, numerical approximations methods, the structure of the general solution of linear equations, phase plane analysis, equilibria categorization and stability analysis, bifurcation theory, and series methods of approximation.

Chapter 1 begins with a fundamental existence and uniqueness theorem for initial value problems associated with a single first order equation. This general theorem is followed by graphical and numerical approximation procedures for solutions. These approximation methods appear early in the book because they are general (i.e., are not limited to specialized types of equations) and are therefore available for use throughout the book.

Chapters 2 and 3 study solution formula methods, for selected special types of first order equations, and some basic qualitative and approximation methods of analyzing solutions appropriate when formulas are unavailable (or unnecessary). Chapter 2 treats linear equations and the integrating factor method, which results in the fundamentally important Variation of Constants formula in the one dimensional (scalar) case. After a discussion of some shortcut solution methods (e.g., Undetermined Coefficients), Chapter 2 closes with a brief look at autonomous linear equations. Although rather trivial from the point of view of solution formulas, these equations introduce (in the simplest setting) some fundamental concepts of dynamical systems (equilibria, stability, phase line portraits, etc.) that anticipate some of the main themes of Chapter 3.

Nonlinear equations are the subject of Chapter 3. The chapter begins (Section 1) with a fairly thorough introduction to the qualitative analysis of phase line portraits for autonomous equations, including an introduction to bifurcation theory. This analysis illustrates how one can obtain great deal of information about solutions of a differential equation without use of solution formulas. Autonomous equations are, however, a special type of separable equation, a category of equations for which we can in principle obtain solution formulas (provided the necessary integrals can be calculated). Separable equations are covered in Section 2. Section 3 deals with a selection of other special types of equations and the methods for finding solution formulas. The main theme of this section is how an appropriate change of variable can transform an equation to a type for which a solution method is known. Section 4 presents some analytic approximation methods for first order equations, including Taylor polynomial and Picard iteration methods. Two forms of Taylor polynomial approximations are given. It is common in an introductory course to cover solution approximation formulas using Taylor polynomials in the independent variable (and the limiting case of power series formulas for solutions). Section 4 also includes a method of approximation based on Taylor polynomials in a parameter that appears in the equation. This is not so commonly done in introductory courses, but I find quite natural and straightforward in my course to introduce and use this example of a "perturbation" method. (Perturbation methods are of great importance in both classical and modern applied mathematics.)

Chapters 4 through 8 cover systems of first order equations (and higher order equations). The development parallels that of single first order equations in Chapters 1 through 3. The fundamental existence and uniqueness theorem for initial values problems, and the graphical and numerical approximations studied in Chapter 1, are extended to systems of equations in Chapter 4. Chapters 5 and 6 cover linear systems. Homogeneous linear systems are studied in Chapter 5, with an emphasis placed on the solution of autonomous systems and their phase plane portraits. Solution procedures for single linear equations studied in Chapter 2 are extended to nonhomogeneous systems of linear equations in Chapter 6 (specifically, the Variation of Constants formula and the method of Undetermined Coefficients). The approximation techniques introduced for single first order equations in Chapter 3 (Taylor polynomial, Picard, and perturbation approximations) are extended to systems in Chapter 7. Higher order equations are given special treatment in Chapters 5, 6 and 7 in their own sections. Nonlinear systems (and higher order equations) are the subject of Chapter 8, which focuses primarily on planar autonomous systems and the problem of constructing phase plane portraits. The basic notions of equilibria, stability, linearization, and bifurcation first introduced in Chapter 3 are extended to systems in Chapter 7 (which also includes an introduction to Poincar-Bendixson theory). The focus throughout is on two dimensional systems, although closing sections briefly discuss extensions to higher dimensional systems. The book concludes with an study of Laplace transforms for linear equations (first order, higher order, and systems).

A typical syllabus I use in my course is:

Chapter 1, Chapter 2, Chapter 3.1, 3.4, Chapter 4, Chapter 5.1–5.5, Chapter 6, Chapter 8.

Supplements

The book introduces numerical and graphical approximation methods early so that they can be used throughout. If an instructor desires, it is possible to use the book in a way that de-emphasizes the use of computers and hand calculators. However, to utilize the book fully, readers should have access to software that draws slope (direction) fields and numerically approximates solutions of initial value problems by using the Euler algorithm, Heun (modified or improved Euler) algorithm, and a higher order algorithm (e.g., the fourth order Runge-Kutte algorithm). The user should have control over the step size in the algorithms and should be able to view both numerical and graphical output. I have used a variety of software over the years, both freeware (e.g., Winplot) and commercial software (e.g., Phaser, MATLAB and Maple). Some of my students have succeeded quite well using only programmable graphics hand calculators! (Calculators are also useful for graphing and solving algebraic equations that occasionally arise, particularly in the extended applications.)

Instructors who wish to use either a MATLAB, Maple, or Mathematica manual that contains basic syntax and various projects/problem sets can order any of the following from Prentice Hall (which are free when shrinkwrapped with this text):

Ordinary Differential Equations Using MATLAB 3/e by John Polking(ISBN 0-13-145679-2 for text and manual)

Maple Projects for Differential Equations by Robert Gilbert and George Hsaia (ISBN 0-13-047974-8 for text and manual)

Using Mathematica in Differential Equations by Selwyn Hollis (ISBN 0-13-046329-9 for text and manual)

Mathematica for Differential Equations: *Projects, Insights, Syntax, and Animations* by David Calvis (ISBN 0-13-143976-6 for text and manual)

Online programs constitute another option which is becoming increasingly available. For example, the reader will find Java programs online at prenhall.com/cushing. John Polking's DField and PPlane that can be used with the book. That webpage also contains other material relevant (such as errata, updates, new applications, and so on).

Answers are provided for approximately half of the more than 1700 exercises in the book. A solutions manual for these selected exercises is available from Prentice Hall. Also available is an instructor's solution manual for all exercises in the book.

Acknowledgments

I owe thanks to a large number of people who helped, in one way or another, in the preparation of this text. Several people have class tested various parts of the book: Shandelle Henson (at the University of Arizona, The College of William and Mary, and Andrews University), Jeffrey Edmunds and Suzanne Sumner (at Mary Washington College), Linda Allen (at Texas Tech University), and Guadalupe Lozano, Nakul Chitnis, and Sheree Levarge (at the University of Arizona).

I have also received valuable suggestions and critiques from many other people, including

Michael Colvin, California Polytechnic State University
Saber Elaydi, Trinity University
Richard Elderkin, Pomona College
Heidi Fuchs, Woods Hole Oceanographic Institute
Sophia Jang, University of Louisiana
Daniel Kemp, South Dakota State University
Aaron King, University of Tennessee
Nathan Kutz, University of Washington
Steven Levandosky, Stanford University
David Lomen, University of Arizona
David Lovelock, University of Arizona
Don Meyers, University of Arizona
Jamison Moeser, University of Colorado at Boulder
Michael Neubert, Woods Hole Oceanographic Institute
E. Arthur Robinson, George Washington University
Henri Schurz, Southern Illinois University
Patrick Sullivan, Valpraiso University
Hans Volkmer, University of Wisconsin at Milwaukee

Several students at the University of Arizona helped with proof reading and editing: Ivan Barrientos, Tyler Byers, Nakul Chitnis, Hanees Haniffa, Sheree LeVarge, Doug Owen and Dave Schumann. I, however, accept responsibility for any errors.

I extend my deepest appreciation and thanks to the production staff at Prentice Hall and in particular to Jeanne Audino (production editor), Jennifer Brady (editorial assistant), Patricia M. Daly (copy editor), and Dennis Kletzing (compositor) who, despite having to deal with an eccentric professor, managed to put this book together. Finally, special thanks go to my editor, George Lobell, whose vision for the book made it all possible.

J. M. Cushing
cushing@math.arizona.edu

DIFFERENTIAL EQUATIONS
AN APPLIED APPROACH

Introduction

Preliminaries

Mathematical applications typically involve one or more equations to be solved for unknown quantities. Often applications involve rates of change and therefore lead to equations containing derivatives. Such equations are called *differential* equations.

A student's first encounter with differential equations is usually in a calculus course where antiderivatives (or indefinite integrals) are studied. For example, consider the problem of finding the antiderivative of t^2. This problem can be formulated as follows: Find a function $x = x(t)$ whose derivative is t^2, or, in other words, find a function $x = x(t)$ that satisfies the equation

$$x' = t^2. \tag{1.1}$$

(Here we have used the notation x' for the derivative of x with respect to t. We will also occasionally use the notation dx/dt.) Equation (1.1) is a differential equation for the unknown function $x = x(t)$. Notice what it means to "solve" this equation: Find a function $x = x(t)$ that, when substituted into both sides of the equation, makes the left-hand side *identically* equal to the right-hand side. That is to say, a solution is a function that, upon substitution into the equation, reduces the equation to a mathematical identity in t. Also notice it is not accurate to speak of *the* solution of this differential equation. This is because it has many solutions, namely $x(t) = t^3/3 + c$, where c is any constant (the so-called constant of integration).

It is not always as easy to find formulas for solutions of a differential equation as it is for the equation (1.1). For example, consider the differential equation

$$x' = x. \tag{1.2}$$

This equation is fundamentally different from (1.1) because the unknown function x appears on the right-hand side. This equation cannot be solved by an anti-differentiation of the right-hand side, because the right-hand side is not a known function of t. Later we will learn how to solve this equation, but for now notice that $x(t) = e^t$ is a solution (i.e., a substitution of e^t for x into the left- and the right-hand sides of the equation yields the same result, namely e^t). Similarly, $x(t) = ce^t$ is a solution of this equation for *any* constant c (including $c = 0$). Notice, however, that $x(t) = e^t + c$ is *not* a solution (unless $c = 0$). To see this, we calculate $x'(t) = e^t$ and note that it is not equal to $x(t) = e^t + c$ (unless $c = 0$). This shows that constants of integration do not always appear additively in formulas for solutions of differential equations.

As another example, consider the differential equation

$$x' = x^2. \tag{1.3}$$

The function $x(t) = 1/(1-t)$ is a solution of this equation so long as $t \neq 1$, because the derivative $x' = 1/(1-t)^2$ is identical to x^2 for $t \neq 1$. We say this function is a solution on the interval $-\infty < t < 1$ or on the interval $1 < t < +\infty$ (or on any interval not containing $t = 1$). Similarly, for a constant c, the function $x(t) = 1/(c-t)$ is a solution on any interval that does not contain $t = c$. Notice each solution obtained by assigning a numerical value to c has a different singular point $t = c$ and hence is associated with a different interval of existence. [Incidentally, the constant function $x \equiv 0$ is also a solution that is not included in the formula $x(t) = 1/(c-t)$.]

A solution of a differential equation is associated with an *interval of existence*. The solutions $x(t) = 1/(c-t)$ of equation (1.3) show that there is not necessarily a common interval of existence for all solutions of a differential equation. This example also illustrates that the differential equation itself might give little or no clue about the intervals of existence of its solutions.

For differential equations (1.1), (1.2), and (1.3) it is possible, as we have seen, to write down formulas for solutions. For other equations, it is not possible to calculate solution formulas. In the latter case, we must use other methods to study equations and their solutions. In this book we will study some types of equations for which we can derive solution formulas, but we will also study many methods of analysis that do not require solution formulas. These methods are of particular importance since it is not possible to calculate solution formulas for the differential equations that arise in many, if not most, scientific and engineering applications.

The equations (1.1), (1.2), and (1.3) are examples of a general class of ordinary differential equations of the form

$$x' = f(t, x).$$

Here all terms in the equation not involving the derivative have been placed on the right-hand side. In general, both the independent variable t and the dependent variable x can appear on the right-hand side. Letters or symbols representing unspecified numerical constants called "coefficients" or "parameters" might also appear. Here are some further examples:

$$x' = x^2 + t^2$$
$$x' = -2x$$
$$x' = px, \quad \text{where } p \text{ is a constant}$$
$$x' = r\left(1 - \frac{x}{K}\right)x, \quad \text{where } r > 0, K > 0 \text{ are constants.}$$

It is important to recognize those letters and symbols that represent independent variables, those that represent dependent variables, and those that represent coefficients or parameters. The independent variable is, of course, the variable with respect to which the derivative is being taken. In the above equations we use the letter t for the independent variable; this will be done throughout the book. This choice is motivated by the many applications in which the independent variable represents time. (Other letters can, of course, be used.) On the other hand, throughout the book we use a variety of letters for the *dependent* variable (sometimes referred to as the "state variable"). In applications, a letter suggestive of the meaning of the variable in that application is usually chosen. For example, we will encounter differential equations involving symbols such as x', y', N', and P' for the derivatives of the dependent variables x, y, N, and P with respect to t. If it is necessary to emphasize the role of the independent variable t, we sometimes write derivatives as

$$\frac{dx}{dt}, \frac{dy}{dt}, \frac{dN}{dt}, \frac{dP}{dt}.$$

Applications often involve several differential equations for several unknown functions (i.e., a *system of differential equations*). Some examples are

$$\begin{cases} x' = y \\ y' = -\sin x \end{cases}$$

$$\begin{cases} x' = -r_1 x - r_2 y \\ y' = r_1 x - (r_1 + r_2) y \end{cases}$$

$$\begin{cases} x' = y \\ y' = -\dfrac{k}{m} x - \dfrac{c}{m} y \end{cases}$$

$$\begin{cases} x' = r\left(1 - \dfrac{x}{K}\right) x - cxy \\ y' = -dy + xy \end{cases}$$

$$\begin{cases} x' = y \\ y' = -x - \alpha\left(x^2 - 1\right) y. \end{cases}$$

In each of these examples there are two differential equations for two unknown functions x and y. All other letters represent coefficients (or parameters).

A solution of a system of two equations is a *pair* of functions $x = x(t)$, $y = y(t)$. For example, the pair $x(t) = 2e^{2t}$, $y(t) = -e^{2t}$ is a solution of the system

$$x' = 5x + 6y$$
$$y' = x + 4y.$$

To see this, we note that $x' = 4e^{2t}$ is identical to

$$5x + 6y = 5\left(2e^{2t}\right) + 6\left(-e^{2t}\right) = 4e^{2t}$$

(i.e., the *first* equation is satisfied for all t) and *also* that $y' = -2e^{2t}$ is identical to

$$x + 4y = \left(2e^{2t}\right) + 4\left(-e^{2t}\right) = -2e^{2t}$$

(i.e., the *second* equation is also satisfied for all t). The reader can check that $x(t) = 3e^{7t}$, $y(t) = e^{7t}$ is another solution pair of this same system.

Applications also arise in which higher-order derivatives appear in the equation. Here are some examples of higher order differential equations:

$$x'' + x = 0$$
$$mx'' + cx' + kx = a \sin \beta t$$
$$x''' + 3x'' + 3x' + 2x = 0$$
$$\begin{cases} m_1 x'' + (k_1 + k_2)x - k_2 y = 0 \\ m_2 y'' - k_2 x + k_2 y = 0. \end{cases}$$

The *order* of a differential equation is that of the highest-order derivative appearing in the equation. Thus, the equation $x' = x$ is a *first-order* equation. The first two equations above are *second order* and the third equation is *third order*. The last pair of equations constitute a *second-order system* of equations.

Solutions of higher-order equations must reduce the equation(s) to identities upon substitution. For example, $x(t) = \sin t$ is a solution of the second-order equation $x'' + x = 0$ for all t [as is $x(t) = \cos t$]. The exponential function $x(t) = e^{-2t}$ is a solution of the third-order equation $x''' + 3x'' + 3x' + 2x = 0$ for all t.

Any higher-order equation (or system of higher-order equations) can be associated with an equivalent system of first-order equations. The following example illustrates the most common way to convert a higher-order equation to an equivalent first-order system. The function $x(t) = \sin t$ is a solution (for all t) of the second-order equation

$$x'' + x = 0. \tag{1.4}$$

Define y to be the derivative of x (i.e., $y = x'$). Then $y(t) = \cos t$ and the pair $x(t) = \sin t$, $y(t) = \cos t$ solves the first-order system

$$x' = y \tag{1.5}$$
$$y' = -x.$$

This shows how a particular solution of the second-order equation (1.4) can be used to construct a solution of the first-order system (1.5).

More generally, suppose $x = x(t)$ is *any* solution of the second-order equation (1.4) [i.e., $x''(t) + x(t) = 0$]. Define $y = x'(t)$. The calculations

$$x'(t) = y(t)$$
$$y'(t) = x''(t) = -x(t)$$

show that the pair $x(t)$, $x'(t)$ solves the system (1.5). This shows that *any* solution of the second-order equation (1.4) gives rise to a solution pair for the first-order system (1.5). Is the converse true? Can a solution of the first-order system (1.5) be used to obtain a solution of the second-order equation (1.4)? If so, then we could say that the second-order equation (1.4) is "equivalent" to the first-order system (1.5) in the sense that solving one is the same as solving the other.

Suppose $x = x(t)$, $y = y(t)$ is a solution pair of the first-order system (1.5). Then

$$x'(t) = y(t)$$
$$y'(t) = -x(t). \tag{1.6}$$

We need to show how we can obtain a solution of the second-order equation (1.4) from the solution pair of the system. The way to do this is simply to choose the first component x of the solution pair. We can show that the first component $x = x(t)$ satisfies the second-order equation by differentiating both sides of the first equation in the system (1.6), to obtain $x''(t) = y'(t)$, and then use the second equation in the system to obtain $x''(t) = -x(t)$, or, in other words, $x'' + x = 0$.

The procedure we used to derive the system (1.5) equivalent to the equation (1.4) is not peculiar to that second-order equation. For example, by the same method, we can show that the second-order equation

$$x'' + \sin x = 0$$

is equivalent to the first-order system

$$x' = y$$
$$y' = -\sin x.$$

In general, we can show (by a similar procedure) that *any* second-order differential equation of the general form

$$x'' = f(t, x, x')$$

is equivalent to the first-order system

$$x' = y$$
$$y' = f(t, x, y).$$

An extension of the method also applies to equations of order higher than two. For example, we can obtain an equivalent first-order system for the third-order equation

$$x''' + 3x'' + 3x' + 2x = 0$$

by defining *two* new dependent variables

$$y = x', \quad z = x''.$$

As previously, we can show solutions of this equation give rise to solutions of the system

$$x' = y$$
$$y' = z$$
$$z' = -2x - 3y - 3z$$

and vice versa.

A further extension of the method can be used for higher-order systems as well. For example, consider the second-order system

$$x'' + 2x - z = 0$$
$$2z'' - x' + z = 0$$

for two unknowns x and z. We apply the procedure twice, once on each equation, by defining two new dependent variables

$$y = x', \quad w = z'$$

and obtaining the equivalent first-order system of four equations

$$x' = y$$
$$y' = -2x + z$$
$$z' = w$$
$$w' = \tfrac{1}{2}y - \tfrac{1}{2}z.$$

The ability to convert higher-order equations to a first-order system is required by many (if not most) computer programs available for the study of differential equations.

One way to classify differential equations is by their order. Another way to classify equations is based on the notion of linearity. *A differential equation is linear if the dependent variable and all of its derivatives appear linearly.* Thus, in a linear first-order equation, both x and x' appear linearly. This means that

$$x' = 3x + 1$$
$$2x' - x = 2 + \sin t$$
$$x' = tx + a$$
$$e^t x' = \frac{x}{t} + \ln t$$

are all linear (first-order) differential equations. Note that the independent variable plays no role in the definition of linearity. For example, the second equation is linear even though the independent variable t appears in a nonlinear way (in the $\sin t$ term). We can write each of these equations in the form

$$x' = p(t)x + q(t)$$

for appropriate coefficients $p(t)$ and $q(t)$. By definition, an equation is linear if it has this form (or can be rewritten in this form).

The equations

$$x' = x^2 - 1$$
$$xx' = x + t$$
$$\left(x'\right)^2 = tx - 4$$
$$x' = r\left(1 - \frac{x}{K}\right)x$$

are nonlinear. The first and fourth equations are nonlinear because of the term x^2. The second equation is nonlinear because of the term xx' and the third equation is nonlinear because of the term $\left(x'\right)^2$ (not because of the term tx).

A second- or higher-order equation is linear if the dependent variable and all of its derivatives appear linearly in the equation. The second-order equations

$$x'' + x = 0$$
$$x'' + x' + x = \sin t$$
$$x'' + (\sin t)x = 0$$

are linear because x, x' and x'' appear linearly. The equations

$$x'' + \alpha(1 - x)x' + x = 0$$
$$x'' + \sin x = 0$$

are nonlinear (the first because of the term xx' and the second because of the term $\sin x$).

Systems of equations are linear if (and only if) *all* of the equations are linear in *all* of the *dependent* variables and their derivatives. Thus,

$$x' = y$$
$$y' = -x$$

$$x' = -rx + ry$$
$$y' = rx - 2ry$$

are linear systems and

$$x' = \left(1 - x - \tfrac{1}{2}y\right)x$$
$$y' = \left(1 - \tfrac{1}{2}y - x\right)y$$

$$x' = (x_{in} - x)d - \frac{1}{\gamma}\frac{mx}{a+x}y$$
$$y' = \left(\frac{mx}{a+x} - d\right)y$$

are nonlinear systems [because of the terms x^2, xy, and y^2 in the first system and the term $mxy/(a+x)$ in the second].

Exercises

What are the orders of the following equations? Explain your answers.

1. $t^2x' + x^3 = 0$

2. $3x' - 2x^2 = 0$

3. $e^t (x')^2 + x^3 = 0$

4. $3(x'')^3 - 2x^5(x')^2 = 0$

5. $x'x^3x'' - t^7x^{1/2} = 0$

6. $x' + t^2x^2 + x''' = 2$

7. $x' + t^{1/2}x = \ln t$

8. $x'(x'')^2 - 5t^{1/2}x^3 = 2$

9. $tx = e^x + (x')^2$

10. $x'' + a\sin x = 0$

11. $t^2x' + x^3 = t\cos t$

12. $x' = p(t)x + q(t)$

13. $2xx' + x''' + (x'')^3 - x^4 = 0$

14. $(x''')^2 + (x'')^5 + 3(x')^7 - \sin x = 0$

Which of the following are solutions and which are not solutions of the equation $x' + 3x = 0$? Explain your answers.

15. e^{-3t}

16. e^{3t}

17. $-e^{-3t}$

18. $3e^{-t}$

Which of the following are solutions and which are not solutions of the equation $x' - 2tx = 0$? Explain your answers.

19. e^{2t}

20. $2e^{-2t}$

21. $-7e^{t^2}$

22. $1 + e^{t^2}$

Which of the following are solutions and which are not solutions of the equation $2x' + 3x^{5/3} = 0$? Explain your answers.

23. $t^{-3/2}$

24. $-t$

25. $(t - 1)^{-3/2}$

26. $(1 - t)^{-3/2}$

27. $t^{3/2}$

28. $-(t + 3)^{-3/2}$

29. $(t - 2)^{-2/3}$

30. $(t - c)^{-3/2}$ (where c is any constant)

Which of the following are solutions and which are not solutions of the equation $x'' - 5x' + 6x = 0$? Explain your answers.

31. e^{-2t}

32. e^{2t}

33. e^{3t}

34. e^{-3t}

35. $5e^{2t}$

36. $-7e^{3t}$

37. $e^{2t} + e^{3t}$

38. $c_1e^{2t} + c_2e^{3t}$ for constants c_1 and c_2

In Exercises 39–43, determine which of the functions are solutions of the given differential equation and which are not.

39. For the equation $x' + 5x = 0$:

    (a) $x = e^{-5t}$ **(b)** $x = 3e^{-5t}$ **(c)** $x = 5e^{-3t}$

40. For the equation $x' = 2x$:

    (a) $x = e^{3t}$ **(b)** $x = -3e^{2t}$ **(c)** $x = e^{2t}$

41. For the equation $x' + x^2 = 0$:

    (a) $x = \dfrac{1}{t}$ **(b)** $x = \dfrac{2}{t}$ **(c)** $x = \dfrac{1}{t-2}$

42. For the equation $x' = x + e^t$:

    (a) $x = e^t$ **(b)** $x = te^t$ **(c)** $x = e^t + te^t$

43. For the equation $tx'' + x' = 0$.

    (a) $x = \ln t$ **(b)** $x = 1$ **(c)** $x = t$

Which of the following are solutions and which are not solutions of the equation $x''' - 4x'' - 4x' + 16x = 0$? Explain your answers.

44. $x = e^{4t}$ **45.** $x = -2e^{4t}$

46. $x = ce^{4t}$, where c is any constant

47. $x = e^{2t}$ **48.** $x = \frac{1}{2}e^{-2t}$

49. $x = c_1 e^{4t} + c_2 e^{2t} + c_3 e^{-2t}$ for any constants c_1, c_2, c_3

50. $x = e^{4t} e^{2t}$

Which of the following are solutions and which are not solutions of the equation $x'' + x' - 2x = 0$? Explain your answers.

51. $x = e^t$ **52.** $x = e^{-2t}$

53. $x = e^t e^{-2t}$ **54.** $x = e^t + 2e^{-2t}$

55. Do $x = e^{4t}$ and $y = -2e^{4t}$ form a solution pair for the two equations $x' = 2x - y$, $y' = -6x + y$?

56. Do $x = e^{3t} \sin 5t$ and $y = e^{3t} \cos 5t$ form a solution pair for the equations $x' = 3x + 5y$, $y' = -5x + 3y$?

Which of the following are solution pairs of the system below? Which are not solution pairs? Explain your answers.

$$x' = 4x + 3y$$
$$y' = -2x - y$$

57. $x = e^t$, $y = -e^t$ **58.** $x = -e^t$, $y = e^t$

59. $x = e^t$, $y = e^t$ **60.** $x = -e^t$, $y = -e^t$

61. $x = 3e^{2t}$, $y = -2e^{2t}$ **62.** $x = e^{2t}$, $y = -e^{2t}$

63. $x = e^t + 3e^{2t}$, $y = -e^t - 2e^{2t}$

64. $x = -2e^t + 6e^{2t}$, $y = 2e^t - 4e^{2t}$

65. $x = c_1 e^t + 3c_2 e^{2t}$, $y = -c_1 e^t - 2c_2 e^{2t}$ for constants c_1 and c_2

66. For each function that is a solution in Exercises 15–18, identify the interval on which it is a solution.

67. For each function that is a solution in Exercises 23–30, identify the interval on which it is a solution.

Convert the following equations to equivalent first-order systems.

68. $x'' + x' - 3x = 0$ **69.** $x'' - 6x' + 4x = 0$

70. $3x'' - 6xx' + 12x^2 = 1$

71. $5x'' + 10x'x = 5e^t$

72. $2x''' - 6x'' + 4x' + x = -3$

73. $x''' + 2x'' - x' + x = 1$

74. $x'' + 2x' + 4x = \cos t$

75. $2x'' + 3x' + 9x = 0$

76. $t^2 x'' + (x')^2 + \cos x = 0$

77. $xx'' + (x')^2 + x^{1/2} = e^t$

78. $x'' = -2x' - x + z$, $z'' = -z' + 2x - z$

79. $x''' + x'' - 2x' + 7x = t$

80. Convert the second-order system

$$2x'' - x' + 2z' + 4x - 8z = 0$$
$$z'' + 2x' - z' - x + 3z = \sin t$$

to an equivalent first-order system.

81. Convert the second-order system

$$x'' - 5x' - 6z' + x - z = 0$$
$$3z'' - 6x' - z' + 12x + 3z = 21e^{-3t}$$

to an equivalent first-order system.

Which of the following first-order equations are linear? If an equation is nonlinear, explain why.

82. $x' = 2x + 1$ **83.** $3x' + 4x = \frac{1}{2}$

84. $x' = tx^2 - 1$ **85.** $x' = t^2 x - 1$

86. $t^2 x' = x$ **87.** $x' = x \sin t$

88. $x' = t \sin x$ **89.** $x' = e^x$

Which of the following second-order equations are linear? If an equation is nonlinear, explain why. (a is a constant.)

90. $x'' + xx' + x = 0$ **91.** $x'' + tx' + x = 0$

92. $t^2 x'' + tx' + x = 1$ **93.** $x^2 x'' + tx' + x = 1$

94. $x'' + a(1 - x)x = 0$ **95.** $x'' + a(1 - t)x = t$

96. $x'' + e^{-x} t = \sin t$ **97.** $x'' + e^{-t} x = \sin t$

Which of the following systems are linear? If a system is nonlinear, explain why.

98. $\begin{cases} x' = x + y \\ y' = x - y \end{cases}$ **99.** $\begin{cases} x' = (1 - x)x - xy \\ y' = -y + xy \end{cases}$

100. $\begin{cases} x' = x - y \\ y' = xy \end{cases}$

101. $\begin{cases} x' = ax + by \\ y' = cx + dy \\ \text{where } a, b, c \text{ are constants} \end{cases}$

Which of the following equations (or systems of equations) are linear?

102. $x' = a(r - x)$, where a and r are constants

103. $x' = a(r - x)$, where $a = a(x)$ is a decreasing function of x

104. $x'' + f(x)x = 0$, where $f = f(x)$ is a function of x satisfying $\frac{df(x)}{dx} > 0$ (for all x)

105. $mx'' + cx' + kx = a \sin \beta t$, where m, c, k, a and β are positive constants

106. $mx'' + c \sin x = 0$, where m and c are positive constants

107. $\begin{cases} x' + y' = x + y \\ x' - y' = 2x + y \end{cases}$

108. $\begin{cases} x' = \ln(ty) \\ y' = x \end{cases}$

109. $\begin{cases} x' = y \sin t \\ y'x' = x + y \end{cases}$

110. $\begin{cases} x' - 2x = y + \cos t \\ y - e^{2t}x = y' - 1 \end{cases}$

Determine whether the following equations can be rewritten as linear equations or not.

111. $x' = \ln(2^x)$

112. $x' = \begin{cases} \dfrac{x^2 - 1}{x - 1} & \text{if } x \neq 1 \\ 2 & \text{if } x = 1 \end{cases}$

Mathematical Models

Mathematical models are descriptions of phenomena that involve mathematical equations, symbols, and concepts. These descriptions express laws, assumptions and/or hypotheses relevant to questions arising within some scientific discipline. One wishes to obtain answers to these questions by using the model. This involves "solving" the equations appearing in the model, where *solving* might mean the usual process of calculating a formula for the solution, or it might instead mean obtaining an approximation to the solution or even applying some other means of analysis that derives information about the solution. Even after the solution step is completed, further work might be needed to answer the scientific questions. Information about the mathematical solution needs to be interpreted and applied in the original scientific context. Thus, a modeling exercise involves three major steps that we can term the derivation ("setup") step, the solution step, and the interpretation step. Often the interpretation step reveals deficiencies in the model (e.g., implications and predictions of the model solution might not compare well with data). Such shortcomings can provide feedback to the derivation step by means of which a modified (presumably improved) model is constructed. Figure 1 shows a schematic representation of these stages of a modeling effort. While these are idealized procedural steps, they can often help to orient and guide oneself while embedded in the details of an elaborate model.

The model derivation step in the modeling cycle involves translating the statement of a problem from the language and jargon of a particular discipline (e.g., physics, chemistry, biology, engineering, economics, etc.) into mathematical terminology, symbols, and equations. The statement of the problem involves laws, principles, and/or assumptions that are to be used in the application. The first task is to identify the relevant unknown quantity or quantities and assign symbols to them. It may be necessary to choose symbols for other quantities as well (time, length, mass, growth and decay rates, coefficients of friction, etc.). These symbols must then be related to each other according to the statement of the problem, utilizing the stated laws and assumptions. The result will be a mathematical problem, usually one or more mathematical equations, to be solved for the unknown quantity (or quantities).

A law (or assumption) often used in mathematical modeling states that one quantity is proportional to another. This means one quantity is a constant multiple of the other. The mathematical expression of the assumption requires the designation of a

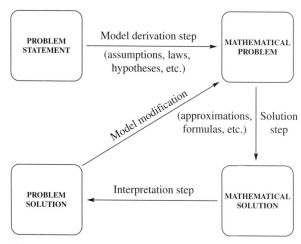

Figure 1 The modeling cycle.

symbol for the constant multiple, called the constant of proportionality. For example, if the force F exerted by a spring is proportional to its elongation s, then we write $F = ks$, where k is a constant of proportionality.

If rates of change are involved in an application, then the mathematical model usually involves a differential equation. For example, suppose the problem is to determine the velocity of an object with mass m subject to Newton's law of motion $F = ma$. This law states that the force F exerted on the object equals its mass m times its acceleration a. Denoting time by t and velocity by $v = v(t)$ and recalling that $a = v'$, we obtain the differential equation $mv' = F$ for the velocity v. For this differential equation to be fully specified we need more information (assumptions, laws, etc.) about the forces acting on the object so that we can write a mathematical expression for F.

In many applications the time rate of change of a quantity x is proportional to the quantity itself, in which case $x' = rx$ for a constant of proportionality r. This case is often described in another way, namely that the per unit rate of change x'/x is constant. In some applications, the rate of change of a quantity x is proportional to other functions of the quantity. For example, if the rate of change of x is proportional to the square of x, then $x' = px^2$ for a constant of proportionality p.

Another modeling assumption is that a quantity is jointly proportional to other quantities. This means the quantity is a constant multiple of the product of the other quantities. Thus, if a is jointly proportional to b and c, then we write $a = kbc$, where k is the constant of proportionality. For example, suppose the reaction rate of a chemical substrate is jointly proportional to its own concentration and to the concentration of an enzyme that catalyzes the reaction. If c and e denote the concentrations of the substrate and the enzyme respectively, then $c' = kce$ for a constant of proportionality k. In this case, the constant k is negative, since the reaction decreases the concentration of the substrate c. To emphasize this, a more convenient notation is to write the constant of proportionality as $k = -m$, where $m > 0$. Thus $c' = -mce$. As a general rule, when defining symbols in an application it is useful to let letters stand for positive quantities.

In many applications involving rates of change a useful model derivation procedure is called *compartmental modeling*. In compartmental models the unknown quantities move into and out of designated compartments at certain rates. A compartment may be a physically well-defined entity, such as a reaction tank in which chemical reactions occur. In other applications a compartment may be more loosely defined, such as the soil bank in a forest or the body tissues in a physiological problem involving the absorption of a medicinal drug. The *balance law (or balance equation)*

$$x' = \text{inflow rate} - \text{outflow rate} \qquad (2.1)$$

applies to the amount of the quantity $x = x(t)$ contained in a compartment at time t. The inflow and outflow rates are those of the quantity into and out of the designated compartment. Coupled with information about these rates that allows us to write mathematical expressions for each of them, the balance law (2.1) becomes a differential equation for x. For example, if the quantity flows into the compartment at a constant rate r and flows out of the compartment at a rate proportional to the amount present, then the inflow rate is r and the outflow rate is px (where p is a constant of proportionality p). Then the balance law (2.1) yields the differential equation $x' = r - px$ for x.

The solution step of the modeling cycle in Fig. 1 focuses on the mathematical problem of solving the equation(s) in the model. A basic mathematical question is whether the equations even have a solution or not. If not, the problem is ill posed, and we must reassess the original statement of the problem and/or the derivation step. Another fundamental question concerns the number of solutions and, if there is more than one, which solution is relevant to the application. A problem is usually called well posed if it has a solution and only one solution. Assuming the mathematical problem in the model is well posed, we would then like to solve the equation(s). One way to do this is to find a formula for the solution. However, it is not possible to find solution formulas for most differential equations. In such cases, we can seek a formula that approximates the solution, or use a computer to calculate numerical approximations to the solution $x(t)$ at selected values of t. From these approximations we can draw approximate graphs of the solution. Another approach is to approximate the differential equation by a simpler equation, where by *simpler* we mean one for which we are able to calculate a solution formula. Yet another approach is to obtain the desired information about the solution directly from the differential equation itself, without the aid of solution formulas or approximations.

Finally, in the interpretation step of the modeling cycle the mathematical results from the solution step are utilized and interpreted so as to provide an answer to the original question. The mathematical solution may not immediately provide the answer and further use and manipulations of the solution, as well as additional information, might be necessary.

The following examples illustrate the modeling cycle.

EXAMPLE 1

Assume the number of bacteria in a culture grows at a constant per capita rate. If an initial population of one thousand bacteria doubles in thirty minutes, in how many minutes will there be one million bacteria present? ∎

To derive the model equations (model derivation step) we let t be time measured in minutes, $t = 0$ be the initial time, and $x = x(t)$ be the number of bacteria at time t. With these symbols the given initial condition becomes $x(0) = 10^3$ and the growth rate assumption yields the differential equation $x' = px$, where p is a constant of proportionality. A formula for the solution of these equations is $x = 10^3 e^{pt}$ (model solution step). To answer the question (model interpretation step) we determine the time t at which $x(t) = 10^6$; that is, we solve the equation $10^3 e^{pt} = 10^6$ for

$$t = \frac{1}{p} \ln 10^3 = \frac{3}{p} \ln 10.$$

To obtain a numerical answer in minutes we need a numerical value for p. This value is found from the stated fact that the population doubles in thirty minutes; that is, in symbols, from $x(30) = 2 \times 10^3$. Solving the resulting equation

$$10^3 e^{30p} = 2 \times 10^3$$

for

$$p = \frac{1}{30} \ln 2$$

we obtain our final answer of

$$t = \frac{90}{\ln 2} \ln 10$$

or approximately 298.97 minutes.

EXAMPLE 2

A chemical pesticide is applied to a stand of trees. This pesticide is absorbed into the tissues of the trees and, because of the natural exchange of material between the trees and the soil, the pesticide is transferred from the soil to the trees and vice versa. Assume these transitions take place at a per unit (pesticide) rate of 2 per year. In addition, the pesticide decomposes in the soil at a per unit rate of 3 per year. No amount of pesticide is initially present in the soil at which time a pesticide dosage of $d > 0$ units is applied to the trees. Determine the maximum amount of pesticide that will occur in the soil and the time at which this maximum occurs. In the long run, what fraction of the pesticide is in the soil? ∎

As part of the model derivation step we consider two compartments: trees and soil. Let $x = x(t)$ denote the amount of pesticide in the trees at time t and let $y = y(t)$ denote the amount in the soil. The "compartmental diagram" in Fig. 2 shows how the pesticide moves between the two compartments. According to the stated assumptions, the outflow rate from the trees to the soil is $2x$ (units of pesticide per year) and the inflow rate is $2y$. The balance equation (2.1) yields the differential equation

$$x' = 2y - 2x. \tag{2.2}$$

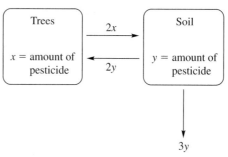

Figure 2

The flow rate into the soil is $2x$. There are two outflow rates from the soil, the absorption rate $2y$ into the trees and the decomposition rate $3y$, yielding a total outflow rate from the tree compartment of $5y$. The balance equation (2.1) yields the differential equation

$$y' = 2x - 5y. \tag{2.3}$$

Together these equations form a system of two first-order linear differential equations for the two unknowns x and y. These unknowns are subject to the given initial conditions

$$x(0) = d, \qquad y(0) = 0. \tag{2.4}$$

As part of the model solution step, we begin with some approximate solutions of these equations, which we use to form some tentative answers to the questions. Figure 3 shows some computer drawn plots of the $y = y(t)$ component of the solution for a selection of initial doses d. In Fig. 4 the fraction of pesticide in the soil, $y/(x+y)$, is plotted for each of these cases.

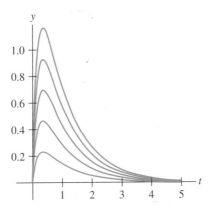

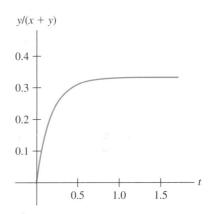

Figure 3 The amount of pesticide in the soil, $y(t)$, is plotted as a function of time t for initial doses $d = 1, 2, 3, 4, 5$. The maximum of y occurs at approximately $t = 1/3$ year and appears to be independent of the initial dose.

Figure 4 The fraction of pesticide in the soil $y/(x + y)$ is plotted as a function of time for initial doses $d = 1, 2, 3, 4, 5$. The result, which is the same for all doses, shows an increase from 0 to approximately $1/3$.

A visual inspection of these graphs suggests the following conclusions. The maximum amount of pesticide occurring in the soil is proportional to the initial dosage (i.e., if d is doubled, tripled, etc., the maximum amount is doubled, tripled, etc.). On the other hand, the time at which the maximum occurs is independent of the initial dosage d and is approximately equal to $1/3$ year. The graph in Fig. 4 suggests the fraction of pesticide in the soil approaches approximately $1/3$ as time goes on; that is,

$$\lim_{t \to +\infty} \frac{y(t)}{x(t) + y(t)} = \frac{1}{3}.$$

These conjectures are formulated from the plots shown in Figs. 3 and 4 (which were obtained from computer-generated approximations to the solution of the differential equations).

It turns out there are formulas for the solution of the system of differential equations (2.2)–(2.3) and the initial conditions (2.4), namely

$$x(t) = d\left(\frac{1}{5}e^{-6t} + \frac{4}{5}e^{-t}\right)$$

$$y(t) = d\left(\frac{2}{5}e^{-t} - \frac{2}{5}e^{-6t}\right).$$

(See Exercise 19, Sec. 5.2 in Chapter 5.) From these formulas we can obtain more accurate and general answers to our questions than we can get from computer experiments.

For example, using calculus methods, we find the maximum of $y(t)$ to occur at $t = (\ln 6)/5$, the root of the derivative

$$y'(t) = -\frac{2}{5}e^{-t} + \frac{12}{5}e^{-6t}.$$

Our conjecture was accurate, since $(\ln 6)/5 = 0.35835 \approx 1/3$. Our conjecture that the maximum is proportional to d was also correct, as we see from the calculation

$$y\left(\frac{1}{5}\ln 6\right) = \frac{1}{3}6^{-1/5}d \approx 0.2329d.$$

Finally, we calculate the long-term fraction of pesticide in the soil by taking the limit

$$\lim_{t\to+\infty} \frac{y(t)}{x(t) + y(t)} = \lim_{t\to+\infty} 2\frac{1 - e^{-5t}}{6 - e^{-5t}} = \frac{1}{3}.$$

This also agrees with our conjecture.

It is important to remember that mathematical models are built from assumptions. Thus, models take the form of "what if" questions: What are the logical conclusions if the assumptions are valid? The assumptions state not only which laws and principles are used in the model, but also (by exclusion) what phenomena and mechanisms are ignored. Some effects are small (relative to those included in the model) and can presumably be safely excluded while still obtaining useful and accurate answers. For example, in some circumstances friction may be negligible compared to other forces acting on an object in motion. However, if the predictions of the model turn out to be unacceptably inaccurate and if this is due to frictional forces, then we must return to the derivation step, include frictional forces in the model, and derive new mathematical equations. Usually this cycle results in more complicated equations to solve, and thus there is a relentless trade-off between the accuracy or realism of models and their mathematical tractability.

Here is an illustration. In a laboratory experiment an object was dropped and left to fall vertically to the ground under the influence of gravity. We would like a model that predicts the distance fallen $x = x(t)$ at each instant of time $t > 0$. In the model derivation step we assume Newton's law of motion $F = ma$. We also assume that near the surface of the earth objects fall with constant acceleration g (this is an approximation to another gravitational law of Newton). This constant is known to be approximately $g = 9.8$ m/s^2 (32 ft/s^2). Under these assumptions we have $F = mg = mv'$, where $v = x'$ is the object's velocity, and the mathematical model becomes, after the cancellation of a factor m,

$$v' = 9.8, \quad v(0) = 0.$$

The initial condition $v(0) = 0$ results from the object being dropped (i.e., it is not given any initial velocity). We can carry out the model solution step in this example by performing a straightforward integration of $v' = 9.8$. The result, when $(0) = 0$ is taken into account, is the solution formula $v(t) = 9.8t$. Since $x' = v$ this formula for v implies $x(t) = 4.9t^2$. [The constant of integration is 0 because at the initial instant the distance fallen equals 0, i.e., $x(0) = 0$.] This formula predicts the distance fallen at each time $t > 0$. For example, it predicts the object falls 4.9 meters in $= 1$ second and $4.9(2)^2 = 19.6$ meters in $t = 2$ seconds.

To test the accuracy of the model, its predictions can be compared observational data obtained from the experiment. In the experiment it turned out that the object fell 0.61 meter in $t = 0.347$ second and 7.0 meters in $t = 1.501$ seconds. The model prediction $x(0.347) = 4.9(0.347)^2 = 0.59$ for $t = 0.347$ is fairly accurate, making an error of approximately 3%. However, at $t = 1.501$ seconds the model predicts $x(1.501) = 4.9(1.501)^2 = 11.04$ meters whereas the object actually fell only 7 meters, an error of over 50%. The model interpretation step reveals deficiency. To obtain more accurate predictions we reconsider the model and its derivation (model modification step).

The falling object in the experiment, it turns out, was a shuttlecock used in the game of badminton. A shuttlecock is a designed to experience considerable air resistance (it is essentially a small ball with cometlike tail of feathers). Therefore, in addition to gravity, the force of friction should be included in the total force F acting on the shuttlecock. To do this requires modeling assumption about how friction acts on the shuttlecock.

Since friction is related to the motion of the shuttlecock, the force due to friction is a function of the velocity v. This function equals 0 when $v = 0$ (there is no friction when the shuttlecock is rest) and increases as v increases (friction increases as velocity increases). A simple relationship of this kind is direct proportionality (i.e., the force due to friction equals $-kv$ for a constant k of proportionality). (We assume $k > 0$ and the minus sign occurs because friction works against the motion of the object.) The constant k is called the coefficient of friction. Under this assumption we have $F = mg - kv$. and Newton's law $F = ma$ lead to the modified model and differential equation $mg - kv = mv'$ for the object's velocity. After dividing by m and letting $g = 9.8$, we obtain the differential equation

$$v' = 9.8 - k_0v.$$

Here k_0 stands for k/m (the per unit mass coefficient of friction).

In order to make numerical predictions, k_0 needs to be assigned a numerical value. Data yields an estimated value of $k_0 = 1.128$ (see Sec. 3.6.2 Chapter 3). This yields the equations

$$v' = 9.8 - 1.128v, \quad v(0) = 0 \tag{2.5}$$

for the velocity v of the falling shuttlecock. After these equations are solved (either by formula or approximation), the distance fallen is found by integrating v (i.e., $x = \int_0^t v\,dt$).

It turns out that this modified model, with friction included, makes more accurate predictions at later times than the original model without the frictional force does. For example, $x(1.501) = 6.75$ meters, an error of only 3.5%.

For more details about this application see Sec. 3.6.2, Chapter 3.

Mathematical books naturally focus on the solution step of the modeling cycle. In this book we will study methods for approximating solutions of differential equations (graphically, numerically, and analytically), methods for obtaining solution formulas, and methods for analyzing properties of solution. Throughout the book we will, however, use model equations arising from applications to illustrate these methods.

Exercises

In the exercises below you are asked to derive differential equations for unknown functions and, in some cases, initial conditions for the equations. Except when explicitly asked to do so, do not solve the equations.

1. Suppose the balance in a savings account grows, at each instant of time, at a rate proportional to the balance present at that time. (This is called continuous compounding.) Suppose an initial deposit of d dollars is made.

   (a) Write a differential equation and initial condition for the balance $x = x(t)$ as a function of time t.

   (b) Suppose withdrawals are made at a constant rate w. Write a differential equation and initial condition for the balance $x = x(t)$ as a function of time t.

2. Samples of radioactive isotopes decay at a rate proportional to the amount present. Suppose an initial amount x_0 is present.

   (a) Write a differential equation and initial condition for the amount of radioactive isotope $x = x(t)$ present at time t.

   (b) Suppose an amount of radioactive material is added at a constant rate a. Write a differential equation and initial condition for the amount of radioactive isotope $x = x(t)$ present at time t.

3. A container holds a volume v of fluid. Suppose a fluid is pumped into the container at a rate $d > 0$ (volume per unit time) and is rapidly mixed. To keep the volume constant at v, the well-mixed fluid is pumped out at a rate d. Suppose the incoming fluid contains a concentration c_{in} of a chemical substrate, which is initially absent.

   (a) Write a differential equation and initial condition for the concentration $c = c(t)$ of the substrate in the container at time t.

   (b) An enzyme is added to the container in such a way that its concentration e is held constant. Suppose this enzyme reacts with the substrate at a rate (per unit volume) jointly proportional to both concentrations. Write a differential equation and initial condition for the concentration $c = c(t)$.

4. (a) Write a differential equation and initial condition for the velocity of a dropped shuttlecock under the assumption that the force of friction is proportional to the square of its velocity.

   (b) Show that

   $$v(t) = \sqrt{\frac{9.8}{k_0} \frac{1 - \exp\left(-2t\sqrt{9.8k_0}\right)}{1 + \exp\left(-2t\sqrt{9.8k_0}\right)}}$$

   solves the equations in (a), where k_0 is the per unit mass coefficient of friction.

   (c) Show $v(t)$ has a limit as $t \to +\infty$. What does this mean about the motion of the falling shuttlecock?

5. Suppose a population has a constant per unit death rate $d > 0$ and a constant per unit birth rate $b > 0$.

   (a) Using the balance law (2.1), write a differential equation for the population concentration $x(t)$.

   (b) Suppose the per unit death rate d is not constant but is instead proportional to population concentration. Write a differential equation for the population concentration $x(t)$.

6. Suppose a population has a constant per unit death rate $d > 0$ and a per unit birth rate that is proportional to population concentration x (with constant of proportionality denoted by $a > 0$). Using the balance law (2.1), write a differential equation for the population concentration $x(t)$.

7. A basic model of the growth of a tumor is based upon the assumption that the per unit volume growth rate of the tumor decreases exponentially with time (i.e., proportionally to an exponential of the form e^{-bt} for a constant $b > 0$). Assume the tumor initially has volume $v_0 > 0$. Write a differential equation and initial condition for the tumor volume $v = v(t)$ as a function of time.

8. Assume a population with constant per unit birth and death rates b and d is subjected to an influx of immigrants at a constant rate I (not a per unit rate, but a *constant* rate!). Write a differential equation for the size of the population.

9. Newton's law of cooling states that the rate at which a body cools is proportional to the difference in temperature between the body and its surrounding environment. Write a differential equation for the temperature of the body.

10. Modify the system of equations for the tree and soil compartmental Example 2 to account for the application of pesticide to the trees continuously at a constant rate p. Assume initially there is no pesticide in either the trees or the soil.

11. Suppose two distinct cultural groups live in the same city. Each grows at a rate proportional to its numbers. However, each group loses population numbers at a rate proportional to the size of the other (since the two groups do not get along). Write a compartmental model (two differential equations) for the rate of change of each group's size. Is the system of differential equations linear or nonlinear?

12. Consider two populations that grow exponentially in the absence of the other (i.e., their per unit birth and death rates are constants). When placed together in a common habitat, the two populations compete for a vital resource. Because of this competition each population's per unit death rate is increased, but the per unit birth rates remain constant. Specifically, assume each population's per capita death rate is proportional to the other population's density. Write down a system of differential equations for the population densities x and y of each population. Is the system linear or nonlinear? (This model is a famous system in theoretical ecology called the Lotka-Volterra competition model.)

13. A salt-water concentration of 2 lb per gallon is added to a 150-gallon tank full of initially pure water at a rate of 5 gallons per minute. Suppose the mixture is well stirred so that the tank always has a uniform concentration of salt throughout. The well-stirred mixture is drained out the bottom of the tank at a rate of 5 gallons per minute (so that the tank remains full at 150 gallons). Write a differential equation for the number of lbs of salt in the tank. Is the equation linear or nonlinear?

14. Suppose a container initially contains an amount $x_0 > 0$ of a chemical substance. Suppose the chemical flows into the container at a constant rate $r > 0$ and flows out at a rate proportional to the elapsed time (with proportionality constant $k > 0$).

    (a) Using the balance law (2.1), write a differential equation for the amount of the chemical $x = x(t)$ as a function of time t. Is the equation linear or nonlinear?

    (b) Obtain a formula for the solution x.

    (c) Show the amount of the chemical in the container initially increases to a maximum level x_m at a time $t_m > 0$ before decreasing to 0 in a finite amount of time t_f. Find formulas for x_m, t_m, and t_f.

15. Write a differential equation and initial condition for the concentration $x = x(t)$ of a population under the following assumptions: The initial population concentration is x_0; the birth rate is proportional to population concentration; the death rate is proportional to the square of population concentration; and the population is harvested at a constant rate $h > 0$. Is the differential equation linear or nonlinear?

16. Suppose a drug is to be injected into the blood of a patient. Consider the circulatory system as one compartment and all of the tissues that serve to eliminate the drug from the circulatory system (such as the kidney) as another compartment. Let x and y be the mass of the drug in these two compartments, respectively. Suppose the per unit mass rates at which the drug flows out of a compartment (and into the other) is proportional to the mass present in the compartment. Also assume the drug is removed from a compartment, through elimination or degradation, at a (per unit mass) rate that is proportional to the amount of mass in the compartment. Finally, assume no drug is present in the patient until an initial injection of amount $x_0 > 0$ is given into the bloodstream (after which no further drug is added to the system). Using the balance law (2.1), write differential equations and initial conditions for $x = x(t)$ and $y = y(t)$.

17. A manufactured item is given an initial set price of p_0. However, as supply and demand for the item changes with time t, so does the price $p = p(t)$. To complicate matters, the supply and demand in turn depend on the price. Write a differential equation for the price p under the following assumptions: The rate of change of the price p is proportional to the difference between the supply s and demand d (price increases if demand exceeds supply); the supply s is proportional to the price (if the price goes up the supply goes up proportionally); and the demand d is *inversely* proportional to the price. Is the equation linear or nonlinear?

18. Consider two armies engaged in a battle. Let x and y represent the strengths of the two armies (measured, for example, as the number of troops and/or armaments). Assume the strength of each army decreases at a rate proportional to the strength of the other army. In addition, each army receives reinforcements at a constant rate. Write differential equations for x and y. Are these equations linear or nonlinear?

19. Pure water is pumped into a tank of volume V (liters) initially filled with salt water. Suppose the pure water is pumped in at a rate of r liters per hour and the well-stirred salt-water mixture is pumped out at the same rate (so that the volume in the tank remains constant at V).

    (a) Write a differential equation for the salt concentration $x = x(t)$ in the tank as a function of time t.

(b) If the water pumped into the tank is not pure but instead has a salt concentration of s grams/liter, modify your equation in (a) accordingly.

20. In biology, allometry is the study of the relative size and growth of different parts of an organism. Suppose $x = x(t)$ and $y = y(t)$ are measures of the sizes of two different parts of a particular organism as functions of time. A simple model of allometry is based on the assumption that the per unit growth rates of the different parts are proportional to each other. Treating y as a function of x, derive a differential equation for y. Is this equation linear or nonlinear?

21. The birth rate of a population is proportional to its size $x = x(t)$ with constant of proportionality $b > 0$. The death rate of the population is proportional the population size of a deadly virus $y = y(t)$ with constant of propor-

tionality $c > 0$. The virus population has a negative per unit growth rate $-r$.

(a) Write a differential equation and initial condition for the population sizes $x = x(t)$ and $y = y(t)$, assuming initial sizes of x_0 and y_0.

(b) Show that

$$x = \left(x_0 - \frac{c}{r+b}y_0\right)e^{bt} + \frac{c}{r+b}y_0e^{-rt},$$

$$y = y_0e^{-rt}$$

solve the differential equations you obtained in (a).

(c) Only positive values of x and y are of relevance. Keeping this in mind describe what happens to x as $t \to +\infty$ and explain the implications with regard to the survival of the population x.

C H A P T E R

1

First-Order Equations

In this chapter we consider first-order differential equations of the form

$$x' = f(t, x).$$

A fundamental question concerns the existence of solutions to such an equation. Under what conditions [i.e., for what kind of expressions $f(t, x)$] can we be assured that solutions exist? Another question concerns the number of solutions. We know from calculus that integration problems have infinitely many solutions and, therefore, we anticipate that this is also true for a first-order differential equation. On the other hand, in applications there are often requirements (in addition to the differential equation) that serve to select exactly one solution. For a first-order differential equation the most common requirement is that the solution $x(t)$ equal a specified value x_0 for a specified value of t, that is, that $x(t_0) = x_0$ for a given t_0 and x_0. A fundamental mathematical question is whether the resulting *initial value problem*

$$x' = f(t, x), \quad x(t_0) = x_0$$

has a solution. In this chapter we learn conditions which, when placed on $f(t, x)$, guarantee that this initial value problem has one and only one solution (i.e., has a "unique" solution).

For specialized equations [i.e., for $f(t, x)$ with special properties] we can calculate formulas for solutions. We study some examples in Chapters 2 and 3. However, for most differential equations it is not possible to find solution formulas. Nonetheless, it is possible to obtain useful approximations to solutions of any first-order equation, especially with the aid of a computer. In this chapter we study some basic methods for approximating solutions, both graphically and quantitatively. In applications these methods are often sufficient to obtain the desired answers. Other approximation methods appear in Chapter 3.

1.1 The Fundamental Existence Theorem

We begin with a definition.

DEFINITION 1. A **solution** of a differential equation $x' = f(t, x)$ on an interval $a < t < b$ is a differentiable function $x = x(t)$ that reduces the equation to an identity on the interval, i.e., $x'(t) = f(t, x(t))$ for all values of t from the interval.[1] The interval $a < t < b$ may be the whole real line, in which case we say the function is a solution for all t.

For the differential equation

$$x' = t^2 \tag{1.1}$$

we have $f(t, x) = t^2$. The function $x(t) = t^3/3 + 1$ is a solution of this equation for all t because $x'(t) = t^2$ equals $f(t, x(t)) = t^2$ for all t.

More generally, the unknown x might appear in $f(t, x)$. For example, for the equation $x' = tx$ we have $f(t, x) = tx$. The function $x(t) = e^{t^2/2}$ is a solution of this equation for all t because $x'(t) = te^{t^2/2}$ and $f(t, x(t)) = tx(t) = te^{t^2/2}$ are identical for all t.

From calculus we know the differential equation (1.1) has infinitely many solutions and the set of all solutions is given by the formula

$$x = \frac{1}{3}t^3 + c, \tag{1.2}$$

where c is an arbitrary constant. This is an example of a general solution of a differential equation.

DEFINITION 2. The collection of all solutions of the differential equation $x' = f(t, x)$ is called the **general solution** (or the **solution set**).

An initial condition $x(t_0) = x_0$ selects a particular solution from the general solution. For example, suppose we require that a solution of the equation (1.1) satisfy the initial condition $x(0) = 1$. From the general solution (1.2) we obtain $x(0) = c$ and therefore this initial condition is satisfied by choosing (and only by choosing) $c = 1$. That is, there is a unique solution of the initial value problem

$$x' = t^2, \quad x(0) = 1$$

namely, $x = t^3/3 + 1$.

In an initial value problem the "initial" time need not be $t_0 = 0$. For example, we can use the general solution (1.2) to find the unique solution

$$x(t) = \frac{1}{3}t^3 - \frac{11}{3}$$

of the initial value problem

$$x' = t^2, \quad x(2) = -1.$$

---

[1] As a mathematical function $f(t, x)$ has a domain of t- and x-values. It is assumed, in this definition, that all values of t taken from the interval $a < t < b$ and the corresponding values of $x(t)$ [i.e., the range of the function $x(t)$] lie in the domain of f. Otherwise $f(t, x(t))$ makes no sense.

In fact, we can solve the general initial value problem

$$x' = t^2, \quad x(t_0) = x_0.$$

using the general solution (1.2) by setting

$$x(t_0) = \frac{1}{3}t_0^3 + c$$

equal to the desired initial value x_0 and solving for

$$c = x_0 - \frac{1}{3}t_0^3.$$

This results in the unique solution

$$x(t) = \frac{1}{3}t^3 + x_0 - \frac{1}{3}t_0^3.$$

EXAMPLE 1

A differential equation for the velocity $v = v(t)$ of a falling object subject to the force of gravity and air resistance is $v' = f(t, v)$, where $f(t, v) = g - k_0 v$. Here g and k_0 are constants (the acceleration due to gravity and the per unit mass coefficient of friction, respectively). The function

$$v(t) = e^{-k_0 t} + \frac{g}{k_0}$$

is a solution for all t. To see this, note that

$$v'(t) = -k_0 e^{-k_0 t}$$

is equal to

$$f(t, v(t)) = g - k_0 \left(e^{-k_0 t} + \frac{g}{k_0} \right)$$

for all t.

For any constant c the function

$$v(t) = c e^{-k_0 t} + \frac{g}{k_0}$$

is also solution for all t since

$$v'(t) = -k_0 c e^{-k_0 t}$$

is equal to

$$f(t, v(t)) = g - k_0 \left(c e^{-k_0 t} + \frac{g}{k_0} \right)$$

for all t. In Chapter 2 it is shown that this formula is in fact the general solution.

The solution satisfying the initial condition $v(0) = 0$ (which describes an object that is initially dropped) is found from the general solution by solving

$$v(0) = c + \frac{g}{k_0} = 0$$

for

$$c = -\frac{g}{k_0}.$$

This yields the solution

$$v(t) = -\frac{g}{k_0}e^{-k_0 t} + \frac{g}{k_0}.$$

In applications solutions are not always defined for all t. Here is an example.

EXAMPLE 2

An equation describing the growth of the world's human population $x(t)$ in billions as a function of time t (in years) is

$$x' = kx^{p+1},$$

where $k > 0$ and $p > 0$ are positive constants estimated from data (see Chapter 3, Sec. 3.6.) The function

$$x(t) = \frac{1}{(1 - pkt)^{\frac{1}{p}}}$$

is defined on the interval $t < 1/pk$. (The denominator vanishes at $t = 1/pk$.) This function is a solution for $t < 1/pk$ since

$$x'(t) = k\frac{1}{(1 - pkt)^{\frac{p+1}{p}}}$$

and

$$f(t, x(t)) = k\left(\frac{1}{(1 - pkt)^{\frac{1}{p}}}\right)^{p+1} = k\frac{1}{(1 - pkt)^{\frac{p+1}{p}}}$$

are identically equal for all $t < 1/pk$.

Similar calculations show that the function

$$x(t) = \frac{x_0}{\left(1 - pkx_0^p t\right)^{\frac{1}{p}}}$$

is a solution on the interval $t < 1/pkx_0^p$ for any constant $x_0 > 0$. This solution satisfies the initial condition $x(0) = x_0$.

A formula for the general solution of an equation $x' = f(t, x)$ cannot always be found. The right-hand side of the equation $f(t, x)$ involves the unknown solution x and therefore is not a known function of t that we can integrate. Nonetheless, the initial value problem

$$x' = f(t, x), \quad x(t_0) = x_0 \tag{1.3}$$

has one and only one solution under appropriate conditions placed on $f(t, x)$ as a function of t and x. The derivative of $f(t, x)$ with respect to x is denoted by $\partial f(t, x)/\partial x$ (and called the partial derivative of f with respect to x).

THEOREM 1

Fundamental Existence and Uniqueness Theorem Suppose $f(t, x)$ and its derivative $\partial f(t, x)/\partial x$ with respect to x are continuous for x near x_0 and t near t_0.[2] Then the initial value problem (1.3) has a solution on an interval containing t_0. Moreover, there is no other solution of the initial value problem on this interval.

---

[2]By "continuous for x near x_0" we mean continuous on an interval $a < x < b$ containing x_0. Similarly, by "continuous for t near t_0" we mean continuous on an interval $c < t < d$ containing t_0.

For example, consider the initial value problem

$$x' = tx, \quad x(0) = \frac{1}{2}.$$

The function

$$f(t, x) = tx$$

and its derivative

$$\frac{\partial f(t, x)}{\partial x} = t$$

are continuous for all x and t (and therefore, certainly for x near $x_0 = 1/2$ and t near $t_0 = 0$). Therefore, by Theorem 1 this initial value problem has a unique solution on an interval containing $t_0 = 0$. [From the formula $x(t) = e^{t^2/2}$ for the solution it is seen that the solution is defined for all t, a fact not obtainable from Theorem 1.]

EXAMPLE 3

An initial value problem describing the growth of a population in a periodically fluctuating environment is

$$x' = rx \left(1 - \frac{x}{K + a \sin t} \right), \quad x(0) = x_0,$$

where x_0 is the initial population size and r, K and $a < K$ are positive constants. Since the denominator never vanishes, the function

$$f(t, x) = r \left(x - \frac{x^2}{K + a \sin t} \right)$$

and its derivative with respect to x

$$\frac{\partial f(t, x)}{\partial x} = r \left(1 - 2 \frac{x}{K + a \sin t} \right)$$

are continuous for all x and t. Therefore, the initial value problem has a unique solution on an interval containing $t_0 = 0$. No algebraic formula is available for the general solution of this equation, nor for the solution of initial value problems. ∎

If one or both of the conditions on $f(t, x)$ in the existence and uniqueness Theorem 1 fail to hold, then we can draw no conclusions from this theorem. In particular, we *cannot* conclude in this case that there is not a solution. For example, for the initial value problem

$$x' = x^{1/3}, \quad x(0) = 0 \tag{1.4}$$

the function

$$f(t, x) = x^{1/3}$$

fails to satisfy the conditions in Theorem 1 because the derivative

$$\frac{\partial f(t, x)}{\partial x} = \frac{1}{3} x^{-2/3}$$

is not continuous at $x = 0$ (it is not even defined there). Yet this initial value problem does have a solution: $x(t) = 0$. For an example of an initial value problem that has no solution, see Exercise 26.

The initial value problem (1.4) also provides an example of nonuniqueness since $x(t) = 0$ and

$$x(t) = \left(\frac{2}{3}t\right)^{3/2}$$

are two different solutions. This does not contradict Theorem 1 because the theorem does not apply to this initial value problem.

The fundamental existence and uniqueness theorem (Theorem 1) provides criteria under which an initial value problem has a solution on an interval containing the initial point $t = t_0$. The *maximal interval* of the solution is the largest interval containing t_0 on which it solves the differential equation. Theorem 1 gives no information about the maximal interval of a solution. In fact, without a solution formula it is usually difficult to determine the maximal interval. The function $f(t, x)$ may satisfy the criteria of Theorem 1 for all values of t and x and yet solutions may not be defined for all t.

EXAMPLE 4

Consider the initial value problem

$$x' = 2tx^2, \quad x(0) = 1.$$

The function

$$f(t, x) = 2tx^2$$

and its derivative

$$\frac{\partial f(t, x)}{\partial x} = 4tx$$

are continuous for all x and t. Theorem 1 implies that there exists a unique solution on an interval containing $t_0 = 0$. The solution formula

$$x(t) = \frac{1}{1 - t^2}$$

shows that the maximal interval is $-1 < t < 1$. See Fig. 1.1. ∎

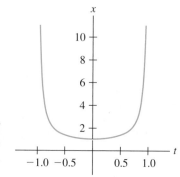

Figure 1.1 The solution of the initial value problem $x' = 2tx^2$, $x(0) = 1$ has vertical asymptotes at $t = \pm 1$.

The importance of the interval of existence of a solution can sometimes be overlooked. Here is an example.

| EXAMPLE 5 | A popular computer program gives the formula $x(t) = \sin t$ for the solution of the initial value problem |

$$x' = \sqrt{1 - x^2}, \quad x(0) = 0$$

without indicating the solution interval. Since $\sin t$ is defined for all t, the implication is that $\sin t$ is a solution for all t. This is false, however, since

$$x'(t) = \cos t$$

and

$$f(t, x(t)) = \sqrt{1 - \sin^2 t}$$

are equal only on intervals where $\cos t$ is positive. Thus, the formula $x(t) = \sin t$ defines a solution on the interval $-\pi/2 < t < \pi/2$, but on no larger interval containing $t_0 = 0$. (However, this interval is not the maximal interval of the solution of the initial value problem! See Exercise 27.) ∎

Exercises

Find the general solution of the following differential equations.

1. $x' = 1 + t^2$

2. $x' = \cos \pi t$

3. $x' = e^{2t}$

4. $x' = te^{-t}$

Find the unique solution of the following initial value problems.

5. $x' = t^2$, $x(1) = 2$

6. $x' = e^{-3t}$, $x(0) = 1$

7. $x' = te^{-t}$, $x(0) = 1$

8. $x' = \sin 3t$, $x\left(\frac{\pi}{6}\right) = 0$

For which initial value problems can the fundamental existence and uniqueness theorem (Theorem 1) be applied? Explain your answer. In each case, what do you conclude from this theorem?

9. $x' = t^2 + x^2$, $x(0) = 0$

10. $x' = \dfrac{t^2}{x^2}$, $x(0) = 0$

11. $x' = \tan x$, $x\left(\frac{\pi}{2}\right) = 0$

12. $x' = \tan x$, $x(0) = 0$

13. $x' = \tan x$, $x(0) = \frac{\pi}{2}$

14. $x' = \ln(tx)$, $x(1) = 2$

15. $x' = \dfrac{1}{\sin x}$, $x(0) = \frac{\pi}{2}$

16. $x' = \dfrac{1}{t - x}$, $x(-1) = 2$

17. For what values of the constant a can the fundamental existence and uniqueness theorem be applied to the initial value problems below? Explain your answer. What do you conclude from this theorem for such values of a? What do you conclude from this theorem for other values of a?

18. $x' = \ln(a - x)$, $x(0) = 0$

19. $x' = \tan ax$, $x(0) = \frac{\pi}{2}$

20. $x' = \sqrt{a^2 - x^2}$, $x(1) = 2$

21. $x' = \dfrac{1}{a - x}$, $x(1) = 2$

For which t_0 and x_0 does the fundamental existence and uniqueness theorem apply to the initial value problem $x' = f(t, x)$, $x(t_0) = x_0$, with the functions $f(t, x)$ below? Explain your answer. What do you conclude from this theorem for such initial points? What do you conclude from this theorem for other initial points?

22. $f = \ln(t^2 + x^2)$

23. $f = \dfrac{t^2}{x}$

24. $f = \tan bx$, $b = $ constant

25. $f = \sqrt{t^2 + x^2 - b^2}$, $0 < b = $ constant

26. Consider the initial value problem $x' = f(t, x)$, $x(0) = 0$, where

$$f(t, x) = \begin{cases} 1, & \text{for } t \geq 0 \text{ and all } x \\ -1, & \text{for } t < 0 \text{ and all } x. \end{cases}$$

(a) Show that the existence and uniqueness Theorem 1 does not apply. What do you conclude?

(b) Show that this initial value problem does not have a solution on any interval containing $t_0 = 0$.

27. Consider the equation $x' = \sqrt{1 - x^2}$.

(a) Show that the constant functions $x(t) = 1$ and $x(t) = -1$ are solutions for all t.

(b) Show that the function

$$x(t) = \begin{cases} 1 & \text{for } t \geq \frac{\pi}{2} \\ \sin t & -\frac{\pi}{2} < t < \frac{\pi}{2} \\ -1 & \text{for } t \leq -\frac{\pi}{2} \end{cases}$$

is a solution for all t. Thus, the maximal interval for the solution of the initial value problem $x(0) = 0$ is the whole real line.

(c)   The solution $x(t) = 1$ and the solution in (b) both satisfy the same initial value problem $x(\pi/2) = 1$ for all t. Why does this not contradict Theorem 1?

1.2 Approximation of Solutions

Formulas for solutions of differential equations are not in general available. For this reason we need other methods for studying equations and their solutions. For some applications it is sufficient to obtain approximations to solutions. For example, roughly sketched graphs of solutions are sometimes adequate. In other applications, more accurate graphs or even numerical approximations are necessary. We can also obtain algebraic formulas for approximations to solutions. In this section we study some graphical and numerical approximation methods. Analytic approximation methods are studied in Chapter 3. We begin with a procedure for making sketches of solution graphs.

1.2.1 Slope Fields

From algebra and calculus we learn that graphs are a useful way to study functions. The derivative of a function is the slope of its graph. A differential equation therefore tells us something about the slopes of the graphs of its solutions.

Specifically, if the graph of a solution $x = x(t)$ of

$$x' = f(t, x) \tag{2.1}$$

passes through a point (t, x), then the slope $x'(t)$ of its graph at this point equals $f(t, x(t))$. In other words, each point (t, x) in the domain of f is associated with a slope equal to the number $f(t, x)$.

For example, the graph of a solution of $x' = t^2 + x^2$ that passes through the point $(t, x) = (1, 1)$ necessarily has slope $1^2 + 1^2 = 2$ at this point. Similarly, the solution whose graph passes through the point $(-2, 1/3)$ must have slope $(-2)^2 + (1/3)^2 = 37/9$ at this point.

The association of a slope $f(t, x)$ with each point (t, x) defines the *slope field* of the differential equation (2.1). Solutions of differential equation must "fit" its slope field. This means at each point on a solution's graph the slope (of the tangent) must equal the slope associated with that point.

One way to obtain a picture of a slope field is to draw, through each of several points in the (t, x)-plane, a short straight line segment that has the slope associated with that point. By drawing such line segments through a sufficient number of points in the plane, we can get a good approximation to the overall slope field and hence the graphs of solutions.

Rather than randomly choosing points in the plane, it is better to proceed in a systematic manner. We discuss two ways to do this: the grid and the isocline methods. The grid method is particularly well suited for computer use. The isocline method is sometimes a convenient way to obtain a sketch of the slope field by hand.

The Grid Method   One way to approximate a slope field is to draw a short line segment with the appropriate slope at points lying on a rectangular grid in the (t, x)-plane. This *grid method* can be done by hand; however, most computer programs that solve differential equations will also draw slope fields using this grid method and display the results graphically.

When sketching a slope field by the grid method, we must chose a grid fine enough so that the essential features of the slope field are apparent but coarse enough so as not to be visually cluttered. It usually takes a several attempts to find a suitable grid size. Sample slope fields for several differential equations, drawn using the grid method, appear in Fig. 1.2.

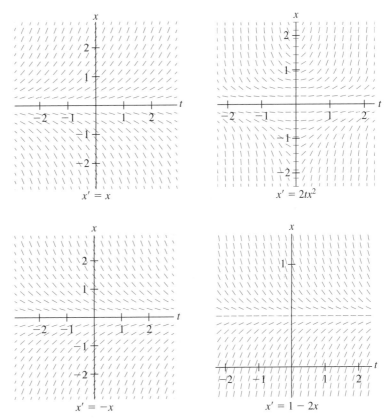

Figure 1.2 Slope fields are shown for four different differential equations.

We can sketch the solution graph of an initial value problem $x(t_0) = x_0$ by drawing a curve that both fits the slope field and passes through the point (t_0, x_0). Such a sketch can often suggest important properties of solutions. For example, the slope field and solution sketched in Fig. 1.3 suggest that the solution is monotonically increasing without bound as $t \to +\infty$ and that the x-axis is a horizontal asymptote as $t \to -\infty$.

The next example shows how a slope field can yield important general properties of solutions.

EXAMPLE 1

Figure 1.4 shows the slope fields of the logistic equation

$$x' = rx\left(1 - \frac{x}{K}\right)$$

for several choices of the parameters r and K. These slope fields, together with the sample solution graphs, suggest that solutions with positive initial conditions $x(0) =$

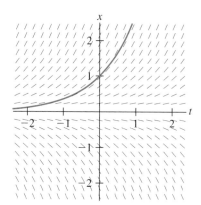

Figure 1.3 The slope field for $x' = x$ and the solution satisfying the initial condition $x(0) = 1$.

$x_0 > 0$ tend monotonically to a horizontal asymptote at $x = K$ as $t \to +\infty$. This important fact about the logistic equation will be proved in Chapter 3. Note that $x(t) = K$ is a solution.   ∎

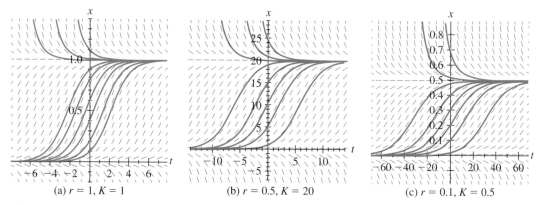

(a) $r = 1$, $K = 1$ (b) $r = 0.5$, $K = 20$ (c) $r = 0.1$, $K = 0.5$

FIGURE 1.4 Selected slope fields and solutions for the logistic equation $x' = r(1 - x/K)x$.

The Isocline Method   In Fig. 1.4 it is interesting to note that the points lying on a horizontal straight line appear to be associated with the same slope. The reason for this is that $f(t, x) = rx(1 - x/K)$, and hence the slope at a point (t, x), does not depend on t. This observation in fact applies to any equation whose right-hand side f does not depend on the independent variable t, i.e., to any so-called *autonomous* equation (Chapter 3).

A curve all of whose points are associated with the same slopes in the slope field of a differential equation is called an *isocline* (*iso* means "same" and *cline* means "slope"). The isoclines of an autonomous equation $x' = f(x)$ are horizontal straight lines. Points on a horizontal line $x = a$ are associated with slope $f(a)$. This fact can be a useful aid in sketching the slope field of an autonomous equation. Figure 1.5 shows a sketch of the slope field for the equation $x' = x(1 - x)$ obtained using this isocline method.

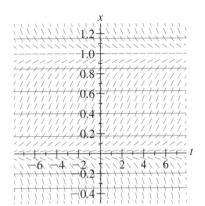

Figure 1.5 Some isoclines for $x' = x\,(1 - x)$ are shown.

The concept of an isocline is not restricted to autonomous equations. For any equation $x' = f(t, x)$ we can find isoclines by determining those points in the plane that are associated with a common slope m. These points satisfy the *isocline equation*

$$f(t, x) = m.$$

The graph of this equation is, in general, a curve in the plane called the *isocline* associated with slope m.

For nonautonomous equations isoclines are not necessarily horizontal lines. If they can be conveniently graphed, isoclines can be used to sketch slope fields for nonautonomous equations in the same way they were used for autonomous equations. On an isocline we draw several short line segments each having the slope associated with that isocline. Doing this for a collection of isoclines we obtain a sketch of the slope field. The following example illustrates the method.

EXAMPLE 2

What are the isoclines associated with the equation

$$x' = t^2 + x^2?$$

Suppose we find the isocline associated with slope $m = 1$. The equation for this isocline is $t^2 + x^2 = 1$, which we recognize as the equation the circle with radius 1 and center at the origin $(0, 0)$. Drawing this circle and placing on it several short line segments with slope 1, we obtain part of the slope field. This procedure can be repeated using other slopes m. Points associated with slope $m = 2$ lie on the circle of radius $\sqrt{2}$ while points associated with slope $m = 0.25$ lie on the circle of radius $\sqrt{0.25}$ and so on. The typical isocline equation $t^2 + x^2 = m$ yields the circle of radius $\sqrt{m}$, provided $m > 0$. A degenerate isocline is obtained for slope $m = 0$, namely the single point $(0, 0)$. There are no isoclines associated with negative slopes $m < 0$. See Fig. 1.6(a). ∎

Isoclines are not necessarily easy to identify or graph. Their usefulness for slope field sketching depends on the right hand side $f(t, x)$ of the differential equation. If we can easily identify and graph isoclines, then this method for drawing slope fields is convenient. Otherwise it is not.

Caution: A common mistake is to confuse isoclines with the solution graphs. Isoclines are *not* graphs of solutions. For example, compare the solution graph in Fig. 1.6(b) to the circular isoclines in Fig. 1.6(a).

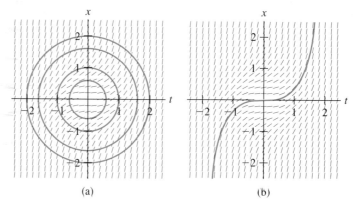

Figure 1.6 (a) Selected circular isoclines of $x' = t^2 + x^2$. (b) The solution satisfying the initial condition $x(0) = 0$.

Exercises

Use a computer to obtain sketches of the slope fields for the differential equations in the exercises below. Using the slope field, sketch (by hand) the graph of the solutions satisfying each of the given initial conditions.

1. $x' = 1 - x$, $x(0) = 3$, $x(0) = 0$, $x(-1) = 2$

2. $x' = 2 - 3x$, $x(0) = 1$, $x(0) = \frac{2}{3}$, $x(0) = -1$

3. $x' = (1-x)^2$, $x(0) = 1$, $x(-2) = 1$, $x(0) = 0$, $x(1) = 1.2$

4. $x' = x\left(1 - \dfrac{x}{2 + \cos t}\right)$, $x(0) = 0$, $x(0) = 1$, $x(0) = -0.1$, $x(-2) = 2$

5. $x' = x \cos t$, $x(0) = 0$, $x(1) = 4$, $x(0) = 1$, $x(0) = -1$

6. $x' = -\frac{1}{2}x + \sin t$, $x(0) = 0$, $x(0) = 1$, $x(-2) = -1$, $x(0) = -1$

7. $x' = x \sin x$, $x(0) = 0$, $x(0) = \frac{\pi}{2}$, $x(0) = -3$, $x(0) = 4$

8. $x' = \left(1 + t^2 + x^2\right)^{-1/2}$, $x(0) = 0$, $x(-1) = -1.5$

9. $x' = (1 - x) x \sin^2 t$, $x(0) = -0.25$, $x(0) = 2$, $x(-2) = 0.5$

10. $x' = (1 - x^2)(\sin t - x)$, $x(0) = 0$, $x(0) = -0.5$, $x(1) = 1.5$, $x(0) = 1$

11. $x' = x(1 - x)(x + 1)$, $x(0) = 0.5$, $x(0) = -0.5$, $x(0) = 1.5$, $x(0) = -1.5$

12. Consider the differential equation in Example 3:

$$x' = rx\left(1 - \frac{x}{K + a \sin t}\right).$$

(a) Use a computer to sketch the slope fields of the equation for the cases below.

(i) $r = 1, K = 2, a = 1$

(ii) $r = 1, K = 5, a = 1$

(iii) $r = 0.5, K = 5, a = 2$

(iv) $r = 0.5, K = 5, a = 4$

(b) For each case in (a), use the slope field to sketch (by hand) the graphs of the solutions satisfying the initial condition $x(0) = 1$.

(c) What do all the solutions graphed in (b) seem to have in common?

Use a computer to obtain a sketch of the slope field for the equations below. Do this for a selection of values for the constant a. How are the slope fields for $a > 1$ different from those for $a < 1$?

13. $x' = -a + 2x - x^2$

14. $x' = x(x - a)(1 - x)$

Describe (geometrically) and sketch the isoclines for the differential equations below and use them to obtain a sketch of the slope fields.

15. $x' = 1 - x$

16. $x' = 4 - 2x$

17. $x' = \left(1 + t^2 + x^2\right)^{-1/2}$

18. $x' = -x + \sin t$

Find first-order differential equations whose isoclines are as described below. Here m denotes the slope in the field slope.

19. The family of lines $x = t + m$, where m allowed to be any constant.

20. The family of parabolas $x = t^2 + m$, where m is allowed to be any constant.

21. The family of lines $x = t + \frac{1}{m}$, where m is allowed to be any nonzero constant.

22. The family of parabolas $x = t^2 + \frac{1}{m}$, where m is allowed to be any nonzero constant.

23. The family of ellipses $2x^2 + 3t^2 = m^{1/3}$, where m is allowed to be any positive constant.

24. The family of circles $x^2 + t^2 = 1 - 2m^2$, where m is allowed to be any positive constant satisfying $0 < m < 1/\sqrt{2}$.

1.2.2 Numerical and Graphical Approximations

Slope fields provide approximate graphs of solutions of differential equations. However, it is often desirable to have a more accurate approximation to a solution and its graph than can be obtained from a slope field. Another way to obtain an approximate graph of a solution on an interval $t_0 \leq t \leq T$ is to calculate numerical approximations x_i to the solution $x(t_i)$ at $t = t_i$, where

$$t_0 < t_1 < t_2 < \cdots < t_{n-1} < t_n = T$$

and, in the (t, x)-plane, connect the points $(t_0, x_0), (t_1, x_1), \ldots, (t_n, x_n)$ by straight line segments. See Fig. 1.7.

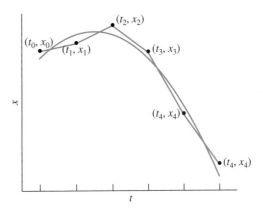

Figure 1.7 A "broken line" approximate graph obtained from approximations x_i to the solution at t_i.

We want to obtain the approximations $x_i \approx x(t_i)$ in such a way that if the number of points t_i increases (and the distances between them tend to zero), then the approximations x_i become more accurate and the approximate ("broken line") graphs approach the (smooth) graph of the solution $x = x(t)$.

In this section we study a basic method for approximating the solution of the initial value problem

$$x' = f(t, x), \quad x(t_0) = x_0 \tag{2.2}$$

at specified values of $t > t_0$. The method, called the *Euler algorithm*, is a fundamental method that serves as an introduction to the numerical approximation of solutions of differential equations. It is, however, rarely used for other than pedagogical reasons because it "converges" too slowly. Section 1.3 gives some methods that converge more quickly (and hence are more commonly used). Nonetheless, Euler's algorithm, by providing a basis for understanding how solutions are numerically approximated, is a good starting point for the study of more efficient (and hence complicated) algorithms.

Consider the problem of approximating the solution $x = x(t)$ of (2.2) at $t = t_1 > t_0$. Since $x(t)$ is a solution, we can integrate both sides of the equation $x'(t) = f(t, x(t))$ from $t = t_0$ to $t = t_1$ to obtain

$$x(t_1) - x(t_0) = \int_{t_0}^{t_1} f(t, x(t)) \, dt$$

or, using the initial condition,

$$x(t_1) = x_0 + \int_{t_0}^{t_1} f(t, x(t)) \, dt. \tag{2.3}$$

The right-hand side of this equation does not give a formula for $x(t_1)$ because it involves the unknown solution $x(t)$. However, we can use (2.3) to approximate $x(t_1)$ by making an approximation to the integral on the right-hand side. For example, we can use integration approximation methods studied in calculus, such as the rectangle rule, the trapezoid rule, or Simpson's rule.

The Euler algorithm is obtained by using the (left-hand) rectangle rule to approximate the integral:

$$\int_{t_0}^{t_1} f(t, x(t)) \, dt \approx (t_1 - t_0) f(t_0, x(t_0)).$$

Defining the first step size by $s_0 = t_1 - t_0$ and recalling the initial condition $x(t_0) = x_0$, we have

$$\int_{t_0}^{t_1} f(t, x(t)) \, dt \approx s_0 f(t_0, x_0)$$

and consequently from (2.3) we have the approximation

$$x(t_1) \approx x_0 + s_0 f(t_0, x_0).$$

Denote this approximation by x_1; that is, we define x_1 by

$$x_1 = x_0 + s_0 f(t_0, x_0).$$

To obtain an approximation x_2 to the solution value $x(t_2)$ at the next point t_2, we proceed in a similar manner. Integrate both sides of the equation $x'(t) = f(t, x(t))$ from $t = t_1$ to $t = t_2$. Using the fundamental theorem of calculus, the (left-hand) rectangle rule to approximate the integral, and the approximation $x_1 \approx x(t_1)$, we obtain

$$x(t_2) = x(t_1) + \int_{t_1}^{t_2} f(t, x(t)) \, dt \approx x_1 + (t_2 - t_1) f(t_1, x_1).$$

We denote this approximation to the solution at $t = t_2$ by

$$x_2 = x_1 + s_1 f(t_1, x_1), \qquad s_1 = t_2 - t_1.$$

In calculating the approximation x_2, we introduced *two* sources of error. First, there is the error made in using the rectangle rule to approximate the integral (called the truncation error) and, second, there is the error in using the approximation x_1 to $x(t_1)$. Together these errors account for the accumulation error at the point $t = t_2$.

If this procedure is repeated, we obtain the following formulas:

$$x_0 = x_0$$
$$x_{i+1} = x_i + s_i f(t_i, x_i), \quad s_i = t_{i+1} - t_i, \quad i = 0, 1, 2, \ldots$$

of the Euler algorithm. The number x_i is an approximation to the solution $x = x(t)$ of the initial value problem (2.2) at the point $t = t_i$. Usually equally spaced points are chosen, in which case $s_i = s$ for all i and the algorithm reduces to

$$x_0 = x_0 \qquad\qquad (2.4)$$
$$x_{i+1} = x_i + sf(t_i, x_i) \quad \text{for } i = 0, 1, 2, \ldots, n.$$

The common distance s is called the *step size* of the algorithm.

The formulas (2.4) are recursive. That is, we utilize the same formula sequentially to calculate the approximations at each of the points $t_1, t_2, \ldots, t_n$, using at each step the approximation made at the previous step. This makes the method ideally suited for programming on a computer or calculator.

The accuracy of the integral approximation obtained by the rectangle rule increases if the step size s decreases. For this reason we expect the accuracy of the approximations obtained from the Euler algorithm (2.4) to increase if the step size s decreases. There is a cost for this increased accuracy, however, because decreasing the step size s will increase the number n of steps necessary to get from the initial condition t_0 to the end point T. This means more repetitions of the algorithm (2.4) are required, and consequently more arithmetic work is necessary to reach the endpoint $t_n = T$. (This also means more round-off errors!)

EXAMPLE 3

In this example we use the Euler algorithm (2.4) to approximate the solution $x = x(t)$ of the initial value problem

$$x' = x, \quad x(0) = 1$$

at $T = 1$ using step size $s = 0.2$. The Euler algorithm (2.4) for this problem is

$$x_{i+1} = x_i + sx_i \quad \text{for } i = 0, 1, 2, \ldots$$

with $x_0 = 1$. Using step size $s = 0.2$, we need to calculate approximations at the five points $t = 0.2, 0.4, 0.6, 0.8, 1.0$. The calculations are

$$x_1 = x_0 + sx_0 = 1 + 0.2 \times 1 = 1.2$$
$$x_2 = x_1 + sx_1 = 1.2 + 0.2 \times 1.2 = 1.44$$
$$x_3 = x_2 + sx_2 = 1.44 + 0.2 \times 1.44 = 1.728$$
$$x_4 = x_3 + sx_3 = 1.728 + 0.2 \times 1.728 = 2.0736$$
$$x_5 = x_4 + sx_4 = 2.0736 + 0.2 \times 2.0736 = 2.48832.$$

The Euler algorithm with step size $s = 0.2$ yields the approximation $x(1) \approx x_5 = 2.48832$. ∎

How good is the approximation x_5 in the previous example? More generally, how accurate are the approximations (2.4) of the Euler algorithm? Can we estimate the size of the error, and if not how can we have any confidence in the numerical approximations obtained from the formulas (2.4)?

An accurate estimate of the error resulting from approximation methods such as the Euler algorithm is usually not possible. However, we expect that the numerical approximations will get more accurate as the step size s decreases and that they will tend to the exact solution in the limit as $s \to 0$. This turns out to be true for the Euler algorithm, on the solution's interval of existence, under the assumptions of the fundamental existence and uniqueness theorem (Theorem 1).

One useful way to study the accuracy of the Euler algorithm (and of other algorithms as well) is to consider the *rate* at which the approximations converge to the exact solution. The Euler algorithm is said to be first-order or of order 1. What this means is that the magnitude of the error at $t = T$ is no larger than constant multiple of the first power of s. That is, there exists a constant $c > 0$ such that $| x(T) - x_n | \le cs$. This inequality guarantees the Euler approximations converge to the value of the solution at least as fast as s decreases to 0. Thus, roughly speaking, if the step size s is halved, then in general we expect the error to be (at least) halved. If the step size is decreased by a factor of $1/10$, then in general we expect the error to decrease by a factor of $1/10$ and so on. (For an example, see Table 1.2 on page 35.) We summarize this by saying that the Euler algorithm is "$O(s)$" (pronounced "Oh of s").

We can gain confidence in the accuracy of numerical approximations by observing their changes as the step size s decreases. This is commonly done by decreasing s by a fixed fraction. For example, if s is decreased by one half several times, we expect the error to be cut in half each time. Since the approximations at a fixed t approach the solution value $x(t)$, the leading digits in the resulting sequence of approximations should eventually "stabilize" (i.e., remain unchanged as s decreases further). As a practical matter, we accept these digits as correct. However, none of these digits may be accurate, since we cannot be sure that they will remain unchanged if the step size s decreases further.

EXAMPLE 4

In this example we repeat Example 3 by halving the step size s six consecutive times and observe the resulting change in the approximation to $x(1)$. The number of calculations necessary to perform the approximation increases as s decreases. For example, the algorithm (2.4) must be used 320 times for the step size $s = 0.003125$.

We use a computer to perform the calculations, and the results appear in Table 1.1. We expect the approximation $x(1) \approx 2.714047$ obtained from the smallest step size $s = 0.003125$ to be the most accurate, but how many of these digits are correct? We know the sequence of approximations converges to the exact value of the solution at $T = 1$. Since only two digits appear to have stabilized in Table 1.1, we accept only the two digit approximation 2.7 as accurate. ∎

There is a formula for the solution of the initial value problem in Examples 3 and 4, namely $x(t) = e^t$. Therefore, the exact value of the solution at $t = 1$ is $x(1) = e$ (recall $e \approx 2.718282$). Using this formula, we can investigate how accurate the approximations in Table 1.1 really are.

The percent error of each approximation is given in Table 1.2. Notice that the percent error decreases by a factor of (approximately) $1/2$ at each consecutive step. This is what we expect, since the step size s decreases by a factor of $1/2$ at each step and the Euler algorithm is $O(s)$.

We approximate the graph of solution of the initial value problem $x' = x$, $x(0) = 1$ by connecting the points (t_i, x_i) with straight line segments. This is done in Fig. 1.8 for decreasing step sizes on the interval $0 \le t \le 5$. The convergence, as s decreases, of these approximate graphs to the graph of the solution $x = e^t$ is apparent.

Table 1.1 The Euler algorithm approximations to the solution at $t = 1$ of the initial value problem $x' = x$, $x(0) = 1$ obtained by repeatedly halving the step size.	
Step Size s	**Approximation to $x(1)$**
0.200000	2.488320
0.100000	2.593742
0.050000	2.653298
0.025000	2.685064
0.012500	2.701485
0.006250	2.709836
0.003125	2.714047

Table 1.2 The percent errors of the approximations in Table 1.1.		
Step Size s	**Approximation to $x(1)$**	**% Error**
0.200000	2.488320	8.4598
0.100000	2.593742	4.5816
0.050000	2.653298	2.3906
0.025000	2.685064	1.2220
0.012500	2.701485	0.6179
0.006250	2.709836	0.3107
0.003125	2.714047	0.1558

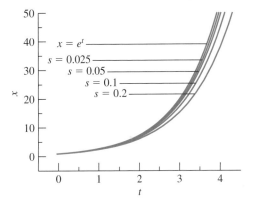

Figure 1.8 The broken line graphs calculated from approximations using Euler's algorithm converge to the solution of the initial value problem $x' = x$, $x(0) = 1$ as the step size s decreases.

We should not accept a graphical approximation to a solution obtained from a single step size s alone (e.g., the default step size in a computer program). Instead, before accepting a graphical approximation, we should decrease the step size until little change occurs in two consecutive graphical approximations.

EXAMPLE 5

The equation $x' = ae^{-bt}x$ describes the growth of a tumor, where $x = x(t)$ is a measure of its size (e.g., weight or number of cells) and t is time. Figure 1.9 shows approximate graphs of the solution of the initial value problem with $x_0 = 5$ and parameter values $a = 20$ and $b = 15$. These graphs result from the Euler algorithm using a decreasing sequence of step sizes starting with $s = 0.1$. Little change occurs in the graphs for the last two steps sizes $s = 0.003125$ and 0.0015625 and therefore we accept the final graph as an accurate approximation. All of the graphs indicate that the tumor size x approaches a maximal size as $t \to +\infty$. However, the inaccurate graphs obtained from the larger steps sizes considerably over estimate the maximal size of the tumor. ∎

The convergence rate $O(s)$ of the Euler algorithm is sometimes too slow for practical purposes. In Table 1.2 only two digits of accuracy for $x(1)$ are obtained with a step size $s = 0.003125$. To obtain more accuracy a smaller step size is needed. However, there are more intermediate steps with each decrease in step size and it takes longer

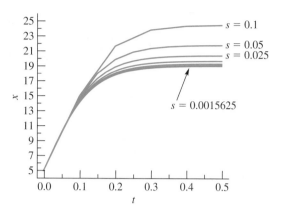

Figure 1.9 The Euler algorithm with a decreasing sequence of steps sizes yields converging approximate graphs for the solution of the initial value problem $x' = 20e^{-15t}x$, $x(0) = 5$.

to perform all of the necessary calculations. Furthermore, other sources of error, such as round-off errors at each step, might eventually prevent increased accuracy if the number of steps (and hence calculations) becomes too large.

Table 1.3 shows an example that dramatically illustrates the slow convergence of the Euler algorithm. In this example no accurate digits are found with a step size as small as $s = 0.000391$.

Table 1.3 Euler algorithm estimates to the solution of the initial value problem $x' = x^2$, $x(0) = 0.99$, at $t = 1$ for a decreasing sequence of step sizes. The solution formula $x(t) = 99/(100 - 99t)$ for this initial value problem gives the exact value $x(1) = 99$.[3]

Step Size s	Approximation to $x(1)$
0.100000	5.862897
0.050000	8.905711
0.025000	13.766320
0.012500	21.242856
0.006250	31.967263
0.003125	45.709606
0.001563	60.736659
0.000781	74.330963
0.000391	84.517375

Fortunately, practical algorithms with faster rates of convergent are available. In the following section we discuss algorithms of orders two and four. An algorithm has *order of convergence p* (or more succinctly is of *order p*), written $O(s^p)$, if the accumulative error is bounded in magnitude by a constant multiple of s^p [i.e., if $|x(T) - x_n| \leq cs^p$].

---

[3]The interval of existence for the solution is $-\infty < t < 100/99 \approx 1.0101$. It is interesting to note that the Euler algorithm will calculate "approximations" at t values outside of this interval. For example, with step size $s = 0.1$, eleven repetitions of the algorithm produce the number $x_{11} = 9.30025$. However, this number cannot be taken as an approximation to the solution at $t = 1.1$ because the solution is not defined at this value of $t > 100/99$.

To see the advantage of a convergence rate of order greater than $p = 1$ consider an algorithm of order $p = 2$, for which the error satisfies $| x(T) - x_n | \le cs^2$. We can expect the error to decrease by a factor of $(1/2)^2 = 1/4$ if the step size s is decreased by a factor of $1/2$, or by a factor of $(1/10)^2 = 1/100$ if the step size s is decreased by a factor of $1/10$, and so on. For an algorithm of order 4 the error decreases even faster [e.g., by a factor of $(1/10)^4 = 1/10000$ if the step size is decreased by a factor of $1/10$].

1.3   Another Numerical Algorithm

In deriving the Euler algorithm we used the rectangle rule to approximate the integral

$$\int_{t_i}^{t_{i+1}} f(t, x(t)) \, dt.$$

More accurate approximations to this integral lead to algorithms that converge faster than the Euler algorithm. For example, we could use the trapezoid rule. (Another choice is Simpson's rule; see Exercise 10.) Integrating both sides of the equation $x'(t) = f(t, x(t))$ from $t = t_i$ to $t = t_{i+1}$, we obtain

$$x(t_{i+1}) = x(t_i) + \int_{t_i}^{t_{i+1}} f(t, x(t)) \, dt.$$

From the trapezoid rule approximation

$$\int_{t_i}^{t_{i+1}} f(t, x(t)) \, dt \approx \frac{s_i}{2} \left[f(t_{i+1}, x(t_{i+1})) + f(t_i, x(t_i)) \right]$$

we get

$$x(t_{i+1}) \approx x(t_i) + \frac{s_i}{2} \left[f(t_{i+1}, x(t_{i+1})) + f(t_i, x(t_i)) \right].$$

Assuming that we already have an approximation $x(t_i) \approx x_i$ to the solution at the point $t = t_i$, we can write

$$x(t_{i+1}) \approx x_i + \frac{s_i}{2} \left[f(t_{i+1}, x(t_{i+1})) + f(t_i, x_i) \right].$$

Unfortunately, we cannot use the right-hand side to calculate an approximation x_{i+1} to $x(t_{i+1})$ because it involves $x(t_{i+1})$. This is an example of what is called an *implicit* algorithm because the equation

$$x_{i+1} = x_i + \frac{s_i}{2} \left[f(t_{i+1}, x_{i+1}) + f(t_i, x_i) \right]$$

is not explicitly solved for the approximation x_{i+1}. (The Euler algorithm is an example of an *explicit* algorithm.) To find the approximation x_{i+1}, we have to solve this equation. To do this at each step results in a highly complicated algorithm. One way to deal with this difficulty is to perform another approximation. For example, we can use the Euler approximation for the x_{i+1} on the right-hand side. Thus, at each step we use the formulas

$$x_{i+1}^* = x_i + s_i f(t_i, x_i)$$

$$x_{i+1} = x_i + \frac{s_i}{2} \left[f(t_{i+1}, x_{i+1}^*) + f(t_i, x_i) \right], \quad i = 0, 1, 2, \ldots$$

to calculate the approximation x_{i+1}. This algorithm is called *Heun's algorithm* (sometimes the *improved Euler algorithm* or the *modified Euler algorithm*). It is an example of a "predictor-corrector" algorithm. At each step the Euler approximation x_{i+1}^* is the prediction and x_{i+1} is the correction.

If equal step sizes $s_i = s$ are used Heun's algorithm is

$$x_{i+1}^* = x_i + sf(t_i, x_i) \tag{3.1}$$

$$x_{i+1} = x_i + \frac{s}{2}\left[f(t_{i+1}, x_{i+1}^*) + f(t_i, x_i)\right], \quad i = 0, 1, 2, \ldots.$$

The initial condition x_0 starts the algorithm. It turns out that Heun's algorithm of order $O(s^2)$.

Compare the results in Table 1.4 with those in Table 1.2. Note that the error in Table 1.4 decreases approximately by a factor of $1/4$ as the step size is decreased by a factor of $1/2$. Heun's algorithm is a popular procedure; for example, it is often used with programmable hand calculators.

Table 1.4   **The Heun algorithm approximations to the solution of the initial value problem $x' = x$, $x(0) = 1$, at $t = 1$ obtained by repeatedly halving the step size.**

Step Size s	Approximation to $x(1)$	% Error
0.200000	2.702708	0.5729
0.100000	2.714081	0.1545
0.050000	2.717191	0.0401
0.025000	2.718004	0.0102
0.012500	2.718212	0.0026
0.006250	2.718264	0.0007

We saw in Table 1.3 an example of an initial value problem for which the Euler algorithm converges too slowly to be practical. Table 1.5 shows the results of applying Heun's method to the same initial value problem. The estimates obtained from the two numerical algorithms differ considerably. At each step size Heun's algorithm provides a more accurate approximation to $x(1) = 99$ than does the Euler algorithm.

Even higher-order algorithms are available, although they involve more complicated formulas at each step. A widely used class of algorithms is the class of Runge-Kutta algorithms. These algorithms are available for any order of convergence. A popular algorithm is the fourth-order Runge-Kutta algorithm. You can see the complicated formulas for this algorithm in Exercise 1. Table 1.6 shows the results applying this algorithm to the same initial value problem in Tables 1.3 and 1.5. This faster converging algorithm provides an accurate approximation to $x(1) = 99$.

Step Size s	Approximation to $x(1)$
0.100000	19.346653
0.050000	33.073325
0.025000	52.217973
0.012500	72.662362
0.006250	87.787581
0.003125	95.273334
0.001563	97.719807
0.000781	98.719804
0.000391	98.928245

Table 1.5 Heun's algorithm estimates to the solution of the initial value problem $x' = x^2$, $x(0) = 0.99$, at $t = 1$ for a decreasing sequence of step sizes. The solution formula $x(t) = 99/(100 - 99t)$ for this initial value problem gives the exact value $x(1) = 99$.

Step Size s	Approximation to $x(1)$
0.100000	53.355933
0.050000	75.881773
0.025000	91.639594
0.012500	97.671604
0.006250	98.856123
0.003125	98.988718
0.001563	98.999238
0.000781	98.999951
0.000391	98.999997

Table 1.6 Fourth-order Runge-Kutta algorithm estimates to the solution of the initial value problem $x' = x^2$, $x(0) = 0.99$, at $t = 1$ for a decreasing sequence of step sizes. The solution formula $x(t) = 99/(100 - 99t)$ for this initial value problem gives the exact value $x(1) = 99$.

Exercises

1. The following formulas constitute the *fourth-order Runge-Kutta algorithm*:

$$x_0 = x_0$$
$$x_{i+1} = x_i + s\frac{L_1 + 2L_2 + 2L_3 + L_4}{6} \quad \text{for } i = 0, 1, 2, \ldots,$$

where

$$L_1 = f(t_i, x_i)$$
$$L_2 = f\left(t_i + \frac{s}{2}, x_i + \frac{s}{2}L_1\right)$$
$$L_3 = f\left(t_i + \frac{s}{2}, x_i + \frac{s}{2}L_2\right)$$
$$L_4 = f(t_i + s, x_i + sL_3).$$

At each step we must calculate, in order, the four numbers L_1, L_2, L_3, and L_4 before calculating x_{i+1}.

(a) Use the fourth-order Runge-Kutta method to approximate the solution of $x' = x$, $x(0) = 1$ at $t = 1$. Start with step size $s = 0.2$ and calculate a sequence of approximations by repeated step size halving.

(b) Use the solution formula $x = e^t$ to calculate percent errors. Do the errors decrease at the expected rate?

(c) Compare the results in (a) and (b) with those of the Euler and Heun's algorithms in Tables 1.2 and 1.4.

2. Let $x = x(t)$ denote the solution of the initial value problem $x' = x^3$, $x(0) = 0.6$. It turns out that $x(1) \approx 1.1338934190$.

(a) Use the Euler algorithm to obtain an approximation to $x(1)$ with step size $s = 0.1$. How many correct significant digits does this approximation have?

(b) Obtain Euler approximations by repeatedly halving the step size (starting at $s = 0.1$). At which step size s is the Euler approximation first correct to 2 decimal places? To 3 decimal places?

(c) Compute the absolute error at each step size, starting from $s = 0.1$ and halving four times. Is the fractional decrease in the error correct for the Euler Algorithm?

3. Repeat Exercise 2 using Heun's algorithm.

4. Repeat Exercise 2 using the Runge-Kutta algorithm.

5. Repeat Exercise 2 using any other algorithm available on your computer.

6. Let $x = x(t)$ denote the solution of the initial value problem $x' = e^x$, $x(0) = 0$. It turns out that $x(0.8) \approx 1.6094379124$.

(a) Use the Euler algorithm to obtain an approximation to $x(0.8)$ with step size $s = 0.1$. How many correct significant digits does this approximation have?

(b) Obtain Euler approximations by repeatedly halving the step size. At which step size s is the Euler approximation first correct to 2 decimal places? To 3 decimal places?

(c) Compute the absolute error at each step size, starting from $s = 0.1$ and halving four times. Is the fractional decrease in the error correct for the Euler algorithm?

7. Repeat Exercise 6 using Heun's algorithm.

8. Repeat Exercise 6 using the Runge-Kutta algorithm.

9. Repeat Exercise 6 using any other algorithm available on your computer.

10. Euler's algorithm was derived by using the rectangle rule to approximate the integral $\int_{t_i}^{t_{i+1}} f(t, x(t))\, dt$ and Heun's algorithm was derived by using the trapezoid rule. In this exercise you derive an algorithm by using Simpson's rule to approximate this integral. Simpson's rule for approximating an integral $\int_a^b g(t)\, dt$ is

$$\int_a^b g(t)\, dt \approx \frac{1}{3}\left[g(b) + 4g\left(\frac{a+b}{2} \right) + g(a) \right].$$

   (a) Use Simpson's rule to obtain a predictor-corrector algorithm for an approximation x_{i+1} of the solution $x = x(t)$ of the initial value problem $x' = f(t, x)$, $x(t_0) = x_0$ at $t = t_{i+1}$. Use equal step sizes of length s.

   (b) Why is the algorithm derived in (a) called a "two-step" algorithm? What problem does this cause at the start (i.e., at t_1) and how might this be solved?

   (c) Use you answers from (a) and (b) to obtain an approximation to the solution of $x' = x$, $x(0) = 1$ at $T = 0.2$ using a step size of $s = 0.1$. (The exact solution is $e^{0.2} = 1.2214027582$.)

   (d) Compare your answer in (c) to the approximations obtained by the Euler and Heun's. If your computer has the fourth-order Runge-Kutta algorithm, compare its approximations also. Which algorithm gives the best approximation at $T = 0.2$?

11. Approximate the solution of the initial value problem $x' = t^2 + x^2$, $x(0) = 0$ at $T = 0.5$ using the Euler algorithm,

Heun's algorithm, and the Runge-Kutta algorithm. Start with step size $s = 0.1$ and repeat by halving the step size four times. What are the accurate digits obtained from each algorithm? What is the best approximation obtained from all methods?

12. Use a computer obtain an accurate graphical solution of the initial value problem $x' = t^2 + x^2$, $x(0) = 0$ on the interval from $t = 0$ to $T = 1$ using the Euler algorithm. Repeatedly halve the step size s starting with $s = 0.1$. What step size did you stop with and why?

13. Repeat Exercise 12 using Heun's algorithm.

14. Repeat Exercise 12 using the Runge-Kutta algorithm.

15. Repeat Exercise 12 using any other algorithm available on your computer.

16. Use a computer obtain an accurate graphical solution of the initial value problem $x' = \frac{x^3}{x-t}$, $x(0) = 1$ on the interval from $t = 0$ to $T = 1$ using the Euler Algorithm. Use a window size of $-20 < x < 20$. Repeatedly decrease the step size s by a factor of one tenth, starting with $s = 0.1$. What step size did you stop with and why?

17. Repeat Exercise 16 using Heun's algorithm.

18. Repeat Exercise 16 using the Runge-Kutta algorithm.

19. Repeat Exercise 16 using any other algorithm available on your computer.

20. **(a)** Use any algorithm you wish to obtain a graphical solution of the initial value problem $x' = 500\cos(200t)$, $x(0) = 0$. Start with step size $s = 0.1$ and decrease until the graph has stabilized. What do you conclude about the solution?

   (b) Obtain a formula for the solution and use it to explain the graphical solution.

1.4 Chapter Summary and Exercises

A solution $x = x(t)$ of the differential equation $x' = f(t, x)$ is a differentiable function for which $x'(t) = f(t, x(t))$ holds for all t on an interval. In general, a differential equation has infinitely many solutions. The general solution is the set of all solutions. We need an additional requirement in order to specify a unique solution. For a given point (t_0, x_0), the initial condition $x(t_0) = x_0$ is such a requirement. Theorem 1 gives conditions under which an initial value problem $x' = f(t, x)$, $x(t_0) = x_0$ has one and only one solution. Specifically, if $f(t, x)$ and its derivative $\partial f(t, x)/\partial x$ with respect to x are both continuous for t near t_0 and x near x_0, then there is one and only one solution. Although formulas for the solution cannot always be calculated, many kinds of approximation methods are available. The slope field associated with the differential equation helps in to sketching a graph of the solutions. A computers is useful for plotting the slope fields by the grid method; this method associates the slope $f(t, x)$ with each point (t, x) on from a chosen grid of points in the (t, x)-plane. Also useful for sketching slope fields are isoclines, which are curves in the (t, x)-plane made up of those points associated with a common slope. Numerical approximations to solution values $x(t)$ yield more accurate graphs of the solution. If $x_1, x_2, \ldots, x_n$ approximate the solution values $x(t_1), x(t_2), \ldots, x(t_n)$ for $t_1 < t_2 < \cdots < t_n$, then by connecting the points (t_i, x_i) with straight line segments we construct an approximate (broken line segment) solution graph. Usually equally spaced points t_i are chosen and the common distance between them is the step size s of the method. If the approximations converge to the solution values as s tends to 0, then the broken line graph tends to the solution graph as s tends to 0. The Euler algorithm is one method for calculating such approximations. It is based on the left hand rectangle rule for approximating an integral. Under the conditions on $f(t, x)$ in Theorem 1 the Euler approximations converge to the solution values as the step size s decreases to 0. The Euler algorithm is of order 1, which means the errors tend to 0 at the same rate that s tends to 0. Under the same conditions on $f(t, x)$, faster converging algorithms are available. Heun's algorithm is of order 2, which means the error tends to 0 at the same rate that s^2 tends to 0. A fourth-order method called the Runge-Kutta algorithm is commonly used.

Exercises

Find formulas for the general solutions of the differential equations below.

1. $x' = \dfrac{1}{(1 - t)t}$

2. $x' = \dfrac{2t}{1 + t^2}$

Find solution formulas for the following initial value problems.

3. $x' = \dfrac{1}{1 + t^2}, x(1) = \dfrac{\pi}{2}$

4. $x' = \dfrac{1 + t + t^2}{(1 + t^2)t}, x(1) = 1$

5. Does existence and uniqueness Theorem 1 apply to the initial value problem $x' = \sqrt{1 - x}, x(1) = 0$? Explain your answer. What do you conclude?

6. Does the existence and uniqueness Theorem 1 apply to the initial value problem $x' = (4 - x^2)^{-1}, x(2) = 0$. Explain your answer. What do you conclude?

For which initial values t_0 and x_0 does the existence and uniqueness Theorem 1 apply to the problems below? Explain your answer. What do you conclude? What can you conclude about initial value problems for other t_0 and x_0?

7. $x' = \ln|x - t|, x(t_0) = x_0$

8. $x' = \sqrt{9 - x^2 - t^2}, x(t_0) = x_0$

9. $x' = |x|, x(t_0) = x_0$

10. $x' = t^{\frac{1}{3}}x, x(t_0) = x_0$

Explain why Theorem 1 does not apply to the initial value problems below. What do you conclude?

11. $x' = \sqrt{x^2 + t^2}, x(0) = 0$

12. $x' = \sqrt{\sin(x^2 + t^2)}, x(0) = 0$

13. Apply the existence and uniqueness Theorem 1 to the initial value problem $x' = \sqrt{1 - x^2}, x(0) = 0$ in Example 5 of Sec. 1.1. What do you conclude?

14. Let $f(t, x)$ be a polynomial in t and x. Prove that any initial value problem $x' = f(t, x), x(t_0) = x_0$ has a unique solution on an interval containing t_0.

15. Let $p(z, w)$ be a polynomial in z and w and let $f(t, x) = p(\sin t, \sin x)$. Prove that any initial value problem $x' = f(t, x), x(t_0) = x_0$, has a unique solution on an interval containing t_0.

Use a computer to obtain sketches of the slope fields associated with the following differential equations. By hand, sketch graphs of the solutions satisfying each of the given initial conditions.

16. $x' = t^2 + 4x^2, x(0) = 0, x(0.5) = 0.5$

17. $x' = -\dfrac{t}{x}, x(0) = 1, x(1) = -1$

18. $x' = \dfrac{t^2 - x^2}{t^2 + x^2}, x(0) = 1, x(-1) = -1$

19. $x' = \ln(t^2 + x^2), x(1) = 0, x(0) = 0.1$

20. Match each equation with its slope field.

(a) $x' = x - t$ (b) $x' = t - x$

(c) $x' = x^2 - t$ (d) $x' = x - t^2$

(e) $x' = tx^2 - x$ (f) $x' = tx^2 + x$

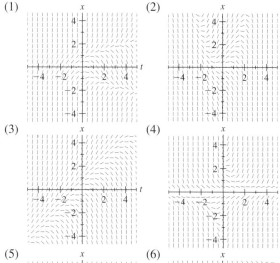

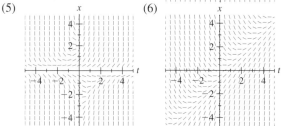

21. Find the isocline equation for the differential equations in Exercises 16–19 and graph several typical isoclines. Use your results to sketch the slope field of the equation.

Use a computer to obtain sketches the slope fields associated with the equations in the following initial value problems. Hand sketch a graph of the solution satisfying each of the given initial conditions.

22. $x' = 1 - x$, $x(0) = 0$, $x(0) = 1.5$

23. $x' = x - 1$, $x(0) = 0$, $x(0) = 1.5$

24. $x' = 1 - x^2$, $x(0) = 0$, $x(0) = 1.5$

25. $x' = \sin(x^2 + t^2)$, $x(0) = 0$, $x(0) = -0.5$

26. Consider the initial value problem $x' = x^3 e^{-t}$, $x(0) = 1$. Apply the Euler algorithm to approximate the solution at $T = 0.6$.

   (a) Start with step size $s = 0.1$ and halve it four times. Which digits in the resulting approximations do you think are accurate? Explain your answer.

   (b) Halve the step size four more times. Now which digits in the resulting approximations do you think are accurate? Explain your answer.

27. Consider the initial value problem $x' = x^3 e^{-t}$, $x(0) = 1$. Apply Heun's algorithm to approximate the solution at $T = 0.6$.

   (a) Start with step size $s = 0.1$ and halve it four times. Which digits in the resulting four approximations do you think are accurate? Explain your answer.

   (b) Halve the step size four more times. Now which digits in the resulting four approximations do you think are accurate? Explain your answer.

28. Consider the initial value problem $x' = x^3 e^{-t}$, $x(0) = 1$. Apply the Runge-Kutta algorithm to approximate the solution at $T = 0.6$. (See Exercise 1 of Sec. 1.3.)

   (a) Start with step size $s = 0.1$ and halve it four times. Which digits in the resulting four approximations do you think are accurate? Explain your answer.

   (b) Halve the step size four more times. Now which digits in the resulting four approximations do you think are accurate? Explain your answer.

29. Use the formula $x(t) = \left(2e^{-t} - 1\right)^{-1/2}$ for the solution of the initial value problem in Exercises 26, 27, and 28 to calculate the error and the per cent error of the approximations in these exercises for step size $s = 0.00625$. Round all numbers to 6 significant digits.

30. Use the Euler algorithm and a computer program to obtain an accurate graph of the solution of the initial value problem $x' = 1.5x^3 \sin 10t$, $x(0) = 1$ on the interval from $t = 0$ to $T = 1$. Use a window size of $-2 < x < 2$. Repeatedly halve the step size s starting with $s = 0.2$. At what step size did you stop and why?

31. Repeat Exercise 30 using Heun's algorithm.

32. Repeat Exercise 30 using the Runge-Kutta algorithm.

33. Suppose the decay rate of a radioactive isotope is $r = -0.35$ per year. The differential equation for the amount $x(t)$ at time t is $x' = -0.35x$.

   (a) Use a computer to study the graphs of solutions with many different initial conditions $x_0 > 0$ and formulate a conjecture about the length of time it takes a sample amount of the isotope to decay to one-half of its initial amount.

   (b) Use the solution formula $x(t) = x_0 e^{-0.35t}$ to verify or disprove your conjecture.

34. Let $x = x(t)$ be the dollars in an investment account that is compounded continuously at a rate of 4.5%.

   (a) Perform numerical experiments on the model equations $x' = 0.045x$, $x(0) = s$ to formulate a conjecture about how long will it take for the initial investment of s dollars to triple.

   (b) Use the solution formula $x(t) = se^{0.045t}$ to prove or disprove your conjecture.

35. Suppose a population has a per capita death rate $d > 0$ and a per capita birth rate that is proportional to population size x (with constant of proportionality denoted by $a > 0$).

(a) Use the balance law (2.1) in the Introduction to write down a model differential equation for the population size $x = x(t)$.

(b) Perform numerical experiments and formulate a conjecture about the fate of the population. [*Hint*: Choose a pair of model parameter values, such as

$a = 1$ and $d = 1$, and compute solution graphs for many initial population sizes $x(0) = x_0$. Then repeat for other values for a and d.]

(c) Use the solution formula

$$x(t) = \frac{dx_0}{x_0 a + e^{dt}(d - x_0 a)}$$

to verify or disprove your conjectures in (b).

1.5 APPLICATIONS

1.5.1 Bacterial Cell Growth

When placed in an environment of abundant resources (nutrients, space, etc.), cell cultures typically grow in such a way that their per capita rate of change is constant. Mathematically, this means the number of cells $x = x(t)$ at time t satisfies the differential equation

$$x' = rx,$$

where the constant $r > 0$ is the per capita growth rate. Often a particular microorganism's growth rate is described by the time it takes the number of cells in the culture to double. This time δ is called the doubling time (or generation time) and it is related to the growth rate according to the formula

$$r = \frac{\ln 2}{\delta}.$$

For more detailed discussion of these topics and of population growth models, see Sec. 3.6, Chapter 3.

As an example, the doubling time of the bacterium *Staphylococcus aureus* is approximately $\delta = 30$ minutes, which corresponds to a per capita growth rate of

$$r = \frac{\ln 2}{30} = 0.02310 \text{ (per minute)}.$$

The growth of a culture of *S. aureus* initially consisting of 10^6 cells is described by the initial value problem

$$x' = 0.02310x \qquad (5.1)$$
$$x(0) = 1.$$

Here we x is measured in units of 10^6 cells.

According to Theorem 1, this initial value problem has a unique solution $x = x(t)$. A slope field and a solution graph (drawn using Heun's algorithm with step size $s = 0.05$) appear in Fig. 1.10. Notice that the number of cells grows rapidly, following a seemingly exponential-like curve. Indeed, the solution formula for the initial value problem

$$x = e^{0.02310t}$$

shows the growth is indeed exponential.

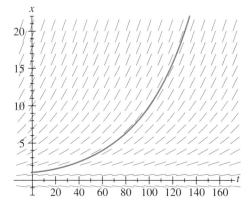

Figure 1.10 The slope field of the differential equation $x' = 0.02310x$ and the solution of the initial value problem (5.1) drawn using Heun's algorithm with step size $s = 0.05$.

S. *aureus* is a common cause of bacterial skin infection (particularly in patients with HIV). The rapid exponential growth of a staph infection can be a serious problem if left untreated. Our modeling application involves determining the effect of a medical treatment that removes staph cells from the patient at a certain rate $h > 0$ (cells/minute). We set up our mathematical model (i.e., perform the model derivation step in Fig. 0.1 of the Introduction) by applying the balance law (2.1) of the Introduction to the staph cell population numbers. This leads to the differential equation

$$x' = 0.02310x - h. \tag{5.2}$$

More specifically, suppose a milligram (mg) of antibiotic in a particular patients kills staph cells at a rate of 10^4 per minute. Then a dosage of d mg kills a total of $10^4 d$ staph cells per minute. In units of 10^6 cells, we have

$$h = \frac{10^4}{10^6}d = 0.01d \text{ (per minute).} \tag{5.3}$$

Suppose, for the moment, that this removal rate h remains constant in time, as might be the case for example if the antibiotic were continuously administered intravenously. We want to know what dosages d, if any, will eliminate the staph infection from the patient, and if so in what amount of time.

The antibiotic kill rate h in (5.3) leads to the initial value problem

$$x' = 0.02310x - 0.01d \tag{5.4}$$
$$x(0) = 1$$

for the number of staph cell $x = x(t)$. Our next goal is to perform the model solution step in the modeling cycle. What we want to learn from the solution $x = x(t)$ is whether or not it continues to increase or whether it decreases and eventually equals 0. The answer will presumably depend on the dosage d.

One way to obtain answers to our questions would be from a formula for the solution $x(t)$. We will learn how to find such a formula in Chapter 2. Here, however, we will investigate the solution by means of the methods developed in Sec. 1.2.

Figure 1.11 shows slope fields and solution graphs, for a selection of dosages d, obtained by a computer. These graphs indicate the existence of a critical dosage level d_{cr} above which the staph infection is eliminated and below which it is not. From

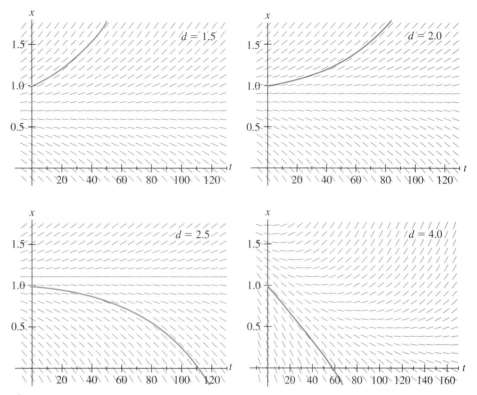

Figure 1.11 The slope field of the differential equation $x' = 0.02310x - 0.01d$ and the solution of the initial value problem (5.4) for selected values of the antibiotic dose d.

Fig. 1.11 this critical dose lies between 2.0 g and 2.5 g. Further computer explorations, using other values of d, suggest this critical value is approximately $d_{\mathrm{cr}} = 2.31$ g.

Another way to determine the critical value is to reason as follows. For $d < d_{\mathrm{cr}}$, the staph infection increases ($x' > 0$) and for $d > d_{\mathrm{cr}}$ it decreases ($x' < 0$). Therefore, at the critical dose $d = d_{\mathrm{cr}}$ the infection should do neither, but instead remain constant. From the initial value problem (5.4), we see that x remains at $x(0) = 1$, and hence $x' = 0$, means

$$0.02310 - 0.01d_{\mathrm{cr}} = 0$$

or

$$d_{\mathrm{cr}} = 2.31.$$

At the critical dose d_{cr} the staph infection remains constant, but at a higher dose $d > d_{\mathrm{cr}}$ our computer studies indicate that $x(t) = 0$ at some finite time t_c. This ("cured") time $t_c = t_c(d)$ when the infection is eliminated depends on d, as Fig. 1.11 shows. The higher the dose, the quicker the staph is eliminated; that is, $t_c(d)$ is a decreasing function.

We emphasize that computer explorations do not prove our conclusions about the existence of a critical dosage and the dependence of t_c on d. This is because, when doing computer studies, we can calculate only a finite number of solutions for only a finite selection of dosages d. An advantage of a solution formula, if available (or, if not, other methods of analysis) is that these conclusions can be rigorously established. (See Exercise 34 in Sec. 2.6 of Chapter 2.)

Often antibiotics are not continuously administered to a patient, but a dose is applied by pill or injection. In this case, the effect of the antibiotic is not constant, but decreases over time. To account for this change we return to the model equation (5.2) to see what adjustments must be made (this is the model modification step of the modeling cycle). To proceed we need information concerning how the effectiveness of the antibiotic changes over time, so that we can derive a formula for the staph removal rate h.

Suppose, for example, the effectiveness of the antibiotic decreases exponentially so that

$$h = 0.01de^{-at}.$$

Under this model assumption, the initial effectiveness of the antibiotic is $0.01d$ (cells/minute), but the effectiveness decreases over time with an exponential decay rate of $a > 0$. Suppose it is observed that the effectiveness decreases by 50% every hour. This allows us to calculate a. In 60 minutes, h is decreased by a fraction of $1/2$ and therefore

$$e^{-a60} = 0.5$$

or

$$a = 0.01155.$$

These assumptions lead us to a new initial problem for a staph infection starting with 10^6 cells:

$$x' = 0.02310x - 0.01de^{-0.01155t} \tag{5.5}$$
$$x(0) = 1.$$

(Recall that x is measured in units of 10^6.)

Again we ask, What dosages d, if any, will eliminate the staph infection?

Figure 1.12 shows the slope field and the solution of the initial value problem (5.5) for some selected values of the dose d. These samples suggest that this initial value problem also has a critical dosage d_{cr} below which the treatment does not eliminate the staph infection. The particular examples in Fig. 1.12 indicate that d_{cr} lies between 3.3 g and 3.5 g. (See Exercise 35 in Sec. 2.6 of Chapter 2.)

An interesting difference between the intravenous treatment modeled by (5.4) and the pill or injection treatment modeled by (5.5) occurs for doses below the critical level d_{cr}. Unlike the intravenous treatment, the pill or injection treatment can show an initial improvement (x initially decreases in Fig. 1.12 for $d = 2.5$ and 3.3) even though the infection ultimately "bounces back" and grows unabated. Thus, we must guard against a mistaken conclusion, based on its early effectiveness, that the treatment will result in a cure.

1.5.2 Running a Curve

One of the most famous laws of physics is Newton's law of motion given by the equation $F = ma$. Here m is the mass of a moving object and a is its acceleration. The letter F represents the force (or a collection of many forces $F = F_1 + F_2 + \cdots$) acting on the object. Since

$$a = \frac{dv}{dt} = \frac{d^2x}{dt},$$

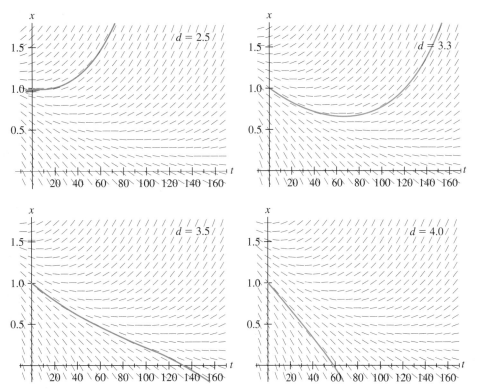

Figure 1.12 The slope field of the differential equation $x' = 0.02310x - 0.01de^{-0.01155t}$ and the solution of the initial value problem (5.5) for selected values of the antibiotic dose d.

where v is the object's velocity and x is its position (measured from some reference point), the application of Newton's law usually results in a differential equation that describes the motion of the object when subject to the force F. The modeling derivation step of the modeling cycle involves describing the relevant forces acting on the object so as to obtain a mathematical expression for F. We will utilize this law in a variety of applications throughout the book. In this section, we will use Newton's law to study a sprinter running a race of fixed distance.

One model of a sprinter running in a straight line assumes two forces are involved: the propulsive force exerted by the runner and a resistive force (due mainly to air resistance).[4] Thus, $F = F_p + F_r$ in Newton's law. In this model it is assumed that the runner exerts a constant propulsive force throughout the race, an assumption that seems reasonable for sprint of short distance. Thus,

$$F_p = mf,$$

where $f > 0$ is the per unit mass force characteristic of a particular individual runner. The resistive force, on the other hand, depends on the runner's velocity. It is absent when the runner is not moving and it increases with the runner's speed. The simplest law assumes the resistive force is proportional to velocity v; that is,

$$F_r = -cv,$$

---

[4]J. B. Keller, 1973. *Physics Today*, 26(9), p. 42.

where the coefficient of friction $c > 0$ is another characteristic of each particular runner. The reason for the negative sign is that the resistive force works against the runner.

In the absence of other forces, Newton's law yields the differential equation

$$m\frac{dv}{dt} = mf - cv$$

for the runner's velocity v. If we divide both sides by m and denote the per unit mass coefficient of friction c/m by σ, this equation becomes

$$\frac{dv}{dt} = f - \sigma v. \tag{5.6}$$

The model parameters f and σ can be approximated from performance records. For example, the parameters for 1968 Mexico City Olympics gold medalist Tommie Smith have been estimated to be $f = 13.46$ (Newtons/kg) and $\sigma = 1.252$ (per second).[5]

In races one is usually interested in the time it takes to run a fixed distance x_d from a starting line at $x = 0$ from which the runner's begin from a standing start ($v(0) = 0$). To determine this time from the initial value problem

$$\frac{dv}{dt} = f - \sigma v, \quad v(0) = 0$$

we calculate the runner's position from $v = dx/dt$, that is,

$$x(t) = \int_0^t v(s)\,ds,$$

and then solve the equation $x(t) = x_d$ for t.

For example, consider gold medalist Tommie Smith running a 100-m race. We can use a computer to approximate the solution of the initial value problem

$$\frac{dv}{dt} = 13.46 - 1.252v \tag{5.7}$$
$$v(0) = 0.$$

The algorithms studied in Secs. 1.2 and 1.3 produce approximations to the velocity $v(t)$ at points t_i (depending on the chosen step size s) lying in, say, the interval $0 \leq t \leq 12$. The resulting table of approximations for $v(t_i)$ permits us to approximate the distance

$$x(t_i) = \int_0^{t_i} v(s)\,dx$$

run at each point in time t_i by using an numerical integration procedure (for example, the trapezoid rule). Table 1.7 shows some of the results.

We make two observations from the numerical solution of the initial value problem (5.7). First, the model predicts that Tommie Smith could run 100 m in approximately 10.1 seconds (Table 1.7). Second, the model predicts that after about 6 seconds (55 m), Smith's velocity $v(t)$ is very nearly constant at 10.75 (m/s) for the rest of the race.

---

[5]A. Armenti, Jr., 1993. *The Physics of Sports*, American Institute of Physics, New York, pp. 105–108.

Table 1.7 Some results of applying Heun's algorithm (Sec. 1.3) with step size $s = 0.01$ to the initial value problem (5.7).

t	$x(t)$	$v(t)$
10.00	98.92	10.75
10.01	99.03	10.75
10.02	99.14	10.75
10.03	99.24	10.75
10.04	99.35	10.75
10.05	99.46	10.75
10.06	99.57	10.75
10.07	99.67	10.75
10.08	99.78	10.75
10.09	99.89	10.75
10.10	99.99	10.75
10.11	100.10	10.75
10.12	100.21	10.75

Some sprints are not run in a straight line but involve running a curve at the beginning of the race, with staggered starting positions for the racers. For example, this is the case for most 200-m races in which the course lies on a (circular) curve for 100 m before straightening out for the last 100 m.

When running along the curve, the sprinter's propulsive force must supply an additional centripetal acceleration, which depends on the radius of curvature of the curve. Furthermore, the radius of curvature is different for each lane. Lanes are typically 1.22 meters wide, and the inner radius of the nth lane is given by the formula

$$R(n) = \frac{100}{\pi} + 1.22\,(n - 1).$$

We will not delve into the physics of the derivation here, but leave it to say that the new equation motion that results when the additional force due to the centripetal acceleration is taken into account leads to the initial value problem[6]

$$\frac{dv}{dt} = \left(f^2 - \left(\frac{v^2}{R(n)}\right)^2\right)^{1/2} - \sigma v \tag{5.8}$$

$$v(0) = 0.$$

These are the equations of motion during the first 100 m of the 200-m race.

During the second 100 m of the race equation (5.6) is applicable. The initial condition associated with (5.6) would be the runner's velocity v_c at the end of the first 100 m (along the curve portion) of the race.

Thus, the model solution step of the modeling cycle involves, in this application, the numerical solution of the initial value problem (5.8) until the time t_c is reached at which $x(t_c) = 100$. At this time the runner's velocity is $v_c = v(t_c)$, which constitutes the initial condition for equation (5.6). This initial value problem is solved until the

---

[6]A. Armenti, Jr., 1993. *The Physics of Sports*, American Institute of Physics, New York, pp. 105–108.

finishing time for the runner is reached; that is, the time t_f at which $x\left(t_f\right) = 100$ (the second 100 m of the race).

We can, however, simplify the second step of the solution procedure as follows. It will turn out (as in the 100-m sprint example above) that the runner's velocity will reach a constant by the time the final 100-m portion of the race is reached. Therefore, rather than solve a second initial value problem using equation (5.6), we can obtain a good approximation to the time for the last 100 m by assuming a constant velocity v_f is maintained, in which case the final 100-m time is given by the formula $100/v_c$. The model predicted sprint time for the 200-m race is then

$$t_f = t_c + \frac{100}{v_c}. \tag{5.9}$$

Notice that all that is needed by the model to make a prediction for a sprinter's time in a 200-m race are the parameter values f and σ (obtained from the sprinter's performance data on straight courses) and the lane assignment n.

As an example we consider gold medalist Tommie Smith's performance in the 1968 Mexico City Olympics. In the 200-m finals Smith was assigned lane $n = 3$. Using $f = 13.46$, $\sigma = 1.252$ and $R(3) = 34.27$ we approximate the solution of (5.8) using Heun's algorithm.

From Table 1.8 we see that the model predicts Smith will run the first 100 meters along the curve in approximately $t_c = 10.33$ seconds and at the end of the curve his velocity will be approximately $v_c = 10.45$. From (5.9) we calculate the predicted time for Smith's 200-m sprint in lane $n = 3$ to be approximately

$$t_f = 10.33 + \frac{100}{10.45} = 19.90.$$

In fact, Smith ran the race in 19.83 seconds (at that time a world record).

Table 1.8 Some results of applying Heun's algorithm with step size $s = 0.01$ (Sec. 1.3) with step size $s = 0.01$ to the initial value problem (5.8) with $f = 13.46$, $\sigma = 1.252$, and $R(3) = 34.27$.

t	$x(t)$	$v(t)$
10.25	99.20	10.45
10.26	99.31	10.45
10.27	99.41	10.45
10.28	99.51	10.45
10.29	99.62	10.45
10.30	99.72	10.45
10.31	99.83	10.45
10.32	99.93	10.45
10.33	100.0	10.45
10.34	100.1	10.45
10.35	100.2	10.45
10.36	100.4	10.45
10.37	100.6	10.45

Table 1.9 Some results of applying Heun's algorithm with step size $s = 0.01$ (Sec. 1.3) with step size $s = 0.01$ to the initial value problem (5.8) with $f = 13.46$, $\sigma = 1.252$, and $R(1) = 31.83$.

t	$x(t)$	$v(t)$
10.25	98.85	10.40
10.26	98.96	10.40
10.27	99.06	10.40
10.28	99.17	10.40
10.29	99.27	10.40
10.30	99.37	10.40
10.31	99.48	10.40
10.32	99.58	10.40
10.33	99.69	10.40
10.34	99.79	10.40
10.35	99.89	10.40
10.36	100.0	10.40
10.37	100.1	10.40

We can use the model (5.8) to predict what might have been the result if Smith been given a different lane assignment. For example, the results in Table 1.9 for lane $n = 1$ show a slower predicted time of

$$t_f = 10.36 + \frac{100}{10.40} = 19.98.$$

Runners dislike lane 1 as being "too tight." The slower time predicted by the model for $n = 1$ bears out this opinion. Had Smith run in lane $n = 8$, however, his world record, according to the model, would have been even lower than 19.83 seconds. See Exercise 10.

Exercises

When we talk of a population's doubling time in the exercises below we imply that the population grows, in the absence of any limiting facts, with a constant per unit growth rate: $x' = rx$.

1. *E. coli* has a doubling time of approximately 20 minutes. Assume h cells per minute are removed from a culture initially at 10^8 cells. Use a computer to solve the initial value problem for the number of cells $x = x(t)$ at time t. Determine the critical value h_{cr} of h above which the culture will die out.

2. *E. coli* has a doubling time of approximately 20 minutes. Assume that he^{-at} cells per minute are removed from a culture initially at 10^8 cells. Use a computer to solve the initial value problem for the number of cells $x = x(t)$ at time t. Explore those values of a and h for which the culture goes extinct. Specifically, for selected values of a, calculate the critical value h_{cr} for h above which the culture goes extinct. Determine a relationship between a and h_{cr}.

3. The bacterium *M. Tuberculosis* has a doubling time of approximately 13 hours. Assume that h cells per hour are removed from a culture initially at 10^7 cells. Use a computer to solve the initial value problem for the number of cells $x = x(t)$ at time t. Determine the critical value h_{cr} of h above which the culture will die out.

4. The bacterium *M. Tuberculosis* has a doubling time of approximately 13 hours. Assume that he^{-at} cells per hour are removed from a culture initially at 10^7 cells. Use a computer to solve the initial value problem for the number of cells $x = x(t)$ at time t. Explore those values of a and h for which the culture goes extinct. Specifically, for selected values of a, calculate the critical value h_{cr} for h above which the culture goes extinct. Determine a relationship between a and h_{cr}.

5. In the model (5.2) suppose the effectiveness of the antibiotic decays more slowly than exponentially. Specifically, assume $h = 200d\,(1 + at)^{-1}$, where $a > 0$ is a constant.

Assume there is a 50% drop in effectiveness after 60 minutes.

(a) Modify the initial value problem (5.5) to account for this new assumption.

(b) Using slope fields and computer calculated solution graphs, determine whether or not this new model has a critical dosage value d_{cr} below which the infection is not controlled and above which the infection is eliminated.

No population can grow exponentially indefinitely. Many populations eventually decrease their rate of growth and level off at a number K appropriate to their environment and available resources. A differential equation often used to model this kind of growth is $x' = rx\,(1 - x/K)$, where r is the exponential growth rate at low population numbers. Suppose such a population is harvested at a constant rate h. Then the equation governing the populations growth is $x' = rx\,(1 - x/K) - h$. As an example, suppose the number of fish in a large lake grows according to this law. It is estimated that the lake can support $K = 10^4$ fish, and it is known that during the exponential growth phase (i.e., low population numbers) the fish population will double in two years. Use a computer to investigate the following question: At what maximal annual rate h_{cr} can the fish be harvested without causing extinction if initially there are the following numbers in the lake?

6. $x(0) = 10^2$ 7. $x(0) = 10^3$ 8. $x(0) = 10^4$

9. Investigate many initial conditions $x(0) > K/2 = 0.5 \times 10^4$. What do you notice about h_{cr}?

10. Calculate the model (5.8) predicted time for gold medalist Tommie Smith had he run in lane $n = 8$ of the 200-m finals in the 1968 Mexico City Olympics.

11. The current world record for 200 m of 19.32 seconds, set at the 1996 Atlanta Olympics, is held by Michael Johnson. In order for Tommie Smith to equal this time on a straight

course what higher value of the per unit mass propulsive form f would he have to attain?

12. The current world record for 200 m of 19.32 seconds, set at the 1996 Atlanta Olympics, is held by Michael Johnson. In order for Tommie Smith to better this time on a curved course in lane $n = 3$, what higher value of the per unit mass propulsive form f would he have to attain?

13. Estimated parameter values for sprinter Jim Hines are $f = 7.10$ (N/kg) and $\sigma = 0.581$ (per second). In a 100-m (straight line) race, who would win, Jim Hines or Tommie Smith?

14. Estimated parameter values for sprinter Jim Hines are $f = 7.10$ (N/kg) and $\sigma = 0.581$ (per second). In a 200-m (curve course), race who would win, Jim Hines in lane $n = 5$ or Tommie Smith in lane $n = 4$?

15. Who would win if Hines and Smith switched lanes in Exercise 14?

2

Linear First-Order Equations

There is no method that will succeed in calculating a formula for the general solution of every first-order differential equation $x' = f(t, x)$. We can find solution formulas only for certain kinds of equations [i.e., for equations with specialized right-hand sides $f(t, x)$]. Nonetheless, for several reasons it is important to learn solution methods for some specialized types equations, even though they are necessarily limited in scope. First, some types of equations arise often enough in applications that formulas for their general solutions are very useful. Second, certain types of equations serve as approximations to more complicated equations. Third, a study of various types of equations increases one's general understanding of differential equations. In this chapter we study one of the most important special types of differential equation, namely, linear differential equations.

A first-order equation $x' = f(t, x)$ is linear if the right-hand side $f(t, x)$ is linear as a function of x; that is, if

$$f(t, x) = p(t)x + q(t).$$

Notice that it is irrelevant how the *independent* variable t appears. What matters is that the *dependent* variable x, and its derivative x', appear linearly (i.e., to the first power and the first power only).

> **DEFINITION 1.** A **linear differential equation of first-order** has the form
>
> $$x' = p(t)x + q(t),$$
>
> where the **coefficients** $p(t)$ and $q(t)$ are defined on an interval $a < t < b$.

If the coefficients $p(t)$ and $q(t)$ are continuous on an interval $a < t < b$, then $f(t, x) = p(t)x + q(t)$ and its partial derivative $\partial f(t, x)/\partial x = p(t)$ are both continuous for t on the interval $a < t < b$ and for all x. Thus, the existence and uniqueness theorem (Theorem 1) in Chapter 1 applies to the initial value problem

$$x' = p(t)x + q(t), \quad x(t_0) = x_0$$

for any initial time t_0 from the interval $a < t < b$ and for any initial condition x_0. As a result, we know there exists a unique solution of this initial value problem on an interval containing t_0. As we will see (Sec. 2.1), it turns out that a stronger existence and uniqueness result holds for linear equations, namely that the solution exists on the *whole* interval $a < t < b$.

It is important to distinguish between linear equations in which the term $q(t)$ is present and those in which it is not.

DEFINITION 2. The linear equation $x' = p(t)x$, in which $q(t)$ is identically equal to 0, is **homogeneous**. If the term $q(t)$ is not identically equal to 0, the linear equation $x' = p(t)x + q(t)$ is **nonhomogeneous**. The term $q(t)$ is called the **nonhomogeneous term** (or **forcing function**).

In many applications a quantity changes at a rate proportional to the amount present at each moment of time. For example, this is true for a radioactive sample as it decays over time. Another example is the account balance of a continuously compounded investment. Yet another example is the growth of a bacterial culture in an environment of abundant resources. In all of these cases the rate of change x' is proportional to x (i.e., $x' = px$ for a constant of proportionality p). This is an example of a homogeneous, linear differential equation. Because the coefficient p is constant, the equation is also called *autonomous*. An *autonomous* linear equation is one in which both p and q are constants (so that the independent variable t does not appear in $f = px + q$). Otherwise, the equation is *nonautonomous*. Below is a list of linear equation that arise in applications that we will encounter in examples and exercises.

$$
\begin{aligned}
&\textbf{(a)}\ \ x' = -rx \\
&\textbf{(b)}\ \ x' = g - cx \\
&\textbf{(c)}\ \ x' = ae^{-bt}x \\
&\textbf{(d)}\ \ x' = a\left(b_{\mathrm{av}} + \alpha \sin\left(\tfrac{2\pi}{T}t\right) - x\right) \\
&\textbf{(e)}\ \ x' = c_{\mathrm{in}}\,r_{\mathrm{in}} - r_{\mathrm{out}}\tfrac{1}{rt+V_0}x,\ \text{where}\ r = r_{\mathrm{in}} - r_{\mathrm{out}}.
\end{aligned}
\tag{0.1}
$$

All coefficients are positive constants. Equation (a) is homogeneous and autonomous since $q(t) = 0$ and $p(t) = p$ are both constants. Since $p(t) = -c$ and $q(t) = g$ are constants, equation (b) is also autonomous. Equation (c) is homogeneous and nonautonomous since $p(t) = ae^{-bt}$ and $q(t) = 0$. In equation (d) $p(t) = -a$ and $q(t) = a(b_{\mathrm{av}} + \alpha \sin(2\pi t/T))$, and therefore this equation is nonhomogeneous and nonautonomous. Equation (e) is also nonhomogeneous and nonautonomous.

2.1 The Solution of Linear Equations

We learn in this section how to calculate formulas for solutions of linear equations. We can use the solution method to find formulas for general solutions and for solutions of initial value problems.

First we consider homogeneous linear equations. As a motivating example, we take the equation $x' = 2x$. As a first step we rewrite the equation as

$$x' = 2x$$
$$x' - 2x = 0$$

and then multiply by the exponential e^{-2t} to get

$$e^{-2t}x' - 2e^{-2t}x = 0.$$

Putting aside for the moment the reason why we chose this particular exponential, observe what results from this multiplication step. We can rewrite the left-hand side of the equation and obtain

$$\left(e^{-2t}x\right)' = 0.$$

It might not have immediately occurred to the reader to rewrite the left-hand side of the equation this way, but to check that it is correct is a straightforward application of the product rule from differential calculus. From this last equation, we conclude that the term $e^{-2t}x$ in the parentheses is a constant, so that

$$e^{-2t}x = c$$

and

$$x = ce^{2t}.$$

Since all of the equations in this list are equivalent in the sense that a solution x of one is a solution of the others, we have obtained a formula for all solutions (the general solution) of the differential equation.

We can apply the method used to find the general solution of the example above to obtain a formula for the general solution of any homogeneous equation. The key step is to determine an appropriate multiple to use in place of the exponential e^{-2t} that we used in the example.

We rewrite the homogeneous equation

$$x' = p(t)x$$

as

$$x' - p(t)x = 0$$

and multiply by a (yet to be determined) factor $e^{-P(t)}$

$$e^{-P(t)}x' - p(t)e^{-P(t)}x = 0 \tag{1.1}$$

so that the left-hand side can be rewritten as

$$\left(e^{-P(t)}x\right)' = 0. \tag{1.2}$$

Then $e^{-P(t)}x = c$ and $x = ce^{P(t)}$ is the general solution. However, how do we find such a factor? That is, how do we find $P(t)$ so that the left-hand sides of equations (1.1) and (1.2) are identical. We must find a $P(t)$ so that

$$e^{-P(t)}x' - p(t)e^{-P(t)}x = \left(e^{-P(t)}x\right)'.$$

Using the product rule for differential, we see this is the same as

$$e^{-P(t)}x' - p(t)e^{-P(t)}x = e^{-P(t)}x' - P'(t)e^{-P(t)}x.$$

After some algebra, we find that $P'(t) = p(t)$ and consequently that

$$P(t) = \int p(t)\,dt.$$

As a result we have found the formula

$$x = ce^{P(t)}$$

for the general solution of the homogeneous equation $x' = p(t)x$.

| EXAMPLE 1 | |

The autonomous homogeneous equation

$$x' = \left(-1.245 \times 10^{-4}\right) x$$

describes the radioactive decay of a sample $x = x(t)$ of radioactive carbon-14 (C^{14}). (See Exercises 20–23.) An integral of the coefficient $p(t) = -1.245 \times 10^{-4}$ is $P(t) = \left(-1.245 \times 10^{-4}\right) t$ and the general solution is given by the formula

$$x = ce^{\left(-1.245 \times 10^{-4}\right) t}.$$ ■

We can also apply the same procedure, used above for homogeneous equations, to nonhomogeneous linear equations. The steps are as follows.

$$x' = p(t)x + q(t)$$

$$x' - p(t)x = q(t)$$

$$e^{-P(t)}x' - p(t)e^{-P(t)}x = e^{-P(t)}q(t) \tag{1.3}$$

$$\left(e^{-P(t)}x\right)' = e^{-P(t)}q(t)$$

$$e^{-P(t)}x = c + \int e^{-P(t)}q(t)\,dt$$

$$x = ce^{P(t)} + e^{P(t)}\int e^{-P(t)}q(t)\,dt,$$

where c is an arbitrary constant. The factor

$$e^{-P(t)}, \quad P(t) = \int p(t)\,dt \tag{1.4}$$

is called an *integrating factor* for the differential equation. The procedure presented in (1.3) is called the *integrating factor method* for calculating the general solution of a linear differential equation.

It is worth pointing out that when applying the integrating factor method (1.3) one must be sure (in the second step) to multiply *both* sides of the equation by the integrating factor. It is a common mistake to forget to multiply the right-hand side $q(t)$ of the equation by the integrating factor.

Notice that, in the third step of the method, the term $e^{-P(t)}x$ differentiated on the left-hand side is the integrating factor times x. Keeping this in mind when using the method, we can sometimes spot an error made in calculating the integrating factor.

| EXAMPLE 2 | |

The integrating factor for the linear differential equation

$$x' = -3x + 2e^{-t}$$

is

$$e^{-P(t)} = e^{-\int(-3)\,dt} = e^{3t}.$$

The integrating factor method (1.3) for obtaining the general solution of the equation consists of the steps

$$x' + 3x = 2e^{-t}$$
$$e^{3t}x' + 3e^{3t}x = 2e^{3t}e^{-t}$$
$$\left(e^{3t}x\right)' = 2e^{2t}$$
$$e^{3t}x = c + e^{2t}$$
$$x = ce^{-3t} + e^{-t}.$$ ∎

There is more than one integrating factor for a differential equation. This is because the exponent $P(t)$ used to define the integrating factor (1.4) is not unique. An arbitrary constant arises when calculating the integral of $p(t)$, and we can use any $P(t) = \int p(t)\,dt + k$, for any constant k, to construct an integrating factor. However, all these integrating factors produce the same general solution. See Exercise 50.

For example, suppose in Example 2 that we choose $P(t) = \int(-3)\,dt + 2 = -3t + 2$ and use the integrating factor $e^{-P(t)} = e^{3t-2}$. The method (1.3) then yields

$$x' + 3x = 2e^{-t}$$
$$e^{3t-2}x' + 3e^{3t-2}x = 2e^{3t-2}e^{-t}$$
$$\left(e^{3t-2}x\right)' = 2e^{2t-2}$$
$$e^{3t-2}x = d + e^{2t-2}$$
$$x = de^{-3t+2} + e^{-t}$$
$$x = \left(de^2\right)e^{-3t} + e^{-t},$$

where d is an arbitrary constant. This formula for the general solution describes the same set of solutions as that found in Example 2, once we relabel the constant de^2 as c.

EXAMPLE 3

The velocity $v = v(t)$ of an object falling near the surface of the earth, subject to gravity and friction due to air resistance, satisfies the equation

$$v' = g - kv,$$

where k is the (per unit mass) coefficient of friction and g is the (constant) acceleration due to gravity. This is a linear (nonhomogeneous and autonomous) equation with coefficients $p(t) = -k$ and $q(t) = g$. An integral of $p(t)$ is $P(t) = -kt$ and an integrating factor is $e^{-P(t)} = e^{kt}$. Applying the integrating factor method (1.3), we obtain a formula for the general solution as follows:

$$v' + kv = g$$
$$e^{kt}v' + ke^{kt}v = ge^{kt}$$
$$\left(e^{kt}v\right)' = ge^{kt}$$
$$e^{kt}v = c + \frac{g}{k}e^{kt}$$
$$v = ce^{-kt} + \frac{g}{k}.$$

It is interesting to note that this formula implies $\lim_{t\to+\infty} v = g/k$, that is, the object approaches a limiting velocity g/k (until it hits the ground, of course). ∎

One way to solve an initial value problem

$$x' = p(t)x + q(t), \quad x(t_0) = x_0 \tag{1.5}$$

is to first find the general solution by using the integrating factor method (1.3) and then use the initial condition $x(t_0) = x_0$ to determine the arbitrary constant c in the general solution.

EXAMPLE 4   To find a formula for the solution of the initial value problem

$$x' = -3x + 2e^{-t}, \quad x(0) = 5$$

we can use the general solution

$$x = ce^{-3t} + e^{-t}$$

calculated by the integrating factor method (1.3) in Example 2. From this formula we find that

$$x(0) = c + 1$$

and therefore the initial condition requires $c + 1 = 5$ or $c = 4$. The solution of the initial value problem is

$$x = 4e^{-3t} + e^{-t}. \qquad\blacksquare$$

Another way to solve the initial value problem (1.5) is to incorporate the initial condition into the integrating factor method. In this way, the method finds the solution formula directly, without first calculating the general solution of the differential equation. This is done by using definite integrals in the method instead of general integrations (antidifferentiations) that involve arbitrary constants. Specifically, we use the specific integrating factor

$$e^{-P(t)}, \quad P(t) = \int_{t_0}^{t} p(s)\,ds$$

in the method.[1] Note that $P(t_0) = 0$ and $e^{-P(t_0)} = 1$. With this integrating factor we perform the calculations [just as in (1.3)]

$$x' = p(t)x + q(t)$$
$$x' - p(t)x = q(t) \tag{1.6}$$
$$e^{-P(t)}x' - p(t)e^{-P(t)}x = e^{-P(t)}q(t)$$
$$\left(e^{-P(t)}x\right)' = e^{-P(t)}q(t).$$

At this step we calculate definite integrals and use the fundamental theorem of calculus to get

$$e^{-P(t)}x(t) - e^{-P(t_0)}x_0 = \int_{t_0}^{t} e^{-P(u)}q(u)\,du \tag{1.7}$$

$$e^{-P(t)}x(t) = x_0 + \int_{t_0}^{t} e^{-P(u)}q(u)\,du.$$

---

[1]The integration variable in this formula for $P(t)$ is named s in order to distinguish it from the upper limit t.

This gives the formula

$$x(t) = x_0 e^{P(t)} + e^{P(t)} \int_{t_0}^{t} e^{-P(u)} q(u) \, du$$

for the solution of the initial value problem (1.5). This formula is called the *variation of constants formula.* It gives the solution of an initial value problem in terms of the coefficients $p(t)$ and $q(t)$ and the initial conditions t_0 and x_0.

THEOREM 1   If the coefficient $p(t)$ and nonhomogeneous term $q(t)$ are continuous on an interval $a < t < b$ containing t_0,[2] then the unique solution of the initial value problem

$$x' = p(t)x + q(t), \quad x(t_0) = x_0$$

is given by the variation of constants formula

$$x(t) = x_0 e^{P(t)} + e^{P(t)} \int_{t_0}^{t} e^{-P(u)} q(u) \, du, \qquad (1.8)$$

where

$$P(t) = \int_{t_0}^{t} p(s) \, ds.$$

We can use the formula (1.8) to calculate a solution formula for an initial value problem. However, most people prefer instead to calculate solutions using the integrating factor method [i.e., by performing the steps (1.6)–(1.7) used to arrive at the formula]. Here is an example.

EXAMPLE 5   Consider the initial value problem

$$v' = 9.8 - 1.128v, \quad v(0) = 0. \qquad (1.9)$$

The linear differential equation is a special case of the equation in Example 3.
    This initial value problem arises in a laboratory experiment involving an object falling under the influence of gravity. See Sec. 3.6.2 in Chapter 3.
        Using

$$P(t) = \int_{t_0}^{t} p(s) \, ds = \int_{0}^{t} (-1.128) \, ds = -1.128t$$

and the integrating factor

$$e^{-P(t)} = e^{1.128t},$$

we perform the steps

$$v' + 1.128v = 9.8$$
$$e^{1.128t} v' + 1.128 e^{1.128t} v = 9.8 e^{1.128t}$$
$$\left(e^{1.128t} v \right)' = 9.8 e^{1.128t}.$$

---

[2]The continuity of p and q guarantees the differentiability of $P(t)$ and $x(t)$ on the interval $a < t < b$.

Next we calculate *definite* integrals from $t_0 = 0$ to t and use the fundamental theorem of calculus.

$$e^{1.128t} v(t) - 1 \cdot v(0) = \int_0^t 9.8 e^{1.128s} \, ds$$

$$e^{1.128t} v(t) - 0 = \frac{9.8}{1.128} e^{1.128s} \Big|_0^t$$

$$e^{1.128t} v(t) = \frac{9.8}{1.128} e^{1.128t} - \frac{9.8}{1.128}$$

Finally, we arrive at the solution formula

$$v(t) = \frac{9.8}{1.128} - \frac{9.8}{1.128} e^{-1.128t}.$$

Using this formula for the velocity, we can also calculate a formula for the distance $x(t)$ fallen in time t:

$$x(t) = \int_0^t v(s) \, ds$$

$$= \int_0^t \left(\frac{9.8}{1.128} - \frac{9.8}{1.128} e^{-1.128s} \right) ds$$

$$= \frac{9.8}{1.128} t + \frac{9.8}{1.128^2} e^{-1.128t} - \frac{9.8}{1.128^2}$$

$$\approx 8.688t + 7.702 e^{-1.128t} - 7.702. \qquad \blacksquare$$

EXAMPLE 6

According to Newton's law of cooling, the temperature $x = x(t)$ of an object, residing in an environment of temperature b, satisfies the nonhomogeneous linear equation

$$x' = a (b - x),$$

where coefficient a is positive. Suppose x_0 is the initial temperature of the object. To solve the initial value problem $x(0) = x_0$ we can use either the variation of constants formula (1.8) or the integrating factor method. In either case, $p(t) = -a$ and $P(t) = \int_0^t (-a) \, ds = -at$.

Using formula (1.8), we obtain the solution formula by substituting $P(t) = -at$ and $q(t) = ab$:

$$x = x_0 e^{-at} + e^{-at} \int_0^t e^{as} ab \, ds$$

$$= (x_0 - b) e^{-at} + b.$$

If instead we use the integrating factor method, the solution steps are

$$x' + ax = ab$$
$$x'e^{at} + axe^{at} = abe^{at}$$
$$\left(e^{at}x\right)' = abe^{at}$$
$$e^{at}x(t) - 1 \cdot x(0) = \int_0^t abe^{as}\, ds$$
$$e^{at}x(t) - x_0 = be^{as}\Big|_0^t$$
$$e^{at}x(t) - x_0 = be^{at} - b \cdot 1$$
$$x(t) = (x_0 - b)\, e^{-at} + b.$$

Note $\lim_{t \to +\infty} x(t) = b$ (since $a > 0$). Thus, Newton's law of cooling predicts, in the long run, that the temperature of the object will (exponentially) approach that of its environment. ∎

For homogeneous linear equations ($q(t) = 0$), the variation of constants formula (1.8) gives the formula

$$x(t) = x_0 e^{P(t)}, \qquad P(t) = \int_{t_0}^t p(s)\, ds \qquad (1.10)$$

for the solution of the initial value problem

$$x' = p(t)x, \qquad x(t_0) = x_0.$$

EXAMPLE 7   The initial value problem

$$x' = ae^{-bt}x, \qquad x(0) = x_0 > 0$$

arises from a model of tumor growth (see Exercise 32). The constants a and b are positive and the size of the tumor, $x = x(t)$, is initially equal to x_0. From $p(t) = ae^{-bt}$ we calculate

$$P(t) = \int_0^t ae^{-bs}\, ds = \frac{a}{b}\left(1 - e^{-bt}\right).$$

Using formula (1.10), we obtain the solution formula

$$x = x_0 \exp\left(\frac{a}{b}\left(1 - e^{-bt}\right)\right).$$

Note this model implies that the tumor monotonically approaches a limiting size

$$\lim_{t \to +\infty} x(t) = x_0 \exp\left(\frac{a}{b}\right).$$ ∎

As a final observation, we note that the variation of constants formula (1.8) shows that the solution of a linear differential equation exists on the whole interval on which the coefficients $p(t)$ and $q(t)$ of the equation are continuous. This is because the continuity of $p(t)$ and $q(t)$ on an interval $a < t < b$ guarantees the integrals appearing in the formula define differentiable functions on the interval $a < t < b$.

◆ **Corollary 1**   If the coefficient $p(t)$ and nonhomogeneous term $q(t)$ are continuous on an interval $a < t < b$ containing t_0, then the solution of the initial value problem $x' = p(t)x + q(t)$, $x(t_0) = x_0$, exists on the whole interval $a < t < b$.

In equations (a)–(d) in (0.1), the coefficients $p(t)$ and nonhomogeneous terms $q(t)$ are continuous for all t and, by this corollary, solutions of initial value problems are defined for all t (i.e., on the interval interval $-\infty < t < +\infty$). In equation (e), $p(t) = (rt + V_0)^{-1}$ is continuous except at $t = -V_0/r$ and, as a result, solutions are defined for all $t > -V_0/r$ (if $t_0 > -V_0/r$) or for all $t < -V_0/r$ (if $t_0 < -V_0/r$).

Exercises

Which of the following equations are linear and which are non-linear? For those equations that are linear, identify the coefficient $p(t)$ and nonhomogeneous term $q(t)$.

1. $x' - t^2 x = t$

2. $x' + t^2 x = t^2$

3. $x' - tx^2 = t^2$

4. $x' + x = \sin t$

5. $x' = tx + \sin x$

6. $x' = -e^t x + \sqrt{x}$

7. $xx' = 2x + 1$

8. $x \sin t + \dfrac{1}{t} - x' = 1$

9. $t^2 x - \cos 3t + x \sin t - 5x' = \dfrac{1}{t^2 + 1}$

10. $t^2 x - \cos 3t + t \sin x - 5x' = \dfrac{1}{t^2 + 1}$

Which of the following equations are linear homogeneous? Which are linear nonhomogeneous? Which are nonlinear? For the linear equations, identify the coefficient $p(t)$ and nonhomogeneous term $q(t)$.

11. $x' = t^2 x - 1$

12. $x' = \dfrac{t + x}{t - 1}$

13. $x' - t^2 x = 0$

14. $x' = \dfrac{t + x}{t - x}$

15. $e^t x' = 3x - t$

16. $2tx' = -e^t x$

On what interval $a < t < b$ do the solutions of the following initial value problems exist? (Do not find formulas for solutions. Use Corollary 1.)

17. $x' = \dfrac{1}{t} x$, $x(1) = 1$ and $x(-1) = 1$

18. $x' = \dfrac{1}{t(t - 1)} x$, $x(0.5) = x_0$, $x(-1) = x_0$ and $x(2) = x_0$

19. $x' = (\tan t)x + t^2$, $x(0) = 0$ and $x\left(\frac{3\pi}{4}\right) = \frac{\pi}{2}$

20. $x' = \dfrac{1}{\sin 2\pi t} x + \dfrac{1}{\cos 2\pi t}$, $x\left(\dfrac{1}{8}\right) = 0$ and $x\left(\dfrac{15}{16}\right) = \pi$?

21. $x' = \dfrac{1}{\alpha^2 + t^2} x + \dfrac{1}{\alpha^2 - t^2}$, $x(t_0) = x_0$, where α is a positive constant and $t_0 \neq \alpha$.

22. $x' = (\ln t)x + \ln(t - \alpha)$, $x(t_0) = x_0$, where α and t_0 are positive constants.

23. Use a computer program to plot the graph of the solution of the initial value problem $x' = e^{t^2} x$, $x(0) = 10$. On what interval does the solution appear to exist? Apply Corollary 1 to determine on what interval the solution exists.

24. Use a computer program to plot the graph of the solution of the initial value problem $x' = tx + e^t$, $x(0) = 10$. On what interval does the solution appear to exist? Apply Corollary 1 to determine on what interval the solution exists.

25. Identify the coefficient $p(t)$ and nonhomogeneous term $q(t)$ for the equations of (0.1). Which equations are homogeneous and which are nonhomogeneous? Which equations are autonomous?

Find a formula for the general solution of the following linear homogeneous equations.

26. $x' = -3x$

27. $x' = -\frac{1}{2} x$

28. $x' = \dfrac{1}{t} x$

29. $x' = tx$

30. $x' = e^{-3t} x$

31. $x' = x \sin 2t$

32. $x' = \dfrac{t}{1 + t^2} x$

33. $x' = \dfrac{1}{t - t^2} x$

34. $\dfrac{x'}{x} = t \sin t$

35. $tx - e^t x' = 0$

Find a formula for the general solution of the following linear homogeneous equations. (a and b are constants.)

36. $x' = \frac{1}{a} x$, $a \neq 0$

37. $x' = bx \cos at$

38. $x' = e^{at} x$

39. $x' = \dfrac{a}{b + t} x$

Find a formula for the general solution of the linear nonhomogeneous equations below. (a and b are constants.)

40. $x' = -2x + 12$

41. $x' = 3x - 4$

42. $x' - tx = t$

43. $x' = \left(-\dfrac{1}{t}\right) x + \sin t$

44. $x' = ax + \cos bt$ **45.** $x' = ax + \sin bt$

46. $tx' = -x + t^{1/3}$ **47.** $tx' = -x - t^2$

48. $x' = tx - 1$ **49.** $x' = x \cos t + \sin t$

50. Prove that all integrals $P(t)$ lead to the same general solution x of $x' = p(t)x + q(t)$.

Find formulas for the solutions of the following initial value problems. (a and b are constants.)

51. $x' = \pi x$, $x(1) = -2$ **52.** $x' = -\dfrac{3x}{2}$, $x(-2) = 3$

53. $x' = \dfrac{x}{1+t^2}$, $x(1) = e^\pi$

54. $x' - (\sec^2 t)x$, $x\left(\frac{\pi}{3}\right) - e^{\sqrt{3}}$

55. $x' = (\sin at)x$, $x(0) = 1$, $a \neq 0$

56. $x' = \left(a + \dfrac{b}{t}\right)x$, $x(1) = 1$

Find formulas for the solutions of the following initial value problems. (a and b are constants.)

57. $x' = 3x - 2$, $x(0) = 5$ **58.** $x' = -2x + 6$, $x(0) = -1$

59. $x' = \frac{x}{t} + t^3$, $x(2) = 0$ **60.** $x' = -\frac{x}{t} + \sqrt{t}$, $x(1) = 1$

61. $x' = x \cos at + b \cos at$, $x(0) = 0$

62. $x' = 2atx + bt$, $x(0) = 0$

63. $x' = -\dfrac{x}{t} + \dfrac{2}{1+t^2}$, $x(1) = \ln 8$

64. Solve the initial value problem $x' = (b - d)x$, $x(0) = p_0$, for a population of size $x = x(t)$ that has a per capita birth rate $b > 0$ and a per capita death rate $d > 0$.

65. Show that the length of time it takes the solution of $x' = rx$, $x(t_0) = x_0 > 0$ (where $r > 0$) to double its initial size x_0 is independent of the initial condition x_0. Find a formula for this "doubling" time.

66. Show that the length of time it takes the solution of $x' = -rx$, $x(t_0) = x_0 > 0$ (where $r > 0$) to decrease by 50% is independent of the initial condition x_0. Find a formula for this "halving" time.

67. Suppose a population $x = x(t)$ naturally grows at an exponential rate r, i.e., $x' = rx$. However, suppose the population also is subject to harvesting (removal of individuals) at a constant rate $h > 0$. Then $x' = rx - h$. Find a formula for the solution of the initial value problem

$$x' = rx - h, \quad x(0) = x_0.$$

68. Suppose in Exercise 67 that the population is harvested periodically at the rate $h = h_{av} + \alpha \sin(2\pi t/T)$. Here $h_{av} > 0$ is the average rate over one harvesting period of length $T > 0$. Find a formula for the solution of the resulting initial value problem.

69. Suppose the temperature $x = x(t)$ of an object satisfies the nonhomogeneous (nonautonomous) linear equation

$$x' = a\left(b_{av} + \alpha \sin\left(\frac{2\pi}{T}t\right) - x\right), \quad a > 0.$$

This equation arises from Newton's law of cooling when the environmental temperature $b_{av} + \alpha \sin\left(\frac{2\pi}{T}t\right)$ oscillates sinusoidally with period T, average b_{av}, and amplitude α. Suppose x_0 is the initial temperature of the object.

(a) Find a formula for the solution $x(t)$ of the initial value problem $x(0) = x_0$.

(b) For large $t > 0$ and some constant θ, show

$$x(t) \approx b_{av} + \frac{Ta}{\sqrt{a^2T^2 + 4\pi^2}}\alpha \sin\left(\frac{2\pi}{T}t - \theta\right).$$

(c) Use the result in (b) to discuss the relationship between the oscillating temperature of the environment and that of the object.

70. (a) Use a computer to investigate the solutions of the equation $x' = e^{-0.5t}x$. What properties do all solutions have in common as $t \to +\infty$? What differences do solutions have as $t \to +\infty$?

(b) Find a formula for the solution of the initial value problem $x(0) = x_0$ and it to verify your observations in (a).

71. Consider the initial value problem

$$x' = (500 \sin 600\pi t)\, x, \quad x(0) = 1.$$

(a) Use a computer program to approximate $x(0.2)$.

(b) Use a computer program to graph the solution on the interval $0 \leq t \leq 0.2$. Describe the important features of the graph.

(c) Find a formula for the solution of the initial value problem and use it to calculate $x(0.2)$. Compare your answer with that obtained in (a).

(d) Describe the important features of the graph of the solution found in (c) and compare your answer with your description in (b).

72. Consider the initial value problem

$$x' = \left(\frac{1}{(t-1)^2}\tan\left(\frac{1}{t-1}\right)\right)x, \quad x(0) = 1.$$

(a) Use a computer program to approximate $x(0.35)$.

(b) Use a computer program to graph the solution on the interval $0 \leq t \leq 1$. Describe the important features of the graph. (The graph near $t = 1$ will be difficult to get, even with small step sizes, with the fourth-order Runge-Kutta algorithm. Try an adaptive algorithm [or other algorithms] if you have one available.)

(c) Find a formula for the solution of the initial value problem and use it to calculate $x(0.35)$. Compare your answer with that obtained in (a).

(d) Describe the important features of the graph of the solution found in (c) and compare your answer with your description in (b).

73. Consider the initial value problem

$$x' = \frac{1}{t}x + te^{-t}, \quad x(-1) = 0.$$

(a) Notice that the right-hand side of the differential equation is not defined for $t = 0$. Use a computer to obtain a graph of the solution on the interval $-1 < t < 0$. Describe the graph. In particular, what happens as $t \to 0-$?

(b) Use a computer to obtain approximations to the solution at $t = -0.1, -0.01, -0.001,$ and -0.0001. How do these approximations compare to your answer in (a)?

(c) Find a formula for the solution of the initial value problem and use it to compute $\lim_{t \to 0-} x(t)$. How does your answer compare with your answers in (a) and (b)?

74. Consider the initial value problem

$$x' = (\cos 60\pi t)x + 100 \cos 60\pi t, \quad x(0) = 0.$$

(a) Use a computer program to approximate the solution x at $t = 0.99$.

(b) Use a computer program to graph the solution on the interval $0 \le t \le 1$. Describe the important features of the graph.

(c) Find a formula for the solution of the initial value problem and use it to calculate $x(0.99)$. How does your answer compare with that in (a)?

(d) Use the formula obtained in (c) to explain the features of the graph found in (b).

75. Consider the equation $x' = -x - 15 \sin 5t + 3 \cos 5t$.

(a) Use a computer program to find a periodic solution. Do this by investigating solution graphs for many initial conditions $x(0) = x_0$. What is the approximate period and amplitude of this periodic solution? What relationship do all other solutions have to this periodic solution?

(b) Find a formula for the solution of the initial value problem $x(0) = x_0$.

(c) Use the formula found in (b) to show there is exactly one periodic solution. (*Hint:* Show that the formula gives a periodic function if and only if one special value of x_0 is chosen.) What is the period and amplitude of this periodic solution? Do your answers compare favorably to your answers in (a)?

2.2 Properties of Solutions

Suppose $x_p(t)$ is a particular solution of the nonhomogeneous equation

$$x' = p(t)x + q(t)$$

and suppose $x(t)$ is any other solution of this same equation. Let y denote the difference between these two solutions (i.e., $y = x - x_p$). Then

$$
\begin{aligned}
y' &= x' - x_p' \\
&= p(t)x + q(t) - \left[p(t)x_p + q(t)\right] \\
&= p(t)\left(x - x_p\right) \\
&= p(t)y.
\end{aligned}
$$

In other words, the difference y is a solution of the homogeneous equation

$$x' = p(t)x.$$

It follows that $y = x - x_p$ must be found in the general solution $ce^{P(t)}$, $P'(t) = p(t)$, of this equation. We conclude that *any* solution of the nonhomogeneous equation can be written in the form $x(t) = ce^{P(t)} + x_p(t)$ for some constant c.

THEOREM 2 The general solution of the nonhomogeneous linear equation

$$x' = p(t)x + q(t)$$

has the additive decomposition

$$x = x_h + x_p,$$

where

$$x_h = ce^{P(t)}, \quad P'(t) = p(t)$$

is the general solution of the associated homogeneous equation

$$x' = p(t)x$$

and x_p is **any** particular solution of the nonhomogeneous equation.

The additive decomposition $x = x_h + x_p$ of the general solution often provides a shortcut for its calculation.

STEP 1: Calculate the general solution $x_h = ce^{P(t)}$ of the homogeneous equation $x' = p(t)x$.

STEP 2: Find *any* particular solution x_p of the *nonhomogeneous* equation $x' = p(t)x + q(t)$.

STEP 3: Add the answers from Step 1 and Step 2 to get the *general* solution $x = x_h + x_p$.

The integrating factor method (1.3) provides a formula for x_p in Step 2:

$$x_p = e^{P(t)} \int e^{-P(t)} q(t)\, dt.$$

However, frequently there are shortcuts for finding a particular solution x_p. Sometimes there is even an "obvious" solution that can be found by inspection.

EXAMPLE 1 Consider the nonhomogeneous equation

$$v' = g - kv$$

for the velocity v of a falling object subject to gravity and a frictional force. (See Example 3 in Section 2.1.) The general solution of the associated homogeneous equation $v' = -kv$ is $v_h = ce^{-kt}$. The constant solution $v_p = g/k$ is found by inspection. Therefore, the general solution is the sum

$$v = ce^{-kt} + \frac{g}{k}. \qquad \blacksquare$$

Perhaps the constant solution in the preceding example ("found by inspection") would not have immediately occurred to the reader. Sometimes there are shortcuts for finding a particular solution x_p that are more systematic than simply guessing. In the next example we illustrate one such method. The method (which involves only algebraic calculations and thereby avoids having to calculate integrals) starts by making a "reasonable guess" for x_p.

EXAMPLE 2

Suppose a population $x = x(t)$ grows exponentially according to the equation $x' = x$. If we harvest this population at a rate $h(t)$, then

$$x' = x - h(t).$$

For example, if $h(t) = e^{-t}$, then we harvest the population at an exponentially decreasing rate, as time goes by. In this case we have the nonhomogeneous equation

$$x' = x - e^{-t}.$$

The homogeneous equation $x' = x$ associated with this equation has the general solution $x_h = ce^t$. To find a particular solution x_p of the nonhomogeneous equation, we reason as follows. If $x'_p - x_p$ is equal to e^{-t}, then x_p must somehow involve the exponential function e^{-t}. Suppose, for example, we try to find a solution x_p that is a constant multiple of e^{-t}. That is, suppose we try to find a constant k such that $x_p = ke^{-t}$ solves the differential equation. Since

$$x'_p = -ke^{-t}$$

and

$$x_p - e^{-t} = (k-1)e^{-t},$$

this exponential function is a solution if and only if $-k = k - 1$, i.e., $k = 1/2$. This choice for k yields the particular solution $x_p = e^{-t}/2$ and consequently the general solution

$$x = ce^t + \frac{1}{2}e^{-t}. \qquad \blacksquare$$

See Sec. 2.3 for a further discussion of shortcuts like the one used in this example.

If each function from a set of functions is multiplied by a (possibly different) constant and the resulting products added, the sum is called a *linear combination* of the functions. Thus, $k_1x_1(t) + k_2x_2(t)$, where k_1 and k_2 are constants, is a linear combination of the two functions $x_1(t)$ and $x_2(t)$. The sum $k_1x_1(t) + k_2x_2(t) + k_3x_3(t)$ is a linear combinations of the three functions $x_1(t)$, $x_2(t)$ and $x_3(t)$.

A fundamental fact about solutions of linear homogeneous equations is that a linear combination of solutions is itself a solution.[3] This fact is the "superposition principle" for homogeneous equation. To see why this is true, consider a linear combination $x(t) = k_1x_1(t) + k_2x_2(t)$ of two solutions $x_1(t)$ and $x_2(t)$ of the linear homogeneous equation

$$x' = p(t)x.$$

That $x_1(t)$ and $x_2(t)$ are solutions means

$$x'_1 = p(t)x_1, \quad x'_2 = p(t)x_2.$$

To show that the linear combination x is also a solution we calculate

$$\begin{aligned} x' &= k_1x'_1 + k_2x'_2 \\ &= k_1p(t)x_1 + k_2p(t)x_2 \\ &= p(t)(k_1x_1 + k_2x_2) \\ &= p(t)x. \end{aligned}$$

---

[3]In the terminology of linear algebra, the set of solutions of a homogeneous linear equation is a linear vector space.

A similar calculation shows that a linear combination of any number of solutions is a solution (Exercise 9).

The superposition principle for homogeneous equations does *not* hold for nonhomogeneous equations. Here is an example. The two functions

$$x_1(t) = 1, \quad x_2(t) = e^t + 1$$

are solutions of the nonhomogeneous equation $x' = x - 1$. However, the linear combination

$$x = x_1(t) + x_2(t) = e^t + 2$$

is not a solution (since $x' = e^t$ does not equal $x - 1 = e^t + 1$).

There is, however, a superposition principle for nonhomogeneous equations.

THEOREM 3   **Superposition Principle** Suppose, on an interval $a < t < b$, $x = x_1(t)$ is a solution of the equation

$$x' = p(t)x + q_1(t)$$

and $x = x_2(t)$ is a solution of the equation

$$x' = p(t)x + q_2(t).$$

Then the linear combination $x = k_1 x_1(t) + k_2 x_2(t)$ solves the equation

$$x' = p(t)x + [k_1 q_1(t) + k_2 q_2(t)]$$

on the interval $a < t < b$ for any constants k_1 and k_2.

The student is asked to prove this theorem in Exercise 10.

Notice that all three nonhomogeneous equations in this theorem have the same associated homogeneous equation $x' = p(t)x$ and differ only in their nonhomogeneous terms $q_1(t)$, $q_2(t)$ and $k_1 q_1(t) + k_2 q_2(t)$. If $q_1(t) = q_2(t) = 0$ in Theorem 3, we obtain the superposition principle for homogeneous equations.

Theorem 3 concerns linear combinations of two solutions of two equations. An analogous principle holds for any number of solutions of any number of equations. For example, if $x_1(t)$, $x_2(t)$, and $x_3(t)$ are solutions of the three linear nonhomogeneous equations

$$\begin{aligned} x' &= p(t)x + q_1(t) \\ x' &= p(t)x + q_2(t) \\ x' &= p(t)x + q_3(t), \end{aligned} \tag{2.1}$$

respectively, then the linear combination

$$x = k_1 x_1(t) + k_2 x_2(t) + k_3 x_3(t)$$

solves the linear nonhomogeneous equation

$$x' = p(t)x + [k_1 q_1(t) + k_2 q_2(t) + k_3 q_3(t)]. \tag{2.2}$$

The same linear combination is made of the solutions as is made of the nonhomogeneous terms q_1, q_2, q_3. See Exercises 11 and 12.

The superposition principle is often useful when searching for a particular solution x_p of the nonhomogeneous equation. This is because it allows the equation to be "broken up" into a set of simpler nonhomogeneous equations, whose solutions can be "put back together again" (as a linear combination) to obtain a solution of the original nonhomogeneous equation. Here is an example.

EXAMPLE 3

Suppose in Example 2 the population is harvested at the rate

$$h(t) = 1 + 9e^{-t}.$$

Thus, the harvesting rate starts at $h(0) = 10$ and exponentially decreases over time to 1. This yields the nonhomogeneous equation

$$x' = x - 1 - 9e^{-t}$$

for the population size x.

The general solution of the associated homogeneous equation $x' = x$ is $x_h = ce^t$. To find the general solution of the nonhomogeneous equation we need only a particular solution x_p.

We begin by noting that the nonhomogeneous term

$$q(t) = -1 - 9e^{-t}$$

is as a linear combination $k_1 q_1(t) + k_2 q_2(t)$ of

$$q_1(t) = -1 \qquad \text{and} \qquad q_2(t) = -e^{-t}.$$

Specifically,

$$q = q_1 + 9q_2$$

(i.e., $k_1 = 1$ and $k_2 = 9$). We can therefore use the superposition principle (Theorem 3) to construct a particular solution x_p by forming the same linear combination of solutions of the equations

$$x' = x - 1$$
$$x' = x - e^{-t}.$$

The first equation has the constant solution $x_1 = 1$. From Example 2, $x_2 = e^{-t}/2$ is a solution of the second equation. Using the superposition principle, we construct a particular solution as the linear combination

$$x_p = x_1 + 9x_2,$$

that is,

$$x_p = 1 + 9\left(\frac{1}{2}e^{-t}\right).$$

Finally, the general solution is the sum $x = x_h + x_p$, or

$$x = ce^t + 1 + \frac{9}{2}e^{-t}.$$

■

Exercises

1. Given that $x_p = t^{100} e^t$ solves $x' = x + 100 t^{99} e^t$, find the general solution of this equation.

2. Given that $x_p = \frac{t}{1+t}$ solves $x' = (\cos t) x + \frac{1}{1+2t+t^2}(1 - t \cos t - t^2 \cos t)$, find the general solution of this equation.

3. Given that $x_1 = 2e^{-3t}$ solves $x' = -x - 4e^{-3t}$ and that $x_2 = e^t$ solves $x' = -x + 2e^t$, find the general solution of the equation $x' = -x - 4e^{-3t} + 2e^t$.

4. Given that $x_1 = \sin t$ solves $x' = 2x + \cos t - 2\sin t$ and that $x_2 = \ln t$ solves $x' = 2x - 2\ln t + \frac{1}{t}$, find the general solution of the equation $x' = 2x + 2\cos t - 4\sin t + 2\ln t - \frac{1}{t}$.

5. Given that $x_p = 10$ solves the differential equation, find the solution of the initial value problem $x' = x - 10$, $x(0) = 5$.

6. Given that $x_p = e^t$ solves the differential equation, find the solution of the initial value problem $x' = e^{-t} x + e^t - 1$, $x(0) = 0$.

7. Given that $x_p = \frac{h}{r}$ solves the differential equation, find the solution of the initial value problem $x' = rx - h$, $x(0) = 0$.

8. Given that $x_p = 2 + \frac{a}{a^2+1}(a \sin t - \cos t)$ solves the differential equation, find the solution of the initial value problem $x' = -a(x - 2 - \sin t)$, $x(0) = 0$.

9. Suppose $x_1, x_2, \ldots, x_m$ are m solutions of the linear homogeneous equation $x' = p(t)x$. Show the linear combination $x = k_1 x_1 + k_2 x_2 + \cdots + k_m x_m$ is also a solution for any constants $k_1, k_2, \ldots, k_m$.

10. Prove Theorem 3

11. If $x_1(t)$, $x_2(t)$, and $x_3(t)$ are solutions of the three linear nonhomogeneous equations (2.1), respectively, show $x = k_1 x_1(t) + k_2 x_2(t) + k_3 x_3(t)$ is a solution of (2.2) for any constants k_1, k_2, k_3.

12. If $x_i(t)$ solves $x' = p(t)x + q_i(t)$ for $i = 1, 2, \ldots, n$, show that a linear combination $x = \sum_{i=1}^{n} k_i x_i(t)$ solves the equation $x' = p(t)x + \sum_{i=1}^{n} k_i q_i(t)$.

2.3 The Method of Undetermined Coefficients

In this section we study a shortcut method for finding a particular solution of nonhomogeneous equations with a constant coefficient $p(t) = p$ and special types of nonhomogeneous terms $q(t)$. The shortcut involves no antidifferentiation and is therefore easier to apply than the variation of constants formula. The method is also applicable to higher order equations and to systems of first-order equations (see Chapter 6).

The general solution of the linear equation

$$x' = px + q(t), \tag{3.1}$$

where p is a constant, has the form

$$x = ce^{pt} + x_p(t),$$

where c is an arbitrary constant and x_p is *any* particular solution of the equation. One way to find a particular solution x_p is to make a reasonable guess based on the nonhomogeneous term $q(t)$.

In Example 2, we sought an exponential solution x_p because $q(t)$ is an exponential function. Here is another example. Suppose $q(t) = q$ is a constant. If we seek a constant solution $x_p = k$ of equation (3.1), we find that

$$x_p' = 0 = pk + q$$

implies $k = -q/p$, that is, there is a constant solution $x_p = -q/p$ (provided $p \neq 0$). In these examples, the particular solution x_p is the same type of function as the nonhomogeneous term $q(t)$. The *method of undetermined coefficients* is a shortcut procedure based on this idea. This methods finds a particular solution x_p by making a "reasonable guess" according to the type of functions appearing in $q(t)$. (Some authors call it the *method of judicious guessing*.)

Here is another example of the method.

EXAMPLE 1

Suppose a population $x = x(t)$ grows exponentially according to the equation $x' = x$. If individuals immigrate or emigrate at a rate $q(t)$, then the population changes according to the equation

$$x' = x + q(t).$$

If $q(t) > 0$, then at time t individuals are being added to the population at the rate $q(t)$. If $q(t) < 0$, then at time t individuals are removed from the population at the rate $q(t)$.

Consider the case when immigration and emigration alternate periodically at the rate $q(t) = \cos t$. Then we have the linear nonhomogeneous equation

$$x' = x + \cos t$$

for x. The general solution has the form $x = ce^t + x_p$, where x_p is any particular solution. To find a particular solution x_p we first note that since the nonhomogeneous term $q(t)$ is the trigonometric function $\cos t$ it is natural to look for a particular solution that is similar, perhaps a multiple of $\cos t$. However, a quick look at the differential equation shows that a multiple of $\cos t$ is not a reasonable guess for a solution. To see this, consider what happens if a multiple of $\cos t$ is substituted into the equation: The left-hand side would be a multiple of $\sin t$ while the right-hand side would be a multiple of $\cos t$ and hence the two cannot be equal.

However, if a linear combination of $\sin t$ and $\cos t$ were substituted into the equation, then both sides would be linear combinations of $\sin t$ and $\cos t$ and their identity is possible. Specifically, consider the "guess"

$$x_p = k_1 \sin t + k_2 \cos t, \tag{3.2}$$

where k_1 and k_2 are "undetermined coefficients." The goal is to find k_1 and k_2 so that this linear combination is a solution of the nonhomogeneous equation; that is, so that

$$x_p' = k_1 \cos t - k_2 \sin t$$

is identical to

$$x_p + \cos t = k_1 \sin t + (k_2 + 1) \cos t.$$

Equating the coefficients of sine and cosine terms, we obtain

$$k_1 = k_2 + 1$$
$$-k_2 = k_1$$

an algebraic system of equations for k_1 and k_2 whose solution is

$$k_1 = \frac{1}{2}, \qquad k_2 = -\frac{1}{2}.$$

Using these coefficients in (3.2), we obtain the particular

$$x_p = \frac{1}{2} \sin t - \frac{1}{2} \cos t$$

and the general solution

$$x = ce^t + \frac{1}{2} \sin t - \frac{1}{2} \cos t.$$

∎

In the preceding example a guess for x_p based solely on $\cos t$ failed because, upon differentiation, it produces a term, $-\sin t$, on the left-hand side of the differential equation that is not present on the right-hand side. However, when we used a linear combination of $\cos t$ and its derivative $\sin t$ for x_p the method succeeded. The reason is that differentiation of such a linear combination introduces no new type of function. This is the basic idea behind the construction of a reasonable guess when using the method of undetermined coefficients.

The method proceeds as follows. A linear combination, with undetermined coefficients, is formed from $q(t)$ together with functions that are produced by repeated differentiations of $q(t)$. A requirement of the procedure is that only a finite number of different functions arise from repeated differentiations of $q(t)$. It turns out that this requirement is met by and only by products of sine or cosine functions, exponential functions, and/or polynomials (and linear combinations of such products).

Examples include t^2, te^{2t}, and $t \sin 3t$. The polynomial t^2 produces (up to constant multiples) the polynomials t and 1 under repeated differentiation. If $q(t)$ is a multiple of t^2, the method of undetermined coefficients suggests the guess $x_p = k_2 t^2 + k_1 t + k_0$ for a particular solution. The function te^{2t} produces (up to linear combinations) only one new function under repeated differentiation, namely e^{2t}, so when $q(t)$ is a multiple of te^{2t} the appropriate guess is $x_p = k_1 te^{2t} + k_2 e^{2t}$. The function $t \sin 3t$ produces (up to linear combinations) three new functions, $t \cos 3t$, $\sin 3t$, and $\cos 3t$, under repeated differentiation, so the appropriate guess is $x_p = k_1 t \sin 3t + k_2 t \cos 3t + k_3 \sin 3t + k_4 \cos 3t$ when $q(t)$ is a multiple of $t \sin 3t$.

The method of constructing a reasonable guess for a particular solution x_p can be formalized into sequence of systematic steps for nonhomogeneous terms $q(t)$ of the type described previously. See Exercises 1–10 and 11–22. However, the method is practical only if $q(t)$ is a product of two or three exponential functions, sines and cosines and/or low degree polynomials. The larger the number of undetermined coefficients, the less useful the method is as a shortcut.

| EXAMPLE 2 | The homogeneous equation $x' = 2x$ associated with the nonhomogeneous equation |

$$x' = 2x - 2te^t \sin t$$

has general solution $x_h = ce^{2t}$. The nonhomogeneous term $q(t) = -2te^t \sin t$ is a multiple of $te^t \sin t$. Using the method of undetermined coefficients, we construct a particular solution x_p from $te^t \sin t$ and all new functions (up to linear combinations) that arise by repeated differentiation of $te^t \sin t$; that is, $te^t \cos t$, $e^t \sin t$, and $e^t \cos t$. Thus, we search for a particular solution of the form

$$x_p = k_1 te^t \sin t + k_2 te^t \cos t + k_3 e^t \sin t + k_4 e^t \cos t.$$

Our goal is to calculate the four undetermined coefficients.

Equating the coefficients of like terms in

$$x_p' = (k_1 - k_2) te^t \sin t + (k_1 + k_2) te^t \cos t$$
$$+ (k_1 + k_3 - k_4) e^t \sin t + (k_2 + k_3 + k_4) e^t \cos t$$

and

$$2x_p - 2te^t \sin t = (2k_1 - 2) te^t \sin t + 2k_2 te^t \cos t$$
$$+ 2k_3 e^t \sin t + 2k_4 e^t \cos t,$$

we obtain the four linear algebraic equations

$$k_1 - k_2 = 2k_1 - 2$$
$$k_1 + k_2 = 2k_2$$
$$k_1 + k_3 - k_4 = 2k_3$$
$$k_2 + k_3 + k_4 = 2k_4$$

to solve for

$$k_1 = 1, \quad k_2 = 1, \quad k_3 = 0, \quad k_4 = 1.$$

Thus, a particular solution is

$$x_p = te^t \sin t + te^t \cos t + e^t \cos t$$

and the general solution $x = x_h + x_p$ is

$$x = ce^{2t} + te^t \sin t + te^t \cos t + e^t \cos t. \qquad \blacksquare$$

In all of the examples so far, the guess for x_p constructed by the method described above has succeeded in determining a particular solution of the nonhomogeneous equation. However, the method of undetermined coefficients *as described previously does not always work*. If one of the functions constructed from $q(t)$ and its derivatives is the exponential e^{pt} (i.e., is a solution of the associated *homogeneous* differential equation), then the method may not succeed in obtaining a solution x_p. In this event, the procedure must be modified as follows: All of the functions in the list derived from $q(t)$ are multiplied by t before they are used in the linear combination for x_p. See Exercises 11–23. Here is an example.

EXAMPLE 3

The general solution of the homogeneous equation $x' = x$ associated with the nonhomogeneous equation

$$x' = x + e^t$$

is $x_h = ce^t$. The nonhomogeneous term $q(t) = e^t$ produces only the exponential e^t under repeated differentiation.

If we were to follow the original procedure described previously, we would construct the guess $x_p = ke^t$. For this guess $x_p' = ke^t$ and $x_p + e^t = (k+1)e^t$, which requires $k = k + 1$, an impossibility. This means there is no solution of the form ke^t.

Following the modified method described previously, since e^t is a solution of the homogeneous equation we instead construct the guess $x_p = kte^t$. Substituting this guess into the differential equation, we obtain

$$x_p' = kte^t + ke^t$$
$$x_p + e^t = kte^t + e^t.$$

These terms are identical if (and only if) $k = 1$. The resulting particular solution $x_p = te^t$ yields the general solution $x = ce^t + te^t$. $\qquad \blacksquare$

The final example uses both the method of undetermined coefficients together with the superposition principle to find a particular solution.

EXAMPLE 4

The general solution of the homogeneous equation $x' = x$ associated with the nonhomogeneous equation

$$x' = x - a + b\cos t$$

is $x_h = ce^t$. Therefore, $x = ce^t + x_p$ is the general solution of this equation if x_p is a particular solution. To find a particular solution x_p we notice that

$$q(t) = -a + b\cos t$$

is a linear combination of $q_1(t) = -1$ and $q_2(t) = \cos t$; that is,

$$q(t) = a(-1) + b\cos t.$$

By the superposition principle the same linear combination of solutions x_1 and x_2 of the two nonhomogeneous equations

$$x' = x - 1$$
$$x' = x + \cos t$$

namely,

$$x_p = ax_1 + bx_2.$$

The first equation has the constant solution $x_1 = 1$. From Example 1 we find that the second equation has solution

$$x_2 = \frac{1}{2}\sin t - \frac{1}{2}\cos t.$$

Thus,

$$x_p = a\,(1) + b\left(\frac{1}{2}\sin t - \frac{1}{2}\cos t\right)$$

and the general solution is

$$x = ce^t + a + \frac{b}{2}\sin t - \frac{b}{2}\cos t. \qquad \blacksquare$$

Exercises

The method of undetermined coefficients applies to a linear equation $x' = px + q(t)$ when p is constant and $q(t)$ is a term of the form

$$t^m e^{at}\sin bt \qquad or \qquad t^m e^{at}\cos bt,$$

where $m \geq 0$ is a nonnegative integer and a, b are constants. [The method actually applies when $q(t)$ is a linear combination of functions of these types.] These functions (and only these kinds of functions) generate a finite number of independent functions when repeatedly differentiated. Notice that a and/or b are allowed to equal 0. Therefore, the method applies when $q(t)$ involves terms of the forms t^m or $t^m e^{at}$ or $e^{at}\sin bt$ and $e^{at}\cos bt$, or $t^m \sin bt$ and $t^m \cos bt$.

For each equation below, determine whether or not the method of undetermined coefficients applies.

1. $x' = 2x + t^2 e^{-t}$

2. $x' = -x + 2t^{-3}\sin 2t$

3. $x' = tx + \frac{1}{2}t\sin t$

4. $x' = 5x + t\cos t$

5. $x' = -\pi x + ae^{2t}$, a is a constant

6. $x' = x - 3e^t$

7. $x' = -2x - te^{-2t}$

8. $x' = e^t x - 5e^{2t}$

9. $x' = x - 3te^{bt}$, b is a constant

10. $x' = px + t^2 e^{bt}$, p and b are constants

When $q(t)$ has the form in the preceding exercise, the method of undetermined coefficients determines a guess for a particular solution x_p of $x' = px + q(t)$ as follows.

1. Construct the list of all independent functions obtained by differentiating $q(t)$ indefinitely.
2. If this list contains a solution of the associated homogeneous equation (i.e., if the list contains a term of the form e^{pt}), then multiply all functions in the list by t.
3. Formulate a "guess" for x_p by constructing a linear combination (with undetermined coefficients) of the functions from the list.

This procedure works for (and only for) those equations in which p is a constant and $q(t)$ is a multiple of terms of the form

$$t^m e^{at} \sin bt \quad or \quad t^m e^{at} \cos bt.$$

The finite list of independent functions obtained from a term of this types is

$$t^m e^{at} \sin bt, \ t^{m-1} e^{at} \sin bt, \dots, \ t e^{at} \sin bt, \ e^{at} \sin bt$$
$$t^m e^{at} \cos bt, \ t^{m-1} e^{at} \cos bt, \dots, \ t e^{at} \cos bt, \ e^{at} \cos bt.$$

Clearly this list is long if the exponent m is large. The method is only practical as a shortcut procedure if m is not too large. For such cases the list is usually more easily found by inspection than by using this formal list.

For the following equations

(a) construct the appropriate "guess" for a particular solution x_p and
(b) use the "guess" to find a particular solution of the equation. (p and a = constants.)

11. $x' = 0.5x - 0.3e^{0.2t}$ 12. $x' = 4x - 3e^{\pi t}$

13. $x' = 3x - 15e^{3t}$ 14. $x' = -2x - 3e^{-2t}$
15. $x' = -\frac{2}{3}x - \frac{15}{16}e^{-t}\sin t$
16. $x' = 0.1x + 2te^{-t}\cos t$
17. $x' = -x + 5t^4 \cos 2t$ 18. $x' = 2x - 3t^3 e^t \sin 5t$
19. $x' = px + \frac{1}{3}t^3 e^{at}$ 20. $x' = px - 14t^2 e^{at}$
21. $x' = x + 2\cos 2t$ 22. $x' = 2x + \sin 3t - \cos 3t$
23. For those equations in Exercises 1–10 to which the method of undetermined coefficients is applicable, find a particular solution $x_p(t)$.
24. Use the method of undetermined coefficients to find the general solution of the equations in Exercises 11–22.
25. Find the general solution of the equation
$$x' = 2x + ae^t + bte^{2t},$$
where a and b are constants.

We can combine the method of undetermined coefficients with the superposition principle when $q(t)$ is a linear combination of functions of the type $t^m e^{at} \sin bt$ and $t^m e^{at} \cos bt$. We use the superposition principle to decompose the differential equation into several new equations whose $q(t)$ terms are of these types. Next we find a particular solution of each of these equations by using the method of undetermined coefficients. We form the appropriate linear combination of these solutions to obtain a particular solution x_p of the original differential equation. Use this procedure to find a particular solution $x_p(t)$ of the following equations.

26. $x' = -x + 2e^t - 3\sin t$
27. $x' = -x - 0.5t + 3e^{2t}$
28. $x' = x + 3e^t - 4\cos t$
29. $x' = 2x - e^{2t} + 0.5e^t$

2.4 Autonomous Linear Equations

A linear equation is called *autonomous* if both $p(t)$ and $q(t)$ are constant functions:

$$x' = px + q, \tag{4.1}$$

where p and q are constants. Although autonomous linear equations are a specialized kind of linear equations, we will nonetheless discuss them briefly in order to introduce several concepts that will play important roles later in our study of nonlinear autonomous equations and systems. We focus on the *asymptotic dynamics* of autonomous equations (i.e., on the behavior of solutions as $t \to \pm\infty$). We need to distinguish between equations (4.1) with $p = 0$ and those with $p \neq 0$.

DEFINITION 3. The autonomous equation $x' = px + q$ is **hyperbolic** if $p \neq 0$. It is **nonhyperbolic** if $p = 0$.

We begin with two hyperbolic examples. Figure 2.1(a) shows graphs of several solutions of the equation

$$x' = -x + 1.$$

These graphs appear to have a horizontal asymptote at $x = 1$. That is, the solutions tend to 1 as $t \to +\infty$. The horizontal asymptote is itself the graph of a solution, namely, the constant solution $x = 1$. As $t \to -\infty$, the solutions in Fig. 2.1(a) appear to be unbounded. The formula $x = (x_0 - 1)\, e^{-t} + 1$ for the solution of the initial value problem $x(0) = x_0$ shows that these observations hold for all solutions of the equation.

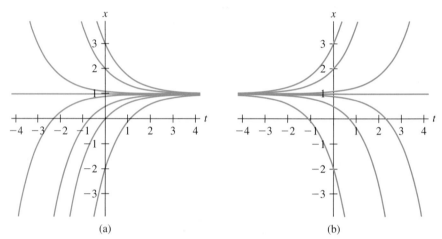

(a)   (b)

Figure 2.1 (a) Selected solutions of $x' = -x + 1$. (b) Selected solutions of $x' = x - 1$.

Figure 2.1(b) shows solution graphs for the equation

$$x' = x - 1,$$

where the asymptotic dynamics are reversed. Solutions appear to have a horizontal asymptote at $x = 1$ as $t \to -\infty$ and are unbounded as $t \to +\infty$. These facts hold for all solutions (except the constant solution $x = 1$), as you can see from the solution formula $x = (x_0 - 1)\, e^t + 1$.

As we will see, the two examples in Fig. 2.1 turn out to be typical for hyperbolic equations.

The general solution of the homogeneous equation $x' = px$ associated with equation (4.1) is $x_h = ce^{pt}$. To find the general solution, all we need is a particular solution. A hyperbolic equation (4.1) has a very special particular solution, namely the constant solution $x = -q/p$.

DEFINITION 4. A constant solution is called an **equilibrium solution** or simply an **equilibrium**.[4]

---

[4] An equilibrium is sometimes called a *rest point*, a *critical point*, or a *singular point*.

We will denote equilibrium solutions x_e. Using the equilibrium $x_e = -q/p$ as a particular solution, we can have the formula

$$x = ce^{pt} - \frac{q}{p}, \quad c = \text{arbitrary constant}$$

for general solution of (4.1). From this general solution we obtain the formula

$$x = \left(x_0 + \frac{q}{p}\right)e^{p(t-t_0)} - \frac{q}{p} \tag{4.2}$$

for the solution of the initial value problem $x(t_0) = x_0$.

Using (4.2), we discover the following general facts about hyperbolic linear equations. If $p < 0$, then all nonequilibrium solutions are strictly monotonic (i.e., strictly increasing or decreasing, depending on the sign of the coefficient $x_0 + q/p$) and tend to the equilibrium $x_e = -q/p$ as $t \to +\infty$. (This is because the exponential term is strictly monotonic and tends to 0 when $p < 0$.) In this case, the equilibrium is called a *sink*.

On the other hand, if $p > 0$, then all nonequilibrium solutions ($x_0 \neq -q/p$) are strictly monotonic and are exponentially unbounded as $t \to +\infty$. (This is because the exponential term is strictly monotonic and unbounded when $p > 0$.) In this case, the equilibrium is called a *source*.

THEOREM 4

If the linear autonomous equation

$$x' = px + q$$

is hyperbolic (i.e., if $p \neq 0$), then all nonequilibrium solutions are strictly monotonic (i.e., strictly increasing or decreasing). The equilibrium $x_e = -q/p$ is a sink if $p < 0$ and a source if $p > 0$.

Note that simply checking the sign of the coefficient p is sufficient to obtain a great deal of information about all solutions. It is not necessary to solve the differential equation in order to determine whether the equilibrium is a sink or a source. Here is an example involving a sink.

EXAMPLE 1

The equation

$$v' = g - kv$$
$$g > 0, \quad k > 0$$

for the velocity $v = v(t)$ of an object falling under the influence of gravity and a frictional force kv is linear and autonomous. Since the coefficient of friction $k > 0$ is positive, the equation coefficient $p = -k$ is negative and the equilibrium $v_e = g/k$ is a sink. Thus, all solutions $v(t)$ tend to v_e as $t \to +\infty$ in a strictly monotonic fashion. If, for example, the object is dropped, so that $v(0) = 0$, its velocity strictly increases to the "limiting velocity" v_e as $t \to +\infty$. ∎

The next example involves a source.

EXAMPLE 2

The linear autonomous equation

$$x' = rx - h$$
$$r > 0, \quad h \geq 0$$

describes the dynamics of a harvested population $x = x(t)$. The constant h is the rate at which the population is harvested. The population has a positive (per unit) growth rate r when unharvested. That is, when $h = 0$ the population grows exponentially according to the equation $x' = rx$. Since the coefficient $p = r$ is positive, the equilibrium $x_e = h/r$ is a source. This means all nonequilibrium populations are exponentially unbounded, growing without bound for initial population densities $x(0) > x_e$ and decreasing without bound for $x(0) < x_e$. In the latter case, $x(t)$ equals 0 and the population becomes extinct at some finite time $t > 0$. ∎

In the nonhyperbolic case, when $p = 0$, the autonomous equation (4.1) becomes

$$x' = q.$$

The solution of the initial value problem $x(t_0) = x_0$ associated with this equation is

$$x = q(t - t_0) + x_0.$$

By inspection of this formula we obtain the following theorem for the nonhyperbolic case.

THEOREM 5   If the linear autonomous equation (4.1) is nonhyperbolic (i.e., if $p = 0$), then all solutions are equilibrium solutions if $q = 0$ or are (linearly) unbounded if $q \neq 0$.

Theorems 4 and 5 completely account for the asymptotic dynamics of the linear autonomous equation (4.1). They show that *the asymptotic dynamics an autonomous linear equation can be determined from an inspection of the coefficient p and nonhomogeneous term q alone.* There is no need to solve the equation (i.e., find a formula for solutions).

The coefficient p plays a determining role. If $p < 0$, then all solutions tend to the equilibrium $x_e = -q/p$; on the other hand, if $p > 0$, then all nonequilibrium solutions tend away from the equilibrium and are unbounded. These facts are conveniently summarized in a graphical manner by a *phase line portrait*, which is drawn as follows. Locate the equilibrium x_e on an x-axis (called the phase space or, in this case, the phase line). The equilibrium point separates the phase line into two half lines. If $p < 0$, place two arrows on these half-lines that point toward the equilibrium. These directed half-lines are called *orbits*. The resulting picture is the phase line portrait of a sink. See Fig. 2.2(a). If $p > 0$, the two arrows are reversed and the resulting phase line portrait is that of a source. See Fig. 2.2(b). Phase line portraits are also important for nonlinear equations and are studied in more detail in Chapter 3.

The phase line portraits for the equations in the Examples 1 and 2 are

$$\longrightarrow \frac{g}{k} \longleftarrow \qquad \text{and} \qquad \longleftarrow \frac{h}{r} \longrightarrow,$$

respectively.

Imagine varying (or "tuning") the coefficient p from $-\infty$ to $+\infty$. For $p < 0$ the equilibrium is a sink. However, as soon as p is increased through 0 and becomes positive, the sink changes into a source and the arrows reverse their directions in the phase line portrait.

A drastic change in asymptotic dynamics as a coefficient passes through a critical value is called a *bifurcation*. The critical value of the coefficient is called a *bifurcation value*. In this context the coefficient is called a *bifurcation parameter*. With respect to

$\longrightarrow x_e \longleftarrow$

(a)

$\longleftarrow x_e \longrightarrow$

(b)

Figure 2.2 The phase line portraits of (a) a sink and (b) a source.

the bifurcation parameter p, the linear autonomous equation $x' = px + q$ undergoes a bifurcation at the bifurcation value $p = 0$ (q being held fixed). Notice that the equation is nonhyperbolic at this bifurcation value. This is not a coincidence; it is typically the case, as we will see in Chapter 3, where these bifurcation concepts are extended to nonlinear equations.

Exercises

Under what conditions (i.e., for what values of the parameter a) are the following equations hyperbolic? Under what conditions are they nonhyperbolic? Explain your answers. In all equations a is a constant.

1. $x' = x \sin a + 5a$ 2. $x' = x \cos a + a$

3. $x' = a^2 x + 5$ 4. $x' = (1 - a^2) x + \sin a$

Without solving the equations, find the equilibrium solutions and determine the asymptotic dynamics of all other solutions as $t \to +\infty$. (a is a constant.)

5. $x' = -5x - 7$ 6. $x' = -2x + 4$

7. $x' = 2x - 10$ 8. $x' = 6x + 7$

9. $x' = (a - 1)x + 2$ 10. $x' = (a^2 - 4)x - 1$

11. $x' = \sin a$ 12. $x' = \ln(a - 1), a > 1$

Describe what solutions do as $t \to -\infty$ when the linear autonomous equation $x' = px + q$ is

13. hyperbolic and the equilibrium is a sink

14. hyperbolic and the equilibrium is a source

15. is non-hyperbolic and $q \neq 0$

16. is non-hyperbolic and $q = 0$

17. A population $x = x(t)$ with per capita birth and death rates $b > 0$ and $d > 0$ satisfies the equation $x' = bx - dx$. Suppose we harvest the population at constant rate $h \geq 0$. Then $x' = bx - dx - h$. With respect to the parameter b, are there any bifurcation points and, if so, what are they?

18. Fix the coefficient $p < 0$ in the linear autonomous equation (4.1). Does the equation undergo a bifurcation (i.e., a qualitative change in the asymptotic dynamics of its solutions) as the nonhomogeneous term q is varied from $-\infty$ to $+\infty$? Explain your answer. Does your answer change if $p > 0$?

19. Use a computer program to study the asymptotic dynamics of the equation $x' = (a - e^{-a}) x + 1$, where $a > 0$ is a constant. Formulate conjectures about the dependence of the asymptotic dynamics on a. Are there any bifurcation points? Solve the equation and prove (or disprove) your conjectures.

20. Use a computer program to study the asymptotic dynamics of the equation $x' = (a^2 - 3a + 2) x + 1$, where $a > 0$ is a constant. Formulate conjectures about the dependence of the asymptotic dynamics on a. Are there any bifurcation points? Solve the equation and prove (or disprove) your conjectures.

2.5   Chapter Summary and Exercises

A linear first-order equation has the form $x' = p(t)x + q(t)$. The integrating factor method [which uses the integrating factor $e^{P(t)}$, $P'(t) = p(t)$] can be used to calculate a formula for the general solution of a linear equation, or for the solution of an initial value problem. In a general setting the method produces the variation of constants formula for solutions. The general solution has the additive decomposition $x = x_h + x_p$, where x_p is any particular solution and $x_h = ce^{P(t)}$ is the general solution of the related homogeneous equation $x' = p(t)x$. In some cases shortcut methods exist for finding a particular solution x_p, including the superposition principle and the method of undetermined coefficients. When p and q are constants, the equation is called autonomous. If $p \neq 0$, the equation is hyperbolic. If $p < 0$, all solutions tend, as $t \to +\infty$, to the equilibrium $x_e = -q/p$, which is then called a sink. If $p > 0$, all nonequilibrium solutions are unbounded and the equilibrium is called a source. These facts are summarized by phase line portraits.

Exercises

Find the general solution of the following equations. (a, b, and r are constants.)

1. $x' - atx = 0$

2. $x' = \left(\dfrac{1}{1+t^2}\right)x$

3. $x' + te^{-t^2/2}x = 0$

4. $x' = a(1 + \cos bt)x$

5. $x' = ae^{rt}x$

6. $x' = ate^{rt^2}x$

7. $x' = x \ln t$

8. $x' = ax \ln bt$

9. $x' = x \tan t$

10. $x' = (r + ae^{bt})x$

Find the general solution of the following equations by means of an integrating factor and by using the variation of constants formula. (a, b, and β are constants.)

11. $x' = x + \cos t$

12. $x' = -x + \sin t$

13. $x' = x \cos \beta t + \cos \beta t$

14. $x' = \dfrac{a}{t}x + b$

15. $x' = x \ln t + t^t$

16. $x' = \left(1 + \dfrac{1}{t}\right)x + t$

17. $x' = x + e^{at}$

18. $x' = x + te^{at}$

Find the general solution of the following equations using the method of undetermined coefficients.

19. $x' = x + \sin 10t$

20. $x' = -2x + te^{2t}$

21. $x' = -2x + te^{-2t}$

22. $x' = x + 2t^2 e^{-2t}$

Write down the appropriate guess for x_p from the method of undetermined coefficients for the following equations. Do not solve for x_p.

23. $x' = x - 2t^3 e^{-t} \cos 2t$

24. $x' = -x - 2t^3 e^{-t} \cos 2t$

Use the superposition principle together with the method of undetermined coefficients to find the general solution of the following equations.

25. $x' = x + 2\sin 10t + 3\cos t$

26. $x' = -2x + 3te^{2t} - 5te^{-2t}$

27. $x' = 3x + 2 + 3t - t^2 + 6t^3$

28. $x' = -x + 2 - 3e^{-t} + \cos t$

29. $x' = -x + 3\sin t + 2\sin 2t$

30. $x' = x - \cos 4t + 2\cos 2t$

Solve the following initial value problems.

31. $x' - tx = 0$, $x(5) = \pi$

32. $x' = \left(\dfrac{1}{1+t^2}\right)x$, $x(1) = 1$

33. $x' = x + \sin 10t$, $x(0) = 0$

34. $x' = x + \sin 10t$, $x(0) = x_0$

35. $x' = -2x + te^{2t}$, $x(0) = x_0$

36. $x' = -2x + te^{2t}$, $x(0) = -1$

Suppose the size of a tumor grows according to the following law: The per unit size rate of growth is a decreasing function of time $p(t)$. Then $x'/x = p(t)$ or $x' = p(t)x$. Solve the initial value problem $x(0) = x_0$ for each of the growth rates below. In all cases $a > 0$, $b > 0$ and p_0 are constants.

37. $p(t) = ate^{-bt}$

38. $p(t) = p_0 + ae^{-bt}$

39. $p(t) = \dfrac{a}{1+t}$

40. $p(t) = p_0 + \dfrac{a}{1+t}$

41. $p(t) = \dfrac{a}{1+t^2}$

42. $p(t) = p_0 + \dfrac{a}{1+t^2}$

43. A chemical substance is dissolved in a fluid contained in a container. Fluid flows into the container at a rate $r_{in} > 0$ and out of the container at rate $r_{out} > 0$. The concentration of the substance in the entering fluid is c_{in}. If we denote the amount of the substance in the container at time t by $x(t)$, then x satisfies the equation

$$x' = c_{in}r_{in} - \frac{x}{rt + V_0}r_{out},$$

where $r = r_{in} - r_{out}$ and $V_0 \geq 0$ is the initial amount of fluid in the container. If the container initially contains no chemical substance, then $x(0) = 0$. Suppose the volume of the container is 10 and the initial volume of fluid is $V_0 = 2$.

(a) Find the general solution when the volume of fluid in the container remains constant (i.e., $r_{in} = r_{out}$). Then solve the initial value problem. Use your answer to find the long term concentration in the container.

(b) Suppose the container is initially empty and the inflow rate is twice the outflow rate (i.e., $r_{in} = 2r_{out}$). Find the general solution. Then solve the initial value problem. What is the concentration at the moment the container is full? Is it more or is it less than the concentration in the incoming fluid?

Suppose a population $x = x(t)$ is grows (decays) exponentially according to the equation $x' = rx$. Suppose this population is subject to immigration (seeding) and emigration (harvesting) at rate $q(t)$. Interpret the immigration/emigration rate $q(t)$ and then find a formula for the solution of the initial value problem $x(0) = x_0$.

44. $q(t) = e^{-t} \sin t$

45. $q(t) = te^{-t}$

46. $q(t) = 1 + 2\cos t$

47. $q(t) = -1 + 2\cos t$

48. Consider the initial value problem

$$x' = \left(100 e^{\sin 40\pi t} \cos 40\pi t\right) x, \quad x(0) = 1.$$

(a) Use a computer program to approximate the solution x at $t = 2/3$.

(b) Use a computer program to graph the solution. Describe the important features of the graph.

(c) Solve the initial value problem and use the resulting formula to calculate x at $t = 2/3$. Compare your answer with the results in (a).

(d) Describe the important features of the graph of the solution found in (c) and compare your answer with your description in (b).

49. Consider the initial value problem

$$x' = 100 t \cos(100 t^2) x, \quad x(0) = 1.$$

(a) Use a computer program to approximate the value $x(1)$ of the solution at $t = 1$.

(b) Use a computer program to graph the solution on the interval $0 \le t \le 1$. Describe the important features of the graph.

(c) Solve the initial value problem and use the resulting formula for $x(t)$ to obtain $x(1)$. Compare your answer with that obtained in (a).

(d) Describe the important features of the graph of the solution found in (c) and compare your answer with your description in (b).

Without solving the equation, determine the asymptotic dynamics of each of the following autonomous equations. Draw the phase line portrait. In Exercises 56–59, in which a is a constant, find all bifurcation values.

50. $x' = -0.5x + 1$ **51.** $x' = x - 3$

52. $x' = -x + 2$ **53.** $x' = 0.01x - 1$

54. $x' = -\pi x + 7$ **55.** $x' = \frac{3}{2}x - e$

56. $x' = (2a - 1)x + 1$ **57.** $x' = -x + e^a$

58. $x' = (a^2 - 1)x + 1 + a$

59. $x' = (\ln a)x - \ln 2, \ a > 0$

60. (a) Use a computer program to investigate the asymptotic dynamics as $t \to +\infty$ of the equation $x' = (e^{-a} - a^2)x - 1, 0 < a < 1$. Describe how the asymptotic dynamics depend on the parameter a. Find (approximately) any bifurcation points for a.

(b) Formulate a conjecture about the phase line portraits as they depend on a.

(c) Prove or disprove your conjectures by utilizing the theorems in the Sec. 2.4.

For each equation find a periodic solution. Discuss the asymptotic dynamics as $t \to +\infty$ of all other solutions.

61. $x' = x + \sin 10t$

62. $x' = x + \sin 10t + \cos 10t$

63. $x' = -x + \sin 10t + \cos t$

64. $x' = -3x + 2 \sin t \cos t$

Find a formula for the general solution of the following equations in which p is a constant.

65. $x' = px + 2 \sin 2\pi t$ **66.** $x' = px - 3 \cos t$

67. Consider the initial value problem equation $x' = -ax + 0.5 \sin 2\pi t, x(0) = 0$, where $a > 0$ is a constant.

(a) Use a computer program to study the solution for selected values of a, ranging from $a = 0.1$ to 50. The solutions will approach periodic oscillations as $t \to +\infty$. Formulate conjectures about how the period, amplitude and phase of this oscillation depend on a. Relate these properties of the oscillation to the nonhomogeneous nonhomogeneous term $q(t) = 0.5 \sin 2\pi t$.

(b) Find a formula for the solution of the initial value problem.

(c) Use the formula in (b) to prove (or disprove) your conjectures in (a).

68. In this exercise you are asked to prove the linear equation $x' = px + q(t)$ has exactly one periodic solution of period T when $p \ne 0$ and $q(t)$ is a periodic solution of period T.

(a) Prove that $x(t)$ is a periodic solution if $x(0) = x(T)$. [*Hint*: Show that the function $y(t) = x(t + T)$ is a solution.]

(b) Use the variation of constants formula to prove that there exists exactly one initial condition x_0 for which the condition in (a) holds.

(c) Prove that $p < 0$ implies all other solutions tend to the unique periodic solution as $t \to +\infty$. Prove that $p > 0$ implies that no other solution tends to the unique periodic solution as $t \to +\infty$ (and are in fact exponentially unbounded).

(d) Apply these results to the equation $x' = (-0.3)x + (2 + \cos 2\pi t)^{-1}$. Use a computer program to study and then describe the periodic solution. Can you find a formula for the periodic solution?

69. A function $h(t)$ is called "*bounded*" for $t \ge t_0$ if there is a constant $M > 0$ such that

$$|h(t)| \le M \quad \text{for all } t \ge 0.$$

The constant M is called *a bound for* $h(t)$. Consider the initial value problem

$$x' = -x + q(t), \quad x(0) = x_0,$$

where the nonhomogeneous term $q(t)$ is bounded for $t \geq 0$. Show that the solution of this initial value problem is bounded for $t \geq 0$. (*Hint*: Use the variation of constants formula.)

2.6 APPLICATIONS

The three applications in this section involve linear, first-order differential equations. The first application concerns the onset of AIDS in HIV-infected individuals and utilizes a homogeneous linear equation. The second application, which deals with temperature changes of an object, involves a nonhomogeneous linear equation. The third application deals with the flow of a substance dissolved in a fluid and also involves a nonhomogeneous linear equation.

2.6.1 A Model of HIV/AIDS

Differential equations are widely used to model problems related to the spread of diseases in biological populations. In this section we consider a problem concerning the human immunodeficiency virus HIV. When antibodies to HIV are detected, a patient is said to be HIV positive. An infected patient eventually acquires AIDS (immunodeficiency syndrome). One significant feature of AIDS with regard to its spread through a population is its incubation period (i.e., the time between being diagnosed as HIV positive and exhibiting symptoms of AIDS). We consider a model for the course of the disease during this period.

Consider an initial population of x_0 individuals all of whom are HIV positive at time $t = 0$ but none of whom have symptoms of AIDS. Let $x = x(t)$ denote the number of individuals from this population who still do not have symptoms of AIDS at time t. Then $x(0) = x_0$ and we wish to determine $x = x(t)$ for $t > 0$.

Table 2.1

Years	Number Developing Symptoms of AIDS
1	4
2	15
3	21
4	26
5	23
6	12
7	2

Table 2.1 shows data from a population of 103 patients who obtained the HIV virus through blood transfusions. The number of patients showing symptoms of AIDS after 1, 2, 3, 4, 5, and 6 years are given (after 7 years all 103 patients had contracted AIDS).[5] From these data we can calculate, for each year, the number x of HIV-positive patients who have not yet acquired AIDS. In addition, we can calculate the average (per year per capita) rate of change of x. The results of these calculations appear in Table 2.2.

---

[5]Thomas A. Peterman et al., *Epidemiologic Reviews* 7 (1985), pp. 7–21.

Table 2.2

Years	Patients without AIDS	Average Per Capita Change
t_i	x_i	$\dfrac{x_{i+1} - x_i}{x_i}$
0	103	*
1	99	−0.0388
2	84	−0.1515
3	63	−0.2500
4	37	−0.4127
5	14	−0.6216
6	2	−0.8571

In Table 2.2 notice that the average per capita rate of change is not constant. In fact, it appears that the (negative) per capita rate of change increases in magnitude over time. This seems reasonable when it is remembered that the HIV virus progressively weakens the immune system, making a patient more and more susceptible to opportunistic diseases as time goes on. A model based on the assumption that the per capita rate of change is a constant, that is,

$$\frac{x'}{x} = -r,$$

where r is a positive constant, would fail to account for this observation. Instead, these data suggest we should build a model that assumes the per capita rate of change increases in magnitude with time. For example, we could write

$$\frac{x'}{x} = -r(t), \tag{6.1}$$

where $r(t) \geq 0$, the "rate of conversion" from HIV infection to AIDS, is an increasing function of t. To complete the model derivation step of the model cycle (Fig. 0.1 of the Introduction) we need a mathematical expression for $r(t)$.

We begin with the simple assumption that $r(t)$ is proportional to time t. We will investigate the resulting model and see how well its solution fits the data in Table 2.2. If the fit is unsatisfactory in some way, we can return to this assumption and modify it in some appropriate way (following the model modification step in the modeling cycle).

With $r(t) = kt$, where $k > 0$ is a constant of proportionality yet to be determined, we obtain from equation (6.1) the initial value problem

$$x' = -ktx \tag{6.2}$$
$$x(0) = x_0$$

to be solved for x. The differential equation in this initial value problem is linear, homogeneous, and nonautonomous [with $p(t) = -kt$ and $q(t) = 0$]. Using the solution method of Sec. 2.1, we obtain the formula

$$x(t) = x_0 \exp\left(-\frac{k}{2}t^2\right) \tag{6.3}$$

for the solution of the initial value problem (6.2). However, to use this formula we need a numerical values for the x_0 and the constant k. We will obtain these from the data in Table 2.2.

It is beyond the scope of the book to delve into the extensive, and very important, subject of parameter estimation from data. We will make use here, and throughout the rest of the book, of two methods: interpolation or least squares fit ("regression"). Which method we use will depend on convenience or on pedagogical reasons.

To estimate k from the data in Table 2.2, we can use a linear regression procedure. From the solution formula (6.3) we find

$$\ln \frac{x(t)}{x_0} = -\frac{k}{2}t^2.$$

This expression implies a linear relationship (with slope $m = -k/2$) between the variables

$$y = \ln \frac{x}{x_0} \quad \text{and} \quad s = t^2.$$

To estimate k we use the data in Table 2.2 to plot the points

$$y_i = \ln \frac{x_i}{103} \quad \text{and} \quad s_i = t_i^2$$

and fit them with a straight line. A standard way to find the best straight line fit is based on the so-called method of least squares. The procedure, called linear regression, is available on most computers and hand calculators. When applied to the data in Table 2.2, this procedure produces the straight line $y = ms$ with slope $m = -9.3825 \times 10^{-2}$. See Fig. 2.3.

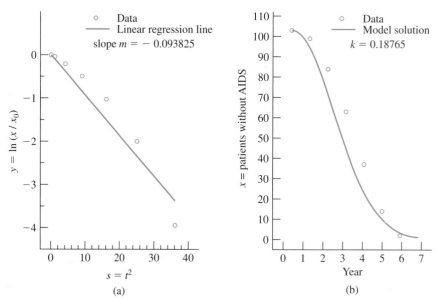

Figure 2.3 (a) A linear regression from the data in Table 2.2. (b) A fit of the model equation (6.2) to the data in Table 2.2.

Using the estimated value of $k = -2m = 0.18764$ in (6.3), we obtain the solution formula

$$x(t) = 103e^{-\left(9.3825 \times 10^{-2}\right)t^2}.$$

This formula gives the model predicted HIV population levels $x(t)$ for $t > 0$.

How well does the model solution fit the data in Table 2.2? You can use the plots in Fig. 2.3(b) to compare the model solution $x(t)$ with the data from Table 2.2. As you can see, the model fit is good near the beginning and the end of the seven-year time interval. The model fit does less well in the middle of the time interval where it consistently under predicts the number of HIV patients who develop AIDS. [Also notice that the data in Fig. 2.3(a) lie more often above, than below, the least squares linear fit.] Errors that are consistently too high or too low usually indicate that there is room for improvement in a model. (Errors that are uniformly distributed above and below the data might indicate sampling errors or other random errors and not a deficiency in the model.)

We will make an attempt to improve our model. As a generalization of the model considered previously, suppose $r(t)$ is proportional to an unspecified power of t. If we write $r(t) = -kt^p$ for some constant $k > 0$ and some power $p > 0$, then our model equations become

$$x' = -kt^p x \tag{6.4}$$
$$x(0) = x_0.$$

If the initial HIV population x_0 is known (as it is in the case of the data in Table 2.2), then there are two parameters k and p for which numerical values are needed. With this extra degree of freedom we expect to obtain a better fit to the data in Table 2.2.

The differential equation in the initial value problem (6.4) is linear, homogeneous, and nonautonomous [with $p(t) = -kt^p$ and $q(t) = 0$]. Using the solution methods of Sec. 2.1 we obtain the formula

$$x(t) = x_0 \exp\left(-\frac{k}{p+1}t^{p+1}\right) \tag{6.5}$$

for the solution. We will again use the data in Table 2.2 to determine values for the parameters in the solution, in this case the quantities x_0, k, and p. As before, we have $x_0 = 103$ directly from Table 2.2. We will determine values of k and p by interpolating two other data points from Table 2.2; we do this (rather than a regression procedure used above) in order to illustrate the interpolation method.

In general, there are several schemes for selecting those data points to be interpolated. For example, one could choose the data points at $t = 1$ and 2 [with $x_0 = 103$ thereby interpolating the first three years of data in Table 2.2 by the solution (6.5)]. The idea is to see how well the resulting solution predicts the future data for $t > 2$. Another scheme is to interpolate a selection of data points distributed over the whole time interval—for example, a point from the middle and end of the data set—in order to better fit (hopefully) the data throughout the time interval. We will opt here for the latter scheme and ask the reader to carry out the former scheme in Exercise 3.

Suppose we choose the data points at $t = 3$ and 6 for interpolation. That is, we require that the solution of the initial value problem (6.4) satisfy $x(3) = 63$ and $x(6) = 2$. The solution formula (6.5) yields the two equations

$$103 \exp\left(-\frac{k}{p+1}3^{p+1}\right) = 63$$

$$103 \exp\left(-\frac{k}{p+1}6^{p+1}\right) = 2$$

for the two unknowns k and p. These equations can be solved using a computer or, as it turns out, algebraically by hand. Rewriting the equations as

$$\frac{k}{p+1}3^{p+1} = -\ln\left(\frac{63}{103}\right)$$

$$\left(\frac{k}{p+1}3^{p+1}\right)2^{p+1} = -\ln\left(\frac{2}{103}\right),$$

(6.6)

we can substitute the first equation into the second equation to obtain

$$-\ln\left(\frac{63}{103}\right)2^{p+1} = -\ln\left(\frac{2}{103}\right)$$

and hence

$$p = -1 + \frac{\ln\left(\frac{\ln 2/103}{\ln 63/100}\right)}{\ln 2} \approx 2.093.$$

Using this number for p, we can determine k from the first equation in (6.6):

$$k = -\left(\frac{p+1}{3^{p+1}}\right)\ln\left(\frac{63}{103}\right) \approx 0.0509.$$

These values for p and k, placed into the formula (6.5), produce the formula

$$x(t) = 103\exp\left(-0.01644t^{3.093}\right)$$

(6.7)

for the solution that interpolates the three selected data points. See Fig. 2.4 for a graph of this improved fit of the data in Table 2.2.

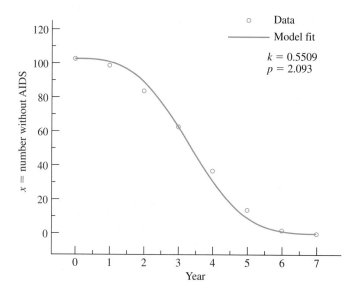

Figure 2.4 A fit of the model equation (6.2) to the data in Table 2.2 that interpolates the data at $t = 0, 3,$ and 6. The solution is given by the formula (6.7).

2.6.2 Newton's Law of Cooling

The temperature of an object placed in a cooler environment will, of course, decrease. Our goal is to predict the future temperature of the object from a knowledge of its initial temperature and that of its environment. To do this (i.e., to accomplish the model derivation step in the modeling cycle), we must accept some assumptions or laws about how temperature change occurs. *Newton's law of cooling* states that the rate of change of an object's temperature is proportional to the temperature difference between the object and the surrounding environment. (Although this is referred to as Newton's law of *cooling* it also applies to the *warming* of an object placed in a warmer environment.)

To translate Newton's law into mathematical equations, we let $x = x(t)$ denote the object's temperature at time t and let b denote the temperature of the surrounding environment. Then Newton's law states that $x' = k|x - b|$, where k is a constant of proportionality. The sign of x' must be the opposite that of $x - b$ because the object's temperature decreases when the object is warmer than the environment. Thus, k is negative if the object is cooling ($x - b > 0$) and positive if the object is warming ($x - b < 0$). A way to write both cases is

$$x' = -a(x - b), \quad a > 0. \tag{6.8}$$

The constant of proportionality a is determined experimentally from data. Its numerical value depends on various properties of the object; for example, its thermal conductivity (i.e., how well the material out of which the object is made conducts heat), its geometric shape, how uniformly it cools, and so on.

To determine the object's temperature as a function of time, using Newton's law of cooling (6.8), we also need to know the temperature x_0 of the object at an initial time t_0. The initial value problem

$$x' = -a (x - b), \quad x(t_0) = x_0 \tag{6.9}$$

is the mathematical description of our modeling assumptions. The differential equation is linear and nonhomogeneous. If a and b are constants, then the equation is also autonomous. We will assume a is a constant and consider two different assumptions for the environmental temperature, the case when it is held constant and the case when it fluctuates periodically. The first case holds, for example, in a temperature controlled room. The second case might be appropriate if the environmental temperature fluctuated on, say, a daily basis.

Constant Environment   Suppose both a and b are constants in the initial value problem (6.9). Using any one of the methods in this chapter (undetermined coefficients is perhaps the easiest), we carry out the solution step of the modeling cycle by calculating the solution formula

$$x(t) = (x_0 - b) e^{-a(t-t_0)} + b. \tag{6.10}$$

Once numerical values for t_0, x_0, a, and b are prescribed, this formula provides a prediction for the temperature of the object at any time t. Even without such numerical values, one interesting result follows from the formula. No matter what the parameter values are, we see that

$$\lim_{t \to +\infty} x(t) = b.$$

This conclusion of the model (as part of the interpretation step in the modeling cycle) is not so surprising. It states that the object's temperature approaches, in the long run, the temperature of its environment.

How well does our model describe real data? To examine the accuracy of Newton's law of cooling, consider the data in Table 3.3. These data are temperature readings taken from a cooking thermometer that the author heated in an oven to $100°\mathrm{F}$ and then left to cool on a counter top in a kitchen of constant temperature. Since $t_0 = 0$ and $x_0 = 100$, the solution (6.10) in this case is

$$x(t) = (100 - b) e^{-at} + b.$$

To use this solution to "fit" the data in Table 2.3, we need numerical values for the parameters a and b.

If the room temperature was measured during the experiment, then b would be known and we would have to estimate only a from the data. If the room temperature was not measured, which we assume is the case, then we must estimate both a and b from the data in Table 2.3. One way to do this is to interpolate two selected data points [i.e., we choose a and b so that the solution (6.10) passes through two data points from Table 2.3].

Table 2.3	
Time (minutes)	**Data °F**
0.0	100.0
0.5	94.5
1.0	90.0
1.5	87.0
2.0	84.5
2.5	83.5
3.0	82.0
3.5	81.0
4.0	80.5
4.5	80.0
5.0	80.0
5.5	79.5
6.0	79.5
6.5	79.0
7.0	79.0

In order to obtain a good fit over the duration of the experiment, we choose the last data point $x = 79$ at $t = 7$ and the middle data point $x = 81$ at $t = 3.5$ as the two points to be interpolated. Thus, we require $x(3.5) = 81$ and $x(7) = 79$. This means a and b must satisfy the two (nonlinear) algebraic equations

$$(100 - b) e^{-3.5a} + b = 81$$
$$(100 - b) e^{-7a} + b = 79.$$

With the help of a computer or calculator program we obtain the solutions $a \approx 0.6432$ and $b \approx 78.77$. Thus, Newton's law of cooling (when the three data points at $t = 0$,

3.5, and 7 are interpolated) yields the solution formula

$$x(t) = 21.23e^{-0.6432t} + 78.77.$$

The temperature predictions made by this formula, and the resulting errors, are tabulated in Table 2.4. A graph of this solution together with the data appears in Fig. 2.5. The fit is remarkably good.

Table 2.4

Time (minutes)	Data °F	Prediction °F	% Error
0.0	100.0	100	0
0.5	94.5	94.2	0.4
1.0	90.0	89.9	0.08
1.5	87.0	86.9	0.02
2.0	84.5	84.6	−0.2
2.5	83.5	83.0	−0.02
3.0	82.0	81.8	0.02
3.5	81.0	81.0	0
4.0	80.5	80.4	0.1
4.5	80.0	79.9	0.08
5.0	80.0	79.6	0.5
5.5	79.5	79.4	0.01
6.0	79.5	79.2	0.4
6.5	79.0	79.1	−0.1
7.0	79.0	79.0	0

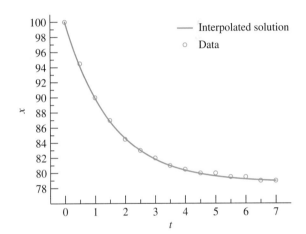

Figure 2.5

Incidentally, the estimated value of b in Newton's law of cooling provides a prediction of the kitchen's temperature during the experiment, namely $b = 78.77°F$. It turns out that the kitchen temperature was measured during this experiment and found to be $78.5°F$!

We can obtain another fit to the data in Table 2.2 using the fact that the kitchen temperature was measured to be $b = 78.5$ and estimating a by the interpolation of only one data point. See Exercise 7.

Fluctuating Environment Suppose the temperature of an object's environment is not constant. Suppose instead that the environmental temperature oscillates periodically, as might be the case, for example, of an object subject to daily temperature fluctuations. Then $b = b(t)$ is a periodic function of time in the equation (6.8) expressing Newton's law of cooling. This leads to the nonautonomous linear equation

$$x' = -a(x - b(t)) \tag{6.11}$$

for the object's temperature $x = x(t)$. Unlike for the case of constant environmental temperature b, this equation does not have an equilibrium solution. Our goal is to determine the long-term properties of solutions x when $b = b(t)$ is periodic.

We consider a special case only. We suppose the environmental temperature has regularly fluctuations that are reasonably well described by a sine function. Specifically, we take b to have the form

$$b(t) = b_{av} + \alpha \sin\left(\frac{2\pi}{T}t\right). \tag{6.12}$$

This expression for b represents a periodic oscillation around an average temperature of $b_{av} > 0$ with a period $T > 0$ and an amplitude α. Equation (6.11) now becomes the linear, nonhomogeneous equation

$$x' = -ax + a\left(b_{av} + \alpha \sin\left(\frac{2\pi}{T}t\right)\right). \tag{6.13}$$

The general solution of (6.13) has the form $x = ce^{-at} + x_p$, where c is an arbitrary constant and x_p is any particular solution. One way to find a particular solution is to use the method of undetermined coefficients. This method yields a solution that is a linear combination of 1, $\sin(2\pi t/T)$, and $\cos(2\pi t/T)$, specifically,

$$x_p(t) = b_{av} + \alpha \frac{aT^2}{a^2T^2 + 4\pi^2}\left(a \sin\left(\frac{2\pi}{T}t\right) - \frac{2\pi}{T}\cos\left(\frac{2\pi}{T}t\right)\right). \tag{6.14}$$

Using this particular solution, we have the formula

$$x(t) = ce^{-at} + b_{av} + \alpha \frac{aT^2}{a^2T^2 + 4\pi^2}\left(a \sin\left(\frac{2\pi}{T}t\right) - \frac{2\pi}{T}\cos\left(\frac{2\pi}{T}t\right)\right)$$

for the general solution of (6.13). The arbitrary constant c is determined by the object's initial temperature x_0 at $t_0 = 0$, namely,

$$c = x_0 - b_{av} + \alpha\frac{2\pi aT}{a^2T^2 + 4\pi^2}.$$

Having completed the model solution step of the modeling cycle, we can proceed to the model interpretation step and, using the solution formula, draw conclusions concerning the object's temperature.

Notice that $x - x_p = e^{-at}$ tends to 0 as $t \to +\infty$. As a result every solution of (6.13) approaches x_p as $t \to +\infty$, regardless of its initial condition x_0. See Fig. 2.6. We conclude that, in the long run, the object's temperature is given (approximately) by the formula (6.14).

Note that $x_p(t)$ is a periodic function of period T, just as the environment temperature $b(t)$ is. However, $x_p(t)$ and $b(t)$ are not identical, as the formulas (6.14) and (6.12) show. We conclude that the object's temperature, even in the long run, does not exactly match that of the environment.

To investigate the relationship between these two fluctuations more closely, it is convenient to rewrite the formula (6.14) for $x_p(t)$. Trigonometric terms in $x_p(t)$ can be written as a phase shifted sine function as follows:

$$\alpha\frac{aT^2}{a^2T^2 + 4\pi^2}\left(a \sin\left(\frac{2\pi}{T}t\right) - \frac{2\pi}{T}\cos\left(\frac{2\pi}{T}t\right)\right)$$

$$= \alpha\frac{aT}{\left(a^2T^2 + 4\pi^2\right)^{1/2}}\sin\left(\frac{2\pi}{T}t - \theta\right),$$

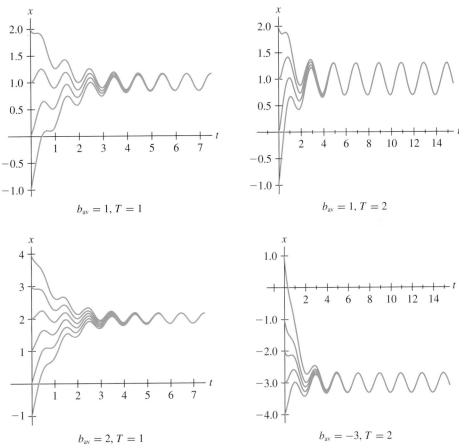

Figure 2.6   Some sample solutions of (6.13) with $a = 1$ and $\alpha = 1$.

where

$$\theta = \arctan\left(\frac{2\pi}{aT}\right).$$

Using this trigonometric identity, we rewrite the long-term temperature of the object as

$$x_p(t) = b_{av} + \alpha \frac{aT}{\left(a^2T^2 + 4\pi^2\right)^{1/2}} \sin\left(\frac{2\pi}{T}t - \theta\right).$$ (6.15)

From this formula we see, first of all, that $x_p(t)$ is a sinusoidal oscillation around the average b_{av}. Thus, the object's long-term temperature periodically fluctuates with the same period and average as the environmental temperature. These implications of the model probably seem reasonable and intuitive. Perhaps less intuitive, however, is the fact that the object's temperature fluctuations have a different amplitude and phase from that of the environmental temperature fluctuations. The amplitude of $x_p(t)$ in (6.15) is

$$\alpha \frac{aT}{\left(a^2T^2 + 4\pi^2\right)^{1/2}}.$$

Since

$$\frac{aT}{\left(a^2T^2 + 4\pi^2\right)^{1/2}} < 1,$$

we conclude that the amplitude of $x_p(t)$ is *less than* α, the amplitude of the environmental oscillation.

Moreover, the oscillation in $x_p(t)$ is out of phase with $b(t)$ by

$$\frac{T}{2\pi}\theta = \frac{T}{2\pi}\arctan\left(\frac{2\pi}{aT}\right)$$

time units. Since this phase difference satisfies

$$\frac{T}{2\pi}\arctan\left(\frac{2\pi}{aT}\right) \approx \begin{cases} 0 & \text{for } aT \text{ large} \\ \frac{T}{4} & \text{for } aT \text{ small}, \end{cases}$$

we see that the temperature oscillations of the object closely track those of the environment if aT is large. However, if aT is small, then the object's temperature oscillations are out-of-phase with the environmental oscillations by one-quarter of a period.

It is also interesting to note that the amplitude of $x_p(t)$ satisfies

$$\alpha\frac{aT}{\left(a^2T^2 + 4\pi^2\right)^{1/2}} \approx \begin{cases} \alpha & \text{for } aT \text{ large} \\ \alpha\frac{aT}{2\pi} & \text{for } aT \text{ small}. \end{cases}$$

See Figure 2.7.

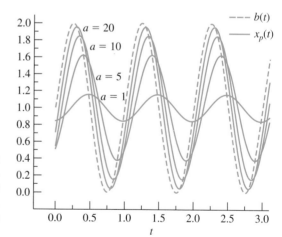

Figure 2.7 Plots of $b(t)$ given by (6.12) and the periodic solution $x_p(t)$ given by (6.14) with $b_{av} = \alpha = T = 1$. As a increases, $x_p(t)$ and $b(t)$ look more and more alike.

2.6.3 Flow of Dissolved Substances

Many important problems in science and engineering involve the movement or flow of a substance into and/or out of one or more compartments. A compartment may be a real physical container, such as a small beaker in a laboratory, a large vat in a brewery, a septic tank in a sewer system, or a natural lake. A compartment may also be a more physically complicated and diffuse assemblage, such as a soil bank, an underground water reservoir, or the circulatory system in the human body. The substance of interest may be, for example, a simple chemical compound (such as a dissolved salt or mineral), a large molecular biochemical (such as DNA, a medicinal drug, or polluting insecticide), or a biological organism (such as bacteria or unicellular animals, zooplankton, phytoplankton). In addition to the physical movement of the substance into and out of the compartment, chemical reactions or other processes (e.g., biological processes including reproduction, mortality) may occur in the compartment that add to or subtract from the quantity of substance. In general, we want to predict the amount (or concentration) $x = x(t)$ of the substance in the compartment as a function of time t. For example, we might want to know, Will the substance ultimately disappear from the compartment or will it persist indefinitely? If the substances persists, does it equilibrate or oscillate? What will the maximum and minimum amounts be and at what time(s) will these extrema occur?

Compartmental modeling is primarily based on the balance law

$$x' = \text{inflow rate} - \text{outflow rate.} \tag{6.16}$$

This law applies to each compartment involved in an application. When applying compartmental modeling procedures, the primary concern in the modeling derivation step (Fig. 0.1 of the Introduction) is the mathematical specification of all the inflow and outflow rates. Inflow and outflow mean not only the physical import of the substance into and out of each compartment but also the possible creation or destruction of the substance within a compartment by processes occurring inside that compartment (chemical reactions, birth and death processes, etc.).

In this section we consider applications involving only one compartment and, therefore, a single differential equation. We will consider problems in which a substance is only transported into and out of a compartment. See Chapter 3, Sec. 3.6.3, for some applications in which the substance also enters (or leaves) the compartment through chemical or biological reactions that take place in the compartment.

Consider a fluid, containing a dissolved substance, that flows into a container at a constant rate r_{in}. The contents are rapidly stirred and the mixture is pumped out of the container at a constant rate r_{out}. No processes occur in the container that either destroy or create the dissolved substance. The balance law (6.16) yields the equation

$$x' = \left\{ \begin{array}{l} \text{rate at which } x \text{ flows} \\ \text{into the container} \end{array} \right\} - \left\{ \begin{array}{l} \text{rate at which } x \text{ flows} \\ \text{out of the container} \end{array} \right\} \tag{6.17}$$

for the amount $x(t)$ of substance in the container at time t. To complete the model, we must mathematically describe the two rates on the right-hand side of the equation.

Suppose the inflowing fluid contains a concentration c_{in} of the dissolved substance (amount per unit volume). Then

$$\left\{ \begin{array}{l} \text{rate at which } x \text{ flows} \\ \text{into the container} \end{array} \right\} = \frac{\text{amount}}{\text{volume}} \times \frac{\text{volume}}{\text{time}} = c_{in} \, r_{in}.$$

Next we turn our attention to the outflow rate. Denote the fluid volume in the container by $V = V(t)$. Then the concentration in the container at time t is $x(t)/V(t)$ and

$$\left\{ \begin{array}{l} \text{rate at which } x \text{ flows} \\ \text{out of the container} \end{array} \right\} = \frac{\text{amount}}{\text{volume}} \times \frac{\text{volume}}{\text{time}} = \frac{x(t)}{V(t)} r_{out}.$$

From the balance law (6.17) we obtain the linear differential equation

$$x' = c_{in} r_{in} - \frac{r_{out}}{V(t)} x \tag{6.18}$$

for $x(t)$. The model equation will be complete when the volume $V(t)$ is determined. We consider two cases: the case when the volume is constant and a case when the volume changes over time.

Constant Volume   The volume V in the container will remain unchanged if the rate at which the fluid is removed r_{out} equals the rate at which it is added r_{in}. Let $r = r_{in} = r_{out}$ denote this common rate. Then equation (6.18)

$$x' = c_{in} r - \frac{r}{V} x \tag{6.19}$$

is linear and autonomous. Since the coefficient $-r/V$ is negative, the equilibrium $x_e = c_{in} V$ is a sink (Theorem 4). Thus, all solutions tend monotonically to x_e as $t \to +\infty$. As a result, the concentration in the container tends to $x_e/V = c_{in}$, which not surprisingly is the concentration of the inflowing fluid. Here is a specific example.

EXAMPLE 1

A salt-water solution with a concentration of 10 grams of salt per liter is added at a rate of 5 liters per minute to a 100-liter tank. The tank initially contains 50 liters of salt water with a concentration of 0.2 grams of salt per liter. The solution in the tank is rapidly mixed so that the solution has a uniform concentration at all times. The uniform mixture in the tank is also drained out at a rate of 5 liters per minute so as to maintain a constant volume of 50 liters of salt water in the tank at all time.

Let $x(t)$ be the amount of salt dissolved in the tank at time t. In the equation (6.19) for x we have the numerical values

$$c_{in} = 10 \text{ (grams per liter)}$$
$$r = r_{in} = r_{out} = 5 \text{ (liters per minute)}$$
$$V = 50 \text{ (liters)}$$

for the coefficients and the governing equation for x is

$$x' = -0.1x + 50.$$

All solutions of this linear equation tend to the equilibrium $x_e = c_{in} V = 500$ (grams). Thus, the concentration of salt water in the container tends to $x_e/V = c_{in} = 10$ (grams per liter).

The initial amount of salt in the container is $0.2 \times 50 = 1$. The solution of the resulting initial value problem

$$x' = 50 - 0.1x, \quad x(0) = 1$$

is

$$x(t) = -499 e^{-0.1t} + 500.$$

Using this formula, we can calculate the amount of salt in the tank at any time in the future.

For example, after one minute there are

$$x(1) = -499 e^{-\frac{1}{10}} + 500 = 48.49$$

grams of salt present [i.e., a concentration of $x(1)/50 = 0.9697$ grams per liter]. After ten minutes there are

$$x(10) = -499e^{-\frac{1}{10}10} + 500 = 316.4$$

grams present [i.e., a concentration of $x(10)/50 = 6.329$ grams per liter]. ∎

Nonconstant Volume   If the constant rates at which the fluid is pumped into and out of the container are *not* the same, that is, if

$$r_{in} \neq r_{out},$$

then the volume of fluid $V = V(t)$ in the container will not remain constant and equation (6.18) becomes nonautonomous. If $r_{in} < r_{out}$ the container will eventually empty. If $r_{in} > r_{out}$ the container will eventually fill.

The balance law (6.16) applied to the volume of fluid in the container yields the equation

$$V' = r_{in} - r_{out}$$

for V. This equation and equation (6.18) constitute a system of two equations

$$V' = r_{in} - r_{out}$$

$$x' = c_{in}r_{in} - \frac{r_{out}}{V}x$$

for two unknowns V and x. The system is uncoupled, however, in that the first equation does not involve x. Therefore, we can solve the first equation for V separately and use the answer in the second equation for x. Specifically, if V_0 is the initial amount of fluid in the tank, then the first equation yields

$$V(t) = rt + V_0, \tag{6.20}$$

where $r = r_{in} - r_{out} \neq 0$. Using this formula for V in the second equation of x, we obtain

$$x' = c_{in}r_{in} - \frac{r_{out}}{rt + V_0}x. \tag{6.21}$$

Since $r \neq 0$, this is a nonautonomous linear equation.

Suppose we consider the case $r < 0$ (i.e., $r_{in} < r_{out}$) when the container volume is decreasing. (See Exercise 31 for the case $r > 0$ when the container volume is increasing.) Solving

$$V(t) = rt + V_0 = 0$$

for t, we find that the container will be empty at time

$$t_{empty} = -V_0/r > 0.$$

Our problem is to determine the concentration of the substance at the time that the container becomes empty.

Some graphs of a few typical solutions of (6.21) for an example case appear in Fig. 2.8. Also plotted is the concentration $c(t) = x(t)/V(t)$. From these graphs it appears, in this example case anyway, that at the time that the container becomes empty its concentration is that of the inflowing fluid c_{in}. Is this in general true for solutions of the model equation (6.21)?

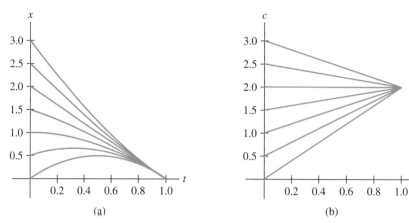

Figure 2.8 (a) Selected solutions of equation (6.21) with $c_{in} = 2$, $r_{in} = 1$, $r_{out} = 2$, $V_0 = 1$. (b) The concentration $x(t)/V(t)$ for the solutions in (a) is plotted.

Since equation (6.21) is linear, we can find a formula for the solution of the initial value problem $x(0) = x_0$ using the integrating factor (or the variation of constants formula). A formula for the solution is

$$x(t) = c_{in}(rt + V_0) + \left(\frac{rt + V_0}{V_0}\right)^{-r_{out}/r}(x_0 - c_{in}V_0). \qquad (6.22)$$

(See Exercise 29.) A formula for the concentration $c(t) = x(t)/V(t)$ is therefore

$$c(t) = c_{in} + \left(\frac{rt + V_0}{V_0}\right)^{-r_{in}/r}(x_0 - c_{in}V_0).$$

Since $-r_{in}/r > 0$, the second term on the right-hand side tends to 0 as $t \to t_{empty}$ and therefore, as we saw in the sample example in Fig. 2.8,

$$\lim_{t \to t_{empty}} c(t) = c_{in}.$$

Exercises

1. The rate $r(t)x$ at which HIV-positive patients become AIDS patients is the rate at which the AIDS epidemic sweeps through the HIV population and can be called the intensity of the epidemic. Using model (6.2), find formulas for the maximal intensity and the time at which it occurs. Show that the maximal intensity increases with k while the time at which it occurs decreases with k.

2. Show that the model (6.4) predicts a unique time $t_{max} > 0$ at which the rate of the AIDS epidemic peaks and this time is a decreasing function of k. Also show that the maximal rate which occurs at this time is an increasing function of k.

3. Determine values of k and p in (6.5), with $x_0 = 103$, so

that this solution interpolates the data at times $t = 1$ and 2. Make a table showing the model predicted values at each year and the percent error. Discuss your results.

4. In the models given in the text for the conversion of HIV-positive patients to AIDS patients, the population was assumed closed. That is, no new HIV-positive patients entered the population at any time. Suppose instead that new HIV-positive patients do enter the HIV-positive population and they do so at a rate proportional to the number of HIV-positive patients present at any given time. (This assumption is meant to model the infectiousness of HIV-positive individuals.)

Assume that the rate of conversion from HIV positive to AIDS is the same as in model (6.4) (i.e., is given by $-kt^px$ for positive constants $k > 0$ and $p > 0$). Let $r > 0$ be the constant of proportionality for the rate at which new HIV positives enter the population. Call r the infection rate. Under these assumptions we have the initial value problem

$$x' = -kt^px + rx, \quad x(0) = x_0$$

for the number $x = x(t)$ of HIV-positive patients at time $t > 0$, assuming there are $x_0 > 0$ present initially.

(a) Explore the solutions of these equations using a computer program for selected values of k, p, r, and x_0. Describe the features of x as a function of t (e.g., its maximum and minimum, its asymptotic dynamics, etc.). Based on your explorations, formulate conjectures about how these features depend on the infection rate r.

(b) What kind of differential equation appears in this model? Find the general solution and use it to solve the initial value problem.

(c) Use your solution from (b) to address your conjectures in (a). Are they true or false? Why?

(d) The intensity of the AIDS epidemic can be studied by means of the rate kt^px at which HIV patients become AIDS patients. Compute this rate, using your answer in (b), and graphically study it using a computer graphics program. In particular, study how the time $t_{max} > 0$ at which this rate is maximal depends on the infection rate r.

(e) Find an algebraic formula for $t_{max} > 0$ using the formula for kt^px obtained from your answer in (b). Use this formula to prove (or disprove) the your conclusion in (d) about how t_{max} depends on r.

5. Consider the model (6.4) for the conversion rate of HIV-positive patients to AIDS patients. Take $p = 3$.

(a) Using linear least squares, determine a value of k from the data in Table 2.2. (*Hint:* Use least squares on $\ln x_i$ versus t_i^4.)

(b) Use a computer program to compare the solution graph of the resulting initial value problem with the data in Table 2.2. Does the result appear to fit the data as well as the fits in Figs. 2.3 and 2.4?

(c) What is the formula for the solution of resulting initial value problem with $x_0 = 103$?

6. An object initially at $25°C$ is placed in a room where the temperature is held fixed at $20°C$. After 2 minutes the temperature of the object is measured at $24°C$.

(a) What will the temperature of the object be after 4 minutes?

(b) How long will it take the temperature of the object to reach $22°C$?

(c) What temperature will the object have in the long run?

7. Fit the data in Table 2.2 using Newton's law of cooling with $b = 78.5$ (°F) and the initial condition $x(0) = 100$. Determine a by interpolating the final data point $x = 79$ at $t = 7$. Compute the percent errors.

8. Find a and b so that the solution of Newton's law of cooling $x' = -a(x - b)$ interpolates the three data points $(t, x) = (t_0, x_0)$, $(t_0 + \delta, x_1)$, and $(t_0 + 2\delta, x_2)$, where x_0, x_1, and x_2 are three different numbers and $\delta > 0$. Is this possible for all such data points x_0, x_1, and x_2? If not, what constraints must be placed on these data points?

9. Assume b is known. Find a formula for a so that the solution of Newton's law of cooling $x' = -a(x - b)$ interpolates the two data points (t_0, x_0) and $(t_0 + \delta, x_1)$. Is this possible for all such data points x_0 and x_1? If not, what constraints must be placed on these data points?

10. The data in the following table were obtained by the author using a kitchen thermometer heated in an oven to $150°F$ and then observed to cool.

Time (minutes)	Data °F
0	150
1	109
2	94
3	85
4	83
5	81
6	80

(a) Use Newton's law of cooling to fit these temperature readings at $t = 0$, 3, and 6 minutes. What is the predicted room temperature? Compute the temperature predictions and their percent errors at the times in the table.

(b) Given that the room temperature during cooling was constant at $b = 79°F$ use Newton's law of cooling to fit these temperature readings at $t = 0$ and 6 minutes. Compute the temperature predictions and their percent errors at the times in the table.

(c) Compare the results in (a) and (b).

11. The data in the table below were taken by the author using a kitchen thermometer chilled in a freezer to $50°F$ and then observed to warm.

Time (minutes)	Data °F
0.0	50.0
0.5	56.0
1.0	60.5
1.5	64.5
2.0	68.0
2.5	71.0
3.0	72.5
3.5	74.0
4.0	75.0
4.5	75.5
5.0	76.0
5.5	76.5
6.0	77.0
6.5	77.5
7.0	77.5

(a) Use Newton's law of cooling to fit these temperature readings at $t = 0$, 3.5, and 7 minutes. What is the predicted room temperature? Compute the temperature predictions, their errors, and their percent errors at the times in the table.

(b) Given that the room temperature during cooling was constant at $b = 78.5°F$ use Newton's law of cooling to fit these temperature readings at $t = 0$ and 7 minutes. Compute the temperature predictions, their errors, and their percent errors at the times in the table.

(c) Compare the results in (a) and (b).

12. Consider Newton's law of cooling $x' = -a(x - b)$ for constants $a > 0$ and b. Suppose an object is initially hotter than its environment so that $x_0 > b$. As we know, the object's temperature $x(t)$ approaches the environmental temperature b as $t \to +\infty$. Therefore, in the limit the object cools a total of $x_0 - b$ degrees.

(a) Use a computer program to formulate a conjecture about how long it takes the object to cool 50% of the initial temperature difference $x_0 - b$. How does this time depend on the initial temperature x_0? On b? On a?

(b) Find a formula for the time it takes the object to cool a fraction $\varphi \in (0, 1)$ of the initial difference $x_0 - b$.

(c) Use your answer in (b) to validate or disprove your conjectures in (a).

13. Repeat Exercise 12 when $x_0 < b$ and the object warms to the environmental temperature b.

14. Derive the solution (6.14).

15. Suppose an object is placed in an environment whose temperature is initially $b_0 > 0$ but is cooling exponentially to temperature 0 (with rate $r > 0$).

(a) Use Newton's law of cooling to obtain an initial value problem for the temperature $x(t)$ of the object as a function of time t.

(b) Use a computer program to experiment with the initial value problem in (a) and formulate conjectures about the long-term temperature of the object. Specifically, does the object's temperature approach a limit as $t \to +\infty$? If so, what is the limit and at what rate does the temperature of the object approach it? Does the temperature of the object change in a monotonic fashion?

(c) Find a formula for the solution of your initial value problem in (a).

(d) Use your answer in (c) to verify or disprove your conjectures in (b).

16. The table below gives the average temperature rise of the earth (above that of the year 1860) for the first five decades of the twentieth century.[6]

Year	Temperature Rise (°C) (above that of 1860)
1900	0.03
1910	0.04
1920	0.06
1930	0.08
1940	0.10
1950	0.13

(a) Assume the temperature rise is growing exponentially and determine the per unit rate of change r by interpolating the first and last data points (at years 1900 and 1950). Calculate the percent errors that your solution makes at the intervening years.

(b) What does your answer in (a) predict the rise in temperature was in 1960 and 1970? It turns out that these rises were 0.18 and 0.24, respectively. Compute the errors in these model predictions. What do you conclude about this exponential model's accuracy?

17. A model for the velocity of a sprinter is described by the equation

$$v' = F_0 - k_0 v. \qquad (6.23)$$

---

[6]From *Applying Mathematics*, by D. N. Burghes, I. Huntley, and J. McDonald, Ellis Horwood Publishers, 1982.

Here F_0 is the (assumed constant) force exerted per unit mass of the runner and k_0 is the coefficient of friction (per unit mass) experience by the runner due to air resistance. For a race that begins with a standing start, we have the initial condition

$$v(0) = 0. \qquad (6.24)$$

(a) Find a formula for the solution of the initial value problem.

(b) Find a formula for the distance $x(t)$ run by the sprinter at time t.

(c) The following table shows time t and distance x measurements for a college sprinter during a 100-meter dash. Determine values for F_0 and k_0 so that $x(t)$ interpolates the data at times $t = 6$ and 11.92.

Time t_i (s)	Distance x_i (m)
0	0.0
1	2.0
2	8.5
3	15.5
4	25.5
5	34.0
6	44.5
7	53.0
8	64.0
9	72.5
10	83.0
11	91.5
11.92	100.0

(d) Graph the solution you obtain from your answers to (b) and (c). Include a plot of the data. Based on the visual evidence in this plot, how well does the solution appear to fit the data?

(e) Using the solution from (b) and (c), make a table of the predicted distances at the those times appearing in the table of data. Compare the model predictions to the data by including percentage errors in your table. At what times does the model make the most accurate predictions and at what times does it make the least accurate predictions? Can you offer a physical explanation for the times when the model is the least accurate?

18. Repeat Exercise 17, interpolating the data at $t = 3$ and $t = 6$. Compare the results with those from the model in Exercise 17.

19. Another way to obtain numerical estimates for the model parameters F_0 and k_0 from the race data in Exercise 17

is to use a linear least squares procedure based on the observation that equation (6.23) asserts a linear relationship between acceleration $a = v'$ and velocity v, namely $a = F_0 - k_0 v$. Thus, F_0 and k_0 could be estimated by linear least squares if velocity and acceleration data were available. In this exercise a and v are estimated from the data in Exercise 17.

(a) A popular difference quotient approximation for velocity is the so-called centered difference

$$v(t_i) = \frac{x(t_i + \Delta t) - x(t_i - \Delta t)}{2\Delta t}.$$

Use this formula to estimate the velocity at times $t = 1$ through 11 from the data in Exercise 17.

(b) Use the answers in (a) and the centered difference quotient estimate to obtain estimates for the acceleration $a = v'$ at times $t = 2$ through 10 for the data in Exercise 17.

(c) Use linear least squares to estimate F_0 and k_0 from the estimated velocities and accelerations obtained in (a) and (b).

(d) Obtain a formula for the distance $x = x(t)$ from the initial value problem (6.23) and (6.24) using your answers in (c).

(e) Use the formula in (d) to calculate predicted distances x at the data times in Exercise 17. Calculate the percent errors made by these predictions. Round your answers to the nearest tenth of a meter.

20. The amount of a radioactive isotope $x(t)$ decays are a rate proportional to the amount present. Denote the constant of proportionality by $-r$. Show the time it takes for a sample to decay to 50% of its initial amount is independent of the initial size x_0 of the sample. This time, denote H, is called the half-life of the isotope. Find the relation between r and H.

21. At time t_0 an amount x_0 of a radioactive isotope was present in a sample. Under the decay law in Exercise 20, find a formula for the age $t - t_0$ of the sample

(a) in terms of the half-life H, the initial amount x_0, and the amount present at time t

(b) in terms of the half-life H and the decay rates at time t_0 and t

22. The half-life of carbon-14 (C^{14}) is $H = 5568$ years. A burnt wooden beam taken in 1950 from the roof of a building in the ancient Babylonian city of Nippur gave a C^{14} count of 4.09 disintegrations (per minute per gram). Use the living wood disintegration rate of 6.68 (per minute per gram) to obtain an approximate date of the building. *Hint:* Use the formula from Exercise 21(b).

23. The half-life of carbon-14 (C^{14}) is $H = 5568$ years. In 1950 a piece of burnt wood from a fire pit in the Lascaux cave in France gave an average count of 0.97 disintegrations of C^{14} per minute per gram while a sample of living wood gave a count of 6.68 disintegrations per minute per gram. Approximately when was the fire built in the cave? *Hint*: Use the formula from Exercise 21(b).

24. The half-life of radium is approximately 600 years. What fraction of a 1-gram sample will remain after 1000 years? Of a 2-gram sample? Of a half-gram sample?

25. Show that the fraction of a radioactive sample remaining after T years is independent of the initial size of the sample.

26. The half-life of uranium-238 is approximately 4.51 billion years. How long will it take the number of disintegrations per minute measured at a contaminated site to drop 10%?

27. Suppose a quantity x decays exponentially at a rate r. Let f be a number between 0 and 1 (but not equal to 0 or 1). Show that the elapsed time necessary for x to decay from its original amount x_0 to the fraction f of its original amount is inversely proportional to r and is independent of x_0.

28. Cell populations growing in uninhibited circumstances grow in such a way that the per capita growth rate is a constant $r > 0$.

  (a) Write an initial value problem for the number of cells $x = x(t)$ at time t.

  (b) Find a formula for the solution of the initial value problem. What does the formula predict in the long run (i.e., as $t \to +\infty$)?

  (c) Use the formula calculated in (b) to show that the time it takes the cell population to double its initial size x_0 is independent of the initial size.

29. Derive the solution formula (6.22) for the initial value problem

$$x' = c_{in} r_{in} - \frac{r_{out}}{rt + V_0} x, \quad x(0) = x_0.$$

30. Suppose in Example 1 the uniform mixture in the container is drained out at a rate of 3 liters per minute (instead of 5 liters per minute).

  (a) Find the time it takes for the 100 liter tank to fill.

  (b) Find the concentration of salt when the tank is full.

31. In equation (6.21) suppose the inflow rate is larger than the outflow rate ($r_{in} > r_{out}$) so that the volume in the tank is increasing.

  (a) Write down the initial value problem for the salt concentration x.

  (b) Use a computer program to investigate how the amount of salt $x(t)$ changes in the short term. Based on your experiments, formulate a conjecture about the short-term change in the amount of salt in the tank depends on the initial amount of salt x_0 in the tank. (*Hint*: Study whether the amount of salt in the tank initially decreases or increases.)

  (c) Solve the initial value problem in (a).

  (d) Use the differential equation in (a) itself to prove (or disprove) your conjecture in (b).

  (e) Use your solution in (c) to prove (or disprove) your conjecture in (b).

32. It has been observed that the volume $x = x(t)$ of a solid tumor does not grow in the same way as freely growing cells (Exercise 28). Instead, the per capita growth rate decreases with time t. Suppose the per capita growth rate decreases exponential according to the formula ae^{-bt}, where a and b are positive constants.

  (a) Write an initial value problem for the volume of the tumor $x = x(t)$.

  (b) Use a computer to plot sample solutions of the initial value problem. Include the cases $a = b = 1$; $a = 2$, $b = 1$; and $a = 1$, $b = 0.5$. Based on these plots, what do you conjecture happens in general (that is, for all values of a and b) to the size of the tumor as $t \to +\infty$?

  (c) Find a formula for the solution of the initial value problem. Use the formula to corroborate (or disprove) your conjecture in (b).

33. Consider a company whose management goal is to produce the maximum annual dividend for its shareholders. If the total profit is paid out as dividends, then the company does not grow. On the other hand, if the total profit is reinvested in capital, then the company grows and profit increases but no dividends are paid out. Management's problem is to determine what fraction of the profit should be reinvested so as to maximize dividends.

  (a) Assume the rate of change of profit $x = x(t)$ is proportional to the fraction c of profit reinvested in capital ($0 \le c \le 1$). Write a differential equation for x and solve the initial value problem $x(0) = x_0$, where x_0 is the initial capital available.

  (b) Let $d = d(t)$ denote the total dividend paid to shareholders over the time interval 0 to t is $d = d(t)$. Assume the dividend is paid out at a rate proportional to the profit not reinvested in capital. Write a differential equation for d and solve the initial value problem $d(0) = 0$.

(c) Show graphically that if $k \leq 2$, then $c = 0$ produces the maximum dividend, but if $k > 2$, then there is a unique value of $c > 0$ that produces the maximum dividend in one year.

34. Use your result from (a) to answer (b) and (c).

(a) Find a solution formula for the initial value problem (5.4) in Sec. 1.5.1, Chapter 1, for a staph infection treated by an intravenously administered antibiotic of dosage d.

(b) Show that there is a critical dosage level d_{cr} above which the staph infection is cured.

(c) Show that the time $t_c = t_c(d)$ it takes to eliminate the infection is a decreasing function of $d > d_{cr}$.

35. Use your result from (a) to answer (b)–(d).

(a) Find a solution formula for the initial value problem (5.5) in Sec. 1.5.1, Chapter 1, for a staph infection treated by an antibiotic of dosage d administered by a pill or an injection.

(b) Show that there is a critical dosage level d_{cr} above which the staph infection is cured and below which it is not cured.

(c) Show that there is another critical value $d^* < d_{cr}$ such that if d satisfies $d^* < d < d_{cr}$, then $x(t)$ will initially decrease before increasing without bound.

(d) Show that the time $t_c = t_c(d)$ it takes to eliminate the infection is a decreasing function of $d > d_{cr}$.

3

Nonlinear First-Order Equations

In Chapter 2 we learned to solve and analyze linear first-order differential equations. We now turn our attention to nonlinear equations. In general, nonlinear equations are more difficult to solve and analyze than are linear equations. The fundamental existence and uniqueness theorem (Theorem 1) in Chapter 1 tells us, under very general conditions, that nonlinear equations do have solutions. For specialized types of nonlinear equations mathematicians have developed methods for obtaining formulas for the solutions. Although such methods can be useful, applications frequently do not involve equations of these specialized types. In these cases, we must turn to other methods of analysis.

In an application that involves a differential equation we generally want to answer specific questions about solutions. For example, we may want to know whether the solution has zeros or not; whether it is increasing or decreasing; whether it has maxima or minima; whether it is periodic; whether its graph has an asymptote; and so on. If we can "solve" the equation, in the sense of obtaining a formula for solutions, then we can use the solution formula to answer such questions. This approach is possible only for those special types of equations for which solution methods are available and tractable. Otherwise we will have to use other methods to obtain answers to our questions. It turns out, in fact, that methods are available for certain types of analysis that are often much easier to use even when solution formulas are readily obtainable. This chapter begins with a study of a very important class of nonlinear equations, called autonomous equations, for which this is the case. In this chapter we will see that a great deal can be learned about solutions of autonomous equations directly from the equation itself, without the need of a solution formula. These qualitative methods of analysis serve as a basic introduction to the modern theory of dynamical systems.

Of course solution formulas, when available, can be useful. One reason is that we can use them in those applications that happen to involve the specialized kinds of equations for which solution methods are available. Another reason is that specialized types of equations can serve as prototypical and insightful examples for the study of more difficult equations. Finally, some special types of equations, and their solution formulas (e.g., linear equations), can serve as approximations to more complicated equations and their solutions.

Autonomous equations (studied in Sec. 3.1) are a special case of so-called separable equations for which a method is available to calculate solution formulas. The solution method for separable equations is covered in Sec. 3.2. In Sec. 3.3 two other special types of equations and their solution methods are studied; these methods illustrate the important procedure of changing variables in a mathematical problem in order to simplify it in a way that allows a solution formula to be found. Finally, in Sec. 3.4 we study methods that produce formulas for approximations to solutions, rather than formulas for the solutions themselves.

3.1 Autonomous Equations

An important class of first-order differential equations are those in which the independent variable t does not explicitly appear on the right-hand side:

$$x' = f(x).$$

Equations of this kind are called *autonomous*. (See Exercise 1.) In Chapter 2 we studied *linear* autonomous equations [when $f(x) = px + q$ for constants p and q]. Examples of *nonlinear* autonomous equation are

$$x' = r\left(1 - \frac{x}{K}\right)x$$
$$x' = -a\,(b - x)\,|b - x|^p$$
$$v' = 9.8 - k_0 v^2,$$

provided all the parameters (r, K, a, b, p, and k_0) are constants. The first equation arises in applications from many disciplines, including population dynamics, and is called the logistic equation. The second equation arises in the study of the heating and cooling of objects. The third equations describes the motion of an object falling near the surface of the earth under the influence of gravity and frictional forces. For applications of these and other nonlinear autonomous equations see Sec. 3.6.

An equation that is not autonomous is called *nonautonomous*. The independent variable t appears explicitly in a nonautonomous equation. For example, the linear equation $x' = x + \sin t$ is nonautonomous. Another example arises from the logistic equation when one of its constants r or K is replaced by a function of t. For example, the nonautonomous equation

$$x' = r\left(1 - \frac{x}{K_0 + a\sin\left(\frac{2\pi}{T}t\right)}\right)x$$

arises in an application to population dynamics studied in Sec. 3.6.

3.1.1 Basic Properties of Solutions

Consider an autonomous differential equation

$$x' = f(x). \tag{1.1}$$

We assume $f(x)$ is defined and continuously differentiable on some interval I of x values. Then, for each initial conditions x_0 from the interval I, the fundamental existence and uniqueness theorem implies the initial value problem $x(t_0) = x_0$ has a unique solution (for any t_0). For most equations we study, $f(x)$ is continuously differentiable for all x or, in other words, the interval I is the entire real number line.

It is possible for an autonomous equation (1.1) to have a constant solution, that is to say, to have a solution $x(t) = x_e$ where x_e is a real number. (The graph of such a solution is a horizontal straight line.) Since the derivative of a constant equals zero, a constant solution $x(t) = x_e$ must satisfy $f(x_e) = 0$. In other words, constant solutions correspond to the roots of $f(x)$.

> **DEFINITION 1.** A constant solution of (1.1) is called an **equilibrium** (or a rest point or a critical point). Equilibria are the roots of $f(x)$.

To find the equilibria of an autonomous equation (1.1) we must solve the *equilibrium equation*

$$f(x) = 0.$$

EXAMPLE 1

The equilibrium equation of the autonomous differential equation

$$x' = \frac{1}{3}\left(1 - x^3\right)$$

is

$$\frac{1}{3}\left(1 - x^3\right) = 0.$$

This polynomial has one (real) root $x = 1$ and therefore the differential equation has one equilibrium $x_e = 1$.

Another example is the equation

$$x' = x^2 - 1$$

whose equilibrium equation

$$x^2 - 1 = 0$$

has two roots ± 1. This differential equation has two equilibria

$$x_e = 1 \quad \text{and} \quad x_e = -1. \qquad \blacksquare$$

The equilibrium equations of some differential equations are not easily solved by hand. Graphic methods or numerical approximation methods (using a computer or calculator) often help determine the equilibria of a differential equation.

EXAMPLE 2

Consider the autonomous equation

$$x' = \frac{a^2}{b^2 + x^4} - x$$

in which a and b are nonzero constants. It is not possible to solve the equilibrium equation

$$\frac{a^2}{b^2 + x^4} - x = 0$$

for x algebraically. However, rewriting the equation as

$$\frac{a^2}{b^2 + x^4} = x$$

and plotting the right-hand side and the left-hand side on the same graph, we see that these graphs intersect in exactly one point (no matter what the nonzero values of a and b are). See Fig. 3.1.

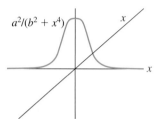

Figure 3.1

For particular cases in which a and b have specified numerical values, we can use a computer or hand calculator to obtain accurate approximations to the root of the equilibrium equation. For example, with $a = b = 1$ the equilibrium equation becomes

$$\frac{1}{1 + x^4} - x = 0.$$

Using a computer or calculator, we find the root of this equation to be approximately $x_e \approx 0.7549$. ∎

We now turn our attention to nonequilibrium solutions of autonomous equations. Unlike equilibrium solutions, nonequilibrium solutions might not be defined for all t. We saw examples in Chapter 1 (Examples 2 and 5). We denote the *maximal interval* on which a solution $x = x(t)$ is defined by $\alpha < t < \beta$. This means there is no larger interval containing $\alpha < t < \beta$ on which the solution $x = x(t)$ is defined. For equilibria $\alpha = -\infty$ and $\beta = +\infty$. For nonequilibrium solutions one or both of the endpoints α and β may be finite.

We begin our look at nonequilibrium solutions with two motivating examples.

EXAMPLE 3

Consider the autonomous equation

$$x' = \frac{1}{3}\left(1 - x^3\right). \tag{1.2}$$

Figure 3.2 shows computer generated graphs of solutions for a selection of initial values $x(0) = x_0$. Further computer experimentation, using a wider selection of initial values, will show these graphs are typical. The nonequilibrium graphs in Fig. 3.2 all appear to approach the limit 1 as $t \to +\infty$ and to do so in a strictly monotonic fashion (i.e., the solutions are either strictly increasing or strictly decreasing). Note that $x_e = 1$ is the equilibrium solution of equation (1.2). These numerical examples and observations encourage us to conjecture that all nonequilibrium solutions monotonically approach the equilibrium $x_e = 1$ as $t \to +\infty$. We must remain cautious in making this conjecture, however, since computer examples cannot prove general statements like this. This is because it is possible to investigate only a finite number of

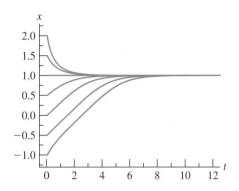

Figure 3.2 Graphs of se-lected solutions of the equation $x' = \left(1 - x^3\right)/3$.

examples and also because one cannot be sure what the graphs of solution are like outside the display window. ■

In Example 3 we conjectured, on the basis of some computer explorations, that all nonequilibrium solutions of the equation (1.2) monotonically approach an equilibrium as $t \to +\infty$. This is one possibility for nonequilibrium solutions of autonomous equations; there are others, however. For example, all nonequilibrium solutions $x(t) = x_0 e^t$ of the linear autonomous equation $x' = x$ are monotonic but do not approach the equilibrium $x_e = 0$. Instead they are unbounded, either increasing without bound ("blowing up") if $x_0 > 0$ or decreasing without bound ("blowing down") if $x_0 < 0$ as $t \to +\infty$. For nonlinear autonomous equations some nonequilibrium solutions approach an equilibrium while others are unbounded. Here is an example.

EXAMPLE 4

The differential equation

$$x' = x^2 - 1$$

has equilibria $x_e = -1$ and $x_e = 1$. We can use a computer to explore the graphs of selected nonequilibrium solutions. Figure 3.3 shows the slope field and several solution graphs. The graphs of the equilibria are horizontal straight lines. The nonequilibrium solutions displayed in Fig. 3.3 are monotonic. Those with initial values $x(0) = x_0$

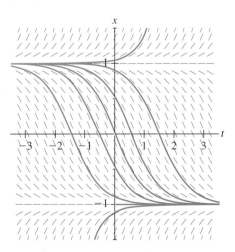

Figure 3.3 The slope field associated with equation $x' = x^2 - 1$ and graphs of selected solutions.

lying between the equilibria -1 and 1 are decreasing and those with initial values outside this interval are increasing. From these computer explorations we conjecture that all nonequilibrium solutions with initial values between -1 and 1 are decreasing (approaching the equilibrium $x_e = -1$ as $t \to +\infty$). We also conjecture that all other nonequilibrium solutions are increasing; those with initial conditions $x_0 < -1$ appear to approach the equilibrium $x_e = -1$ and those with $x_0 > 1$ appear to be unbounded as $t \to +\infty$. ∎

While the computer explorations in Examples 4 and 3 do not prove the conjectured monotonicity of nonequilibrium solutions for those nonlinear autonomous equations, the conjectures turn out to be true. In fact the conjecture is true for all autonomous differential equations, as the following theorem asserts.

THEOREM 1

Assume $f(x)$ is continuously differentiable. All nonequilibrium solutions of the autonomous equation $x' = f(x)$ are strictly monotonic (i.e., nonequilibrium solutions are either strictly increasing or strictly decreasing). The solution of the initial value problem

$$x' = f(x), \quad x(t_0) = x_0$$

is an equilibrium if $f(x_0) = 0$, is strictly increasing if $f(x_0) > 0$ or is strictly decreasing if $f(x_0) < 0$.

To see why this theorem is true, consider a nonequilibrium solution $x = x(t)$ of the autonomous equation (1.1) on the (maximal) interval $\alpha < t < \beta$. First, note that this solution can never, for any value of t, equal an equilibrium x_e. [In other words, the graph of the solution $x = x(t)$ in the x, t-plane can never intersect the horizontal straight line graph of an equilibrium x_e.] The reason for this is as follows. Suppose $x(t)$ did equal x_e at some value of t, say at $t = t^*$. Then there would exist two different solutions of the initial value problem

$$x' = f(x), \quad x(t^*) = x_e$$

namely, the nonequilibrium solution $x(t)$ and the equilibrium solution x_e itself. This would contradict the existence and uniqueness theorem (specifically, the uniqueness assertion of the theorem). From this contradiction we conclude that $x(t)$ cannot equal an equilibrium, that is, a root of $f(x)$, for any value of t. In other words, we have shown that $f(x(t)) \neq 0$ for all t. This means either $x'(t) = f(x(t)) > 0$ or $x'(t) = f(x(t)) < 0$ on the whole interval $\alpha < t < \beta$. In the first case the solution $x = x(t)$ is strictly increasing and in the second case it is strictly decreasing. This is the conclusion of Theorem 1.

EXAMPLE 5

For the equation $x' = (1 - x^3)/3$ in Example 3 an initial value $x_0 < 1$ implies $f(x_0) = (1 - x_0^3)/3 > 0$ and Theorem 1 implies the solution is strictly increasing. On the other hand, an initial value $x_0 > 1$ implies $f(x_0) < 0$ and the solution is therefore strictly decreasing. This application of Theorem 1 proves the monotonicity conjecture concerning nonequilibrium made in Example 3.

For the equation $x' = x^2 - 1$ in Example 4 we see $f(x_0) = x_0^2 - 1 < 0$ for $-1 < x_0 < 1$. By Theorem 1 solutions associated with these initial conditions are strictly decreasing. For initial conditions satisfying $x_0 < -1$ or $x_0 > 1$ we see that $f(x_0) > 0$ and the solutions are strictly increasing. ∎

In Examples 3 and 4 we conjectured that nonequilibrium solutions either approach an equilibrium or increase (decrease) without bound as t increases. We now investigate these alternatives for solutions of general autonomous equation.

A function $x = x(t)$, defined on an interval $\alpha < t < \beta$, is said to be *bounded above* if there is a number M such that $x(t) \leq M$ for all t from the interval. It is *bounded below* if there is a number m such that $m \leq x(t)$ holds on the interval.

If $x = x(t)$ is increasing (or decreasing) on an interval $\alpha < t < \beta$ and is bounded above (or below), then the limit

$$\lim_{t \to \beta} x(t) = x_L$$

exists. If $x = x(t)$ is increasing (or decreasing) on an interval $\alpha < t < \beta$ and is not bounded above (or below), then we write

$$\lim_{t \to \beta} x(t) = +\infty \ (\text{or} \ -\infty).$$

Here we allow the possibility that $\beta = +\infty$.

The exponential $x = e^t$ is an example of a solution of an autonomous equation (namely, $x' = x$) that is not bounded above on its interval of definition $-\infty < t < +\infty$. It is, however, bounded below (with $m = 0$, for example). On the other hand, solution $x = -e^t$ is bounded above (with $M = 0$), but not below.

It is also possible that a solution is not be bounded above (i.e., it may "blow up") even when its interval of definition is finite. For example, $x = \tan t$ is a solution of the equation $x' = x^2 + 1$ on the interval $-\pi/2 < t < \pi/2$. This solution is neither bounded above (since $\lim_{t \to \pi/2} \tan t = +\infty$) nor bounded below (since $\lim_{t \to -\pi/2} \tan t = -\infty$). Graphically the solution has vertical asymptotes at $t = \pm\pi/2$.

Now we turn our attention to *bounded* solutions of an autonomous equation $x' = f(x)$. Since nonequilibrium solutions are either strictly increasing or decreasing there are two cases to consider as $t \to \beta$, namely, increasing solutions that are bounded above and decreasing solutions that are bounded below. We consider increasing solutions bounded above. Decreasing solutions bounded below can be treated in a similar way.

Let $x = x(t)$ be an increasing solution of $x' = f(x)$ that is bounded above on its interval of definition $\alpha < t < \beta$. Exercise 11 shows that such a solution must exist for all $t > \alpha$ (i.e., $\beta = +\infty$). Let x_L denote the limit

$$\lim_{t \to +\infty} x(t) = x_L.$$

Because $x(t)$ is increasing and approaches a limit (in other words, its graph approaches a horizontal asymptote) the derivative $x'(t)$ must approach 0. (See Exercise 12.) Thus,

$$\lim_{t \to +\infty} x'(t) = \lim_{t \to +\infty} f(x(t)) = 0.$$

On the other hand, because $f(x)$ is a continuous function of x, we have that

$$\lim_{t \to +\infty} x'(t) = \lim_{t \to +\infty} f(x(t)) = f\left(\lim_{t \to +\infty} x(t)\right) = f(x_L).$$

We conclude that $f(x_L) = 0$. In other words, the limit x_L is a root of $f(x)$ and hence is an equilibrium of the differential equation.

We have shown that an increasing solution of an autonomous equation, bounded above, must approach an equilibrium as $t \to +\infty$. Conversely, if there is an equilibrium x_e greater than the initial value x_0 of an increasing solution, then the solution is certainly bounded above (take $M = x_e$). In summary, an increasing solution is bounded above if and only if there is an equilibrium greater than its initial value x_0.

Similar facts about decreasing functions that are bounded below can be derived in an analogous way. We summarize these findings in the following theorem.

THEOREM 2   Assume $f(x)$ is continuously differentiable for all x and consider the initial value problem

$$x' = f(x)$$
$$x(t_0) = x_0.$$

Let $x = x(t)$ be a nonequilibrium solution [i.e., assume $f(x_0) \neq 0$] and let $\alpha < t < \beta$ be its maximal interval of existence.

If $f(x_0) > 0$, then $x = x(t)$ is strictly increasing and one of the following alternatives holds:

(a) If there is no equilibrium greater than x_0, then $\lim_{t \to \beta} x(t) = +\infty$;

(b) If there is equilibrium greater than x_0, then $\beta = +\infty$ and $\lim_{t \to +\infty} x(t) = x_e$, where x_e is the smallest such equilibrium.

If $f(x_0) < 0$, then $x(t)$ is strictly decreasing and one of the following alternatives holds:

(c) If there is no equilibrium smaller than x_0, then $\lim_{t \to \beta} x(t) = -\infty$;

(d) If there is equilibrium x_e smaller than x_0, then $\beta = +\infty$ and $\lim_{t \to +\infty} x(t) = x_e$, where x_e is the largest such equilibrium.

In cases (a) and (c), β may be finite or $+\infty$. In these cases, the solution "blows up" (or "blows down") in either a finite amount of time or as $t \to +\infty$, respectively.

Using Theorem 2, we can prove the conjectures made in Examples 3 and 4.

EXAMPLE 6   For the equation

$$x' = \frac{1}{3}\left(1 - x^3\right)$$

in Example 3, $f(x) = \left(1 - x^3\right)/3$ has only one root, namely $x_e = 1$. Since $f(x_0) < 0$ for an initial value $x_0 > 1$ the solution, by part (d) of Theorem 2, decreases to $x_e = 1$ as $t \to +\infty$. Since $f(x_0) > 0$ for an initial value $x_0 < 1$, the solution, by part (b) of Theorem 2, increases to $x_e = 1$ as $t \to +\infty$.

For the equation

$$x' = x^2 - 1$$

in Example 4, $f(x) = x^2 - 1$ has two roots, namely -1 and 1. Moreover, $f(x)$ is negative between these roots and positive elsewhere. Therefore, a solution with initial value $x_0 < -1$, by part (b) of Theorem 2, increases to $x_e = -1$ as $t \to +\infty$. The solution for an initial value x_0 between -1 and 1, by part (d) of Theorem 2, decreases to $x_e = -1$ as $t \to +\infty$. Finally, the solution for an initial value $x_0 > 1$, by part (a) of Theorem 2, increases without bound as $t \to \beta$. (Exercise 19 shows β is finite in this case.)

A similar investigation can also be made of nonequilibrium solutions for decreasing t (i.e., as $t \to \alpha$). The result is that nonequilibrium solutions either approach an equilibrium as $t \to -\infty$ or "blow up" (or "down") as $t \to \alpha$.

The monotonicity and limiting properties of solutions described in Theorem 2 are called the *asymptotic dynamics* of the equation $x' = f(x)$. Using this theorem, *all we need to do in order to determine the asymptotic dynamics of an autonomous equation is find the roots of $f(x)$ and determine the sign of $f(x)$ between the roots.*

Exercises

1. If $x = x(t)$ is a solution of an autonomous equation $x' = f(x)$ for all t, show $y = x(t+k)$ is also a solution for any constant k. This shows that "translations" $x(t + k)$ of solutions of autonomous equations are also solutions. This is another defining characteristic of autonomous differential equations.

Find all equilibria of the following equations.

2. $x' = x^2 + 2x - 3$

3. $x' = 4x^3 - 4x^2 - x + 1$

4. $x' = \ln \dfrac{2x}{1 + x}$

5. $x' = \dfrac{3x}{1 + x^2} - 1$

6. $x' = x^2 - e^{-x}$

7. $x' = 1 - x - \ln(1 + x)$

How many equilibria do the following equations have?

8. $x' = \dfrac{1}{a^2 + x} - \ln x$

9. $x' = ax - e^{-x^2}, a \neq 0$

10. $x' = a - x - \dfrac{x^2}{1 + x^2}, a > 1$

11. $x' = x^4 - xe^{-ax}, a > 0$

For which initial values x_0 are the solutions of the following equations strictly increasing and for which are they strictly decreasing?

12. $x' = x(1 - x^4)$

13. $x' = \ln\left(x^2 + \frac{1}{4}\right)$

14. $mx' = g - cx^2$, where m, g, c are positive constants

15. $x' = r\left(1 - \dfrac{x}{K}\right)x$, where r, K are positive constants

16. $x' = 6x^2 - 5x + 1$

17. $x' = (1 - x)\left(1 - e^{-x}\right)$

3.1.2 Phase Line Portraits

In this section we seek a convenient graphical way to summarize the asymptotic dynamics of an autonomous equation. We do this by means of the *phase line portrait*. We begin with an example.

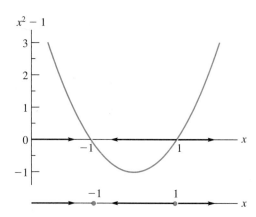

Figure 3.4 The phase line portrait for the equation $x' = x^2 - 1$.

Consider the equation $x' = x^2 - 1$ in Example 4, for which $f(x) = x^2 - 1$. From the graph of $f(x)$ in Fig. 3.4 we see that the two roots -1 and 1 divide the x-axis into three disjoint intervals:

$$x < -1, \quad -1 < x < 1 \quad \text{and} \quad 1 < x.$$

On the interval $-1 < x < 1$, where $f(x)$ is negative (i.e., where its graph is below the x-axis), place an arrow pointing to the left. This arrow indicates that solutions with initial values in this interval are decreasing. On the two half-line intervals $x < -1$ and $x > 1$ where $f(x)$ is positive (i.e., where its graph of $f(x)$ is above the x-axis) place an arrow pointing to the right. This arrow indicates that solutions with initial values in these intervals are increasing. The result appears in Fig. 3.4. Extracting the x-axis from this graph, we obtain a line divided into subintervals with orientations indicated by arrows. This line is the phase line portrait associated with the equation $x' = x^2 - 1$. It summarizes the asymptotic dynamics of this equation.

We can apply the procedure used to obtain the phase line portrait of the equation $x' = x^2 - 1$ to any autonomous equation $x' = f(x)$. All that we need know are the roots of $f(x)$ and the signs of $f(x)$ between the roots. This even includes the case when $f(x)$ has infinitely many roots, provided they are isolated. A root is *isolated* if it can be placed at the center of an interval in which there are no other roots. The function $f(x) = \sin x$, with roots at integer multiples of π, is an example. *We assume from now on that the roots of $f(x)$ are isolated.*

> **DEFINITION 2.** The **phase line portrait** associated with an autonomous equation $x' = f(x)$ consists of those subintervals of the x-axis created by the equilibria [i.e., the roots of $f(x)$] together with arrows pointing to the right if $f(x)$ is positive on a subinterval or to the left if $f(x)$ is negative on a subinterval.

The phase line portrait summarizes the asymptotic dynamics of an autonomous equation. If the initial value x_0 lies in a subinterval with an arrow pointing to the right, then the solution with this initial value strictly increases and approaches the right-hand endpoint of the subinterval (which may be $+\infty$) as t increases. If the initial value x_0 lies in a subinterval with an arrow pointing to the left, then the solution with this initial condition strictly decreases and approaches the left hand end point of the interval as t increases (which may be $-\infty$). (*Note*: The asymptotic dynamics for decreasing t are summarized in the phase portrait obtained by reversing the orientation arrows.)

EXAMPLE 7

The roots of $f(x) = x^2(1 - x^2)$ are -1, 0, and 1. These three roots are the equilibria of the equation

$$x' = x^2(1 - x^2)$$

and they determine four subintervals of the x-axis. The sign of $f(x)$ on each of these intervals can be deduced from the formula $f(x) = x^2(1 - x^2)$ or from the graph shown in Fig. 3.5.[1] Either way we obtain the phase line portrait shown in Fig. 3.5.

From the phase line portrait we see, for example, that an initial condition $x_0 > 1$ implies the solution $x(t)$ is decreases to 1 as $t \to +\infty$. Or, an initial condition satisfying $0 < x_0 < 1$ implies $x(t)$ is increases to 1 as $t \to +\infty$ and approaches 0 as $t \to -\infty$. ∎

---

[1] Sometimes test points selected from the subintervals are useful. For example, in this case we calculate $f(1/2) = 3/16 > 0$ and conclude that $f(x)$ is positive on the subinterval between 0 and 1.

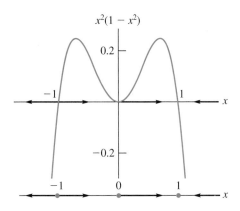

Figure 3.5 The phase line portrait for the equation $x' = x^2 \left(1 - x^2\right)$.

There is a connection between the phase line portrait of the equation $x' = f(x)$ and the graphs of solutions $x = x(t)$ in the t, x-plane. Equilibrium points are the (horizontal) projections of the horizontal line plots of the equilibria onto the (vertical) x-axis. The subintervals of the phase line portrait are the projections of nonequilibrium solution graphs onto the x-axis. Thus, the subintervals in the phase line portrait are the ranges of nonequilibrium solutions. This is illustrated in Fig. 3.6 for the equation $x' = x^2 - 1$. In Fig. 3.6 arrows have been added to the graphs of the solutions, indicating the direction of increasing t, so that one can see how the monotonicities of solutions determine the orientations in the phase line portrait.

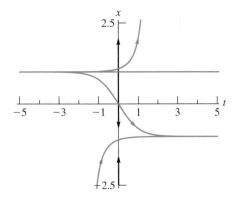

Figure 3.6 Horizontal projections of solution graphs of an equation $x' = f(x)$ produce the phase line portrait on the (vertical) x-axis. The oriented subintervals in the phase line portrait are the orbits of the equation.

DEFINITION 3. The range of a solution $x = x(t)$ of an autonomous equation $x' = f(x)$, together with an orientation in the direction of increasing t, is called the **orbit** associated with the solution.

The roots of $f(x)$, which are points on the phase line portrait, are the orbits associated with equilibrium solutions. Notice, however, that many different nonequilibrium solutions project to the same orbit. In general, infinitely many solutions share the same orbit.

An example is seen in Fig. 3.6, where all solutions with initial values between -1 and 1 project onto the same orbit, namely the subinterval $-1 < x < 1$. The equation

$x' = x^2 - 1$ has infinitely many solutions, but only five orbits: two equilibrium (point) orbits and three nonequilibrium (subinterval) orbits.

EXAMPLE 8   Orbits of the equation

$$x' = x^2(1 - x^2)$$

are displayed the phase line portrait in Fig. 3.5. There are four nonequilibrium orbits, namely the four intervals

$$x < -1, \quad -1 < x < 0, \quad 0 < x < 1, \quad 1 < x$$

oriented to the left, right, right, and left, respectively. There are also three equilibrium orbits at the equilibria $x_e = -1, 0$, and 1 for a total of seven orbits. ■

In the following example there are infinitely many equilibria and nonequilibrium orbits.

EXAMPLE 9   The roots of $f(x) = \sin x$ are

$$x_e = n\pi, \quad n = 0, \pm 1, \pm 2, \dots.$$

These equilibrium orbits of the equation $x' = \sin x$ determine infinitely many nonequilibrium orbits, namely the intervals

$$n\pi < x < (n + 1)\pi, \quad n = 0, \pm 1, \pm 2, \dots.$$

The phase line portrait appears in Fig. 3.7. ■

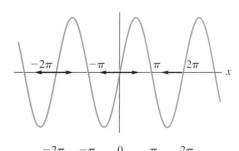

Figure 3.7 The phase line portrait of the equation $x' = \sin x$.

Phase line portraits for linear autonomous equations are particularly simple, as we see in the next example.

EXAMPLE 10   If $p \neq 0$, the linear equation

$$x' = px + q$$

has one equilibrium $x_e = -q/p$ and two nonequilibrium orbits, namely the intervals $x < -q/p$ and $-q/p < x$. If $p = 0$ and $q \neq 0$, there is no equilibrium. As a result there is only one orbit, namely the whole x-axis, with a right orientation if $q > 0$ and a left orientation if $q < 0$. (If both $p = 0$ and $q = 0$, then every point is an equilibrium and the equilibria are not isolated.)

$x' = px + q$	Phase Line Portrait
$p < 0$	$\longrightarrow \quad -\frac{q}{p} \quad \longleftarrow$
$p > 0$	$\longleftarrow \quad -\frac{q}{p} \quad \longrightarrow$
$p = 0, q > 0$	$\longrightarrow$
$p = 0, q < 0$	$\longleftarrow$

An (isolated) equilibrium separates two nonequilibrium orbits in a phase line portrait. As a result there are a limited number of possible orbit configurations near this equilibrium (in fact, only three). The possibilities are listed in Fig. 3.8.

Equilibrium Type	Phase Line Portrait
Sink (or attractor)	$\longrightarrow \quad x_e \quad \longleftarrow$
Source (or repellor)	$\longleftarrow \quad x_e \quad \longrightarrow$
Shunt	$\begin{cases} \longrightarrow \quad x_e \quad \longrightarrow \\ \longleftarrow \quad x_e \quad \longleftarrow \end{cases}$

Figure 3.8 The phase line portraits in the neighborhood of an isolated equilibrium.

DEFINITION 4. Consider an autonomous equation $x' = f(x)$, where $f(x)$ is continuously differentiable for all x. Assume the roots of $f(x)$ are isolated.

An equilibrium is called a **sink** (or an attractor) if the orientation arrows of both adjacent orbits point toward it.

An equilibrium is called a **source** (or a repellor) if the orientation arrows of both adjacent orbits point away from it.

An equilibrium is called a **shunt** if the orientation arrows of both adjacent orbits point in the same direction.

For linear equations $x' = px + q$ these definitions of sink and source are the same are those given in Chapter 2. However, sinks and sources of linear equations have a property that sinks and sources of a nonlinear equation might not have. For linear equations *all* orbits move toward a sink and *all* nonequilibrium orbits move away from a source. This is not necessarily true for nonlinear equations. For example, from the phase line portrait of the equation $x' = x^2(1-x^2)$ in Fig. 3.5 we see that the equilibrium $x_e = -1$ is a source, the equilibrium $x_e = 0$ is a shunt and the equilibrium $x_e = -1$ is a sink.

EXAMPLE 11

The parabolic graph of the quadratic function

$$f(x) = r\left(1 - \frac{x}{K}\right)x$$

appears in Fig. 3.9 for positive constants r and K. From this graph we obtain the phase line portrait in Fig. 3.9 for the logistic equation

$$x' = r\left(1 - \frac{x}{K}\right)x.$$

The equilibrium $x_e = 0$ is a source and the equilibrium $x_e = K$ is a sink. ∎

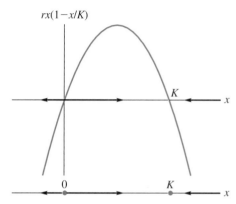

Figure 3.9 The graph of $f(x) = rx(1-x/K)$ together with the phase line portrait of logistic equation $x' = rx(1 - x/K)$.

From the geometric way by which the graph of $f(x)$ is used to construct phase line portraits, we obtain straightforward geometric tests for the three different types of equilibria in Definition 4.

THEOREM 3

Geometric Test Suppose that $x = x_e$ is an isolated equilibrium of $x' = f(x)$. Then
x_e is a **sink** if and only if the graph of $f(x)$ **decreases** through x_e;
x_e is a **source** if and only if the graph of $f(x)$ **increases** through x_e;
x_e is a **shunt** if and only if the graph of $f(x)$ has a (local) **extremum** at x_e.
See Figure 3.10.

The graph of $f(x) = x^2(1 - x^2)$ shown in Fig. 3.5 illustrates this theorem. The graph increases through the root -1, has a (local) minimum at the root 0 and decreases through the root 1. For the differential equation $x' = x^2(1 - x^2)$ this implies the equilibrium $x_e = -1$ is a source, $x_e = 0$ is a shunt, and $x_e = 1$ is a sink.

Another example is provided by the graph of $f(x) = rx(1 - x/K)$ in Fig. 3.9, which is seen to increase through the root 0 and decrease through the root K. Theorem 3 implies $x_e = 0$ is a source and $x_e = K$ is a sink for the logistic equation $x' = rx(1 - x/K)$.

The monotonicity of a function $f(x)$ at a point $x = x_e$ is related to its derivative df/dx evaluated at this point, which we denote by

$$\left.\frac{df}{dx}\right|_{x_e}.$$

If $df/dx|_{x_e} \neq 0$, then the function $f(x)$ cannot have an extremum at x_e and it either decreases through x_e (if $df/dx|_{x_e} < 0$) or increases through x_e (if $df/dx|_{x_e} > 0$). Thus, we can use the derivative of $f(x)$, evaluated at an equilibrium x_e, to determine whether x_e is a sink or a source, provided this derivative does not vanish.

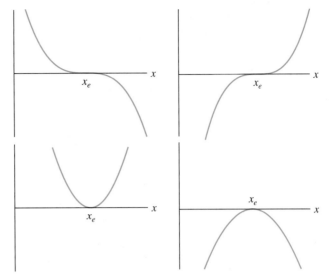

Figure 3.10 If the graph of $f(x)$ decreases or increases through the equilibrium x_e, as in the upper two graphs, then the equilibrium is a sink or a source, respectively. If the graph of $f(x)$ has an extremum at the equilibrium x_e, as in the lower two graphs, then the equilibrium is a shunt.

DEFINITION 5. Suppose x_e is an equilibrium of $x' = f(x)$. If

$$\left.\frac{df}{dx}\right|_{x_e} \neq 0,$$

then x_e is called **hyperbolic.**

This definition is consistent with that given in Chapter 2 for a linear autonomous equation $x' = px + q$, since in this case $f(x) = px + q$ and $df/dx = p$. From the geometric test in Theorem 3 we obtain a derivative test for the equilibrium type.

THEOREM 4 **Derivative Test** If x_e is a hyperbolic equilibrium of $x' = f(x)$, then

$$\left.\frac{df}{dx}\right|_{x_e} < 0 \quad \text{implies } x_e \text{ is a sink and}$$

$$\left.\frac{df}{dx}\right|_{x_e} > 0 \quad \text{implies } x_e \text{ is a source.}$$

If the derivative of $f(x)$ evaluated at an equilibrium x_e equals 0 (that is to say, if the equilibrium is nonhyperbolic), then nothing can be deduced from this theorem. For example, for $f(x) = x^2(1 - x^2)$ we find

$$\frac{df}{dx} = 2x - 4x^3$$

and consequently

$$\frac{df}{dx}\bigg|_{-1} = 2, \qquad \frac{df}{dx}\bigg|_{0} = 0 \qquad \frac{df}{dx}\bigg|_{1} = -2.$$

From Theorem 4 we conclude that $x_e = -1$ is a source and $x_e = 1$ is a sink. The equilibrium $x_e = 0$ is nonhyperbolic, however, and we cannot conclude anything about it from this theorem. (From the geometric test we can see, however, that $x_e = 0$ is a shunt.)

Do not make the mistake of concluding that a nonhyperbolic equilibrium must necessarily be a shunt. Although $df/dx|_{x_e} = 0$ is a necessary condition, it is not sufficient to imply that $f(x)$ has an extremum at x_e. Here is an example.

EXAMPLE 12

The equilibria of the equation

$$x' = x^3 - x^4$$

are the roots of

$$f(x) = x^3 - x^4,$$

namely 0 and 1. Since

$$\frac{df}{dx}\bigg|_{1} = -1 < 0,$$

the equilibrium $x_e = 1$ is a hyperbolic sink. Since

$$\frac{df}{dx}\bigg|_{0} = 0,$$

the equilibrium $x_e = 0$ is nonhyperbolic (and Theorem 4 does not apply). However, $f(x)$ does not have an extremum at 0 and hence $x_e = 0$ is not a shunt. The graph in Fig. 3.11 shows that $f(x)$ is increasing through 0 and therefore by Theorem 3 the nonhyperbolic equilibrium $x_e = 0$ is a source. ∎

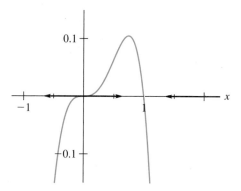

Figure 3.11 The graph of $f(x) = x^3 - x^4$ together with the phase line portrait of $x' = x^3 - x^4$.

Both adjacent orbits approach a sink as $t \to +\infty$. For this reason a sink is called *asymptotically stable* or, more commonly, simply *stable*. In applications a variable located at a stable equilibrium returns to that same equilibrium when slightly perturbed away. This is in contrast to a source for which such a perturbation results in motion away from the equilibrium. For this reason a source is called *unstable*. A shunt is also called unstable (because both adjacent orbits do not approach it as $t \to +\infty$) or semistable (because one adjacent orbit approaches it and the other does not). For more on stability see Exercises 13 and 14 of Sec. 3.5.

Exercises

For each of the following equations, find all equilibria. Sketch a graph of $f(x)$ and use it to draw the phase line portrait. Identify the type of each equilibrium.

1. $x' = x^3 - 1$ **2.** $x' = x^3 - x$

3. $x' = x^3 - x^2$ **4.** $x' = x - e^{-x}$

For each of the following equations, find all equilibria. Use the derivative test (if applicable) to determine which are sinks and which are sources. Otherwise use the geometric test. Draw the phase line portrait. (a is a constant.)

5. $x' = -x + \cos x$

6. $x' = x^2(x - 1)^3(2 - x)$

7. $x' = x(x + 1)(x - 0.5)^4$

8. $x' = x(1 - e^x)$

9. $x' = (1 - x^2)(1 - e^{1-x})$

10. $x' = x^2 - a$

11. $x' = -x^3 + (1 + a)x^2 - ax$

12. $x' = a - x^2(1 + x^2)^{-1}$

13. Find all equilibria of the equation $x' = x \sin x$ and determine which are hyperbolic. Identify the type of all equilibria. Draw the phase line portrait.

14. Find all equilibria of the equation $x' = \sin^2 x$ and determine which are hyperbolic. Identify the type of all equilibria. Draw the phase line portrait.

For each of the phase line portraits drawn below, write down a first-order differential equation of the form $x' = f(x)$.

15. $\longrightarrow -3 \longleftarrow 3 \longrightarrow$ **16.** $\longleftarrow -3 \longrightarrow 3 \longleftarrow$

17. $\longrightarrow 0 \longrightarrow 2 \longleftarrow$ **18.** $\longleftarrow 1 \longleftarrow 10 \longrightarrow$

19. $\longrightarrow 0 \longrightarrow 1 \longrightarrow$ **20.** $\longleftarrow -2 \longleftarrow 5 \longleftarrow$

21. $\longrightarrow a \longleftarrow b \longrightarrow$ **22.** $\longleftarrow a \longrightarrow b \longleftarrow$

23. $\longrightarrow 1 \longleftarrow 2 \longrightarrow 3 \longleftarrow$

24. $\longleftarrow 1 \longrightarrow 2 \longrightarrow 3 \longleftarrow$

25. $\longrightarrow a \longleftarrow b \longrightarrow c \longleftarrow d \longleftarrow$

26. $\longrightarrow a \longrightarrow b \longrightarrow c \longleftarrow d \longleftarrow$

27. The velocity $v \geq 0$ of a falling object satisfies the equation

$$v' = 9.8 - k_0 v^2,$$

where $k_0 > 0$ is the (per unit mass) coefficient of friction. Draw the phase line portrait for $v \geq 0$. Classify the positive equilibrium. Is this equilibrium hyperbolic?

28. Let $x = x(t)$ denote the temperature of an object. According to a modified Newton's law of cooling x satisfies the equation

$$x' = -a\,(b - x)\,|b - x|^p,$$

where b is the constant environment temperature and a and p are positive constants. Draw the phase line portrait. Classify the equilibrium. Is the equilibrium hyperbolic?

29. The concentration $c = c(t)$ of a substrate in a container in which an enzyme is present (in constant concentration $e > 0$) satisfies the equation

$$c' = d\,(c_{\text{in}} - c) - \frac{mc}{a + c}e.$$

All coefficients d, c_{in}, m, and a are positive constants. In this exercise let $d = c_{\text{in}} = a = e = 1$ and $m = 2$. Draw the phase line portrait for $c \geq 0$. Classify the equilibrium. Is the equilibrium hyperbolic?

3.1.3 The Linearization Principle

The derivative test in Theorem 4 is related to the linearization principle, one of the most important principles in applied mathematics. Linearization is a procedure for studying solutions (or orbits) of an equation $x' = f(x)$ near a hyperbolic equilibrium x_e by approximating $f(x)$ with its tangent line at x_e.

Since $f(x_e) = 0$ at an equilibrium x_e, the equation of the tangent line to $f(x)$ at x_e is $y = \lambda\,(x - x_e)$, where

$$\lambda = \left.\frac{df}{dx}\right|_{x_e}.$$

The graph of this tangent line approximates the graph of $f(x)$ near the tangent point and therefore

$$f(x) \approx \lambda(x - x_e)$$

for x near x_e. This suggests that we may learn about solutions of

$$x' = f(x) \tag{1.3}$$

near an equilibrium x_e from the solutions of the linear equation

$$x' = \lambda(x - x_e). \tag{1.4}$$

In fact, the phase portraits near x_e of these two equations are identical if $\lambda \neq 0$. See Fig. 3.12. The equilibrium x_e is a sink for both equations if $\lambda < 0$; it is a source for both equation if $\lambda > 0$.

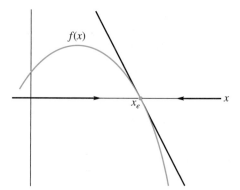

Figure 3.12 The graph of $f(x)$ and that of its tangent line $\lambda(x - x_e)$, $\lambda = df/dx|_{x_e}$, produce the same phase line portrait near a hyperbolic equilibrium x_e.

We call linear differential equation (1.4) the *linearization of the equation (1.3) at the equilibrium* x_e. We can simplify the linearized equation by a change of variables. Let $u = x - x_e$. Then $u' = x'$ and the linearization becomes

$$u' = \lambda u, \quad \text{where} \quad \lambda = \left.\frac{df}{dx}\right|_{x_e}. \tag{1.5}$$

Referring to Example 10, we see that the linear equation $u' = \lambda u$ has a sink at the equilibrium $u_e = 0$ if $\lambda < 0$ and a source at $u_e = 0$ if $\lambda > 0$. This fact and Theorem 4 imply that a hyperbolic equilibrium x_e of $x' = f(x)$ has the same type as that of the equilibrium $u_e = 0$ of its linearization. This is the *linearization principle*.

THEOREM 5   **Linearization Principle** An autonomous equation $x' = f(x)$ has a sink (or source) at a hyperbolic equilibrium x_e if its linearization (1.5) at x_e has a sink (or source) at $u_e = 0$, that is, if $\lambda < 0$ (or $\lambda > 0$).

EXAMPLE 13   The logistic equation

$$x' = r\left(1 - \frac{x}{K}\right)x, \ r > 0, \ K > 0$$

has two equilibria $x_e = 0$ and K. To apply the linearization principle at each of these equilibria, we evaluate the derivative

$$\frac{df}{dx} = r\left(1 - 2\frac{x}{K}\right)$$

of $f(x) = rx\,(1 - x/K)$ at each equilibrium. Since $df/dx|_0 = r > 0$, the linearization at $x_e = 0$ has a source at 0. Since $df/dx|_K = -r < 0$, the linearization at $x_e = K$ has a sink at 0. By the linearization principle the logistic equation has a source at $x_e = 0$ and a sink at $x_e = K$. ∎

The linearization principle does not hold for a nonhyperbolic equilibrium x_e (i.e., when $\lambda = 0$). That is, the linearization at a nonhyperbolic equation cannot be used in general to determine the phase portrait near the equilibrium. For example, graphs of x^2, x^3 and $-x^3$ show the equilibrium $x_e = 0$ is a shunt for $x' = x^2$, a source for $x' = x^3$ and a sink for $x' = -x^3$. Yet all three equations have the same linearization $u' = 0$ at 0.

Exercises

Find the linearization of the following equations at each of their equilibria.

1. $x' = x^3 - x$
2. $x' = \sin x$
3. $x' = rx\left(1 - \dfrac{x}{K}\right)$, where r, K are positive constants
4. $mx' = mg - cx^2$, where m, g, c are positive constants
5. $x' = \dfrac{x^3}{1 + x^2}$
6. $x' = \sqrt{2}x - 4\dfrac{x^2}{1 + x^2}$
7. $x' = (1 - x)x - h$ where $0 \le h \le 1/4$
8. The velocity $v \ge 0$ of a falling object satisfies the equation

$$v' = 9.8 - k_0 v^2,$$

where $k_0 > 0$ is the (per unit mass) coefficient of friction. Find this equation's positive equilibrium, and then find the linearization at this equilibrium. What kind of equilibrium does the linearization have? If applicable, use the linearization principle (Theorem 5) to classify the positive equilibrium.

9. Let $x = x(t)$ denote the temperature of an object. According to a modified Newton's law of cooling, x satisfies the equation

$$x' = -a\,(b - x)\,|b - x|^p,$$

where b is the constant environment temperature and a and p are positive constants. Find this equation's equilibrium, and then find the linearization at this equilibrium. What kind of equilibrium does the linearization have? If applicable, use the linearization principle (Theorem 5) to classify the equilibrium.

10. The concentration $c = c(t)$ of a substrate in a container in which an enzyme is present (in constant concentration $e > 0$) satisfies the equation

$$c' = d\,(c_{\text{in}} - c) - \frac{mc}{a + c}e.$$

All coefficients d, c_{in}, m and a are positive constants. In this exercise let $d = c_{\text{in}} = a = e = 1$ and $m = 2$. Find this equation's positive equilibrium, and then find the linearization at this equilibrium. What kind of equilibrium does the linearization have? If applicable, use the linearization principle (Theorem 5) to classify the positive equilibrium.

3.1.4 Bifurcations

Differential equations that arise in applications often contain unspecified numerical constants called parameters or coefficients. The radioactive decay equation

$$x' = -rx$$

has one parameter, r. The logistic equation

$$x' = r\left(1 - \frac{x}{K}\right)x$$

has two parameters, r and K. The "spruce budworm" equation

$$x' = r\left(1 - \frac{x}{K}\right)x - c\frac{x^2}{a + x^2}$$

has four parameters r, K, c, and a.

The graph of $f(x)$ and consequently the phase line portrait of an autonomous equation $x' = f(x)$ depend on the values assigned to the parameters appearing in $f(x)$. In applications it is often important to understand how changes in parameter values alter the phase line portrait. Parameters may change, for example, from naturally occurring events or from deliberate (or inadvertent) manipulations by humans. Moreover, in applications parameters have to be estimated numerically (e.g., from data) and therefore we must investigate the phase line portrait throughout a statistical confidence interval for these estimates. *Bifurcation theory* is the study of how changes in parameters alter the phase line portrait and the asymptotic dynamics of an equation.

To introduce some basic ideas, consider the homogeneous linear autonomous equation

$$x' = px, \tag{1.6}$$

where p is a constant. The phase line portrait is depends on the sign of p. Specifically, the equilibrium $x_e = 0$ is a sink if $p < 0$ and it is a source if $p > 0$. (See Fig. 3.8.) Thus, the phase line portrait changes in a significant way when the parameter p is increased (or decreased) through 0. Such a radical change in the phase line portrait of an equation is called a bifurcation. Thus, in the linear equation (1.6) a bifurcation occurs when p passes through 0. This critical value 0 of the parameter p is called a bifurcation value.

The phase line portrait of the linear equation $x' = -x + q$, however, is unaltered if q is changed. The equilibrium $x_e = q$ is a sink for all q. In this case, we say that the phase portraits remain qualitatively equivalent and that no bifurcation occurs.

To make the concept of bifurcation more precise we need a definition. We have been considering autonomous equations $x' = f(x)$ for which the roots of $f(x)$ (the equilibria) are isolated. From now on we assume more, namely that $f(x)$ has at most a finite number of roots (in which case they are necessarily isolated). Between each pair of consecutive roots, the function $f(x)$ is either positive or negative. The sign of $f(x)$ determines the orientation direction of the orbit between two consecutive roots in the phase line portrait. The set of roots of $f(x)$ and the sequence of signs of $f(x)$ between consecutive roots characterize the phase line portrait.

DEFINITION 6. Assume $f(x)$ is continuously differentiable for all x and has at most a finite number of roots. The number of equilibria of $x' = f(x)$ and the orientation directions of the nonequilibrium orbits (more precisely, the sequence of signs of $f(x)$ between consecutive equilibria) determine the "**structure of the phase line portrait**" (or the "**orbit structure**") of the equation.

Two differential equations might differ considerably but still have the same orbit structure. For this situation we have the following terminology.

> **DEFINITION 7.** Two phase line portraits are said to be **qualitatively equivalent** if they have the same orbit structure as defined in Definition 6.

EXAMPLE 14

The function $f(x) = (1 - x)e^{-x}$ has only the root 1. Since

$$\left. \frac{df}{dx} \right|_1 = -e^{-1} < 0,$$

the equilibrium $x_e = 1$ of the differential equation

$$x' = (1 - x)e^{-x}$$

is a sink. Figure 3.13 shows the phase line portrait for this equation. The phase line portrait of the equation $x' = -x$ also appears in Fig. 3.13. By Definitions 6 and 7, the phase portraits of these two equations are qualitatively equivalent. ∎

Equation	Phase Line Portrait
$x' = (1 - x)\,e^{-x}$	$\longrightarrow$ 1 $\longleftarrow$
$x' = -x$	$\longrightarrow$ 0 $\longleftarrow$

Figure 3.13 The phase line portraits in the neighborhood of an isolated equilibrium.

If two differential equations do not have the same number of equilibria, then their phase portraits cannot have the same orbit structure. Thus, for the qualitative equivalence of two phase line portraits it is *necessary* that they have the *same number* of equilibria. Having the same number of equilibria is *not sufficient* for qualitative equivalence, however, because the nonequilibrium orbit orientations might not be identical. The next example illustrates this.

EXAMPLE 15

Figure 3.14 shows the phase line portraits of the logistic equation

$$x' = rx(1 - \frac{x}{K})$$

and the equation

$$x' = rx^2(1 - \frac{x}{K}),$$

where $r > 0$ and $K > 0$. Both have equilibria $x_e = 0$ and K. However, the orientation of the nonequilibrium orbits differ and therefore the phase line portraits are not qualitatively equivalent. ∎

Equation	Phase Line Portrait
$x' = rx(1 - \frac{x}{K})$	$\longleftarrow$ 0 $\longrightarrow$ K $\longleftarrow$
$x' = rx^2(1 - \frac{x}{K})$	$\longrightarrow$ 0 $\longrightarrow$ K $\longleftarrow$

Figure 3.14

Sometimes phase portraits remain qualitatively equivalent as a parameter in an equation is changed. For example, the phase line portraits of the linear equation $x' = -x + q$ remain qualitatively equivalent for all values of q. Bifurcation theory, on the other hand, is concerned with the loss of qualitative equivalence as a parameter is changed. Here is an example.

EXAMPLE 16   The equation

$$x' = x^2 - p \tag{1.7}$$

has no equilibria if $p < 0$. If $p > 0$, this equation has two equilibria $x_e = \pm\sqrt{p}$. The phase line portraits for both cases appear in Fig. 3.15. The phase line portraits have the same orbit structure and therefore are qualitatively equivalent for all negative values of p. The same is true for all positive values of p. However, the phase line portraits for negative p are not qualitatively equivalent to those for positive p. They do not have the same number of equilibria. Thus, there is a change in orbit structure as p passes through 0. ∎

$x' = x^2 - p$	Phase Line Portrait
$p < 0$	$\longrightarrow$
$p > 0$	$\longrightarrow$ $-\sqrt{p}$ $\longleftarrow$ $\sqrt{p}$ $\longrightarrow$

Figure 3.15

In the preceding example a bifurcation occurs at the bifurcation value $p = 0$ because the orbit structure of the equation changes as the parameter p passes through 0. This motivates a general definition of a bifurcation value for an autonomous differential equation with a parameter.

Consider the equation

$$x' = f(x, p), \tag{1.8}$$

where a parameter p is included in the variable list of the function f. There may be other parameters in an equation, but we specify as p only the parameter whose effect on the phase line portrait we want to study. This parameter we designate as the bifurcation parameter. As the bifurcation parameter p is allowed to vary over a designated interval, we require that our basic assumptions hold: f is continuously differentiable (with respect to x) and has (at most) a finite number of roots.

DEFINITION 8. The phase line portrait of equation (1.8) is **stable at** p_0 if its orbit structure remains unchanged for all values of p in an interval centered on p_0. If the phase portrait of (1.8) is not stable at p_0, then a **bifurcation** occurs at $p = p_0$ and p_0 is called a **bifurcation value**.

A bifurcation occurs at $p_0 = 0$ for the linear equation $x' = px$ because the orbit structure is a sink for $p < 0$ and a source for $p > 0$. Any interval centered on $p_0 = 0$ contains both negative and positive values of p and therefore, by Definition 8, the phase portrait is unstable at $p_0 = 0$. In Example 16 $p_0 = 0$ is a bifurcation value for equation (1.7) for the same reason.

One graphical way to describe bifurcations of an equation (1.8) is to plot the equilibria as a function of the parameter p. This is the same as plotting the graph described by the equation

$$f(x, p) = 0$$

in the p, x-plane. The resulting graph is called an *orbit diagram* or, if bifurcations occur, a *bifurcation diagram*.

If, in addition, the equilibria type is indicated on the bifurcation diagram graph, then phase line portraits can be constructed from the graph at a selected value of p. For example, we might simply label the graph with letters or words; or one might indicate sinks (the stable equilibria) by solid lines and sources and shunts (the unstable equilibria) by dashed lines. For example, Fig. 3.16 shows the bifurcation diagram for the equation (1.7) in the Example 16. This is the graph of the equation $x^2 - p = 0$ in the p, x-plane. For $p > 0$ the source $x_e = \sqrt{p}$ is plotted as a dashed line and the sink $x_e = -\sqrt{p}$ is plotted as a solid line.

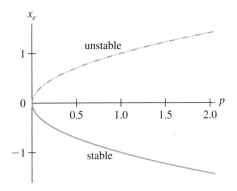

Figure 3.16 The bifurcation diagram for $x' = x^2 - p$ in which the equilibria $x_e = \pm\sqrt{p}$ are plotted against p.

One goal of bifurcation theory is to categorize different kinds of bifurcations. We consider only a few basic types.

The bifurcation in Example 16 is called a *saddle-node bifurcation*.[2] This basic type of bifurcation is characterized by the following features. On one side of the bifurcation value p_0 there are two equilibria (a source and a sink). These two equilibria merge to a single equilibrium x_e as p approaches p_0 in the limit. For p on the other side of p_0 there are no equilibria (at least near x_e). Thus, these two equilibria collide and annihilate each other as p passes through p_0. The bifurcation diagram of a saddle-node bifurcation has a parabolic shape that opens either to the right or the left and has its "nose" at the bifurcation value p_0. See Fig. 3.17.

Another fundamental type of bifurcation is illustrated in the next example.

EXAMPLE 17

The equation

$$x' = px - x^3$$

---

[2]We will learn the reason for this odd name in Chapter 8 (Sec. 6). We could also refer to this type of bifurcation as a *blue-sky bifurcation*, a term that colorfully captures the fact that the two equilibria involved suddenly appear, as if out of nowhere, as p passes through the critical value p_0.

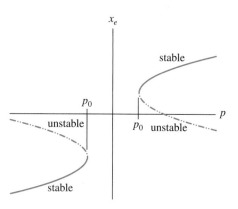

Figure 3.17 Saddle-node bifurcations.

has equilibrium $x_e = 0$ for all values of p and the equilibria $x_e = \pm\sqrt{p}$ for $p > 0$. Since $f(x, p) = px - x^3$ and

$$\left.\frac{\partial f}{\partial x}\right|_0 = p$$

the equilibrium $x_e = 0$ is a sink for $p < 0$ and a source for $p > 0$. For $p > 0$

$$\left.\frac{\partial f}{\partial x}\right|_{\pm\sqrt{p}} = -2p < 0$$

and both equilibria $x_e = \pm\sqrt{p}$ are sinks. For $p = 0$, the equation reduces to $x' = -x^3$ whose only equilibrium is the sink $x_e = 0$. A bifurcation occurs at $p_0 = 0$ because the orbit structure for $p < 0$, consisting of a single sink, is different from that for $p > 0$, which consists of two sinks and a source. All these facts are summarized by the bifurcation diagram in Fig. 3.18. This graph is found by solving the equation $px - x^3 = 0$ and plotting the solutions $x = 0$ and $p = x^2$. Notice the pitchfork shape of the graph. ∎

Figure 3.18 shows a typical *pitchfork bifurcation*. There are three equilibria for p on one side of p_0 and only one on the other side of p_0. The three equilibria merge to a single equilibrium x_e as p approaches p_0. A pitchfork bifurcation diagram may open to the right (as in Fig. 3.18) or to the left. (See Exercises 38, 39, and 40.) Typically the equilibria on the upper and lower branches of the pitchfork have the same stability properties (i.e., are all stable or unstable), and the equilibria on the middle branch have the opposite stability property from those on the outer branches.

The following example illustrates a third type of bifurcation.

EXAMPLE 18   The equation

$$x' = px - x^2$$

has equilibria $x_e = 0$ and $x_e = p$ for all values of p. Since $f(x, p) = px - x^2$ and

$$\left.\frac{\partial f}{\partial x}\right|_0 = p$$

the equilibrium $x_e = 0$ changes from a sink when $p < 0$ to a source for $p > 0$. Since

$$\left.\frac{\partial f}{\partial x}\right|_p = -p$$

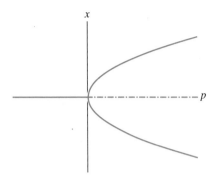

$x' = px - x^3$	**Phase Line Portrait**
$p < 0$	$\longrightarrow \quad 0 \quad \longleftarrow$
$p = 0$	$\longrightarrow \quad 0 \quad \longleftarrow$
$p > 0$	$\longrightarrow \quad -\sqrt{p} \quad \longleftarrow \quad 0 \quad \longrightarrow \quad \sqrt{p} \quad \longleftarrow$

Figure 3.18 The pitchfork bifurcation of $px - x^3$.

the equilibrium $x_e = p$ is a source for $p < 0$ and a sink for $p > 0$ (exactly the opposite of the situation for the equilibrium $x_e = 0$). For $p = 0$ the equation reduces to $x' = -x^2$ whose only equilibrium is a shunt at $x_e = 0$.

The orbit structure for $p \neq 0$ is different from the orbit structure for $p = 0$ and therefore a bifurcation occurs at $p_0 = 0$. The bifurcation diagram in Fig. 3.19 summarizes all these facts. ∎

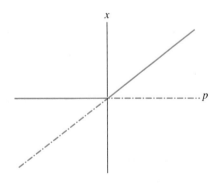

	Phase Line Portrait
$p < 0$	$\longleftarrow \quad p \quad \longrightarrow \quad 0 \quad \longleftarrow$
$p = 0$	$\longleftarrow \quad 0 \quad \longleftarrow$
$p > 0$	$\longleftarrow \quad 0 \quad \longrightarrow \quad p \quad \longleftarrow$

Figure 3.19 The transcritical bifurcation of $x' = px - x^2$.

The bifurcation in the preceding example is called *transcritical*. Transcritical bifurcations are characterized by the crossing of two equilibrium branches in the bifurcation diagram. There are two equilibria on each side of the bifurcation value p_0 which merge to a single equilibrium x_e at p_0. Moreover, the equilibrium type on each branch changes as p passes through p_0. For example, in Fig. 3.19 the equilibria on one branch change from sinks to sources and on the other branch from sources to sinks. This is called an *exchange of stability* and it is a typical feature of transcritical bifurcations.

It is possible for some equations to have more than one bifurcation.

EXAMPLE 19

Consider the equation

$$x' = 3x - x^3 - p.$$

The roots of the cubic polynomial $3x - x^3 - p$ and hence the equilibria of this equation are not easily found algebraically. However, as we will see, it is not necessary to calculate the roots in order to draw the bifurcation diagram.

The bifurcation diagram is the graph in the p, x-plane associated with the (equilibrium) equation

$$3x - x^3 - p = 0.$$

In principle, we want to solve this equation for x and graph the result as a function of p. A simpler way to obtain this graph, however, is to do the just opposite: Solve the equation for p in terms of x and graph p as a function of x. We can then obtain the desired bifurcation diagram by reflecting this graph through the $45°$ line $p = x$. Thus, we solve for

$$p = 3x - x^3$$

and graph this cubic polynomial in Fig. 3.20(a). The bifurcation diagram, obtained by reflecting this graph through the line $p = x$, is shown in Fig. 3.20(b).

From the bifurcation diagram we observe that there are two saddle-node bifurcations. One is located at $p = 2$ and the other at $p = -2$. We also see that there is one equilibrium for $p < -2$ and one equilibrium for $p > 2$. For p between -2 and 2, however, there are three equilibria.

To determine the type of equilibria in the bifurcation diagram we can apply the derivative test. From $f(x) = 3x - x^3 - p$ we obtain

$$\frac{df}{dx} = 3\left(1 - x^2\right).$$

It follows that

$$\left.\frac{df}{dx}\right|_{x_e} = 3\left(1 - x_e^2\right) \quad \begin{cases} < 0 & \text{for those equilibria } x_e < -1 \text{ and } x_e > 1 \\ > 0 & \text{for those equilibria } -1 < x_e < 1. \end{cases}$$

From this we have the following phase line portraits:

For $p < -2$: $\longrightarrow$ $\circ$ $\longleftarrow$

For $-2 < p < 2$: $\longrightarrow$ $\circ$ $\longleftarrow$ $\circ$ $\longrightarrow$ $\circ$ $\longleftarrow$

For $p > 2$: $\longrightarrow$ $\circ$ $\longleftarrow$

The bifurcation diagram in Fig. 3.20 graphically summarizes this information.   ∎

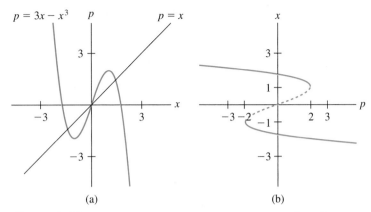

Figure 3.20 The bifurcation diagram for $x' = 3x - x^3 - p$ in (b) is obtained by reflecting the graph of the cubic $p = x^3 - 3x$ in (a) through the line $p = x$. The solid line consists of stable equilibria (sinks). The dashed line consists of unstable equilibria (sources).

The method we used to obtain the bifurcation diagram in Example 19 is often a convenient one. The problem of drawing the bifurcation diagram associated with a first-order equation $x' = f(x, p)$ is the problem of drawing the graph defined by the equilibrium equation

$$f(x, p) = 0$$

in the p, x-plane. Ideally we can do this by algebraically solving the equation for x and graphing the answer as a function of p. However, since we are concerned with nonlinear equations, it is usually not easy to solve this equation for x. However, it is frequently the case that the parameter p appears in the equation in a simpler algebraic way than x does. In such a case we can usually more easily solve the equation for p, instead of x. If we do this and plot the answer p as a function of x in the x, p-plane, we obtain the sought after bifurcation diagram by reflecting the resulting graph through the line $p = x$.[3]

In applications bifurcations often play an important and crucial role. Here is an example. The equation

$$x' = r\left(1 - \frac{x}{K}\right)x - c\frac{x^2}{a + x^2} \qquad (1.9)$$

has been used to describe the dynamics of spruce budworm populations. The variable x denotes the number or density of the budworm population. Outbreaks of this defoliating insect have caused major deforestations in Canada and the United States. One explanation that has been given for the occurrence of outbreaks is based on the multiple bifurcations that occur in the equation (1.9).

---

[3] Another way to accomplish the same thing is to reflect the graph through the vertical p-axis and then rotate the result $90°$ clockwise.

As an example, consider equation (1.9) with $a = 0.01$, $c = 1$, and $K = 1$ and $p = r$ as a parameter:

$$x' = rx\,(1 - x) - \frac{x^2}{0.01 + x^2}. \tag{1.10}$$

To obtain the bifurcation diagram for this equation, we use the procedure described above. That is, we solve the equilibrium equation

$$rx\,(1 - x) - \frac{x^2}{0.01 + x^2} = 0$$

for the parameter

$$r = \frac{1}{x\,(1 - x)}\frac{x^2}{0.01 + x^2}$$

and plot the result in the x, r-plane. This plot is shown in Fig. 3.21(a). We obtain the bifurcation diagram in Fig 3.21(b) by reflecting the graph through the $r = x$ line.

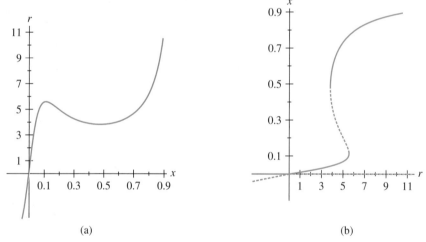

(a)   (b)

Figure 3.21 The bifurcation diagram for the spruce budworm equation (1.10) in (b) is obtained by reflecting the graph of

$$r = \frac{1}{x\,(1 - x)}\frac{x^2}{0.01 + x^2}$$

in (a) through the $r = x$ line. Dashed lines indicate unstable equilibria. There is a transcritical bifurcation at $r = 0$ and there are saddle-node bifurcations located at $r \approx 3.84$ and 5.55. Selected phase line portraits are shown below.

$r = 3$	$r = 5$	$r = 7$
← 0 → ∘ ←	← 0 → ∘ ← ∘ → ∘ ←	← 0 → ∘ ←

In Fig. 3.21(b) we observe three bifurcations associated with equation (1.10). A transcritical bifurcation occurs at $r = 0$. For $r < 0$ the equilibrium $x_e = 0$ is a sink (and the budworm population goes extinct), whereas for $r > 0$ an exchange of stability occurs and there is a positive sink (and the budworm persists). The remaining two bifurcations in Fig. 3.21(b) are saddle-node bifurcations similar to those in Fig. 3.20 in Example 19. Because of these saddle-node bifurcations, the sink undergoes discontinuous changes as r passes through the two bifurcation values located approximately at $r \approx 3.84$ and 5.55. For example, if r is increased from small values (where the sink is also small) to a value larger than 5.55, then the sink discontinuously jumps to a higher level. This indicates a spruce budworm outbreak.

The bifurcation diagram in Fig. 3.21(b) contains another important feature. If some kind of control measures are put into effect to decrease r, in an attempt to reverse the spruce budworm outbreak and infestation, the budworm population (now at the higher equilibrium level with $r > 5.55$) will not return to the lower equilibrium level until r is decreased below the smaller critical value 3.84. At that point there is a collapse of the population to the lower sink. Interestingly, $r = 3.84$ (at which the outbreak is eradicated) is less than $r = 5.55$ (at which the outbreak occurs). This phenomenon is called *hysteresis*. It occurs in many other applications as well.

Exercises

Which pairs of equations have qualitatively equivalent phase line portraits? Which do not and why?

1. $\begin{cases} x' = 2 - 3x \\ x' = 3 - 2x \end{cases}$

2. $\begin{cases} x' = 2 + 3x \\ x' = 3 + 2x \end{cases}$

3. $\begin{cases} x' = px + q, \, p < 0 \\ x' = -a^2x + 1, \, a \neq 0 \end{cases}$

4. $\begin{cases} x' = -x + q \\ x' = px + q, \, p > 0 \end{cases}$

5. $\begin{cases} x' = x - e^{-x} \\ x' = -x \end{cases}$

6. $\begin{cases} x' = x - e^{-x} \\ x' = x \end{cases}$

7. $\begin{cases} x' = (x - 1)(x + 2)^2 \\ x' = (x - 2)^3(x + 1)^{10} \end{cases}$

8. $\begin{cases} x' = (2x - 1)(x - 2)^3 \\ x' = (2x - 1)^2(x - 2)^3 \end{cases}$

9. $\begin{cases} x' = 1 - 2e^{-x^2} \\ x' = 2 - 2e^{-x^2} \end{cases}$

10. $\begin{cases} x' = 1 + bx + x^2, \, b > 2 \\ x' = px + q, \, p > 0 \end{cases}$

11. $\begin{cases} x' = -1 + 5x - x^2 \\ x' = x(1 - x) \end{cases}$

12. $\begin{cases} x' = e^x \dfrac{x^2 - 1}{x^2 + 1} \\ x' = x^4 - 1 \end{cases}$

13. $\begin{cases} x' = 1 - x + x^2 - x^3 \\ x' = -x \end{cases}$

14. $\begin{cases} x' = 1 - x^4 \\ x' = 1 - x^3 \end{cases}$

15. $\begin{cases} x' = x(1 + x^2) \\ x' = x\left(1 - x^2\right) \end{cases}$

16. $\begin{cases} x' = -2 + \sin 2x \\ x' = \dfrac{1}{1 + e^x} \end{cases}$

17. $\begin{cases} x' = 1 \\ x' = e^{-x} \end{cases}$

18. $\begin{cases} x' = (1 + \sin^2 x)(e^x - 1) \\ x' = x \end{cases}$

Describe the phase line portraits of the following equations and how they depend on the value of p. Classify all bifurcations (if possible).

19. $x' = p - x^2, \, -\infty < p < +\infty$

20. $x' = px + (x - 1)(x - 4), \, p > 0$

21. $x' = x\left(x^2 - 1 - p\right), \, -\infty < p < +\infty$

22. $x' = x\left(p - x^2\right), \, -\infty < p < +\infty$

23. $x' = x^2 - p^2, \, -\infty < p < +\infty$

24. $x' = x^2 - px, \, -\infty < p < +\infty$

25. $x' = px - \sin x, \, p > 2/\pi$

26. $x' = (x - p)^2, \, -\infty < p < +\infty$

27. $x' = px^2, \, -\infty < p < +\infty$

28. $x' = px^3, \, -\infty < p < +\infty$

29. $x' = p + x(1 - x)(x - 2), \, -\infty < p < +\infty$

30. $x' = p - x^3, \, -\infty < p < +\infty$

31. $x' = p - x^4, \, -\infty < p < +\infty$

32. $x' = x^2p - 1, \, -\infty < p < +\infty$

33. $x' = p - e^{-x^2}, \, p > 0$

34. $x' = x^2\left(p - e^{-x^2}\right), \, p > 0$

35. Draw the bifurcation diagram for each equation in Exercises 19–34 as a function of p on the indicated interval.

36. Show that all phase line portraits of the logistic equation $x' = rx(1 - x/K)$, $K > 0, r > 0$, are qualitatively equivalent.

37. A logistically growing population $x = x(t)$ satisfies the equation $x' = rx(1 - x/K)$, $r > 0$, $K > 0$. If the population is harvested at a constant rate $h > 0$, then $x' = rx(1 - x/K) - h$. Draw a bifurcation diagram using $h > 0$ as a parameter. Identify any bifurcations that occur. Sketch the relevant phase line portraits. What are the biological implications of the bifurcation diagram?

38. Show that the bifurcation diagram of the equation $x' = -px - x^3$ is a pitchfork that opens to the left. Describe the stability properties of the branches.

39. Show that the bifurcation diagram of the equation $x' = -px + x^3$ is a pitchfork that opens to the right. Describe the stability properties of the branches.

40. Write down a differential equation whose bifurcation diagram is a pitchfork that opens to the left in which a source is exchanged between the branches.

41. Definition 7 of qualitatively equivalent phase portraits is based upon the geometry of the portraits. It turns out that although this definition works well for first-order equations, it is too simplistic for systems of equation (and higher-order equations). For these higher-dimensional problems a more general definition is needed. One common definition is based on the topological notion that two phase portraits are equivalent if one can be continuously distorted into the other. Mathematicians make this notion precise by using homeomorphisms. A homeomorphism is a continuous function $h: R \to R$ that has a continuous inverse. Two equations are said to have qualitatively equivalent (or topologically equivalent) phase portraits if by making a change of variables from x to $h(x)$ one phase portrait is mapped to the other such that orbits go to orbits with their orientations preserved. Analytically this can be tested as follows. Consider two autonomous first-order equations

$$x' = f(x)$$
$$x' = g(x)$$

and let $x = x(t, x_0)$ and $x = \psi(t, x_0)$ be the solutions with initial value $x(0) = x_0$. The phase portraits are qualitatively equivalent if there exists a homeomorphism $h(x)$ such that

$$h(x(t, x_0)) = \psi(t, h(x_0))$$

for all t in the domains of the solutions.

(a) Show that

$$h(x) = \begin{cases} x^2 & \text{if } x > 0 \\ 0 & \text{if } x = 0 \\ -x^2 & \text{if } x < 0 \end{cases}$$

defines a homeomorphism.

(b) Use the homeomorphism in (a) to show that the phase portraits of the equations

$$x' = -x$$
$$x' = -2x$$

are qualitatively equivalent.

42. Using the homeomorphism in Exercise 41 as a guide, construct a homeomorphism $h(x)$ and use it to show that the phase portraits of the two linear equations

$$x' = ax, \quad a \neq 0$$
$$x' = bx, \quad b \neq 0$$

are qualitatively equivalent if a and b have the same sign.

3.2 Separable Equations

In Sec. 3.1 we learned how to study the solutions of an autonomous differential equation $x' = f(x)$ by constructing phase line portraits. It is also sometimes useful to have formulas for solutions. In this section we learn a method that (at least in principle) can produce formulas for the general solution of an autonomous equation and, in fact, for a more general class of equations called *separable*.

To motivate the method we consider the linear autonomous equation

$$x' = x.$$

We can find a solution formula using a method different from that used in Chapter 2—a method that can also be applied to nonlinear equations. If we divide both sides of the equation by x and notice from the chain rule that

$$\frac{d}{dt}\ln|x| = \frac{1}{x}x',$$

we arrive at the equation

$$\frac{d}{dt}\ln|x| = 1.$$

[In dividing by x we have ignore the possibility that a solution $x = x(t)$ might equal zero for some t. We return to this point below.] If the derivative of an expression identically equals 1, then that expression must, by the fundamental theorem of calculus, equal $t + k$ for some constant k. In other words, integrating both sides of this equation, we obtain

$$\ln|x| = t + k,$$

where k is an arbitrary constant. This equation defines an *implicit solution* of the differential equation. It is called an "implicit" solution because the equation has not been solved explicitly for x. We can solve for x by exponentiating both sides of the equation to obtain

$$|x| = e^{t+k}$$

or

$$x = \pm e^{t+k},$$

which we rewrite as

$$x = \left(\pm e^{k}\right) e^{t}.$$

Because k is an arbitrary real number, e^k is an arbitrary positive real number. Thus, $\pm e^k$ represents an arbitrary positive or negative real number. If, to simplify notation, we give this expression a new name

$$c = \pm e^{k},$$

then c is an arbitrary *nonzero* real number.

We have obtained a set of solutions of the equation $x' = x$ described by the formula

$$x = ce^{t}, \quad \text{where } c \text{ is an arbitrary nonzero constant.}$$

However, this formula does not describe the general solution because it is missing (at least) the solution $x = 0$. The method missed finding the solution $x = 0$ because of the division by x carried out in the initial step. However, we can include this exceptional solution in the solution formula above by dropping the word nonzero. Even so, the question remains: Have we missed any other solutions? In other words, is

$$x = ce^{t}, \quad \text{where } c \text{ is an arbitrary constant,}$$

the general solution of the equation $x' = x$? From our study of linear equations in Chapter 2 we know the answer is yes. However, if we apply this solution method to other types of equations (and in particular to nonlinear equations) the answer might not be so obvious. The next example illustrates this.

EXAMPLE 1   Consider the first-order autonomous equation

$$x' = -x^2. \tag{2.1}$$

Ignoring the possibility that $x = x(t)$, and hence $x^2(t)$, might equal zero for some value of t, we divide both sides of the equation by $-x^2$ and write

$$-\frac{1}{x^2}x' = 1$$

$$\frac{d}{dt}\left(\frac{1}{x}\right) = 1.$$

After an antidifferentiation we arrive at the implicit solution

$$\frac{1}{x} = t + c,$$

where c is an arbitrary constant.[4] Solving for x, we obtain a set of explicit solutions

$$x = \frac{1}{t+c}.$$

However, this set of solutions is missing the solution $x = 0$. If we include this solution we get a set of solutions

$$x = \begin{cases} \dfrac{1}{t+c} & c \text{ is an arbitrary constant.} \\ 0 \end{cases} \tag{2.2}$$

∎

Is (2.2) the general solution of the equation (2.1)? Or are there other solutions? This question about the completeness of the solution set arises whenever the method used above is applied to an equation

$$x' = f(x).$$

Formally, we ignore the possibility that $f(x(t))$ might equal 0 for some value of t and write

$$\frac{1}{f(x)}x' = 1.$$

Next we recognize (using the chain rule) the left-hand side of this equation as the derivative of

$$\int^x \frac{1}{f(s)}\,ds$$

and write the equation as

$$\frac{d}{dt}\int^x \frac{1}{f(s)}\,ds = 1.$$

---

[4]The reader might wonder why a constant of integration is introduced for the integration on the right-hand side but not for the integration the left-hand side. If this were done, we would have $x^{-1}+c_1 = t+c_2$ for two arbitrary constants of integration. However, this formula can be rewritten $x^{-1} = t + c$, where $c = c_2 - c_1$.

After integrating both sides we obtain

$$\int^x \frac{1}{f(s)}\, ds = t + c, \tag{2.3}$$

where c is an arbitrary constant. These steps lead to a set of implicitly defined solutions of the original differential equation. The formula can only be made explicit in specific examples if we can carry out the integration and solve the resulting equation for x. We have seen in the examples above, however, that the formula (2.3) may not be the general solution because some solutions may be missing.

How can we find missing solutions? The clue lies in the original assumption, namely that $f(x(t))$ is never equal to zero. Because of this assumption the method will not find any solution $x = x(t)$ of the differential equation for which $f(x(t))$ equals zero for some value of t [i.e., any solution that equals a root of $f(x)$ for some value of t]. Recall that the roots of $f(x)$ are equilibria (constant solutions). Also recall that nonequilibrium solutions $x(t)$ can never equal an equilibrium x_e for any value of t (see the proof of Theorem 1). It follows that the method will find all nonequilibrium solutions. Formally, the four steps of the method produce four equivalent equations

$$x'(t) = f(x(t))$$

$$\frac{1}{f(x(t))} x'(t) = 1$$

$$\frac{d}{dt} \int^{x(t)} \frac{1}{f(x(s))}\, ds = 1$$

$$\int^{x(t)} \frac{1}{f(x(s))}\, ds = t + c$$

for nonequilibrium solutions. By adding all equilibria to this set of solutions, we obtain the general solution.

A more concise, shorthand way of writing the solution method is as follows.

$$\frac{dx}{dt} = f(x)$$

$$\frac{1}{f(x)} dx = dt$$

$$\int \frac{1}{f(x)} dx = \int dt$$

$$\int \frac{1}{f(x)} dx = t + c$$

We can also use the solution method above on some nonautonomous equations. Here is an example. The equation

$$x' = -2tx \tag{2.4}$$

is a particular case of the equation

$$x' = -kt^p x, \quad k > 0, \quad p > 0, \tag{2.5}$$

which arises in a model for the spread of AIDS in a population infected with the human immunodeficiency virus HIV. See Sec. 6.1 of Chapter 2. We can find solution formulas for equation (2.4) as follows:

$$\frac{dx}{dt} = -2tx$$

$$\frac{1}{x} dx = -2t \, dt$$

$$\int \frac{1}{x} dx = -\int 2t \, dt + k$$

$$\ln |x| = -t^2 + k$$

$$|x| = e^{-t^2 + k}$$

$$x = \pm e^k e^{-t^2}$$

$$x = ce^{-t^2},$$

where $c = \pm e^k$ is an arbitrary, nonzero constant. The solution $x = 0$ is not included in this formula (because of the division by x in the first step of the solution method). We can include this solution in the formula, however, by allowing c to equal 0, with the result that the general solution is given by the formula

$$x = ce^{-t^2}, \quad \text{where } c \text{ is an arbitrary constant.} \tag{2.6}$$

The procedure used to solve equation (2.4) is called the *separation of variables method*. This name derives from the first step in which the dependent variable x and the independent variable t are separated to opposite sides of the equation (including the differentials dx and dt). This key step in the method is possible for equation (2.4) because the right-hand side of the equation $f(t, x) = -2tx$ is multiplicatively separable [i.e., it is a product of two factors, one depending only on t and the other only on x]. The method of separating variables can be applied to any equation $x' = f(t, x)$ for which $f(t, x)$ is multiplicatively separable in this way. This suggests the following definition.

DEFINITION 9. The first-order equation $x' = f(t, x)$ is called **separable** if $f(t, x)$ is multiplicatively separable in the variables t and x, that is, if it can be written in the factored form $f(t, x) = g(t)h(x)$.

We can find solution formulas for a separable equation

$$\frac{dx}{dt} = g(t)h(x)$$

as follows. Ignoring for the moment the possibility that $h(x)$ might equal 0, we write

$$\frac{dx}{dt} = g(t)h(x)$$

$$\frac{1}{h(x)} dx = g(t) \, dt$$

$$\int \frac{1}{h(x)}\,dx + c_1 = \int g(t)\,dt + c_2$$

$$\int \frac{1}{h(x)}\,dx = \int g(t)\,dt + c,$$

where the two constants of integration c_1 and c_2 have been put together as the single arbitrary constant $c = c_2 - c_1$. If the two integrals can be calculated (by hand and/or with the help of integral tables or a computer), the resulting equation in x and t defines a set of solutions. Does this set include *all* solutions of the equation?

The four equations in the steps above are equivalent for solutions such that $h(x(t)) \neq 0$ (for all t on the interval of definition). Therefore, the method of separating of variables finds all such solutions. Note that that a root x_e of $h(x)$ defines an equilibrium (a constant solution). In Exercise 37 the reader is asked to show the *only* solutions for which $h(x(t))$ equals 0 at some value of t are equilibrium solutions. It follows that the general solution is given by

$$\int \frac{1}{h(x)}\,dx = \int g(t)\,dt + c, \quad c = \text{arbitrary constant}$$

together with the set of all equilibrium solutions [i.e., the roots of $h(x)$]. To find the general solution of a particular separable equation we can carry out the method, by separating variables and integrating, or just use this formula.

With a formula for the general solution in hand, we can solve any initial value problem. The arbitrary constant is chosen so that the initial condition is met. Alternatively, we can solve an initial value problem directly by the separation of variables method by incorporating the initial condition into the method. This is done by calculating definite integrals (instead of indefinite integrals). To solve the initial value problem

$$x' = g(t)h(x)$$
$$x(t_0) = x_0$$

we proceed as follows. If x_0 is an equilibrium [i.e., a root of $h(x)$], the solution is this equilibrium: $x(t) = x_0$. If x_0 is not an equilibrium, then we calculate

$$\frac{dx}{dt} = g(t)h(x)$$

$$\frac{1}{h(x)}\,dx = g(t)\,dt$$

$$\int_{x_0}^{x} \frac{1}{h(s)}\,ds = \int_{t_0}^{t} g(s)\,ds.$$

By carrying out the indicated definite integrals, we arrive can arrive at a formula for the solution of the initial value problem.

For example, we can find a formula for the solution of the initial value problem

$$x' = -x^2$$
$$x(1) = 10$$

by the steps

$$\frac{dx}{dt} = -x^2$$

$$-\frac{1}{x^2}\,dx = dt$$

$$-\int_{10}^{x}\frac{1}{s^2}\,ds = \int_{1}^{t} ds$$

$$\frac{1}{x} - \frac{1}{10} = t - 1$$

$$x = \frac{10}{10t - 9}.$$

The next example uses separation of variables to find a formula for the solution of the initial value problem for the famous logistic equation.

EXAMPLE 2   Consider the initial value problem

$$x' = r\left(1 - \frac{x}{K}\right)x$$

$$x(t_0) = x_0.$$

The roots of $f(x) = rx(1 - x/K)$ are 0 and K. If $x_0 = 0$, the solution is the equilibrium solution $x_e = 0$. Similarly, if $x_0 = K$, then the solution is the equilibrium solution $x_e = K$. If x_0 does not equal 0 or K, then the solution is a nonequilibrium solution. To find a formula for this solution we use the separation of variables method as follows.

$$\frac{dx}{dt} = rx\left(1 - \frac{x}{K}\right)$$

$$\frac{1}{x\left(1 - \frac{x}{K}\right)}\,dx = r\,dt$$

$$\int_{x_0}^{x}\frac{1}{s\left(1 - \frac{s}{K}\right)}\,ds = \int_{t_0}^{t} r\,ds$$

$$\int_{x_0}^{x}\left(\frac{1}{s} + \frac{\frac{1}{K}}{1 - \frac{s}{K}}\right)ds = r(t - t_0)$$

$$\ln\left|\frac{x}{1 - \frac{x}{K}}\frac{1 - \frac{x_0}{K}}{x_0}\right| = r(t - t_0)$$

$$\left|\frac{x}{1 - \frac{x}{K}}\right| = \left|\frac{x_0}{1 - \frac{x_0}{K}}\right|e^{r(t-t_0)}$$

This equation defines the solution x implicitly. To find an explicit formula we must solve this equation for x. Eliminating the absolute value signs, we obtain

$$\frac{x}{1 - \frac{x}{K}} = \pm \frac{x_0}{1 - \frac{x_0}{K}} e^{r(t-t_0)}.$$

Which sign, $+$ or $-$, should we use? Setting $t = t_0$ and $x = x_0$ shows that the $+$ sign is required. Solving for x we get, after some algebraic manipulations, the explicit solution

$$x(t) = \frac{x_0 K}{x_0 + (K - x_0) e^{-r(t-t_0)}}. \tag{2.7}$$

∎

The solution formula (2.7) shows that solutions with positive initial conditions tend to K as $t \to +\infty$, while solutions with negative initial conditions tend to $-\infty$. We can obtain these conclusions in a simpler way from the phase line portrait of the equation (see Example 15 of Sec. 3.1 and Fig. 3.14). On the other hand, more details about the solutions are available from the solution formula (for example, numerical values for x at specific numerical values of t).

The initial value problem in Example 2 involves an autonomous equation. The next example involves a nonautonomous equation.

EXAMPLE 3   For all constant growth rates $r > 0$ all solutions of the logistic equation

$$x' = rx \left(1 - \frac{x}{K}\right)$$

with positive initial conditions $x_0 > 0$ tend to the carrying capacity $K > 0$ as $t \to +\infty$. The equation

$$x' = r (1 + \cos t) x \left(1 - \frac{x}{K}\right) \tag{2.8}$$

is a modification of the logistic equation in which the growth rate oscillates between $2r$ and 0 with period 2π and an average of r. This modification accounts for oscillations in birth and death rates that might be due, for example, to seasonal fluctuations in life cycles, food and water supplies, temperature, and so on.

Figure 3.22 shows the graphs of some numerically computed solutions for the case $r = 0.25$ and $K = 1$. These graphs indicate that solutions with positive initial conditions tend to $K = 1$ as $t \to +\infty$. This suggests a general conjecture: Solutions of the modified logistic equation (2.8) with positive initial conditions $x_0 > 0$ tend to K as $t \to +\infty$. We can prove this conjecture using the solution formula obtained by separating variables.

First note $x = 0$ and K are equilibrium solutions. For $x_0 \neq 0$ and $x_0 \neq K$ we separate variables as follows:

$$\int_{x_0}^{x} \frac{1}{\left(1 - \frac{s}{K}\right) s} \, ds = \int_{0}^{t} r (1 + \cos s) \, ds$$

$$\int_{x_0}^{x} \left(\frac{1}{s} + \frac{1}{K - s}\right) ds = r (s + \sin s)|_0^t$$

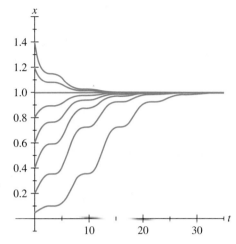

Figure 3.22 Solutions of (2.8) for $r = 0.25$ and $K = 1$ and a selection of initial conditions.

$$\ln\left|\frac{s}{K-s}\right|\Big|_{x_0}^{x} = r\,(t + \sin t)$$

$$\ln\left(\left|\frac{x}{K-x}\right|\,\left|\frac{x_0}{K-x_0}\right|^{-1}\right) = r\,(t + \sin t).$$

The last equation implicitly defines the solution x. To find an explicit formula we solve this equation for x:

$$\left|\frac{x}{K-x}\right|\,\left|\frac{x_0}{K-x_0}\right|^{-1} = e^{r(t+\sin t)}$$

$$\left|\frac{x}{K-x}\right| = \left|\frac{x_0}{K-x_0}\right|e^{r(t+\sin t)}$$

$$\frac{x}{K-x} = \pm\frac{x_0}{K-x_0}e^{r(t+\sin t)}.$$

By setting $t = 0$, we see that $+$ is the appropriate choice of sign. Solving for x, we obtain the explicit solution formula[5]

$$x = K\frac{x_0}{x_0 + (K - x_0)\,e^{-r(t+\sin t)}}. \tag{2.9}$$

This formula also gives the equilibrium solutions when $x_0 = 0$ and $x_0 = K$.
    From (2.9) we find

$$\lim_{t\to+\infty} x = \lim_{t\to+\infty} K\frac{x_0}{x_0 + (K - x_0)\cdot 0} = K$$

when $x_0 > 0$, as conjectured above.   ∎

---

[5]A word of caution: Some computer programs that perform symbolic calculus do not successfully obtain this general solution. This is because they use $\int \frac{1}{x}\,dx = \ln x + c$ instead of $\int \frac{1}{x}\,dx = \ln|x| + c$.

We give one more example to illustrate a final point. Whether or not we need a solution formula for a differential equation depends on the questions we want to answer. In fact, solution formulas do not always provide an easy way to answer some questions.

EXAMPLE 4

We can find a formula for the general solution of the autonomous equation

$$x' = \frac{1}{3}\left(1 - x^3\right)$$

by separating variables:

$$\frac{3}{1 - x^3}dx = dt$$

$$\int \frac{3}{1 - x^3}\,dx = \int dt + c.$$

The integral on the left-hand side can be calculated using a computer program, a table of integrals, or by hand (using partial fraction decomposition). From this calculation we obtain

$$\ln\left|\frac{\sqrt{x^2 + x + 1}}{x - 1}\right| + \sqrt{3}\tan^{-1}\left(\frac{2x + 1}{\sqrt{3}}\right) = t + c. \tag{2.10}$$

This equation implicitly defines all nonequilibrium solutions. (The only equilibrium is $x_e = 1$.) The equation cannot be algebraically solved explicitly for x. Moreover, because of the complexity of equation (2.10), we cannot easily use it to determine what happens to x as $t \to +\infty$. The answer, however, is rather easily obtained by applying the methods of Sec. 3.1.2 to $f(x) = \left(1 - x^3\right)/3$ to obtain the phase line portrait

$$\longrightarrow \quad 1 \quad \longleftarrow$$

All solutions approach the equilibrium $x_e = 1$ as $t \to +\infty$ and do so in a strictly monotonic fashion. ∎

Exercises

Find formulas for the general solution of the following equations. Implicit formulas are acceptable.

1. $x' = 1 + x^2$

2. $x' = 1 - x^2$

3. $x' = x - \dfrac{1}{x}$

4. $x' = 1 - x^4$

5. $x' = \cot x$

6. $x' = x^2(1 - x)$

Find formulas for the solutions of the following initial value problems.

7. $x' = -x^4, x(1) = 1$

8. $x' = 1 + x^2, x(\pi) = 1$

9. $x' = -\dfrac{1}{x}e^{-x}, x(0) = -1$

10. $x' = \dfrac{x^4 - 1}{x^3}, x(0) = \sqrt{2}$

11. $x' = x - \dfrac{1}{x}, x(0) = \dfrac{1}{2}$

12. $x' = x - \dfrac{1}{x}, x(0) = 2$

Find explicit formulas for the solutions of the following initial value problems.

13. $x' = x - \dfrac{1}{x}, x(0) = -\dfrac{1}{2}$

14. $x' = x - \dfrac{1}{x}, x(0) = -3$

15. The velocity $v = v(t) \geq 0$ of a falling object subject to constant force of gravity and a quadratic law for the frictional force of air resistance satisfies the equation

$$v' = 9.8 - k_0 v^2.$$

Here 9.8 (meters/s^2) is acceleration due to gravity and k_0 is the (per unit mass) coefficient of friction.

(a) Find an explicit formula for the general solution of this equation.

(b) If the object is dropped, then $v(0) = 0$. Find a formula for the solution of this initial value problem.

(c) Use your answer in (b) to calculate the limiting velocity $\lim_{t \to +\infty} v(t)$.

16. A modified Newton's law of cooling yields the equation

$$x' = -a\,(b - x)\,|b - x|^p$$

for the temperature of an object. Here the constant b is the environmental temperature. The positive constants a and p depend on properties of the object (its material, geometry, etc.).

(a) Find a formula for the general solution when $p = 1/3$.

(b) Let $x_0 > b$ be the initial temperature of the object. Find a formula for the solution of the initial value problem.

(c) Use your answer in (b) to calculate the long-term temperature of the object (i.e., $\lim_{t \to +\infty} x$).

17. Suppose $f(x)$ is a continuously differentiable function of x and suppose $x = y(t)$ is a solution of the autonomous equation $x' = f(x)$. Show that if $f(y(t^*)) = 0$ for some value of $t = t^*$, then $y(t)$ must be an equilibrium. [*Hint:* Note that $y(t^*)$ is a root of $f(x)$ and apply Theorem 1 of Chapter 1 to the initial value problem $x' = f(x)$, $x(t^*) = y(t^*)$.]

18. Find the solution of the initial value problem $x' = x^2 - x$, $x(0) = x_0$ and determine its maximal interval of existence $\alpha < t < \beta$. Show that if $x_0 > 1$, then $\beta < +\infty$.

19. Find the solution of the initial value problem $x' = x^2 - 1$, $x(0) = x_0$ and determine its maximal interval of existence $\alpha < t < \beta$. Show that if $x_0 > 1$ then $\beta < +\infty$.

Find formulas for the general solutions of the following equations.

20. $x' = t^2 x$

21. $x' = \dfrac{x}{t}$

22. $x' = \dfrac{x^2}{t^2}$

23. $x' = e^{t+x}$

24. $x' = t - tx^2$

25. $x' = t^2 \tan x$

26. $x' = (x^2 - 3x + 2)e^t$

27. $x' = (t + 1)x^{-4}$

28. $x' = (a^2 - x^2)\cos t$

29. $x' = x^a t^b$

Find formulas for the solutions of the following initial value problems.

30. $x' = t^2 x,\ x(0) = 1$

31. $tx' + \sqrt{x} = 0,\ x(1) = 2$

32. $x' = (t + 1)x^{-4},\ x(0) = -1$

33. $x' = e^{t+x},\ x(0) = 0$

34. $x' = 2tx^{-a},\ x(1) = b$

35. $x' = 2at(x^2 + a^2),\ x(0) = 0$

36. Find a formula for the solution of the initial value problem $x(0) = x_0 > 0$ for equation (2.5). Use your formula to find $\lim_{t \to +\infty} x(t)$.

37. Suppose $x = x(t)$ is a solution of a separable equation $x' = g(t)h(x)$ and suppose $h(x(t^*)) = 0$ for some value of $t = t^*$. Show $x(t)$ must be an equilibrium. [*Hint:* Note that $x(t^*)$ is a root of $h(x)$ and apply Theorem 1 of Chapter 1 to the initial value problem $x' = g(t)h(x),\ x(t^*) = x(t^*)$.]

3.3 Change of Variables

A common method, used throughout mathematics, for solving equations is to change variables. Often a difficult mathematical equation can be transformed into a simpler equation by an appropriate change of variables. This is true for many specialized kinds of differential equations. We give two examples: Bernoulli and Ricatti equations.

An equation of the form

$$x' = p_1(t)x + p_2(t)x^n, \quad n \neq 0,\ 1 \tag{3.1}$$

is called a *Bernoulli equation*.

For example, the logistic equation

$$x' = r\left(1 - \frac{x}{K}\right)x = rx + \left(-\frac{r}{K}\right)x^2$$

is a Bernoulli equation with $p_1(t) = r$, $p_2(t) = -r/K$ and $n = 2$. The modified logistic in Example 3 of Sec. 3.2 is also a Bernoulli equation.

When $n = 0$ or $n = 1$ equation (3.1) is linear and hence solvable by the methods of Chapter 2. We therefore consider the Bernoulli equation (3.1) when $n \neq 0$ or 1.

Equation (3.1) has the equilibrium solution $x = 0$. We can find other solutions by making a change of dependent variable, from x to y, where

$$y = x^{1-n}. \tag{3.2}$$

We will see below that this change of variables results in a linear equation for y. We know how to calculate solution formulas for linear equations (Chapter 2).

To change variables in equation (3.1) means to derive a differential equation for y. We calculate the derivative y' using the chain rule:

$$y' = (1 - n)x^{-n}x'.$$

Using Bernoulli's equation (3.1), we substitute $p_1(t)x + p_2(t)x^n$ for x' on the right-hand side of this expression and obtain

$$y' = (1 - n)x^{-n}\left[p_1(t)x + p_2(t)x^n\right]$$
$$= (1 - n)p_1(t)x^{1-n} + (1 - n)p_2(t)$$

or, recalling that $y = x^{1-n}$,

$$y' = (1 - n)p_1(t)y + (1 - n)p_2(t). \tag{3.3}$$

This differential equation for y is linear and solvable by the methods of Chapter 2. After we have calculated a formula for y, we can obtain a formula for the solution $x = y^{1/(1-n)}$ of Bernoulli's equation.

For example, the equation

$$x' = r\,(1 + \cos t)\left(1 - \frac{x}{K}\right)x \tag{3.4}$$
$$= r\,(1 + \cos t)\,x - \frac{1}{K}r\,(1 + \cos t)\,x^2$$

from Example 3 of Sec. 3.2 is a Bernoulli equation with $n = 2$. We can change this equation for x to a linear equation for $y = x^{1-n} = x^{-1}$. From a calculation of y' [or using (3.3)] we obtain the differential equation

$$y' = -r\,(1 + \cos t)\,y + \frac{1}{K}r\,(1 + \cos t)$$

for y. The general solution of this linear equations is

$$y = ce^{-r(t+\sin t)} + \frac{1}{K}.$$

(The reader should derive this solution formula.) From $x = y^{-1}$ we obtain the formula

$$x = \frac{1}{ce^{-r(t+\sin t)} + \frac{1}{K}}$$
$$= \frac{K}{Kce^{-r(t+\sin t)} + 1},$$

which, together with the equilibrium $x = 0$, gives the general solution of the equation (3.4).

The nonlinear equations

$$v' = g - k_0 v^2$$

$$x' = r\left(1 - \frac{x}{K}\right)(x - aK)$$

have quadratic expressions on their right-hand sides but are not Bernoulli equations. The first equation describes the velocity $v \geq 0$ of an object falling vertically under the influence of gravity and friction caused by air resistance (see Sec. 3.6.2). The second equation is a modification of the logistic equation that incorporates a minimum threshold population level for survival (called an *Allee effect*; see Exercises 19 and 9, Sec. 3.6). These equations are examples of a second type of equation, called a Ricatti equation, that is solvable by means of a change of variables.

A first order differential equations of the form

$$x' = p_0(t) + p_1(t)x + p_2(t)x^2$$

is called a *Ricatti equation*. If one solution of a Ricatti equation can be found (e.g., an equilibrium), then the general solution can be calculated by means of a change of variables. Namely, if $x = x_p(t)$ is a known solution of the equation, then the change of variable from x to y defined by

$$y = \frac{1}{x_p - x},$$

or, equivalently,

$$x = x_p - \frac{1}{y}, \tag{3.5}$$

yields the linear differential equation

$$y' = -\bigl(p_1(t) + 2p_2(t)x_p(t)\bigr)y + p_2(t) \tag{3.6}$$

for y (see Exercise 18). From (3.5) a solution formula for the general solution y of this linear equation produces a formula for the solution x of the Riccati equation.

For example, the equation

$$v' = g - k_0 v^2$$

for the velocity of a falling object has the equilibrium solution

$$v_e = \left(\frac{g}{k_0}\right)^{1/2}.$$

With $x_p(t) = v_e$, $p_1(t) = 0$, and $p_2(t) = -k_0$, the transformed Ricatti equation (3.6) becomes

$$y' = (2k_0 v_e)y - k_0.$$

The general solution of this linear equation is

$$y = ce^{2k_0 v_e t} + \frac{1}{2v_e}$$

(the reader should derive this formula) from which (3.5) yields

$$v = v_e - \frac{2v_e}{2v_e c e^{2(gk_0)^{1/2}t} + 1}, \tag{3.7}$$

where c is an arbitrary constant.

To solve an initial value problem, we choose the constant c in the formula for the general solution accordingly. For example, for $v(0) = 0$ (which is the initial condition of a dropped object) we choose c so that

$$v(0) = v_e - \frac{2v_e}{2v_e c + 1} = 0$$

that is,

$$c = \frac{1}{2v_e}.$$

With this c the general solution gives the formula

$$v = v_e \frac{1 - e^{-2k_0 v_e t}}{1 + e^{-2k_0 v_e t}}$$

for the solution of the initial value problem. Notice

$$\lim_{t \to +\infty} v = v_e,$$

which is consistent with the phase line portrait shown in Fig. 3.23. The equilibrium v_e is called the *terminal velocity* of the object. For the case when $v_0 \neq 0$ see Exercise 20.

Figure 3.23 The phase line portrait for $v' = g - k_0 v^2$.

Exercises

Find a formula for the general solution of the equations below.

1. $x' = x - x^2 \sin t$

2. $x' = ax - e^t x^3$, $a = $ constant

3. $x' = -\frac{x}{t} + te^{-2t}x^2$

4. $x' = r(1 - (1 + \cos t)x)x$

5. $x' = x - x^3$

6. $x' = 2x + x^4$

Find a formula for the solution of the initial value problems.

7. $x' = \frac{x}{t} + x^3$, $x(1) = \sqrt{3}$

8. $x' = \frac{t^2 + x^2}{tx}$, $x(1) = 2$

9. $x' = -x + e^{-t}x^3$, $x(0) = \frac{1}{2}$

10. $x' = x + tx^4$, $x(0) = 1$

11. Treat the logistic equation $x' = rx(1 - x/K)$, $r > 0$, $K > 0$ as a Bernoulli equation and find a formula for the general solution. Then find a formula for the solution of the initial value problem $x(0) = x_0 > 0$ and use it to show $\lim_{t \to +\infty} x(t) = K$.

12. The logistic equation $x' = rx(1 - x/K)$, $r > 0$, $K > 0$, describes the growth of a population. The equilibrium $K > 0$, called the environmental carrying capacity, is a sink and all populations with positive initial conditions $x(0) > 0$ tend to K as $t \to +\infty$. Suppose the environmental carrying capacity is not constant but instead degrades with time. Specifically, suppose the carrying capacity tends exponentially to 0 as $t \to +\infty$. Then we have the Bernoulli equation

$$x' = rx \left(1 - \frac{x}{Ke^{-t}} \right).$$

Find a formula for the general solution and a formula for the solution of the initial value problem $x(0) = x_0$. If $x_0 > 0$, what is the long-term fate of the population?

Find a formula for the general solution of the following Ricatti equations.

13. $x' = (x^2 - 5x + 6) \sin t$

14. $x' = 1 - t^2 + (t^2 - 1)x^2$

15. $x' = -\dfrac{1}{t} - \dfrac{1}{t^2} + tx^2$ given that $x_p = \frac{1}{t}$ is a solution.

16. $x' = -t^2 + \dfrac{1}{t}x + x^2$ given that $x_p = t$ is a solution.

17. Find a formula for the general solution of the equation $x' = r(1 + \sin t)(1 - x/K)(x - aK), 0 < a < 1, 0 < r, 0 < K$, treating it as a Ricatti equation.

18. Derive equation (3.6).

19. (a) Draw the phase line portrait for the modified logistic equation $x' = rx(1 - x/K)(x - aK), 0 < a < 1, 0 < r, 0 < K$.

(b) Describe the asymptotic dynamics of solutions with initial conditions $x_0 > 0$. What is the threshold level for initial conditions that must be exceeded for survival?

(c) Find a formula for the general solution, treating the equation as a Ricatti equation with known solution $x_p(t) = K$.

(d) Find a formula for the solution of the initial value problem $x(0) = x_0$.

(e) Show that $0 < x_0 < aK$ implies that there is a $t_1 > 0$ such that $x(t_1) = 0$.

(f) Show that $0 < x_0 < aK$ implies that there is a $t_2 > t_1$ such that $\lim_{t \to t_2} x(t) = -\infty$.

(g) Show that $\lim_{t \to +\infty} x(t) = K$ if $x_0 > aK$.

20. Use the general solution formula (3.7) to show all solutions of the equation $v' = g - k_0 v^2$ for the velocity v of a falling object approach $v_e = (g/k_0)^{1/2}$ (the terminal velocity) as $t \to +\infty$ when $v_0 > -v_e$.

3.4 Approximation Formulas

Consider the initial value problem

$$x' = f(t, x) \tag{4.1}$$
$$x(t_0) = x_0.$$

Under the assumptions of the fundamental existence and uniqueness theorem (i.e., when f and $\partial f/\partial x$ are continuous) there is a unique solution $x(t)$. When formulas for the solution $x(t)$ cannot be calculated, we can attempt to approximate solutions. For example, in Chapter 1 we learned how to approximate the solution numerically and also how to approximate its graph. Another approach is to obtain formulas for approximations to the solution $x(t)$.

In this section we look at several methods for generating approximation formulas. Each method is algorithmic and repeated application produces a sequence of approximations $x_n(t)$ which approaches the exact solution in the limit, that is,

$$\lim_{n \to +\infty} x_n(t) = x(t)$$

for each t. Thus, we can obtain an approximation of any desired accuracy after a finite number of applications of the method.

3.4.1 Taylor Polynomials and Power Series

The Taylor polynomial of order n, centered at $t = t_0$, for the solution $x(t)$ of the initial value problem (4.1) is

$$x_n(t) = k_0 + k_1(t - t_0) + k_2(t - t_0)^2 + \cdots + k_n(t - t_0)^n, \tag{4.2}$$

where the coefficients are given by the formulas[6]

$$k_0 = x(t_0), \quad k_1 = \left.\frac{dx}{dt}\right|_{t_0}, \quad k_2 = \frac{1}{2}\left.\frac{d^2x}{dt^2}\right|_{t_0}, \quad \dots, \quad k_n = \frac{1}{n!}\left.\frac{d^n x}{dt^n}\right|_{t_0}. \tag{4.3}$$

The polynomial $x_n(t)$ matches the solution $x(t)$ and its first n derivatives at $t = t_0$ and, as a result, it approximates the solution $x(t)$ for t near $t = t_0$.

The initial condition $x(t_0) = x_0$ in (4.1) determines the first coefficient

$$k_0 = x_0.$$

To calculate coefficients k_1 we let $t = t_0$ in the differential equation

$$x'(t) = f(t, x(t)) \tag{4.4}$$

and obtain

$$k_1 = \left.\frac{dx}{dt}\right|_{t_0} = f(t_0, x_0).$$

To obtain further coefficients, we differentiate both sides of the differential equation before letting $t = t_0$.

For example, using the chain rule, we differentiate both sides of equation (4.4) with respect to t to find

$$\frac{d^2x}{dt^2} = \frac{\partial f}{\partial t} + \frac{\partial f}{\partial x}\frac{dx}{dt}$$

or

$$\frac{d^2x}{dt^2} = \frac{\partial f}{\partial t} + \frac{\partial f}{\partial x}f, \tag{4.5}$$

which, when evaluated at $t = t_0$, yields

$$k_2 = \frac{1}{2}\left(\left.\frac{\partial f}{\partial t}\right|_{(t_0, x_0)} + \left.\frac{\partial f}{\partial x}\right|_{(t_0, x_0)} f(t_0, x_0) \right). \tag{4.6}$$

To calculate more coefficients, we continue to differentiate both sides of the differential equation (4.5) before evaluating at $t = t_0$. For example, from (4.5) we obtain (using the chain rule and the product rule)

$$\frac{d^3x}{dt^3} = \frac{\partial^2 f}{\partial t^2} + 2\frac{\partial^2 f}{\partial t \partial x}f + \frac{\partial f}{\partial x}\left(\frac{\partial f}{\partial t} + \frac{\partial f}{\partial x}f\right) + \frac{\partial^2 f}{\partial x^2}f^2.$$

We can use this rather complicated formula to calculate

$$k_3 = \frac{1}{3!}\left.\frac{d^3x}{dt^3}\right|_{t_0}.$$

It is easier, however, to carry out these calculations in specific examples rather than derive and use these kinds of general formulas.

---

[6]The notation $dx/dt|_{t_0}$ means the derivative dx/dt is evaluated at $t = t_0$. The notation $d^n x/dt^n|_{t_0}$ means the nth-order derivative $d^n x/dt^n$ is evaluated at $t = t_0$.

EXAMPLE 1   Consider the initial value problem

$$x' = -x + \frac{1}{2}x^2, \quad x(0) = 1. \tag{4.7}$$

We seek the sequence of Taylor polynomial approximations to the solution up to order three, that is,

$$
\begin{aligned}
x_0(t) &= k_0 \\
x_1(t) &= k_0 + k_1 t \\
x_2(t) &= k_0 + k_1 t + k_2 t^2 \\
x_3(t) &= k_0 + k_1 t + k_2 t^2 + k_3 t^3,
\end{aligned} \tag{4.8}
$$

where the coefficients are

$$
\begin{aligned}
k_0 &= x(0) \\
k_1 &= \left.\frac{dx}{dt}\right|_0 \\
k_2 &= \left.\frac{1}{2}\frac{d^2x}{dt^2}\right|_0 \\
k_3 &= \left.\frac{1}{3!}\frac{d^3x}{dt^3}\right|_0.
\end{aligned} \tag{4.9}
$$

Differentiating the differential equation twice, we obtain

$$
\begin{aligned}
\frac{dx}{dt} &= -x + \frac{1}{2}x^2 \\
\frac{d^2x}{dt^2} &= -\frac{dx}{dt} + x\frac{dx}{dt} \\
\frac{d^3x}{dt^3} &= -\frac{d^2x}{dt^2} + x\frac{d^2x}{dt^2} + \left(\frac{dx}{dt}\right)^2.
\end{aligned}
$$

An evaluation at $t = 0$ yields

$$
\begin{aligned}
x(0) &= 1 \\
\left.\frac{dx}{dt}\right|_0 &= -x(0) + \frac{1}{2}x(0)^2 = -\frac{1}{2} \\
\left.\frac{d^2x}{dt^2}\right|_0 &= -\left(-\frac{1}{2}\right) + 1 \times \left(-\frac{1}{2}\right) = 0 \\
\left.\frac{d^3x}{dt^3}\right|_0 &= -0 + 1 \times 0 + \left(-\frac{1}{2}\right)^2 = \frac{1}{4}
\end{aligned}
$$

and, from (4.9), we get the coefficients

$$k_0 = 1$$
$$k_1 = -\frac{1}{2}$$
$$k_2 = 0$$
$$k_3 = \frac{1}{24}.$$

These coefficients give us the sequence of Taylor polynomial approximations

$$x_0(t) = 1$$
$$x_1(t) = 1 - \frac{1}{2}t$$
$$x_2(t) = 1 - \frac{1}{2}t$$
$$x_3(t) = 1 - \frac{1}{2}t + \frac{1}{24}t^3.$$

The differential equation in the initial value problem (4.7) is separable and we can calculate a solution formula:

$$x(t) = \frac{2}{e^t + 1}.$$

Graphs of this solution formula and of Taylor polynomial approximations appear in Fig. 3.24. Notice how the sequence of approximations improve for t near $t_0 = 0$. Also notice that the approximations are not accurate for large values of t. ■

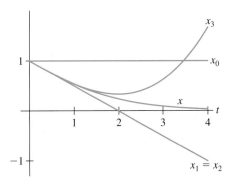

Figure 3.24 The graphs of the Taylor polynomial approximations $x_0(t)$, $x_1(t)$, $x_2(t)$, $x_3(t)$ to the solution $x(t)$ of the initial value problem (4.7).

We can also calculate Taylor polynomials for equations containing coefficients (parameters). When this is done, the coefficients of the Taylor polynomials will in general depend on the coefficients appearing in the equation. For example, if a coefficient denoted by the letter ε appears in the differential equation, then the Taylor polynomial coefficients k_0, k_1, k_2, ... will depend on ε (i.e., become functions of ε). To express this fact we can write $k_0 = k_0(\varepsilon)$, $k_1 = k_1(\varepsilon)$, and so on. Here is an example.

EXAMPLE 2

Consider the initial value problem

$$x' = -x + \varepsilon x^2, \quad x(0) = 1,$$

where ε is a real number. We seek Taylor polynomial approximations (4.8) to the solution up to order three. Because the coefficients in these polynomials depend on ε we write

$$
\begin{aligned}
x_0 &= k_0(\varepsilon) \\
x_1 &= k_0(\varepsilon) + k_1(\varepsilon)t \\
x_2 &= k_0(\varepsilon) + k_1(\varepsilon)t + k_2(\varepsilon)t^2 \\
x_3 &= k_0(\varepsilon) + k_1(\varepsilon)t + k_2(\varepsilon)t^2 + k_3(\varepsilon)t^3.
\end{aligned}
\tag{4.10}
$$

From the differential equation we calculate

$$\frac{dx}{dt} = -x + \varepsilon x^2$$

$$\frac{d^2x}{dt^2} = -\frac{dx}{dt} + 2\varepsilon x\frac{dx}{dt}$$

$$\frac{d^3x}{dt^3} = -\frac{d^2x}{dt^2} + 2\varepsilon x\frac{d^2x}{dt^2} + 2\varepsilon \left(\frac{dx}{dt}\right)^2$$

and

$$x(0) = 1$$

$$\left.\frac{dx}{dt}\right|_0 = -x(0) + \varepsilon x^2(0) = -1 + \varepsilon$$

$$\left.\frac{d^2x}{dt^2}\right|_0 = (-1 + \varepsilon)(-1 + 2\varepsilon)$$

$$\left.\frac{d^3x}{dt^3}\right|_0 = (-1 + \varepsilon)\left(6\varepsilon^2 - 6\varepsilon + 1\right).$$

From the coefficient formulas (4.9) we have

$$k_0(\varepsilon) = 1$$
$$k_1(\varepsilon) = -1 + \varepsilon$$

$$k_2(\varepsilon) = \frac{1}{2}(-1 + \varepsilon)(-1 + 2\varepsilon)$$

$$k_3(\varepsilon) = \frac{1}{6}(-1 + \varepsilon)\left(6\varepsilon^2 - 6\varepsilon + 1\right)$$

and the sequence of Taylor polynomial approximations

$$x_0(t) = 1$$

$$x_1(t) = 1 + (-1 + \varepsilon)\,t$$

$$x_2(t) = 1 + (-1 + \varepsilon)\,t + \frac{1}{2}(-1 + \varepsilon)\,(-1 + 2\varepsilon)\,t^2$$

$$x_3(t) = 1 + (-1 + \varepsilon)\,t + \frac{1}{2}(-1 + \varepsilon)\,(-1 + 2\varepsilon)\,t^2$$
$$+ \frac{1}{6}(-1 + \varepsilon)\left(6\varepsilon^2 - 6\varepsilon + 1\right)t^3.$$

With $\varepsilon = 1/2$ we obtain the initial value problem and the Taylor polynomial approximations in the Example 1. ∎

One way to find the coefficients in a Taylor polynomial approximation (4.2) is to differentiate the differential equation repeatedly and use formula (4.3), as we did in the examples above. Another, sometimes simpler, way to find the coefficients is to substitute the Taylor polynomial (4.2) into both sides of the differential equation and equate coefficients of like powers of $t - t_0$ obtained from each side.

For example, if we substitute

$$x_3(t) = k_0(\varepsilon) + k_1(\varepsilon)t + k_2(\varepsilon)t^2 + k_3(\varepsilon)t^3$$

into the right-hand side of the differential equation

$$x' = -x + \varepsilon x^2$$

we have

$$-x_3(t) + \varepsilon\,(x_3(t))^2 = -\left(k_0(\varepsilon) + k_1(\varepsilon)t + k_2(\varepsilon)t^2 + k_3(\varepsilon)t^3\right)$$
$$+ \varepsilon\left(k_0(\varepsilon) + k_1(\varepsilon)t + k_2(\varepsilon)t^2 + k_3(\varepsilon)t^3\right)^2,$$

which, after some algebra and after gathering like terms (in powers of t), becomes

$$-x_3(t) + \varepsilon\,(x_3(t))^2 = \left(\varepsilon k_0^2 - k_0\right) + (2\varepsilon k_0 k_1 - k_1)\,t$$
$$+ \left(\varepsilon k_1^2 + 2\varepsilon k_0 k_2 - k_2\right)t^2 + \cdots.$$

(For the moment, we have dropped ε for notational simplicity.) Here the dots "$\cdots$" denote terms of degree three and higher in t. If we substitute $x_3(t)$ into the left-hand side of the differential equation, we have

$$x_3'(t) = k_1 + 2k_2 t + 3k_3 t^2 + \cdots.$$

If we equate these expressions obtained from the right- and left-hand sides of the differential equation

$$x_3'(t) = -x_3(t) + \varepsilon\,(x_3(t))^2,$$

we obtain

$$k_1 + 2k_2 t + 3k_3 t^2 + \cdots = \left(\varepsilon k_0^2 - k_0\right) + (2\varepsilon k_0 k_1 - k_1)\,t$$
$$+ \left(\varepsilon k_1^2 + 2\varepsilon k_0 k_2 - k_2\right)t^2 + \cdots.$$

Polynomials in t are identical if and only if their coefficients match, that is, if and only if the coefficients of like powers of t are the same. Matching the coefficients from both sides of the equation, we get

$$k_1 = \varepsilon k_0^2 - k_0$$
$$2k_2 = 2\varepsilon k_0 k_1 - k_1$$
$$3k_3 = \varepsilon k_1^2 + 2\varepsilon k_0 k_2 - k_2.$$

Since initial condition $x(t_0) = k_0$ determines k_0, we can use these equations to calculate the three coefficients k_1, k_2 and k_3 sequentially.

For example, $x(0) = 1$ implies $k_0 = 1$ and as a result

$$k_1(\varepsilon) = \varepsilon - 1$$
$$k_2(\varepsilon) = \frac{1}{2}(\varepsilon - 1)(2\varepsilon - 1)$$
$$k_3(\varepsilon) = \frac{1}{6}(\varepsilon - 1)\left(6\varepsilon^2 - 6\varepsilon + 1\right).$$

We arrive at the same cubic Taylor polynomial approximation obtained in Example 2.

If the sequence of Taylor polynomials

$$x_n(t) = k_0 + k_1(t - t_0) + k_2(t - t_0)^2 + \cdots + k_n(t - t_0)^n$$
$$k_n = \frac{1}{n!}\left.\frac{d^n x}{dt^n}\right|_{t_0}$$

converges as n increases without bound, we obtain the power series representation of $x(t)$:

$$\lim_{t \to \infty} x_n(t) = k_0 + k_1(t - t_0) + k_2(t - t_0)^2 + \cdots + k_n(t - t_0)^n + \cdots ,$$

where the dots "$\cdots$" on the right-hand side indicate the limit process. Alternatively, we can also write

$$\lim_{t \to \infty} x_n(t) = \sum_{n=0}^{\infty} k_n(t - t_0)^n.$$

In calculus we learn that power series converge for t in an interval

$$t_0 - R < t < t_0 + R$$

with t_0 at its center and "radius" $R \geq 0$. The power series

$$x(t) = k_0 + k_1(t - t_0) + k_2(t - t_0)^2 + \cdots = \sum_{i=0}^{\infty} k_i(t - t_0)^i \qquad (4.11)$$

is a meaningful representation of the solution $x(t)$ only if its radius of convergence R is positive. (If $R = 0$ the series converges only for $t = t_0$ and the power series cannot be a solution of a differential equation on an interval containing t_0.)

In summary, if for an initial value problem we can find a general formula for all the coefficients k_n and if the radius of converge is positive, then the power series (4.11) is a formula for the solution. The Taylor polynomial approximation procedure can, by passing to the limit in this way, lead to formulas for solutions.

It is not always true, unfortunately, that the radius of converge of (4.11) is positive. However, mathematicians have established conditions on f that guarantee $R > 0$ for solutions of the differential equation.

For example, if $f = f(x)$ has a power series in x, centered at x_0, with positive radius of convergence, then the solution of the initial value problem $x' = f(x), x(t_0) = x_0$, has a power series, centered at t_0, with a positive radius of convergence.

In the most successful applications of the power series method, we are able to find a formula for k_n, use the formula in (4.11) to obtain a power series formula for the solution, and sum up the power series (which usually means to recognize the power series as that of some familiar function, one involving, for example, exponential or trigonometric functions). All three of these steps are not always possible, however.

Here is an example. We can solve the initial value problem

$$x' = x \tag{4.12}$$
$$x(0) = 1$$

using the methods in Chapter 2. Nonetheless, let us use this problem pedagogically to illustrate the power series solution method. The procedure begins with the substitution of the power series

$$x(t) = k_0 + k_1 t + k_2 t^2 + \cdots = \sum_{i=0}^{\infty} k_i t^i \tag{4.13}$$

into both sides of the differential equation. This will create power series on both sides of the equation, whose coefficients of like powers of t we will equate to one another. Since[7]

$$x'(t) = k_1 + 2k_2 t + 3k_3 t^2 + \cdots = \sum_{i=0}^{\infty} (i+1) k_{i+1} t^i \tag{4.14}$$

the differential equation yields

$$\sum_{i=0}^{\infty} (i+1) k_{i+1} t^i = \sum_{i=0}^{\infty} k_i t^i .$$

Equating coefficients of like powers of t, we have

$$(i+1) k_{i+1} = k_i$$

or

$$k_{i+1} = \frac{1}{i+1} k_i, \quad i = 0, 1, 2, \ldots . \tag{4.15}$$

This formula permits us to calculate the coefficients of the power series sequentially or recursively. For that reason, it is called the *recursive formula* for the coefficients.

---

[7]Differentiating the power series (4.13) termwise, we get $\sum_{i=0}^{\infty} i k_i t^{i-1} = \sum_{i=1}^{\infty} i k_i t^{i-1}$. When working with several power series and using the sigma notation, it is convenient to write the series so that they all have the same power on the variable t (or $t - t_0$). If, in this series for the derivative, we shift the index i to a new index $n = i - 1$ [so that the power on t becomes the index, as in (4.13)] we get $\sum_{n=0}^{\infty} (n+1) k_{n+1} t^n$. This is the same series as (4.14).

Since the initial condition $x(0) = 1$ implies $k_0 = x(0) = 1$, this formula produces

$$k_1 = k_0 = 1$$

$$k_2 = \frac{1}{2}k_1 = \frac{1}{2}$$

$$k_3 = \frac{1}{3}k_2 = \frac{1}{3}\frac{1}{2} = \frac{1}{3!}$$

$$k_4 = \frac{1}{4}k_3 = \frac{1}{4}\frac{1}{3}\frac{1}{2} = \frac{1}{4!}$$

and so on. In this example we see a distinctive pattern in these coefficients. For example, we suspect (without actually making the calculation) that

$$k_5 = \frac{1}{5}\frac{1}{4}\frac{1}{3}\frac{1}{2} = \frac{1}{5!},$$

a conjecture that is borne out by the recursive formula with $i = 4$, namely,

$$k_5 = \frac{1}{5}k_4.$$

Using the factorial notation, we can simply these formulas by writing

$$k_1 = \frac{1}{1!}$$

$$k_2 = \frac{1}{2!}$$

$$k_3 = \frac{1}{3!}$$

$$k_4 = \frac{1}{4!}$$

$$k_5 = \frac{1}{5!}$$

and conjecture that in general

$$k_n = \frac{1}{n!}. \tag{4.16}$$

(A formal justification of this conjecture for k_n involves an induction proof.) This formula, which is the solution of the recursive formula (4.15), gives the power series representation

$$x(t) = \sum_{n=0}^{\infty} \frac{1}{n!}t^n$$

for the solution of the initial value problem (4.12). This power series should be familiar. In calculus we learn that it is power series representation of e^t, and hence $x(t) = e^t$.

For most initial value problems, we usually cannot carry out all of the steps that led to the solution $x = e^t$ in the previous example. Even if we successfully find the recursive formula for the power series coefficients, it might be very difficult or even impossible to solve this formula for k_n [as we did in going from (4.15) to (4.16) in the preceeding example]. Even if we can solve for k_n the resulting power series most likely cannot be summed (i.e., recognized as some combination of elementary functions).

EXAMPLE 3

Consider the initial value problem

$$x' = x^2 + t^2$$
$$x(0) = 0.$$

We substitute the power series (4.13) into both sides of the equation and equate coefficients of like powers of t. To do this for the right-hand side involves some thoughtful calculation. We first note that

$$x^2 = \left(\sum_{n=0}^{\infty} k_n t^n \right)^2$$
$$= \left(k_0 + k_1 t + k_2 t^2 + k_3 t^3 + \cdots \right) \left(k_0 + k_1 t + k_2 t^2 + k_3 t^3 + \cdots \right)$$
$$= k_0 k_0 + (k_0 k_1 + k_1 k_0) t + (k_0 k_2 + k_1 k_1 + k_2 k_0) t^2 + \cdots.$$

Observe that each coefficient involves the sum of products in which the subscripts sum to the power on t. Thus,

$$x^2 = \sum_{i=0}^{\infty} \left(\sum_{p+q=i} k_p k_q \right) t^i.$$

This observation, together with (4.14), implies

$$k_1 = k_0 k_0$$

$$k_2 = \frac{1}{2} (k_0 k_1 + k_1 k_0)$$

$$k_3 = \frac{1}{3} (k_0 k_2 + k_1 k_1 + k_2 k_0 + 1)$$

$$k_{i+1} = \frac{1}{i+1} \sum_{p+q=i} k_p k_q, \quad i = 3, 4, 5, \ldots$$

These are the recursive formulas for the coefficients of the power series solution.

The initial condition $x(0) = 0$ implies $k_0 = 0$. We can use the recursive formulas to compute sequentially as many coefficients as we wish. For example,

$$k_1 = 0$$
$$k_2 = 0$$
$$k_3 = \frac{1}{3}$$
$$k_4 = k_5 = k_6 = 0$$
$$k_7 = \frac{1}{7} \frac{1}{3} \frac{1}{3} = \frac{1}{63}$$

and so on. In this example, it is difficult to solve the recursive formula, although we can use the recursive formula to calculate Taylor polynomials of any degree. The preceeding coefficients yield

$$x(t) = \frac{1}{3}t^3 + \frac{1}{63}t^7 + \cdots .$$

∎

Power series solution methods have also found many applications to higher order differential equations. We will revisit these methods in Chapter 5.

3.4.2 A Perturbation Method

Taylor polynomials are designed to approximate the solution of an initial value problem near the initial point $t = t_0$. In Example 1 the solution of the initial value problem

$$x' = -x + \frac{1}{2}x^2, \quad x(0) = 1 \tag{4.17}$$

is approximated by the cubic polynomial

$$x_3(t) = 1 - \frac{1}{2}t + \frac{1}{24}t^3$$

for t near $t_0 = 0$. This can be seen graphically in Fig. 3.24. For large values of t, however, this cubic polynomial is a poor approximation to the solution. This can also be seen in Fig. 3.24. The graph in Fig. 3.24 indicates that $x(t) \rightarrow 0$ as $t \rightarrow +\infty$. However, this property of the solution cannot be discovered from the cubic polynomial approximation $x_3(t)$ (since this cubic polynomial grows without bound as $t \rightarrow +\infty$). Indeed, we cannot determine this asymptotic property of the solution by using a Taylor polynomial approximation of *any* order n because no polynomial tends to 0 as $t \rightarrow +\infty$. In this section we consider a modification of the Taylor polynomial method that often leads to approximations accurate for t far from t_0.

The initial value problem (4.17) is a particular case of the initial value problem

$$\begin{aligned} x' &= -x + \varepsilon x^2 \\ x(0) &= 1. \end{aligned} \tag{4.18}$$

The solution of this problem depends on the parameter ε.

A modification of the approximation method we used in the previous section is based on the construction of Taylor polynomials in ε (rather than t). Such a polynomial is then an approximation of the solution for small ε. We view the initial value problem (4.18) for small ε as a perturbation of the initial value problem when $\varepsilon = 0$. This method is therefore called a *perturbation method*.

In applying the perturbation method we seek an approximation to solutions of the differential equation

$$x' = f(t, x, \varepsilon)$$

of the form

$$x_n(t) = k_0(t) + k_1(t)\varepsilon + k_2(t)\varepsilon^2 + \cdots + k_n(t)\varepsilon^n \tag{4.19}$$

(a Taylor polynomial in ε, instead of t). Here $k_i(t)$ are undetermined coefficients that we can calculate by either the differentiation or the substitution method used in the previous section. When this is done, $x_n(t)$ is called an *nth-order perturbation approx-imation*.

To illustrate the procedure we find a first-order ($n = 1$) perturbation approximation to the solution of the initial value problem (4.18) by using the substitution method. (The student is asked to use the differentiation method in Exercise 25.) That is to say, we seek a Taylor polynomial approximation (in ε) of the form

$$x_1(t) = k_0(t) + k_1(t)\varepsilon. \tag{4.20}$$

We calculate the coefficients $k_0(t)$ and $k_1(t)$ by substituting this expression into the differential equation (4.18a) and equating coefficients of like powers of ε. From the left-hand side

$$x_1'(t) = k_0'(t) + k_1'(t)\varepsilon$$

and from the right-hand side

$$-x_1(t) + \varepsilon x_1^2(t) = -k_0(t) + \left(-k_1(t) + k_0^2(t)\right)\varepsilon + \cdots,$$

where the dots "$\cdots$" denote terms of order two and higher in ε. Equating coefficients of like powers of ε, we obtain the equations

$$k_0' = -k_0$$
$$k_1' = -k_1 + k_0^2.$$

To obtain the Taylor polynomial approximation (4.20) we need to solve these equations for the coefficients k_0 and k_1. We can do this as follows. The first equation involves only k_0 (and not k_1) and is uncoupled from the second equation. Moreover, the first equation is linear and we can calculate formulas for its solutions (the general solution is $k_0(t) = ce^{-t}$). Once we have a solution $k_0(t)$ of the first equation, we can substitute it into the second equation. With $k_0^2(t)$ a known function, the second equation is linear and we can calculate formulas for its solutions as well. To be more specific about this procedure we make use of the initial condition $x(0) = 1$ in (4.18).

A substitution of (4.20) into the initial condition yields $k_0(0) + k_1(0)\varepsilon = 1$ or

$$k_0(0) + k_1(0)\varepsilon = 1 + 0 \cdot \varepsilon.$$

Equating coefficients of like powers of ε from both sides of this equation, we find that

$$k_0(0) = 1$$
$$k_1(0) = 0.$$

These give us two linear initial value problems

$$k_0' = -k_0 \tag{4.21}$$
$$k_0(0) = 1$$

and

$$k_1' = -k_1 + k_0^2 \tag{4.22}$$
$$k_1(0) = 0$$

to solve for the coefficients k_0 and k_1, respectively. [Notice that the lowest-order initial value problem (4.21) for k_0 is the same as the original initial value problem (4.18) with $\varepsilon = 0$.]

The (linear homogeneous) initial value problem (4.21) has the solution

$$k_0(t) = e^{-t}.$$

Following the procedure outlined previously, we substitute this solution formula for k_0 into the initial value problem (4.22) for k_1. This yields the *linear* (nonhomogeneous) initial value problem

$$k_1' = -k_1 + e^{-2t}$$
$$k_1(0) = 0$$

whose solution is

$$k_1(t) = e^{-t} - e^{-2t}.$$

Using these coefficients k_0 and k_1, we obtain the first-order approximation

$$x_1(t) = e^{-t} + \left(e^{-t} - e^{-2t}\right)\varepsilon \tag{4.23}$$

to the solution of the initial value problem (4.18).

Figure 3.25 shows the accuracy of both the first-order approximation $x_1(t)$ (with $\varepsilon = 1/2$) and the zeroth-order approximation $x_0(t) = k_0 = e^{-t}$ as approximations to the solution of the initial value problem (4.17). Notice the approximations are fairly accurate for all t, not just for t near $t_0 = 0$. In fact, the approximation $x_1(t)$ tends to 0 as $t \to +\infty$, as does the solution $x(t)$ of the initial value problem (which is evident from the phase line portrait associated with the differential equation).

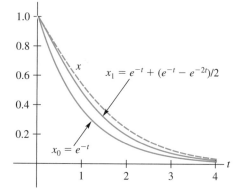

Figure 3.25 The zeroth- and first-order perturbation approximations to the solution of the initial value problem (4.18).

We can also use the perturbation method to approximate a solution of differential equation that has a desired property (rather than satisfying an initial condition). For example, we might be interested in approximating a periodic solution of a differential equation. We can use the perturbation method to do this by requiring that the coefficients k_i in (4.19) to be periodic functions. This requires solving the linear differential equations for periodic solutions $k_i = k_i(t)$, rather than for specified initial conditions. See Exercises 28 and 29 for examples.

3.4.3 Picard Iterates

In this section we study an important approximation method that is useful for a variety of problems in mathematics. The procedure involves the calculation of a sequence approximations, called Picard iterates, that converges to a solution. We will apply this iteration procedure to initial value problems for differential equations. The Picard iteration method has been generalized, however, to so many other types of equations that it has become a fundamental tool in mathematics.

Consider the initial value problem (4.1). We begin with a crude approximation to the solution x obtained from the initial condition, namely the constant function

$$x_0(t) \equiv x_0.$$

This approximation is accurate for t close to t_0 because the solution $x(t)$ is continuous at t_0.

To obtain improved approximations to the solution, we proceed as follows. For t close to t_0, the right-hand side $f(t, x)$ of the differential equation is approximated by $f(t, x_0(t))$. We define the second Picard iterate $x_1(t)$ as the solution of the initial value problem

$$x' = f(t, x_0(t))$$
$$x(t_0) = x_0$$

that is,

$$x_1(t) = x_0 + \int_{t_0}^{t} f(s, x_0(s)) \, ds.$$

If this procedure is repeated, approximating the right-hand side $f(t, x)$ of the differential equation is approximated by $f(t, x_1(t))$, we obtain a third Picard iterate

$$x_2(t) = x_0 + \int_{t_0}^{t} f(s, x_1(s)) \, ds.$$

If this procedure is applied repeatedly, we obtain a sequence $x_n(t)$, $n = 0, 1, 2, \ldots$, defined recursively by the formula

$$x_{i+1}(t) = x_0 + \int_{t_0}^{t} f(s, x_i(s)) \, ds. \tag{4.24}$$

The terms $x_n(t)$ are called *Picard iterates*. Under the conditions of the fundamental existence and uniqueness theorem, the Picard iterates $x_n(t)$ converge to the solution (4.1) for t close to t_0.

EXAMPLE 4

Consider the initial value problem

$$x' = -x + \frac{1}{2}x^2 \tag{4.25}$$
$$x(0) = 1.$$

Using the first Picard iterate $x_0(t) = 1$ and the formula (4.24), we obtain the second Picard iterate

$$x_1(t) = x_0 + \int_0^t \left(-x_0(s) + \frac{1}{2}x_0^2(s) \right) ds$$

$$= 1 + \int_0^t \left(-\frac{1}{2} \right) ds$$

$$= 1 - \frac{1}{2}t.$$

The third Picard iterate, by (4.24), is

$$x_2(t) = x_0 + \int_0^t \left(-x_1(s) + \frac{1}{2}x_1^2(s) \right) ds$$

$$= 1 + \int_0^t \left(-\frac{1}{2} + \frac{1}{8}s^2 \right) ds$$

$$= 1 - \frac{1}{2}t + \frac{1}{24}t^3.$$

See Fig. 3.26.   ■

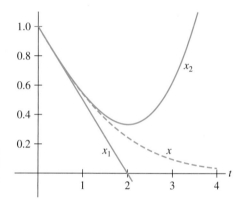

Figure 3.26 The Picard iterate approximations x_1 and x_2 to the solution x of the initial value problem (4.25).

Exercises

Find the quadratic Taylor polynomial approximation to the solution of the initial value problems below.

1. $x' = \sin x$, $x(\pi) = 1$ **2.** $x' = xe^{-x}$, $x(2) = -1$

3. $x' = rx\,(1 - x/K)$, $x(t_0) = x_0$

4. $x' = ae^{bt}x$, $x(0) = x_0$

Use a computer program to plot both the graph of the solution and the quadratic Taylor polynomial approximation of the initial value problems below. Do this for intervals of different lengths. How does the accuracy of the approximation depend on the interval?

5. $x' = \sin x$, $x(\pi) = 1$ **6.** $x' = xe^{-x}$, $x(2) = -1$

Find the recursive formulas for the power series solutions of the initial value problems below. Solve the recursive formulas and sum the series solution.

7. $x' = tx$, $x(0) = 2$ **8.** $x' = -2x$, $x(0) = 1$

9. $x' = x + \cos t - \sin t$, $x(0) = 0$

10. $x' = x + e^{2t}$, $x(0) = 2$

Find the first four nonzero terms in the power series expansion of the solution to the following initial value problems.

11. $x' = x + \dfrac{1}{1-t}$, $x(0) = 0$

12. $x' = x^2 + \dfrac{1}{1-t^2}$, $x(0) = 0$

13. $x' = 2x - \ln(1-t)$, $x(0) = 1$

14. $x' = \dfrac{t}{1-t}x$, $x(0) = 3$

Find the first-order perturbation expansion for the solution of the following initial value problems.

15. $x' = (1 + \varepsilon e^{-t})x$, $x(0) = 2$

16. $x' = (\sin t + \varepsilon \cos t)x$, $x(0) = -1$

17. $x' = 2x - \varepsilon \sin t$, $x(0) = 1$

18. $x' = -x + \varepsilon e^{-2t}$, $x(0) = -7$

19. $x' = x - x^2 + \varepsilon \sin t$, $x(0) = 1$

20. $x' = x - x^2 + \varepsilon e^{-t}$, $x(0) = 1$

21. Consider the initial value problem

$$x' = 4 - \frac{1}{4}x^2 + \varepsilon \sin(2t)$$

$$x(0) = 4.$$

(a) Find the quadratic Taylor polynomial approximation.

(b) Find the coefficients $k_0(t)$, $k_1(t)$ in the first-order perturbation approximation.

(c) Use a computer to graph the solution with $\varepsilon = 1$, together with both approximations obtained in (a) and (b). Discuss the results.

Find the initial value problems for the coefficients $k_0(t)$, $k_1(t)$, $k_2(t)$ in the second-order perturbation approximation $x_2(t) = k_0(t) + k_1(t)\varepsilon + k_2(t)\varepsilon^2$ to the solution of the following initial value problems. Solve these initial value problems and construct the approximation $x_2(t)$.

22. $x' = -(1 + \varepsilon \sin t)x$, $x(t_0) = x_0$

23. $x' = x - \varepsilon x^2$, $x(t_0) = x_0$

24. Find the recursive formulas for the coefficients $k_i(t)$ of the nth-order perturbation approximation $x_n(t) = \sum_{i=0}^{n} k_i(t)\varepsilon^i$ of the solution of the initial value problem

$$x' = -x + \varepsilon \sin t, \ x(0) = 1.$$

Show that all coefficients $k_i(t)$ except $k_0(t)$ and $k_1(t)$ are equal to 0 and that therefore the solution has the form $x = k_0(t) + k_1(t)\varepsilon$.

25. Derive the second-order approximation (4.23) to the solution of the initial value problem (4.18) using the differentiation method. *Hint:* Use

$$c_i = \frac{1}{i!}\frac{d^i x}{d\varepsilon^i}\bigg|_{\varepsilon=0}.$$

26. Consider the initial value problem

$$x' = a\left(b + \varepsilon \sin\left(\frac{2\pi}{T}t\right) - x\right), \ x(0) = x_0$$

obtained from Newton's law of cooling for an object placed in an environment with a sinusoidally fluctuating temperature $b + \varepsilon \sin(2\pi t/T)$.

(a) Using the amplitude ε of the environmental temperature oscillations as a small parameter, find the equations for the coefficients $c_i(t)$ in the perturbation approximation $x(t) \approx c_0(t) + c_1(t)\varepsilon + c_2(t)\varepsilon^2$ of the solution.

(b) Compute $c_1(t)$ and $c_2(t)$.

(c) Show that all remaining coefficients $c_i(t) \equiv 0$, $i \geq 2$, and therefore that the perturbation approximation to the first-order in ε in fact yields the exact solution.

27. Consider an object falling under the influence of gravity. If we assume the acceleration due to gravity is a constant g and air resistance exerts a frictional force proportional to the object's velocity $v = v(t)$, then $v' = g - cv$, where $c > 0$ is the coefficient of friction. Suppose also that an additional small frictional force is present that is proportional to v^2. Then $v' = g - cv - \varepsilon v^2$, where $\varepsilon > 0$ is a small constant. Assume the object is dropped, so that $v(0) = 0$.

(a) Find the first-order perturbation approximation $k_0(t) + \varepsilon k_1(t)$ to the initial value problem for v.

(b) An approximation to the terminal velocity v_e of the object is the limit as $t \to +\infty$ of the first-order perturbation approximation found in (a). Find this approximation.

(c) Draw the phase line portrait associated with the equation $v' = g - cv - \varepsilon v^2$.

(d) Find a formula for the terminal velocity v_e.

(e) Compare your answers in (b) and (d). [*Hint:* Find the first terms in the Taylor series of your answer in (d), with respect to ε, and compare it to your answer in (b).]

28. Consider the differential equation

$$x' = x\left(1 - \frac{2}{9}x\right) - (1 + \varepsilon \sin 2\pi t).$$

(a) Show that the coefficients of the first-order perturbation approximation $x_1(t) = k_0(t) + k_1(t)\varepsilon$ satisfy the differential equations

$$k_0' = k_0\left(1 - \frac{2}{9}k_0\right) - 1$$

$$k_1' = k_1\left(1 - \frac{4}{9}k_0\right) - \sin 2\pi t.$$

(b) In order to approximate periodic solutions x we require that k_0 and k_1 be periodic solutions of the equations in (a). Find all periodic solutions of the equation for k_0. (*Hint:* Equilibria are periodic solutions.)

(c) For each periodic solution k_0 from (b) find a periodic solution of the equation for k_1 in (a).

(d) Use your answers in (b) and (c) to construct first-order perturbation approximations to periodic solutions x.

(e) Use a computer program to obtain graphs of the periodic solutions when $\varepsilon = 0.5$ and compare them with the graphs of the approximations obtained in (d).

29. Consider the differential equation

$$x' = rx \left(1 - \frac{x}{K}\right) + \varepsilon \sin 2\pi t$$

with $r > 0$, $K > 0$ and $\varepsilon \geq 0$. This equation models a population that normally grows according to the logistic equation $x' = rx \left(1 - x/K\right)$ but that is periodically harvested and seeded with sinusoidal rate $\varepsilon \sin 2\pi t$.

(a) When $\varepsilon = 0$ there are two equilibria, $x = 0$ and K. Draw the phase line portrait. What do solutions with $x(0) > 0$ do as $t \to +\infty$?

(b) For small $\varepsilon > 0$, find the first-order approximation to a periodic solution near K.

(c) Show that the first-order approximation

 (i) has period 1;

 (ii) has average K;

 (iii) has amplitude proportional to, but less than, ε.

(d) Use the first-order approximation in (b) show the population survives indefinitely for small ε but goes extinct for large enough ε.

Calculate the first three Picard iterates for the initial value problems below.

30. $x' = px$, $x(0) = 1$

31. $x' = px + q$, $x(t_0) = x_0$

32. $x' = \sin x$, $x(1) = \frac{1}{2}\pi$

33. $x' = ae^{bt}x$, $x(0) = 1$

Use a computer program to plot both the graph of the solution and the first three Picard iterates for the initial value problems.

35. $x' = \sin x$, $x(1) = \pi/2$

36. $x' = e^{-t}x$, $x(0) = 1$

3.5 Chapter Summary and Exercises

The asymptotic dynamics of autonomous equation $x' = f(x)$ can be analyzed, by means of phase line portraits, without having to calculate a formulas for solutions. The phase line portrait summarizes the monotonicity properties of solutions and the classification of its equilibria (as sinks, sources, and shunts). We learned several ways to construct phase line portraits, including a graphical method based on a plot of $f(x)$ and a method based on the derivative test for equilibria. Phase line portraits form the basis of bifurcation theory, which is the study of how phase line portraits depend on a parameter p that appears in the differential equation. A bifurcation point p_0 is a value of p at which the phase line portrait significantly changes in a way made precise by the notion of the qualitative equivalence of phase line portraits. We studied three basic types of bifurcations (saddle-node, pitchfork, and transcritical) and how they can be graphically represented in a bifurcation diagram. Autonomous equations are an important special case of separable equations. It is possible to find solution formulas for separable equations (including autonomous equations) by the separation of variables method provided appropriate integrals can be calculated. In general, methods are available for calculating solution formulas only for special types of differential equations, such as linear and separable equations. We studied two other special types of equations in this chapter, Ricatti and Bernoulli equations, and how to find solution formulas for them by change-of-variable methods. In place of solution formulas, we can instead calculate formulas for approximations to solutions. In this chapter we looked at several approximation methods, including two methods based on Taylor polynomials. One method calculates Taylor polynomials in t for solutions $x = x(t)$ and another (the perturbation method) calculates Taylor polynomial approximations in terms of a parameter appearing in the differential equation. Picard iteration is a third method that generates a sequence of approximations that converge to a solution of an initial value problem.

Exercises

For which initial values x_0 are the solutions of the following equations strictly increasing and for which are they strictly decreasing?

1. $x' = 1 - e^{-ax}, a > 0$

2. $x' = (x - a)(b - x), a, b = $ constants satisfying $a \le b$

For each of the following equations, find all equilibria and determine which are hyperbolic. Use the derivative test in Theorem 4 to determine which are sinks and which are sources. Sketch a graph of the right hand side $f(x)$ and use it to obtain the phase line portrait. Identify the type of all equilibria. (a is a constant.)

3. $x' = \cos^2 x$

4. $x' = (ax - 1)(a - x), a > 0$

5. $x' = \dfrac{a}{x} - x, a = $ constant

6. $x' = \tan x$

Find the linearization of the following equations at each of their equilibria.

7. $x' = x^2(1 - x)$ 8. $x' = x(1 - x)^2$

9. $x' = b - e^{-ax}, a \ne 0, b > 0$

10. $x' = ae^x - be^{-x}, a > 0, b > 0$

11. Prove $\beta = +\infty$ for an increasing solution $x = x(t)$ of an autonomous first-order equation $x' = f(x)$ that is bounded above. [*Hint:* Suppose $\beta < +\infty$. Apply the fundamental existence and uniqueness theorem to the initial value problem $x' = f(x), x(\beta) = \lim_{t \to \beta} x(t)$.]

12. Suppose an increasing solution $x = x(t), x'(t) > 0$, of an autonomous first-order equation $x' = f(x)$ approaches a (finite) limit $x_L < +\infty$ as $t \to +\infty$.

   (a) Show that $x'(t)$ approaches a (finite) limit $x'_L \ge 0$ as $t \to +\infty$.

   (b) Show that $x'_L = 0$. [*Hint:* Suppose that $x'_L > 0$, i.e., suppose that $x'(t) > 0$ does *not* tend to 0. Then since $x'(t)$ approaches the positive number $x'_L > 0$ as $t \to +\infty$, it follows that for sufficiently large t it must be true that $x'(t) \ge x'_L/2 > 0$. From $x(t) - x(t_0) = \int_{t_0}^{t} x'(s)\, ds$ deduce a contradiction. Since $x'_L \ge 0$ and since the assumption $x'_L > 0$ lead to a contradiction, the only possibility is $x'_L = 0$.]

13. An equilibrium x_e of $x' = f(x)$ is called *stable* if any solution $x = x(t)$ that starts close to x_e remains close for all $t \ge 0$. Formally, x_e is stable if for any $\varepsilon > 0$ there exists a $\delta > 0$ such that $|x_0 - x_e| < \delta$ implies that $|x(t) - x_e| < \varepsilon$ for all $t > 0$. Prove $x_e = 0$ is stable as an equilibrium solution of the linear equation $x' = px$ if $p \le 0$.

14. An equilibrium x_e of $x' = f(x)$ is called *asymptotically stable* if it is stable and, in addition, there is a $\delta_0 > 0$ such that $|x_0 - x_e| < \delta_0$ implies that $|x(t) - x_e| \to 0$ as $t \to +\infty$. Prove that $x_e = 0$ is asymptotically stable as an equilibrium solution of the linear equation $x' = px$ if $p < 0$.

15. Prove that the equilibrium $x_e = 0$ is stable as a solution of the equation $x' = -ax^3$ provided $a \ge 0$. (*Hint:* Find a formula for the general solution.)

16. Prove the equilibrium $x_e = 0$ is asymptotically stable as a solution of the equation $x' = -ax^3$ provided $a > 0$. (*Hint:* Find a formula for the general solution.)

17. Consider the function f defined by

$$f(x) = \begin{cases} x^2 \sin\left(\frac{1}{x}\right) & \text{for } x \ne 0 \\ 0 & \text{for } x = 0. \end{cases}$$

This function is continuous and differentiable for all x, including $x = 0$.

   (a) Show that $x_e = 0$ is not an isolated equilibrium of the equation $x' = f(x)$. It follows that Theorem 3 does not apply to this equilibrium.

   (b) Show that $x_e = 0$ is not hyperbolic. It follows that Theorem 4 does not apply to this equilibrium.

   (c) Prove that $x_e = 0$ is stable.

   (d) Prove that $x_e = 0$ is not asymptotically stable. Because $x_e = 0$ is not isolated it is not included in the equilibrium classification scheme given in this chapter (i.e., it is neither a source nor a sink nor a shunt).

For each of the equations,

   (a) *find all equilibria and draw the phase line portrait;*

   (b) *classify all equilibria;*

   (c) *determine which equilibria are hyperbolic and which are nonhyperbolic;*

   (d) *find the linearization at each equilibrium and classify its equilibrium;*

   (e) *apply the linearization principle (Theorem 5), if possible, to classify the equilibria.*

18. $x' = x^3 \dfrac{2 - x^2}{1 + x^2}$ 19. $x' = \dfrac{x(1 - x)}{1 + x^4} e^x$

20. $x' = p + x^4$ 21. $x' = p + e^{-x}$

Write down autonomous first-order differential equations that have the following phase portraits.

22. $\longrightarrow 0 \longrightarrow 1 \longrightarrow 2 \longleftarrow$

23. $\longleftarrow -2 \longrightarrow -1 \longleftarrow 0 \longrightarrow 1 \longrightarrow$

24. $\longrightarrow p \longleftarrow$ for $p > 0$ and $\longleftarrow p \longrightarrow$ for $p < 0$

25. $\longrightarrow p \longrightarrow$ for $p > 0$ and $\longleftarrow p \longleftarrow$ for $p < 0$

Which pairs of equations have qualitatively equivalent phase portraits?

26. $\begin{cases} x' = x^7 \dfrac{1 - x^2}{1 + x^2} \\ x' = (x + 1)^5(1.5 - x)^3(2 + x) \end{cases}$

27. $\begin{cases} x' = -x \\ x' = xe^{-x} \end{cases}$

28. $\begin{cases} x' = 1 + x \\ x' = -xe^x \end{cases}$ **29.** $\begin{cases} x' = 1 - x \\ x' = -xe^x \end{cases}$

Determine the bifurcation values for the following equations. Determine the type of bifurcation that occurs. Draw a bifurcation diagram.

30. $x' = x^3(x - p)$

31. $x' = p + (x - 1)^2(x + 1)^2$

32. $x' = p + e^{-x^2}$ **33.** $x' = x\left(p + e^{-x^2}\right)$

34. $x' = 1 + (x - 1)^2 - p$

35. $x' = px - \dfrac{x}{1 + x^2},\ p > 0$

36. Find all the equilibria of the equation $x' = -x + px^2$ where p is a constant. Show all equilibria are hyperbolic. Using phase line portraits, explain why a bifurcation occurs at $p = 0$.

Find formulas for the general solutions of the following equations.

37. $x' + x = x^2$ **38.** $x' = \dfrac{1 + x^3}{x^2}$

39. $x' = 1 - t - x + tx$

40. $x' = \exp(ax - bt),\ a \neq 0,\ b \neq 0$

41. $x' = (t - ax)^2 - t^2$ **42.** $x' = a^2 + x^2$

Find formulas for the solutions of the following initial value problems.

43. $x' = ax + x^3,\ x(0) = x_0,\ a \neq 0$

44. $x' = (a + \sin bt)x^2,\ x(0) = x_0,\ b \neq 0$

Find formulas for the general solution of the following equations.

45. $x' = t^{3/2}x$ for $t > 0$ **46.** $x' = 2x + x^2 \cos t$

47. $x' = -x + (\sin t + 1)x^3$ **48.** $x' = \dfrac{x^2}{t}$ for $t \neq 0$

49. $x' = x + tx^3$ **50.** $x' = te^{ax}$

51. $x' = \dfrac{t}{\sin at},\ a \neq 0$ **52.** $x' = t^{-1}x - x^{-1}$

53. $x' = -x^2 t + x^2 + 3xt - 3x - 2t + 2$

Find formulas for the solutions of the following initial value problems.

54. $x' = te^{ax},\ x(0) = 0$ and $x(0) = -\frac{1}{a}$ (assuming $a \neq 0$)

55. $x' = -x + x^4 t,\ x(0) = 1$

56. $x' = -t^{-1}x + x^2,\ x(1) = -1$ and $x(-1) = 1$

57. $x' = \dfrac{t}{\sin ax},\ x(0) = 1$ and $x(0) = -\frac{\pi}{a}$ (assuming $a \neq 0$)

58. $x' = 1 - t - x + xt,\ x(0) = 0$

59. $x' = 1 - t^2 + x - xt^2,\ x\left(\sqrt{3}\right) = 1$

(a) *Find the third-order Taylor polynomial approximation to the solution of the initial value problems below.*

(b) *Use a computer program to obtain a graphical approximation to these initial value problems on the interval $0 \le t \le 1$. Also plot the first-, second-, and third-order Taylor polynomial approximations to the solutions. Compare the graphs. Do the approximations appear to improve as the order of the approximating polynomial increases?*

60. $x' = x + te^t,\ x(0) = 0$ and 1

61. $x' = 2x + \cos t,\ x(0) = 0$ and 1

Find the recursive formula for the coefficients of the power series solution of the initial value problems below. Use the formula to write out the first few terms of the power series.

62. $x' = t(x - 1),\ x(0) = x_0$

63. $x' = -x + \dfrac{1}{1 - 2t},\ x(0) = x_0$

64. $x' = x,\ x(1) = 1$

65. $x' = -x + \ln t,\ x(1) = 2$

66. Find the recursive formulas for the coefficients $k_i(t)$ of the series representation

$$x = \sum_{i=0}^{\infty} k_i(t)\varepsilon^i$$

of the solution x of the initial value problem

$$x' = a(T(t) - x),\ x(0) = x_0$$

with $T(t) = \tau(1 + \varepsilon \sin t)$. Here $a > 0$, $\tau > 0$, $a > 0$ are positive constants and ε is a small constant. Show that all coefficients $k_i(t)$ except $k_0(t)$ and $k_1(t)$ are equal to 0.

67. Find the recursive formulas for the coefficients $c_i(t)$ of the perturbation approximation

$$x_n(t) = \sum_{i=0}^{n} k_i(t)\varepsilon^i$$

to the solution of the initial value problem

$$x' = (1 + \varepsilon \cos t)x,\ x(0) = 1,$$

where ε is a small constant.

(a) *Compute the first three Picard iterates $x_0(t)$, $x_1(t)$, and $x_2(t)$ of the initial value problems below.*

(b) *Use a computer program to plot the graphs of the first three Picard iterates on the interval $0 \le t \le 1$, together with the numerical solution obtained by use of a computer. Which iterate provides the best approximation? Repeat for the intervals $0 \le t \le 2$ and $0 \le t \le 3$. How do the approximations depend on the length of the intervals?*

68. $x' = r\left(1 - \dfrac{x}{K}\right)x$, $x(0) = \frac{1}{2}K$

69. $x' = r\left(1 - \dfrac{x}{K}\right)x$, $x(0) = K$

70. $x' = x^2 + t^2$, $x(0) = 0$

71. $x' = \cos(x + t)$, $x(0) = 0$

72. A tank full of water has the shape of an inverted circular cone of height H with a very small hole at the bottom out which water is draining. The circular top of the tank has radius R. Suppose the rate at which water drains out the bottom hole at any time is proportional to the square root of the depth of the water in the tank at that time.

(a) Derive an initial value problem for the depth $x = x(t)$ of water in the tank at time t. Let $k > 0$ denote the constant of proportionality and $x_0 > 0$ the initial depth of water. [*Hint:* The volume of water at time t is $v(t) = \pi r^2(t)x(t)/3$, where $r(t)$ is the radius of the circular surface of the water at time t. Apply the balance law, $v' = $ inflow rate $-$ outflow rate, to the volume of water.]

(b) Classify the differential equation derived in (a). Discuss the application to this equation of the fundamental existence and uniqueness theorem.

(c) Determine the dynamics of the equation derived in (a) as time t increases. Discuss your answer with respect to the emptying of the water out of the tank.

(d) Using a computer program, explore the behavior of the solutions of the model derived in (a). Formulate conjectures about the amount of time t_{empty} it will take for the tank to empty of water and how this length of time depends on the initial water depth x_0. (For example: Is t_{empty} proportional to x_0? If not, what are some properties of the dependence?)

(e) Solve the initial value problem in (a). (An explicit solution is not necessary.)

(f) Use your answer in (d) to determine a formula for t_{empty}.

(g) Use your answer in (e) to address your conjectures in (c).

(h) A conical tank of height 9 meters is initially completely full with 100 cubic meters of water. After one half hour the depth of water in the tank is 8 meters. How long will it take for the tank to be half full? To be empty?

(i) Calculate the first- and second-order Taylor polynomial approximations $x_1(t)$ and $x_2(t)$ to the solution.

(j) Use the Taylor polynomial approximations in (h) to approximate t_{empty}. Use your answer in (e) to calculate the percent error made by each approximation.

(k) Calculate the first- and second-order Picard iterate approximations $x_1(t)$ and $x_2(t)$ to the solution.

(l) Use the Picard iterate approximations in (h) to approximate t_{empty}. Use your answer in (e) to calculate the percent error made by each approximation.

3.6 APPLICATIONS

This section contains applications from three different disciplines (population dynamics, the dynamics of motion, and chemical reactions) that involve autonomous and nonautonomous first-order equations. Numerical and graphical approximations, phase line portraits, bifurcation diagrams, perturbation approximations, and solution formulas are among the methods used.

3.6.1 Population Dynamics

Let $b > 0$ and $d > 0$ denote the per capita birth and death rates of individuals in a biological population. This means that the total number of births and deaths per unit time in a population of size x are bx and dx, respectively. If births account for all entries into the population and if deaths account for all exits from the population (that is, there is no immigration, emigration, seeding or harvesting, etc.), then the balance law

$$x' = \text{inflow rate} - \text{outflow rate} \tag{6.1}$$

yields the differential equation

$$x' = bx - dx$$

for the population size $x = x(t)$ at time t. We can rewrite this first-order, linear homogeneous differential equations as

$$x' = rx, \tag{6.2}$$

where $r = b - d$ (the per capita net growth rate). Assuming r is a constant, we can solve the initial condition $x(t_0) = x_0$ for this equation and obtain the formula

$$x(t) = x_0 e^{r(t-t_0)} \tag{6.3}$$

for the population size at future times $t > t_0$. The assumption of a constant per capita net growth rate r leads to exponential growth if $r > 0$ and exponential decay if $r < 0$.

EXAMPLE 1

Suppose we assume the world population of humans has a constant per capita growth rate r. In 1900 the human population of the world is estimated to have been 1.65 billion. In 1910 it is estimated to have been 1.75 billion. Let t be measured in years, x measured in billions of humans, and take

$$t_0 = 1900, \ x_0 = 1.65.$$

Then from (6.3) we have

$$x(t) = 1.65 e^{r(t-1900)}.$$

Using the census data for 1910, we find from this formula that

$$1.65 e^{10r} = 1.75$$

and solving for r that

$$r = \frac{1}{10} \ln \left(\frac{1.75}{1.65} \right) \approx 5.884 \times 10^{-3} \ (\text{per year}). \qquad \blacksquare$$

The value of r obtained in the preceding example gives the formula

$$x(t) = 1.65 e^{5.884 \times 10^{-3}(t-1900)} \tag{6.4}$$

for the world population since 1900. For example, the model predicts the world population in 1920 was

$$x(1920) = 1.65 e^{5.884 \times 10^{-3}(1920-1900)} \approx 1.86 \ \text{billion}.$$

It turns out that the estimated world population was 1.86 billion! World population census estimates up to 1990, as compiled by the United Nations,[8] appear in Table 3.1 along with the predictions of the exponential growth law (6.4) and the errors made by these predictions. Figure 3.27 shows a plot of this data and solution.

The exponential model (6.4), by construction, passes exactly through (i.e., interpolates) the first two data points at years 1900 and 1910 and, as already observed,

---

[8]United Nations, Population Division, Department of Economic and Social Information and Policy Analysis, 1994.

Table 3.1   Percent errors were calculated by the formula $100(x - x(t))/x$, where x is the population data value at year t and $x(t)$ is the model predicted population size. The answer was rounded to the nearest percent.

Year	World Population (billion)	Model (6.4) Prediction (billion)	Error %
1900	1.65	1.65	0
1910	1.75	1.75	0
1920	1.86	1.86	0
1930	2.07	1.97	5
1940	2.30	2.09	9
1950	2.52	2.21	12
1960	3.02	2.35	22
1970	3.70	2.49	33
1980	4.45	2.64	41
1990	5.30	2.80	47

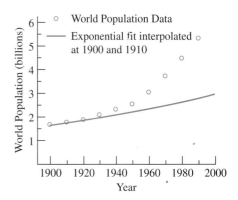

Figure 3.27

predicts the third data point at 1920 accurately. However, the model gives increasingly poor predictions after 1920. During the twentieth century, the world population has grown considerably faster than this exponential model predicts (it was nearly double the prediction in 1990)!

We can try to obtain a better fit of the exponential model (6.4) using other methods. For example, we could interpolate other pairs of data points to determine the values of x_0 and r. We might choose to interpolate the first and last data points (at 1900 and 1990) in order to obtain a better fit (hopefully) of the model to the data over the entire century. However, there is a feature of the data in Table 3.1 that suggests the exponential growth model (6.4) is not an appropriate model for this world population data. This feature concerns the so-called doubling time of an exponentially growing population.

A fundamental fact about solutions of the differential equation (6.2) is that the time it takes a population to double in size is independent of its size. To see this, notice that a population of size x_0 at time t_0 reaches size $2x_0$ when

$$x_0 e^{r(t-t_0)} = 2x_0$$

or

$$t - t_0 = \frac{1}{r} \ln 2. \tag{6.5}$$

This formula shows that the doubling time is independent of x_0 and depends only on r.

Approximate doubling times, calculated from the world population data in Table 3.1, appear in Table 3.2. These approximations seem to indicate that the doubling time of the world population is not independent of its size. In fact, it seems the doubling time *decreases* as the population size increases. This suggests the world population is not growing exponentially but instead growing *faster* than exponentially!

	Table 3.2	
Year	World Population (billion)	Approximate Doubling Time (years)
1900	1.65	60.4
1910	1.75	50.8
1920	1.86	50.0
1930	2.07	40.6
1940	2.30	40.2
1950	2.52	30.7

Population Explosion   We modeled the world population of humans by means of the differential equation (6.4), used its solution (6.2) to compare the model predictions with data, and found the result to be unsatisfactory. In keeping with the modeling cycle discussed in the Introduction), we will proceed to the model modification step and attempt to obtain a new model whose solution fits the data more accurately.

In Table 3.2 we find that the doubling time of the world population decreases as the population increases. This suggests that r increases with population size x [see (6.5)]. To capture this feature of the data, let us assume that r is proportional to some power of x, that is,

$$r = kx^p$$

for constants $k > 0$ and $p > 0$. Under this assumption our new model for the growth of the world population is

$$x' = kx^{p+1} \qquad (6.6)$$
$$x(t_0) = x_0.$$

The differential equation in (6.6) is autonomous (separable). Using the solution method in Sec. 3.2, we calculate the solution formula

$$x(t) = \frac{x_0}{\left(1 - x_0^p\, pk(t - t_0)\right)^{1/p}} \qquad (6.7)$$

for the initial value problem (6.6). If we use the initial condition $x_0 = 1.65$ at $t_0 = 1900$ from Table 3.1, this formula becomes

$$x(t) = \frac{1.65}{(1 - 1.65^p\, pk(t - 1900))^{1/p}}. \qquad (6.8)$$

In this formula there remain two parameters, p and k, that need to be determined before we can use the formula for numerical predictions.

We can use the data in Table 3.1 and an interpolation method to obtain estimates for p and k. Since there are two parameters to estimate, we can interpolate two data points. Let us choose a point at the end and in the middle of the data time period, say the data points at $t = 1940$ and 1990. To interpolate these particular points, the solution (6.8) must satisfy

$$x(1940) = 2.30$$
$$x(1990) = 5.30,$$

or, in other words, the two parameters p and k must satisfy the two equations

$$\frac{1.65}{(1 - 40 \times 1.65^p\, pk)^{\frac{1}{p}}} = 2.30$$

$$\frac{1.65}{(1 - 90 \times 1.65^p\, pk)^{\frac{1}{p}}} = 5.30.$$

The solutions of these equations can be obtained with the aid of a computer program or a calculator:

$$p \approx 1.2633, \quad k \approx 3.6022 \times 10^{-3}.$$

Using these values of p and k in (6.8), we arrive at the formula

$$x(t) = \frac{1.65}{\left(17.277 - 8.5669 \times 10^{-3}t\right)^{0.79158}} \tag{6.9}$$

for the model predicted world population at time t.

Table 3.3 shows the errors made by the formula (6.9) in comparison to the data throughout the twentieth century. These errors indicate that the model (6.9) fits the world population during that time period reasonably well. The graph in Fig. 3.28 also illustrates the accuracy of the fit.

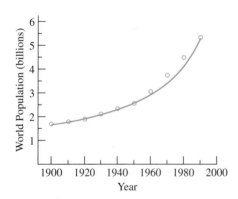

Table 3.3

Year	World Population (billion)	Model (6.9) Prediction (billion)	% Error
1900	1.65	1.65	0
1910	1.75	1.77	−1
1920	1.86	1.91	3
1930	2.07	2.09	−1
1940	2.30	2.30	0
1950	2.52	2.57	−2
1960	3.02	2.92	3
1970	3.70	3.41	8
1980	4.45	4.12	7
1990	5.30	5.30	0

Figure 3.28 The predictions and errors of the solution (6.9). Percent errors were calculated by the formula $100\,(x - x(t))\,/x$, where x is the population data value at year t and $x(t)$ is the model predicted population size. The answer was rounded to the nearest percent.

What does the model (6.9) predict for the future of the world population? The graph in Fig. 3.29 shows a very rapid increase in population numbers during the first two decades of the twenty-first century. In fact, the graph suggests that the solution has a vertical asymptote (approximately at $t = 2016$). Indeed, the solution (6.9) does have a vertical asymptote, located at that value of t for which its denominator equals 0, which turns out to be $t \approx 2016.7$. This model predicts the human population will

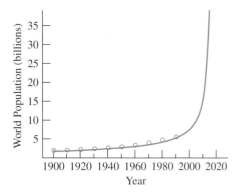

Figure 3.29 The world population data and the model prediction (6.9).

tend to infinity on Monday, September 12, 2016, at about 4:48 A.M.! For this reason, the model is called a doomsday model.

The doomsday date predicted by the solution formula (6.8) depends on the numerical values used for the parameters p and k. A different estimation of these parameters (obtained, for example, by interpolating different data points or using an entirely different method, such as least squares) will change the doomsday date. For example, see Exercises 4 and 5. Therefore, predicted doomsday dates using the model should not be taken seriously. However, an important fact is that no matter what values are assigned to p and k, the model equations (6.6) always predict a doomsday,[9] since the solution (6.7) has a vertical asymptote where its denominator equals 0, that is, at

$$t = 1900 + \frac{1}{1.65^p \, pk}.$$

No population can increase beyond all bounds, of course, as is predicted by both the linear model (6.2) and by the nonlinear model (6.6). A model predicts the inevitable consequences of the modeling assumptions on which it is based, provided these assumptions remain valid in the future. The models serve to illustrate the seriousness of exponential and supraexponential growth patterns and that population numbers must—will be—regulated in some manner or other. In the following sections we consider some models of regulated population growth.

Regulated Population Growth   The assumption that the per capita growth rate x'/x is a positive constant leads to unbounded exponential growth. We have also seen how the assumption that x'/x increases with population size x, proportionally to a power of x, can lead to a population explosion in finite time. Realistically it is clear that no population can grow so as to exceed all bounds. Population growth must, therefore, be regulated in some way and population numbers must eventually level off or decrease.

An example appears in Table 3.4. A major insect pest that thrives in stored grain products (e.g., the flour in your kitchen cupboard) is the so-called flour beetle. The census numbers of the adult beetles that appear in Table 3.4 are from a culture of flour beetles maintained in a container with a constantly replenished supply of fresh flour. These numbers show that the beetle population, after an initial period of rapid

---

[9]For another example of a doomsday model for the world population (one that predicted as doomsday date of November 13, 2026), see Heinz von Foerster et al., *Science* 132 (1960): 1291–1295.

Table 3.4

Census Time (unit = 2 wks)	Number of Beetles (per 20 gram of flour)	Average per Capita Growth Rate (6.10)
0	112	0.3571
1	152	0.3947
2	212	0.2170
3	258	0.1860
4	306	0.0098
5	309	0.0194
6	315	−0.0159
7	310	−0.0387
8	298	−0.0268
9	290	0.0448
10	303	−0.0264
11	295	0.0542
12	311	−0.0096
13	308	−0.0292
14	299	0.0334
15	309	−0.0162

increase, eventually level off at approximately 300 individuals (per 20 g of flour); see Fig. 3.30(a). The population is self-regulated so as to prevent numbers that exceed the available resources. (In this case, the beetles regulate their numbers by cannibalism—a solution to the problem of excessive population growth that humans hopefully will not adopt.) Table 3.4 also gives the (average) per capita growth rate

$$\frac{x(t+1) - x(t)}{x(t)}. \tag{6.10}$$

From the graph in Fig. 3.30(b) we see that the per capita growth rate (generally) decreases with population size. In fact, from the graph we see that the relationship between these two variables is nearly linear, that is,

$$\frac{x(t+1) - x(t)}{x(t)} \approx b + mx,$$

where the slope m is negative. Also shown in Fig. 3.30(b) is the least squares straight line fit (linear regression) to these data. This best linear fit has slope $m = -0.002184$ and intercept $b = 0.6709$. If we approximate the average per capita growth rate (6.10) by x'/x, we arrive at the differential equation

$$\frac{x'}{x} = 0.6709 - 0.002184x$$

or

$$x' = x\,(0.6709 - 0.002184x) \tag{6.11}$$

as a model for the population growth of the flour beetles.

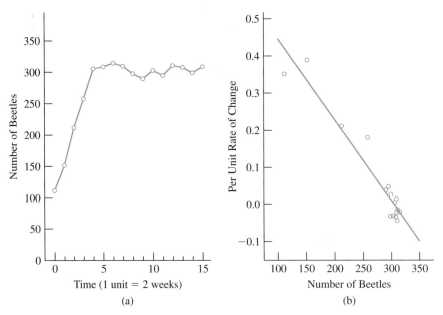

Figure 3.30 (a) The flour beetle census data from Table 3.4. (b) A plot of the per capita growth rate and population size from Table 3.4. Also shown is the least squares best linear fit to these data.

The differential equation (6.11) is an example of the *logistic equation* (or the *Pearl-Verhulst equation*). The logistic equation derives from the balance law (6.1) under the assumption that the per capita birth and death rates are linearly related to population size. Specifically, if we assume the per capita birth rate decreases and the per capita death rate increases with increased population size, so that we can write these two rates as $b - c_1 x$ and $d + c_2 x$, respectively, then from (6.1) we have

$$x' = (b - c_1 x) x - (d + c_2 x) x.$$

Letting $r = b - d$ and $c = c_1 + c_2$ for notational simplicity, we have the equation

$$x' = x(r - cx). \tag{6.12}$$

Equation (6.11), which we obtained from the data on flour beetles, has this form with $r = 0.6709$ and $c = 0.002184$.

The logistic equation (6.12) is autonomous. When $c \neq 0$, there are two equilibria $x_e = 0$ and r/c. From the two possible phase line portraits shown in Fig. 3.31, we see that a transcritical bifurcation and an exchange of stability between the two equilibria occur at the critical value $r = 0$.

The coefficient r has the following biological interpretation. When population numbers are low, $x(r - cx) \approx rx$ and $x' \approx rx$. Therefore, at low numbers the population grows (approximately) exponentially with rate r. For this reason r is called the *inherent (per capita) growth rate.*

According to Fig. 3.31, if the inherent growth rate r is negative, all solutions x [with positive initial conditions $x(0) > 0$, which are the only initial conditions of interest in the application] tend to 0 as $t \to +\infty$. Consequently, the model predicts

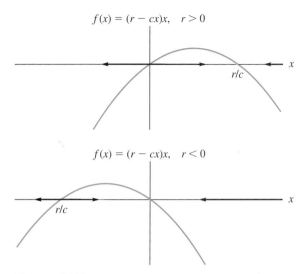

Figure 3.31 The two possible phase line portraits of equation (6.12).

that the population ultimately goes extinct, no matter what its initial size is. On the other hand, if $r > 0$, then the population approaches the equilibrium r/c as $t \to +\infty$. Therefore, the logistic model does not predict unbounded population growth but instead predicts a long-term equilibrium state. This equilibrium, traditionally denoted by

$$K = \frac{r}{c},$$

is called the (environmental) *carrying capacity*.

We can place the carrying capacity K explicitly into equation (6.12) by making the substitution $c = r/K$. The logistic equation is then in its traditional form

$$x' = rx\left(1 - \frac{x}{K}\right). \tag{6.13}$$

We can calculate a formula for the solution of the initial value problem $x(t_0) = x_0$ using the method of separating variables (see Example 2 of Sec. 3.2):

$$x(t) = \frac{x_0 K}{(K - x_0)e^{-r(t-t_0)} + x_0}. \tag{6.14}$$

(The logistic is also a Bernoulli and a Ricatti equation and the solution methods for these equations also yield this formula. See Exercise 8.)

Returning to the equation (6.11) derived from the beetle data in Table 3.4, we see that $r = 0.6709$ and $c = 0.002184$, or

$$r = 0.6709, \quad K = 307.2.$$

With these coefficients and with the initial condition $x_0 = 112$ and $t_0 = 0$, the formula (6.14) provides predictions for the beetle population size at any time t. Figure 3.32 and Table 3.5 show how well this formula fits the data from Table 3.4, as well as how well it predicts five new data points not included in that table.

Table 3.5   Percent errors were calculated by the formula 100(x − x(t))/ x, where x is the population data value at year t and x(t) is the model predicted population size. The answer was rounded to the nearest percent.

Census Time (unit = 2 wks)	Number of Beetles (per 20 gram of flour)	Logistic (6.11)	% Error
0	112	112.00	0
1	152	162.45	−7
2	212	211.06	0
3	258	249.17	3
4	306	274.51	10
5	309	289.59	6
6	315	297.93	5
7	310	302.39	2
8	298	304.72	−2
9	290	305.93	−5
10	303	306.55	−1
11	295	306.87	−4
12	311	307.03	1
13	308	307.11	0
14	299	307.16	−3
15	309	307.18	1
16	304	307.19	−1
17	311	307.19	1
18	306	307.20	−0
19	310	307.20	1
20	308	307.20	0

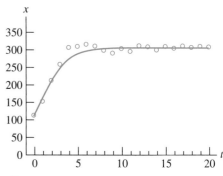

Figure 3.32 The logistic equation (6.11) fit to the data in Table 3.5, plus 5 additional points.

Based on the accuracy of the logistic fit to the data in Table 3.5, we conclude that the flour beetle populations will, in the long run, maintain approximately 307 beetles per 20 gram of flour.

Fluctuating Environments   In the logistic equation (6.13) the parameters r and K are positive constants. The carrying capacity K can be considered a property of the population's environment and habitat. It is not unusual, however, for important properties and characteristics of a population's habitat to fluctuate in time. For example, a common situation would be seasonal fluctuations in food resource availability, temperature, and so on. Important fluctuations can occur on a daily basis as well, or even a monthly basis. Our problem is to determine how the dynamics of a logistically growing population change when it is placed in a habitat with a fluctuating carrying capacity.

When $K = K(t)$ is a function of t, the logistic equation (6.13) becomes nonautonomous. We consider here the case when the habitat fluctuations are regular over time, a situation we will model by assuming they are exactly periodic. Then $K = K(t) > 0$ is a periodic function of time with period denoted by $T > 0$. We assume the inherent growth rate $r > 0$ remains constant. Under these assumptions we have a nonautonomous (periodically forced) version

$$x' = rx \left(1 - \frac{x}{K(t)} \right) \tag{6.15}$$

of the logistic equation for our model of the growth of a population in a periodically fluctuating habitat.

In order to address the question we asked above, we need to determine the properties of solutions to the equation (6.15). In particular, we are interested in the long-term dynamics of solutions of the initial value problem $x(0) = x_0 > 0$. When K is a constant, we know $\lim_{t \to +\infty} x(t) = K$. What does the solution do as $t \to +\infty$ when $K(t)$ is nonconstant but oscillates periodically? To carry out the solution step of the modeling cycle we have several choices. We can try to find useful solution formulas or, if that is not possible, we can calculate approximations to solutions.

We will confine our investigation to the special case when the periodically varying carrying capacity is given:

$$K(t) = K_0 + a \sin \left(\frac{2\pi}{T} t \right). \tag{6.16}$$

Here $K(t)$ oscillates around the average $K_0 > 0$ in a sinusoidal fashion with amplitude a and period $T > 0$. Since $K(t)$ must remain positive, it is required that $|a| < K_0$.

With the model (6.16) for the environmental carrying capacity, the logistic equation becomes

$$x' = rx \left(1 - \frac{x}{K_0 + a \sin \left(\frac{2\pi}{T} t \right)} \right). \tag{6.17}$$

Before studying the solutions of this equation, it is helpful to perform some computer explorations of solutions for selected values of r, K_0 and the period T. These will help focus our analysis.

Figure 3.33 shows graphs of selected solutions of (6.17) with $r = 1$, $K_0 = 1$, $a = 0.5$, and $T = 1$. Notice that all the displayed solution plots tend to the same periodic oscillation as t increases. A close observation of the plots shows that this limiting periodic oscillation has period 1, which is that of the carrying capacity $K(t)$ in this example. Moreover, it turns out that if we use the initial condition $x_0 = 0.80162$,

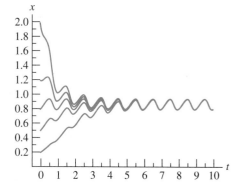

Figure 3.33 Selected solutions of the periodic logistic equation (6.15)–(6.16) with $r = K_0 = T = 1$ and $a = 0.5$.

the resulting solution appears identical to the limiting periodic oscillation. We conclude that the limiting periodic oscillation is apparently itself a solution of the periodic logistic.

The observation that there exists a periodic solution to which all other (positive) solutions tend as $t \to +\infty$ is not, it turns out, peculiar to the sinusoidal carrying capacity (6.16) or the special parameter choices in Fig. 3.33, but in fact holds true in general for the periodic logistic (6.15). This can be proved by noting that the periodic logistic is Bernoulli equation (or a Ricatti equation) and by calculating a solution formula (Sec. 3.3). These details are left as Exercise 10. Our goal is to describe the basic features of the attracting periodic solution $x_p(t)$ of the logistic (6.16), and hence what the basic features of all positive solutions are in the long run.

For example, we would like to know the amplitude of $x_p(t)$. We would also like to know its "phase" relative to that of the environmental carrying capacity $K(t)$, that is, whether the oscillation in population size $x_p(t)$ lags behind the oscillations in the environment $K(t)$ or whether it closely tracks that of $K(t)$. Presumably the answers to these questions depend on the numerical values of the four parameters r, K_0, a, and T. In order to focus our study we will hold fixed the parameters related to the environmental carrying capacity $K(t)$, namely K_0, a, and T, and determine how the characteristics of the population's (long-term) ocillations $x_p(t)$ depend on its inherent (exponential) growth rate r.

It turns out that answers to these questions are difficult to obtain from the solution formula for $x_p(t)$ obtained by the Bernoulli or Ricatti method (see Exercise 11). Therefore, we will instead study $x_p(t)$ using a perturbation approximation. To do this we will consider small amplitude environmental oscillations, which means mathematically that $|a|$ is a small number. We seek periodic coefficients in the a-expansion

$$x_p(t) = c_0(t) + c_1(t)a + \cdots \tag{6.18}$$

of the periodic solution. We will calculate only the first-order perturbation approximation

$$x_p(t) \approx c_0(t) + c_1(t)a \tag{6.19}$$

and use it to carry out our study of $x_p(t)$.

Following the method in Sec. 3.4.2, we obtain recursive formulas for the coefficients c_i in (6.18) by substituting the expansion (6.18) into both sides of the periodic logistic equation (6.17) and equating coefficients of like powers of a.

From the left-hand side of the periodic logistic we have

$$x_p' = c_0' + c_1'a + \cdots .$$

To find the expansion of the right-hand side of the equation we use the expansion[10]

$$\frac{1}{K_0 + a \sin\left(\frac{2\pi}{T}t\right)} = \frac{1}{K_0} + \left(-\frac{1}{K_0^2}\sin\left(\frac{2\pi}{T}t\right)\right)a + \cdots$$

---

[10]Recall the geometric series $\frac{1}{1-r} = 1 + r + r^2 + \cdots$, for $|r| < 1$.

in a to obtain

$$rx_p \left(1 - \frac{x_p}{K_0 + a \sin\left(\frac{2\pi}{T}t\right)} \right) = rc_0 \left(1 - \frac{c_0}{K_0} \right)$$

$$+ \left\{ rc_1 \left(1 - 2\frac{c_0}{K_0} \right) + r\frac{c_0^2}{K_0^2} \sin\left(\frac{2\pi}{T}t\right) \right\} a + \cdots.$$

Equating coefficients of like powers of a on both sides, we get the differential equations

(a) $c_0' = rc_0 \left(1 - \dfrac{c_0}{K_0} \right)$

(b) $c_1' = rc_1 \left(1 - 2\dfrac{c_0}{K_0} \right) + r\dfrac{c_0^2}{K_0^2} \sin\left(\dfrac{2\pi}{T}t\right)$

(6.20)

for the periodic coefficients c_0 and c_1, respectively. Notice that the second equation (b) for c_1 involves the solution c_0 from the first equation (a). The first equation (a), however, does not involve c_1. In fact, equation (6.20a) is the logistic equation for c_0. Remember that we seek periodic coefficients c_0 and c_1.

The only periodic solutions of the (autonomous) logistic (6.20a) are the equilibrium solutions 0 and K_0. The first choice $c_0 = 0$, when used in (6.20), leads to the equation $c_1' = rc_1$ whose only periodic solution is $c_1 = 0$ (all other solutions are exponentials). This choice leads to the perturbation approximation $c_0 + c_1 a = 0$, a choice that does not approximate the positive periodic solution in which we are interested. The second choice

$$c_0(t) = K_0$$

for a periodic solution of the the logistic (6.20a), when used in (6.20b), gives the linear nonhomogeneous equation

$$c_1' = -rc_1 + r\sin\left(\frac{2\pi}{T}t\right)$$

(6.21)

for c_1. Again, remember that we seek a periodic solution $c_1(t)$.

The method of undetermined coefficients (Sec. 2.3, Chapter 2) is useful here, since it calculates a periodic solution as a linear combination of $\sin(2\pi t/T)$ and $\cos(2\pi t/T)$. This method yields the periodic solution

$$c_1(t) = \left(\frac{r^2 T^2}{r^2 T^2 + 4\pi^2} \right) \sin\left(\frac{2\pi}{T}t\right) + \left(\frac{-2\pi rT}{r^2 T^2 + 4\pi^2} \right) \cos\left(\frac{2\pi}{T}t\right).$$

Since the general solution of (6.21) is the sum of this particular solution and the exponential ce^{-rt}, we see that $c_1(t)$ is the only periodic solution of (6.21).

In order to address the questions raised above, it is convenient to rewrite $c_1(t)$. Using the sum angle trigonometric identity for the sine function, we write

$$c_1(t) = \frac{rT}{\sqrt{r^2 T^2 + 4\pi^2}} \sin\left(\frac{2\pi}{T}t - \theta\right),$$

where

$$\theta = \arctan\left(\frac{2\pi}{rT}\right).$$

In summary, we have calculated the first-order perturbation approximation

$$x_p(t) \approx K_0 + a\frac{rT}{\sqrt{r^2T^2 + 4\pi^2}} \sin\left(\frac{2\pi}{T}t - \theta\right) \tag{6.22}$$

to the periodic solution of the periodic logistic (6.17), which we expect to be accurate for small environmental amplitudes $|a| < K_0$.

We can use the approximation (6.22) to address the questions raised above concerning the periodic solution $x_p(t)$. Our goal was to compare the population oscillation $x_p(t)$ to the environmental carrying capacity oscillation $K(t)$. Both the approximation (6.22) and the carrying capacity $K(t)$ are sinusoidal oscillations of period T with average K_0. However, the first-order perturbation approximation (6.22) and $K(t)$ have different amplitudes and phases.

First, notice that because

$$\frac{rT}{\sqrt{r^2T^2 + 4\pi^2}} < 1$$

the amplitude of the approximation (6.22) is less than a. That is,

$$\text{amplitude of } x_p(t) \approx a\frac{rT}{\sqrt{r^2T^2 + 4\pi^2}} < a = \text{amplitude of } K(t).$$

We conclude that the oscillation in population numbers is less in magnitude than that of the carrying capacity.

The difference in the two amplitudes is least for r large and greatest for r small. To see this, note that (for fixed period T)

$$\lim_{r\to+\infty} a\frac{rT}{\sqrt{r^2T^2 + 4\pi^2}} = a$$

$$\frac{rT}{\sqrt{r^2T^2 + 4\pi^2}} = \frac{rT}{2\pi} + O(r^3) \approx 0 \quad \text{for } r \text{ small.}$$

If we investigate the approximation (6.22) for large and for small r, we find, after making use of

$$\lim_{r\to+\infty} \arctan\left(\frac{2\pi}{rT}\right) = 0$$

$$\arctan\left(\frac{2\pi}{rT}\right) \approx \frac{\pi}{2} \quad \text{for } r \text{ small,}$$

that

$$x_p(t) \approx K_0 + a\sin\left(\frac{2\pi}{T}t\right) = K(t) \quad \text{for large } r$$

and

$$x_p(t) \approx K_0 + a\frac{rT}{2\pi}\sin\left(\frac{2\pi}{T}\left(t - \frac{T}{4}\right)\right) \quad \text{for small } r.$$

In summary, the population tracks the carrying capacity very closely for large inherent growth rates r. For small r, however, the population has a low-amplitude oscillation that lags behind the carrying capacity by a quarter cycle. These properties of the periodic logistic are illustrated in Fig. 3.34.

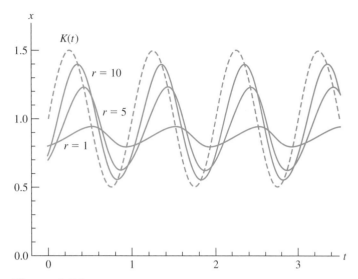

Figure 3.34 The solid lines show three typical periodic solutions $x_p(t)$ of the periodic logistic (6.17) with $K_0 = 1$, $a = 0.5$, $T = 1$ and three selected values of r. For comparison purposes, the dashed line shows the periodic carrying capacity.

With regard to a comparison between the average of the population's oscillation and that of its environment, note that the first-order perturbation approximation (6.22) is a sinusoidal oscillation around the average K_0, which is the average of the carrying capacity $K(t)$. To first-order approximation, the average population size equals the average carrying capacity. However, it turns out that the second-order perturbation approximation has an average less than K_0! (See Exercise 12.) Thus, one sometimes needs higher-order approximations to obtain accurate conclusions.

3.6.2 Objects in Motion

Newton's laws of motion are widely used in science and engineering to derive equations for the motion of objects subject to various forces. One of Newton's laws is expressed by the famous equation $F = ma$ in which m is the mass of the object, a is its acceleration, and F is the sum total of all forces acting on the object. Since acceleration is the derivative of velocity v (or the second derivative of position x) with respect to time, this equation is a differential equation for velocity (or position). In general, the quantities x, v, a, and F are vector functions of time t. We will restrict our attention to motion constrained to a straight line (rectilinear motion). In this case, these quantities are scalar functions of time t.

The use of Newton's law[11] $F = ma$ requires an accounting of all forces acting on the object in motion (gravity, friction, engine propulsion, etc.). The properties of these forces require modeling assumptions for their description. Some forces, being small in comparison to others, can often be neglected. Others require mathematical expressions or models to describe them. The models for these forces are necessarily based upon

---

[11]This form of Newton's famous law assumes that the mass m of the object remains constant in time. More generally, the law is $F = (mv)'$. The product mv is called the momentum of the object.

the specifics of the situation. They also usually require some simplifying assumptions that take advantage of particular circumstances.

A common force for many situations is the frictional force caused by the medium through which the object is moving. Air resistance is an example. Frictional forces caused by fluids such as water or oil are other examples. We denote the frictional force exerted on the moving object by F_f. In a specific application a model that describes this force must be chosen. A general feature of a frictional force is that it depends on the velocity v of the object. If the object is at rest, then $F_f = 0$. As the velocity increases, so does the frictional force F_f. Thus, we write

$$F_f = F_f(v), \qquad F_f(0) = 0.$$

Since the force of friction acts in the opposite direction from that of the objects motion, F_f should be positive when v is negative and negative when v is positive. A convenient way to describe these facts mathematically is to assume that F_f satisfies

$$v F_f(v) < 0 \quad \text{for} \quad v \neq 0. \tag{6.23}$$

Of course, F_f might also depend on the object's position and/or on time [i.e., $F_f = F_f(v, x, t)$]. This would occur, for example, if the medium through which the object is moving is not uniform is space or time. We assume in our application, however, that the medium is uniform in space and time and write $F_f = F_f(v)$.

If no other force acts on the object, then $F = F_v(v)$ in Newton's law $F = ma$ and (since $a = v'$) we have

$$mv' = F_f(v). \tag{6.24}$$

This is a first-order autonomous differential equation for the velocity $v = v(t)$. (See Exercise 15.)

A *linear law* for the frictional force assumes that F_f is proportional to the velocity v, that is,

$$F_f = -kv. \tag{6.25}$$

Here $k > 0$ must be positive so that the constant of proportionality $-k$ is negative; this guarantees (6.23) holds. The coefficient k is called the *coefficient of friction*. The differential equation (6.24) becomes, in this case, the linear autonomous equation

$$mv' = -kv, \quad k > 0.$$

Given an initial velocity of v_0, it follows from this model that the object's velocity is an exponentially decreasing function

$$v(t) = v_0 e^{-kt/m}$$

of time. By integrating this exponential, we find that the object's position $x = x(t)$ (measured from a selected reference point) is

$$x(t) = -\frac{m}{k} v_0 e^{-kt/m} + c,$$

where c is a constant of integration determined by the object's initial position $x(0) = x_0$. Thus, $c = x_0 + mv_0/k$ and

$$x(t) = -\frac{m}{k} v_0 e^{-kt/m} + x_0 + \frac{m}{k} v_0.$$

As $t \to +\infty$ this model implies the object's velocity $v(t)$ approaches 0 and its position approaches $x_0 + mv_0/k$. Thus, in the long run, the object shifts from its initial positive x_0 by an amount mv_0/k. Note the shift in position is proportional to its mass m and to its initial velocity v_0 but is inversely proportional to the coefficient of friction k.

Another friction law is the *quadratic law*. This law assumes the force of friction has a magnitude proportional to the *square* of the object's velocity v. Thus,

$$F_f = \begin{cases} -kv^2 & \text{if } v \geq 0 \\ kv^2 & \text{if } v \leq 0 \end{cases} = -kv\,|v|\,. \tag{6.26}$$

(See Exercise 16.) Other laws, such as $F_f = \pm kv^p$, for other powers v are often used. An appropriate model for the force due to friction depends on such things as the shape of the object (e.g., is it blunt or pointed and streamlined?), the material out of which it is made (e.g., is it solid, pliable, porous?), and the properties of the medium in which it is moving.

In general, moving objects are subject to several forces. For example, an object falling near the surface of the earth is subject not only to air resistance F_f but to the force of gravity F_g. In this case, $F = F_g + F_f$ and Newton's law is

$$mv' = F_g + F_f\,.$$

We have already discussed the modeling of the frictional force F_f. To complete the model in the presence of a gravitational force we need a model for F_g.

Another fundamental law of Newton asserts that the gravitational force between two objects is jointly proportional to their masses and inversely proportional to the distance between them (i.e., between their centers of mass). An accurate approximation can be made for small objects moving near the surface of the earth. In this case, the distance moved by the object is small compared to the distance to the center of the earth. Therefore, the distance between (the centers) of the object and the earth is nearly constant in time and Newton's law of gravity implies the gravitational force exerted on the object remains nearly constant. Since this constant force is proportional to the mass m of the object, we can write $F_g = mg$ for some constant of proportionality g. For an object in free fall and subject to no other forces, Newton's law of motion $F = ma$ with $F = F_g$ becomes $mg = ma$ and we see that the constant of proportionality $g = a$ is the acceleration of the object. This constant acceleration due to gravity g is approximately 9.8 meters/s^2 (or 32 feet/s^2) near the surface of the earth. Using this approximation for the force of gravity and using the friction law discussed above, we arrive at the autonomous, first-order differential equation

$$mv' = mg + F_f(v), \tag{6.27}$$

where $F_f(v)$ satisfies (6.23). In the following sections we give an application of both the linear friction law (6.25) and the quadratic friction law (6.26).

A Falling Object: The Linear Friction Law   The game of badminton uses a shuttlecock or birdie, which is a ball with feathers (these days usually made of plastic) attached in a cometlike fashion. The intent of these feathers is to provide additional air resistance and thereby significantly affect the shuttlecock's flight. We consider the motion of a badminton shuttlecock when it is dropped and allowed to fall vertically to the ground under the influence of gravity. Data taken in such an experiment appear in Table 3.6.[12]

Table 3.6

Time t (seconds)	Distance Fallen x (meters)
0.000	0.00
0.347	0.61
0.470	1.00
0.519	1.22
0.582	1.52
0.650	1.83
0.674	2.00
0.717	2.13
0.766	2.44
0.823	2.74
0.870	3.00
1.031	4.00
1.193	5.00
1.354	6.00
1.501	7.00
1.727	8.50
1.873	9.50

We assume that the only forces acting on the shuttlecock are a constant force of gravity $F_g = mg$ and a frictional force $F_f = F_f(v)$ due to air resistance. In the law of motion described by (6.27), we first consider the linear frictional law (6.25)

$$mv' = mg - kv,$$

where m is the mass of the shuttlecock. We can simplify this differential equation by dividing both sides by m and letting $k_0 = k/m$ (the coefficient of friction per unit mass)

$$v' = g - k_0 v$$

or, since our units are meters and seconds,

$$v' = 9.8 - k_0 v. \tag{6.28}$$

In the experiment from which the data in Table 3.6 was taken, the shuttlecock was simply dropped from a height above the floor. Therefore, we have the initial condition

$$v(0) = 0. \tag{6.29}$$

---

[12]From "Terminal Velocity of a Shuttlecock in Vertical Fall" by Mark Peastrel, Rosemary Lynch, and Angelo Armenti, Jr., in *The Physics of Sports*, Angelo Armenti, Jr. (editor), American Institute of Physics, New York, 1993, pp. 301–303.

A formula for the solution of this (linear, nonhomogeneous) initial value problem is

$$v(t) = -\frac{9.8}{k_0}e^{-k_0 t} + \frac{9.8}{k_0}. \tag{6.30}$$

An antidifferentiation of $x' = v$ yields the formula

$$x(t) = \frac{9.8}{k_0^2}e^{-k_0 t} + \frac{9.8}{k_0}t + c$$

for the distance fallen. The arbitrary constant c is determined by the initial condition $x(0) = 0$ and yields

$$x(t) = \frac{9.8}{k_0^2}e^{-k_0 t} + \frac{9.8}{k_0}t - \frac{9.8}{k_0^2}. \tag{6.31}$$

To see how well this model fits the observed data in Table 3.6, we need a numerical value for the coefficient k_0. We will estimate k_0 by an interpolation procedure. That is, we will determine k_0 by requiring that the solution of the initial value problem (6.28)–(6.29) reproduce a selected data point from Table 3.6.

There are many choices for the interpolating point. For example, we can choose the final point for interpolation [i.e., require that $x(1.873) = 9.5$] with the idea that this will increase the chance that all the intermediate data points will be accurately approximated by the model. The reader is asked to do this in Exercise 18. Another strategy is to interpolate a middle data point. The idea then is to obtain a reasonable approximation to the data over the first half of the data set. Then we can assess the accuracy of the model approximation over the second half of the data to see how well the model predicts into the future. We will take this strategy here and interpolate the data point at time $t = 0.717$.

Specifically, we want to determine a value of k_0 so that $x(1.873) = 9.5$ or, using the formula (6.31), so that

$$\frac{9.8}{k_0^2}e^{-0.717 k_0} + \frac{7.027}{k_0} - \frac{9.8}{k_0^2} = 2.13.$$

Using a computer or calculator, we solve this equation for $k_0 \approx 0.7333$. With this approximate value of k_0 in the formula (6.31) the linear friction law model predicts that the distance fallen in t seconds is

$$x(t) \approx 18.23 \exp(-0.7333t) + 13.37t - 18.23. \tag{6.32}$$

How well this model fits the data can be judged by the errors in Table 3.7. Also see Fig. 3.35. In general the errors are not small. A striking observation is the steadily increasing magnitude of error that occurs after the interpolating point at the midpoint $t = 0.717$. The model based on the linear friction law predicts the shuttlecock will fall further than was actually observed. Presumably this friction law does not adequately account for the air resistance that results from the design of the shuttlecock. Apparently there is more force due to friction than is accounted for by the linear law, since the shuttlecock falls slower than the law predicts. In the next section we investigate the quadratic friction law as a model for the falling shuttlecock.

Table 3.7 The predicted distances (rounded to the nearest hundreth) of an interpolating solution of the linear equation (6.28) and their percent errors. Percent errors were calculated by the formula $100(x - x(t))/x$, where x is the population data value at year t and $x(t)$ is the model predicted population size. The answer was rounded to the nearest percent.

Time t (seconds)	Distance Fallen x (meters)	Model Prediction (to nearest 100th)	Error %
0.000	0.00	0.00	0
0.347	0.61	0.54	11
0.470	1.00	0.97	3
0.519	1.22	1.17	4
0.582	1.52	1.45	5
0.650	1.83	1.78	3
0.674	2.00	1.90	5
0.717	2.13	2.13	0
0.766	2.44	2.41	1
0.823	2.74	2.74	0
0.870	3.00	3.03	−1
1.031	4.00	4.11	−3
1.193	5.00	5.32	−6
1.354	6.00	6.63	−10
1.501	7.00	7.90	−13
1.727	8.50	10.00	−18
1.873	9.50	11.43	−20

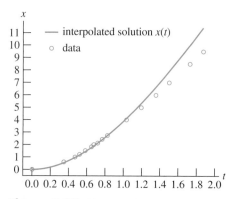

Figure 3.35 The data from Table 3.7 are plotted together with an interpolated solution of the linear equation (6.32).

A Falling Object: A Nonlinear Friction Law   For the falling shuttlecock the velocity is always positive (distance is measured downward from the drop point) and so the quadratic friction law (6.26) is

$$F_f = -kv^2 \text{ for } v \geq 0.$$

Newton's law $F = ma$ now yields the initial value problem

$$v' = 9.8 - k_0 v^2, \qquad v(0) = 0, \tag{6.33}$$

where $k_0 = k/m$. The nonlinear differential equation appearing in this initial value problem is autonomous and separable. The method of separating variables leads to the formula[13]

$$v(t) = \sqrt{\frac{9.8}{k_0}} \tanh \sqrt{9.8k_0}t. \tag{6.34}$$

Antidifferentiating $x' = v$ and using $x(0) = 0$, we obtain the formula

$$x(t) = \frac{1}{k_0} \ln \left(\cosh \sqrt{9.8k_0}t \right) \tag{6.35}$$

for the distance fallen as a function of time t.

---

[13] $\sinh x = (e^x - e^{-x})/2$, $\cosh x = (e^x + e^{-x})/2$, and $\tanh x = (e^x - e^{-x})/(e^x + e^{-x})$.

As we did for the linear model (6.28), we can obtain a numerical estimate for k_0 from the data in Table 3.6 by interpolation. Specifically, we will interpolate the data point at $t = 0.717$ by requiring $x(0.717) = 2.13$. (The reader is asked to interpolate the final data point in Table 3.6 in Exercise 18.) From (6.35) we obtain the equation

$$\frac{1}{k_0} \ln\left(\cosh 2.245\sqrt{k_0}\right) = 2.13$$

to solve for k_0. With the aid of a computer or calculator we find the solution $k_0 \approx 0.2415$, which together with (6.35) yields

$$x(t) \approx 4.142 \ln(\cosh 1.538\,t).$$

How well this quadratic law model fits the data can be judged by the errors in Table 3.8; see also Fig. 3.36. These errors show that this model better fits the shuttlecock data than does the linear friction law model.[14] This is particularly true during the second half of the fall. The quadratic model does, however, have a bias toward underpredicting the distance fallen.

Table 3.8 The predicted distances (rounded to the nearest hundreth) of an interpolating solution of the nonlinear equation (6.33) and their percent errors. Percent errors were calculated by the formula $100(x - x(t))/x$, where x is the population data value at year t and $x(t)$ is the model predicted population size. The answer was rounded to the nearest percent.

Time t (seconds)	Distance Fallen x (meters)	Model Prediction (to nearest 100th)	Error %
0.000	0.00	0.00	0
0.347	0.61	0.56	8
0.470	1.00	1.00	0
0.519	1.22	1.20	2
0.582	1.52	1.48	3
0.650	1.83	1.80	2
0.674	2.00	1.91	4
0.717	2.13	2.13	0
0.766	2.44	2.38	2
0.823	2.74	2.69	2
0.870	3.00	2.95	2
1.031	4.00	3.87	3
1.193	5.00	4.83	3
1.354	6.00	5.82	3
1.501	7.00	6.73	4
1.727	8.50	8.15	4
1.873	9.50	9.07	4

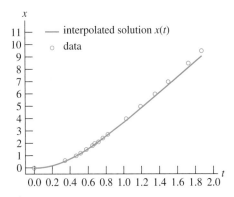

Figure 3.36 The data from Table 3.8 are plotted together with an interpolated solution of the quadratic equation (6.33).

[14]This same conclusion is reached by Peastrel and colleagues (see note 12) by using a different method of estimating k_0 based on a least squares preocedure.

A Falling Object: Terminal Velocity   The phase line portrait for the differential equation (6.28) expressing the linear friction law for the velocity of the falling shuttlecock is

$$\longrightarrow \frac{9.8}{k_0} \longleftarrow .$$

Thus, this law predicts a *terminal velocity* of $9.8/k_0$, that is,

$$\lim_{t \to +\infty} v(t) = \frac{9.8}{k_0}.$$

This can also be seen from the solution formula (6.30) for $v(t)$ from which we find

$$\lim_{t \to +\infty} v(t) = \lim_{t \to +\infty} \left(-\frac{9.8}{k_0} e^{-k_0 t} + \frac{9.8}{k_0} \right) = \frac{9.8}{k_0}.$$

From the estimated value $k_0 \approx 0.7333$ for the experimental data in Table 3.6 the terminal velocity of the shuttlecock predicted by the linear friction law is therefore

$$v_l \approx 13.36 \text{ meters/s}.$$

We found, however, that the quadratic friction law gives a better fit to the data in Table 3.6. The phase line portrait for the differential equation (6.33) expressing the quadratic friction law for the velocity $v \geq 0$ of the falling shuttlecock is

$$\longrightarrow \sqrt{\frac{9.8}{k_0}} \longleftarrow .$$

Thus, when the shuttlecock is dropped ($v(0) = 0$) the quadratic friction law predicts a *terminal velocity* of $\sqrt{9.8/k_0}$. This can also be seen from the formula (6.34) for $v(t)$ from which we calculate

$$\lim_{t \to +\infty} v(t) = \lim_{t \to +\infty} \sqrt{\frac{9.8}{k_0}} \tanh \sqrt{9.8 k_0} t = \sqrt{\frac{9.8}{k_0}}.$$

From the estimated value $k_0 \approx 0.2415$ for the experimental data in Table 3.6 the terminal velocity predicted by the quadratic friction law is

$$v_q \approx 6.37 \text{ meters/s},$$

which is slower than the terminal velocity predicted by the linear law. For more on terminal velocity, see the Exercises.

3.6.3 Flow and Reaction Problems

In Chapter 2 we considered some problems involving dissolved substances flowing into and out of a container (Sec. 2.6.3). In those problems we assumed there was no creation or destruction of the substance within the container, such as might occur if chemical reactions are involved (or, in the case of biological organisms, if birth and death processes are involved). In this section we return to the case considered in Sec. 2.6.3 in which the fluid volume in the container remains constant (by equal inflow and outflow rates), but we now assume a chemical reaction takes place in the container that consume the substance. This chemical reaction contributes a term to the outflow rate term in the balance law

$$x' = \text{inflow rate} - \text{outflow rate} \tag{6.36}$$

used to derive the governing differential equation.

In Sec. 2.6.3 we saw that if a fluid with concentration c_{in} of the substance flows into the container with a rate r and if the (rapidly) mixed fluid in the container is pumped out at the same rate r (so that the volume V in the container remains constant), then— in the absence of chemical reactions in the container—the equation for the amount of substance in present $x(t)$ is

$$x' = c_{in} r - \frac{r}{V} x. \tag{6.37}$$

Chemical reaction rates generally depend on the concentrations of the chemicals involved. Therefore, it is convenient to work with the concentration

$$c(t) = \frac{x(t)}{V}$$

of the substance in the container (amount per unit volume). Dividing both sides of differential equation (6.37) by the volume V, we obtain the differential equation

$$c' = d\,(c_{in} - c) \tag{6.38}$$

for the concentration $c(t)$. In this equation

$$d = \frac{r}{V} \tag{6.39}$$

is called the *dilution rate* (it has units of $1/\text{time}$).

If a reaction takes place that consumes the chemical at a rate ρ (amount per unit time per unit volume), then the outflow rate in the balance law (6.36) has the additional term ρ and hence equation (6.38) has the additional term $-\rho$ on the right-hand side:

$$c' = d\,(c_{in} - c) - \rho. \tag{6.40}$$

The final step in the model derivation step of the modeling cycle involves a specification of the *reaction rate* ρ. Usually the reaction rate ρ is a function of the concentration c, as well as the concentrations of other chemical compounds involved in the reaction. To complete the model equation (6.40) we need a mathematical expression for ρ appropriate for the chemical reaction involved.

A *catalyst* is a substance that facilitates a chemical reaction but that undergoes no chemical change itself. Enzymes in biological reactions are examples of catalysts. Virtually all chemical reactions in cells involve the participation of enzymes. A basic enzyme reaction involves another compound called a *substrate* with which the enzyme reacts to produce a complex. This complex, in turn, produces a product compound and the enzyme. In effect, the enzyme has transformed the substrate into a product while leaving itself unchanged. For example, hemoglobin in red blood cells is an enzyme that combines in the lungs with oxygen as a substrate to form oxyhemoglobin, which then releases oxygen in the tissues to form hemoglobin again.

Suppose a fluid containing a concentration c_{in} of substrate is added to a container of volume V (mils) at a rate r (mils per second) in which an enzyme is present at a fixed concentration $z > 0$. The result is well stirred and pumped out at the same rate

r. To determine the substrate concentration c in the container from equation (6.40) we require a mathematical expression for the reaction rate ρ between the enzyme and the substrate.

Under some circumstances the reaction between the substrate and the enzyme occurs at a rate *jointly proportional* to the concentration of both substances (i.e., at a rate proportional to the product of both concentrations). With $\rho = kzc$, where $k \geq 0$ is a constant of proportionality, equation (6.40) becomes the linear, autonomous equation

$$c' = -(d + kz)c + dc_{\text{in}} \tag{6.41}$$

for the substrate concentration c. Since the coefficient $-(d + kz) < 0$ of c in this equation is negative, the equilibrium concentration

$$c_e = \frac{d}{d + kz}c_{\text{in}}$$

is a sink. Therefore,

$$\lim_{t \to +\infty} c(t) \to c_e$$

for all initial conditions and all positive values of the model parameters c_{in}, d, z, k. Notice that if $k > 0$ (i.e., the enzyme is present), then the long-term concentration c_e in the container is less than the inflow concentration c_{in}. On the other hand, in the absence of the enzyme ($k = 0$) the limiting concentration is simply c_{in}.

Michaelis-Menten (Monod) Reaction Rates

In the model equation (6.41), the reaction rate $\rho = kzc$ is proportional to the substrate concentration c. However, this modeling assumption is usually not accurate for high substrate concentrations. It is often the case that the *rate* at which ρ increases *lessens* as c grows large and, as a result, ρ levels off and approaches a maximum value as $c \to +\infty$. Therefore, a more accurate model at high substrate concentrations requires a different submodel for ρ as a function of c. It requires a submodel for ρ that increases with c, but approaches a horizontal asymptote as $c \to +\infty$. The *Michaelis-Menten* (or *Monod*) expression

$$\rho = \frac{mc}{a + c}z, \tag{6.42}$$

where $m > 0$ and $a > 0$ are constants, is a widely used submodel of this type. The quotient

$$\frac{mc}{a + c} \tag{6.43}$$

is the reaction (or uptake) rate of the substrate *per unit* enzyme concentration (i.e., ρ/z). This per unit rate is an increasing function of c and satisfies

$$\lim_{c \to +\infty} \frac{mc}{a + c} = m.$$

For this reason the constant $m > 0$ is the *maximal reaction* or *maximal uptake rate*. The constant $a > 0$ is the *half saturation level* (or the *Michaelis-Menten constant*). This is because from (6.43) we find that the per unit reaction rate (6.43) attains half its maximal value (namely, $m/2$) when the concentration $c = a$. See Fig. 3.37. The constants m and a are often experimentally measured.

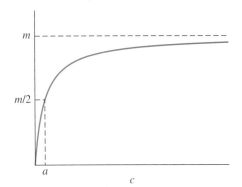

Figure 3.37 The graph of the Michaelis-Menten uptake rate (6.42) as a function of substrate concentration c.

Using the Michaelis-Menten the reaction rate (6.42) in equation (6.40), we obtain the nonlinear, autonomous equation

$$c' = d\,(c_{\text{in}} - c) - \frac{mc}{a+c}z \tag{6.44}$$

for the substrate concentration c. Our goal is to determine what long-term substrate concentration is predicted by this kind of reaction rate model.

We can answer this question by determining the phase line portrait associated with the autonomous equation (6.44). To do this we must study the equilibria (only positive equilibria are relevant to the application). We can study the roots of the equilibrium equation

$$d\,(c_{\text{in}} - c) = \frac{mc}{a+c}z$$

algebraically (see Exercise 24), or geometrically by considering the intersection points of the straight line $d\,(c_{\text{in}} - c)$ with the graph of $mcz/(a+c)$ (as functions of c). From the graphs in Fig. 3.38 we see, for all positive values of the parameters, that there is one and only one intersection point of these two graphs. Thus, there is one and only one root $c_e > 0$ of equation (6.44).

Moreover, from the graphs in Fig. 3.38 we see that the straight line lies above the curve $mcz/(a+c)$ for $c < c_e$ and below it for $c > c_e$. This observation tells us that the phase line portrait association with (6.44), for all model parameter values, is

$$\longrightarrow c_e \longleftarrow .$$

From Fig. 3.38 it is clear that the long-term substrate concentration in the container c_e is less than the input concentration c_{in}.

Growth-Inhibiting Substrates Both the linear flow reaction equation (6.41) and the nonlinear Michaelis-Menten equation (6.44) have a unique positive sink for all (positive) values of the equation parameters (their phase line portraits are qualitatively equivalent). Bifurcations do not occur in either these models as parameters change. For other types of chemical reactions, however, bifurcations can occur.

For example, the reaction rate

$$\rho = \frac{mc}{a+c^2}z \tag{6.45}$$

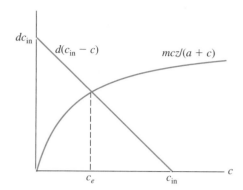

Figure 3.38

is appropriate for a so-called *growth-inhibiting substrate*—a substrate whose reaction rates *decrease* at *large* concentrations. The graph of this kind of reaction rate appears in Fig. 3.39. The governing differential equation for the concentration of this kind of substrate is

$$c' = d\,(c_{\text{in}} - c) - \frac{mc}{a + c^2}z.$$

(6.46)

Our goal is, once again, to determine the long-term substrate concentration in the container.

We can study the equilibria of this equation in a geometric way similar to that we used for the Michaelis-Menten equation (6.44) in Fig. 3.38. In this case, however, there are several intersection possibilities, including more than one intersection points of the graph of (6.45) with the straight line

$$\rho = d\,(c_{\text{in}} - c)\,.$$

(6.47)

From Fig. 3.39(a) we find, as d increases, the sequence of phase line portraits shown in Fig. 3.39. From the graphs in Fig. 3.39 you will observe two saddle-node bifurcations [which occur for those values of d when the straight line (6.47) is tangent to the graph of (6.45)]. A bifurcation diagram appears in Fig. 3.39(b).

The bifurcation diagram in Fig. 3.39(b) has an interesting implication. If d is small [less than d_1 in Fig. 3.39(b)], then the substrate concentration equilibrates at the low level located on the lower stable branch of the bifurcation diagram. If d increases slightly, the equilibrium changes only slightly and the substrate concentration will reequilibrate at a new equilibrium on the lower branch. In this way the equilibrium concentration follows the lower stable branch on the bifurcation diagram in Fig. 3.39(b) as d increases. This occurs until d exceeds the saddle-node bifurcation point d_2 and the lower stable branch of equilibria abruptly disappears (it merges with the middle equilibrium branch). As d increases through d_2 the substrate concentration, therefore, undergoes a discontinuous jump to a higher level—the equilibrium concentration located on the upper stable branch of the bifurcation diagram. If we now decrease d, the equilibrium concentration will, in a similar manner, follow the *upper* stable branch. Thus, the equilibrium concentration will not retrace its steps along the lower equilibrium concentration branch, an occurrence that might well puzzle an experimenter. The substrate equilibrium concentration will follow the higher branch as d decreases until d is less than the saddle-node bifurcation point d_1, at which point it will undergo a discontinuous jump down to the lower stable branch. In the scenario we just described,

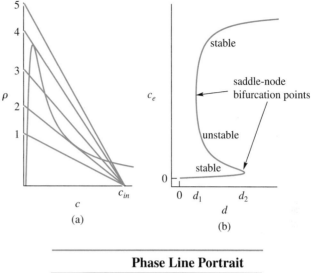

(a)

(b)

Phase Line Portrait

Line 1 : ⟶ ∘ ⟵
Line 2 : ⟶ ∘ ⟵ ∘ ⟵
Line 3 : ⟶ ∘ ⟵ ∘ ⟶ ∘ ⟵
Line 4 : ⟶ ∘ ⟶ ∘ ⟵
Line 5 : ⟶ ∘ ⟵

Figure 3.39 In (a) the graph of (6.45) intersects the straight line (6.47) for five increasing values of d. For line 1 (the smallest value of d) there is one intersection point. For line 3 there are three intersection points. For line 5 there is again only one intersection point. The intermediate lines 2 and 4 have a tangency with the graph of (6.45) and yield exactly two intersection points. These tangent points result in saddle-node bifurcations, as two intersection points coalesce and disappear. The phase line portraits corresponding to these lines appear in the table. Graph (b) shows the bifurcation diagram.

the two critical points d_1 and d_2 at which the equilibrium concentration leaves and returns to the lower stable branch are different. This is called a *hysteresis* effect.

Exercises

Assume the world population is growing exponentially according to the model equations $x' = rx$, $x(t_0) = x_0$. Use the data in Table 3.1, with $t_0 = 1900$ and $x_0 = 1.65$, to do the following problems.

1. (a) Determine the value of r so that the solution interpolates the world population at $t = 1950$ and obtain a formula for $x(t)$.

   (b) Construct tables of model estimates at each census time and calculate their percent errors.

2. (a) Determine the value of r so that the solution interpo-

lates the world population at $t = 1990$ and obtain a formula for $x(t)$.

   (b) Construct tables of model estimates at each census time and calculate their percent errors.

3. The United Nations estimate of world population numbers in the year 2000 was 6.06 billion. This is less than the prediction of 7.7 billion from the doomsday model (6.9). Moreover, 6.06 is approximately double the world population of 3.02 billion in 1960, suggesting that the steady decrease in the doubling time seen during the first half of the twentieth century (Table 3.2) has ceased. These facts

suggest a slowing of world population growth. The United Nations estimates of world population numbers from 1950 to 2000 and the United Nations predictions of future population numbers until 2050 appear in the table below.

World Population	
Year	(billion)
1950	2.52
1960	3.02
1970	3.70
1980	4.45
1990	5.30
2000	6.06
2015	7.21
2025	7.94
2050	9.32

(a) Use the logistic equation (6.13) to interpolate the population data at 1950, 2000, and 2050. What does this solution predict for the world population in the year 2100?

(b) What carrying capacity for the world population does this interpolated logistic model predict?

4. Obtain the solution of (6.6) that interpolates the first three data points at $t = 1900$, 1910, and 1920 in Table 3.1. Calculate the errors and % errors and draw a graph of the solution together with the data. Discuss the results and compare them to the errors in Table 3.3. What doomsday date does this interpolation predict?

5. (a) Use the data in Table 3.1 to approximate the average (yearly) rate of change per billion individuals of the world population from $t = 1900$ to 1980. That is, calculate

$$\frac{1}{x(\Delta t)} \frac{x(t + \Delta t) - x(t)}{\Delta t} \qquad (6.48)$$

with $\Delta t = 10$ for $t = 1900$ to 1980.

(b) Find the least squares best fit (linear regression) $mx + b$ to these rates. Using (6.48) as an approximation to x'/x we arrive, after these calculations, at the differential equation $x'/x = mx + b$ as a model for world population growth.

(c) Solve the differential equation in (b) using the initial condition $x(1900) = 1.65$ from Table 3.1.

(d) Calculate the errors and % errors made by the solution in (c).

(e) Sketch a graph of the solution in (c), together with the data from Table 3.1.

(f) What doomsday date (if any) does the model in (b) predict?

6. When completely unrestricted a population of microorganisms grows at an exponential rate of 0.25 per minute. When placed in a culture dish with limited food resources the population grows logistically with a carrying capacity of 10,000 per cm^2. If the initial density is 4000 per cm^2 how long will it take the population to double its density? Compare this with the doubling time under unrestricted exponential growth. The doubling times under logistic and exponential growth laws are compared in the next exercise.

7. Suppose a population grows logistically, $x' = rx(1 - x/K)$, from an initial size $x_0 > 0$. When $x_0 < K/2$ it makes sense to consider the population's doubling time t_d. Use a computer program to study t_d. Formulate conjectures about how t_d depends on the parameters r and K and the initial condition x_0. Formulate a conjecture about how t_d compares to the doubling time for exponential growth with rate r. Find a formula for t_d and use it to verify or disprove your conjectures.

8. (a) Find a formula for the solution of the initial value problem $x' = rx(1 - x/K)$, $x(t_0) = x_0$ treating the equation as a Bernoulli equation. You should obtain the answer (6.14).

(b) Repeat (a) treating the equation as a Ricatti equation.

9. (a) A modification of the logistic equation is $x' = r(x - aK)x(1 - x/K)$, $r > 0$, $K > 0$. Letting a range over *all* real numbers $-\infty < a < \infty$ draw a bifurcation diagram. Where do the bifurcations occur and of what type are they?

(b) Consider a biological population $x = x(t)$ whose dynamics are described by the modified logistic in (a). The existence of a threshold such that populations initially below the threshold go extinct, while those above the threshold survive, is called an Allee effect. For what values of the parameter a does the modified logistic equation in (a) have an Allee effect?

10. Consider the periodic logistic equation (6.15) under the assumption that $r > 0$ and $K(t) > 0$ is periodic with period $T > 0$.

(a) Show there exists a positive, periodic solution $x_p(t)$ (and only one) with period T. [*Hint*: Use the fact that (6.15) is a Bernoulli equation and the result about periodic linear equations from Exercise 68 in Sec. 2.5 of Chapter 2.]

(b) Show that all solutions with initial conditions $x_0 > 0$ tend to $x_p(t)$ as $t \rightarrow +\infty$ [i.e., $\lim_{t \to +\infty} (x(t) - x_p(t)) = 0$].

11. Find a formula for the periodic solution $x_p(t)$ of the periodic logistic equation (6.17). [*Hint*: The equation is a Bernoulli equation.]

12. Find the differential equation for the coefficient $c_2(t)$ in the perturbation expansion (6.18) for the periodic solution of the periodic logistic (6.17). Use your answer to show the average population size $T^{-1} \int_0^T x_p(t)\, dt$ is *smaller* than the average K_{av} of the carrying capacity $K(t)$ (at least for small amplitude environmental oscillations).

13. Consider an object subject to a constant force $F_p > 0$ and a force of resistance $F_f = F_f(v)$, $F_f(0) = 0$, for which $dF_f(v)/dv < 0$ for all v. Show (6.23) holds.

14. Consider the friction law (6.24) assuming $F_f(v) = -kv^{1/3}$ and an initial condition $v(0) = v_0 > 0$.

  (a) Use a computer program to explore the numerical solution of this initial value problem using a selection of values for m, k, and v_0. Formulate conjectures about the domain of existence of solutions $v(t)$. Describe the behavior of $v(t)$ on its domain of existence for $t \geq 0$. How do your conjectures depend on m, k, and v_0?

  (b) Solve this initial value problem and determine the maximal interval of existence $\alpha < t < \beta$ of the solution. Relate your answer to the conjectures you made in (a).

  (c) What happens to the velocity v as $t \to \beta-$? Relate your answer to the conjectures you made in (a). What is the physical meaning of your answer?

  (d) From $v = x'$ find the position $x = x(t)$ of the object if its initial position is $x(0) = x_0$.

  (e) Compute $\lim_{t \to \beta-} x(t)$. Interpret your answer.

  (f) How far does the object move? What happens to the distance moved if the mass m of the object is doubled? If the initial velocity v_0 is doubled? The coefficient of friction k doubled?

15. Draw the phase line portrait of the general friction law (6.24) under the assumption that the friction model is described by a function $F_f(v)$ satisfying (6.23). What can you conclude from this phase line portrait about the long-term dynamics of the object?

16. Consider the general friction law (6.24) under the assumption that the friction model is described by the quadratic law

$$F_f(v) = \begin{cases} -kv^2 & \text{for } v \geq 0 \\ kv^2 & \text{for } v \leq 0. \end{cases}$$

  (a) Find a formula for the solution of the initial value problem $mv' = F_f(v)$, $v(0) = v_0$.

(b) Compute the terminal velocity $\lim_{t \to +\infty} v(t)$. Interpret your answer.

(c) From $v = x'$ find the position $x = x(t)$ of the object if its initial position is $x(0) = x_0$.

(d) Compute $\lim_{t \to +\infty} x(t)$. Interpret your answer and compare it to the same limit in the case of a linear friction law. How do you explain the difference in these limits?

17. Interpolate the last point in Table 3.6 to determine a value for k_0 in the linear friction model (6.28)–(6.29). Find a formula for the distance fallen $x(t)$ and use it to construct a table of errors like Table 3.7. Discuss the errors.

18. Repeat Exercise 17 using the quadratic friction law model (6.33).

19. (a) Calculate the value of k_0 in the linear model (6.28), with $v(0) = 0$, so that the distance fallen $x(t)$ interpolates the data point at $t = 0.766$ in Table 3.6.

  (b) Use the value of k_0 obtained in (a) to calculate predicted distances $x(t)$ from the linear model at the data times in Table 3.6. Calculate the percent errors made by these predictions.

20. Repeat Exercise 19 using the quadratic friction law model (6.33).

21. A bullet is shot through a block of wood 25 centimeters in thickness. It enters the block at a velocity of 250 meters/second and leaves at 10 meters/second. Assume the bullet travels in a straight line and that its deceleration is proportional to the square of its velocity.

  (a) Derive an initial value problem for the velocity $v = v(t)$ as a function of time t.

  (b) Find a formula for the solution of the initial value problem in (a).

  (c) Use your answer in (b) to find the position of the bullet $x = x(t)$ as a function of time t (using the entering point as the origin of reference).

  (d) Determine the constant of proportionality in the model.

  (e) How long does it take the bullet to pass through the block of wood?

  (f) If a bullet traveling 1 meter/second is consider harmless, how thick should the block of wood be to serve as an effective shield?

  (g) According to this model, how thick should the block of wood be in order to bring the bullet to rest?

22. In this exercise you are asked to derive and analyze a model for weight gain and loss by using the balance law (6.1). Let $w = w(t)$ be an individual's weight at time t. Denote by f the rate of weight gain due to food consumption and denote by p the rate of weight loss due to physical activity such as exercise. Finally, let m denote the loss of weight due to body metabolism, which is a function of body weight w.

   (a) Derive a differential equation for w under the assumptions that $f = f(t)$ and $p = p(t)$ are functions of time and metabolism is proportional to w^q, where q is a positive constant.

   (b) Assume that both f and p are constants. Determine the phase line portrait of the resulting autonomous equation in (a) and draw a bifurcation diagram using p as a parameter.

   (c) Discuss your results in (b) with regard to long-term weight gain or loss.

   (d) Suppose food consumption is not constant, but is more realistically periodic. Specifically, let $f(t) = f_{av} + \varepsilon \sin 2\pi t$. With ε constant and $q = 3/4$ (which, it turns out, is a reasonable value suggested by data) use a computer to study solutions of the equation in (a). Specifically, choose numerical values for f_{av}, c, and $\varepsilon < 1$ and study the solutions of the initial value problem $x_0 = 1$ as they depend on p. What conclusions do you reach with regard to long-term weight gain or loss?

23. Suppose a concentration $z > 0$ of an enzyme is maintained in a container that initially contains a concentration $c_0 > 0$ of substrate. Suppose no fluid is added or removed from the container.

   (a) Write down an initial value problem for the substrate concentration $c = c(t)$ assuming a Michaelis-Menten uptake rate (6.42).

   (b) Draw the phase line portrait of the differential equation from your answer in (a) and show $\lim_{t \to +\infty} c(t) = 0$ when $c_0 > 0$.

   (c) Use a computer program to study the time t_{half} it takes for the substrate c to reach half of its initial concentration c_0. Specifically, formulate a conjecture about how t_{half} is related to c_0. What does the halving time t_{half} do as $c_0 \to 0$?

   (d) Find a formula for the solution of the initial value problem in (a). You need not find an explicit solution.

   (e) Use your solution in (c) to prove or disprove your conjecture in (c).

   (f) Repeat (c)–(e) replacing c_0 by the enzyme concentration z.

   (g) Repeat (c)–(e) replacing c_0 by the maximal uptake rate m.

   (h) Repeat (c)–(e) replacing c_0 by the half saturation constant a.

24. Find a formula for the equilibria of equation

$$c' = d\,(c_{in} - c) - \frac{mc}{a+c}z.$$

Show that there is one and only one positive equilibrium for all positive values of the five parameters.

25. It was shown in Exercise 24 that the equation

$$c' = d\,(c_{in} - c) - \frac{mc}{a+c}z$$

has a unique positive equilibrium for all positive values of the five parameters.

   (a) Explore the solutions of this equation using a computer program to see if you think that the equilibrium level always increases or always decreases as the enzyme concentration $z > 0$ is increased.

   (b) Prove or disprove your conjecture in (a) by using a graphical argument based on the graphs of the straight line $d(c_{in} - c)$ and the Michaelis-Menten expression $mcz/(a+c)$ (as functions of c).

   (c) Prove or disprove your conjecture in (a) by an implicit differentiation of the equilibrium equation.

   (d) Prove or disprove your conjecture in (a) using the formula for the equilibrium found in Exercise 24.

26. It has been found in many enzyme reactions that a plot of the reaction rate ρ against the concentration c has an inflection point (i.e., a change in concavity) unlike that of the Michaelis-Menten function (6.42) sketched in Fig. 3.37, which is concave down for all $c \geq 0$. The explanation of phenomenon is that an enzyme can possess several active binding centers for the substrate and the rates of binding activity at one center depends on the activity at other centers. For this reason, such a enzymatic reaction is called cooperative, as opposed to one described by the Michaelis-Menten function, which is called noncooperative. A commonly used expression for such a reaction rate ρ is Hill's equation:

$$\rho = \frac{mc^n}{a+c^n}z,$$

where the exponent $n > 0$ is determined from data.

   (a) Show that Hill's equation has a positive inflection point as a function of c if and only if $n > 1$. This is why an enzyme/substrate reaction is called cooperative if data shows n is greater than one.

   (b) Determine the phase line portraits of equation (6.40) when ρ is taken as Hill's equation. Are there any bifurcations for positive values of the parameters?

4

Systems and Higher-Order Equations

Many applications involve more than one first-order differential equation for more than one unknown function. For example, suppose we are interested in two quantities x and y that change over time. If these quantities effect each others rates of change, then we could have two differential equations of the form

$$x' = f(x, y) \qquad (0.1)$$
$$y' = g(x, y)$$

that describe how x and y effect their rates of change x' and y'. Examples include the interaction of a predator species with a prey, a reaction between two chemicals, and the motions of two planetary bodies.

A common situation that gives rise to a system of equations involves the amounts of a substance, x and y, present in two different locations or compartments. If the movement of the substance into and out of the compartments includes exchanges between between the compartments, then a system of the form (0.1) arises from the balance laws

$$x' = \text{inflow rate} \ - \ \text{outflow rate}$$
$$y' = \text{inflow rate} \ - \ \text{outflow rate}.$$

For example, if the flow rates for each compartment are proportional to the amounts present in the compartment, then a typical compartment model diagram appears as in Fig. 4.1. This diagram, together with the balance laws, yields the system

$$x' = -r_1 x + r_2 y \qquad (0.2)$$
$$y' = r_1 x - (r_2 + r_3) y$$

of differential equations. The pesticide application in the Introduction, (Sec. 0.2)

$$x' = -2x + 2y \qquad (0.3)$$
$$y' = 2x - 5y$$

is a specific example in which $r_1 = r_2 = 2$ and $r_3 = 3$.

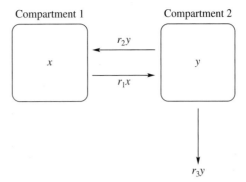

Figure 4.1

Problems involving three or more quantities changing in time will lead to systems of three or more differential equations. Examples include ecological communities involving three or more species, reactions involving several chemical compounds, the motions of many planetary objects, and compartmental systems with three or more compartments.

In this chapter we turn our attention to systems of first-order equations. Although systems of equations are in general more difficult to analyze than single equations, we learn in this chapter how to extend the fundamental existence and uniqueness theorem to systems of equations and how to adapt numerical approximation methods for single equations to systems of equations. We will also look in some detail at systems of linear equation and discover that many fundamental properties of single linear equations, and their solutions, also hold for systems of linear equations.

In the Introduction we saw how to rewrite a higher-order equation as an equivalent first-order system. For example, the second-order equation

$$x'' + px' + qx = 0$$

is equivalent to the system

$$x' = y \tag{0.4}$$
$$y' = -qx - py$$

of first-order equations. Objects in motion, as described by Newton's laws of motion, often give rise to second-order differential equations. For example, the equation (called the simple harmonic oscillator equation)

$$x'' + \frac{k}{m}x = 0 \tag{0.5}$$

arises in the study of the oscillatory motion of a mass m attached to a spring. The system

$$x' = y \tag{0.6}$$
$$y' = -\frac{k}{m}x$$

is equivalent to this second order equation. The association between higher equations and systems of first-order equations allows us to apply to higher-order equations any solution formulas, approximation techniques, methods of analysis, and general results derived for systems of equations. Treating higher equations within the context of systems provides a unified approach and an efficiency in presentation. It also provides a natural setting for several important concepts, such as phase space and the dimension of a dynamical system. (As a practical matter, most programs for solving differential equations require the user to write a higher equation as an equivalent system.) On the other hand, we will see in Chapters 5 and 6 that in some circumstances there are short-cut methods for treating higher equations that do not utilize an equivalent first-order system.

4.1   The Fundamental Existence Theorem

A solution of a system of first-order equations

$$x' = f(t, x, y) \tag{1.1}$$
$$y' = g(t, x, y)$$

is a *pair* of functions $x = x(t)$, $y = y(t)$. More precisely, we have the following definition.

> **DEFINITION 1.** A **solution pair** of the system (1.1) is a pair of differentiable functions $x = x(t)$, $y = y(t)$ that reduces both equations to identities on an interval $\alpha < t < \beta$; that is, $x'(t) = f(t, x(t), y(t))$ and $y'(t) = g(t, x(t), y(t))$ for all values of t from the interval.

Note that for a second-order equation

$$x'' = g(t, x, x').$$

Definition 1, when applied to the equivalent first order system

$$x' = y$$
$$y' = g(t, x, y),$$

implies that a solution $x(t)$ and its derivative $y(t) = x'(t)$ are differentiable, that is, that $x(t)$ is twice differentiable on the interval $\alpha < t < \beta$.

For example, the formulas

$$x = 2e^{-t}$$
$$y = e^{-t}$$

define a solution pair for the system (0.3) for all t. To see this we first calculate the derivative

$$x' = -2e^{-t}$$

and find that it is identical to

$$-2x + 2y = -2\left(2e^{-t}\right) + 2\left(e^{-t}\right)$$

for all t, and second we calculate the derivative

$$y' = -e^{-t}$$

and find that it is identical to

$$2x - 5y = 2\left(2e^{-t}\right) - 5\left(e^{-t}\right)$$

for all t.

As another example, the simple harmonic oscillator system

$$
\begin{aligned}
x' &= y \\
y' &= -x
\end{aligned}
\tag{1.2}
$$

is equivalent to the simple harmonic equation

$$x'' + x = 0. \tag{1.3}$$

The two functions

$$
\begin{aligned}
x &= \cos t \\
y &= -\sin t
\end{aligned}
$$

constitute a solution pair for all t. This is true because $x' = -\sin t$ identically equals y, and $y' = -\cos t$ identically equals $-x$, for all t. Or, equivalently, $x'' + x = -\cos t + \cos t = 0$ for all t.

The system (0.2) describes the rates at which a pesticide is exchanged between a stand of trees and its soil bed. Knowing these rates is not sufficient, however, to determine the amounts of pesticide in the trees and soil at future times. We must also know the initial amounts present [i.e., $x(0) = x_0$ and $y(0) = y_0$]. The equations

$$
\begin{aligned}
x' &= -r_1 x + r_2 y \\
y' &= r_1 x - (r_2 + r_3)\, y \\
x(0) &= x_0, \qquad y(0) = y_0
\end{aligned}
$$

constitute an initial value problem for the first-order system (0.2).

It is probably not surprising that two initial conditions are necessary in order to specify a unique solution of a system of two first-order differential equations. This is because, roughly speaking, two integrations are needed to solve the two differential equations, and as a result two constants of integration arise in the general solution. Or, put another way, the differential equations specify the rates of change of x and y, but to determine (predict) future values of x and y initial conditions are required.

An initial value problem for a second-order equation $x'' + px' + qx = 0$, or its equivalent first-order system (0.4), consists of the two initial conditions $x(0) = x_0$ and $y(0) = x'(0) = x_0$. For example, the initial value problem for the simple harmonic oscillator system is

$$
\begin{aligned}
x' &= y \\
y' &= -x \\
x(0) &= x_0, \qquad y(0) = y_0.
\end{aligned}
\tag{1.4}
$$

The equations

$$
\begin{aligned}
x' &= f(t, x, y) \\
y' &= g(t, x, y) \\
x(t_0) &= x_0, \qquad y(t_0) = y_0
\end{aligned}
\tag{1.5}
$$

describe an *initial value problem* for the general first-order system (1.1). The following theorem is an extension to systems of the basic existence and uniqueness theorem for single equations (see Sec. 1.1 of Chapter 1).

THEOREM 1

Fundamental Existence and Uniqueness Theorem Suppose $f(t, x, y)$ and $g(t, x, y)$ and all the (partial) derivatives

$$\frac{\partial f}{\partial x}, \quad \frac{\partial f}{\partial y}, \quad \frac{\partial g}{\partial x}, \quad \frac{\partial g}{\partial y} \tag{1.6}$$

are continuous for x near x_0, y near y_0, and t near t_0.[1] Then the initial value problem (1.5) has a unique solution pair $x(t)$, $y(t)$ on an interval containing t_0.

As an example, for system (0.3) the functions $f(t, x, y) = -2x + 2y$ and $g(t, x, y) = 2x - 5y$ are linear in x and y and therefore continuous for all x and y (and t). Moreover, their partial derivatives

$$\frac{\partial f}{\partial x} = -2, \quad \frac{\partial f}{\partial y} = 2,$$

$$\frac{\partial g}{\partial x} = 2, \quad \frac{\partial g}{\partial y} = -5$$

are all constant and therefore continuous for all x and y (and t). It follows from Theorem 1 that any initial value problem $x(t_0) = x_0$, $y(t_0) = y_0$ for system (0.3) has a unique solution $x = x(t)$, $y = y(t)$ on an interval containing t_0.

Theorem 1 applies to the equivalent system of a second-order equation and therefore provides the existence and uniqueness of solutions for initial value problems associated with second-order equations.

For example, by Theorem 1 the initial problem (1.4) for the simple harmonic oscillator has a unique solution. This is because $f(t, x, y) = y$, $g(t, x, y) = -x$ and the derivatives

$$\frac{\partial f}{\partial x} = 0, \quad \frac{\partial f}{\partial y} = 1,$$

$$\frac{\partial g}{\partial x} = -1, \quad \frac{\partial g}{\partial y} = 0$$

are continuous functions for all t, x, and y.

As a second example, consider the system

$$x' = y$$
$$y' = -x - \alpha \left(x^2 - 1\right) y,$$

where α is a constant. This system is equivalent to the second-order equation

$$x'' + \alpha \left(x^2 - 1\right) x' + x = 0. \tag{1.7}$$

---

[1] By "continuous for t near t_0" we mean continuous on an interval $a_1 < t < b_1$ containing t_0 (i.e., for which $a_1 < t_0 < b_1$). Thus, the requirements in this theorem are that all of the partial derivatives (1.6) be continuous functions of their three arguments t, x, and y on a region in three-dimensional space described by the intervals $a_1 < t < b_1$, $a_2 < x < b_2$, and $a_3 < y < b_3$ that contains the initial conditions (i.e., such that $a_1 < t_0 < b_1$, $a_2 < x_0 < b_2$ and $a_3 < y_0 < b_3$). The conclusion is that a unique solution pair exists on an interval $\alpha < t < \beta$ such that $\alpha < t_0 < \beta$.

This famous equation, called the *van der Pol equation*, arises in the theory of electric circuits. Since $f(t, x, y) = y$, $g(t, x, y) = -x - \alpha (x^2 - 1) y$ and the derivatives

$$\frac{\partial f}{\partial x} = 0, \quad \frac{\partial f}{\partial y} = 1,$$

$$\frac{\partial g}{\partial x} = -1 - 2\alpha x y, \quad \frac{\partial g}{\partial y} = -\alpha (x^2 - 1)$$

are continuous for all x and y, all initial value problems $x(0) = x_0$, $x'(0) = y_0$ have unique solutions (on an interval containing $t_0 = 0$).

In order to apply the fundamental existence and uniqueness theorem it is necessary to verify that the functions f and g and all of their partial derivatives (1.6) are continuous at the initial conditions. However, it is often not necessary to calculate these partial derivatives. Instead, we can rely on theorems from calculus for this purpose. For example, we know from calculus that sums, differences, products, and composites of continuous and differentiable functions are themselves continuous and differentiable (and so are quotients, at least where the denominator does not equal 0).

For a system of three first-order equations, which involves a triple of solutions, an initial value problem specifies three initial conditions, one for each component of the triple. In general, an initial value problem for n first-order differential equations

$$x_1' = f_1(t, x_1, x_2, \ldots, x_n)$$
$$x_2' = f_2(t, x_1, x_2, \ldots, x_n)$$
$$\vdots$$
$$x_n' = f_n(t, x_1, x_2, \ldots, x_n)$$

for n unknowns $x_1, x_2, \ldots, x_n$ has the form

$$x_1(t_0) = x_1^0, \quad x_2(t_0) = x_2^0, \ldots, \quad x_n(t_0) = x_n^0,$$

where t_0 and the n initial conditions $x_1^0, x_2^0, \ldots, x_n^0$ are specified real numbers. Theorem 1 has a straightforward extension to systems of any number n of equations. This extension states that the initial value problem has a unique solution on an interval containing t_0 provided each of the functions f_i and all of its partial derivatives with respect to each x_i are continuous for $x_1, x_2, \ldots, x_n$ near $x_1^0, x_2^0, \ldots, x_n^0$ and t near t_0.

Exercises

Which of the following are solution pairs of the system (0.3) and which are not? Justify your answers.

1. $\begin{cases} x = e^{-6t} \\ y = -2e^{-6t} \end{cases}$

2. $\begin{cases} x = e^{-6t} \\ y = 2e^{-6t} \end{cases}$

3. $\begin{cases} x = 4e^{-t} + 4e^{-6t} \\ y = 2e^{-t} + 8e^{-6t} \end{cases}$

4. $\begin{cases} x = 2e^{-t} + 3e^{-6t} \\ y = e^{-t} - 6e^{-6t} \end{cases}$

Which of the following are solution pairs of the harmonic oscillator system (1.2) and which are not? Justify your answers. (c, c_1, and c_2 are constants.)

5. $x = 2\cos t$, $y = -2\sin t$

6. $x = \sin t$, $y = \cos t$

7. $x = \sin 2t$, $y = \cos 2t$

8. $x = 2\sin t$, $y = 2\cos t$

9. $x = \cos t + \sin t$, $y = -\sin t + \cos t$

10. $x = \cos t - \sin t$, $y = \sin t - \cos t$

11. $x = c\sin t$, $y = c\cos t$

12. $x = c_1\sin t + c_2\cos t$, $y = c_1\cos t - c_2\sin t$

13. **(a)** Derive an equivalent first-order system for the second-order equation $mx'' + kx = 0$, where $m > 0$ and $k > 0$ are positive constants.

(b) Show that $x = \cos \omega t$, $y = -\omega \sin \omega t$ is a solution pair for all t where $\omega = \sqrt{k/m}$.

For each system below, determine those initial conditions t_0, x_0, and y_0 for which Theorem 1 applies. Explain your answer. What do you conclude for these initial conditions? What do you conclude for other initial conditions? (a, b, c, d, r_1, and r_2 are constants.)

14. $\begin{cases} x' = x(1-x) - \dfrac{xy}{1+x} \\ y' = -y + \dfrac{xy}{1+x} \end{cases}$

15. $\begin{cases} x' = x(1 - \dfrac{xy}{y+x}) \\ y' = (-1+x)y \end{cases}$ $x(t_0) = x_0$
$y(t_0) = y_0$

16. $\begin{cases} x' = ax + by \\ y' = cx + dy \end{cases}$

17. $\begin{cases} x' = r_1 (1 - ax - by)x \\ y' = r_2 (1 - cx - dy)y \end{cases}$

18. $\begin{cases} x' = x(1 - \dfrac{x}{2+\sin t}) - xy \\ y' = -y + xy \end{cases}$

19. $\begin{cases} x' = (2 + \cos t)(1 - x) - \dfrac{x^2 y}{1+x^2} \\ y' = -y + \dfrac{x^2 y}{1+x^2} \end{cases}$

For each second-order equation below, determine those initial conditions t_0, x_0, and y_0 for which Theorem 1 applies. (First find an equivalent first-order system.) Explain your answer. What do you conclude for these initial conditions? What do you conclude for other initial conditions?

20. $x'' + x = \sin t$ (a forced simple harmonic oscillator)

21. $x'' + x' + x - \cos t$ (a forced oscillator with friction)

22. $t^2 x'' + tx' + x = 0$ (a Legendre equation)

23. $x'' + px' + qx^3 = 0$, where p and q are constants (Duffing equation)

24. $x'' + \alpha(x^2 - 1)x' + x = \beta \sin \theta t$, where α, β, and θ are constants (a forced van der Pol equation)

25. $mx'' + k \sin x = 0$, where $m \neq 0$ and k are constants (the frictionless pendulum equation)

4.2   Approximating Solutions of Systems

In this section we extend the numerical approximation methods and graphical techniques in Chapter 1 to systems of first-order equations.

4.2.1   Numerical and Graphical Approximations

Consider the problem of numerically approximating the solution $x = x(t)$, $y = y(t)$ of the initial value problem

$$x' = f(t, x, y)$$
$$y' = g(t, x, y)$$
$$x(t_0) = x_0, \quad y(t_0) = y_0,$$

at points t_i between the initial time t_0 and a chosen endpoint T

$$t_0 < t_1 < t_2 < \cdots < t_{n-1} < t_n = T.$$

To do this we follow the method used in Chapter 1 (Sec. 1.2) to approximate the solution of a single first-order equation. To get an approximation at the first point t_1 we integrate both of the equations

$$x'(t) = f(t, x(t), y(t)) \tag{2.1}$$
$$y'(t) = g(t, x(t), y(t))$$

from $t = t_0$ to $t = t_1$. By the fundamental theorem of calculus, together with the initial conditions $x(t_0) = x_0$, $y(t_0) = y_0$, we obtain

$$x(t_1) = x_0 + \int_{t_0}^{t_1} f(t, x(t), y(t)) \, dt$$

$$y(t_1) = y_0 + \int_{t_0}^{t_1} g(t, x(t), y(t)) \, dt. \tag{2.2}$$

We can use methods for approximating integrals to obtain numerical estimates for $x(t_1)$ and $y(t_1)$. For example, the left-hand rectangle rule applied to both integrals yields the approximations

$$x(t_1) \approx x_0 + (t_1 - t_0) f(t_0, x_0, y_0)$$
$$y(t_1) \approx y_0 + (t_1 - t_0) g(t_0, x_0, y_0).$$

If we denote these approximations by x_1 and y_1, that is,

$$x_1 = x_0 + (t_1 - t_0) f(t_0, x_0, y_0)$$
$$y_1 = y_0 + (t_1 - t_0) g(t_0, x_0, y_0),$$

then we have the first step of the Euler algorithm for systems.

The left-hand rectangle rule yields approximations x_{i+1}, y_{i+1} to the solution values $x(t_{i+1})$, $y(t_{i+1})$ at the time t_{i+1}, assuming we have approximations x_i, y_i at time t_i. Specifically, integrating the equations in (2.1) from $t = t_i$ to t_{i+1} we have

$$x(t_{i+1}) = x(t_i) + \int_{t_i}^{t_{i+1}} f(t, x(t), y(t)) \, dt \tag{2.3}$$

$$y(t_{i+1}) = x(t_i) + \int_{t_i}^{t_{i+1}} g(t, x(t), y(t)) \, dt$$

and from the left-hand rectangle rule

$$x(t_{i+1}) \approx x_i + (t_{i+1} - t_i) f(t_i, x_i, y_i)$$
$$y(t_{i+1}) \approx y_i + (t_{i+1} - t_i) g(t_i, x_i, y_i).$$

Thus, approximations to the solutions at $t = t_{i+1}$ are given by the quantities

$$x_{i+1} = x_i + s_i f(t_i, x_i, y_i)$$
$$y_{i+1} = y_i + s_i g(t_i, x_i, y_i),$$

where $s_i = t_{i+1} - t_i$ are the step sizes. In practice, equally spaced step sizes are usually chosen. If $s = t_{i+1} - t_i$ denotes a fixed step size, we obtain the *Euler algorithm* for systems

$$x_0 = x_0, \quad y_0 = y_0 \tag{2.4}$$
$$x_{i+1} = x_i + s f(t_i, x_i, y_i)$$
$$y_{i+1} = y_i + s g(t_i, x_i, y_i) \quad \text{for } i = 0, 1, 2, 3, \ldots.$$

As an example, the Euler algorithm formulas for the initial value problem

$$x' = -2x + 2y$$
$$y' = 2x - 5y \tag{2.5}$$
$$x(0) = 1, \quad y(0) = 0$$

are

$$x_0 = 1, \quad y_0 = 0$$
$$x_{i+1} = x_i + s\,(-2x_i + 2y_i) \tag{2.6}$$
$$y_{i+1} = y_i + s\,(2x_i - 5y_i)\,.$$

To approximate the solution pair at $T = 2$ using step size $s = 0.4$, the Euler algorithm sequentially generates approximations to the solution at the four points

$$t_1 = 0.4, \quad t_2 = 0.8, \, t_3 = 1.6, \, t_4 = T = 2.$$

We obtain the approximations x_1, y_1 at $t_1 = 0.4$ from (2.6) using $i = 0$ as follows:

$$x_1 = x_0 + s\,(-2x_0 + 2y_0) = 1 + 0.4 \times (-2 + 0) = 0.2$$
$$y_1 = y_0 + s\,(2x_0 - 5y_0) = 0 + 0.4 \times (2 - 5 \times 0) = 0.8.$$

Using these approximations for x_1, x_2, we can now calculate the approximations x_2, y_2 at $t_2 = 0.8$ using $i = 1$ in (2.6):

$$x_2 = x_1 + s\,(-2x_1 + 2y_1) = 0.2 + 0.4 \times (-0.4 + 1.6) = -0.68$$
$$y_2 = y_1 + s\,(2x_1 - 5y_1) = 0.8 + 0.4 \times (0.4 - 4) = -0.64.$$

Continuing this process two more times, we obtain approximations at $t_3 = 1.6$ and $t_4 = 2$. The results are (to 6 significant digits)

$$x_3 = x_2 + s\,(-2x_2 + 2y_2) = -0.376000$$
$$y_3 = y_2 + s\,(2_2 - 5y_2) = 1.18400$$

$$x_4 = x_3 + s\,(-2x_3 + 2y_3) = 0.872000$$
$$y_4 = y_3 + s\,(2x_3 - 5y_3) = -1.48480$$

$$x_5 = x_4 + s\,(-2x_4 + 2y_4) = -1.01344$$
$$y_5 = y_4 + s\,(2x_4 - 5y_4) = 2.18240.$$

Thus, the Euler approximations to the solution pair at time $t = 2$ with step size $s = 0.4$ are (to 6 significant digits)

$$x(2) \approx -1.01344, \quad y(2) \approx 2.18240.$$

Reducing the step size s will increase the accuracy of the Euler approximations but at the same time will increase the number of steps and hence the amount of numerical calculations we must perform. Table 4.1 shows the approximations obtained by halving the step size several times.

The Euler algorithm for systems has the same order of convergence for the Euler algorithm for a single equation, namely $O(s)$. Recall that this means the errors ($|x(T) - x_n|$ and $|y(T) - y_n|$) are bounded by a constant multiple of the step size s and that they tend to zero at the same rate that s tends to zero. Thus, if the step size s is decreased by a certain factor (one-half, one-tenth, etc.), then we can expect the errors to decrease by (roughly speaking) this same factor. For example, in Table 4.1 the errors are approximately halved with each halving of the step size s; see Exercise 1.

Table 4.1 Euler algorithm approximations to the solution of (2.5).		
s	$x(2) \approx$	$y(2) \approx$
0.400	−1.01344	2.18240
0.200	0.0858994	0.0429496
0.100	0.0972613	0.0486307
0.050	0.102810	0.0514046
0.025	0.105551	0.0527742

In Chapter 1 (Sec. 1.3) we used the trapezoid rule for approximating integrals to derive Heun's algorithm for a first-order equation. If the trapezoid rule is used to approximate the integrals in (2.3), we obtain *Heun's algorithm* for systems:

$$x_0 = x_0, \qquad y_0 = y_0$$
$$x_{i+1} = x_i + \frac{s}{2}\left(f(t_{i+1}, x_{i+1}^*, y_{i+1}^*) + f(t_i, x_i, y_i)\right)$$
$$y_{i+1} = y_i + \frac{s}{2}\left(g(t_{i+1}, x_{i+1}^*, y_{i+1}^*) + g(t_i, x_i, y_i)\right),$$

where

$$x_{i+1}^* = x_i + sf(t_i, x_i, y_i)$$
$$y_{i+1}^* = y_i + sg(t_i, x_i, y_i).$$

This algorithm is second-order [i.e., converges at a rate $O(s^2)$].

Table 4.2 contains the Heun's algorithm approximations to the solution of the initial value problem (2.5) using step size $s = 0.4$. Table 4.3 shows the rate at which the approximations converge as the step size s decreases. In comparison with the Euler algorithm approximations in Table 4.1, this rate is considerably faster. (See Exercise 2.)

Table 4.2 Heun's algorithm approximations to the solution (2.5) (step size $s = 0.4$).		
t_i	x_i	y_i
0.0	1.00000	0.00000
0.4	0.84000	−0.32000
0.8	0.80800	−0.69120
1.2	0.89990	−1.17094
1.6	1.13062	−1.83362
2.0	1.53648	−2.78217

Table 4.3 Heun's algorithm approximations to the solution (2.6).		
s	$x(2) \approx$	$y(2) \approx$
0.400	1.536480	−2.78217
0.200	0.110248	0.0544010
0.100	0.108662	0.0543216
0.050	0.108363	0.0541779
0.025	0.108293	0.0541430

Runge-Kutta algorithms, discussed in Chapter 1 (Sec. 1.3), are also available for systems of equations. For example, the fourth-order Runge-Kutta algorithm is widely used.

As we did for first-order equations, we can use numerical approximations to draw approximate graphs of the solution pairs. For the solution component $x = x(t)$ this is done by drawing straight line segments between the pairs of points (t_0, x_0) and (t_1, x_1),

(t_1, x_1) and (t_2, x_2), and so on. Similarly, for the solution component $y = y(t)$ this is done by drawing straight line segments between the points (t_i, y_i). As the step size s is decreased, these broken line graphs will converge to the graphs $x = x(t)$ and $y = y(t)$, respectively.

Figure 4.2 shows graphical approximations to the solutions x and y of the initial value problem (2.5). Notice how these broken line graphs get smoother in appearance and converge to a smooth curve as the step size s decreases.

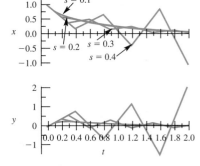

Figure 4.2 Broken line approximations to the graphs of the solution components x and y of the initial value problem (2.5) using the Euler algorithm with a decreasing sequence of step sizes s.

In practice we should compute approximate solution graphs for several decreasing step sizes until the graphs appear unchanged upon any further decreases. Then we have some confidence that the approximate graphs have sufficiently converged so as to provide an accurate approximation to the graph of the solution.

Figure 4.3 shows computer-drawn graphs of the solutions x and y of the initial value problem

$$x' = y$$

$$y' = -x - \frac{1}{2}(x^2 - 1)y \tag{2.7}$$

$$x(0) = 2, \quad x'(0) = 0.$$

[This first-order system is equivalent to the van der Pol equation (1.7) with $\alpha = 1/2$.] The graphs were constructed using Heun's algorithm for two steps sizes. Since the step sizes result in indistinguishable graphs, we conclude that the graphs are accurate approximations.

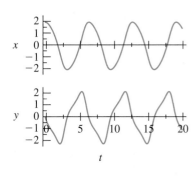

Figure 4.3 Graphs of the solution components x and y of the initial value problem (2.7) using Heun's method with step sizes $s = 0.05$ and 0.025. The two step sizes yield indistinguishable graphs.

So far we have graphically represented solutions of initial value problems (1.5) by drawing two graphs, one for each component x and y of the solution pair. Another way is to draw a three-dimensional graph using a (t, x, y)-coordinate axis and plotting points $(t, x(t), y(t))$. See Fig. 4.4. While sometimes useful, such three-dimensional graphs are often difficult to draw and use. For certain kinds of differential equations— namely autonomous equations—another useful graphical representation of solutions is available. For systems of autonomous equations a powerful method is to plot the pair $(x(t), y(t))$ in the x, y-plane. This amounts to projecting the three-dimensional pictures into the x, y-plane (parallel to the t-axis). We study this method in the next section.

Figure 4.4 Three-dimensional graphs of the solution the initial value problems (2.5) (left) and (2.7) (right).

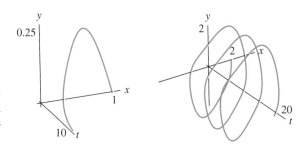

Exercises

1. Formulas for the solution pair of the initial value problem (2.5) are

$$x(t) = \frac{1}{5}e^{-6t} + \frac{4}{5}e^{-t} \qquad y(t) = \frac{2}{5}e^{-t} - \frac{2}{5}e^{-6t}.$$

Use these formulas to calculate the (absolute value of) the errors of the Euler approximations to $x(2)$ and $y(2)$ using step sizes $s = 0.4, 0.2, 0.1, 0.05, 0.0025$. At what rate do the errors decrease? Is this rate appropriate for the Euler algorithm?

2. Repeat Exercise 1 using Heun's algorithm.

3. Formulas for the solution pair of the initial value problem

$$x' = -x + y$$
$$y' = x - 2y$$
$$x(0) = 1, \quad y(0) = 0$$

are

$$x(t) = \frac{1}{10}\left(5 + \sqrt{5}\right)e^{-\frac{1}{2}(3-\sqrt{5})t}$$
$$+ \frac{1}{10}\left(5 - \sqrt{5}\right)e^{-\frac{1}{2}(3+\sqrt{5})t}$$
$$y(t) = \frac{1}{5}\sqrt{5}e^{-\frac{1}{2}(3-\sqrt{5})t} - \frac{1}{5}\sqrt{5}e^{-\frac{1}{2}(3+\sqrt{5})t}.$$

Use these formulas to calculate the (absolute value of) the errors made by the Euler approximations to $x(1)$ and $y(1)$ using step sizes $s = 0.25, 0.125, 0.0625, 0.03125, 0.015625$. At what rate do the errors decrease and is this rate appropriate for the Euler algorithm?

4. Repeat Exercise 3 using Heun's algorithm.

5. (a) Use a computer to graph each component $x(t)$ and $y(t)$ of the solution pair of the initial value problem

$$x' = -3x + y$$
$$y' = \frac{1}{2}x - y$$
$$x(0) = \frac{1}{2}, \quad y(0) = 0$$

for $0 \le t \le 5$. What are the differences and similarities between these two graphs?

   (b) What changes occur in the graphs of $x(t)$ and $y(t)$ if the initial condition $y(0) = 0$ is changed to $y(0) = -\frac{1}{2}$?

6. (a) Derive an equivalent system for the general van der Pol equation

$$x'' + \alpha\left(x^2 - 1\right)x' + x = 0$$

in which α is a constant.

(b) Use a computer to graph the solution $x(t)$ of the initial value problem $x(0) = 2$, $x'(0) = 0$ for $0 \leq t \leq 50$ when $\alpha = -1$.

(c) In what fundamental way does the graph change if $\alpha = 1$?

(d) Draw the graph of $x(t)$ for some other values of α. At what value of α does the change (c) appear to occur?

4.2.2 The Phase Plane for Autonomous Systems

Consider the system

$$x' = y \tag{2.8}$$
$$y' = -x.$$

(This first-order system is equivalent to the second-order equation $x'' + x = 0$.) The trigonometric functions

$$x = \sin t, \quad y = \cos t \tag{2.9}$$

form a solution pair of this system. The graphs of these trigonometric functions should be familiar to the reader. However, instead of graphing each function individually, suppose we plot the points $(x, y) = (\sin t, \cos t)$ in the x, y-plane for each t. The resulting set of points is the circle of radius 1 centered at the origin. This is because

$$x^2 + y^2 = \sin^2 t + \cos^2 t = 1$$

for all t. As t increases, the point (x, y) given by (2.9) moves continuously around the unit circle in a clockwise manner. Therefore, we place arrows pointing clockwise on the unit circle to indicate the direction of the motion along this circular path as t *increases*. See Fig. 4.5. The (clockwise-oriented) unit circle is another graphical way to represent the solution (2.9). This circle is an example of an "orbit" associated with the solution of a system of differential equations.

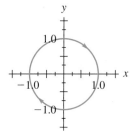

Figure 4.5 The phase plane orbit associated with the solution (2.9) of the system (2.8).

DEFINITION 2. If $x = x(t)$, $y = y(t)$ is a solution pair (on an interval $\alpha < t < \beta$) of an autonomous system

$$x' = f(x, y) \tag{2.10}$$
$$y' = g(x, y),$$

then the set of points $(x(t), y(t))$ in the x, y-plane is called the **orbit** associated with this solution. An orbit is assigned an **orientation** in the direction of increasing t. The x, y-plane is called the **phase plane** and the set of all orbits, together with their orientations, is called the **phase plane portrait** (or simply its **phase portrait**) of the system (2.10).

Another solution of the system (2.8) is the constant pair $(x, y) = (0, 0)$. The orbit associated with this solution is a single point, namely the origin in the phase plane. A constant solution is called an *equilibrium* and its orbit is called an *equilibrium point* (or a *rest point* or a *critical point*).

Geometrically the orbit of a solution is the projection onto the x, y-plane of the three-dimensional graph of the solution obtained by plotting the points $(t, x(t), y(t))$ (as in Fig. 4.4). This is the analog of the one-dimensional case of a single autonomous equation $x' = f(x)$, where the orbit of a solution is obtained by projecting the two dimensional graph of the solution $(t, x(t))$ onto the x-axis (see Chapter 1). Mathematically, the orbit is the range of the solution pair considered as a function: $t \rightarrow (x(t), y(t))$.

Autonomous systems (2.10) arise often in applications, and a major goal is to determine and sketch their phase portraits.

EXAMPLE 1

The formulas

$$x = c_1 \cos t + c_2 \sin t$$
$$y = -c_1 \sin t + c_2 \cos t$$
$$c_1, \ c_2 \ = \text{any constants}$$

turn out to define the general solution of the system (2.8). Note that

$$
\begin{aligned}
x^2 + y^2 &= (c_1 \cos t + c_2 \sin t)^2 + (-c_1 \sin t + c_2 \cos t)^2 \\
&= \left(c_1^2 + c_2^2\right) \left(\cos^2 t + \sin^2 t\right) \\
&= c_1^2 + c_2^2.
\end{aligned}
$$

Thus, every orbit associated with this system is a circle centered at the origin and the phase portrait consists of the collection of all such circles. See Fig. 4.6. ∎

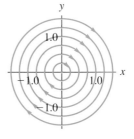

Figure 4.6 The phase portrait of the system (2.8) consists of all circles centered at the origin, oriented clockwise.

As in the previous example, one approach to drawing phase portraits is to find formulas for solutions and plot a selection of orbits using the formulas. This approach is feasible only for specialized systems for which solution methods are available (linear systems are an example, as we will see in Chapter 5).

Another method for sketching phase portraits, a method that does not require finding formulas for solutions, uses slope fields. At a point (x, y) on an orbit

$$\frac{dy}{dx} = \frac{dy/dt}{dx/dt} = \frac{g(x, y)}{f(x, y)}$$

gives the slope of the tangent line to the orbit curve. See Fig. 4.7. (We exclude equilibria, at which the ratio is undefined. If f vanishes at the point, and g doesn't, then the tangent is vertical and the slope is "infinite.") In other words, each point in the (x, y)-plane is associated with a unique slope $f(x, y)/g(x, y)$ and an orbit associated with the system (2.10) must have that slope at each of points through which it passes. The association of a slope with each point in the plane is called a *slope field*. Graphically, we represent the slope field by drawing a small line segment through each point with the assigned slope.

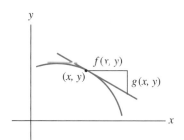

Figure 4.7

As t increases the orbit passes through a point (x, y) in one of two directions. The direction is that of the vector with components $f(x, y)$ and $g(x, y)$; that is, the vector

$$\begin{pmatrix} f(x, y) \\ g(x, y) \end{pmatrix}$$

We denote this direction by an arrow on the slope field line segment. The resulting association of *directed* line segments with each point (x, y) is called the *vector field* (or *direction field*) associated with the system (2.10).[2]

The vectors usually are not drawn to scale. That is, the arrows indicate the direction, but not the length of the vector $(f(x, y), g(x, y))$.

The orbits of an autonomous system must fit the vector field in the sense that when an orbit passes through a point it must do so in the direction and with the slope assigned to that point by the vector field. Thus, from a sufficiently detailed sketch of the vector field we can usually visualize typical orbits and hence the phase portrait.

For example, Fig. 4.8 shows the vector field associated with the system (2.8) whose orbits are circles centered at the origin, as we saw in Fig. 4.6.

EXAMPLE 2   Figure 4.9 shows a sketch of the vector field of the system

$$x' = -x + y$$
$$y' = -2x - y$$

together with a selection of orbits. Notice that the orbits appear to tend toward the origin as $t \to +\infty$. From these selected graphs it seems reasonable to conjecture that all orbits of this system tend toward the origin as $t \to +\infty$. The reader can test this conjecture by further computer exploration of orbits. (Note, however, that no amount

---

[2]The vectors are usually not drawn to scale. That is, the arrows indicate the direction but not the length of the vector $(f(x, y), g(x, y))$.

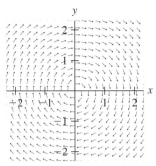

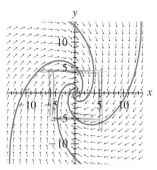

Figure 4.8 The vector field associated with the system (2.8).

Figure 4.9

of computer exploration will rigorously prove this conjecture, since there are infinitely many orbits and only a finite number can be numerically calculated.) ■

EXAMPLE 3

Figure 4.10 shows the vector field of the system in (2.7) and some typical orbits. From these graphs it appears that orbits beginning near the origin spiral outward and orbits beginning far from the origin spiral inward and that both types of orbits approach a closed loop as $t \to +\infty$. ■

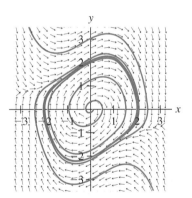

Figure 4.10

Exercises

For each system below, draw by hand the vector field at the indicated points.

1. $\begin{cases} x' = y \\ y' = -x \end{cases}$
   at $(x, y) = (1, 1), (1, 0), (0, -2), (-1, -1), \left(-\frac{1}{2}, 0\right)$

2. $\begin{cases} x' = -x + y \\ y' = -2x - y \end{cases}$
   at $(x, y) = (1, 1), (-1, -1), (-2, 1), \left(1, \frac{1}{2}\right), \left(-\frac{1}{2}, \frac{1}{2}\right)$

For each system below, use a computer to obtain the vector field for the indicated rectangle in the x, y-plane. Sketch by hand the orbit through the given points.

3. $\begin{cases} x' = -2x + 2y \\ y' = 2x - 5y \end{cases}$
   for $-1 < x < 1, -1 < y < 1$. Sketch the orbits passing through the points $(x, y) = (0, 1)$ and $(-1, 0)$.

4. $\begin{cases} x' = 2x - y \\ y' = -3x - 2y \end{cases}$

for $-2 < x < 2$, $-2 < y < 2$. Sketch the orbits passing through the points $(x, y) = (0, 1)$ and $(0.5, 1)$.

5. $\begin{cases} x' = -y \\ y' = -x + \left(1 - \frac{1}{3}y^2\right) y \end{cases}$

for $-2.5 < x < 2.5$, $-2.5 < y < 2.5$. Sketch the orbits passing through the points $(x, y) = (2, 1)$ and $(2, 2)$ This system is equivalent to the Rayleigh equation $x'' - \left(1 - \frac{1}{3}(x')^2\right) x' + x = 0$.

6. $\begin{cases} x' = -y \\ y' = -x + \left(1 - \frac{1}{3}y^2\right) y \end{cases}$

for $-5 < x < 5$, $-5 < y < 5$. Sketch the orbits passing through the points $(x, y) = (-2, -0.5)$ and $(-2, -1.5)$.

7. Match each system with its vector field. (Do not use a computer.)

(a) $\begin{cases} x' = -2x - y \\ y' = -x - y \end{cases}$ (b) $\begin{cases} x' = xy \\ y' = x - y \end{cases}$

(c) $\begin{cases} x' = x - y \\ y' = xy \end{cases}$ (d) $\begin{cases} x' = x^2 + y^2 \\ y' = x^2 - y^2 \end{cases}$

(e) $\begin{cases} x' = y - 1 \\ y' = x - 1 \end{cases}$ (f) $\begin{cases} x' = x^2 - y + 1 \\ y' = y - x \end{cases}$

(1)

(2)

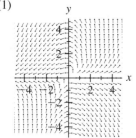

(3)

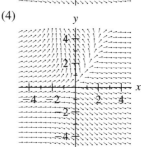

(4)

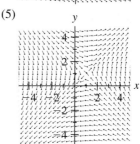

(5)

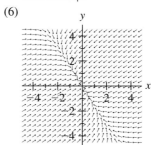

(6)

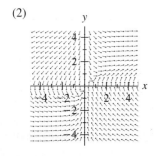

8. The x-**nullcline** of the autonomous system

$$x' = f(x, y)$$
$$y' = g(x, y)$$

is the curve in the x, y-plane defined by the equation $f(x, y) = 0$. The y-**nullcline** is the curve in the x, y-plane defined by the equation $g(x, y) = 0$. These two nullclines divide the x, y-plane into regions on which neither f nor g change sign. Thus, for example, in a region where both f and g are positive, the vector field points to the northeast. Other possibilities are regions in which the vector field is northwest, southeast or southwest. These can be useful aids in drawing a rough sketch of a vector field.

(a) Show that a point (x_0, y_0) lies on both nullclines if and only if it is an equilibrium point.

(b) Show that the slope of the vector field equals zero at each nonequilibrium point on the y-nullcline (and consequently the vector field is horizontal at these points). Show that the slope of the vector field is undefined at each nonequilibrium point on the x-nullcline (and consequently the vector field is vertical at these points).

Identify and sketch the x- and y-nullclines for the following systems. Sketch the vector field by using the nullclines and the regions they define in the x, y-plane (or the portion of the plane indicated). Using your sketch, draw some typical orbits.

9. $\begin{cases} x' = -y \\ y' = x \end{cases}$ 10. $\begin{cases} x' = y \\ y' = x \end{cases}$

11. $\begin{cases} x' = -x + y \\ y' = x + y \end{cases}$ 12. $\begin{cases} x' = x - y \\ y' = x + y \end{cases}$

13. $\begin{cases} x' = x(1 - y) \\ y' = y(1 - x) \end{cases}$ for $x \geq 0, y \geq 0$

14. $\begin{cases} x' = x(1 - y) \\ y' = y(-1 + x) \end{cases}$ for $x \geq 0, y \geq 0$

15. The equations

$$x' = x(r_1 - a_{11}x - a_{12}y)$$
$$y' = y(r_2 - a_{21}x - a_{22}y)$$

with positive coefficients r_i, a_{ij} are called the Lotka-Volterra competition equations. They describe the dynamics of two populations x and y in competition with one another. Only solutions in the first quadrant $x \geq 0, y \geq 0$ are of interest.

(a) What kind of curves are the nullclines?

(b) Study the possible configurations of the nullclines in the first quadrant. For each possibility, sketch the vector field.

(c) Use your results in (b) to sketch several typical orbits for each possible nullcline configuration you found in (b).

(d) Use a computer to sketch the vector field of an example for each possible nullcline configuration you found in (b). Also use the computer to graph typical orbits for each example.

(e) Discuss the implications of your results in (c) and (d) with regard to the long-term fate of both populations.

16. The equations

$$x' = x(r_1 - a_{11}x - a_{12}y)$$
$$y' = y(-r_2 + a_{21}x)$$

with positive coefficients r_i, a_{ij} are called the Lotka-Volterra predator-prey equations. They describe the dynamics of a predator population y and its prey population x. Only solutions in the first quadrant $x \geq 0, y \geq 0$ are of interest. Repeat Exercise 15 for this system of equations.

17. The system

$$x' = y$$
$$y' = -\alpha(x^2 - 1)y - x$$

is equivalent to the second-order equation $x'' + \alpha(x^2 - 1)x' + x = 0$ (the van der Pol equation).

(a) Find and sketch the nullclines for $\alpha > 0$. Use the nullclines to sketch the vector field.

(b) Repeat (a) for $\alpha < 0$.

18. Find and sketch the nullclines of the system

$$x' = y$$
$$y' = -x^3 - x.$$

This system is equivalent to the second-order equation $x'' + x^3 + x = 0$ (the Duffing equation). Use the nullclines to sketch the vector field.

4.3 Linear Systems of Equations

Systems of linear equations (or linear systems) constitute an important class of systems. This is because linear systems often arise in applications and because a thorough knowledge of linear systems is required for the study of nonlinear systems. As we have seen, linear systems also arise from applications involving linear equations of second and higher order. The equation for a simple harmonic oscillator is an example.

Linear systems are those in which both dependent variables x and y appear linearly. That is, $f(x, y)$ and $g(x, y)$ in the system (1.5) are linear functions of x and y.

Thus, a linear system has the form

$$x' = a(t)x + b(t)y + h_1(t) \tag{3.1}$$
$$y' = c(t)x + d(t)y + h_2(t).$$

Examples of linear systems include the systems (0.2) and (0.3) (which arise from a pesticide application problem), the system (0.4) (which is equivalent to the second-order equation $x'' + px' + qx = 0$), and the simple harmonic oscillator system (1.2).

We call $a(t)$, $b(t)$, $c(t)$, and $d(t)$ the *coefficients* of the system (3.1) and $h_1(t)$ and $h_2(t)$ the *nonhomogeneous terms (or forcing functions)*.

We begin by applying the fundamental existence and uniqueness theorem (Theorem 1) to the linear system (3.1). The initial value problem

$$x' = a(t)x + b(t)y + h_1(t)$$
$$y' = c(t)x + d(t)y + h_2(t) \tag{3.2}$$
$$x(t_0) = x_0, \quad y(t_0) = y_0$$

is a special case of the initial value problem (1.5) considered in the Sec. 4.1 in which

$$f(t, x, y) = a(t)x + b(t)y + h_1(t)$$
$$g(t, x, y) = c(t)x + d(t)y + h_2(t).$$

Since f and g are linear in x and y, they are continuous as functions of x and y. If all the coefficients $a(t)$, $b(t)$, $c(t)$, and $d(t)$ are continuous near $t = t_0$, then $f(t, x, y)$ and $g(t, x, y)$ and the derivatives

$$\frac{\partial f}{\partial x} = a(t), \qquad \frac{\partial f}{\partial y} = b(t)$$

$$\frac{\partial g}{\partial x} = c(t), \qquad \frac{\partial g}{\partial y} = d(t)$$

are continuous for t, x, and y near t_0, x_0 and y_0. It follows from Theorem 1 that the initial value problem (3.2) has a unique solution pair $x(t)$, $y(t)$ on an interval containing t_0.

It turns out that a stronger result holds for *linear* systems.

THEOREM 2   Suppose all coefficients $a(t)$, $b(t)$, $c(t)$, $d(t)$ and nonhomogeneous terms $h_1(t)$ and $h_2(t)$ in the linear system (3.1) are continuous on an interval $\alpha < t < \beta$. Then for any initial time t_0 in this interval, and for any x_0 and y_0, the initial value problem (3.2) has a unique solution pair $x(t)$, $y(t)$ defined on the whole interval $\alpha < t < \beta$.

In particular, if the coefficients and nonhomogeneous terms of a linear system are continuous *for all t*, as is often the case in applications, then solutions of initial value problems exist *for all t*. An important case is when the coefficients and nonhomogeneous terms are constants. Thus, for example, solutions of the initial value problem for the pesticide system (0.2) exist for all t, and the same is true for the harmonic oscillator system (1.2).

EXAMPLE 1   The initial value problem system

$$x' = y$$
$$y' = -x + \sin t \tag{3.3}$$
$$x(0) = x_0, \quad y(0) = y_0$$

is equivalent to the initial value problem

$$x'' + x = \sin t$$
$$x(0) = x_0, \quad x'(0) = y_0.$$

This equation is an example of the forced harmonic oscillator. Since $\sin t$ is continuous for all t, the coefficients and nonhomogeneous terms

$$a(t) = 0, \quad b(t) = 1, \quad c(t) = -1, \quad d(t) = 0$$
$$h_1(t) = 0, \quad h_2(t) = \sin t$$

are continuous for all t, x, and y. It follows that the initial value problem has a (unique) solution pair that is defined for all t. ∎

The main goal of this section is to study the general solution of the linear system (3.1). The *general solution* of a system is the set of all solution pairs.

A linear system (3.1) in which $h_1(t) = 0$ and $h_2(t) = 0$, that is, a system of the form

$$x' = a(t)x + b(t)y \qquad\qquad (3.4)$$
$$y' = c(t)x + d(t)y$$

is called *homogeneous*. The systems (0.2) and (1.2) are examples of homogeneous linear systems. A homogeneous linear system is completely specified by the four coefficients $a(t)$, $b(t)$, $c(t)$, and $d(t)$, which we can organize into a matrix

$$\begin{pmatrix} a(t) & b(t) \\ c(t) & d(t) \end{pmatrix}$$

called the *coefficient matrix* of the system. For example, the coefficient matrix of (0.3) is

$$\begin{pmatrix} -2 & 2 \\ 2 & -5 \end{pmatrix}$$

and the coefficient matrix of (0.5) is

$$\begin{pmatrix} 0 & 1 \\ -\frac{k}{m} & 0 \end{pmatrix}.$$

Note that

$$x = 0$$
$$y = 0$$

is a solution pair of any linear *homogeneous* system. This solution is called the *trivial solution pair*. All other solution pairs are called *nontrivial*.

The system (3.3) for a forced harmonic oscillator is an example of a linear system that is not homogeneous (since $h_2(t) = \sin t$) or, in other words, *nonhomogeneous*.

Recall from Chapter 2 that the general solution of the single linear equation $x' = p(t)x + q(t)$ has the form

$$x = x_h(t) + x_p(t),$$

where $x_h(t)$ is the general solution of the associated homogeneous equation $x' = p(t)x$ and $x_p(t)$ is any particular solution of the nonhomogeneous equation $x' = p(t)x + q(t)$. A similar fact is true for linear systems, as we will now see.

Our goal is to show that the general solution of a *nonhomogeneous* system (3.1) has the form

$$x = x_h(t) + x_p(t) \qquad (3.5)$$
$$y = y_h(t) + y_p(t),$$

where

$$x = x_p(t)$$
$$y = y_p(t)$$

is a particular solution pair and

$$x = x_h(t)$$
$$y = y_h(t)$$

is the general solution of the *homogeneous* system (3.4). That is, we want to show that the collection of solution pairs defined by (3.5) is in fact the entire set of solution pairs of the nonhomogeneous system (3.1).

Another way of saying the same thing is that we want to show that if x, y is a solution pair of the system (3.1), then the differences $x - x_p$, $y - y_p$ form a solution pair of the homogeneous system (3.4). To see that this is true, we calculate

$$(x - x_p)' = x' - x_p'$$
$$= a(t)x + b(t)y + h_1(t) - \left(a(t)x_p + b(t)y_p + h_1(t) \right)$$
$$= a(t) \left(x - x_p \right) + b(t) \left(y - y_p \right)$$

and

$$(y - y_p)' = y' - y_p'$$
$$= c(t)x + d(t)y + h_2(t) - \left(c(t)x_p + d(t)y_p + h_2(t) \right)$$
$$= c(t) \left(x - x_p \right) + d(t) \left(y - y_p \right).$$

These two equations imply that $x - x_p$ and $y - y_p$ solve the two equations of the homogeneous system (3.4).

THEOREM 3

The general solution of the linear nonhomogeneous system

$$x' = a(t)x + b(t)y + h_1(t)$$
$$y' = c(t)x + d(t)y + h_2(t)$$

has the additive decomposition

$$x = x_h(t) + x_p(t)$$
$$y = y_h(t) + y_p(t),$$

THEOREM 3
(Continued)

where

$$x = x_p(t), \qquad y = y_p(t)$$

is a particular solution pair of the nonhomogeneous system and

$$x = x_h(t), \qquad y = y_h(t)$$

is the general solution of the associated homogeneous system

$$x' = a(t)x + b(t)y$$
$$y' = c(t)x + d(t)y.$$

We conclude this section with a study of the general solution x_h, y_h of homogenous systems.

EXAMPLE 2

As we saw in the preceding section,

$$x = \cos t \qquad\qquad (3.6)$$
$$y = -\sin t$$

is a solution pair of the simple harmonic oscillator system

$$x' = y$$
$$y' = -x.$$

The multiples

$$x = 2\cos t$$
$$y = -2\sin t$$

also form a solution pair, since

$$x' = -2\sin t = y$$
$$y' = -2\cos t = -x.$$

In fact, any constant multiple c of the solution pair (3.6)

$$x = c\cos t$$
$$y = -c\sin t$$

forms a solution pair, since

$$x' = -c\sin t = y$$
$$y' = -c\cos t = -x. \qquad \blacksquare$$

The preceding example illustrates a general fact about homogeneous systems. Namely, a constant multiple of a solution pair is also a solution pair. This is proved by the following calculations:

$$(cx)' = cx'$$
$$= c\,[a(t)x + b(t)y]$$
$$= a(t)\,(cx) + b(t)\,(cy)$$

and

$$(cy)' = cy'$$
$$= c\,[c(t)x + d(t)y]$$
$$= c(t)\,(cx) + d(t)\,(cy),$$

which show that the pair cx, cy satisfies both equations of the homogeneous system

$$x' = a(t)x + b(t)y$$
$$y' = c(t)x + d(t)y$$

provided that the pair x, y does.

From a single (nontrivial) solution pair x, y we can obtain an infinite set of other solution pairs, namely the set of all constant multiples cx and cy. The next example shows, however, that this infinite set of solution pairs is not the general solution of the homogeneous system.

EXAMPLE 3

The pair

$$x = \sin t \tag{3.7}$$
$$y = \cos t$$

is a solution pair of the simple harmonic oscillator system

$$x' = y$$
$$y' = -x$$

since

$$x' = \cos t = y$$
$$y' = -\sin t = -x.$$

However, this solution pair is not a constant multiple of the solution pair (3.6), since there is no constant c for which $\sin t = c\cos t$ and $\cos t = -c\sin t$.

Another solution pair is given by the sums of the two solution pairs (3.6) and (3.7), that is, by

$$x = \cos t + \sin t$$
$$y = -\sin t + \cos t.$$

This follows from

$$x' = -\sin t + \cos t = y$$
$$y' = -\cos t - \sin t = -x.$$

In fact, any linear combination

$$x = c_1 \cos t + c_2 \sin t$$
$$y = -c_1 \sin t + c_2 \cos t$$

of the two solution pairs (3.6) and (3.7) is a solution pair, since

$$x' = -c_1 \sin t + c_2 \cos t = y$$
$$y' = -c_1 \cos t - c_2 \sin t = x.$$

∎

The preceding example illustrates the following general fact about solutions of linear homogeneous systems.

THEOREM 4

The Superposition Principle If

$$x = x_1(t) \qquad x = x_2(t)$$
$$y = y_1(t) \quad \text{and} \quad y = y_2(t)$$

are two solution pairs of a linear homogenous system (3.4), then for any constants c_1 and c_2 the linear combination

$$x = c_1 x_1(t) + c_2 x_2(t)$$
$$y = c_1 y_1(t) + c_2 y_2(t)$$

is a solution pair.[3]

This theorem is a result of the calculations

$$
\begin{aligned}
x' &= c_1 x_1' + c_2 x_2' \\
&= c_1 \left(a(t)x_1 + b(t)y_1 \right) + c_2 \left(a(t)x_2 + b(t)y_2 \right) \\
&= a(t) \left(c_1 x_1 + c_2 x_2 \right) + b(t) \left(c_1 y_1 + c_2 y_2 \right) \\
&= a(t)x + b(t)y
\end{aligned}
$$

and

$$
\begin{aligned}
y' &= c_1 y_1' + c_2 y_2' \\
&= c_1 \left(c(t)x_1 + d(t)y_1 \right) + c_2 \left(c(t)x_2 + d(t)y_2 \right) \\
&= c(t) \left(c_1 x_1 + c_2 x_2 \right) + d(t) \left(c_1 y_1 + c_2 y_2 \right) \\
&= c(t)x + d(t)y.
\end{aligned}
$$

From the two solution pairs (3.6) and (3.7) of the simple harmonic oscillator system we obtain a collection of solutions by forming all possible linear combinations, that is,

$$x = c_1 \cos t + c_2 \sin t$$
$$y = -c_1 \sin t + c_2 \cos t.$$

Are there any other solutions? Or is this the general solution of the harmonic oscillator system?

More generally, we have the following question: If

$$x = x_1(t) \qquad x = x_2(t)$$
$$y = y_1(t) \quad \text{and} \quad y = y_2(t)$$

are two solution pairs of a linear homogeneous system, do the linear combinations

$$x = c_1 x_1(t) + c_2 x_2(t)$$
$$y = c_1 y_1(t) + c_2 y_2(t)$$

form the general solution?

---

[3] Students familiar with linear algebra will note that this theorem implies that the general solution of a linear homogeneous system is a linear vector space (of function pairs).

The answer might be no. For example, consider the two solution pairs

$$x = \cos t \tag{3.8}$$
$$y = -\sin t$$

and

$$x = 2\cos t \tag{3.9}$$
$$y = -2\sin t.$$

The set of all linear combinations is

$$x = c_1 \cos t + c_2 (2\cos t)$$
$$y = -c_1 \sin t + c_2 (-2\sin t),$$

which, when rewritten as

$$x = (c_1 + 2c_2)\cos t$$
$$y = -(c_1 + 2c_2)\sin t,$$

we see is included in the set

$$x = c\cos t$$
$$y = -c\sin t$$

of solution pairs. We saw in Example 3 that this set is *not* the general solution. Note that the solution pair (3.9) is a multiple of the solution pair (3.8), and vice versa.

> **DEFINITION 3.** Two solution pairs are **dependent** if one pair is a constant multiple of the other. If two solution pairs are not dependent (i.e., are not constant multiples of each other), they are **independent**.

For example, the solution pairs (3.8) and (3.9) are dependent and the solution pairs (3.8) and (3.7), are independent solution pairs.
Another example is the solution pair

$$x_1 = 2e^{-t}$$
$$y_1 = e^{-t}$$

of the system (0.3) and the solution pair

$$x_2 = e^{-6t}$$
$$y_2 = -2e^{-6t}.$$

Since $2e^{-t}$ is not a constant multiple of e^{-6t}, these pairs are independent. By Theorem 4 the linear combinations

$$x = 2c_1 e^{-t} + c_2 e^{-6t}$$
$$y = c_1 e^{-t} - 2c_2 e^{-6t} \tag{3.10}$$

are solutions for any constants c_1 and c_2.
We can seek solutions of initial value problems among the set of linear combinations of two solution pairs. To do this we attempt to choose the two constants c_1 and c_2 so that the linear combination satisfies the given initial conditions.

For example, consider the initial value problem

$$x' = -2x + 2y$$
$$y' = 2x - 5y$$
$$x(0) = 1, \quad y(0) = 0$$

for the pesticide model (0.3). This corresponds to an initial pesticide dosage equal to one unit $x_0 = 1$ in a system in which the soil has initially no pesticide present ($y_0 = 0$). Can the solution of this initial value problem be found in the set of linear combinations (3.10)? That is, can c_1 and c_2 be chosen so that

$$x(0) = 2c_1 + c_2 = 1$$
$$y(0) = c_1 - 2c_2 = 0?$$

These two linear algebraic equations have the (unique) solution

$$c_1 = \frac{2}{5} \quad \text{and} \quad c_2 = \frac{1}{5}$$

and therefore the answer is yes. The solution of the initial value problem is then by the formulas

$$x = \frac{4}{5}e^{-t} + \frac{1}{5}e^{-6t}$$

$$y = \frac{2}{5}e^{-t} - \frac{2}{5}e^{-6t}.$$

But can *any* initial value problem

$$x' = -2x + 2y$$
$$y' = 2x - 5y$$
$$x(0) = x_0, \quad y(0) = y_0$$

be found in the set of linear combinations (3.10)? That is, can the linear algebraic equations

$$x(0) = 2c_1 + c_2 = x_0 \qquad (3.11)$$
$$y(0) = c_1 - 2c_2 = y_0$$

be solved for c_1 and c_2? The answer is yes and the solutions are

$$c_1 = \frac{1}{5}(2x_0 + y_0), \quad c_2 = \frac{1}{5}(x_0 - 2y_0).$$

These yield the solution pair

$$x = \frac{2}{5}(2x_0 + y_0)e^{-t} + \frac{1}{5}(x_0 - 2y_0)e^{-6t}$$

$$y = \frac{1}{5}(2x_0 + y_0)e^{-t} - \frac{2}{5}(x_0 - 2y_0)e^{-6t}.$$

The reason the algebraic equations (3.11) have a unique solution, for all choices of x_0 and y_0, is that the matrix of coefficients

$$\begin{pmatrix} 2 & 1 \\ 1 & -2 \end{pmatrix}$$

is nonsingular (i.e., invertible) or, in other words, its determinant

$$\det \begin{pmatrix} 2 & 1 \\ 1 & -2 \end{pmatrix} = -5$$

is nonzero.

In general, the solution of an initial value problem

$$x' = a(t)x + b(t)y$$
$$y' = c(t)x + d(t)y$$
$$x(t_0) = x_0, \quad y(t_0) = y_0$$

is found among the linear combinations

$$x = c_1 x_1(t) + c_2 x_2(t)$$
$$y = c_1 y_1(t) + c_2 y_2(t)$$

of two solution pairs if the two linear algebraic equations

$$c_1 x_1(t_0) + c_2 x_2(t_0) = x_0$$
$$c_1 y_1(t_0) + c_2 y_2(t_0) = y_0$$

have solutions c_1 and c_2. A unique solution of these two algebraic equations exists if (and only if) the matrix of coefficients is nonsingular or, in other words, the determinant

$$\det \begin{pmatrix} x_1(t_0) & x_2(t_0) \\ y_1(t_0) & y_2(t_0) \end{pmatrix} = x_1(t_0)y_2(t_0) - y_1(t_0)x_2(t_0) \tag{3.12}$$

is not equal to 0.

The condition that the determinant (3.12) be nonzero turns out to be equivalent to the condition that the two solutions be independent. The reader is asked to prove this in Exercises 24–25.

THEOREM 5

Two solution pairs of a linear homogeneous system on an interval $\alpha < t < \beta$ are independent if and only if

$$\det \begin{pmatrix} x_1(t_0) & x_2(t_0) \\ y_1(t_0) & y_2(t_0) \end{pmatrix} = x_1(t_0)y_2(t_0) - y_1(t_0)x_2(t_0) \neq 0$$

for some t_0 in the interval $\alpha < t_0 < \beta$.

An interesting and important fact about solutions of linear systems is that the determinant in Theorem 5 is either equal to 0 for all t on the interval $\alpha < t < \beta$ or is never equal to 0 on this interval. See Exercise 26. Consequently, we can just as well say that two solution pairs are independent on an interval $\alpha < t < \beta$ if and only the determinant (3.12) is nonzero for all t in the interval.

EXAMPLE 4

For the two solution pairs

$$\begin{aligned} x_1 &= \cos t \\ y_1 &= -\sin t \end{aligned} \quad \text{and} \quad \begin{aligned} x_2 &= \sin t \\ y_2 &= \cos t \end{aligned}$$

of the simple harmonic oscillator system

$$x' = y$$
$$y' = -x$$

we find that the determinant

$$\det\begin{pmatrix} x_1(t) & x_2(t) \\ y_1(t) & y_2(t) \end{pmatrix} = \det\begin{pmatrix} \cos t & \sin t \\ -\sin t & \cos t \end{pmatrix}$$

$$= \cos^2 t + \sin^2 t = 1$$

is nonzero for all t, and hence the solution pairs are independent. ∎

As pointed out previously, the determinant (3.12) in Theorem 5 is never equal to 0 or always equal to 0 on the interval $\alpha < t < \beta$. Thus, to show independence of a solution pair it is (conveniently) enough to show that the determinant is nonzero at only one value of t.

EXAMPLE 5

For the two solution pairs

$$\begin{aligned} x_1 &= 2e^{-t} \\ y_1 &= e^{-t} \end{aligned} \qquad \text{and} \qquad \begin{aligned} x_2 &= e^{-6t} \\ y_2 &= -2e^{-6t} \end{aligned}$$

of the system

$$\begin{aligned} x' &= -2x + 2y \\ y' &= 2x - 5y \end{aligned}$$

we find, at $t = 0$, that the determinant

$$\det\begin{pmatrix} x_1(0) & x_2(0) \\ y_1(0) & y_2(0) \end{pmatrix} = \det\begin{pmatrix} 2 & 1 \\ 1 & -2 \end{pmatrix} = -5$$

is nonzero and hence the solution pairs are independent. ∎

We have seen that solutions of initial value problems are found in the set of linear combinations of two solution pairs if the determinant (3.12) is nonzero. By Theorem 5 it follows that this determinant is nonzero if and only if the solution pairs are independent. *Thus, all initial value problems have a solution contained in the set of linear combinations of two independent solution pairs.* From this, we can deduce that the set of linear combinations of two independent solution pairs is the general solution of the system. This is because any solution pair $x(t)$, $y(t)$ whatsoever is a solution of an initial value problem; just pick a t_0 (any will do) and let $x_0 = x(t_0)$, $y_0 = y(t_0)$. The solution pair satisfies the resulting initial value problem (by the way we constructed it). By the italicized statement above, we conclude that any solution pair whatsoever can be found in the collection of linear combinations of two independent solution pairs. We have proved the following theorem.[4]

THEOREM 6

If

$$\begin{aligned} x &= x_1(t) \\ y &= y_1(t) \end{aligned} \qquad \text{and} \qquad \begin{aligned} x &= x_2(t) \\ y &= y_2(t) \end{aligned}$$

are independent solution pairs of a linear homogeneous system (3.4), then the general solution is

$$\begin{aligned} x &= c_1 x_1(t) + c_2 x_2(t) \\ y &= c_1 y_1(t) + c_2 y_2(t). \end{aligned} \qquad (3.13)$$

[4]Students familiar with linear algebra will notice that Theorem 6 implies that the general solution of linear homogeneous system of two equations is a two-dimensional vector space.

By Theorem 6 we can conclude that the linear combinations (3.10) form the general solution of the system

$$x' = -2x + 2y$$
$$y' = 2x - 5y.$$

Similarly, the general solution of the simple harmonic oscillation system

$$x' = y$$
$$y' = -x$$

is

$$x = c_1 \cos t + c_2 \sin t$$
$$y = -c_1 \sin t + c_2 \cos t.$$

Theorem 6 reduces the problem of solving a homogenous system of two linear equations (i.e., finding a formula for the general solution) to a search for only two independent solution pairs.

Unfortunately, for general homogeneous systems there is no universal solution method that will always produce two independent solution pairs. However, for an important restricted class of homogeneous systems, namely those with constant coefficients (i.e., autonomous systems), there is such a method. We study this method in the next chapter.

As a last remark, note that

$$\det \begin{pmatrix} x_1(t) & 0 \\ y_1(t) & 0 \end{pmatrix} = 0$$

for all t. Consequently, no solution pair is independent of the trivial solution pair $x = 0$, $y = 0$. Only nontrivial solution pairs can be independent. This means we can never use the trivial solution pair to construct the general solution of a linear system.

Exercises

Consider the first-order system

$$x' = \left(\frac{3}{2t} \right) x + \left(-\frac{1}{2} \right) y$$

$$y' = \left(-\frac{1}{2t^2} \right) x + \left(\frac{1}{2t} \right) y.$$

1. Show that

$$\begin{array}{cc} x_1 = t & x_2 = t^2 \\ y_1 = 1 & \quad \text{and} \quad \quad y_2 = -t \end{array}$$

are both solution pairs for $t > 0$.

2. Give a formula for the general solution.

3. Solve the initial value problem $x(1) = 2$, $y(1) = 0$.

Consider the first-order system

$$x' = 2ty$$
$$y' = -2tx.$$

4. Show

$$\begin{array}{cc} x_1 = \cos(t^2) & x_2 = \sin(t^2) \\ y_1 = -\sin(t^2) & \quad \text{and} \quad \quad y_2 = \cos(t^2) \end{array}$$

are both solution pairs for all t.

5. Give a formula for the general solution.

6. Find a formula for the solution of the initial value problem $x(0) = 1$, $y(0) = -2$.

In the exercises below, solution pairs are given for each system. If possible, find the general solution. If not possible, explain why.

7. $\begin{cases} x' = -x + 6y \\ y' = 2x + 3y \end{cases}$

$x_1 = -\frac{1}{2}e^{5t}$ $x_2 = 3e^{5t}$ $x_3 = -3e^{-3t}$
$y_1 = -\frac{1}{2}e^{5t}$ $y_2 = 3e^{5t}$ $y_3 = e^{-3t}$

8. $\begin{cases} x' = -x + 6y \\ y' = 2x + 3y \end{cases}$

$x_1 = e^{5t} - 3e^{-3t}$ $x_2 = -2e^{5t} + 6e^{-3t}$
$y_1 = e^{5t} + e^{-3t}$ $y_2 = -2e^{5t} - 2e^{-3t}$

$x_3 = \frac{1}{3}e^{5t} - e^{-3t}$

$y_3 = \frac{1}{3}e^{5t} + \frac{1}{3}e^{-3t}$

9. $\begin{cases} x' = -3x + 4y \\ y' = -x + 2y \end{cases}$

$x_1 = 16e^{-2t} - 2e^{t}$ $x_2 = -8e^{-2t} + e^{t}$
$y_1 = 4e^{-2t} - 2e^{t}$ $y_2 = -2e^{-2t} + e^{t}$

$x_3 = 4e^{-2t} - \frac{1}{2}e^{t}$

$y_3 = e^{-2t} - \frac{1}{2}e^{t}$

10. $\begin{cases} x' = -3x + 4y \\ y' = -x + 2y \end{cases}$

$x_1 = e^{t}$ $x_2 = 4e^{-2t}$ $x_3 = e^{t} + 4e^{-2t}$
$y_1 = e^{t}$ $y_2 = e^{-2t}$ $y_3 = e^{t} + e^{-2t}$

11. $\begin{cases} x' = -3x - 2y \\ y' = 4x + 3y \end{cases}$

$x_1 = -e^{-t}$ $x_2 = e^{t} - e^{-t}$ $x_3 = e^{t}$
$y_1 = e^{-t}$ $y_2 = -2e^{t} + e^{-t}$ $y_3 = -2e^{t}$

12. $\begin{cases} x' = -3x - 2y \\ y' = 4x + 3y \end{cases}$

$x_1 = e^{t} - e^{-t}$ $x_2 = e^{t} + e^{-t}$
$y_1 = -2e^{t} + e^{-t}$ $y_2 = -2e^{t} - e^{-t}$

13. $\begin{cases} x' = x - 2y \\ y' = 2x + y \end{cases}$

$x_1 = e^{t} \cos 2t$ $x_2 = e^{t}(\cos 2t + \sin 2t)$
$y_1 = e^{t} \sin 2t$ $y_2 = e^{t}(\sin 2t - \cos 2t)$

$x_3 = e^{t} \sin 2t$

$y_3 = -e^{t} \cos 2t$

14. $\begin{cases} x' = x - 2y \\ y' = 2x + y \end{cases}$

$x_1 = 6\cos 5t - 4\sin 5t$ $x_2 = -3\cos 5t + 2\sin 5t$
$y_1 = 2\cos 5t + 2\sin 5t$ $y_2 = -\cos 5t - \sin 5t$

$x_3 = 5\cos 5t + \sin 5t$

$y_3 = 2\sin 5t$

15. (a) Show that

$$x_1 = e^{4t} \qquad x_2 = -3e^{-3t}$$
$$y_1 = 2e^{4t} \qquad y_2 = e^{-3t}$$

are both solution pairs of the system

$$x' = -2x + 3y$$
$$y' = 2x + 3y.$$

(b) Show that the solution pairs in (a) are independent.

(c) Write the general solution of the system.

(d) Show that

$$x = 2e^{4t} + 3e^{-3t}$$
$$y = 4e^{4t} - e^{-3t}$$

is also a solution pair of the system.

(e) Write the solution pair in (d) as a linear combination of the independent pairs in (a).

16. (a) Show that

$$x_1 = 2e^{-t} \cos 3t$$
$$y_1 = e^{-t}(\cos 3t - \sin 3t)$$
$$x_2 = 2e^{-t} \sin 3t$$
$$y_2 = e^{-t}(\cos 3t + \sin 3t)$$

are both solution pairs of the system

$$x' = -4x + 6y$$
$$y' = -3x + 2y.$$

(b) Show that the solution pairs in (a) are independent.

(c) Write the general solution of the system.

(d) Show that

$$x = -2e^{-t}(2\cos 3t + \sin 3t)$$
$$y = e^{-t}(-3\cos 3t + \sin 3t)$$

is also a solution pair of the system.

(e) Write the solution pair in (d) as a linear combination of the independent pairs in (a).

17. Use the solution pairs

$$x_1 = -2e^{-5t} \qquad x_2 = e^{5t}$$
$$y_1 = e^{-5t} \qquad y_2 = 2e^{5t}$$

of the system

$$x' = -3x + 4y$$
$$y' = 4x + 3y$$

to solve the following initial value problems.

(a) $x(0) = -1$, $y(0) = 2$

(b) $x(0) = 3$, $y(0) = 1$

18. Use the solution pairs

$$x_1 = 7\cos\sqrt{3}t$$
$$y_1 = 2\cos\sqrt{3}t - \sqrt{3}\sin\sqrt{3}t$$

$$x_2 = 2\cos\sqrt{3}t + \sqrt{3}\sin\sqrt{3}t$$
$$y_2 = \cos\sqrt{3}t$$

of the system

$$x' = -2x + 7y$$
$$y' = -x + 2y$$

to solve the following initial value problems.

(a) $x(0) = 1$, $y(0) = 0$

(b) $x(0) = 0$, $y(0) = 1$

Find the general solution of the nonhomogeneous system

$$x' = y + h_1(t)$$
$$y' = -x + h_2(t)$$

with the nonhomogeneous terms $h_1(t)$ and $h_2(t)$ below. (Hint: Guess a particular solution pair x_p, y_p using the method of undetermined coefficients.)

19. $h_1(t) = 0$, $h_2(t) = \sin 2t$

20. $h_1(t) = 0$, $h_2(t) = \cos 2t$

21. $h_1(t) = 1$, $h_2(t) = 2$

22. $h_1(t) = -1$, $h_2(t) = e^t$

23. Let $\beta = \sqrt{k/m}$. Show that

$$x = c_1\cos\beta t + c_2\sin\beta t$$
$$y = -c_1\beta\sin\beta t + c_2\beta\cos\beta t$$

is the general solution of the system (0.6). This shows $x = c_1\cos\beta t + c_2\sin\beta t$ is the general solution of the simple harmonic oscillator equation $x'' + kx/m = 0$.

In the following two exercises, suppose that $x_1(t)$, $y_1(t)$ and $x_2(t)$, $y_2(t)$ are both solution pairs of a linear homogeneous system (3.4) on an interval $\alpha < t < \beta$.

24. Suppose that these solution pairs are dependent. Show that

$$\det\begin{pmatrix} x_1(t) & x_2(t) \\ y_1(t) & y_2(t) \end{pmatrix} = 0$$

for all t. It follows that if

$$\det\begin{pmatrix} x_1(t_0) & x_2(t_0) \\ y_1(t_0) & y_2(t_0) \end{pmatrix} \neq 0$$

for some t_0, then the solution pairs are independent.

25. Conversely, suppose that

$$\det\begin{pmatrix} x_1(t_0) & x_2(t_0) \\ y_1(t_0) & y_2(t_0) \end{pmatrix} = 0$$

for some t_0. Show that these solution pairs are dependent.

26. Suppose that $x_1(t)$, $y_1(t)$ and $x_2(t)$, $y_2(t)$ are both solution pairs of a linear homogeneous system (3.4) on an interval $\alpha < t < \beta$. Show that the determinant

$$z(t) = \det\begin{pmatrix} x_1(t) & x_2(t) \\ y_1(t) & y_2(t) \end{pmatrix}$$

is either never equal to 0 for $\alpha < t < \beta$ or is always equal to 0 on $\alpha < t < \beta$. [*Hint:* By direct calculation show that the determinant satisfies the first-order, linear homogeneous equation $z' = p(t)z$, where $p(t) = a(b) + d(t)$, and solve the initial value problem for z.]

4.4 Matrix Notation

A convenient way to work with linear systems and their solutions is to use matrix notation. We use $x = x(t)$ and $y = y(t)$ and their derivatives to construct the matrices

$$\begin{pmatrix} x \\ y \end{pmatrix} = \begin{pmatrix} x(t) \\ y(t) \end{pmatrix}$$

$$\begin{pmatrix} x \\ y \end{pmatrix}' = \begin{pmatrix} x' \\ y' \end{pmatrix}.$$

(Matrices with only one column, or only one row, are called vectors.) Recalling the definition of matrix multiplication, we have

$$\begin{pmatrix} a(t) & b(t) \\ c(t) & d(t) \end{pmatrix}\begin{pmatrix} x \\ y \end{pmatrix} = \begin{pmatrix} a(t)x + b(t)y \\ c(t)x + d(t)y \end{pmatrix}.$$

Here we have used the coefficients of x and y in the linear system (3.1) to construct the *coefficient matrix*

$$\begin{pmatrix} a(t) & b(t) \\ c(t) & d(t) \end{pmatrix}$$

of the system. With this notation we can write the homogeneous linear system

$$\begin{aligned} x' &= a(t)x + b(t)y \\ y' &= c(t)x + d(t)y \end{aligned}$$

in the matrix form

$$\begin{pmatrix} x \\ y \end{pmatrix}' = \begin{pmatrix} a(t) & b(t) \\ c(t) & d(t) \end{pmatrix} \begin{pmatrix} x \\ y \end{pmatrix}. \tag{4.1}$$

Since matrices are added by adding corresponding components, we can also write the nonhomogeneous system

$$\begin{aligned} x' &= a(t)x + b(t)y + h_1(t) \\ y' &= c(t)x + d(t)y + h_2(t) \end{aligned}$$

as

$$\begin{pmatrix} x \\ y \end{pmatrix}' = \begin{pmatrix} a(t) & b(t) \\ c(t) & d(t) \end{pmatrix} \begin{pmatrix} x \\ y \end{pmatrix} + \begin{pmatrix} h_1(t) \\ h_2(t) \end{pmatrix}. \tag{4.2}$$

Initial conditions are written as a vector,

$$\begin{pmatrix} x(t_0) \\ y(t_0) \end{pmatrix} = \begin{pmatrix} x_0 \\ y_0 \end{pmatrix}.$$

For example, in matrix form the system (0.3) is written

$$\begin{pmatrix} x \\ y \end{pmatrix}' = \begin{pmatrix} -2 & 2 \\ 2 & -5 \end{pmatrix} \begin{pmatrix} x \\ y \end{pmatrix} \tag{4.3}$$

and the system (0.2) for the pesticide problem is written

$$\begin{pmatrix} x \\ y \end{pmatrix}' = \begin{pmatrix} -r_1 & r_2 \\ r_1 & -(r_2 + r_3) \end{pmatrix} \begin{pmatrix} x \\ y \end{pmatrix}.$$

The simple harmonic oscillator equation $x'' + x = 0$ is equivalent to the system (1.2) which in matrix form is

$$\begin{pmatrix} x \\ y \end{pmatrix}' = \begin{pmatrix} 0 & 1 \\ -1 & 0 \end{pmatrix} \begin{pmatrix} x \\ y \end{pmatrix}. \tag{4.4}$$

The forced harmonic oscillator system (3.3) is written

$$\begin{pmatrix} x \\ y \end{pmatrix}' = \begin{pmatrix} 0 & 1 \\ -1 & 0 \end{pmatrix} \begin{pmatrix} x \\ y \end{pmatrix} + \begin{pmatrix} 0 \\ \sin t \end{pmatrix}.$$

We write a solution pair $x = x(t)$, $y = y(t)$ of the system (4.2) as a *solution vector*

$$\begin{pmatrix} x \\ y \end{pmatrix} = \begin{pmatrix} x(t) \\ y(t) \end{pmatrix}.$$

A solution vector satisfies

$$\begin{pmatrix} x(t) \\ y(t) \end{pmatrix}' = \begin{pmatrix} a(t) & b(t) \\ c(t) & c(t) \end{pmatrix} \begin{pmatrix} x(t) \\ y(t) \end{pmatrix} + \begin{pmatrix} h_1(t) \\ h_2(t) \end{pmatrix} \tag{4.5}$$

for t on an interval $\alpha < t < \beta$. For example, the solution pair $x = 2e^{-t}$, $y = e^{-t}$ of the system (4.3) is written

$$\begin{pmatrix} x \\ y \end{pmatrix} = \begin{pmatrix} 2e^{-t} \\ e^{-t} \end{pmatrix}.$$

In vector form, this means that

$$\begin{pmatrix} x \\ y \end{pmatrix}' = \begin{pmatrix} -2e^{-t} \\ -e^{-t} \end{pmatrix}$$

and

$$\begin{pmatrix} -2 & 2 \\ 2 & -5 \end{pmatrix} \begin{pmatrix} 2e^{-t} \\ e^{-t} \end{pmatrix} = \begin{pmatrix} -2e^{-t} \\ -e^{-t} \end{pmatrix}$$

are identical.

Similarly, the solution pair $x = \cos t$, $y = -\sin t$ of the simple harmonic oscillator system (4.4) is written

$$\begin{pmatrix} x \\ y \end{pmatrix} = \begin{pmatrix} \cos t \\ -\sin t \end{pmatrix}.$$

Then

$$\begin{pmatrix} x \\ y \end{pmatrix}' = \begin{pmatrix} -\sin t \\ -\cos t \end{pmatrix}$$

and

$$\begin{pmatrix} 0 & 1 \\ -1 & 0 \end{pmatrix} \begin{pmatrix} \cos t \\ -\sin t \end{pmatrix} = \begin{pmatrix} -\sin t \\ -\cos t \end{pmatrix}$$

are identical.

By Theorem 6, the general solution of a homogeneous system (4.1) is

$$x = c_1 x_1(t) + c_2 x_2(t)$$
$$y = c_1 y_1(t) + c_2 y_2(t),$$

where

$$x = x_1(t) \qquad x = x_2(t)$$
$$y = y_1(t) \qquad y = y_2(t)$$

are independent solution pairs. In matrix form the general solution is

$$\begin{pmatrix} x \\ y \end{pmatrix} = \begin{pmatrix} c_1 x_1(t) + c_2 x_2(t) \\ c_1 y_1(t) + c_2 y_2(t) \end{pmatrix},$$

which can be written as the linear combination

$$\begin{pmatrix} x \\ y \end{pmatrix} = c_1 \begin{pmatrix} x_1(t) \\ y_1(t) \end{pmatrix} + c_2 \begin{pmatrix} x_2(t) \\ y_2(t) \end{pmatrix}$$

of the *independent solution vectors*

$$\begin{pmatrix} x_1(t) \\ y_1(t) \end{pmatrix}, \qquad \begin{pmatrix} x_2(t) \\ y_2(t) \end{pmatrix}.$$

Another way to write the general solution is

$$\begin{pmatrix} x \\ y \end{pmatrix} = \begin{pmatrix} x_1(t) & x_2(t) \\ y_1(t) & y_2(t) \end{pmatrix} \begin{pmatrix} c_1 \\ c_2 \end{pmatrix}. \tag{4.6}$$

The matrix

$$\begin{pmatrix} x_1(t) & x_2(t) \\ y_1(t) & y_2(t) \end{pmatrix}$$

is called a *fundamental solution matrix* of the homogeneous system (4.1). Thus, the general solution (4.6) is the product of a fundamental solution matrix and the vector

$$\begin{pmatrix} c_1 \\ c_2 \end{pmatrix}$$

of arbitrary constants. For example, the two independent solution vectors

$$\begin{pmatrix} 2e^{-t} \\ e^{-t} \end{pmatrix} \qquad \text{and} \qquad \begin{pmatrix} e^{-6t} \\ -2e^{-6t} \end{pmatrix}$$

of the homogeneous system (4.3) produce the fundamental solution matrix

$$\begin{pmatrix} 2e^{-t} & e^{-6t} \\ e^{-t} & -2e^{-6t} \end{pmatrix}$$

and the general solution

$$\begin{pmatrix} x \\ y \end{pmatrix} = \begin{pmatrix} 2e^{-t} & e^{-6t} \\ e^{-t} & -2e^{-6t} \end{pmatrix} \begin{pmatrix} c_1 \\ c_2 \end{pmatrix}.$$

As a second example, the two independent solution vectors

$$\begin{pmatrix} \cos t \\ -\sin t \end{pmatrix} \qquad \text{and} \qquad \begin{pmatrix} \sin t \\ \cos t \end{pmatrix}$$

of the simple harmonic oscillator system (4.4) produce the fundamental solution matrix

$$\begin{pmatrix} \cos t & \sin t \\ -\sin t & \cos t \end{pmatrix} \tag{4.7}$$

and the general solution

$$\begin{pmatrix} x \\ y \end{pmatrix} = \begin{pmatrix} \cos t & \sin t \\ -\sin t & \cos t \end{pmatrix} \begin{pmatrix} c_1 \\ c_2 \end{pmatrix}.$$

The solution of the initial value problem

$$\begin{pmatrix} x \\ y \end{pmatrix}' = \begin{pmatrix} a(t) & b(t) \\ c(t) & d(t) \end{pmatrix} \begin{pmatrix} x \\ y \end{pmatrix}$$

$$\begin{pmatrix} x(t_0) \\ y(t_0) \end{pmatrix} = \begin{pmatrix} x_0 \\ y_0 \end{pmatrix}$$

is found from the general solution (4.6) by solving the algebraic equation

$$\begin{pmatrix} x_1(t_0) & x_2(t_0) \\ y_1(t_0) & y_2(t_0) \end{pmatrix} \begin{pmatrix} c_1 \\ c_2 \end{pmatrix} = \begin{pmatrix} x_0 \\ y_0 \end{pmatrix}$$

for

$$\begin{pmatrix} c_1 \\ c_2 \end{pmatrix} = \begin{pmatrix} x_1(t_0) & x_2(t_0) \\ y_1(t_0) & y_2(t_0) \end{pmatrix}^{-1} \begin{pmatrix} x_0 \\ y_0 \end{pmatrix}.$$

Note that Theorem 5 guarantees that a fundamental solution matrix is invertible for each t. A formula for the solution of the initial value problem is therefore

$$\begin{pmatrix} x \\ y \end{pmatrix} = \begin{pmatrix} x_1(t) & x_2(t) \\ y_1(t) & y_2(t) \end{pmatrix} \begin{pmatrix} x_1(t_0) & x_2(t_0) \\ y_1(t_0) & y_2(t_0) \end{pmatrix}^{-1} \begin{pmatrix} x_0 \\ y_0 \end{pmatrix}.$$

An even more concise notation is

$$A(t) = \begin{pmatrix} a(t) & b(t) \\ c(t) & c(t) \end{pmatrix}$$

$$\tilde{x} = \begin{pmatrix} x \\ y \end{pmatrix}, \qquad \tilde{h}(t) = \begin{pmatrix} h_1(t) \\ h_2(t) \end{pmatrix}$$

by means of which the nonhomogeneous system (4.2) is written

$$\tilde{x}' = A(t)\tilde{x} + \tilde{h}(t)$$

and the homogeneous system (4.1) becomes

$$\tilde{x}' = A(t)\tilde{x}.$$

Let

$$\Phi(t) = \begin{pmatrix} x_1(t) & x_2(t) \\ y_1(t) & y_2(t) \end{pmatrix}$$

denote a fundamental solution matrix of the homogeneous system and

$$\tilde{c} = \begin{pmatrix} c_1 \\ c_2 \end{pmatrix}.$$

Then the general solution is

$$\tilde{x}(t) = \Phi(t)\tilde{c}$$

and the solution of the initial value problem

$$\tilde{x}' = A(t)\tilde{x} \qquad\qquad (4.8)$$
$$\tilde{x}(t_0) = \tilde{x}_0,$$

where

$$\tilde{x}_0 = \begin{pmatrix} x_0 \\ y_0 \end{pmatrix},$$

is

$$\tilde{x}(t) = \Phi(t)\Phi^{-1}(t_0)\tilde{x}_0.$$

For the homogeneous system (4.3)

$$\Phi(t) = \begin{pmatrix} 2e^{-t} & e^{-6t} \\ e^{-t} & -2e^{-6t} \end{pmatrix}, \qquad \Phi^{-1}(t) = \frac{1}{5} \begin{pmatrix} 2e^{t} & e^{t} \\ e^{6t} & -2e^{6t} \end{pmatrix}$$

and

$$\Phi(t)\Phi^{-1}(0) = \frac{1}{5} \begin{pmatrix} 4e^{-t} + e^{-6t} & 2e^{-t} - 2e^{-6t} \\ 2e^{-t} - 2e^{-6t} & e^{-t} + 4e^{-6t} \end{pmatrix}.$$

The solution of the initial value problem

$$\tilde{x}(0) = \begin{pmatrix} x_0 \\ y_0 \end{pmatrix}$$

is then

$$\tilde{x}(t) = \Phi(t)\Phi^{-1}(0)\tilde{x}_0$$

$$= \frac{1}{5} \begin{pmatrix} 4e^{-t} + e^{-6t} & 2e^{-t} - 2e^{-6t} \\ 2e^{-t} - 2e^{-6t} & e^{-t} + 4e^{-6t} \end{pmatrix} \begin{pmatrix} x_0 \\ y_0 \end{pmatrix}.$$

For example, the solution of the initial value problem

$$\tilde{x}(0) = \begin{pmatrix} 2 \\ 1 \end{pmatrix}$$

is

$$\tilde{x}(t) = \frac{1}{5} \begin{pmatrix} 4e^{-t} + e^{-6t} & 2e^{-t} - 2e^{-6t} \\ 2e^{-t} - 2e^{-6t} & e^{-t} + 4e^{-6t} \end{pmatrix} \begin{pmatrix} 2 \\ 1 \end{pmatrix}$$

$$= \begin{pmatrix} 2e^{-t} \\ e^{-t} \end{pmatrix}.$$

For the fundamental solution matrix

$$\Phi(t) = \begin{pmatrix} \cos t & \sin t \\ -\sin t & \cos t \end{pmatrix}$$

of the simple harmonic oscillator system (4.4), we have

$$\Phi^{-1}(t) = \begin{pmatrix} \cos t & \sin t \\ -\sin t & \cos t \end{pmatrix}^{-1} = \begin{pmatrix} \cos t & -\sin t \\ \sin t & \cos t \end{pmatrix}$$

and the solution of the initial value problem is

$$\tilde{x}(t) = \Phi(t)\Phi^{-1}(t_0)\tilde{x}_0$$

$$= \begin{pmatrix} \cos t & \sin t \\ -\sin t & \cos t \end{pmatrix} \begin{pmatrix} \cos t_0 & \sin t_0 \\ -\sin t_0 & \cos t_0 \end{pmatrix}^{-1} \begin{pmatrix} x_0 \\ y_0 \end{pmatrix}$$

$$= \begin{pmatrix} \cos t & \sin t \\ -\sin t & \cos t \end{pmatrix} \begin{pmatrix} \cos t_0 & -\sin t_0 \\ \sin t_0 & \cos t_0 \end{pmatrix} \begin{pmatrix} x_0 \\ y_0 \end{pmatrix}$$

$$= \begin{pmatrix} \cos t \cos t_0 + \sin t \sin t_0 & -\cos t \sin t_0 + \sin t \cos t_0 \\ -\sin t \cos t_0 + \cos t \sin t_0 & \sin t \sin t_0 + \cos t \cos t_0 \end{pmatrix} \begin{pmatrix} x_0 \\ y_0 \end{pmatrix}$$

or

$$\begin{pmatrix} x(t) \\ y(t) \end{pmatrix} = \begin{pmatrix} \cos(t-t_0) & \sin(t-t_0) \\ -\sin(t-t_0) & \cos(t-t_0) \end{pmatrix} \begin{pmatrix} x_0 \\ y_0 \end{pmatrix}.$$

For example, the solution of the initial value problem

$$x'' + x = 0$$
$$x(0) = 2, \qquad x'(0) = -1$$

is

$$\begin{pmatrix} x(t) \\ y(t) \end{pmatrix} = \begin{pmatrix} \cos t & \sin t \\ -\sin t & \cos t \end{pmatrix} \begin{pmatrix} 2 \\ -1 \end{pmatrix}$$
$$= \begin{pmatrix} 2\cos t - \sin t \\ -2\sin t - \cos t \end{pmatrix}.$$

This initial value problem corresponds to displacing the simple harmonic oscillator 2 units from its rest position and giving it an initial velocity of -1. The formula $x = 2\cos t - \sin t$ describes the resulting oscillation of the object around its rest position (and its velocity is $y = x' = -2\sin t - \cos t$).

Exercises

Solve the following initial value problems for system (4.3):

$$\begin{pmatrix} x \\ y \end{pmatrix}' = \begin{pmatrix} -2 & 2 \\ 2 & -5 \end{pmatrix} \begin{pmatrix} x \\ y \end{pmatrix}.$$

1. $\tilde{x}_0 = \begin{pmatrix} 1 \\ 1 \end{pmatrix}$ **2.** $\tilde{x}_0 = \begin{pmatrix} \frac{1}{2} \\ -1 \end{pmatrix}$

3. $\tilde{x}_0 = \begin{pmatrix} 10 \\ -5 \end{pmatrix}$ **4.** $\tilde{x}_0 = \begin{pmatrix} -5 \\ 2 \end{pmatrix}$

Solve the following initial value problems for the simple harmonic equation $x'' + x = 0$.

5. $x(0) = -1, x'(0) = 1$ **6.** $x(0) = 2, x'(0) = -3$
7. $x(\pi) = -1, x'(\pi) = 1$ **8.** $x(\pi) = 2, x'(\pi) = -3$
9. Find a fundamental solution matrix $\Phi(t)$ and write the general solution for the systems in Exercises 7–14 of Sec. 4.3,

if possible. If it is not possible to find a fundamental solution matrix using the given solutions, explain why.

Use the following two facts about matrices and vectors

$$A(\tilde{x} + \tilde{y}) = A\tilde{x} + A\tilde{y}$$
$$A(k\tilde{x}) = kA\tilde{x} \quad \text{for any real number } k$$

to do the following two exercises.

10. Prove the superposition principle Theorem 4 for the homogeneous linear system

$$\tilde{x}' = A(t)\tilde{x}.$$

11. Show that a linear combination $k_1\tilde{x}_1 + k_2\tilde{x}_2 + \cdots + k_n\tilde{x}_n$ of any number of solutions $\tilde{x}_1, \tilde{x}_2, \ldots, \tilde{x}_n$ of is a solution.

4.5 Chapter Summary and Exercises

A first-order system consists of two (or more) first-order differential equations for two (or more) solutions. Higher-order equations and systems are equivalent to first-order systems. The fundamental existence and uniqueness theorem (Theorem 1) is a generalization to systems of the fundamental existence and uniqueness theorem for single first-order differential equations (Theorem 1, Sec. 1.1 of Chapter 1). Algorithms (Euler, Heun, and Runge-Kutta) for numerically approximating solutions of initial value problems also extend to systems of equations. We can graph solutions of systems in several ways. Each component of the solution can be graphed separately or, in the case of two equations, both components can simultaneously be graphed as functions of t in three-dimensional space. For the important special case of autonomous systems a solution pair $(x(t), y(t))$, when graphed in the x, y-plane, produces an orbit. The set of all orbits constitute the phase plane portrait of the system. Vector fields are useful for studying phase portrait. Systems of linear equations have special properties. The general solution of a homogeneous system of two equations is a linear combination of two independent solutions. The general solution of a nonhomogeneous system has an additive decomposition consisting of a particular solution plus the general solution of the associated homogeneous system.

Exercises

Show that the pairs below solve the system

$$x' = -2x + y$$
$$y' = x - 2y.$$

1. $x = e^{-t}, \quad y = e^{-t}$

2. $x = e^{-3t}, \quad y = -e^{-3t}$

3. $x = e^{-t} + e^{-3t}, \quad y = e^{-t} - e^{-3t}$

4. $x = c_1 e^{-t} + c_2 e^{-3t}, y = c_1 e^{-t} - c_2 e^{-3t}$ for any constants c_1 and c_2

Show that the following pairs solve the system

$$x' = 4x - 2y$$
$$y' = 7x - 5y.$$

5. $x = 2e^{-3t}, y = 7e^{-3t}$

6. $x = e^{2t}, y = e^{2t}$

7. $x = 2e^{-3t} - e^{2t}, y = 7e^{-3t} - e^{2t}$

8. $x = 2c_1 e^{-3t} + c_2 e^{2t}, y = 7c_1 e^{-3t} + c_2 e^{2t}$

9. Show that the pair $x(t) = 3 + 2\cos t - 2\sin t$, $y(t) = -2\sin t - 2\cos t$ solves the system

$$x' = y$$
$$y' = -x + 3$$

for all t.

10. Show that the pair $x(t) = \dfrac{1}{t}, y(t) = \dfrac{1}{t^2}$ solves the system

$$x' = -x\sqrt{y}$$
$$y' = -2xy$$

for $t > 0$.

Theorem 1 applies to which initial value problems below? To which does it not apply? Explain your answers. What conclusions can you draw in each case?

11. $\begin{cases} x' = \sin(x+y) \\ y' = \sin(x-y) \\ x(0) = 1, \, y(0) = 0 \end{cases}$

12. $\begin{cases} x' = \dfrac{1+x}{1-y} \\ y' = \dfrac{1+y}{1-x} \\ x(0) = 0, \, y(0) = 0 \end{cases}$

13. $\begin{cases} x' = \dfrac{1+x}{1-y} \\ y' = \dfrac{1+y}{1-x} \\ x(0) = 0, \, y(0) = 1 \end{cases}$

14. $\begin{cases} x' = \sqrt{t-x-y} \\ y' = \sqrt{t+x+y} \\ x(1) = 0, \, y(1) = y_0 \\ \text{where } -1 < y_0 < 1 \end{cases}$

For which initial conditions x_0, y_0 and initial times t_0 does Theorem 1 apply for the systems below? For which does this theorem not apply? Explain your answers. What conclusions can you draw in each case?

15. $\begin{cases} x' = \dfrac{1+x}{1-y} \\ y' = \dfrac{1+y}{1-x} \end{cases}$

16. $\begin{cases} x' = ax + by \\ y' = cx + dx \end{cases}$

17. $\begin{cases} x' = \sqrt{t-x-y} \\ y' = \sqrt{t+x+y} \end{cases}$

18. $\begin{cases} x' = \ln(1-x-y) \\ y' = \ln(1+x+y) \end{cases}$

For which initial conditions $x(t_0) = x_0$, $y(t_0) = y_0$ do the following systems have unique solutions? On what intervals do those solutions exist?

19. $\begin{cases} x' = x\cos t - 2y \\ y' = 2x - 3y\sin t \end{cases}$

20. $\begin{cases} x' = \dfrac{1}{t}x + y - e^t \cos t \\ y' = 5x - \dfrac{1}{1-t}y \end{cases}$

21. $\begin{cases} x' = x^2 + y \\ y' = x - y \end{cases}$ **22.** $\begin{cases} x' = x^{-2} + y \\ y' = x - y \end{cases}$

For which initial conditions $x(t_0) = x_0$, $x'(t_0) = y_0$ do the following equations have unique solutions? On what intervals do those solutions exist?

23. the mass-spring equation $mx'' + kx = 0$, where $m > 0$ and $k > 0$ are constants

24. the forced mass-spring equation $mx'' + kx = \sin t$, where $m > 0$ and $k > 0$ are constants

25. the Legendre equation $t^2 x'' + tx' + x = 0$

26. $x'' - \dfrac{1}{1 - t^2} x' - 2x = e^t$

27. Show that the initial value problem $x(t_0) = x_0$, $x'(t_0) = y_0$ for the second-order equation $x'' + p(t)x' + q(t)x = h(t)$ has a unique solution on the whole interval $\alpha < t < \beta$ if the coefficients $p(t)$ and $q(t)$ and the nonhomogeneous term $h(t)$ are continuous on the interval $\alpha < t < \beta$.

Consider the following initial value problem:

$$x' = x + 2y$$
$$y' = -x - y$$
$$x(0) = 1, \quad y(0) = 0.$$

28. Use a computer program to approximate the solution at $T = 1$ using Euler's algorithm with step sizes $s = 0.1$, 0.05, 0.025, 0.0125, and 0.00625. Which digits in these approximations do you think are accurate and why?

29. Repeat (a) using Heun's algorithm.

30. Repeat (a) using the Runge-Kutta algorithm.

31. Use a computer program to plot the vector field for this system.

32. Use the Euler algorithm for $0 \le t \le 20$ with step size $s = 0.1$ to study the orbits in the phase plane. What do you conjecture all orbits do as $t \to +\infty$? Repeat using Heun's algorithm and the Runge-Kutta algorithm.

33. Repeat Exercises 28–32 for the initial value problem

$$x' = x + y - 0.05(x^2 + y^2)x$$
$$y' = -x + y - 0.05(x^2 + y^2)y$$
$$x(0) = 0.1, \quad y(0) = 0.0.$$

Consider the following initial value problem:

$$x' = x + 0.5y$$
$$y' = x - y$$
$$x(0) = 1, \quad y(0) = 0.$$

34. Use a computer program to approximate the solution at $T = 1$ using Euler's algorithm with step sizes $s = 0.1$, 0.05, 0.025, 0.0125, and 0.00625. Which digits in these approximations do you think are accurate and why?

35. Repeat (a) using Heun's algorithm.

36. Repeat (a) using the Runge-Kutta algorithm.

37. Use a computer program to plot the vector field for this system of equations.

38. Use the Euler algorithm for $0 \le t \le 20$ with step size $s = 0.1$ to study the orbits in the phase plane. What do you conjecture all orbits do as $t \to +\infty$? Repeat using Heun's algorithm and the Runge-Kutta algorithm.

Which of the following systems are linear and which are nonlinear? If the system is linear, is it homogeneous or nonhomogeneous?

39. $\begin{cases} x' = -3x + y \\ y' = x + 5y \end{cases}$ **40.** $\begin{cases} x' = (\sin t)x + y \\ y' = x + (\cos t)y \end{cases}$

41. $\begin{cases} x' = 2(1 - x) + 3y \\ y' = -x + y^2 \end{cases}$ **42.** $\begin{cases} x' = x(1 - x) - xy \\ y' = -x + xy \end{cases}$

43. $\begin{cases} x' = x - \sin t + y \\ y' = 2x - y \end{cases}$ **44.** $\begin{cases} x' = -1 + y - 3x \\ y' = 1.5x - y + 4.2 \end{cases}$

45. $\begin{cases} x' = rx - 14y \\ y' = x - sy \end{cases}$ r, s are constants

46. $\begin{cases} x' = \pi y - 3.2(x - 2) + e^{-t} \\ y' = k(x + y) + t \end{cases}$ k is a constant

For the linear systems below, find the coefficient matrix A and the nonhomogeneous term $\tilde{h}(t)$. Write the system in matrix form.

47. $\begin{cases} x' = 3(x - y) + 2(x - y) \\ y' = -7 - x - y \end{cases}$

48. $\begin{cases} x' = 2(0.5 - y) + x - 6 \\ y' = 3(2x - y) \end{cases}$

49. $\begin{cases} x' = c(x + 3y) - 4x + t^2 \\ y' = -x + (d - 1)y - t \end{cases}$ c, d are constants

50. $\begin{cases} x' = (k^2 - 1)(x - y) + \sin t \\ y' = x + ky - \cos t \end{cases}$ k is a constant

51. Use the solution pairs

$$x_1 = \cos \beta t \qquad\qquad x_2 = \sin \beta t$$
$$y_1 = -\beta \sin \beta t \qquad\quad y_2 = \beta \cos \beta t$$

of the system

$$x' = y$$
$$y' = -\beta^2 x$$

to solve the following initial value problems. ($\beta > 0$ is a positive constant.)

(a) $x(0) = -1, y(0) = 1$

(b) $x(0) = \beta$, $y(0) = 1$

Given the general solution $x_h = c_1e^{-3t} + c_2e^t$ of the homogeneous second-order equation $x'' + 2x' - 3x = 0$, find the general solution of the following equations. (Hint: Use undetermined coefficients methods to find a particular solution x_p.)

52. $x'' + 2x' - 3x = 6$ **53.** $x'' + 2x' - 3x = -2$

54. $x'' + 2x' - 3x = 2\sin t$

55. $x'' + 2x' - 3x = -e^{-t}$

56. $x'' + 2x' - 3x = kt$, k is a constant

57. $x'' + 2x' - 3x = k\cos t$, k is a constant

4.6   APPLICATIONS

4.6.1   Chemostats

A chemostat is a laboratory apparatus for the culturing of microorganisms. Its importance derives from the role it plays in both applied problems and theoretical studies. It is used, for example, to study the fermentation of commercial products (e.g., genetically altered species) and ecological interactions among several species. The chemostat can be a laboratory model of a waste water treatment plant or even an alpine lake.

A chemostat consists of a vessel into which nutrients required by the microorganism(s) are pumped at a constant rate. In order to keep a constant volume in the vessel, its contents are pumped out at the same rate. The contents are well stirred so that the nutrient, the microorganisms, and all byproducts are uniformly distributed throughout the vessel at all times. The experimentalist can control the nutrient input into the vessel and can assay the contents in the vessel by using the output volume. Usually all nutrients are supplied in abundance except for one which is kept in limited supply (and is referred to as the limiting nutrient or resource). The goal is to determine the population dynamics of the organism and how they depend on the limiting resource.

To begin the model derivation step of the modeling cycle, let $x = x(t)$ denote the *concentration* (amount per unit volume) of the limiting nutrient in the culture vessel and let $y = y(t)$ denote the *concentration* of organisms. The goal is to describe the dynamics of these two concentrations as functions of time. One fundamental question is, Under what circumstances will the microorganism survive indefinitely in the culturing vessel and under what circumstances will it go extinct?

We derive model equations for x and y from the balance law

$$\text{rate of change} = \text{input rate} - \text{output rate.}$$

Consider first the microorganism. Applying the balance law to the *amount* of the organism in the vessel, we have

$$\text{rate of change in the amount of organism} \tag{6.1}$$
$$= \text{population growth rate} - \text{rate population is pumped out.}$$

To derive an equation for the concentration $y = y(t)$ we need mathematical expressions for the three rates in this equation.

Let the volume of the culture vessel be denoted by $V > 0$ so that the amount of microorganism in the vessel is yV. The left-hand side of the balance equation (6.1) is

$$\text{rate of change in the amount of organism} = (y(t)V)'.$$

Next, consider the output rate on the right hand side of (6.1). Suppose the rate at which the volume is being pumped out (and also into) the vessel is denoted by $F > 0$ (units of volume per unit of time). F is called the flow rate. Since the contents of the vessel are well stirred and hence the microorganism is uniformly distributed throughout, the rate at which microorganisms are pumped out of the vessel is

$$\text{rate population is pumped out} = yF.$$

At this point, the balance law (6.1) for the microoganisms

$$(yV)' = \text{population growth rate} - yF. \tag{6.2}$$

What remains is a formula for the population growth rate. This formula is more complicated because it involves the biological and physical processes that the microorganism uses to obtain, consume, and metabolize (and maybe store) the nutrient and to reproduce itself. We will not derive a formula for the population growth rate from detailed physiological and behavioral characteristics of the microorganism. Instead, we will use a mathematical formula that qualitatively describes the population growth rate.

First, what are some basic properties the population growth rate would have? Certainly, at any instant of time this rate would depend on the amount yV of microorganisms present. A simple modeling assumption would be that the *population growth rate is proportional to the population amount Vy*:

$$\text{population growth rate} = pVy.$$

The constant of proportionality p is the *per capita* growth rate.

Is it reasonable to assume the per capita growth rate p constant over time? Not if the amount of nutrient x changes over time, because the reproduction rate of a single microorganism is less if the nutrient is scarce and greater if the nutrient is abundant. Therefore, we assume that $p = p(x)$ is a function of x, specifically, an *increasing function of x* (i.e., $dp/dx > 0$). Furthermore, if no nutrient is available, then there is no population growth at all (recall that the nutrient is necessary for growth). This means $p(0) = 0$.

There is one more mathematical requirement we need to place on the per capita growth rate $p(x)$. This requirement has to do with $p(x)$ for large values of the nutrient concentration x. Biologically, there is a nutrient level above which an individual organism does not need (or cannot not utilize) any more nutrient. Thus, the mathematical formula for $p(x)$ must not increase without bound (i.e., its graph should approach a horizontal asymptote as $x \to +\infty$).

There are many mathematical functions with the three properties we require for the per capita growth rate $p(x)$, that is,

$$p(0) = 0, \qquad \frac{dp(x)}{dx} > 0, \qquad \lim_{x \to +\infty} p(x) = m.$$

A commonly used function with these properties is the *Michaelis-Menten* expression

$$p(x) = m\frac{x}{a+x} \tag{6.3}$$

$$m > 0, \quad a > 0.$$

(This expression is also called a *Monod* or *Holling II* functional response or uptake rate.) The parameter m is the maximal possible per capita growth rate (the saturation level). Since $p(a) = m/2$, the coefficient a is called the *half-saturation coefficient* (or the Michaelis-Menten constant).

We return now to the balance law (6.2) and use the Michaelis-Menten expression (6.3) for p. We obtain the equation

$$(yV)' = \frac{mx}{a+x}Vy - yF.$$

Let us simplify this equation a little. Since the volume V of the culturing vessel is held constant, we can divide both sides of this equation by V and obtain

$$y' = \left(m\frac{x}{a+x} - d\right)y.$$

Here we have introduced the simplifying notation

$$d \doteq \frac{F}{V}$$

which is called the *dilution* or *washout* rate of the chemostat (d has units of 1/time).

To complete the model for the chemostat we need a differential equation for the concentration $x = x(t)$ of the nutrient. The balance law of the amount of nutrient in the vessel is

rate of change in the amount of nutrient $=$ rate amount of nutrient is pumped in

$\qquad\qquad\qquad\qquad$ $-$ rate amount is pumped out $\qquad$ (6.4)

$\qquad\qquad\qquad\qquad$ $-$ rate nutrient is consumed by organism.

We need mathematical expressions for the four rates in this balance equation.

First, since x is a concentration (amount per unit volume) and V is the total volume of the vessel, the amount of nutrient in the vessel at any time is Vx. The rate of change of nutrient in the vessel is

rate of change in the amount of nutrient $= (Vx(t))'$.

Second, if the nutrient concentration being pumped into the vessel is denoted by x_{in}, then the rate at which the *amount* of nutrient is pumped in is

rate amount of nutrient is pumped in $= x_{\text{in}}F$.

Since the contents in the vessel are well stirred (and hence the nutrient uniformly distributed),

rate amount is pumped out $= x(t)F$.

Finally, the growth rate of the microorganism population is $p(x)Vy$. We assume that this is proportional to the rate at which the nutrient is consumed, that is,

rate nutrient is consumed by organism $= \frac{1}{\gamma}p(x)Vy$.

The constant γ is a conversion factor of nutrient into microorganism. (Specifically, γ is the amount of microorganism formed per unit nutrient consumed. It has no dimensions.)

For example, using the Michaelis-Menten uptake rate (6.3) for $p(x)$ we have

$$\text{rate nutrient is consumed by organism} = \frac{1}{\gamma} m \frac{x}{a+x} V y.$$

We now have mathematical expressions for all four rates in the nutrient balance law (6.4), which becomes

$$(Vx(t))' = Fx_{\text{in}} - \left(Fx(t) + \frac{1}{\gamma} \frac{mx}{a+x} V y \right)$$

or, after dividing both sides by V,

$$x' = (x_{\text{in}} - x)d - \frac{1}{\gamma} \frac{mx}{a+x} y.$$

In summary, our modeling assumptions about the chemostat lead to the equations

$$x' = (x_{\text{in}} - x)d - \frac{1}{\gamma} \frac{mx}{a+x} y \tag{6.5}$$

$$y' = \left(\frac{mx}{a+x} - d \right) y$$

for the concentrations x and y of the limiting nutrient and the microorganism in the vessel.

Equations (6.5) constitute a first-order, nonlinear plane autonomous system. Notice the system has five (positive) parameters a, d, m, γ, and x_{in}, all of which, it turns out, experimentalists can measure in the laboratory.

The fundamental question raised earlier—namely, under what circumstances will the microorganism survive indefinitely in the culturing vessel and under what circumstances will it go extinct—becomes a question about the component $y(t)$ of the solution of the initial value problem

$$x(0) = x_0 \geq 0, \quad y(0) = y_0 > 0.$$

For what values of the five model parameters does $y(t)$ equal 0 at some time $t > 0$ or tend to 0 as $t \to +\infty$? We will study this question in detail in Chapter 8, where we will learn enough mathematical tools to carry out a general analysis. Here, however, we will explore the answer to this question for some example problems by using the approximation techniques discussed in Sec. 4.2. Thus, the solution step of the modeling cycle in this application will involve only the numerical approximation of solutions to selected initial value problems.

Suppose a strain of the bacterium E. coli is placed in a chemostat of volume $V = 200$ milligrams with a washout rate of

$$d = 0.075 \text{ (per hour)}.$$

The inflow rate of a nutrient concentration is set to be

$$x_{\text{in}} = 0.005 \text{ (milligrams per liter)}.$$

Suppose the Michaelis-Menten parameters and the conversion factor for this strain have been estimated to be

$$m = 0.68 \text{ (per hour)}$$
$$a = 0.0016 \text{ (milligrams per liter)}$$
$$\gamma = 63 \text{ ($\times 10^6$ cells per milligram)}.$$

The concentration y of *E. Coli* is measured in units of 10^6 cells per liter and that of the nutrient x in milligrams per liter. We can investigate solutions of the chemostat equations (6.5) with these parameter values using a computer.

Figure 4.11(a) shows typical graphs of selected orbits in the first quadrant of the x, y phase plane. Figure 4.11(b) shows graphs of the individual concentrations x and y plotted against time t. These numerically calculated graphs suggest, for the selected parameter values, that the population of *E. Coli* survives indefinitely in the chemostat (eventually attaining and remaining at a concentration of approximately 0.3×10^6 cells per milligram).

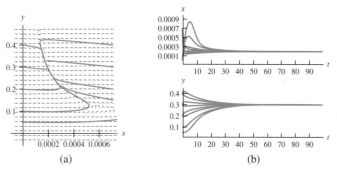

(a)   (b)

Figure 4.11 (a) Selected orbits of the chemostat model (6.5) with parameter values $d = 0.075$, $x_{in} = 0.005$, $a = 0.0016$, $\gamma = 63$, and $m = 0.68$ are shown embedded in the system's slope field. This graph suggests that all orbits (in the positive quadrant) approach a point located at approximately $(x, y) = (0.0002, 0.3)$. The time series plots of x and y along these orbits are shown in (b).

For other values of the parameters in the chemostat equations (6.5) the phase plane portrait can change and the prognosis for the survival of the *E. Coli* culture can be quite different.

For example, consider a different strain of *E. Coli* whose parameter values are identical to those in Fig. 4.11 except that this strain has a different maximal growth rate, namely $m = 0.01$ (per hour). Figure 4.12 shows the new phase plane portrait and orbits that results from this change in m. These graphs indicate that all orbits still approach a point. In this case, however, the limit point lies on the x-axis. A careful examination of the numerical solutions shows that the limit point is $(x, y) = (0.005, 0)$. This point corresponds to the absence of the *E. Coli* culture ($y = 0$) and to the input nutrient concentration $x_{in} = 0.005$. We conclude in this case that this strain of *E. Coli.* goes extinct.

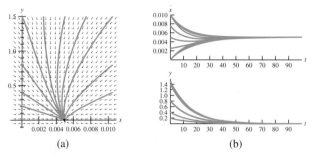

(a)   (b)

Figure 4.12 (a) Selected orbits of the chemostat model (6.5) with parameter values $d = 0.075$, $x_{\text{in}} = 0.005$, $a = 0.0016$, $\gamma = 63$, and $m = 0.01$ are shown embedded in the system's slope field. This graph suggests that all orbits (in the positive quadrant) approach a point located at approximately $(x, y) = (0.005, 0)$. The time series plots of x and y along these orbits are shown in (b).

The two examples shown in Figs. 4.11 and 4.12 are typical for the chemostat model (6.5). That is, typically orbits approach an equilibrium point as $t \to +\infty$. Whether or not the equilibrium lies in the first quadrant and the microorganism survives depends on the values assigned to the model parameters.

4.6.2 Objects in motion

In Chapter 3, Sec. 3.6 we considered the motion of objects along a straight line, as described by Newton's law $F = ma$. For objects not constrained to move in a straight line, Newton's law takes the vector form $\widetilde{F} = m\widetilde{a}$, where $\widetilde{F}$ is the vector of all forces acting on the object and $\widetilde{a} = d\widetilde{v}/dt$ is its acceleration ($\widetilde{v}$ is the velocity vector). In this section we consider motion of objects constrained to remain in a plane, moving near the surface of the earth subject to the forces of gravity and air resistance.

Denote the force acting on an object by the vector

$$\widetilde{F} = \begin{pmatrix} F_x \\ F_y \end{pmatrix}$$

whose components F_x and F_y are measured with respect to a rectangular (x, y)-coordinate system (Fig. 4.13). The position of the object as a function of time t is denoted by the vector

$$\begin{pmatrix} x(t) \\ y(t) \end{pmatrix}$$

and the velocity and acceleration vectors are

$$\text{velocity} = \widetilde{v} = \begin{pmatrix} x'(t) \\ y'(t) \end{pmatrix} = \begin{pmatrix} u(t) \\ v(t) \end{pmatrix}$$

$$\text{acceleration} = \widetilde{a} = \begin{pmatrix} u'(t) \\ v'(t) \end{pmatrix} = \begin{pmatrix} x''(t) \\ y''(t) \end{pmatrix}.$$

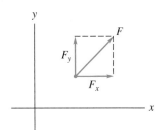

Figure 4.13

Newton's law

$$\begin{pmatrix} F_x \\ F_y \end{pmatrix} = m \begin{pmatrix} u'(t) \\ v'(t) \end{pmatrix}$$

leads to the system of equations

$$mu' = F_x \tag{6.6}$$
$$mv' = F_y.$$

Initial conditions

$$u'(0) = u_0, \qquad v'(0) = v_0 \tag{6.7}$$

are determined by the object's velocity at $t = 0$. Once we specify mathematical expressions for the component forces F_x and F_y in both coordinate directions, the system of differential equations (6.6) for the two velocity components $u = u(t)$ and $v = v(t)$ are the equations of motion of the object.

The Flight of a Golf Ball   Consider the motion of a golf ball, struck by a golfer in such a way that it remains in a vertical plane during its flight. (As golfers know, golf balls do not often remain in a plane but can hook or slice. A more realistic model would have to be three dimensional.) Our goal is to determine the flight path of the ball from a knowledge of the initial velocity and flight angle imparted to the ball by the golfer's swing. Several questions are of interest. For example, how far does the ball travel (when it first hits the ground)? How does the golfer maximize this distance? What is the flight time of the ball? What height does it attain?

Place the origin of a rectangular coordinate system at the point on the ground where the ball is struck. Locate the x-axis in the direction of motion and the y-axis vertically upward, determining the flight plane of the ball. To obtain the equations of motion (6.6) we must specify the two component forces F_x and F_y acting on the golf ball. We consider two cases: the case when only the force of gravity acts on the ball

$$\widetilde{F} = \widetilde{F}_g$$

and the case when both gravity and a frictional force due to air resistance acts on the ball

$$\widetilde{F} = \widetilde{F}_g + \widetilde{F}_f.$$

Here $\widetilde{F}_g$ denotes the force of gravity and $\widetilde{F}_f$ the frictional force due to air resistance. Near the surface of the earth it is accurate to assume the force of gravity is constant and acts vertically downward. Thus,

$$\widetilde{F}_g = \begin{pmatrix} 0 \\ -mg \end{pmatrix},$$

where m is the mass of the ball and g is the acceleration due to gravity (approximately 9.8 meters/s^2 or 32 feet/s^2).

In the first case, when air resistance is ignored, we have from $\widetilde{F}_g = m\widetilde{a}$ that

$$\begin{pmatrix} 0 \\ -mg \end{pmatrix} = m \begin{pmatrix} u'(t) \\ v'(t) \end{pmatrix}.$$

We have completed the model derivation step of the modeling cycle and arrived at the system of differential equations

$$u' = 0$$
$$v' = -g$$

for the velocity components u and v. We now turn to the solution step of the modeling cycle and find solution formulas for u and v, which we will in turn use to determine formulas for the position coordinates x and y.

Antidifferentiating and using the initial conditions (6.7), we obtain

$$u(t) = u_0, \qquad v(t) = -gt + v_0.$$

To calculate the flight path of the ball, we antidifferentiate $x' = u(t)$, $y' = v(t)$, and find

$$x = u_0 t, \qquad y = -\frac{1}{2}gt^2 + v_0 t$$

for the ball's position coordinates at time t. Here we used the fact that the origin of the coordinate system lies at the initial position of the ball so that

$$x(0) = 0, \qquad y(0) = 0. \tag{6.8}$$

Since $t = x/u_0$, we obtain the alternative description

$$y = -\left(\frac{1}{2}\frac{g}{u_0^2}\right)x^2 + \left(\frac{v_0}{u_0}\right)x \tag{6.9}$$

of the flight path. We recognize this to be the equation of a parabola in the (x, y)-plane. We conclude that, in the case of negligible frictional forces, the golf ball will travel along a parabolic arc.

From the formulas above we can determine the distance (measured along the ground) traveled by the golf ball when it first hits the ground,[5] the flight time, and how these depend on how the ball is hit by the golfer. See Exercises 1–5.

It turns out, in the case of no friction just considered, that the distance travelled (along the ground) is maximized by hitting the ball at an initial angle of $\pi/4$ radians (45°). See Exercise 4. However, a careful observation of a good golfer's swing will show that they do not strike the ball at this angle. Indeed, the initial angle they give the ball is usually less than 45°. Why is this?

---

[5]That is, we do not consider any bounces of the ball or with the ball rolling along the ground before it comes to rest.

To obtain a more accurate model of the ball's motion, and hopefully account for this difference between theory and observation, we follow the model modification step of the modeling cycle. We return to the derivation of the model equations and include other forces, neglected above, that act on the golf ball. A more accurate model would include frictional forces due to air resistance. To incorporate frictional forces into the model we need a mathematical expression for $\widetilde{F}_f$.

First, we observe that the frictional force $\widetilde{F}_f$ depends on the ball's velocity (and not, in a uniform atmosphere at least, on its position). This follows from, if nothing else, the observation that there is no air resistance when the ball is at rest and that there is greater air resistance at higher speeds. Thus, we write $\widetilde{F}_f = \widetilde{F}_f(u, v)$ and $\widetilde{F}_f(0, 0) = \widetilde{0}$, where the components of $\widetilde{F}_f$ are increasing functions of the velocity components u and v. Beyond these basic requirements for the modeling of a frictional force, exactly how the force mathematically depends on u and v can be complicated.

In general, the force of friction can depend on many factors, including the size and shape of the object (golf balls are not smooth spheres), the properties of the medium (humidity, air density, etc.), the speeds attained by the object during its flight, and so on. We will consider what is perhaps the simplest modeling assumption that meets the requirements set out above, namely, that each component of the frictional force is proportional to the velocity component. This linear friction law is written mathematically as

$$\widetilde{F}_f = -c \begin{pmatrix} u \\ v \end{pmatrix}, \tag{6.10}$$

where the constant of proportionality $c > 0$ is the called the coefficient of friction. The minus sign appears because the frictional forces at in the opposite direction of the ball's motion. In the following section we consider a different (nonlinear) friction law.

If the frictional force (6.10) is added to the constant force of gravity acting on the golf ball, so that

$$\widetilde{F} = \widetilde{F}_g + \widetilde{F}_f = \begin{pmatrix} -cu \\ -mg - cv \end{pmatrix},$$

we obtain a new system of model equations from Newton's law (6.6), namely,

$$u' = -\frac{c}{m}u \tag{6.11}$$

$$v' = -g - \frac{c}{m}v.$$

This is a linear, nonhomogeneous system for the velocity components u and v. A unique solution is determined by the initial conditions

$$u(0) = u_0, \qquad v(0) = v_0, \tag{6.12}$$

where u_0 and v_0 are the initial velocity components imparted to the ball by the golfer. If we write

$$u_0 = s_0 \cos\theta, \qquad v_0 = s_0 \sin\theta, \qquad 0 < \theta < \frac{\pi}{2},$$

then the initial conditions

$$u(0) = s_0 \cos\theta, \qquad v(0) = s_0 \sin\theta$$

are described by the initial speed

$$s_0 = \left(u_0^2 + v_0^2 \right)^{1/2}$$

and the initial flight angle θ given to the ball by the golfer's swing.

A typical golf ball weighs about $w = 1.62$ oz. or $w = 1.62/16$ lb. Since $g = 32$ ft/s², the ball's mass is $m = 1.62/512 \approx 3.16 \times 10^{-3}$ slug (note: $w = mg$). The coefficient of friction for a golf ball is approximately $c = 7.83 \times 10^{-4}$ lb/ft/s.[6] With these numerical values for m and c, the initial value problem for the flight of the golf ball becomes

$$\begin{aligned} u' &= -0.2478u \\ v' &= -32 - 0.2478v \\ u(0) &= s_0 \cos\theta, \qquad v(0) = s_0 \sin\theta. \end{aligned} \tag{6.13}$$

We can use a computer to explore the solutions of this initial value problem once we assign numerical values to the initial speed s_0 and flight angle θ_0 of the ball. A numerical algorithm (for example, Heun's algorithm) will produce a table of approximations to the velocity components u and v at equally spaced times t_i determined by the choice of step size s. From this table of values we can use a computer to approximate the integrals

$$x(t) = \int_0^{t_i} u(s)\, ds, \qquad y(t) = \int_0^{t_i} v(s)\, ds \tag{6.14}$$

by means of a numerical integration scheme from calculus (for example, the trapezoid rule). The result will be a table of approximations to the positions x and y of the ball at the times t_i. These approximations, when plotted in an (x, y)-plane (and connected by straight lines), provide an approximation to the flight path of the ball.[7] Figure 4.14 shows solution graphs constructed in this way for three selected values of the initial speed s_0 and flight angle θ_0 of the ball.

In Fig. 4.14 notice that the flight paths are nearly, but not exactly, parabolic. If the ground is level, then the ball strikes the ground when the graph crosses the x-axis (i.e., when $y = 0$). It is clear from these examples that the maximal distance traveled by the ball (along the ground) is not attained by hitting the ball at an angle of $\pi/4$ radians (45°), as it is for the frictionless model. The optimal angles, in these examples, are in fact all less than $\pi/4$ radians.

Linear systems are studied in Chapters 5 and 6, where, among other things, methods are developed for calculating solution formulas. However, we can obtain solution formulas for the linear system (6.11) by using methods from Chapter 2 for single linear equations. This is because the equations in this system are uncoupled from one another: The first is a linear homogeneous equation for u and the second is a linear nonhomogeneous equation for v. We could use such formulas, together with formulas for x and y obtained from (6.14), to study properties of the ball's flight path. See Exercises 6–15.

---

[6] Herman Erlichson, "Maximum Projectile Range with Drag and Lift, with Particular Application to Golf," *American Journal of Physics* 51 (1983), pp. 357–362, reprinted in *The Physics of Sports* (Angelo Armenti, Jr., editor), American Institute of Physics, New York, 1993.

[7] An alternative procedure is to numerically solve the system of four differential equations consisting of (6.13) and $x' = u$, $y' = v$ with the added initial conditions $x(0) = 0$, $y(0) = 0$.

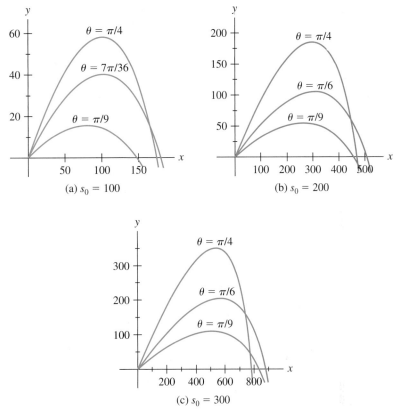

Figure 4.14 (a) Solutions of the initial value problem (6.13) for three different initial flight angles θ at initial speed 100 (ft/s) give rise (through the integrals (6.14)) to near parabolic flight paths. An angle of $\pi/4$ radians (45°) does not result in the farthest hit ball. The same conclusion results from the graphs in (b) and (c) in which the initial speeds are 200 and 300 ft/s.

How Far Can a Baseball Be Hit?   The mathematical description of the flight of a batted baseball would seemingly be the same as that of the golf ball discussed in the preceding section, with appropriate changes in the model parameter values. However, a baseball is larger than a golf ball and a quadratic friction law is usually more accurate than a linear friction law.[8] In this section we use a quadratic friction law to study the flight of a baseball. The resulting equations are nonlinear, and we cannot find solution formulas. We will rely on numerical integrations of the differential equations (using a computer) to draw conclusions about the ball's flight.

We assume the batted baseball remains in a horizontal plane during its flight and subject to a downward, constant force due to gravity:

$$\widetilde{F}_g = \begin{pmatrix} 0 \\ -mg \end{pmatrix}.$$

Let the position coordinates x, y of the ball be measured relative to a Cartesian coordi-

---

[8]Peter J. Brancazio, "Looking into Chapman's Homer: The Physics of Judging a Fly Ball," *American Journal of Physics* 53 (1985), pp. 849–855, reprinted in *The Physics of Sports* (Angelo Armenti, Jr., editor), American Institute of Physics, 1993.

nate system whose origin lies on the ground directly beneath the point where the ball is struck. Thus, the initial coordinates of the ball are

$$\begin{pmatrix} x(0) \\ y(0) \end{pmatrix} = \begin{pmatrix} 0 \\ y_0 \end{pmatrix},$$

where $y_0 \geq 0$ is the height at which the ball is struck. (Unlike a golf ball, a baseball is not typically struck on the ground.)

We assume that the remaining force on the ball $\widetilde{F}_f$ is due to air resistance. Under the quadratic friction law the force vector $\widetilde{F}_f$ points in a direction opposite to that the velocity vector

$$\begin{pmatrix} u \\ v \end{pmatrix} = \begin{pmatrix} x' \\ y' \end{pmatrix}$$

and has a magnitude proportional to the square of the ball's speed, that is, is proportional to

$$s^2 = u^2 + v^2.$$

Since

$$-\frac{1}{s} \begin{pmatrix} u \\ v \end{pmatrix}$$

is a unit vector pointing in the opposite direction of the velocity vector, we have

$$\widetilde{F}_f = -cs^2 \frac{1}{s} \begin{pmatrix} u \\ v \end{pmatrix} = \begin{pmatrix} -csu \\ -csv \end{pmatrix},$$

where $c > 0$ is the constant of proportionality (a nonlinear coefficient of friction).

The total force $\widetilde{F} = \widetilde{F}_g + \widetilde{F}_f$ acting on the base ball is

$$\widetilde{F} = \begin{pmatrix} -csu \\ -mg - csv \end{pmatrix}$$

and Newton's law $\widetilde{F} = m\widetilde{a}$ yields

$$m \begin{pmatrix} u' \\ v' \end{pmatrix} = \begin{pmatrix} -csu \\ -mg - csv \end{pmatrix}.$$

This results in the nonlinear system

$$u' = -\frac{c}{m} u \sqrt{u^2 + v^2} \tag{6.15}$$

$$v' = -g - \frac{c}{m} v \sqrt{u^2 + v^2}$$

of first-order differential equations for the velocity vector components u and v. The initial conditions are the same as for the golf ball in the preceding section, namely,

$$u(0) = s_0 \cos \theta, \qquad v(0) = s_0 \sin \theta, \tag{6.16}$$

where $s_0 > 0$ is the initial speed of the batted ball and θ ($0 < \theta < \pi/2$) is its initial angle of flight.

From the solution of the initial value problem (6.15)–(6.16) we can obtain the position coordinates $x = x(t)$ and $y = y(t)$ from[9]

$$x(t) = \int_0^t u(s)\,ds, \quad y(t) = y_0 + \int_0^t v(s)\,ds.$$

For a typical baseball the coefficient of friction is approximately $c = 4.0723 \times 10^{-6}$ (slugs/ft/s). A baseball weighs about $w = 5$ oz. (or 5/16 lb) so that $m = w/32 = 9.7656 \times 10^{-3}$ (slugs). Using these values for c and m, we obtain the equations of motion

$$u' = -4.17 \times 10^{-4} u \sqrt{u^2 + v^2} \tag{6.17}$$

$$v' = -32 - 4.17 \times 10^{-4} v \sqrt{u^2 + v^2}$$

for the velocity components of the baseball. We can determine the flight path of a batted baseball by numerically solving the nonlinear system (6.17) once the initial speed s_0 and flight angle θ in the initial conditions (6.16) are specified.

One question of interest is how far a batted baseball is hit. This distance is a function of how hard the ball is struck (i.e., the initial velocity s_0) as well as the angle θ and height y_0 at which it is struck.

Here is an example. Suppose the ball is hit $y_0 = 3$ feet above the ground and the ball leaves the bat at $s_0 = 125$ ft/s (≈ 85 mph). At what angle θ will the ball travel farthest? The model derivation step has led us to the initial value problem

$$u' = -4.17 \times 10^{-4} u \sqrt{u^2 + v^2}$$

$$v' = -32 - 4.17 \times 10^{-4} v \sqrt{u^2 + v^2} \tag{6.18}$$

$$u(0) = 125 \cos \theta, \quad v(0) = 125 \sin \theta$$

and the equations

$$x(t) = \int_0^t u(s)\,ds, \quad y(t) = 3 + \int_0^t v(s)\,ds \tag{6.19}$$

for the flight path of the ball. For specified initial flight angles θ we can use a computer to approximate the solutions of these equations numerically and graph the results.

Figure 4.15 shows graphs of some sample flight paths of the ball for selected initial angles θ. These graphs suggest that the optimal angle is about $\pi/4$ radians (45°) and that for this angle the ball travels approximately 425 ft.

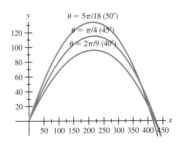

Figure 4.15 The flight paths of a baseball batted at three different angles are shown. Plots were obtained by numerical integration of the initial value problem (6.18) and (6.19).

---

[9]An alternative procedure is to solve numerically the initial value problem consisting of (6.15)–(6.16) together with $x' = u$, $y' = v$ and the added initial conditions $x(0) = 0$, $y(0) = y_0$.

Exercises

Consider the parabolic flight a golf ball, under the assumption of no frictional forces, from the time it is struck by the golfer until it first hits the ground. Assume that the ground is level and described by $y = 0$.

1. Determine the ball's flight time.

2. Determine the distance traveled by the ball.

3. How does the distance traveled by the ball depend on the initial speed $s_0 = \sqrt{u_0^2 + v_0^2}$ of the ball?

4. For a given initial speed s_0 at what angle (from ground level) should the ball be hit in order to maximize the distance traveled?

5. Determine the maximum altitude attained by the ball.

Consider the initial value problem (6.11)–(6.12).

6. Find solution formulas for u and v.

7. Find solution formulas for x and y.

8. Use the answer in Exercise 7 to find an equation $y = f(x)$ for the flight path.

Consider the initial value problem (6.13) for the flight of a golf ball struck with an initial speed of $s_0 = 200$ ft/s.

9. Find solution formulas for the velocity components u and v of the ball.

10. Find solution formulas for the position components x and y of the ball.

11. Use the answer in Exercise 10 to find an equation $y = f(x)$ for the flight path.

12. Calculate the angle θ at which the distance (along the ground) traveled by the ball is maximal. [*Hint*: Use the formula $y = f(x)$ for the flight path found in Exercise 11, set $f(x) = 0$ to obtain an equation for the distance $x = x(\theta)$ as a function of the angel θ, and implicitly differentiate the equation with respect to θ, setting $dx/d\theta = 0$. Solve the resulting equation for $x = x(\theta)$ as a function of the optimal angle θ and substitute the answer into $f(x) = 0$

to obtain an equation for θ. Use a computer to solve this equation for θ.]

13. What is the maximal distance (i.e., the distance obtained when the optimal angle found in Exercise 12 is used)?

14. What is the flight time of the ball when the optimal angle found in Exercise 12 is used?

15. **(a)** Using a typical golf ball weight of $w = 1.62$ oz (or $w = 1.62/16$ lb) and coefficient of friction $c = 7.83 \times 10^{-4}$ lb/ft/s, calculate the optimal angles θ with which to strike the ball so as to maximize its distance traveled. Do this for initial speeds ranging from $s_0 = 100$ ft/s to 300 ft/s at 25 ft/s intervals. Use $g = 32$ ft/s^2.

  (b) Use your answers to calculate the ball's flight time when struck at the optimal angles.

Suppose a head wind, blowing parallel to the level ground surface, exerts a force of magnitude $h > 0$ on a baseball (in addition to gravity and a linear frictional force).

16. Write the equations for the velocity components u and v.

17. Find solution formulas for u and v.

18. Use your answer from Exercise 17 to find formulas for the x and y coordinates of the ball as functions of time t.

19. Let $m = 1.62/512$ slug, $c = 7.83 \times 10^{-4}$ lb/ft/s, and $s_0 = 200$ ft/s. Suppose $h = 0.1$ lb. Use a computer to draw flight paths at selected initial angles θ (as in Fig. 4.14). By experimenting, approximate the optimal angle θ that maximizes the horizontal distance x traveled by the ball. How does this angle and distance differ from those calculated in the text (when no head wind is present)?

20. What is the optimal angle θ and the maximal distance if, in the case studied in the text, the baseball is hit with an initial speed of $s_0 = 100$ ft/s?

21. Using the equations (6.17) with an initial angle of $\theta = 45°$ and an initial speed of $s_0 = 100$ (ft/s) determine, by means of computer explorations, whether the ball will travel farther if the batted pitch is high or low.

CHAPTER

5

Homogeneous Linear Systems and Higher-Order Equations

In Chapter 4 (Sec. 4.3) we learned that the general solution of a linear system of differential equations is additively decomposable into the general solution of the associated homogeneous system and a particular solution of the nonhomogeneous system. As a special case, this is also true of higher-order linear equations. Homogeneous systems are therefore fundamental to the study of general linear systems, and the present chapter is devoted to their study. We consider nonhomogeneous systems in Chapter 6.

Unlike for a single linear equation, there is no general solution procedure applicable to all linear systems of two or more equations (homogeneous or nonhomogeneous). There is, however, a procedure available for an important special category of homogeneous systems, namely, the class of systems with constant coefficients (called *autonomous* systems). In this chapter we learn how to find solution formulas for autonomous, homogeneous linear systems and how to classify the phase portraits associated with these systems. As a special case the procedure applies to higher-order equations with constant coefficients, although shortcut methods are also available for these equations. The focus is on systems consisting of only two differential equations. However, the solution methods used are applicable to systems consisting of more than two equations.

5.1 Introduction

The general solution of the nonhomogeneous system

$$x' = ax + by + h_1(t) \qquad (1.1)$$
$$y' = cx + dy + h_2(t),$$

where the coefficients a, b, c, and d are constants, has the form

$$x = x_h(t) + x_p(t)$$
$$y = y_h(t) + y_p(t),$$

where

$$x = x_p(t), \qquad y = y_p(t)$$

is a particular solution pair of the system and

$$x = x_h(t), \qquad y = y_h(t)$$

is the general solution of the associated homogeneous system

$$x' = ax + by \tag{1.2}$$
$$y' = cx + dy.$$

See Theorem 3 in Sec. 4.3 of Chapter 4. Because the coefficients in this homogeneous system are constants, the system is *autonomous*. By Theorem 6 in Sec. 4.3 of Chapter 4 the general solution of the homogeneous system (1.2) is a linear combination

$$x_h = c_1 x_1(t) + c_2 x_2(t)$$
$$y_h = c_1 y_1(t) + c_2 y_2(t)$$

of any two independent solution pairs

$$x = x_1(t), \qquad x = x_2(t)$$
$$y = y_1(t), \qquad y = y_2(t).$$

We begin with the problem of finding two independent solution pairs for an autonomous homogeneous system (1.2).

The system

$$x' = -2x + 2y \tag{1.3}$$
$$y' = 2x - 5y$$

arose in the pesticide application discussed in Example 2 of the Introduction. Figure 5.1 shows computer calculated graphs of selected solution pairs. These graphs suggest that system (1.3) has (at least some) exponential solutions. Indeed, we saw in Chapter 1 that

$$x_1 = 2e^{-t} \tag{1.4}$$
$$y_1 = e^{-t}$$

is a solution pair of (1.3). The plots of another solution pair appears in Fig. 5.2. The solution components in this case, however, are not exponential functions. Apparently not all solutions of linear systems are simple exponential functions. [This fact is also evident from the trigonometric solution pairs of the simple harmonic oscillator (1.2) in Chapter 1.]

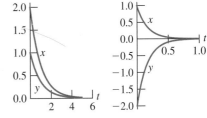

Figure 5.1 Two exponential solution pairs of system (1.3).

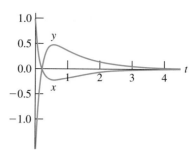

Figure 5.2

Nonetheless, we begin by developing a method to calculate exponential solution pairs for a linear system. As it turns out, the method we develop will also find solutions that are not exponential functions.

To motivate the method we return to (1.3) and attempt to find an exponential solution pair that is independent from the pair (1.4). That is, we look for a solution of the form

$$x = v_1 e^{\lambda t} \tag{1.5}$$
$$y = v_2 e^{\lambda t}.$$

Our task is to choose constants v_1, v_2, and λ in such a way that these expressions solve both equations in the system (1.3). Since we are searching for a solution independent of the pair (1.4) we must find a *nontrivial* solution pair. That is, v_1 *and* v_2 *must not both equal to* 0.

To determine v_1, v_2, and λ we substitute (1.5) into both equations (1.3) and require the left-hand sides to identically equal the right-hand sides by an appropriate choice of the constants v_1, v_2, and λ. This substitution leads to the equations

$$\lambda v_1 e^{\lambda t} = -2v_1 e^{\lambda t} + 2v_2 e^{\lambda t}$$
$$\lambda v_2 e^{\lambda t} = 2v_1 e^{\lambda t} - 5v_2 e^{\lambda t}$$

or upon rearrangement

$$[(-2 - \lambda) v_1 + 2v_2] e^{\lambda t} = 0$$
$$[2v_1 + (-5 - \lambda) v_2] e^{\lambda t} = 0.$$

Since exponential functions are never equal to 0 these identities are satisfied if and only if the terms in both parentheses are equal to 0, that is, if and only if

$$(-2 - \lambda) v_1 + 2v_2 = 0 \tag{1.6}$$
$$2v_1 + (-5 - \lambda) v_2 = 0.$$

These are two linear (algebraic) equations for the two unknowns v_1 and v_2 that have the "trivial" solution $v_1 = 0$, $v_2 = 0$ (the algebraic system is homogeneous). However, we require constants v_1 and v_2 not both equal to 0. Such solutions of (1.6) exist if and only if the two equations are dependent (multiples of each other), which occurs if and only if the determinant of coefficients satisfies

$$\det \begin{pmatrix} -2 - \lambda & 2 \\ 2 & -5 - \lambda \end{pmatrix} = 0$$

or, in other words,

$$\lambda^2 + 7\lambda + 6 = 0. \tag{1.7}$$

By choosing λ as a root of this quadratic equation, we will be able to solve the equations (1.6) for solutions v_1, v_2 not both equal to 0. We then use the resulting numbers λ and v_1, v_2 to obtain a nontrivial solution pair from (1.5).

Since the quadratic (1.7) has two roots

$$\lambda = -1 \quad \text{and} \quad -6$$

we can repeat this procedure twice to obtain two exponential solution pairs. Specifically, for $\lambda = -1$ the algebraic equations (1.6) become

$$-v_1 + 2v_2 = 0$$
$$2v_1 - 4v_2 = 0.$$

Notice that the second equation is a multiple of the first (namely, -2 times the first). Thus, the second equation is satisfied by any solution of the first equation

$$-v_1 + 2v_2 = 0$$

and need not be considered. (This is a useful shortcut to remember!) This equation has infinitely many solutions. For example, we may choose v_2 to be any number (except $v_2 = 0$) and then determine v_1 from the equation. For example, if we choose $v_2 = 1$, then $v_1 = 2$. This choice results in the solution pair (1.4). Other choices for v_2 lead to a multiple of this solution pair and consequently dependent solution pairs. To find an independent solution pair we consider the second root of the quadratic (1.7).

For $\lambda = -6$ the equations (1.6) reduce to

$$4v_1 + 2v_2 = 0$$
$$2v_1 + v_2 = 0.$$

Once again, the second equation is satisfied by any solution of the first equation (since it is a multiple of the first equation, namely, $1/2$ times the first equation). The first equation alone has infinitely many solutions v_1 and v_2. We may, for example, choose v_1 to be any nonzero number and then determine v_2 from the equation. For example, if we choose $v_1 = 1$, then $v_2 = -2$. This choice results in the solution pair

$$x_2 = e^{-6t} \tag{1.8}$$
$$y_2 = -2e^{-6t}.$$

Since the exponential functions e^{-t} and e^{-6t} are not multiples of each other, this solution pair is independent of the solution pair (1.4).

By Theorem 6 of Chapter 4 the general solution of the system (1.3) is a linear combination of the independent solution pairs (1.4) and (1.8), namely

$$x = 2c_1e^{-t} + c_2e^{-6t} \tag{1.9}$$
$$y = c_1e^{-t} - 2c_2e^{-6t}.$$

Is it always possible to find exponential solutions (1.5) for an autonomous homogeneous linear system and use them to construct the general solution? We investigate this question in the next section.

Exercises

Consider the nonhomogeneous system

$$x' = 4x - 2y - 5e^t$$
$$y' = 7x - 5y - 13e^t.$$

1. Show that

$$x_1 = e^{2t} \qquad \text{and} \qquad x_2 = 2e^{-3t}$$
$$y_1 = e^{2t} \qquad\qquad\qquad y_2 = 7e^{-3t}$$

are both solution pairs of the homogeneous system

$$x' = 4x - 2y$$
$$y' = 7x - 5y.$$

2. Find a formula for the general solution of the homogeneous system in Exercise 1.

3. Show that $x_p = e^t$, $y_p = -e^t$ is a solution pair of the nonhomogeneous system.

4. Find a formula for the general solution of the nonhomogeneous system.

5. Find a formula for the solution of the initial value problem $x(0) = 1$, $y(0) = 1$ for the nonhomogeneous system.

Consider the nonhomogeneous system

$$x' = -2x + y + 2$$
$$y' = x - 2y - 1.$$

6. Show that

$$x_1 = -e^{-3t} \qquad \text{and} \qquad x_2 = e^{-t}$$
$$y_1 = e^{-3t} \qquad\qquad\qquad y_2 = e^{-t}$$

are both solution pairs of the homogeneous

$$x' = -2x + y$$
$$y' = x - 2y.$$

7. Find a formula for the general solution of the homogeneous system in Exercise 6.

8. Show that $x_p = 1$, $y_p = 0$ is a solution pair of the nonhomogeneous system.

9. Find a formula for the general solution of the nonhomogeneous system.

10. Find a formula for the solution of the initial value problem $x(0) = 1$, $y(0) = 1$ for the nonhomogeneous system.

5.2 Homogeneous Systems with Constant Coefficients

If we substitute the exponential solution pair

$$x = v_1 e^{\lambda t}$$
$$y = v_2 e^{\lambda t}$$

into the linear homogeneous system

$$x' = ax + by \tag{2.1}$$
$$y' = cx + dy,$$

we obtain

$$\lambda v_1 e^{\lambda t} = a v_1 e^{\lambda t} + b v_2 e^{\lambda t}$$
$$\lambda v_2 e^{\lambda t} = c v_1 e^{\lambda t} + d v_2 e^{\lambda t}$$

or, upon rearrangement,

$$[(a - \lambda) v_1 + b v_2] e^{\lambda t} = 0$$
$$[c v_1 + (d - \lambda) v_2] e^{\lambda t} = 0.$$

To satisfy these two equations for all t, we must (since an exponential function never equal 0) choose λ, v_1, and v_2 so that

$$(a - \lambda) v_1 + b v_2 = 0 \tag{2.2}$$
$$c v_1 + (d - \lambda) v_2 = 0.$$

In order to obtain a nontrivial solution (that is, a solution in which not both v_1 and v_2 are equal to 0), we must choose λ so that the determinant of the coefficients equals 0, that is,

$$\det \begin{pmatrix} a - \lambda & b \\ c & d - \lambda \end{pmatrix} = \lambda^2 - (a + d)\lambda + (ad - bc) = 0.$$

This quadratic equation is called the *characteristic equation* associated with the linear homogeneous system (2.1). The quadratic polynomial $\lambda^2 - (a + d)\lambda + (ad - bc)$ is the *characteristic polynomial* associated with the system.

As in the example in Sec. 5.1, with a (real) root λ of the characteristic equation substituted into the equations (2.2) we can solve for v_1 and v_2 not both equal to 0. If the characteristic equation has two real roots, we obtain two solution pairs. Will these two solution pairs be independent? And what can be done if the characteristic equation does not have real roots or has only one real root (i.e., has a repeated root)? We answer these equations by treating three separate cases.

5.2.1 Case 1: Two Different Real Roots

Suppose the characteristic equation

$$\lambda^2 - (a + d)\lambda + (ad - bc) = 0$$

of the linear homogeneous system (2.1) has two different real roots $\lambda_1 \neq \lambda_2$. For $\lambda = \lambda_1$ the algebraic equations (2.2) have a nontrivial solution that yields the solution pair

$$x_1 = v_1 e^{\lambda_1 t}$$
$$y_1 = v_2 e^{\lambda_1 t}$$

of the system (2.1). For $\lambda = \lambda_2$ the equations (2.2) again have a nontrivial solution that, if we denote this solution by w_1, w_2, yields a second solution pair

$$x_2 = w_1 e^{\lambda_2 t}$$
$$y_2 = w_2 e^{\lambda_2 t}$$

of the system (2.1). Since $\lambda_1 \neq \lambda_2$, the exponential functions $e^{\lambda_1 t}$ and $e^{\lambda_2 t}$ are not constant multiples of each other and these two solution pairs are independent.

The general solution is the linear combination of these two exponential solution pairs

$$x = c_1 v_1 e^{\lambda_1 t} + c_2 w_1 e^{\lambda_2 t}$$
$$y = c_1 v_2 e^{\lambda_1 t} + c_2 w_2 e^{\lambda_2 t}.$$

In the previous section we saw an example of how this method yields the general solution (1.9) of the system (1.3). Here is another example.

EXAMPLE 1

The system

$$x' = -\alpha x - \beta y \qquad (2.3)$$
$$y' = \gamma x - \delta y$$

is a model of the glucose/insulin regulation system in the bloodstream. In these equations x and y are the excess concentrations of glucose and insulin from their equilibrium levels, respectively. Thus, a negative value of x (or y) is a deficiency in the glucose (or insulin) and a positive value is an excess of glucose (or insulin) in the bloodstream. The rate constant $\alpha > 0$ is related to the efficiency that the liver absorbs glucose, $\beta > 0$ to the rate at which glucose is absorbed by muscle, $\gamma > 0$ to the rate that insulin is produced by the pancreas, and $\delta > 0$ the rate at which insulin is degraded by the liver. See Exercise 21.

Some typical values of the rate constants (per hour) are[1]

$$\alpha = 2.92, \qquad \beta = 4.34, \qquad \gamma = 0.208, \qquad \delta = 0.780. \tag{2.4}$$

The characteristic equation for the system with these values of the coefficients

$$\lambda^2 + 3.70\lambda + 3.18 = 0$$

has roots (to three significant digits)

$$\lambda_1 = -2.34, \qquad \lambda_2 = -1.36.$$

The system (2.2) for v_1 and v_2 is

$$(-2.92 - \lambda)\,v_1 - 4.34v_2 = 0 \tag{2.5}$$
$$0.208v_1 + (-0.780 - \lambda)\,v_2 = 0.$$

For the root $\lambda_1 = -2.34$ the first equation becomes

$$-0.58v_1 - 4.34v_2 = 0$$

(the second equation is a multiple of this equation and can be ignored). One solution is

$$v_1 = 4.34, \qquad v_2 = -0.58,$$

which results in the exponential solution pair

$$x_1 = 4.34e^{-2.34t}$$
$$y_1 = -0.58e^{-2.34t}.$$

For the root $\lambda_2 = -1.36$ the first equation in (2.5) becomes

$$-1.56v_1 - 4.34v_2 = 0$$

(the second equation is a multiple of this equation and can be ignored). One solution is

$$v_1 = 4.34, \qquad v_2 = -1.56,$$

which results in the exponential solution pair

$$x_2 = 4.34e^{-1.36t}$$
$$y_2 = -1.56e^{-1.36t}.$$

The general solution is a linear combination of these two exponential solution pairs, that is,

$$x = 4.34c_1e^{-2.34t} + 4.34c_2e^{-1.36t} \tag{2.6}$$
$$y = -0.58c_1e^{-2.34t} - 1.56c_2e^{-1.36t}.$$

---

[1]V. W. Bolie, *Journal of Applied Physiology* 16 (1960), p. 783.

We can solve initial value problems using the general solution by making an appropriate choice of the arbitrary constants c_1 and c_2.

For example, in Example 1 suppose a dose $x_0 > 0$ of glucose is introduced into the bloodstream at $t = 0$. To find the effect of this disturbance from equilibrium we solve the initial value problem

$$x(0) = x_0 > 0, \qquad y(0) = 0$$

for the system (2.3)–(2.4). Setting $t = 0$ in the general solution (2.6), we obtain the equations

$$4.34c_1 + 4.34c_2 = x_0$$
$$-0.58c_1 - 1.56c_2 = 0$$

for c_1 and c_2, which after some algebra yield

$$c_1 = 0.36678x_0, \qquad c_2 = -0.13637x_0.$$

Using these in the general solution we obtain (rounded to three significant digits)

$$x = \left(1.59e^{-2.34t} - 0.59e^{-1.36t}\right) x_0 \qquad (2.7)$$
$$y = \left(-0.213e^{-2.34t} + 0.213e^{-1.36t}\right) x_0.$$

A graph of this solution pair appears in Fig. 5.3. Notice that x decreases to a negative minimum before increasing to 0 (i.e., the glucose concentration in the blood stream drops below and then increases up to the equilibrium level). Also notice that y increases to a maximum and then decreases to 0 (i.e., after the glucose dose is administered the insulin concentration in the blood stream increases above and then returns to the equilibrium level).

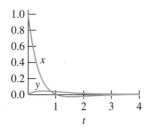

Figure 5.3 The graphs of the solution pair (2.7) with an initial dose $x_0 = 1$ of glucose.

5.2.2 Case 2: Complex Conjugate Roots

Suppose the characteristic equation

$$\lambda^2 - (a + d)\lambda + (ad - bc) = 0$$

of the linear homogeneous system (2.1) has complex roots (which necessarily form a complex conjugate pair). By the quadratic formula

$$\lambda = \frac{a + d \pm \sqrt{(a+d)^2 - 4(ad - bc)}}{2},$$

we see that this occurs when the discriminant $(a + d)^2 - 4(ad - bc)$ is negative, that is,

$$(a + d)^2 < 4(ad - bc).$$

In this case, the exponential $e^{\lambda t}$ is complex and the solutions v_1 and v_2 of the equations (2.2) will consequently be complex numbers. As a result, the exponential solution pair is complex. Our problem is to obtain a real solution pairs from the complex exponential solution pairs.

For example, the equivalent system

$$x' = y$$
$$y' = -x$$

to the simple harmonic oscillator $x'' + x = 0$ has characteristic equation

$$\lambda^2 + 1 = 0$$

with complex roots

$$\lambda_1 = i, \qquad \lambda_2 = -i$$

($i = \sqrt{-1}$). The equations (2.2) for v_1, v_2 are

$$-iv_1 + v_2 = 0$$
$$-v_1 - iv_2 = 0.$$

The second equation is $-i$ times the first. The solution $v_1 = 1$, $v_2 = i$ of the first equation yields the (complex) exponential solution pair

$$x = e^{it}$$
$$y = ie^{it}.$$

Recall that a complex number λ can be written in the form

$$\lambda = \alpha + i\beta,$$

where α and β are real numbers called the *real part* of λ and the *imaginary part* of λ, respectively.[2] Similarly, complex (valued) functions have real and imaginary parts. For example,

$$e^{it} = \cos t + i \sin t$$

and hence the real part of e^{it} is $\cos t$ and the imaginary part is $\sin t$. More generally,

$$e^{i\beta t} = \cos \beta t + i \sin \beta t$$

and from

$$e^{\lambda t} = e^{(\alpha + i\beta)t} = e^{\alpha t} e^{i\beta t}$$

we have

$$e^{\lambda t} = e^{\alpha t} \cos \beta t + ie^{\alpha t} \sin \beta t.$$

---

[2]*Note:* β is the imaginary part of λ, *not $i\beta$*. This is a common mistake. Both the real and imaginary parts of a complex number are *real* numbers.

The key to obtaining real solution pairs from complex solution pairs is the following fact: *If*

$$x = x_1(t) + ix_2(t)$$
$$y = y_1(t) + iy_2(t)$$

is a complex solution pair of the homogenous linear system (1.2), then both the real and the imaginary parts

$$x_1 = x_1(t), \qquad x_2 = x_2(t)$$
$$y_2 = y_1(t), \qquad y_2 = y_2(t)$$

are solution pairs.

To see why, substitute the complex solution into the first equation of the system to obtain

$$x_1' + ix_2' = a\,(x_1 + ix_2) + b\,(y_1 + iy_2)$$
$$= (ax_1 + by_1) + i(ax_2 + by_2).$$

From the second equation we have similarly that

$$y_1' + iy_2' = c\,(x_1 + ix_2) + d\,(y_1 + iy_2)$$
$$= (cx_1 + dy_1) + i(cx_2 + dy_2).$$

Two complex numbers are equal if and only if their real parts and their imaginary parts are equal. From the real parts of these expressions we obtain (remember that a, b, c, and d are real numbers)

$$x_1' = ax_1 + by_1$$
$$y_1' = cx_1 + dy_1,$$

which tells us that the pair x_1, y_1 is a solution pair. From the imaginary parts of the expressions we obtain

$$x_2' = ax_2 + by_2$$
$$y_2' = cx_2 + dy_2,$$

which tells us that the pair x_2, y_2 is also a solution pair.

For example, the real and imaginary parts of the complex solution pair (recall $i^2 = -1$)

$$x = e^{it} = \cos t + i\,\sin t$$
$$y = ie^{it} = -\sin t + i\,\cos t$$

of the simple harmonic oscillator system yield the two real (and independent) solution pairs

$$x_1 = \cos t, \qquad x_2 = \sin t$$
$$y_1 = -\sin t, \qquad y_2 = \cos t.$$

In general, when the root $\lambda = \alpha + i\beta$ of the characteristic equation is complex ($\beta \neq 0$), we obtain a complex solution pair

$$x = v_1 e^{\lambda t}$$
$$y = v_2 e^{\lambda t},$$

where

$$v_1 = u_1 + iw_1$$
$$v_2 = u_2 + iw_2$$

are complex solutions (not both equal to 0) of the algebraic equations (2.2), that is,

$$(a - \lambda)v_1 + bv_2 = 0$$
$$cv_1 + (d - \lambda)v_2 = 0.$$

(These equations are dependent, i.e., one is a multiple of the other.) The real and imaginary parts of this complex solution form two real solution pairs. We need two *independent* solution pairs in order to obtain the general solution. In Exercise 23 it is shown that the two real solution pairs, obtained by taking real and imaginary parts of a nontrivial complex solution pair, will always be independent.

In summary, we can write formulas for the real and imaginary parts of the complex solution pair:

$$x_1 = e^{\alpha t}(u_1 \cos \beta t - w_1 \sin \beta t)$$
$$y_1 = e^{\alpha t}(u_2 \cos \beta t - w_2 \sin \beta t)$$

and

$$x_2 = e^{\alpha t}(u_1 \sin \beta t + w_1 \cos \beta t)$$
$$y_2 = e^{\alpha t}(u_2 \sin \beta t + w_2 \cos \beta t).$$

The general solution is the linear combination

$$x = e^{\alpha t}[(c_1 u_1 + c_2 w_1)\cos \beta t + (-c_1 w_1 + c_2 u_1)\sin \beta t] \qquad (2.8)$$
$$y = e^{\alpha t}[(c_1 u_2 + c_2 w_2)\cos \beta t + (-c_1 w_2 + c_2 u_2)\sin \beta t].$$

For particular systems, it is usually easier to calculate the real and imaginary parts of a complex solution than it is to memorize these formulas.

The second root of the characteristic equation is the complex conjugate $\lambda_2 = \alpha - i\beta$ of the first. Therefore, it turns out that the second complex solution pair obtained from this conjugate root is the complex conjugate of the first solution pair obtained from $\lambda_1 = \alpha + i\beta$. From this, we see that the real part of the second solution pair is the same as the real part of the first solution pair, and the imaginary part of the second solution pair is -1 times the imaginary part of the first solution pair. It follows that the real solution pairs obtained from the conjugate root λ_2 are not independent from those obtained from λ_1. There is no need to consider the solution pair obtained from the second, conjugate root.

EXAMPLE 2

As an example of the complex root case, consider the glucose/insulin regulation system (2.3)

$$x' = -\alpha x - \beta y$$
$$y' = \gamma x - \delta y$$

with coefficients (2.4). Suppose that in a particular patient, the rate at which the liver absorbs glucose and degrades insulin is abnormally low, so that the values of α and δ

drop to one-tenth their former values (i.e., suppose $\alpha = 0.292$ and $\delta = 0.078$). The resulting system

$$x' = -0.292x - 4.34y \tag{2.9}$$
$$y' = 0.208x - 0.078y$$

has characteristic equation

$$\lambda^2 + 0.37\lambda + 0.925\,5 = 0$$

with complex root

$$\lambda = -0.185 + 0.945i.$$

The equations (2.2) for v_1 and v_2 are

$$(-0.107 - 0.945i)\,v_1 - 4.34v_2 = 0$$
$$0.208v_1 + (0.107 - 0.945i)\,v_2 = 0.$$

(The second equation, as usual, is a multiple of the first, in this case the complex multiple $0.0246 + 0.2172i$. Thus, we can ignore the second equation). The choice

$$v_1 = -4.34, \qquad v_2 = 0.107 + 0.945i$$

for a solution of the first equation yields the complex solution pair

$$x = -4.34e^{(-0.185+0.945i)t}$$
$$y = (0.107 + 0.945i)\,e^{(-0.185+0.945i)t}$$

whose real and imaginary parts

$$x_1 = -4.34e^{-0.185t} \cos 0.945t$$
$$y_1 = e^{-0.185t}\,(0.107 \cos 0.945t - 0.945 \sin 0.945t)$$

and

$$x_2 = -4.34e^{-0.185t} \sin 0.945t$$
$$y_2 = e^{-0.185t}\,(0.945 \cos 0.945t + 0.107 \sin 0.945t)$$

are independent solution pairs. The general solution is the linear combination

$$x = e^{-0.185t}\,(-4.34c_1 \cos 0.945t - 4.34c_2 \sin 0.945t)$$
$$y = e^{-0.185t}\,\{(0.107c_1 + 0.945c_2) \cos 0.945t \tag{2.10}$$
$$+ (-0.945c_1 + 0.107c_2) \sin 0.945t\}. \qquad \blacksquare$$

For the system (2.9) we find the solution of the initial value problem

$$x(0) = x_0 > 0, \qquad y(0) = 0$$

by choosing appropriate values for c_1 and c_2 in the general solution. Setting $t = 0$ in the formulas for the general solution (2.10), we obtain equations

$$-4.34c_1 = x_0$$
$$0.107c_1 + 0.945c_2 = 0$$

for c_1 and c_2 whose solutions are

$$c_1 = -0.23041x_0, \qquad c_2 = 0.026089x_0.$$

A substitution of these into the general solution (2.10) yields the solution

$$x = x_0 e^{-0.185t} \left(\cos 0.945t - 0.113 \sin 0.945t \right) \tag{2.11}$$

$$y = 0.220 x_0 e^{-0.185t} \sin 0.945t$$

of the initial value problem. Figure 5.4 shows the graphs of this solution pair. Notice that both glucose and insulin levels return to their equilibrium levels. However, unlike the previous case graphed in Fig. 5.3, both concentrations undergo oscillations, repeatedly dropping below and rising above their equilibrium levels as they return to equilibrium.

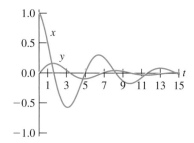

Figure 5.4 The graphs of the solution pair (2.11) with an initial unit dose $x_0 = 1$ of glucose.

5.2.3 Case 3: A Repeated Real Root

From the quadratic formula

$$\lambda = \frac{a + d \pm \sqrt{(a+d)^2 - 4(ad - bc)}}{2}$$

we see that the characteristic equation

$$\lambda^2 - (a+d)\lambda + (ad - bc) = 0$$

has a repeated root if and only if the constant coefficients satisfy

$$(a+d)^2 - 4(ad - bc) = 0$$

in which case the only root is

$$\lambda = \frac{1}{2}(a+d).$$

Using this root and the procedure in Case 1, we can construct an exponential solution pair

$$x_1 = v_1 e^{\lambda t} \tag{2.12}$$

$$y_1 = v_2 e^{\lambda t},$$

where v_1 and v_2 are solutions of the equations

$$(a - \lambda) v_1 + b v_2 = 0 \tag{2.13}$$

$$c v_1 + (d - \lambda) v_2 = 0.$$

Our problem is how to find a second independent solution pair in this case.

In fact, equations (2.13) may yield two independent solution pairs. For example, these equations for the system

$$x' = x$$
$$y' = y$$

are (since $\lambda = 1$ is a repeated root)

$$0 = 0$$
$$0 = 0.$$

In this redundant case, we can choose v_1 and v_2 to be anything we want. For example, the two choices $v_1 = 1$, $v_2 = 0$ and $v_1 = 0$, $v_2 = 1$ yield the two independent solution pairs

$$x_1 = e^t, \qquad x_2 = 0$$
$$y_1 = 0, \qquad y_2 = e^t.$$

This redundant case occurs if and only if $b = c = 0$ and $a = d$. For other cases, equations (2.13) yield only one nontrivial solution pair.

Here is an example. The characteristic equation of the system

$$x' = x + y \tag{2.14}$$
$$y' = y$$

is $\lambda^2 - 2\lambda + 1 = 0$ and therefore has a repeated root $\lambda = 1$. For this example, the

equations (2.13) for v_2 and v_1 are

$$v_2 = 0$$
$$0 = 0.$$

Thus, $v_2 = 0$ and v_1 is arbitrary (but nonzero). The choice $v_1 = 1$ in (2.12) yields the exponential solution pair

$$x_1 = e^t$$
$$y_1 = 0.$$

For this example, we can obtain a second independent solution pair using methods for single first-order equations from Chapter 2. The general solution of the second equation in (2.14) is $y = c_2 e^t$, where c_2 is an arbitrary constant. The first equation then becomes the nonhomogeneous equation $x' = x + c_2 e^t$ for x. The general solution of this equation has the form $x = c_1 e^t + x_p$, where c_1 is an arbitrary constant and x_p is any particular solution. We can use the method of undetermined coefficients to find x_p. Since the nonhomogeneous term is a multiple of e^t, the initial guess for x_p is ae^t. However, this is a solution of the homogeneous equation and therefore the guess must be multiplied by t. Substituting $x_p = ate^t$ into the equation $x' = x + c_2 e^t$ yields $a = c_2$ and $x_p = c_2 te^t$. We have found solution pairs

$$x = c_1 e^t + c_2 te^t$$
$$y = c_2 e^t$$

for arbitrary constants c_1 and c_2. The choice $c_1 = 1$, $c_2 = 0$ produces the solution pair found previously. The choice $c_1 = 0$, $c_2 = 1$ produces the independent solution pair

$$x_2 = te^t$$
$$y_2 = e^t.$$

This example (in particular, the factor of t) shows in the case of repeated roots that it may not be possible to find two independent exponential solution pairs. It turns out in this case that a factor of t (times an exponential) typically (but not always) appears in the general solution.

To summarize, *if $b = c = 0$ and $a = d$, the repeated root is $\lambda = a$ and the general solution is*

$$x = c_1 e^{\lambda t} \tag{2.15}$$
$$y = c_2 e^{\lambda t}.$$

Otherwise there is a solution pair independent from the solution pair (2.12) that has the form

$$x_2 = (w_1 + v_1 t)\, e^{\lambda t} \tag{2.16}$$
$$y_2 = (w_2 + v_2 t)\, e^{\lambda t},$$

where w_1, w_2 solve the equations

$$(a - \lambda)\, w_1 + bw_2 = v_1 \tag{2.17}$$
$$cw_1 + (d - \lambda)\, w_2 = v_2$$

and the general solution is the linear combination

$$x = [c_1 v_1 + c_2\, (w_1 + v_1 t)]\, e^{\lambda t} \tag{2.18}$$
$$y = [c_1 v_2 + c_2\, (w_2 + v_2 t)]\, e^{\lambda t}.$$

Here is an example. The characteristic equation

$$\lambda^2 - 4\lambda + 4 = 0$$

of the system

$$x' = 3x - y$$
$$y' = x + y$$

has the repeated root $\lambda = 2$. From (2.12), the solution $v_1 = 1$, $v_2 = 1$ of the equations (2.13), which in this case are

$$v_1 - v_2 = 0$$
$$v_1 - v_2 = 0,$$

yields the exponential solution pair

$$x_1 = e^{2t}$$
$$y_1 = e^{2t}.$$

The solution $w_1 = 1$, $w_2 = 0$ of the equations (2.17), which in this case are

$$w_1 - w_2 = 1$$
$$w_1 - w_2 = 1,$$

yields, from (2.16), the independent solution pair

$$x_2 = (1 + t)\, e^{2t}$$
$$y_2 = te^{2t}.$$

The general solution is the linear combination

$$x = [c_1 + c_2(1 + t)]\, e^{2t} \tag{2.19}$$
$$y = [c_1 + c_2 t]\, e^{2t}.$$

Exercises

Find a formula for the general solution of each system below.

1. $\begin{cases} x' = 4x - 2y \\ y' = 7x - 5y \end{cases}$

2. $\begin{cases} x' = -2x + y \\ y' = x - 2y \end{cases}$

3. $\begin{cases} x' = \frac{1}{2}x - \frac{3}{2}y \\ y' = \frac{3}{2}x + \frac{1}{2}y \end{cases}$

4. $\begin{cases} x' = x + y \\ y' = x - y \end{cases}$

5. $\begin{cases} x' = -0.012x - 0.45y \\ y' = 2.31x - 3.15y \end{cases}$

6. $\begin{cases} x' = 0.51x - 0.74y \\ y' = 1.42x + 2.67y \end{cases}$

7. $\begin{cases} x' = 3x + 2y \\ y' = -2x - y \end{cases}$

8. $\begin{cases} x' = -x + y \\ y' = -x - 2y \end{cases}$

9. $\begin{cases} x' = -x + y \\ y' = 4x + 2y \end{cases}$

10. $\begin{cases} x' = 3x + 2y \\ y' = -4x - y \end{cases}$

11. $\begin{cases} x' = -6.1x + 0.2y \\ y' = -1.1x - 1.5y \end{cases}$

12. $\begin{cases} x' = 8.3x + 1.2y \\ y' = -1.8x + 0.3y \end{cases}$

13. $\begin{cases} x' = x + 13y \\ y' = -2x - y \end{cases}$

14. $\begin{cases} x' = x + 3y \\ y' = 4x + 2y \end{cases}$

15. $\begin{cases} x' = -\frac{1}{2}x + \frac{3}{4}y \\ y' = -3x + \frac{5}{2}y \end{cases}$

16. $\begin{cases} x' = -5x + 8y \\ y' = -2x + 3y \end{cases}$

17. Find a formula for the solution of the initial value problem $x(0) = 1$, $y(0) = -1$ for the systems in Exercises 1–16.

18. Find a formula for the solution of the initial value problem $x(0) = 2$, $y(0) = 3$ for the systems in Exercises 1–16.

19. Find a formula for the solution of the chemical pesticide problem

$$x' = -2x + 2y$$
$$y' = 2x - 5y$$
$$x(0) = d, \quad y(0) = 0$$

in which an initial dose of pesticide is sprayed in the trees but none is initially present in the soil.

20. **(a)** Use a computer program to sketch the direction field and several typical orbits of the nonsimple system

$$x' = -3x + 6y$$
$$y' = 2x - 4y.$$

   (b) Based on (a), make a conjecture about the asymptotic behavior of solutions as $t \to +\infty$.

   (c) Find a formula for the general solution of the system.

   (d) Find a formula for the solution of the initial value problem $x(0) = x_0$, $y(0) = y_0$.

   (e) Use your answer in (c) to verify (or disprove) your conjecture in (b).

21. Use the fundamental inflow-outflow balance law to derive the equations (2.3) from the following facts. The rate at which the liver absorbs glucose is proportional to the concentration of glucose in the bloodstream. The rate at which skeletal muscle absorbs glucose is proportional to the concentration of insulin in the bloodstream. The rate at which insulin is produced (by the pancreas) is proportional to the concentration of glucose in the bloodstream. All of these rates increase as the concentration of glucose increases. Finally, the rate at which insulin is degraded by the liver is proportional to the concentration of insulin (increasing as the concentration of insulin increases).

22. The system

$$x' = by$$
$$y' = cx$$

has been used as a simplistic model of an arms race. Here b and c are positive constants and x and y are armament budgets of two opposing countries. The equations state that each country increases its armament budget at a rate proportional to the budget of the other country.

(a) Find a formula for the general solution.

(b) Suppose one country (x) establishes an armament budget. Find a formula for the solution of the initial value problem $x(0) = x_0 > 0$, $y(0) = 0$.

(c) What are the long-term consequences of x's armament budget?

23. Show the real and imaginary parts of a complex solution pair $x = v_1 e^{\lambda t}$, $y = v_2 e^{\lambda t}$ of an autonomous linear homogeneous system (1.2) (where $u_1 = u_1 + i w_1$ and $u_2 = u_2 + i w_2$ are not both equal to 0 and $\lambda = \alpha + i\beta$, $\beta \neq 0$) are independent solution pairs. (*Hint*: Assume that they are dependent and derive the contradiction that $v_1 = 0$ and $v_2 = 0$.)

5.3 Homogeneous Second-Order Equations

The homogeneous linear second-order equation

$$ax'' + bx' + cx = 0, \qquad a \neq 0 \tag{3.1}$$

is equivalent to the first-order system

$$x' = y \tag{3.2}$$

$$y' = -\frac{c}{a}x - \frac{b}{a}y.$$

If the coefficients a, b, and c are constants, we can calculate two independent solution pairs of this system, and hence the general solution, using the methods from Sec. 5.2. For second-order equations, however, there is a shortcut.

First, the characteristic equation for the equivalent system is

$$\lambda^2 + \frac{b}{a}\lambda + \frac{c}{a} = 0.$$

The roots of this equation are the same as those of the equation

$$a\lambda^2 + b\lambda + c = 0, \tag{3.3}$$

which we call the *characteristic equation for the second-order equation* (3.1).

Note that the coefficients of the characteristic equation can be read off directly from the differential equation. (They both have the same coefficients.) There is no need to construct the equivalent first-order system.

Suppose the roots λ_1 and λ_2 of the characteristic equation are real and different. We can calculate two independent exponential solution pairs of the equivalent system using the method for Case 1 in Sec. 5.2.

For example, using the root λ_1 we would solve the equations (2.2) for v_1 and v_2 appearing in the exponential solution pair (1.5). It turns out, however, that there is a shortcut to doing this for the system (3.2). Recall that the two equations in (2.2) are dependent and consequently only one need be solved for v_1 and v_2. For the system (3.2) the first equation in (2.2) is

$$-\lambda_1 v_1 + v_2 = 0.$$

The choice $v_1 = 1$, $v_2 = \lambda_1$ yields the solution pair

$$x = e^{\lambda_1 t}$$

$$y = \lambda_1 e^{\lambda_1 t}.$$

For the second-order equation (3.1) we are only interested in the first component of the solution pair: $x_1 = e^{\lambda_1 t}$ (y after all is just x'). Similarly, using the other root λ_2, we obtain the solution $x_2 = e^{\lambda_2 t}$. The general solution is

$$x = c_1 e^{\lambda_1 t} + c_2 e^{\lambda_2 t}.$$

Note that we can construct this general solution immediately from the roots of the characteristic equation (3.3). There is no need to solve the system (2.2) for v_1 and v_2.

For example, the characteristic equation of the homogeneous second-order equation

$$x'' + 2x' - 3x = 0$$

is

$$\lambda^2 + 2\lambda - 3 = 0.$$

The roots of this equation are $\lambda = -3$ and 1, and therefore the general solution is

$$x = c_1 e^{-3t} + c_2 e^t.$$

Similar shortcuts are available for Case 2 (complex conjugate roots) and Case 3 (repeated root). See Exercises 10 and 11. Table 5.1 contains a summary of these shortcuts.

Table 5.1

Roots of Equation $a\lambda^2 + b\lambda + c = 0$	General Solution of $ax'' + bx' + cx = 0, a \neq 0$
Real & different $\lambda_1 \neq \lambda_2$	$x = c_1 e^{\lambda_1 t} + c_2 e^{\lambda_2 t}$
Complex conjugate $\lambda = \alpha \pm \beta i, \beta \neq 0$	$x = e^{\alpha t}(c_1 \cos \beta t + c_2 \sin \beta t)$
Real & repeated $\lambda = \lambda_1 = \lambda_2$	$x = (c_1 + c_2 t) e^{\lambda t}$

To see how to use Table 5.1, consider the two second-order differential equations

$$x'' + 2x' + 2x = 0$$
$$x'' - 2x' + x = 0.$$

The characteristic equation $\lambda^2 + 2\lambda + 2 = 0$ of the first equation has complex roots $\lambda = -1 \pm i$, and therefore this differential equation has general solution $x = e^{-t}(c_1 \cos t + c_2 \sin t)$. (Use Table 5.1 with $\alpha = -1$ and $\beta = 1$.) The characteristic equation $\lambda^2 - 2\lambda + 1 = 0$ of the second equation has a repeated root $\lambda = 1$, and therefore the second differential equation has general solution $x = c_1 e^t + c_2 t e^t$.

EXAMPLE 1

Consider the second-order equation

$$mx'' + cx' + kx = 0$$

with positive coefficients $m > 0$, $c \geq 0$ and $k > 0$. This is a model for an oscillator subjected to a frictional force, such as a mass-spring system attached to the ceiling or the suspension system of an automobile. In the latter example m is the mass of the car frame, k is the stiffness of the suspension, and c is the damping constant of the shock absorbers. The variable $x = x(t)$ is the distance of the frame from its rest position.

Consider the case $m = 1$ and $k = 1$, that is,

$$x'' + cx' + x = 0. \tag{3.4}$$

(For the general case see Exercise 13.) If the shock absorber is removed, or, in other words, if $c = 0$, this equation is the simple harmonic oscillator equation. We are interested in the effects of the shock absorber, so we assume $c > 0$.

When the suspension system is at rest, $x = 0$ and $x' = 0$. Suppose the car suddenly strikes a bump in the road. We can model this occurrence by prescribing an initial velocity $v_0 \neq 0$ to the car frame. Thus, we are interested in the initial value problem

$$x(0) = 0, \quad x'(0) = v_0.$$

Figure 5.5 displays computer-drawn solution graphs for the initial value problem $v_0 = 1$ and two values of the damping constant c. (These were obtained by numerically solving the initial value problem for the equivalent first-order system.) Notice that for the larger value of the damping constant $c = 3$ the car frame returns to the rest position without oscillation. For the smaller value of $c = 0.25$, however, the car frame oscillates as it returns to equilibrium. Assuming that oscillations are undesirable, we are interested in determining the cutoff value of the shock absorber damping constant c_0 that separates oscillations from nonoscillation. To do this we will find a formula for the solution of the initial value problem.

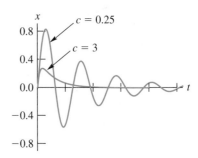

Figure 5.5 The solution of the initial value problem $x'' + cx' + x = 0, x(0) = 0, x'(0) = 1$ for two values of the damping constant c.

The characteristic equation

$$\lambda^2 + c\lambda + 1 = 0$$

has roots

$$\lambda = \frac{1}{2}\left(-c \pm \sqrt{c^2 - 4}\right)$$

and which solution case in Table 5.1 occurs depends on the sign of $c^2 - 4$.

If $c > 2$, the roots

$$\lambda_1 = \frac{1}{2}\left(-c + \sqrt{c^2 - 4}\right), \qquad \lambda_2 = \frac{1}{2}\left(-c - \sqrt{c^2 - 4}\right)$$

are real and different. Notice that both roots are negative (λ_1 is negative because $\sqrt{c^2 - 4} < \sqrt{c^2} = c$.) In this case the general solution is (see Table 5.1)

$$x = c_1 e^{\lambda_1 t} + c_2 e^{\lambda_2 t}.$$

To solve the initial value problem we determine c_1 and c_2 from the equations

$$x(0) = c_1 + c_2 = 0$$
$$x'(0) = \lambda_1 c_1 + \lambda_2 c_2 = v_0.$$

Solving these equations, we obtain (note $\lambda_1 - \lambda_2 = \sqrt{c^2 - 4}$)

$$c_1 = \frac{1}{\sqrt{c^2 - 4}} v_0, \qquad c_2 = -\frac{1}{\sqrt{c^2 - 4}} v_0$$

and the solution formula

$$x(t) = \frac{1}{\sqrt{c^2 - 4}} v_0 \left(e^{\lambda_1 t} - e^{\lambda_2 t} \right).$$

Since both roots are negative, $\lim_{t \to +\infty} x(t) = 0$ and the car frame returns (in the mathematical limit) to its rest position. Since $\lambda_1 \neq \lambda_2$, it follows that $e^{\lambda_1 t} \neq e^{\lambda_2 t}$ for all $t > 0$. Thus, in this case the car frame does *not* oscillate (i.e., does not cross the equilibrium position).

If $c < 2$ the roots of the characteristic equation are complex, namely

$$\lambda = \frac{1}{2} \left(-c \pm i\sqrt{4 - c^2} \right).$$

From Table 5.1 with $\alpha = -c/2$ and $\beta = \sqrt{4 - c^2}/2$ we obtain the general solution

$$x(t) = e^{\alpha t} (c_1 \cos \beta t + c_2 \sin \beta t).$$

From the initial conditions

$$x(0) = c_1 = 0$$
$$x'(0) = \alpha c_1 + \beta c_2 = v_0$$

or $c_1 = 0$ and $c_2 = v_0/\beta$, which yields the solution

$$x(t) = \frac{1}{\beta} v_0 e^{\alpha t} \sin \beta t.$$

Since $\alpha < 0$, once again $\lim_{t \to +\infty} x(t) = 0$ and the car frame returns (in the mathematical limit) to its rest position. However, in this case, $x(t) = 0$ infinitely often, namely for times $t = \pi/\beta, 2\pi/\beta, 3\pi/\beta, 4\pi/\beta, \ldots$.

In conclusion, we find that the car frame does not oscillate (is over damped) if the damping constant is greater than the critical value $c_0 = 2$ and does oscillate (is under damped) if $c < c_0 = 2$. ■

Exercises

Find a formula for the general solution of the following second-order equations.

1. $x'' + x' + x = 0$

2. $x'' - x' + x = 0$

3. $2x'' - x = 0$

4. $x'' + 2x' + x = 0$

5. $x'' + 3x' - 4x = 0$

6. $x'' - 5x' + 4x = 0$

7. $x'' + 5x = 0$

8. $x'' - 6x' + 9x = 0$

9. $x'' + 3.8x' + 3.45x = 0$

10. When the roots of the characteristic equation are complex $\lambda = \alpha \pm \beta i$, $\beta \neq 0$, show that the general solution of the second-order equation $ax'' + bx' + cx = 0$ is

$x = e^{\alpha t}(c_1 \cos \beta t + c_2 \sin \beta t)$.

11. When the characteristic equation has repeated (real) root $\lambda = \lambda_1 = \lambda_2$, verify that the general solution of the second-order equation $ax'' + bx' + cx = 0$ is $x = c_1 e^{\lambda t} + c_2 t e^{\lambda t}$.

12. Consider the equation $mx'' + cx' + kx = 0$ for $m > 0$, $c > 0$, $k > 0$. Show that all solutions tend to 0 as $t \to +\infty$.

13. Consider the initial value problem

$$mx'' + cx' + kx = 0$$
$$x(0) = 0, \qquad x'(0) = v_0 \neq 0$$

for the suspension system of an automobile.

(a) Find a formula for the cutoff value c_0 of the damping constant c such that $c > c_0$ implies that no oscillation occurs and $c < c_0$ implies that oscillations do occur. Justify your answer using formulas for the solutions in each case.

(b) Find a formula for the solution of the initial value problem when $c = c_0$. Does the car frame oscillate in this case? Explain your answer.

Consider the initial value problem

$$Lx'' + Rx' + \frac{1}{C}x = 0$$
$$x(0) = x_0, \qquad x'(0) = 0$$

for the charge $x = x(t)$ on an electric circuit with a resistor (of R Ohms), inductor (of L Henrys), and capacitor (of C Farads). The circuit has an initial charge of x_0 and no initial current $x'(0)$. Determine the phase line portrait type for the circuits with the following parameter values and solve solve the initial value problem.

14. $L = 0.1$, $R = 250$, $C = 10^{-5}$

15. $L = 0.2$, $R = 200$, $C = 10^{-5}$

16. $L = 0.1$, $R = 0$, $C = 10^{-5}$

17. $L = 0.1$, $R = 200$, $C = 10^{-5}$

5.4 Phase Plane Portraits

In Sec. 2.4 of Chapter 2 we saw how phase line portraits summarize basic properties of solutions of the linear equation $x' = px$. When $p \neq 0$, the phase line portrait is, remarkably, one of only two types: a sink (when $p < 0$) or a source (when $p > 0$). When $p = 0$, all solutions are equilibria and hence the equilibria are not isolated points on the phase line portrait. In Sec. 3.1 of Chapter 3 we saw how to construct phase line portraits are for nonlinear equations.

In this section we study phase plane portraits for autonomous systems

$$x' = ax + by \qquad\qquad (4.1)$$
$$y' = cx + dy.$$

We will learn how to construct x, y-phase plane portraits and how to classify them into only a small number of types.

We will consider only the case when the *trivial equilibrium*

$$x = 0, \qquad y = 0$$

is isolated. Since equilibria are solutions of the equations

$$ax + by = 0$$
$$cx + dy = 0,$$

it follows that the trivial equilibrium is the *only* equilibrium provided that these two algebraic equations have no solution other than $x = 0$, $y = 0$. This occurs if and only if $ad - bc \neq 0$. The number $ad - bc$ is the determinant of the matrix

$$\begin{pmatrix} a & b \\ c & d \end{pmatrix}.$$

This matrix is called *coefficient matrix* of the linear system (4.1). (In matrix theory terminology a matrix with a nonzero determinant is *nonsingular* or *invertible*.)

If $ad - bc \neq 0$, we say the homogeneous system (4.1) is simple. Note that the system (4.1) is simple if and only if $\lambda = 0$ is *not* a root of the characteristic equation

$$\lambda^2 - (a + d)\lambda + (ad - bc) = 0.$$

(In matrix theory terminology, a matrix the roots of this equation are the *eigenvalues* of the coefficient matrix.)

The orbit associated with an equilibrium solution pair is a point (x, y) in the phase plane. The orbit associated with the trivial equilibrium is the origin $(0, 0)$. Therefore, all phase portraits of a homogeneous system (4.1) have the origin as one orbit. A simple system has only the trivial equilibrium. All other orbits are curves in the x, y-phase plane. *Our goal is to describe the nonequilibrium orbits and to classify the phase plane portraits of simple systems.*

Since the general solution of the homogeneous system (4.1) falls into the three cases depending on roots of the characteristic equation (Sec. 5.2), it is natural to base our study of the phase plane portraits on these same three cases. Although we begin our study of phase plane portraits using formulas for general solutions, ultimately we will learn how to construct phase portraits without having to find solution formulas.

5.4.1 Case 1: Two Different Real Roots

When the roots λ_1 and λ_2 of the characteristic equation

$$\lambda^2 - (a + d)\lambda + (ad - bc) = 0$$

are real and different ($\lambda_1 \neq \lambda_2$), the general solution of the homogeneous system (4.1) is

$$x = c_1 v_1 e^{\lambda_1 t} + c_2 w_1 e^{\lambda_2 t}$$
$$y = c_1 v_2 e^{\lambda_1 t} + c_2 w_2 e^{\lambda_2 t},$$

where v_1, v_2 and w_1, w_2 are the solutions of equations (2.2) with $\lambda = \lambda_1$ and $\lambda = \lambda_2$, respectively. What kind of curves in the x, y-plane do solution pairs of this type have?

In Sec. 5.1 we found that the system

$$x' = -2x + 2y$$
$$y' = 2x - 5y$$

has the general solution

$$x = 2c_1 e^{-t} + c_2 e^{-6t} \tag{4.2}$$
$$y = c_1 e^{-t} - 2c_2 e^{-6t}.$$

Each choice of the arbitrary constants c_1, c_2 determines a solution pair and an orbit in the phase plane. Consider first some specific choices. Take $c_1 = 1$ and $c_2 = 0$ to obtain the solution pair

$$x = 2e^{-t}$$
$$y = e^{-t}.$$

The set of points (x, y) associated with this solution pair (i.e., the orbit) lies in the first quadrant (since $x > 0$ and $y > 0$ for all t) and lies on the line $y = x/2$. The orbit is a half-line. The orbit's orientation is toward the origin since $\lim_{t \to +\infty} x = 0$ and $\lim_{t \to +\infty} y = 0$. See Fig. 5.6.

The choice $c_1 = -1$, $c_2 = 0$ in the general solution yields the solution pair

$$x = -2e^{-t}$$
$$y = -e^{-t}$$

whose points (x, y) lie on the same line $y = x/2$, but in the third quadrant, with an orientation again toward the origin.

The line $y = x/2$ contains three orbits: the origin (equilibrium) and two half-line orbits. See Fig. 5.6. These three orbits result from the general solution (4.2) with $c_2 = 0$. If c_1 is chosen as a nonzero number other than ± 1, we still obtain points (x, y) lying on the line $y = x/2$. Thus, as is the case with single equations and phase line portraits, infinitely many solutions correspond to an orbit in the phase plane.

In a similar manner, choosing $c_1 = 0$ in the general solution (4.2), we obtain three orbits lying on a straight line, namely the origin and the two half-lines lying on the line $y = -2x$ with orientations toward the origin. See Fig. 5.6.

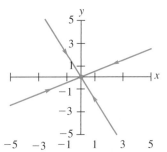

Figure 5.6 Half-line orbits of (4.2).

What kind of orbits do we get from the general solution (4.2) when neither c_1 nor c_2 equal 0? First, note that all orbits approach the origin, since $\lim_{t \to +\infty} x = 0$ and $\lim_{t \to +\infty} y = 0$ for any choice of c_1 and c_2. Second, note that the two straight line orbits found above have constant ratios y/x for all t, namely, $y/x = 1/2$ and $y/x = -2$, respectively. However, for all other orbits (with $c_1 \neq 0$ and $c_2 \neq 0$) this ratio

$$\frac{y}{x} = \frac{c_1 e^{-t} - 2c_2 e^{-6t}}{2c_1 e^{-t} + c_2 e^{-6t}}$$

is not constant. Therefore, none of the other orbits lie on a straight line. Moreover, if both numerator and denominator in this ratio are divided by $c_1 e^{-t}$ (note $c_1 \neq 0$), the result

$$\frac{y}{x} = \frac{1 - 2\frac{c_2}{c_1} e^{-5t}}{2 + \frac{c_2}{c_1} e^{-5t}}$$

shows that

$$\lim_{t \to +\infty} \frac{y}{x} = \frac{1}{2}$$

and hence that all other orbits are tangent to the line $y = x/2$ as they approach the origin $(t \to +\infty)$.

The sketch of the phase plane portrait in Fig. 5.7 summarizes our observations about the orbits of system (4.2). This type of phase portrait is called a *stable node*.

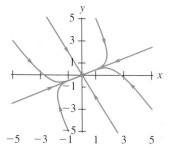

Figure 5.7 The phase plane portrait of the homogeneous system (4.2).

A stable node is characterized by two straight lines through the origin (consisting of half-line orbits approaching the origin as $t \to +\infty$) and by the fact that all other orbits approach the origin tangentially to one of the lines as $t \to +\infty$.

When both roots λ_1 and $\lambda_2 \ne \lambda_1$ are negative, the phase plane portrait of a linear system is always a stable node. To see this, note that the solution pairs

$$x_1 = v_1 e^{\lambda_1 t}, \qquad x_2 = w_1 e^{\lambda_2 t}$$
$$y_1 = v_2 e^{\lambda_1 t}, \qquad y_2 = w_2 e^{\lambda_2 t}$$

yield half-line orbits lying on the two straight lines passing through the origin with slopes v_2/v_1 and w_2/w_1, respectively. Since both roots λ_1 and λ_2 are negative all solutions

$$x = c_1 v_1 e^{\lambda_1 t} + c_2 w_1 e^{\lambda_2 t}$$
$$y = c_1 v_2 e^{\lambda_1 t} + c_2 w_2 e^{\lambda_2 t}$$

satisfy $\lim_{t \to +\infty} x = 0$, $\lim_{t \to +\infty} y = 0$ and their orbits tend to the origin as $t \to +\infty$. Suppose λ_2 is the smaller (more negative) root (i.e., $\lambda_2 < \lambda_1$). Orbits lying on the line with slope w_2/w_1 correspond to solutions with $c_1 = 0$. For all other orbits ($c_1 \ne 0$) the ratio

$$\frac{y}{x} = \frac{c_1 v_2 e^{\lambda_1 t} + c_2 w_2 e^{\lambda_2 t}}{c_1 v_1 e^{\lambda_1 t} + c_2 w_1 e^{\lambda_2 t}}$$

$$= \frac{v_2 + \frac{c_2}{c_1} w_2 e^{(\lambda_2 - \lambda_1)t}}{v_1 + \frac{c_2}{c_1} w_1 e^{(\lambda_2 - \lambda_1)t}}$$

tends to v_2/v_1 as $t \to +\infty$ [since $e^{(\lambda_2 - \lambda_1)t}$ tends to 0 $t \to +\infty$]. This means these orbits tend to the origin tangentially to the half-line orbits with slope v_2/v_1 (the slope associated with the root λ_1 of smallest magnitude).

Figure 5.8 shows a generic stable node, although in specific examples the configuration of the half-line orbits might be different. In principle, the two lines determined by the points (v_1, v_2) and (w_1, w_2) can have any configuration (except they will never be coincident), even lie in the same quadrants.

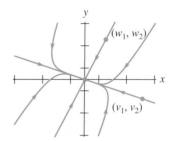

Figure 5.8 When $\lambda_2 <$ $\lambda_1 < 0$ the phase plane portrait is a stable node.

As an example, consider the glucose/insulin regulation system (2.3). The characteristic equation

$$\lambda^2 + (\alpha + \delta)\,\lambda + (\alpha\delta + \beta\gamma)$$

has roots

$$\lambda_1 = \frac{1}{2}\left(-(\delta + \alpha) + \sqrt{(\delta + \alpha)^2 - 4\,(\alpha\delta + \beta\gamma)}\right)$$

$$\lambda_2 = \frac{1}{2}\left(-(\delta + \alpha) - \sqrt{(\delta + \alpha)^2 - 4\,(\alpha\delta + \beta\gamma)}\right),$$

which are real and different provided that the inequality

$$(\delta + \alpha)^2 - 4\,(\alpha\delta + \beta\gamma) > 0$$

holds. In fact, in this case $\lambda_2 < \lambda_1 < 0$ and the phase portrait is a stable node.

For $\lambda = \lambda_1$, equations (2.2), which in this case are

$$(-\alpha - \lambda)\,v_1 - \beta v_2 = 0$$
$$\gamma v_1 + (-\delta - \lambda)\,v_2 = 0,$$

have a solution

$$v_1 = \delta - \alpha + \sqrt{(\delta - \alpha)^2 - 4\beta\gamma} > 0, \qquad v_2 = 2\gamma > 0.$$

Since $v_1 > 0$ and $v_2 > 0$, all (nonlinear) orbits approach the origin tangentially with positive slope v_2/v_1. This means that as both insulin excess and glucose excess above equilibrium levels tend to zero as $t \to +\infty$, the ratio of insulin excess to glucose excess approaches a positive number given by the formula

$$\lim_{t \to +\infty} \frac{y}{x} = \frac{2\gamma}{\delta - \alpha + \sqrt{(\delta - \alpha)^2 - 4\beta\gamma}}.$$

If both roots of the characteristic equation are positive, $0 < \lambda_1 < \lambda_2$, the phase plane portrait of a homogeneous system has the geometric appearance of a stable node, but with the orientation arrows reversed. This reversal in direction occurs because, in this case, both exponential functions $e^{\lambda_1 t}$ and $e^{\lambda_2 t}$ grow without bound as $t \to +\infty$ and consequently orbits move away from the equilibrium at the origin. For this reason, such a phase plane portrait is called an *unstable node*. See Fig. 5.9(a). Note in case of an unstable node that the nonlinear orbits approach the origin tangentially to the half-line orbits with slope v_2/v_1 as $t \to -\infty$ (in reverse time).

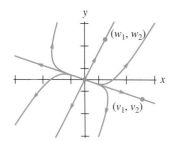

Figure 5.9(a) When $0 <$ $\lambda_1 < \lambda_2$, the phase plane portrait is an unstable node.

EXAMPLE 1

The characteristic equation $\lambda^2 - 12\lambda + 32$ associated with the system

$$x' = 3x + 5y \qquad (4.3)$$
$$y' = -x + 9y$$

has positive, unequal roots $\lambda_1 = 4$ and $\lambda_2 = 8$ and the phase portrait is an unstable node.

For $\lambda_1 = 4$, equations (2.2)

$$-v_1 + 5v_2 = 0$$
$$-v_1 + 5v_2 = 0$$

have solution $v_1 = 5$, $v_2 = 1$ and for $\lambda_2 = 8$ equations (2.2)

$$-5w_1 + 5w_2 = 0$$
$$-w_1 + w_2 = 0$$

have solution $w_1 = 1$, $w_2 = 1$. These two solutions determine the half-line orbits in the phase plane that arise from the solution pairs

$$x_1 = 5c_1 e^{4t}, \qquad x_2 = c_2 e^{8t}$$
$$y_1 = c_1 e^{4t}, \qquad y_2 = c_2 e^{8t}.$$

These independent solution pairs in turn yield the the general solution

$$x = 5c_1 e^{4t} + c_2 e^{8t}$$
$$y = c_1 e^{4t} + c_2 e^{8t}.$$

Because both roots $\lambda_1 = 4$ and $\lambda_2 = 8$ are positive, all orbits [except the origin $(x, y) = (0, 0)$] are unbounded. We see from the calculation

$$\lim_{t \to -\infty} \frac{y}{x} = \lim_{t \to -\infty} \frac{c_1 e^{4t} + c_2 e^{8t}}{5c_1 e^{4t} + c_2 e^{8t}}$$
$$= \lim_{t \to -\infty} \frac{c_1 + c_2 e^{4t}}{5c_1 + c_2 e^{4t}} = \frac{c_1}{5c_1} = \frac{1}{5}$$

that orbits not lying on the first line ($c_1 \neq 0$) approach the origin as $t \to -\infty$ (in reverse time), and they do so tangentially to the straight line orbits associated with $\lambda_2 = 4$. See Fig. 5.9(b). ∎

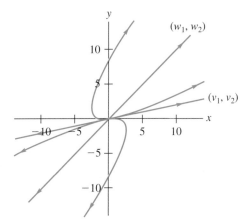

Figure 5.9(b) The phase portrait of the homogeneous system (4.3).

We have seen that when the roots of the characteristic equation have the same sign, the phase plane portrait is a node (either stable, if the roots are negative, or unstable, if the roots are positive). The remaining possibility, in this case of real distinct roots, is that the roots have *different* signs, say $\lambda_1 < 0 < \lambda_2$. In this event, the solution pair x_1, y_1 still approaches the origin along a straight line with slope v_2/v_1, as in the case of a stable node. The second solution pair x_2, y_2 moves away from the origin along a straight line with slope w_2/w_1, as in the case of an unstable node. Thus, in this case, the half-line orbits have opposite orientations.

With regard to the remaining (nonlinear) orbits, note that the ratio

$$\frac{y}{x} = \frac{c_1 v_2 e^{\lambda_1 t} + c_2 w_2 e^{\lambda_2 t}}{c_1 v_1 e^{\lambda_1 t} + c_2 w_1 e^{\lambda_2 t}}$$

$$= \frac{\frac{c_1}{c_2} v_2 e^{(\lambda_1 - \lambda_2)t} + w_2}{\frac{c_1}{c_2} v_1 e^{(\lambda_1 - \lambda_2)t} + w_1}$$

approaches w_2/w_1 as $t \to +\infty$ [the exponential $e^{(\lambda_1 - \lambda_2)t}$ tends to 0 since $\lambda_1 - \lambda_2 < 0$]. Thus, the half-line orbits associated with $\lambda_2 > 0$ are asymptotes of orbits as $t \to +\infty$. (It is left as an exercise to show the half-line orbits associated with $\lambda_1 < 0$ are asymptotes as $t \to -\infty$.) The resulting phase plane portrait, called a *saddle*, appears as in Fig. 5.10.

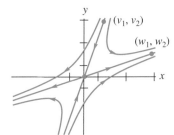

Figure 5.10 When $\lambda_1 < 0 < \lambda_2$, the phase plane portrait is a saddle.

The half-line orbits that approach the origin as $t \to +\infty$ are together called the *stable manifold*; those that approach the origin as $t \to -\infty$ are called the *unstable manifold*. The stable and unstable manifolds determine the geometric framework and shape of a saddle phase portrait.

Here is an example. The system

$$x' = by, \qquad b > 0 \qquad\qquad (4.4)$$
$$y' = cx, \qquad c > 0$$

has been used as a simplistic model of an arms race in which x and y are armament budgets of two opposing countries. See Exercise 22 in Sec. 5.2. Consider the case $b = c = 1$ (which means both countries respond, with budget increase rates, in the same way to the other country's armament budget amount). The characteristic equation

$$\lambda^2 - 1 = 0$$

has real roots

$$\lambda_1 = -1 < 0 < 1 = \lambda_2$$

of opposite signs. The phase plane portrait is a saddle. For $\lambda_1 = -1$, equations (2.2)

$$-v_1 + v_2 = 0$$
$$v_1 - v_2 = 0$$

have solution $v_1 = 1$, $v_2 = 1$ and for $\lambda_2 = 1$ equations (2.2)

$$w_1 + w_2 = 0$$
$$w_1 + w_2 = 0$$

have solution $w_1 = 1$, $w_2 = -1$. The points $(v_1, v_2) = (1, 1)$ and $(w_1, w_2) = (1, -1)$ locate the straight line orbits of the saddle. In Fig. 5.11 we see that the line $y = x$, associated with $(v_1, v_2) = (1, 1)$ and the root $\lambda_2 = 1$, is an asymptote for all orbits with positive initial conditions $x(0) > 0$, $y(0) > 0$ (which are the only initial conditions of interest in the armament model). This means the budgets of both countries grow without bound and an "arms race" ensues.

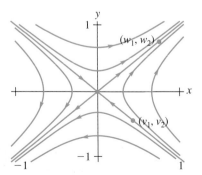

Figure 5.11 The phase plane portrait of the system (4.4) is a saddle, shown here when $a = c = 1$.

5.4.2 Case 2: Complex Conjugate Roots

Suppose the roots of the characteristic equation are complex (conjugates)

$$\lambda = \alpha \pm i\beta, \quad \beta \neq 0.$$

What are the properties of orbits and how does the phase plane portrait look? We begin with an example.

The roots of the characteristic equation

$$\lambda^2 + 1 = 0$$

of the simple harmonic oscillator system

$$x' = y$$
$$y' = -x$$

are the complex conjugate $\lambda = \pm i$ pair. The solution pair

$$x_1 = \cos t$$
$$y_1 = -\sin t$$

defines points lying on a circle. A calculation shows that

$$x_1^2 + y_1^2 = \cos^2 t + \sin^2 t = 1$$

for all t and consequently the orbit lies on the circle of radius 1 centered on the origin. In fact, all solutions have circular orbits. To see this, from the general solution

$$x = c_1 \cos t + c_2 \sin t$$
$$y = -c_1 \sin t + c_2 \cos t$$

we calculate

$$x^2 + y^2 = \left(c_1^2 + c_2^2\right)\left(\cos^2 t + \sin^2 t\right)$$

and hence $x^2 + y^2 = c_1^2 + c_2^2$ for all t. Figure 5.12 shows the resulting phase plane portrait.

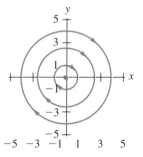

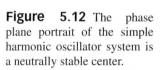

Figure 5.12 The phase plane portrait of the simple harmonic oscillator system is a neutrally stable center.

The phase plane portrait in Fig. 5.12 is called a *center*. A center is said to be neutrally stable. This means that orbits neither approach the origin nor are unbounded as $t \to +\infty$. However, an orbit will remain close (as close as you want) to the origin for all t if it starts close at $t = 0$.

The phase plane portrait of the simple harmonic oscillator is a neutrally stable center. Not all centers consist of circular orbits, however. In general, when the roots of the characteristic equation are purely imaginary (i.e., $\lambda = \pm i\beta$), the orbits are ellipses. For example, the roots of the characteristic equation $\lambda^2 + 25 = 0$ of the system

$$x' = x + 13y \tag{4.5}$$
$$y' = -2x - y$$

are $\lambda = \pm 5i$. Several orbits of this system are plotted in Fig. 5.13.

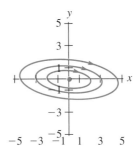

Figure 5.13 The phase plane portrait of the system (4.5) is a center consisting of elliptic orbits.

If the real part of the complex roots are not equal to zero ($\lambda = \alpha \pm i\beta$, $\alpha \neq 0$), then the orbits are no longer ellipses. From the general solution (2.8), we see, for example, that if $\alpha < 0$ all orbits tend to the origin as $t \to +\infty$ (since the exponential term $e^{\alpha t}$ tends to 0 as $t \to +\infty$). The orbits in this case spiral into the origin and the phase plane portrait is a *stable spiral* (or *focus*). If, on the other hand, $\alpha > 0$ the orbits spiral away from the origin and the phase plane portrait is an *unstable spiral* (or *focus*).

For example, the characteristic equation $\lambda^2 + 0.37\lambda + 0.925\,5 = 0$ of the system

$$x' = -0.292x - 4.34y \tag{4.6}$$
$$y' = 0.208x - 0.078y$$

[a particular case of the glucose/insulin model (2.4)] has complex roots $\lambda = -0.185 \pm 0.944i$ with negative real part $\alpha = -0.185$. The phase plane portrait of this system is therefore a stable spiral. See Fig. 5.14.

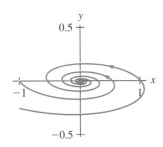

Figure 5.14 The phase plane portrait of the system (4.6) is a stable spiral.

5.4.3 Case 3: A Repeated Root

Suppose the characteristic polynomial

$$\lambda^2 - (a + d)\lambda + (ad - bc)$$

has a repeated root $\lambda \neq 0$. This occurs if (and only if)

$$(a + d)^2 - 4(ad - bc) = 0,$$

in which case the root is

$$\lambda = \frac{1}{2}(a + d)/2.$$

There are two possible general solutions: (2.15) or (2.18).

If $b = c = 0$ and $a = d$ (and hence $\lambda = a$) all orbits associated with the general solution (2.15) are half-lines, as the ratio

$$\frac{y}{x} = \frac{c_2}{c_1}$$

shows. The orbits tend to the origin as $t \to +\infty$ if $\lambda = a < 0$ and tend away from the origin as $t \to +\infty$ if $\lambda > 0$. The phase plane is called a *stable star point* (or a *degenerate node*) in the first case and an *unstable star point* in the second case. See Fig. 5.15.

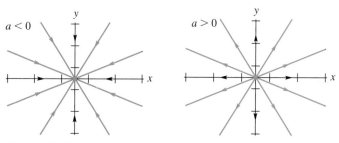

Figure 5.15 The phase plane portrait of the system $x' = ax$, $y' = ay$ is a stable star point if $a < 0$ and an unstable star point if $a > 0$.

If both conditions $b = c = 0$ and $a = d$ are not met (but still assuming $\lambda_1 = \lambda_2$), we have general solution (2.18). What kind of orbits do we have in this case?

Here is an example. The system

$$x' = -x + y \tag{4.7}$$
$$y' = -y$$

has characteristic equation $\lambda^2 + 2\lambda + 1 = (\lambda + 1)^2$ with the repeated root $\lambda = -1$. The general solution (see 2.18) is

$$x = (c_1 + c_2 t)e^{-t}$$
$$y = c_2 e^{-t}.$$

For $c_2 = 0$ we have the solution pairs

$$x_1 = c_1 e^{-t}$$
$$x_2 = 0$$

whose orbits are half-lines lying on the x-axis. These orbits approach the origin as $t \to +\infty$. If we choose $c_2 \neq 0$, we find that the ratio

$$\frac{y}{x} = \frac{c_2}{c_1 + c_2 t}$$

tends to 0 as $t \to +\infty$. This means that all other orbits approach the origin tangentially to the x-axis. See Fig. 5.16. The resulting phase plane portrait is called a *stable improper node* (or *degenerate node*).

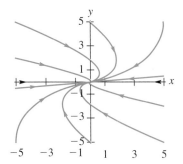

Figure   5.16 The   phase plane portrait of the system (4.7) is a stable improper node.

When $\lambda < 0$ we find from the general solution in this case

$$x = [c_1 v_1 + c_2 (w_1 + v_1 t)] e^{\lambda t}$$
$$y = [c_1 v_2 + c_2 (w_2 + v_2 t)] e^{\lambda t}$$

that all orbits tend to the origin as $t \to +\infty$. There are half-line orbits with slope v_2/v_1 (take $c_2 = 0$) and other orbits (take $c_2 \neq 0$) for which the ratio

$$\frac{y}{x} = \frac{c_1 v_2 + c_2 (w_2 + v_2 t)}{c_1 v_1 + c_2 (w_1 + v_1 t)}$$

tends to the slope v_2/v_1 as $t \to +\infty$. Thus, in general all orbits in a stable improper node tend to the origin with the same tangential direction. Orbits may not be tangential to the x-axis, as in Fig. 5.16. See Fig. 5.17, which shows the phase plane portrait of the system

$$x' = -3x - y \qquad\qquad (4.8)$$
$$y' = x - y.$$

If the repeated root $\lambda > 0$ is positive, the orientation arrows are reversed and the result is an *unstable improper node*.

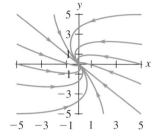

Figure 5.17 The phase plane portrait of the system (4.8) is a stable improper node.

5.4.4 Summary

A summary of the phase plane portraits for (simple) homogeneous systems

$$x' = ax + by$$
$$y' = cx + dy$$
$$ad - bc \neq 0$$

appears in Table 5.2. Although there are infinitely many possible homogeneous systems, there are only a remarkably small number of phase plane types: nodes, saddles, spirals, and centers.

Table 5.2   Phase Portraits of Linear Systems.

Roots of Characteristic Equation	Phase Portrait
Real, different, negative	Stable node
Real, different, positive	Unstable node
Real, opposite signs	Saddle (unstable)
Complex, negative real part	Stable spiral
Complex, positive real part	Unstable spiral
Complex, zero real part	Center (neutrally stable)
Real, equal, negative	Stable (star or improper) node
Real, equal, positive	Unstable (star or improper) node

With the exception of centers, each phase plane type comes in two variants, stable or unstable. A *stable phase plane portrait* is one in which *all* orbits tend to the origin as $t \to +\infty$. An *unstable phase plane portrait* is one in which some orbits move away from the origin and are unbounded as $t \to +\infty$. (Not necessarily all orbits in an unstable phase portrait move away from the origin, as a saddle shows.) From Table 5.2 we see that a phase plane portrait is stable if and only if its characteristic roots are negative, if they are real, or have negative real parts, if they are complex. Since a real number (when viewed as a complex number) is equal to its own real part, this fact is stated more succinctly as follows.

THEOREM 1   The phase plane portrait of a linear homogeneous system is stable if the real parts of all roots of the characteristic equation are negative. The phase plane portrait is unstable if the real part of at least one root is positive.

Note that centers are borderline cases between stable and unstable phase portraits (the roots of their characteristic equation have real part equal to 0). These phase portraits are *neutrally stable*.

EXAMPLE 2

The linear system

$$x' = y$$
$$y' = -x + \alpha y$$

is associated with the nonlinear van der Pol equation

$$x'' + \alpha \left(x^2 - 1 \right) x' + x = 0,$$

which arises in the theory of electric circuits (see Chapter 8). The characteristic equation

$$\lambda^2 - \alpha\lambda + 1 = 0$$

has roots

$$\lambda_1 = \frac{1}{2} \left(\alpha + \sqrt{\alpha^2 - 4} \right), \qquad \lambda_2 = \frac{1}{2} \left(\alpha - \sqrt{\alpha^2 - 4} \right).$$

The roots are complex, with real part α, if $\alpha^2 - 4 < 0$. The roots are real and different if $\alpha^2 - 4 > 0$. If $\alpha^2 = 4$, the root is repeated. From Table 5.2 we obtain the following tabulation of possible phase portraits for this linear system. Note that the system loses stability as α is increased through $\alpha = 0$. ■

$\alpha < -2$	Stable node
$\alpha = -2$	Stable improper node
$-2 < \alpha < 0$	Stable spiral
$\alpha = 0$	Center
$0 < \alpha < 2$	Unstable spiral
$\alpha = 2$	Unstable improper node
$2 < \alpha$	Unstable node

5.4.5 The Trace-Determinant Criteria

The phase plane portrait of a linear homogeneous system

$$x' = ax + by$$
$$y' = cx + dy$$

is determined by the roots of the characteristic equation, as summarized in Table 5.2. The roots of the characteristic equation

$$\lambda^2 - (a + d)\lambda + (ad - bc) = 0$$

seemingly depend on the four coefficients a, b, c, and d in the linear system. However, the roots of a quadratic equation depend only on its coefficients and the coefficients of the characteristic equation are *two* numbers $-(a+d)$ and $ad - bc$. The numbers $a+d$ and $ad - bc$ are called the *trace* and *determinant* of the coefficient matrix

$$A = \begin{pmatrix} a & b \\ c & d \end{pmatrix},$$

respectively. We denote them by

$$\operatorname{tr} A = a + d, \qquad \det A = ad - bc.$$

With this notation the roots of the characteristic equation

$$\lambda^2 - (\operatorname{tr} A)\lambda + \det A = 0$$

are

$$\lambda = \frac{\operatorname{tr} A \pm \sqrt{(\operatorname{tr} A)^2 - 4 \det A}}{2}. \tag{4.9}$$

To determine the phase plane portrait it is only necessary to know the trace and the determinant of the coefficient matrix.

One way to summarize Table 5.2 geometrically is as follows. Consider the two quantities $\operatorname{tr} A$ and $\det A$ as an ordered pair $(\operatorname{tr} A, \det A)$ and hence as the coordinates of a point in a plane. Since knowing $\operatorname{tr} A$ and $\det A$ is sufficient to determine a system's phase portrait, we can associate a unique phase portrait with each point in the plane. (This includes the nonsimple case when $\det A = 0$; see Exercises 37–40.)

For example, the point $(\operatorname{tr} A, \det A) = (2, 2)$ is associated with an unstable spiral because the roots of the quadratic

$$\lambda^2 - (\operatorname{tr} A)\lambda + \det A = \lambda^2 - 2\lambda + 2$$

are the complex conjugates $\lambda = 1 \pm i$ with positive real part (see Table 5.1). The point $(\operatorname{tr} A, \det A) = (2, -2)$ is associated with a saddle because the roots of the quadratic

$$\lambda^2 - (\operatorname{tr} A)\lambda + \det A = \lambda^2 - 2\lambda - 2$$

are $\lambda_1 = 1 - \sqrt{3} < 0$ and $\lambda_2 = 1 + \sqrt{3} > 0$ are of opposite signs.

By associating phase portraits with points in the $(\operatorname{tr} A, \det A)$-plane we obtain a map of phase portraits. For example, the set of points $(\operatorname{tr} A, \det A)$ associated with unstable spirals form a certain region in the plane; the set of points associated with saddles form another region in the plane; and so on. This map is shown in Fig. 5.18.

In Fig. 5.18 the region of points associated with spirals and centers (the phase portraits associated with complex eigenvalues) is separated from the region of points associated with nodes and saddles by a parabola obtained by setting the discriminant under the radical sign in (4.9) equal to 0. This parabola

$$\det A = \frac{1}{4} (\operatorname{tr} A)^2 \tag{4.10}$$

is the set of points associated with stars and improper nodes, because for such points the quadratic formula (4.9) gives a repeated root. A point $(\operatorname{tr} A, \det A)$ lies above the parabola in Fig. 5.18 if $\det A > (\operatorname{tr} A)^2/4$ and below the parabola if $\det A < (\operatorname{tr} A)^2/4$. For points above the parabola, $(\operatorname{tr} A)^2 - 4 \det A < 0$ and the roots (4.9) are complex. By Table 5.2 such points are associated with foci or centers. For points lying below the parabola (4.10) the roots (4.9) are real and are associated with nodes or saddles.

From Fig. 5.18 we notice that a phase portrait is stable (node or spiral) if and only if

$$\operatorname{tr} A < 0 \quad \text{and} \quad \det A > 0.$$

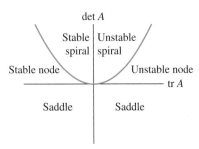

Figure 5.18 The phase portraits of a linear system with coefficient matrix A are determined by the location of the point $(\operatorname{tr} A, \det A)$ on this phase portrait map. The equation of the parabola is $\det A = \frac{1}{4}(\operatorname{tr} A)^2$.

EXAMPLE 3

The equation

$$x'' + cx' + x = 0, \quad c > 0$$

models the suspension system of an automobile. [See Example 1, Sec. 5.3.] This second-order equation has the equivalent first-order system

$$x' = y$$
$$y' = -x - cy$$

whose coefficient matrix

$$A = \begin{pmatrix} 0 & 1 \\ -1 & -c \end{pmatrix}$$

has trace and determinant

$$\operatorname{tr} A = -c, \quad \det A = 1.$$

The point $(\operatorname{tr} A, \det A) = (-c, 1)$ lies in the second quadrant of the phase portrait map in Fig. 5.18. Consequently, the phase portrait is either a stable node or spiral (or improper node or star should the point lie on the parabola). The point lies below the parabola (4.10) if $1 < c^2/4$ and lies above the parabola if $1 > c^2/4$. Therefore, the phase portrait is a stable node if $c > 2$. However, if $c < 2$ the phase portrait is a stable spiral and the solutions are oscillatory. ∎

Exercises

For each of the systems below,

(a) identify the phase portrait type;

(b) sketch the phase portrait;

(c) corroborate your sketch in (b) by using a computer program to draw a direction field and several typical orbits.

1. $\begin{cases} x' = 4x - 2y \\ y' = 7x - 5y \end{cases}$ 2. $\begin{cases} x' = -2x + y \\ y' = x - 2y \end{cases}$

3. $\begin{cases} x' = \frac{1}{2}x - \frac{3}{2}y \\ y' = \frac{3}{2}x + \frac{1}{2}y \end{cases}$ 4. $\begin{cases} x' = x + y \\ y' = x - y \end{cases}$

5. $\begin{cases} x' = -0.012x - 0.45y \\ y' = 2.31x - 3.15y \end{cases}$ 6. $\begin{cases} x' = 0.51x - 0.74y \\ y' = 1.42x + 2.67y \end{cases}$

7. $\begin{cases} x' = 3x + 2y \\ y' = -2x - y \end{cases}$ 8. $\begin{cases} x' = -x + y \\ y' = -x - 2y \end{cases}$

9. $\begin{cases} x' = -x + y \\ y' = 4x + 2y \end{cases}$ 10. $\begin{cases} x' = 3x + 2y \\ y' = -4x - y \end{cases}$

11. $\begin{cases} x' = -6.1x + 0.2y \\ y' = -1.1x - 1.5y \end{cases}$ 12. $\begin{cases} x' = 8.3x + 1.2y \\ y' = -1.8x + 0.3y \end{cases}$

13. $\begin{cases} x' = x + 13y \\ y' = -2x - y \end{cases}$ 14. $\begin{cases} x' = x + 3y \\ y' = 4x + 2y \end{cases}$

15. $\begin{cases} x' = -\frac{1}{2}x + \frac{3}{4}y \\ y' = -3x + \frac{5}{2}y \end{cases}$ 16. $\begin{cases} x' = -5x + 8y \\ y' = -2x + 3y \end{cases}$

For each of the second-order equations below,

(a) identify the phase portrait type;

(b) sketch the phase portrait;

(c) corroborate your sketch in (b) by using a computer program to draw a direction field and several typical orbits.

17. $x'' + x' + x = 0$ 18. $x'' - x' + x = 0$

19. $2x'' - x = 0$ 20. $x'' + 2x' + x = 0$

21. $x'' + 3x' - 4x = 0$ 22. $x'' - 5x' + 4x = 0$

23. $x'' + 5x = 0$ 24. $x'' - 6x' + 9x = 0$

25. $x'' + 3.8x' + 3.45x = 0$

26. Consider two armies engaged in a battle. Let x and y represent the strengths of the two armies (measured, for example, as the number of troops and/or armaments). Assume that the strength of each army decreases at a rate proportional to the strength of the other army. Write a linear system for x and y and determine its phase portrait. What conclusions do you draw from the phase plane portrait about the outcome of the battle?

The system

$$x' = -(0.2 + 0.1)x + 0.5y$$
$$y' = 0.2x - (0.5 + a)y$$

is a model for the amount of drug in the bloodstream x and the amount of drug in the tissues y of a patient. The coefficient a is positive.

27. What type of phase plane portrait does this system have? Show that it is same type for all a and show both x and y tend to 0 in the long run.

28. Sketch a typical phase plane portrait.

29. Suppose $x(0) \geq 0$ and $y(0) \geq 0$ are not both equal to 0. By referring just to the phase plane portrait, show that the ratio y/x of the amount of drug in the tissues to that in the bloodstream approaches a positive limit as $t \to \infty$.

30. Find a formula for $\lim_{t\to\infty} y/x$. Your answer will be in terms of a.

31. Study $\lim_{t\to\infty} y/x$ as a function of $a > 0$. What is the maximum and what is the minimum value of this limit and at which values of a do the occur?

32. Use the map in Fig. 5.18 to determine the phase portrait of the second-order equation $mx'' + cx' + kx = 0$, $m > 0$, $c > 0$, $k > 0$ for the suspension system of an automobile. (See Exercise 13 in Sec. 5.3.)

33. The system

$$x' = -r_1 x + r_2 y$$
$$y' = r_1 x - (r_2 + r_3) y$$

models the amount of pesticide in a stand of trees, x, and its soil bed, y. Use the map in Fig. 5.18 to show that this system has a stable node. All coefficients r_1, r_2, and r_3 are positive.

In the homogeneous linear systems below δ is a real number (positive, negative or zero). Determine the type and stability properties of the phase portrait as they depend on δ.

34. $\begin{cases} x' = -x + (1 - \delta)y \\ y' = x - y \end{cases}$ 35. $\begin{cases} x' = -\delta x + y \\ y' = -x - \delta y \end{cases}$

36. Find the general solution of

$$x' = x + 13y$$
$$y' = -2x - y$$

and show that all (nontrivial) orbits are ellipses.

The following coefficient matrices are for homogeneous linear systems that are not simple. Find the general solution of each system and use it to sketch the phase portrait of the system. It turns out that the phase portrait of any nonsimple linear systems is one of these types (up to a linear change of variables, i.e. to rotations, reflections, and/or rescalings of the coordi- nates axes).

37. $\begin{cases} x' = -x \\ y' = 0 \end{cases}$

38. $\begin{cases} x' = x \\ y' = 0 \end{cases}$

39. $\begin{cases} x' = 0 \\ y' = x \end{cases}$

40. $\begin{cases} x' = 0 \\ y' = 0 \end{cases}$

5.5 Matrices and Eigenvalues

In Chapter 4 (Sec. 4.4) we saw how to represent the homogeneous system

$$x' = ax + by$$
$$y' = cx + dy$$

using matrix notation as

$$\begin{pmatrix} x \\ y \end{pmatrix}' = \begin{pmatrix} a & b \\ c & d \end{pmatrix} \begin{pmatrix} x \\ y \end{pmatrix}$$

or, more concisely, as

$$\tilde{x}' = A\tilde{x}, \tag{5.1}$$

where

$$A = \begin{pmatrix} a & b \\ c & c \end{pmatrix}$$

is the coefficient matrix and

$$\tilde{x} = \begin{pmatrix} x \\ y \end{pmatrix}$$

is the solution vector. The (nontrivial) exponential solution pairs (1.5) are

$$\tilde{x} = \tilde{v}e^{\lambda t}, \qquad \tilde{v} \neq \tilde{0}, \tag{5.2}$$

where

$$\tilde{v} = \begin{pmatrix} v_1 \\ v_2 \end{pmatrix}, \qquad \tilde{0} = \begin{pmatrix} 0 \\ 0 \end{pmatrix}.$$

The general solution is

$$\tilde{x} = \Phi(t)\tilde{c},$$

where

$$\Phi(t) = \begin{pmatrix} x_1(t) & x_2(t) \\ y_1(t) & y_2(t) \end{pmatrix}$$

denotes a fundamental solution matrix of the homogeneous system and

$$\tilde{c} = \begin{pmatrix} c_1 \\ c_2 \end{pmatrix}$$

is a vector of arbitrary constants. The columns of $\Phi(t)$ are two independent solution vectors.

We can calculate a fundamental solution matrix using the procedure in Sec. 5.2. In matrix notation this procedure is as follows. A substitution of (5.2) into the equation (5.1) yields

$$\lambda \tilde{x} e^{\lambda t} = A \tilde{x} e^{\lambda t}.$$

Canceling the exponential $e^{\lambda t}$ and rearranging this equation, we obtain

$$A \tilde{v} - \lambda \tilde{v} = \tilde{0},$$

an equation we can rewrite as

$$(A - \lambda I) \tilde{v} = \tilde{0}, \qquad \tilde{v} \neq \tilde{0}, \tag{5.3}$$

where I denotes the 2×2 identity matrix

$$I = \begin{pmatrix} 1 & 0 \\ 0 & 1 \end{pmatrix}.$$

Equation (5.3) is the matrix form of equations (2.2) in Sec. 5.2. A solution $\tilde{v} \neq \tilde{0}$ of (5.3) yields a nontrivial solution $\tilde{x} = \tilde{v} e^{\lambda t}$ of the homogeneous system (5.1).

A number (real or complex) for which (5.3) has a nonzero solution vector (real or complex) $\tilde{v} \neq \tilde{0}$ is called an *eigenvalue* of the coefficient matrix A. A nonzero solution vector $\tilde{v} \neq \tilde{0}$ is called an *eigenvector*. (*Note*: There is not just one eigenvector associated with an eigenvalue. Any nonzero constant multiple of an eigenvector is also an eigenvector.) The exponential solutions of the homogeneous system (5.1) obtained from eigenvalues and eigenvectors are called *eigensolutions*.

The equation (5.3) has a nontrivial solution vector $\tilde{v} \neq \tilde{0}$ if and only if the matrix $A - \lambda I$ is singular (noninvertible) [i.e., $\det(A - \lambda I) = 0$]. This determinant condition

$$\det \begin{pmatrix} a - \lambda & b \\ c & d - \lambda \end{pmatrix} = \lambda^2 - (a + d)\lambda + (ad - bc) = 0$$

produces the characteristic equation of the coefficient matrix A whose roots are the eigenvalues of A. The characteristic equation can be written

$$\lambda^2 - (\operatorname{tr} A)\lambda + \det A = 0, \tag{5.4}$$

where $\operatorname{tr} A = a + d$ (the sum of the diagonal elements of A) is the *trace* of A. Associated eigenvectors are found by algebraically solving the equation (5.3) for each eigenvalue λ.

There are many computer programs available that will calculate the eigenvalue and eigenvectors of a matrix. Many of these programs will do so not only for matrices A with numerical entries but for matrices with symbolic entries. These programs can be used to find two independent solutions and hence a fundamental solution matrix $\Phi(t)$ and the general solution $\tilde{x} = \Phi(t)\tilde{c}$.

If the eigenvalues $\lambda_1 \neq \lambda_2$ of A are real and different and

$$\tilde{v} = \begin{pmatrix} v_1 \\ v_2 \end{pmatrix}, \qquad \tilde{w} = \begin{pmatrix} w_1 \\ w_2 \end{pmatrix}$$

are associated eigenvectors, then the solutions $\tilde{v} e^{\lambda_1 t}$ and $\tilde{w} e^{\lambda_2 t}$ yield the fundamental solution matrix

$$\Phi(t) = \begin{pmatrix} v_1 e^{\lambda_1 t} & w_1 e^{\lambda_2 t} \\ v_2 e^{\lambda_1 t} & w_2 e^{\lambda_2 t} \end{pmatrix}$$

and the general solution

$$\tilde{x} = \Phi(t)\tilde{c} = \begin{pmatrix} v_1 e^{\lambda_1 t} & w_1 e^{\lambda_2 t} \\ v_2 e^{\lambda_1 t} & w_2 e^{\lambda_2 t} \end{pmatrix} \begin{pmatrix} c_1 \\ c_2 \end{pmatrix}.$$

For example, using a computer we find the eigenvalues of the matrix

$$A = \begin{pmatrix} 1.25 & 1.22 \\ 4.15 & -7.37 \end{pmatrix}$$

to be (rounded to three significant digits)

$$\lambda_1 = -7.92, \qquad \lambda_2 = 1.80$$

and some associated eigenvectors are

$$\tilde{v} = \begin{pmatrix} -0.138 \\ 1.04 \end{pmatrix}, \qquad \tilde{w} = \begin{pmatrix} 0.911 \\ 0.412 \end{pmatrix}.$$

(Remember, eigenvectors are not unique. Your computer may produce multiples of these eigenvectors.) The general solution of the homogeneous system (5.1) is then

$$\tilde{x} = \begin{pmatrix} -0.138 e^{-7.922t} & 0.911 e^{1.802t} \\ 1.04 e^{-7.922t} & 0.412 e^{1.802t} \end{pmatrix} \begin{pmatrix} c_1 \\ c_2 \end{pmatrix}.$$

As an example involving unspecified coefficients, consider the coefficient matrix

$$A = \begin{pmatrix} -\alpha & -\beta \\ \gamma & -\delta \end{pmatrix}$$

of the glucose/insulin regulation system (2.3). A computer program yields the eigenvalues

$$\lambda_1 = \frac{1}{2}\left(-\delta - \alpha + \sqrt{(\delta - \alpha)^2 - 4\beta\gamma} \right)$$

$$\lambda_2 = \frac{1}{2}\left(-\delta - \alpha - \sqrt{(\delta - \alpha)^2 - 4\beta\gamma} \right)$$

and associated eigenvectors

$$\tilde{v} = \begin{pmatrix} \dfrac{\delta - \alpha + \sqrt{(\delta - \alpha)^2 - 4\beta\gamma}}{2\gamma} \end{pmatrix}$$

$$\tilde{w} = \begin{pmatrix} \dfrac{\delta - \alpha - \sqrt{(\delta - \alpha)^2 - 4\beta\gamma}}{2\gamma} \end{pmatrix}.$$

The eigenvalues are real and different provided the coefficients in A satisfy $(\delta - \alpha)^2 - 4\beta\gamma > 0$. In this case, the general solution is

$$\tilde{x} = \begin{pmatrix} \dfrac{\left(\delta - \alpha + \sqrt{(\delta - \alpha)^2 - 4\beta\gamma}\right) e^{\lambda_1 t}}{2\gamma e^{\lambda_1 t}} & \dfrac{\left(\delta - \alpha - \sqrt{(\delta - \alpha)^2 - 4\beta\gamma}\right) e^{\lambda_2 t}}{2\gamma e^{\lambda_2 t}} \end{pmatrix} \begin{pmatrix} c_1 \\ c_2 \end{pmatrix}.$$

If, on the other hand, $(\delta - \alpha)^2 - 4\beta\gamma < 0$ and the eigenvalues and eigenvectors of the coefficient matrix A are complex, then the real and imaginary parts of the complex eigensolution $\tilde{v}e^{\lambda t}$ yield two independent solutions that can be used as the columns in a fundamental solution matrix $\Phi(t)$. From

$$\lambda = \alpha + i\beta, \qquad \tilde{v} = \tilde{u} + i\tilde{w}$$

the complex eigenvalue and eigenvector we have the complex solution

$$\tilde{v}e^{\lambda t} = (\tilde{u} + i\tilde{w})\, e^{\alpha t}\, (\cos\beta t + i\sin\beta t)$$
$$= e^{\alpha t}\, (\tilde{u}\cos\beta t - \tilde{w}\sin\beta t) + ie^{\alpha t}\, (\tilde{w}\cos\beta t + \tilde{u}\sin\beta t)$$

whose real and imaginary parts are

$$e^{\alpha t}\, (\tilde{u}\cos\beta t - \tilde{w}\sin\beta t), \qquad e^{\alpha t}\, (\tilde{w}\cos\beta t + \tilde{u}\sin\beta t).$$

If we denote the components of the real and imaginary parts of the complex eigenvector $\tilde{v}$ by

$$\tilde{u} = \begin{pmatrix} v_1 \\ v_2 \end{pmatrix}, \qquad \tilde{w} = \begin{pmatrix} w_1 \\ w_2 \end{pmatrix},$$

then these two real and independent solutions are

$$e^{\alpha t} \begin{pmatrix} v_1\cos\beta t - w_1\sin\beta t \\ v_2\cos\beta t - w_2\sin\beta t \end{pmatrix}, \qquad e^{\alpha t} \begin{pmatrix} w_1\cos\beta t + v_1\sin\beta t \\ w_2\cos\beta t + v_2\sin\beta t \end{pmatrix}$$

so that a fundamental solution matrix is

$$\Phi(t) = e^{\alpha t} \begin{pmatrix} v_1\cos\beta t - w_1\sin\beta t & w_1\cos\beta t + v_1\sin\beta t \\ v_2\cos\beta t - w_2\sin\beta t & w_2\cos\beta t + v_2\sin\beta t \end{pmatrix}.$$

This gives the general solution

$$\tilde{x} = \Phi(t)\tilde{c} = e^{\alpha t} \begin{pmatrix} (c_1 v_1 + c_2 w_1)\cos\beta t + (-c_1 w_1 + c_2 v_1)\sin\beta t \\ (c_1 v_2 + c_2 w_2)\cos\beta t + (-c_1 w_2 + c_2 v_2)\sin\beta t \end{pmatrix}.$$

Rather than memorize these cumbersome formulas it is generally better in specific problems to calculate a complex eigensolution and extract its real and imaginary parts to use in a fundamental solution matrix. Here is an example.

A complex eigenvalue and eigenvector of the coefficient matrix

$$A = \begin{pmatrix} 1.21 & -4.52 \\ 2.18 & -3.45 \end{pmatrix}$$

are (rounded to three significant digits)

$$\lambda = -1.12 + 2.10i, \qquad \tilde{v} = \begin{pmatrix} -4.52 \\ -2.33 \end{pmatrix} + i\begin{pmatrix} 0 \\ 2.10 \end{pmatrix}.$$

The corresponding complex eigensolution

$$\tilde{v}e^{\lambda t} = \left[\begin{pmatrix} -4.52 \\ -2.33 \end{pmatrix} + i\begin{pmatrix} 0 \\ 2.10 \end{pmatrix}\right] e^{(-1.12+2.10i)t}$$

$$= \left[\begin{pmatrix} -4.52 \\ -2.33 \end{pmatrix} + i\begin{pmatrix} 0 \\ 2.10 \end{pmatrix}\right] e^{-1.12t}\, (\cos 2.10t + i\sin 2.10t)$$

has real part

$$\begin{pmatrix} -4.52 \\ -2.33 \end{pmatrix} e^{-1.12t} \cos 2.10t - \begin{pmatrix} 0 \\ 2.10 \end{pmatrix} e^{-1.12t} \sin 2.10t$$

$$= e^{-1.12t} \begin{pmatrix} -4.52 \cos 2.10t \\ -2.33 \cos 2.10t - 2.10 \sin 2.10t \end{pmatrix}$$

and imaginary part

$$\begin{pmatrix} -4.52 \\ -2.33 \end{pmatrix} e^{-1.12t} \sin 2.10t + \begin{pmatrix} 0 \\ 2.10 \end{pmatrix} e^{-1.12t} \cos 2.10t$$

$$= e^{-1.12t} \begin{pmatrix} -4.52 \sin 2.10t \\ 2.10 \cos 2.10t - 2.33 \sin 2.10t \end{pmatrix}.$$

These solutions, when used as columns, yield a fundamental solution matrix,

$$\Phi(t) = e^{-1.12t} \begin{pmatrix} -4.52 \cos 2.10t & -4.52 \sin 2.10t \\ -2.33 \cos 2.10t - 2.10 \sin 2.10t & 2.10 \cos 2.10t - 2.33 \sin 2.10t \end{pmatrix}.$$

The general $\tilde{x} = \Phi(t)\tilde{c}$ is

$$\tilde{x} = e^{-1.12t} \begin{pmatrix} -4.52 \, (c_1 \cos 2.10t + c_2 \sin 2.10t) \\ (-2.33c_1 + 2.10c_2) \cos 2.10t + (-2.10c_1 - 2.33c_2) \sin 2.10t \end{pmatrix}.$$

Finally, we consider the case when the eigenvalue $\lambda = \lambda_1 = \lambda_2$ of A is repeated. In Sec. 5.2 we saw that there may or may not be a second independent exponential solution pair. If the eigenvalue λ has two independent eigenvectors, then they will produce two independent eigensolutions. However, λ may not have two independent eigenvectors. In either case, a fundamental solution matrix can be found from the formula

$$\Phi(t) = e^{\lambda t} \, (I + (A - \lambda I)t). \tag{5.5}$$

Here is example. The matrix

$$A = \begin{pmatrix} 3 & -1 \\ 1 & 1 \end{pmatrix}$$

has a repeated eigenvalue $\lambda = 2$ with an associated eigenvector

$$\tilde{v} = \begin{pmatrix} 1 \\ 1 \end{pmatrix}.$$

A matrix calculation shows

$$I + (A - \lambda I)t = \begin{pmatrix} 1+t & -t \\ t & 1-t \end{pmatrix}.$$

The formula (5.5) gives the fundamental solution matrix

$$\Phi(t) = e^{2t} \begin{pmatrix} 1+t & -t \\ t & 1-t \end{pmatrix}$$

and the general solution

$$\tilde{x} = e^{2t} \begin{pmatrix} 1+t & -t \\ t & 1-t \end{pmatrix} \begin{pmatrix} c_1 \\ c_2 \end{pmatrix} \tag{5.6}$$
$$= \begin{pmatrix} [(1+t)\,c_1 - c_2 t]\,e^{2t} \\ [c_1 t + (1-t)\,c_2]\,e^{2t} \end{pmatrix}.$$

See Exercise 20.

Exercises

Find a fundamental solution matrix and the general solution for the systems below.

1. $\begin{cases} x' = 4x - 2y \\ y' = 7x - 5y \end{cases}$

2. $\begin{cases} x' = -2x + y \\ y' = x - 2y \end{cases}$

3. $\begin{cases} x' = \frac{1}{2}x - \frac{3}{2}y \\ y' = \frac{3}{2}x + \frac{1}{2}y \end{cases}$

4. $\begin{cases} x' = x + y \\ y' = x - y \end{cases}$

5. $\begin{cases} x' = -0.012x - 0.45y \\ y' = 2.31x - 3.15y \end{cases}$

6. $\begin{cases} x' = 0.51x - 0.74y \\ y' = 1.42x + 2.67y \end{cases}$

7. $\begin{cases} x' = 3x + 2y \\ y' = -2x - y \end{cases}$

8. $\begin{cases} x' = -x + y \\ y' = -x - 2y \end{cases}$

9. $\begin{cases} x' = -x + y \\ y' = 4x + 2y \end{cases}$

10. $\begin{cases} x' = 3x + 2y \\ y' = -4x - y \end{cases}$

11. $\begin{cases} x' = -6.1x + 0.2y \\ y' = -1.1x - 1.5y \end{cases}$

12. $\begin{cases} x' = 8.3x + 1.2y \\ y' = -1.8x + 0.3y \end{cases}$

13. $\begin{cases} x' = x + 13y \\ y' = -2x - y \end{cases}$

14. $\begin{cases} x' = x + 3y \\ y' = 4x + 2y \end{cases}$

15. $\begin{cases} x' = -\frac{1}{2}x + \frac{3}{4}y \\ y' = -3x + \frac{5}{2}y \end{cases}$

16. $\begin{cases} x' = -5x + 8y \\ y' = -2x + 3y \end{cases}$

17. Find a fundamental solution matrix and the general solution of the system

$$x' = -r_1 x + r_2 y$$
$$y' = r_1 x - (r_2 + r_3)\,y$$

in Exercise 33 of Sec. 5.4.

Use a computer program to find the eigensolutions and a fundamental solution matrix of $\tilde{x}' = A\tilde{x}$ for the following coefficient matrices.

18. $A = \begin{pmatrix} 1.256 & -4.500 \\ 2.305 & -0.231 \end{pmatrix}$

19. $A = \begin{pmatrix} -1.2 & 2.4 \\ -1.9 & -2.5 \end{pmatrix}$

20. Show that the formulas (2.19) and (5.6) both yield a general solution of the homogeneous system with coefficient matrix $A = \begin{pmatrix} 3 & -1 \\ 1 & 1 \end{pmatrix}$.

Find a fundamental solution matrix and the general solution for the homogeneous system with coefficient matrices given below.

21. $A = \begin{pmatrix} -7 & -4 \\ 4 & 1 \end{pmatrix}$

22. $A = \begin{pmatrix} \frac{3}{4} & -\frac{1}{2} \\ 2 & \frac{11}{4} \end{pmatrix}$

5.6 Systems of Three or More Equations

In many applications, systems consisting of more than two first-order differential equations arise. For example, ecosystem modelers often use compartmental models to account for quantities transferring into and out of subsystems. Since ecosystems can involve a large number of subsystems, such compartmental models often involve a large number (even hundreds) of differential equations.

Here is an example involving three compartments. Scientists can use a radioactive isotope to trace the flow of nutrients in food chains. Suppose a radioactive isotope is placed into the water of an aquarium in order to trace the flow of nutrients in an aquatic food chain consisting of zooplankton and phytoplankton. Let x_1, x_2, and x_3 denote the concentration of the isotope in the water, phytoplankton, and zooplankton respectively. The compartment diagram in Fig. 5.19 shows the (linear) transfer rates between these subsystems of the food chain. The differential equations for this compartmental model are

$$x_1' = (-0.02 - 0.01)x_1 + 0.06x_2 + 0.05x_3$$
$$x_2' = 0.02x_1 + (-0.06 - 0.06)x_2 \qquad (6.1)$$
$$x_3' = 0.01x_1 + 0.06x_2 - 0.05x_3.$$

This is a linear homogeneous system of equations with constant coefficients.

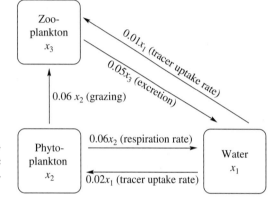

Figure 5.19 The compartmental diagram for an aquatic food chain. (Rates in microcuries per hour.)

The eigenvalue method for constructing solutions of two first-order linear equations applies to systems of three or more first-order equations. Consider a system

$$\tilde{x}' = A\tilde{x}$$

of n first-order equations for n unknowns

$$\tilde{x} = \begin{pmatrix} x_1 \\ x_2 \\ \vdots \\ x_n \end{pmatrix},$$

where

$$A = \begin{pmatrix} a_{11} & a_{12} & \cdots & a_{1n} \\ a_{21} & a_{22} & \cdots & a_{2n} \\ \vdots & \vdots & & \vdots \\ a_{n1} & a_{n2} & \cdots & a_{nn} \end{pmatrix}$$

is an $n \times n$ matrix. An eigenvalue λ is a root of the characteristic (nth-degree polynomial) equation

$$\det(A - \lambda I) = 0,$$

where I is the $n \times n$ identity matrix

$$I = \begin{pmatrix} 1 & 0 & \cdots & 0 \\ 0 & 1 & \cdots & 0 \\ \vdots & \vdots & & \vdots \\ 0 & 0 & \cdots & 1 \end{pmatrix},$$

that is,

$$\det(A - \lambda I) = \det \begin{pmatrix} a_{11} - \lambda & a_{12} & \cdots & a_{1n} \\ a_{21} & a_{22} - \lambda & \cdots & a_{2n} \\ \vdots & \vdots & & \vdots \\ a_{n1} & a_{n2} & \cdots & a_{nn} - \lambda \end{pmatrix}.$$

Eigenvectors are solutions of the algebraic equations

$$(A - \lambda I)\widetilde{v} = \widetilde{0}, \quad \widetilde{v} \neq \widetilde{0}.$$

From the eigenvalue λ and eigenvector $\widetilde{v}$ we construct an eigensolution solution

$$\widetilde{x} = \widetilde{v}e^{\lambda t} \tag{6.2}$$

of the system $\widetilde{x}' = A\widetilde{x}$.

The general solution of an n-dimensional system is a linear combination

$$\widetilde{x} = \sum_{i=1}^{n} c_i \widetilde{x}_i(t),$$

where c_i is an arbitrary constant and

$$\widetilde{x}_i(t) = \begin{pmatrix} x_{1i}(t) \\ x_{2i}(t) \\ \vdots \\ x_{ni}(t) \end{pmatrix}, \quad i = 1, 2, \ldots, n$$

are n independent solutions of the system.[3] Any set of n independent solutions forms an $n \times n$ *fundamental solution matrix*

$$\Phi(t) = \begin{pmatrix} x_{11}(t) & x_{12}(t) & \cdots & x_{1n}(t) \\ x_{21}(t) & x_{22}(t) & \cdots & x_{2n}(t) \\ \vdots & \vdots & & \vdots \\ x_{n1}(t) & x_{n2}(t) & \cdots & x_{nn}(t) \end{pmatrix}$$

whose columns are the independent solutions.[4] In matrix notation the general is

$$\widetilde{x} = \Phi(t)\widetilde{c},$$

---

[3]The solutions $\widetilde{x}_i(t)$ are *dependent* if there are constants c_i not all 0 such that $\sum_{i=1}^{n} c_i \widetilde{x}_i(t) = \widetilde{0}$ for all t. The solutions are *independent* if they are not dependent.
[4]The solutions $\widetilde{x}_i(t)$ are independent if and only if $\det \Phi(t) \neq 0$ for all t.

where

$$
\tilde{c} = \begin{pmatrix} c_1 \\ c_2 \\ \vdots \\ c_2 \end{pmatrix}
$$

is a vector of arbitrary constants c_i.

To find the general solution it is necessary to find n independent solutions. If there are no repeated eigenvalues (i.e., no repeated roots of the characteristic equation), then we can find n independent eigensolutions of the form (6.2). Each real eigenvalue/eigenvector pair contributes a solution $e^{\lambda t}\tilde{v}$ and each complex conjugate pair contributes two real solutions obtained from the real and imaginary parts of $e^{\lambda t}\tilde{v}$. (We ignore here the complicated case of repeated roots.)

EXAMPLE 1

The coefficient matrix

$$
A = \begin{pmatrix} -0.030 & 0.060 & 0.050 \\ 0.020 & -0.120 & 0.000 \\ 0.010 & 0.060 & -0.050 \end{pmatrix}
$$

of the aquatic food chain model (6.1) has eigenvalues and eigenvectors

$$
\lambda_1 \approx -0.07551, \qquad \tilde{v}_1 \approx \begin{pmatrix} 0.5169 \\ 0.2324 \\ -0.7493 \end{pmatrix}
$$

$$
\lambda_2 \approx -0.1245, \qquad \tilde{v}_2 \approx \begin{pmatrix} 0.2677 \\ -1.191 \\ 0.9234 \end{pmatrix}
$$

$$
\lambda_3 = 0, \qquad \tilde{v}_3 \approx \begin{pmatrix} 0.9415 \\ 0.1569 \\ 0.3766 \end{pmatrix}.
$$

These yield the fundamental solution matrix

$$
\Phi(t) \approx \begin{pmatrix} 0.5169e^{-0.07551t} & 0.2677e^{-0.1245t} & 0.9415 \\ 0.2324e^{-0.07551t} & -1.191e^{-0.1245t} & 0.1569 \\ -0.7493e^{-0.07551t} & 0.9234e^{-0.1245t} & 0.3766 \end{pmatrix}
$$

and the general solution $\tilde{x} = \Phi(t)\tilde{c}$ (rounded to two significant digits)

$$
\begin{pmatrix} x_1 \\ x_2 \\ x_3 \end{pmatrix} \approx \begin{pmatrix} 0.52c_1e^{-0.076t} + 0.27c_2e^{-0.12t} + 0.94c_3 \\ 0.23c_1e^{-0.076t} - 1.20c_2e^{-0.12t} + 0.16c_3 \\ -0.75c_1e^{-0.076t} + 0.92c_2e^{-0.12t} + 0.38c_3 \end{pmatrix}.
$$

To solve an initial value problem

$$
\tilde{x}' = A\tilde{x}
$$
$$
\tilde{x}(0) = \tilde{x}_0
$$

we choose the arbitrary constants in the vector $\tilde{c}$ appropriately. Namely, we solve the equation

$$\Phi(0)\tilde{c} = \tilde{x}_0$$

for

$$\tilde{c} = \Phi^{-1}(0)\tilde{x}_0$$

to obtain the solution

$$\tilde{x} = \Phi(t)\Phi^{-1}(0)\tilde{x}_0$$

EXAMPLE 2

Suppose 100 microcuries of tracer are introduced into the water of the aquatic food chain modeled by Fig. 5.19. Suppose no tracer is initially present in either the phytoplankton or the zooplankton. This provides the initial conditions

$$\tilde{x}_0 = \begin{pmatrix} 100 \\ 0 \\ 0 \end{pmatrix} \tag{6.3}$$

for the differential system (6.1). From the fundamental solution matrix found in Example 1 we obtain the solution $\tilde{x} = \Phi(t)\Phi^{-1}(0)\tilde{x}_0$, namely

$$x(t) \approx \begin{pmatrix} 31e^{-0.076t} + 5.5e^{-0.12t} + 64 \\ 14e^{-0.076t} - 25e^{-0.12t} + 11 \\ -44e^{-0.076t} + 19e^{-0.12t} + 25 \end{pmatrix}.$$

As an application, note that the formula implies that the tracer amounts in each compartment tend to constant levels as $t \to +\infty$, namely 64, 11, and 25 (microcuries) for the water, phytoplankton, and zooplankton, respectively. The graphs of each solution component of the initial value problem appear in Fig. 5.20. ∎

Figure 5.20 Shown are computer generated graphs of the solutions of the initial value problem (6.3) for the aquatic food chain system (6.1).

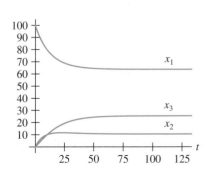

Exercises

Use a computer to find the eigenvalues and eigenvectors of the matrices below. Then find a fundamental solution matrix $\Phi(t)$ for the homogeneous linear system with the coefficient matrix A.

1. $A = \begin{pmatrix} 3 & -6 & -2 \\ 2 & 3 & 2 \\ -2 & 6 & 3 \end{pmatrix}$

2. $A = \begin{pmatrix} -4 & 1 & 2 \\ -22 & 9 & 12 \\ 8 & -4 & -5 \end{pmatrix}$

3. $A = \begin{pmatrix} 7 & 4 & 6 \\ -5 & -3 & -4 \\ -5 & -2 & -5 \end{pmatrix}$

4. $A = \begin{pmatrix} -5 & -10 & 1 \\ 2 & 6 & -2 \\ 6 & 10 & 0 \end{pmatrix}$

5. $A = \begin{pmatrix} -5 & 6 & -3 & -2 \\ -5 & 6 & -1 & -2 \\ 1 & -1 & 2 & 0 \\ -5 & 5 & 0 & -1 \end{pmatrix}$

6. $A = \begin{pmatrix} -13 & -21 & 24 & -15 \\ 11 & 16 & -15 & 11 \\ 4 & 7 & -7 & 4 \\ 5 & 8 & -9 & 7 \end{pmatrix}$

7. Find a formula for the solution of the initial value problem

$$\tilde{x}(0) = \begin{pmatrix} 1 \\ 0 \\ -1 \end{pmatrix}$$

for the systems with coefficient matrices given in Exercises 1–4. (Use a computer as an aid in performing the necessary matrix algebra.)

8. Solve the initial value problem

$$\tilde{x}(0) = \begin{pmatrix} 1 \\ 2 \\ -2 \\ -1 \end{pmatrix}$$

for the systems with coefficient matrices given in Exercises 5–6. (Use a computer as an aid in performing the necessary matrix algebra.)

Find a formula for the general solution of the following higher-order equations.

9. $x''' + 4x'' + x' - 6x = 0$

10. $x''' - 2x'' - \frac{1}{4}x' + \frac{1}{2}x = 0$

11. $x'''' + x''' - 7x'' - x' + 6x = 0$

12. $4x'''' - 5x''' - 39x'' - 22x' + 8x = 0$

13. Find a formula for the solution of the initial value problem $x(0) = 1$, $x'(0) = 0$, $x''(0) = -1$ for the equations in Exercises 9–10.

14. Find a formula for the solution of the initial value problem $x(0) = 1$, $x'(0) = 2$, $x''(0) = 0$, $x'''(0) = -1$ for the equations in Exercises 11–12.

The system below arises from a compartmental model for biomass transfer in a pine-oak forest. The compartments are vegetation (x), litter (y), and humus (z). The unit of time is one year.

$$x' = -\frac{7}{10}x$$
$$y' = \frac{7}{10}x - \frac{3}{10}y$$
$$z' = \frac{3}{10}y - \frac{1}{10}z$$

Suppose the forest is initially free of litter and humus and starts with x_0 units of biomass in vegetation.

15. Use a computer to explore the solution of the resulting initial value problem. How long does it take for the vegetation to decrease by 90%? At what time will the litter biomass be maximum? At what time will the humus biomass be maximum? How do your answers depend on x_0?

16. Find a fundamental solution matrix and the general solution.

17. Find a formula for the solution of the initial value problem.

18. Use your answer in Exercise 17 to corroborate your answers in Exercise 15.

5.7 Chapter Summary and Exercises

We can obtain solution formulas for linear homogeneous systems with constant coefficients

$$x' = ax + by$$
$$y' = cx + dy$$

by using the roots of the characteristic polynomial $\lambda^2 - (a + d)\lambda + ad - bc$. The graph of the points $(x(t), y(t))$ in the x, y-plane (called the phase plane) produces the orbit associated with a solution pair $x = x(t)$, $y = y(t)$. The collection of all orbits is the phase plane portrait of the system. We categorized phase plane portraits into a few basic types: nodes, saddles, spirals, and centers. The roots of the characteristic polynomial determine the phase portrait type (as in Table 5.2). Second-order homogeneous equations with constant coefficients are equivalent to a homogeneous first-order system with constant coefficients. Solution formulas for second-order equations appear in Table 5.1. The solution procedure for systems of two equations also applies to systems of more than two equations.

Exercises

Find a formula for the general solution of each system below.

1. $\begin{cases} x' = -5x \\ y' = -5y \end{cases}$

2. $\begin{cases} x' = 3x - y \\ y' = x + y \end{cases}$

3. $\begin{cases} x' = -3x - y \\ y' = 7x + y \end{cases}$

4. $\begin{cases} x' = 1.33x - 4.31y \\ y' = 8.97x - 1.33y \end{cases}$

5. $\begin{cases} x' = -4.1x - 5.2y \\ y' = 10.1x + 4.1y \end{cases}$

6. $\begin{cases} x' = \sqrt{5}x - y \\ y' = 7x + y \end{cases}$

7. $\begin{cases} x' = -\frac{1}{2}y \\ y' = 5x - \frac{1}{4}\pi y \end{cases}$

8. $\begin{cases} x' = (\sin\theta)x + (\cos\theta)y \\ y' = (\cos\theta)x - (\sin\theta)y, \end{cases}$

where θ is a real number

9. $\begin{cases} x' = 3e^{-a}x - 2e^{-2a}y \\ y' = 2x - e^{-a}y, \end{cases}$

where a is a real number

10. $\begin{cases} x' = x + y \\ y' = x - ay, \end{cases}$

where a is a constant satisfying $-1 < a < 1$

11. $\begin{cases} x' = 0.31x + 4.79y \\ y' = -1.84x - 1.73y \end{cases}$

12. $\begin{cases} x' = -2.3x + .79y \\ y' = 1.84x - 1.73y \end{cases}$

13. Find a formula for the solution of the initial value problem $x(0) = 1$, $y(0) = -1$ for the systems in Exercises 1–12.

14. Find a formula for the solution of the initial value problem $x(0) = 2$, $y(0) = 3$ for the systems in Exercises 1–12.

Find a formula for the general solution of the second-order equations below.

15. $Lx'' + Rx' + \frac{1}{C}x = 0$, where L, R and C are positive constants and $R > 2\sqrt{L/C}$

16. $x'' + (c - d)x - (1 + c)x = 0$, where c and d are positive constants

17. What type of phase plane portrait do the second-order equations appearing in Exercise 15 and Exercise 16 have (stable node, unstable node, saddle, etc.)?

The compartment diagram for the pesticide DDT in an agricultural crop and its soil is as follows:

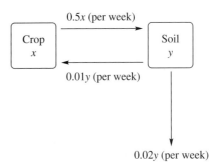

18. Suppose there is initially no DDT in the soil and x_0 units of DDT are sprayed onto the crops. Write an initial value

problem for x (the units of DDT in the crop) and y (the units of DDT in the soil).

19. Use a computer to investigate the solution of the initial value problem in Exercise 18 for a variety of initial dosages $x_0 > 0$. What happens as $t \to +\infty$? What happens to the ratio of DDT in the soil to that in the crop as $t \to +\infty$? How do your answers depend on the initial dosage x_0?

20. Find a formula for the solution of the initial value problem in Exercise 18.

21. Use you answer in Exercise 20 to corroborate your answer in Exercise 19.

The compartmental diagram for the movement of potassium (using a radioactive isotope of potassium as a tracer) between red blood cells and plasma is as follows:

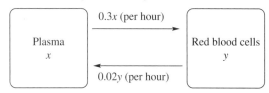

Suppose there is initially no tracer in the red blood cells and x_0 units of tracer are injected into the plasma.

22. Write an initial value problem for x (the units of tracer in the plasma) and y (the units of tracer in the red blood cells).

23. Use a computer to investigate the solution of the initial value problem in Exercise 22 for a variety of initial dosages $x_0 > 0$. What happens as $t \to +\infty$? What happens to the ratio of tracer in the red blood cells to that in the plasma as $t \to +\infty$? How do your answers depend on the initial dose x_0?

24. Find a formula for the solution of the initial value problem in Exercise 22.

25. Use you answer in Exercise 24 to corroborate your answer in Exercise 23.

The following steps describe the Putzer algorithm[5] for finding a fundamental solution matrix $\Phi(t)$ of a linear system $x' = Ax$ with constant coefficients. This method has the advantage that multiple eigenvalues cause no difficulties.

STEP 1: *List the eigenvalues of A with multiplicities:* $\lambda_1, \lambda_2, \lambda_3, \ldots, \lambda_n$.

STEP 2: *Let $P_1 = I$ be the $n \times n$ identity matrix and construct the matrices*

$$P_2 = A - \lambda_1 I$$
$$P_3 = (A - \lambda_2 I) P_2$$
$$\vdots$$
$$P_n = (A - \lambda_{n-1}I) P_{n-1}.$$

STEP 3: *Set $r_1(t) = e^{\lambda_1 t}$ and solve the (linear nonhomogeneous) initial value problems*

$$r_2' = \lambda_2 r_2 + r_1(t), \qquad r_2(0) = 0$$
$$r_3' = \lambda_3 r_3 + r_2(t), \qquad r_2(0) = 0$$
$$\vdots$$
$$r_n' = \lambda_n r_n + r_{n-1}(t), \qquad r_n(0) = 0$$

for $r_2(t), r_3(t), \ldots, r_n(t)$.

STEP 4: *Construct the fundamental solution matrix*

$$\Phi(t) = r_1(t)P_1 + r_2(t)P_2 + \cdots + r_n(t)P_n.$$

Use the Putzer algorithm to find a fundamental solution matrix for systems with the following coefficient matrices.

26. $A = \begin{pmatrix} 4 & 1 & 0 \\ -1 & 2 & 0 \\ 2 & 1 & -3 \end{pmatrix}$ **27.** $A = \begin{pmatrix} -3 & 2 \\ 7 & 2 \end{pmatrix}$

28. $A = \begin{pmatrix} 3 & -6 & -2 \\ 2 & 3 & 2 \\ -2 & 6 & 3 \end{pmatrix}$ **29.** $A = \begin{pmatrix} -4 & 1 & 2 \\ -22 & 9 & 12 \\ 8 & -4 & -5 \end{pmatrix}$

5.8.1 Drug Kinetics

Compartment models are often useful in modeling the administering of pharmaceutical drugs to patients. A compartment in this case might represent a specific physical organ, such as a liver or kidney, or a more diffuse entity such as body tissue or the bloodstream. In this section we consider a two-compartment model for the kinetics of a drug in which one compartment is the bloodstream and the other consists of the

[5]R. A. Horn and C. R. Johnson, *Topics in Matrix Analysis*, Cambridge University Press, Cambridge, 1995.

body tissues into which and out of which a drug is transported. We are interested in the amount of drug in each of these compartment, as a function time, after a dose is introduced into the tissues by an injection. (In Chapter 6, Sec. 6.5.1, we consider the case when the drug is continuously introduced into the bloodstream intravenously.)

In our model we assume there is drug loss from the bloodstream as the blood passes through the kidneys (renal clearance) and that there is an exchange of the drug between the bloodstream and the body tissues. See Fig. 5.21. We carry out the modeling step of the modeling cycle by mathematically describing the flow rates into and out of the two compartments and applying the balance law to each compartment.

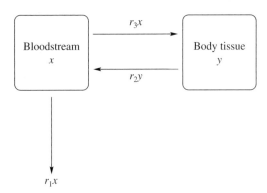

Figure 5.21

Let

$$x = \text{amount of drug in the blood}$$
$$y = \text{amount of drug in body tissue.}$$

These amounts could be measured, for example, in milligrams (mg). (*Note:* These are amounts, not concentrations.) We assume that the drug transfer rates between the blood and tissues, as well the transfer rate out of the blood by renal clearance, are at all times t proportional to the drug amounts present in the compartment at time t. Thus, in the balance equation

$$x' = \left\{ \begin{array}{c} \text{inflow rate of drug} \\ \text{into the bloodstream} \end{array} \right\} - \left\{ \begin{array}{c} \text{outflow rate of drug} \\ \text{from the bloodstream} \end{array} \right\}$$

for the drug amount x in the bloodstream, we have

$$\left\{ \begin{array}{c} \text{inflow rate of drug} \\ \text{into the bloodstream} \end{array} \right\} = \left\{ \begin{array}{c} \text{inflow rate} \\ \text{from tissue} \end{array} \right\} = r_2 y$$

and

$$\left\{ \begin{array}{c} \text{outflow rate} \\ \text{from the bloodstream} \end{array} \right\} = \left\{ \begin{array}{c} \text{renal clearance} \\ \text{rate from bloodstream} \end{array} \right\} + \left\{ \begin{array}{c} \text{outflow rate} \\ \text{to tissues} \end{array} \right\}$$

$$= r_1 x + r_3 x.$$

Here

$$r_1, \quad r_2, \quad r_3 > 0$$

are positive constants of proportionality. Thus, the balance equation yields the differential equation

$$x' = r_2 y - (r_1 x + r_3 x)$$

for the amount of drug x in the bloodstream.

For the tissue compartment we have

$$y' = \left\{ \begin{array}{c} \text{inflow rate of drug} \\ \text{into the tissues} \end{array} \right\} - \left\{ \begin{array}{c} \text{outflow rate of drug} \\ \text{from the tissues} \end{array} \right\}$$

or

$$y' = r_3 x - r_2 y.$$

Suppose by injection an amount y_0 is introduced into the tissues at time $t_0 = 0$ when there is no drug presence in the bloodstream. The initial value problem modeling this situation is

$$\begin{aligned} x' &= -(r_1 + r_3)\, x + r_2 y \\ y' &= r_3 x - r_2 y \\ x(0) &= 0, \qquad y(0) = y_0. \end{aligned} \tag{8.1}$$

In order to be effective, a minimal concentration of the drug must be present in a patient's bloodstream. On the other hand, it will be dangerous (and possibly lethal) to the patient if the drug concentration in the bloodstream gets too high. We want to use the model to answer the following questions: What range of injection doses y_0 will produce an effective concentration in the bloodstream, at least for some interval of time, without causing the lethal concentration level to be reached at any time? For dosages in this range, how long will there be an effective concentration in the patient's bloodstream, that is, how soon will the patient need another injection?

Let's begin with some general observations about the solution of the initial value problem (8.1). The first-order system is linear and homogeneous with coefficient matrix

$$\begin{pmatrix} -(r_1 + r_3) & r_2 \\ r_3 & -r_2 \end{pmatrix}.$$

The roots of the characteristic polynomial (i.e., the eigenvalues) of this matrix are

$$\begin{aligned} \lambda_\pm &= \frac{1}{2}\left(-(r_1 + r_3 + r_2) \pm \sqrt{(r_1 - r_2)^2 + r_3\,(2r_1 + 2r_2 + r_3)} \right) \\ &= \frac{1}{2}\left(-(r_1 + r_3 + r_2) \pm \sqrt{(r_1 + r_3 + r_2)^2 - 4r_1 r_2} \right). \end{aligned}$$

From these formulas we see that both λ_+ and λ_- are real and negative, specifically,

$$\lambda_- < \lambda_+ < 0.$$

It follows that the equilibrium $(x, y) = (0, 0)$ is a stable node. In the context of the application, this means over time that the drug disappears from both the bloodstream and the tissues.

Using the methods of Sec. 5.2, we can obtain formulas for the solution pair of the initial value problem (8.1). See Exercise 1. From these formulas we find that

$$x(t) = \frac{y_0}{\left((r_1 - r_2)^2 + r_3\,(2r_1 + 2r_2 + r_3) \right)^{\frac{1}{2}}} r_2 \left(e^{\lambda_+ t} - e^{\lambda_- t} \right). \tag{8.2}$$

This formula shows that x (which vanishes at $t = 0$) is positive for all $0 < t < +\infty$ and tends to 0 as $t \to +\infty$. A little calculus shows that the term $e^{\lambda_+ t} - e^{\lambda_- t}$—and consequently $x(t)$—increases to its maximum at time

$$t_{\max} = \frac{1}{\lambda_+ - \lambda_-} \ln\left(\frac{\lambda_-}{\lambda_+}\right) \tag{8.3}$$

after which it decreases to 0 as $0 < t < +\infty$.

We can use these mathematical results as follows. In order to attain an effective amount of drug in the bloodstream at some point in time, the maximum value of $x(t)$ must, of course, exceed the effective amount. On the other hand, in order to keep $x(t)$ below the lethal amount for all time, we must keep the maximum value of $x(t)$ less than the lethal amount. Thus, our goal is to administer a dose y_0 so that the maximum of $x(t)$ lies between the effective and lethal levels. Once we have numerical values for the rate coefficients r_i for a particular patient and a specific drug, we can use the formula (8.2) for $x(t)$ to determine the appropriate range of doses y_0 so that this criterion is met.

Here is an example. The drug *lidocaine* is used to treat irregular heartbeats (ventricular arrhythmia) and the simple compartmental model like (8.1) has been used to simulate the treatment. Typical or normal values of the rate coefficients in this model have been estimated to be (approximately)

$$r_1 = 2.40 \times 10^{-2} \text{ (per minute)}$$
$$r_2 = 3.80 \times 10^{-2} \text{ (per minute)} \tag{8.4}$$
$$r_3 = 6.60 \times 10^{-2} \text{ (per minute)}.$$

These parameter values, placed in (8.1), give the initial value problem

$$x' = -\left(9.00 \times 10^{-2}\right)x + \left(3.80 \times 10^{-2}\right)y$$
$$y' = \left(6.60 \times 10^{-2}\right)x - \left(3.80 \times 10^{-2}\right)y \tag{8.5}$$
$$x(0) = 0, \qquad y(0) = y_0$$

for the amounts of lidocaine (in mg) in the bloodstream and tissues. We can obtain formulas for the solution of this initial value problem using the method of Sec. 5.2. Alternatively, to find a formula for x we can simply substitute the values (8.4) for the rates r_i into the general formula (8.2) above. The result is the formula

$$x(t) = 0.3367 y_0 \left(e^{-7.573\,1 \times 10^{-3} t} - e^{-0.12043 t}\right) \tag{8.6}$$

for the amount of lidocaine in the patient's bloodstream at time t after an injection dose of y_0 (mg).

Lidocaine needs to be at a bloodstream *concentration* of 1.5 mg/liter in order to be effective. However, a concentration of 6.0 mg/liter or higher in the bloodstream is dangerous and possibly lethal. The length of time $t = t_e$ (minutes) that it takes for the lidocaine concentration to reach the effective bloodstream level of 1.5 mg/liter depends both on the blood volume of the patient and on the initial dose y_0. To determine what these concentrations imply about the amounts of lidocaine in the bloodstream and body tissues, x and y, we need to know the blood volume of the patient.

Consider a patient with a blood volume of 5 liters. This is typical for a human weighing about 70 kg. For this patient the effective amount of lidocaine in the bloodstream is

$$1.5 \text{ mg/liter} \times 5 \text{ liter} = 7.5 \text{ mg}$$

and the "lethal" level of lidocaine in the bloodstream is

$$6 \text{ mg/liter} \times 5 \text{ liter} = 30 \text{ mg}.$$

We now turn our attention to the dosage criterion described above. An application of the calculus [or formula (8.3)] shows that the maximum of $x(t)$ in (8.6) is attained at time $t_{max} = 24.53$ (minutes). Therefore, the maximum of $x(t)$ is approximately

$$x(t_{max}) = 0.2621 y_0.$$

According to our criterion, we need to determine those injection doses y_0 such that

$$7.5 < x(t_{max}) < 30$$

or, in other words, y_0 must satisfy

$$28.62 < y_0 < 114.5. \qquad (8.7)$$

Injection doses y_0 in the interval (8.7) ensure that the effective amount (concentration) is reached at some time and also that the lethal amount (concentration) is never reached. See Fig. 5.22.

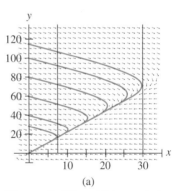

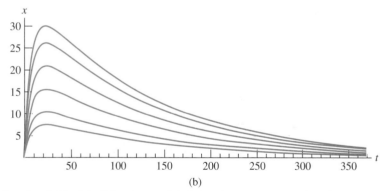

(a)   (b)

Figure 5.22 (a) This graph shows the direction field of the lidocaine model (8.5). Also shown are orbits of the initial value problem associated with the endpoints of the effective, but safe dosage interval (8.7), together with several orbits for doses lying within the interval. The vertical lines located at $x = 7.5$ and 30 indicate the effective and safe levels for the amount x in the bloodstream. (b) This graph shows the x component of the orbits shown in (a).

We would also like to know how long it takes for an effective amount of lidocaine to be reached and how long an effective amount remains in the bloodstream (so that we know when to administer another injection). The answers to these questions depend on the dosage y_0. In Fig. 5.22(b) we see that the effective interval of time is maximal when the maximal safe dose $y_0 = 114.5$ (mg) is administered. Using this dosage in (8.6), we obtain

$$x(t) = 38.55 \left(e^{-7.573 \times 10^{-3} t} - e^{-0.1204 t} \right).$$

Let t_e denote the earliest time at which the effective dose is attained [i.e., at which $x(t) = 37.5$]. Since $x(t)$ ultimately tends to 0, there is also a later time t_l at which $x(t) = 7.5$ [and after which $x(t)$ is always less than 7.5]. See Fig. 5.22(b). These two crucial times are the two roots of the equation $x(t) = 7.5$, that is, of the equation

$$38.55\left(e^{-7.573\,1\times10^{-3}t} - e^{-0.12043t}\right) = 7.5. \tag{8.8}$$

With the aid of a calculator or computer we find these roots to be

$$t_e = 1.949 \ (\text{min}), \quad t_l = 216.1 \ (\text{min}). \tag{8.9}$$

Thus, for the maximal safe dose administered to this patient we expect the drug to take effect in approximately 2 minutes and remain effective for approximately 216 minutes (or about three and one half hours) while at no time exceeding the lethal level.

Although another injection is needed at time t_l, if the drug level in the bloodstream is to be maintained at an effective level, note that the initial value problem (8.1) is not an appropriate model for next injection. This is because of the residual amount of lidocaine present in the tissues and bloodstream at time t_l left over from the first injection. The initial condition for the next injection will have to include these residual amounts, namely, the new initial conditions should be $x(t_l)$ for x and $y(t_l) + y_0$ for y. See Exercise 2.

5.8.2 Oscillations

In many applications the oscillation of a quantity is of primary interest. Problems involving periodic oscillations arise in applications arising from virtually all scientific and engineering disciplines. The mathematical study of periodic solutions of differential equations is often very challenging. However, for linear systems a great deal can be done using the solution methods developed in this chapter. In this application we use these methods to study a linear system (second-order equation) that has proved to be fundamentally important for the study of periodic oscillations. We will study this system in the context of a mechanical model constructed from an object attached to a spring.

Such a mass-spring system is not particularly important in and of itself. Its importance derives from being a basic prototype for oscillators in general. The reason for this is that the mathematical description of the mass-spring system involves a differential equation that often arises in applications from many different disciplines and contexts. The interpretations of the variables and model parameters change from application to application, but from a mathematical point of view the equation is the same. To illustrate this point we will also take a brief look at models for electric circuits.

An advantage of the mass-spring system is that it is conceptually simple, and it is not difficult to study in laboratories and classrooms. The fact that the model equations are mathematically the same as those for other oscillatory systems permits one to think about other oscillatory systems in ways analogous to the mass-spring system. This turns out to be very useful for gaining insight into systems about which one may have less intuition (e.g., electric circuits, vibrating molecules, sound waves, musical tones, water waves, predator-prey interactions).

Consider an object with mass m suspended by a spring from the ceiling. Suppose the object is put into motion by either an initial displacement from its rest position or by striking it so as to give it an initial velocity (or both). Suppose this is done in such a way that the mass moves rectilinearly (i.e., straight up and down, along a vertical straight line only). We wish to know the displacement $x = x(t)$ of the object from its rest position as a function of time t. By assumption, the initial conditions $x(0) = x_0$ and/or $x'(0) = v_0$ are not both equal to 0. Let positive values of x denote displacements below the rest position and negative values denote displacements above the rest position. See Fig. 5.23.

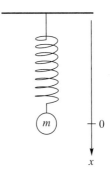

Figure 5.23

We begin the modeling derivation step by recalling Newton's law $F = ma$. Here $a = x''$ is acceleration and F is the sum total of all forces acting on the object. We assume there are only three forces acting on the object in our mass-spring system: the (constant) force of gravity F_g, a force due to the spring F_s (called the restoring force), and a frictional force F_f due to the environmental medium in which the system is immersed (e.g., air, water, oil, etc.). Thus,

$$F = F_g + F_s + F_f.$$

A model equation for x arises from Newton's law as soon as we specify mathematical expressions (submodels) for each of these three forces.

The constant force of gravity F_g gives rise to a constant acceleration g (which is approximately 9.8 meters/s^2 or 32 feet/s^2 near the surface of the earth). Thus,

$$F_g = mg.$$

For the force F_s due to the spring, we use Hooke's law. This law states that the force exerted by a spring is proportional to its elongation (or compression).[6] If, at the rest position, the spring is stretched l units from its natural length, then the force exerted by the spring is kl. The constant of proportionality $k > 0$ is called the *spring constant* (or Hooke's constant). Different springs in general have different spring constants. For two springs, the one with the larger spring constant is a stiffer spring. (The numerical value of the spring constant also depends on the units used.) At the rest position the force of gravity and the force due to the spring balance each other, and consequently

$$kl = mg.$$

Once the object starts in motion, however, the force exerted by the spring will depend on the position of the object. Specifically, the elongation (or compression) of the spring

---

[6]Other restoring force laws are possible, including laws that assume a nonlinear dependence on the elongation of the spring.

when the mass is at position x will be $l + x$. Thus, by Hooke's law

$$F_s = -k(l + x).$$

Notice the minus sign. A negative sign is needed because the force exerted by the spring is up (or negative) when the spring is elongated and down (or positive) when it is compressed.

Finally, we consider the force F_f due to friction. If the mass is not in motion there is, of course, no force due to friction. Only objects in motion experience frictional forces. This means that F_f depends on the velocity x' of the mass. The *linear frictional law* assumes F_f is proportion to x':

$$F_f = -cx'$$

for a constant of proportionality $c \geq 0$ (called the coefficient of friction).[7] The negative sign appears because the force of friction acts in the opposite direction to that of the motion.

Under the aforementioned assumptions, the total of all forces acting on the object is

$$F = mg - k(l + x) - cx'$$

and Newton's law $F = ma$ yields the equation

$$mg - k(l + x) - cx' = mx''.$$

Since $mg = kl$, this second-order, linear differential equation yields the following initial value problem for the motion of the object:

$$mx'' + cx' + kx = 0 \tag{8.10}$$

$$x(0) = x_0, \quad x'(0) = v_0.$$

The goal is to determine the characteristics of the object's motion once it is put into motion and how these characteristics depend on the model parameters (i.e., on the object's mass m, the coefficient of friction c, and the spring constant k, and the initial conditions x_0 and v_0).

Mass-Spring Systems without Friction   If the frictional force is negligible (so that $c = 0$), the differential equation in (8.10) becomes $mx'' + kx = 0$. We can rewrite this equation as $x'' + \beta^2 x = 0$, where $\beta = \sqrt{k/m}$. This leads us to the initial value problem

$$x'' + \beta^2 x = 0, \qquad \beta = \sqrt{\frac{k}{m}} \tag{8.11}$$

$$x(0) = x_0, \qquad x'(0) = v_0$$

for the displacement x (from equilibrium) of the object when friction is negligible. We can also write this second-order equation as a first-order system

$$x' = y$$

$$y' = -\beta^2 x \tag{8.12}$$

$$x(0) = x_0, \qquad y(0) = v_0$$

---

[7]Nonlinear friction laws are more appropriate under some circumstances, depending on the geometric shape of the object, the viscosity ("thickness") of the medium, and so on.

by defining $y = x'$. With the completion of the modeling derivation step of the modeling cycle, we can turn our attention to the solution step. We can mathematically analyze either form of the model equations, (8.11) or (8.12).

Figure 5.24 shows example phase plane portraits of the system (8.12) for selected values of β. We recognize these phase portraits as centers (Sec. 5.4). Solutions of the initial value problem are therefore periodic functions for all initial conditions x_0 and v_0. This means that the object oscillates around its rest point with a certain period p and amplitude a, regardless of how it is put into motion.

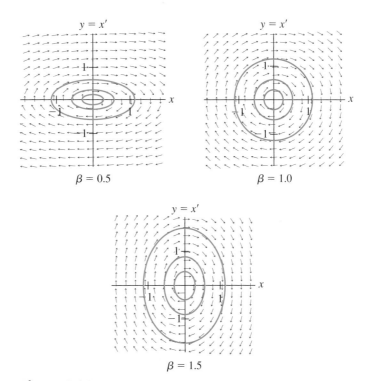

Figure 5.24 The phase plane portraits for the system (8.12) for three selected values of β.

Some interesting questions about the periodic motion of the object are these: How do the amplitude and period of the object's oscillation depend on the manner by which it is put into motion, that is, on the initial conditions x_0 and v_0? How do they depend on β and therefore on the object's mass m and the spring constant k?

We can obtain suggested answers to some of these questions from the phase portraits in Fig. 5.24. In those examples we see an increase in the amplitude a of x when the magnitude of either the initial displacement $|x_0|$ or the initial velocity $|v_0|$ increases. This is because either of these changes in the initial conditions causes the orbit to become a larger ellipse in the phase plane portrait. The phase portraits do not, however, reveal anything about the period of the oscillation. If we plot the solution component x against t obtained from selected orbits, as in Fig. 5.25, we find that the period p of the oscillation, unlike the amplitude, does not seem to depend on the initial conditions.

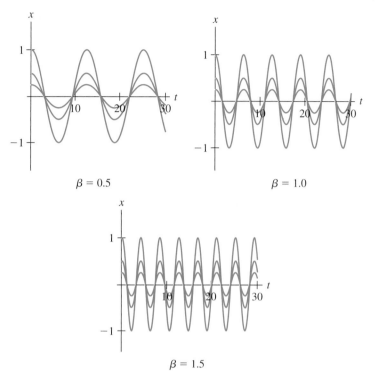

Figure 5.25 Graphs of solutions x of (8.11) [or, equivalently, the x component of the solution of (8.12)] for the same three values of β used in Fig. 5.24. In each case, the three plotted solutions correspond to initial conditions $x_0 = 0.25, 0.5, 1.0$, and $v_0 = 0$.

We cannot rigorously prove these observations concerning the amplitude and period of the oscillations from a selected number of (computer-generated) examples. For all we know, other examples obtained from different choices of β or the initial conditions might contradict our conclusions. One way to validate these observations in a general way and to study other properties of the object's motion—such as how they depend on m and k—is to obtain a solution formula for the initial value problem (8.11) using the method in Sec. 5.3.

The characteristic equation associated with the second-order equation appearing in the initial value problem (8.11) is

$$\lambda^2 + \beta^2 = 0.$$

Its roots

$$\lambda = \pm \beta i, \qquad \beta = \sqrt{\frac{k}{m}}$$

are purely imaginary. This implies that the phase plane portrait is a center (see Table 5.2) and that the general solution is (see Table 5.1)

$$x(t) = c_1 \cos \beta t + c_2 \sin \beta t.$$

To satisfy the initial conditions we choose the arbitrary constants c_1 and c_2 so that

$$x(0) = c_1 = x_0$$
$$x'(0) = \beta c_2 = v_0.$$

Thus, $c_1 = x_0$, $c_2 = v_0/\beta$ and we have the formula

$$x(t) = x_0 \cos \beta t + \frac{v_0}{\beta} \sin \beta t, \qquad \beta = \sqrt{\frac{k}{m}} \tag{8.13}$$

for the solution of the initial value problem (8.11).

Using the solution formula (8.13) we turn to the model interpretation step of the modeling cycle and address the questions about the amplitude and period of the object's motion raised above.

First, the solution formula shows that the object's motion is indeed periodic. This is because the sine and cosine are periodic functions. The period of both $\sin \beta t$ and $\cos \beta t$, and hence of the solution $x(t)$, is

$$p = \frac{2\pi}{\beta} = 2\pi \sqrt{\frac{m}{k}}.$$

Notice that this formula does not contain x_0 or v_0. Thus, as we observed in the selected examples shown in Fig. 5.25, the period of the object's oscillation does not depend on its initial conditions. This exact mathematical relationship between the period of oscillation and the quantities m and k implies that some facts that are perhaps not so intuitive. The period is proportional to the ratio of the *square roots* of the mass and the spring constant. If, for example, the object's mass m is doubled while the spring constant k is held fixed (i.e., the same spring is used), then the period of oscillation does not double but instead increases by a factor of $\sqrt{2} \approx 1.414$. On the other hand, if both m and k are doubled, the period does not change at all!

To study the amplitude of the object's oscillation, we rewrite the solution formula (8.13) (using sum angle identities from trigonometry) as

$$x(t) = a \sin (\beta (t + \varphi))$$

$$a = \sqrt{x_0^2 + \left(\frac{v_0}{\beta} \right)^2} \tag{8.14}$$

$$\beta = \sqrt{\frac{k}{m}}$$

$$\varphi = \frac{1}{\beta} \arctan \left(\frac{\beta}{v_0} x_0 \right).$$

This formula shows that $x(t)$ is a sinusoidal oscillation with amplitude a, period p, and a "phase shift" [i.e., a horizontal translation of the graph of $x(t)$] equal to φ.

The formula for a relates the amplitude of the object's oscillation to the initial conditions x_0 and v_0 and the quantity β. From this formula we see that the amplitude increases with x_0 and v_0, as we conjectured from the selected examples in Fig. 5.24. We also see that the amplitude does not depend on β if $v_0 = 0$ (that is, the object is put into motion by a displacement from the rest position but is given no initial velocity). On the other hand, if $v_0 \neq 0$, then the amplitude is inversely related to β. See Fig. 5.26.

Recalling the formula for β, we see that this in turn implies that the amplitude is larger for heavier masses m and for smaller spring constants k ("weaker" springs).

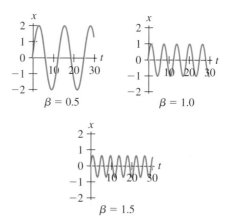

Figure 5.26 Solutions of (8.11) with $x_0 = 0$, $v_0 = 1$ are shown for three values of β.

EXAMPLE 1

Suppose a 2-kilogram mass is placed on a spring and, in the rest position, the spring is elongated 5 centimeters (0.05 meter). The mass is pulled downward 10 centimeters and released. What are the amplitude a and period p of the resulting oscillation?

Since $kl = mg$ and $g = 9.8$ m/s^2, we have $k \times 0.05 = 2 \times 9.8$ or $k = 392$. Thus,

$$\beta = \sqrt{\frac{k}{m}} = 14.$$

From the formulas in (8.14) we obtain (noting that $x_0 = 1/10$ and $v_0 = 0$) the amplitude

$$a = \sqrt{x_0^2 + \left(\frac{v_0}{\beta}\right)^2} = \frac{1}{10} \text{ (meter)}$$

and the period

$$p = \frac{2\pi}{\beta} \approx 0.4488 \text{ s.}$$

The solution formula (8.13) in this case is

$$x(t) = \frac{1}{10} \cos 14t.$$

Suppose, however, the object is not simply released after its displacement but instead is struck from below so as to impart an initial velocity of 2 m/s. What then are the amplitude a and period p of the resulting oscillation?

Since the object is struck from below and the initial velocity is upward, we now have $v_0 = -2$. Thus, the amplitude is

$$a = \sqrt{x_0^2 + \left(\frac{v_0}{\beta}\right)^2} \approx 0.1229 \text{ (meter)}.$$

The period is unaffected by this change in initial condition and therefore remains $p \approx 0.4488$ s. The solution formula (8.13) in this case is

$$x(t) = \frac{1}{10}\cos 14t - \frac{1}{7}\sin 14t.$$

The two solutions are compared in Fig. 5.27. ■

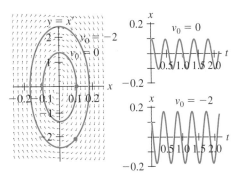

Figure 5.27 The graphs of the solutions in Example 1 are shown. Also shown are the phase plane orbits of these solutions.

The sinusoidal motion of the frictionless mass-spring model is called harmonic motion and the model is called the simple harmonic oscillator. Of course, in reality no mass-spring system will oscillate forever, as predicted by this model. All systems will eventually come to rest, due among other things, to friction experienced by the object during its motion.

Mass-Spring Systems with Friction   To study the motion of the mass-spring system with friction we return to the second-order equation

$$mx'' + cx' + kx = 0 \tag{8.15}$$

with $c > 0$. Figure 5.28 shows graphs of orbits from the equivalent first-order system

$$\begin{aligned} x' &= y \\ y &= -\tfrac{k}{m}x - \tfrac{c}{m}y \end{aligned} \tag{8.16}$$

obtained for selected values of the coefficients k, m and c and initial conditions x_0 and $y_0 = v_0$. These examples indicate that the phase portraits for the small values of c are stable spirals, while the phase portraits for the larger values of c are stable nodes. In all cases orbits tend to the equilibrium $(x, y) = (0, 0)$ point (i.e., the rest position $x = 0$, $x' = 0$ of the object) as $t \to +\infty$. The graphs in Fig. 5.28 suggest that the inclusion of friction into the mass-spring system implies the object will return to rest (at least in the

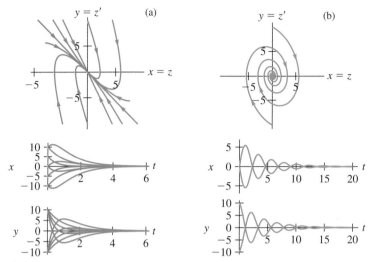

Figure 5.28 The top phase plane graphs show orbits of (8.15) with initial conditions $x_0 = 0$, $v_0 = 1$ with coefficients $m = 1$, $k = 2$, and (a) $c = 3$, (b) $c = \frac{1}{2}$. The lower graphs show the time series plots of the components of selected orbits.

mathematical limit as $t \to +\infty$). For smaller values of c the object oscillates, in the sense that it passes through the rest position infinitely many times. For larger values of c this is not the case.

To investigate the observations made from the sample orbits in Fig. 5.28 in a more general way, we derive a solution formula for the initial value problem $x(0) = x_0$, $x'(0) = v_0$ using the method of Sec. 5.3. The roots of the characteristic equation

$$m\lambda^2 + c\lambda + k = 0$$

associated with the second-order equation in (8.15) are

$$\lambda_1 = \frac{-c + \sqrt{c^2 - 4mk}}{2m}, \qquad \lambda_2 = \frac{-c - \sqrt{c^2 - 4mk}}{2m}.$$

Formulas for the general solution of the differential equation depend on whether these roots are real and different ($c^2 - 4mk > 0$), real and equal ($c^2 - 4mk = 0$), or complex conjugate ($c^2 - 4mk < 0$). In the first two cases, both roots are negative. That is, $c \geq 2\sqrt{mk}$ the phase portrait is a stable node (a stable degenerate node when $c = 2\sqrt{mk}$). When $c < 2\sqrt{mk}$, the roots are complex conjugate with negative real parts

$$\text{Re } \lambda_1 = \text{Re } \lambda_2 = -\frac{c}{2m} < 0$$

and the phase portrait is a stable spiral.

These general results corroborate the phase portrait types observed in Fig. 5.28. For "small" c, namely when $c < 2\sqrt{mk}$ and the phase portrait is a stable spiral, the mass-spring system is said to be underdamped. In this case, the complex root

$$\lambda_1 = \alpha + i\beta$$

$$\alpha = -\frac{c}{2m} < 0, \qquad \beta = \sqrt{\frac{k}{m} - \left(\frac{c}{2m}\right)^2} > 0$$

leads to the general solution

$$x(t) = e^{\alpha t} \left(c_1 \cos \beta t + c_2 \sin \beta t\right).$$

The initial conditions imply

$$c_1 = x_0$$
$$\alpha c_1 + \beta c_2 = v_0$$

and therefore

$$c_1 = x_0, \qquad c_2 = \frac{v_0 - \alpha x_0}{\beta}.$$

We can rewrite the resulting solution formula

$$x(t) = x_0 e^{\alpha t} \cos \beta t + \frac{v_0 - \alpha x_0}{\beta} e^{\alpha t} \sin \beta t$$

as

$$x(t) = e^{\alpha t} a \sin(\beta (t + \varphi)),$$

where

$$a = \sqrt{x_0^2 + \left(\frac{v_0 - \alpha x_0}{\beta}\right)^2}, \qquad \varphi = \frac{1}{\beta} \arctan\left(\frac{x_0 \beta}{v_0 - \alpha x_0}\right).$$

From this solution formula we see that the object oscillates in a sinusoidal fashion with a exponentially decreasing amplitude (remember, $\alpha < 0$). Furthermore, the mass crosses the rest position $x = 0$ infinitely often, or to be more specific, at those times t for which $\sin(\beta t + \varphi) = 0$, namely,

$$\beta(t + \varphi) = n\pi, \qquad n = 0, \pm 1, \pm 2, \ldots$$

or

$$t = \frac{n\pi}{\beta} - \varphi, \qquad n = 0, \pm 1, \pm 2, \ldots .$$

Notice that these times are equally spaced.

It is left as Exercises 18 and 19 to find solution formulas for the initial value problem in the overdamped case when $c > \sqrt{km}$ and in the critically damped case when $c = \sqrt{mk}$. These formulas appear in Table 5.3.

Table 5.3   Solution formulas for the initial value problem $x(0) = x_0$, $x'(0) = v_0$ for the equation (8.15).

Underdamped $c < 2\sqrt{km}$	Critically Damped $c = 2\sqrt{km}$	Overdamped $c > 2\sqrt{km}$
$x(t) = e^{\alpha t}\left(x_0\cos\beta t + \frac{v_0-\alpha x_0}{\beta}\sin\beta t\right)$	$x(t) = x_0 e^{\lambda t} + (v_0 - \lambda x_0)te^{\lambda t}$	$x(t) = \left(\frac{v_0-x_0\lambda_2}{\lambda_1-\lambda_2}\right)e^{\lambda_1 t} + \left(\frac{x_0\lambda_1-v_0}{\lambda_1-\lambda_2}\right)e^{\lambda_2 t}$
$\alpha = -\frac{c}{2m}$	$\lambda = -\frac{c}{2m}$	$\lambda_1 = \frac{-c+\sqrt{c^2-4mk}}{2m}$
$\beta = \sqrt{\frac{k}{m} - \left(\frac{c}{2m}\right)^2}$		$\lambda_2 = \frac{-c-\sqrt{c^2-4mk}}{2m}$

EXAMPLE 2

In Fig. 5.28, $m = 1$ and $k = 2$. In Fig. 5.28(a), $c = 3 > 2\sqrt{2} = 2\sqrt{mk}$ and the mass-spring system is overdamped. From Table 5.3 the solution of the initial value problem is

$$x(t) = (v_0 + 2x_0)e^{-t} - (v_0 + x_0)e^{-2t}.$$

Note that the object is at the rest position (i.e., $x = 0$) if and only if

$$(v_0 + 2x_0)e^{-t} - (v_0 + x_0)e^{-2t} = 0.$$

Solving for t, we obtain

$$t = \ln\left(\frac{v_0 + x_0}{v_0 + 2x_0}\right).$$

Thus, the object crosses the rest position $x = 0$ only once for time $t > 0$, provided that the initial conditions satisfy the inequality

$$\frac{v_0 + x_0}{v_0 + 2x_0} > 1.$$

Otherwise, the object never crosses the rest point.

For example, with no initial velocity $v_0 = 0$ the object never crosses the rest position. The object simply rises (if $x_0 > 0$) or falls (if $x_0 < 0$) to the rest position as $t \to +\infty$.

On the other hand, for $x_0 = 1$ and $v_0 = -3$ the object crosses the rest point once, at time

$$t = \ln\left(\frac{-2}{-1}\right) = 0.6932.$$

In Fig. 5.28(b), $c = 1/2 < 2\sqrt{2} = 2\sqrt{mk}$ and the mass-spring system is underdamped. From Table 5.3 the solution of the initial value problem is

$$x(t) = e^{-\frac{1}{4}t}\left(x_0\cos\left(\frac{\sqrt{31}}{4}t\right) + \frac{\sqrt{31}}{31}(4v_0 + x_0)\sin\left(\frac{\sqrt{31}}{4}t\right)\right).$$

The object crosses the rest position infinitely often, at those times when the parenthetical expression vanishes. For example, for the initial conditions $x_0 = 0$, $v_0 = 1$ the solution

$$x(t) = 4\frac{\sqrt{31}}{31}e^{-\frac{1}{4}t}\sin\left(\frac{\sqrt{31}}{4}t\right)$$

equals 0 when

$$\sin\left(\frac{\sqrt{31}}{4}t\right) = 0$$

or

$$t = \frac{4n\pi}{\sqrt{31}}, \qquad n = 0, \pm 1, \pm 2, \ldots \qquad \blacksquare$$

Electric Circuits A basic electric circuit consists of idealized passive elements that produce changes in voltage that depend on the electric charge or its rates of change. We consider three such elements: capacitors, inductors, and resistors. A capacitor consists of two plates separated by an insulator and causes a voltage difference between the plates. An inductor is a coil in the path of the circuit that builds up a magnetic field within itself, as electrical current flows through it, and hence causes a voltage difference between its ends. A resistor impedes the flow of current through the circuit and thereby causes a voltage drop. We derive a mathematical model for a circuit with these basic elements from assumptions about how the voltage changes (or drops) across these elements depend on the amount of current in, or flowing through, elements. It is also required that the sum of all voltage drops equal the voltage impressed onto the circuit. This requirement is known as Kirchoff's voltage law. (In more elaborate circuits this law applies around any closed loop in the circuit.)

Let $q = q(t)$ denote the electrical charge in the circuit at time t (measured in coulombs). By definition, current $i = i(t)$ is the rate of change of q (i.e., $i = q'$) (measured in amperes). We model simpler circuits using proportionality assumptions for the relationship among the voltage drops across capacitors, inductors, and resistors and the electrical charge and current. Specifically, the voltage drop V_c across a capacitor is proportional to the charge q, the voltage drop V_r across a resistor is proportional to the current i, and the voltage drop V_i across an inductor is proportional to the rate of change of current i'. Therefore, we have

$$V_c = \frac{1}{C}q$$
$$V_r = Ri = Rq'$$
$$V_i = Li' = Lq''.$$

The constants of proportionally have names: $C > 0$ is the capacitance, $L \geq 0$ is the inductance, and $R \geq 0$ is the resistance.

If $V(t)$ denotes the voltage impressed on a simple series circuit (for example, by a battery), then Kirchoff's law is

$$V_i + V_c + V_r = V(t)$$

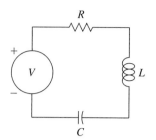

Figure 5.29 A series LCR circuit.

or

$$Lq'' + Rq' + \frac{1}{C}q = V(t). \tag{8.17}$$

We can also write the equation in terms of current i in the circuit:

$$Li' + Ri + \frac{1}{C}q = V(t)$$

or, after a differentiation with respect to t,

$$Li'' + Ri' + \frac{1}{C}i = V'(t). \tag{8.18}$$

Equations (8.17) and (8.18) for q and i are linear differential equations. If $L \neq 0$, they are second-order equations. If $V(t)$ is not identically equal to 0, then the equations are nonhomogeneous. (We postpone this case until the next chapter, Sec. 6.5.3). If $V(t)$ is identically equal to 0, then these equations are homogeneous.

Notice the mathematical similarity between the equations for the electric circuit ((8.17) or (8.18)) when $V(t) = 0$, namely between the equation

$$Lq'' + Rq' + \frac{1}{C}q = 0 \tag{8.19}$$

or

$$Li'' + Ri' + \frac{1}{C}i = 0$$

and the equation for the forced mass-spring system (5.11)

$$mx'' + cx' + kx = 0. \tag{8.20}$$

This observation is the basis of a mechanical-electrical analogy between mechanical systems, such as the mass-spring system, and electric circuits. Table 5.4 summarizes the correspondences in this analogy. In the analogy, resistance plays the role of a coefficient of friction, inductance plays the role of mass, and the reciprocal of the capacitance plays the role of a spring constant.

Table 5.4	The mechanical-electrical analogy.		
Displacement (meters)	x	$\longleftrightarrow$	q	Charge (coulombs)
Velocity (meters/s)	x'	$\longleftrightarrow$	i	Current (amperes)
Mass (kilograms)	m	$\longleftrightarrow$	L	Inductance (Henries)
Coefficient of friction (newtons/velocity)	c	$\longleftrightarrow$	R	Resistance (ohms)
Sprint constant (newtons/meter)	c	$\longleftrightarrow$	$\frac{1}{C}$	1/capacitance (1/farads)
Compliance (meters/newton)	$\frac{1}{k}$	$\longleftrightarrow$	C	Capacitance (farads)

We already analyzed the mass-spring system (8.20) and therefore we need not repeat that analysis for the LCR circuit equations. Instead, we can reinterpret the mass-spring analysis in terms relevant to the electric circuit by making substitutions from Table 5.4.

For example, if no resistor is included in the circuit, then equation (8.19) reduces to the harmonic oscillator equation

$$Lq'' + \frac{1}{C}q = 0 \qquad (8.21)$$

for the charge q in the circuit. Using Table 5.4, together with the solution (8.13) of the harmonic oscillator as a model for the mass-spring system, we find that the charge in an LC circuit oscillates sinusoidally according to the formula

$$q(t) = q_0 \cos \beta t + \frac{i_0}{\beta} \sin \beta t$$

$$\beta = \sqrt{\frac{1}{LC}},$$

where $q(0) = q_0$ is the initial charge on the circuit and $q'(0) = i(0) = i_0$ is the initial amount of current in the circuit.

If we add a resistor to the circuit, then $R > 0$ in (8.19) circuit. In this case, the charge $q(t)$ tends to 0 as $t \to +\infty$. From the mechanical-electrical analogy (see Table 5.3) we can distinguish over- and underdamped oscillations. Thus, in the underdamped case $R < 2\sqrt{L/C}$ the charge in the LCR circuit will oscillate according to the formula

$$q(t) = e^{\alpha t}\left(q_0 \cos \beta t + \frac{i_0 - \alpha q_0}{\beta} \sin \beta t \right)$$

$$\alpha = -\frac{R}{2L} < 0, \qquad \beta = \sqrt{\frac{1}{LC} - \left(\frac{R}{2L}\right)^2} > 0.$$

Similarly, we can obtain formulas for q in the overdamped case $R > 2\sqrt{L/C}$ and the critically damped case $R = 2\sqrt{L/C}$ from Tables 5.3 and 5.4. See Exercise 20.

Exercises

1. Find a formula for the solution of the initial value problem (8.1).

2. In the lidocaine example worked out in the text we found that using the maximal safe dose $y_0 = 114.5$ a second injection is needed after $t_l = 216.1$ minutes. Adjust the initial conditions in the lidocaine model (8.1) to account for the residual amounts and use a computer to determine the effective but safe interval of doses y_0 for a second injection (administered after 216.1 minutes). How long does this second dose remain effective?

3. In the lidocaine example worked out in the text we found an effective but safe time interval of approximately 3.5 hours if the maximal safe dosage was administered. This dose, however, exposes the patient to amounts in the bloodstream that are dangerously close to the lethal level for a short period of time. Suppose, instead, we permit levels no higher than 20 (mg) in order to provide a safety margin.

   (a) Determine the revised range of effective but safe dosages y_0.

   (b) If the revised maximal but safe dose is administered, find the time interval during which there is an effective amount of lidocaine in the patient's bloodstream.

4. For a patient with a blood capacity of 5 liters, we used the model (8.5) to calculate the time interval $t_e < t < t_l$ during which the amount of lidocaine in the bloodstream is at an effective but safe level. See (8.9). Show that this time interval is the same for patients of all blood capacities V.

The total energy E in the mass-spring system is the sum of the potential energy $kx^2/2$ and the kinetic energy $mv^2/2$ $(= my^2/2)$. Thus, $E = kx^2/2 + my^2/2$.

5. For any solution $x = x(t)$ of the mass-spring equation (8.11), show the total energy E remains a constant for all time t.

6. In phase plane coordinates, $E = \frac{1}{2}kx^2 + \frac{1}{2}my^2$. Use Exercise 5 to show all orbits are ellipses.

A mass of 5 kilograms is attached to a spring that, in rest position, is elongated 1.5 meter. Assume no friction. Find the period p and amplitude a of the oscillation under each of the following conditions. Also find the solution x of (8.11).

7. The mass is struck from below so as to impart an initial velocity of -50 cm/s.

8. The mass is struck from above so as to impart an initial velocity of 50 cm/s.

9. The mass is raised 50 centimeters and released.

10. The mass is raised 50 centimeters and struck from above so as to impart an initial velocity of 50 cm/s.

11. The mass is lowered 75 centimeters and struck from below so as to impart an initial velocity of -25 cm/s.

12. The mass is lowered 75 centimeters and struck from above so as to impart an initial velocity of 25 cm/s.

A mass of 2 kilograms is attached to a spring. Find the spring constant k in the following cases. Assume that friction is negligible.

13. The mass is struck from above so that $v_0 = 1.5$ m/s and the mass attains a maximum displacement x of 10 cm.

14. The mass is struck from above so that $v_0 = 1.5$ m/s and the mass attains a maximum displacement x in 1.25 seconds.

15. The mass is pulled down 15 cm, released, and it passes through the rest point in 2.25 seconds.

16. The mass is pulled down 15 cm, released, and it returns in 3 seconds.

17. Show that the amplitude of the frictionless mass-spring oscillator is the same whether the mass is initially struck upward or downward (with the same force).

18. Consider the initial value problem (8.20) for the over-damped case $c > 2\sqrt{km}$.

   (a) Find a formula for the solution.

   (b) Use the formula in (a) to show that the mass crosses the rest position no more than one time for $t \geq 0$.

19. Consider the initial value problem (8.20) for the over-damped case $c = 2\sqrt{km}$.

   (a) Find a formula for the solution.

   (b) Use the formula in (a) to show that the mass crosses the rest position no more than one time.

20. Obtain formulas for the charge q in a series LCR circuit (8.17) in the overdamped $\left(R > 2\sqrt{\frac{L}{C}}\right)$ and critically damped $\left(R = 2\sqrt{\frac{L}{C}}\right)$ cases.

21. The system

$$x' = -r_1 x + r_2 y$$
$$y' = r_1 x - (r_2 + r_3) y$$

models the amount of pesticide in a stand of trees, x, and its soil bed, y. The equations arise from the basic inflow/outflow balance equation

$$\text{rate of change } = \text{ inflow rate } - \text{ outflow rate}$$

applied to each rate of change x' and y'. The assumptions are that the pesticide flows out of the trees and into the soil at a rate $r_1 x$ proportional to the amount present; the pesticide flows out of the soil and into the trees at a rate $r_2 y$ proportional to the amount y present in the soil; and the pesticide degrades in the soil at a rate $r_3 y$ proportional to the amount y present. All coefficients r_1, r_2, and r_3 are positive.

Suppose the initial dosage $p_0 > 0$ is applied to the soil. Suppose there is initially no pesticide in the trees.

(a) Use a computer program to explore the solutions of this initial value problem. In particular, investigate the maximum level reached by the pesticide in the trees and how long it takes this level to be reached. How do these depend on the initial dosage p_0? What happens to the fractions of pesticide in the trees and in the soil in the long run?

(b) Solve the initial value problem and use your answer to prove or disprove your conjectures.

6

Nonhomogeneous Linear Systems

In this chapter we extend to nonhomogeneous linear systems some of the solution methods we developed in Chapter 2 for single, nonhomogeneous linear equations. We will see that the variation of constants formula for the solution of a single linear equations has a generalization to linear systems. This formula permits the calculation of solution formulas, provided the general solution of the associated homogeneous system is known. In the Chapter 5 we learned how to calculate the general solution of a homogeneous linear system with constant coefficients. In this case, we can also apply the shortcut method of undetermined coefficients to those linear systems with special kinds of nonhomogeneous terms. As a special case, these methods for nonhomogeneous linear systems also apply to nonhomogeneous, linear second-order equations.

The homogeneous linear system

$$x' = -\alpha x - \beta y \qquad (0.1)$$
$$y' = \gamma x - \delta y$$

arises as a model of the glucose/insulin regulation system in the bloodstream (see Chapter 5, Sec. 5.2). In these equations x and y are the excess concentrations of glucose and insulin from their equilibrium levels respectively (negative values are deficiencies below equilibrium) and all four coefficients are positive constants. In a glucose tolerance test a patient is given a concentration of glucose intravenously at a constant rate. This input contributes an input rate $g > 0$ in the equation for glucose. The result is the nonhomogeneous linear system

$$x' = -\alpha x - \beta y + g \qquad (0.2)$$
$$y' = \gamma x - \delta y.$$

Under other circumstances glucose may enter the bloodstream at a nonconstant rate $g = g(t)$. For example, if glucose enters at a periodically fluctuating rate (due, for example, to meals taken on a regular schedule), then $g(t)$ would be a periodic function of time t. An example is $g(t) = g_{av} + \alpha \sin \beta t$, where g_{av} is the average, α is the amplitude, and β is the frequency of the oscillating glucose ingestion rate.

Nonhomogeneous systems also arise from nonhomogeneous higher-order equations. If a simple harmonic oscillator (such as a mass-spring system or an inductor/capacitor electric circuit; see Chapter 5, Sec. 5.5.8.2) is subjected to an external force $f(t)$, the resulting second-order equation

$$mx'' + kx = f(t)$$

is equivalent to the nonhomogeneous system

$$x' = y$$
$$y' = -\frac{k}{m}x + \frac{1}{m}f(t).$$

The preceding nonhomogeneous systems are examples of the general nonhomogeneous system

$$x' = a(t)x + b(t)y + h_1(t) \tag{0.3}$$
$$y' = c(t)x + d(t)y + h_2(t).$$

In most applications, such as the preceding, the *nonhomogeneous terms* $h_1(t)$ and $h_2(t)$ arise from forcing or input that is external to the system under consideration (i.e., is independent of the state of the system x and y). For this reason, the nonhomogeneous terms $h_1(t)$ and $h_2(t)$ are often called *forcing functions*.

In Sec. 4.3 of Chapter 4 we learned that the general solution of the nonhomogeneous system (0.3) has the form

$$x = x_h(t) + x_p(t)$$
$$y = y_h(t) + y_p(t)$$

(see Theorem 3), where

$$x = x_p(t)$$
$$y = y_p(t)$$

is a particular solution pair of the system and where

$$x = x_h(t)$$
$$y = y_h(t)$$

is the general solution of the associated homogeneous system

$$x' = a(t)x + b(t)y$$
$$y' = c(t)x + d(t)y.$$

Furthermore, we learned that the general solution of a homogeneous system is a linear combination of two independent solution pairs. In Chapter 5 we studied homogeneous systems with constant coefficients and learned how to find two independent solution pairs and the general solution. In this chapter we consider the problem of finding a particular solution (and the general solution) of nonhomogeneous systems. We will see that two methods used for single linear equations in Chapter 2 have adaptations to systems of linear equations, namely the variation of constants formula and the method of undetermined coefficients. The variation of constants formula applies to any linear system, although it might involve difficult integrations. The method of undetermined coefficients is a shortcut method that avoids integrations, but it applies only to special types of equations.

6.1 The Method of Undetermined Coefficients

The method of undetermined coefficients studied in Sec. 2.3 of Chapter 2 applies to linear equations $x' = px + q(t)$ with a constant coefficient p and nonhomogeneous term $q(t)$ of special type. (See Exercises 1–22, Sec. 2.3 of Chapter 2.) The method is a "shortcut" for obtaining a particular solution x_p that, when added to the general solution $x_h = ce^{pt}$ of the associated homogeneous equation $x' = px$, gives the general solution $x = x_h + x_p$. The method begins with a "guess" for x_p based on the special properties of $q(t)$. We can a use similar method for the nonhomogeneous system

$$x' = ax + by + h_1(t) \tag{1.1}$$
$$y' = cx + dy + h_2(t)$$

with constant coefficients a, b, c, and d and with nonhomogeneous terms $h_1(t)$ and $h_2(t)$ of those same special types. We will not formalize the method for systems but instead will only illustrate its use by means of examples.

In Chapter 5 we used the coefficients

$$\alpha = 2.92, \quad \beta = 4.34, \quad \gamma = 0.208, \quad \delta = 0.780 \tag{1.2}$$

for the glucose/insulin system (0.1). (Glucose is measured in grams, insulin in insulin units and time in hours.) If glucose is intravenously ingested at a constant rate of 18 (grams/hour), the nonhomogeneous system (0.2) becomes

$$x' = -2.92x - 4.34y + 18.0 \tag{1.3}$$
$$y' = 0.208x - 0.780y.$$

Figure 6.1 shows the graphs of two computer calculated solution pairs for this system obtained from two different initial condition pairs. The sample graphs in Fig. 6.1 indicate that the glucose and insulin levels level off as time increases (i.e., both x and y approach finite limits as $t \to +\infty$). Moreover, both examples have the same long term limits for x and y (about 4.5 and 1, respectively). Do all solution pairs approach these limits as $t \to +\infty$? That is, do the glucose and insulin concentrations approach the same long-term levels no matter what their initial concentrations are? We conjecture that this is the case, but we cannot resolve this conjecture by means of (a finite number) of computer examples.

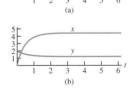

(a)

Figure 6.1 Two solution pairs of (1.3) with initial conditions (a) $x(0) = 0$, $y(0) = 0$ and (b) $x(0) = 0$, $y(0) = 2$.

(b)

One way to address our conjecture is to obtain and study a formula for the general solution of the nonhomogeneous system (1.3). In Chapter 5 we found the general solution of the associated homogeneous system to be

$$x_h = 4.34c_1 e^{-2.34t} + 4.34c_2 e^{-1.36t} \tag{1.4}$$
$$y_h = -0.58c_1 e^{-2.34t} - 1.56c_2 e^{-1.36t}.$$

To find the general solution of the nonhomogeneous system (1.3) we need a particular solution pair x_p, y_p of the nonhomogeneous system.

Noticing that the forcing functions $h_1(t) = 18$, $h_2(t) = 0$ are both constant functions, we look for a constant solution (equilibrium) pair; that is, we formulate the "guess"

$$x_p = k_1, \qquad y_p = k_2,$$

where k_1 and k_2 are undetermined coefficients. A substitution of this guess into (1.3) yields the two equations

$$0 = -2.92k_1 - 4.34k_2 + 18.0$$
$$0 = 0.208k_1 - 0.780k_2$$

for k_1 and k_2. The solutions of these linear, algebraic equations are (rounded to three significant digits)

$$k_1 = 4.41, \qquad k_2 = 1.18.$$

The particular solution pair

$$x_p = 4.41, \qquad y_p = 1.18,$$

together with the general solution of the associated homogeneous system, yields the general solution formula

$$x = 4.34c_1 e^{-2.34t} + 4.34c_2 e^{-1.36t} + 4.41 \tag{1.5}$$
$$y = -0.58c_1 e^{-2.34t} - 1.56c_2 e^{-1.36t} + 1.18.$$

The arbitrary constants c_1, c_2 in the general solution are determined by initial glucose and insulin concentrations. However, independent of their initial concentrations (i.e., for all values of c_1 and c_2), the glucose and insulin concentrations tend in the long run to equilibrium levels

$$\lim_{t \to +\infty} x = 4.41, \qquad \lim_{t \to +\infty} y = 1.18$$

(since both exponential functions $e^{-2.34t}$ and $e^{-1.36t}$ tend to 0 as $t \to +\infty$). This verifies our conjecture based on the examples graphed in Fig. 6.1.

A similar procedure leads to the general solution of the glucose/insulin system (0.2) for any values of its constant coefficients and any constant ingestion rate. We leave this as Exercise 23.

Consider the case of a periodic glucose ingestion rate (which could be administered intravenously or might occur because of a daily intake of food). With the same coefficients (1.2) as above and sinusoidal ingestion rate

$$g(t) = 18 + 18 \sin \frac{2\pi}{24} t \tag{1.6}$$

the glucose/insulin system (0.2) becomes

$$x' = -2.92x - 4.34y + 18 + 18 \sin \frac{2\pi}{24}t \qquad (1.7)$$

$$y' = 0.208x - 0.780y.$$

The formula (1.6) describes a glucose ingestion rate that sinusoidally oscillates around an average of 18 (grams) with a period of 24 (hours), reaching a minimum of 0 and a maximum of 36 (grams).

Two example solution pair graphs appear in Fig. 6.2. Note that both the glucose and insulin concentrations x and y quickly settle into sinusoidal-looking oscillations with the same period as that of the ingestion rate (24 hours). We conjecture that this true for all solution pairs of (0.2). To address this conjecture we will obtain a formula for the general solution of the nonhomogeneous system (1.7).

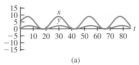

(a)

Figure 6.2 Two solution pairs of (1.7). (a) $x(0) = 0$, $y(0) = 0$ and (b) $x(0) = 0$, $y(0) = 15$.

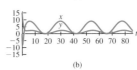

(b)

Since the forcing function (1.6) is a linear combination of a constant and a sine function we look for a solution pair (following the method of undetermined coefficients used in Chapter 2) that has the form of a linear combination of constant and both the sine and cosine functions:

$$x_p = k_1 + k_2 \sin \frac{2\pi}{24}t + k_3 \cos \frac{2\pi}{24}t \qquad (1.8)$$

$$y_p = l_1 + l_2 \sin \frac{2\pi}{24}t + l_3 \cos \frac{2\pi}{24}t.$$

We calculate the undetermined coefficients k_1, k_2, k_3 and l_1, l_2, l_3 by substituting this guess into the system (1.7). This results in the following equations (after gathering constant, sine, and cosine terms together):

$$(2.92k_1 + 4.34l_1 - 18) + \left(2.92k_2 - \tfrac{2\pi}{24}k_3 + 4.34l_2 - 18\right) \sin \tfrac{2\pi}{24}$$
$$+ \left(\tfrac{2\pi}{24}k_2 + 2.92k_3 + 4.34l_3\right) \cos \tfrac{2\pi}{24}t = 0$$

$$(0.208k_1 - 0.780l_1) + \left(0.208k_2 - 0.780l_2 + \tfrac{2\pi}{24}l_3\right) \sin \tfrac{2\pi}{24}t$$
$$+ \left(0.208k_3 - \tfrac{2\pi}{24}l_2 - 0.780l_3\right) \cos \tfrac{2\pi}{24}t = 0,$$

which must hold for all t. This is possible if and only if all six coefficients in parentheses vanish. By hand or using a computer we can solve the following (linear, algebraic) equations:

$$2.92k_1 + 4.34l_1 = 18$$

$$2.92k_2 - \frac{2\pi}{24}k_3 + 4.34l_2 = 18$$

$$\frac{2\pi}{24}k_2 + 2.92k_3 + 4.34l_3 = 0$$

$$0.208k_1 - 0.780l_1 = 0$$

$$0.208k_2 - 0.780l_2 + \frac{2\pi}{24}l_3 = 0$$

$$0.208k_3 - \frac{2\pi}{24}l_2 - 0.780l_3 = 0$$

for the six coefficients (rounded to three significant digits)

$$k_1 = 4.41, \quad k_2 = 4.54, \quad k_3 = 0.100$$
$$l_1 = 1.18, \quad l_2 = 1.10, \quad l_3 = -0.341,$$

which in (1.8) yield the particular solution pair

$$x_p = 4.41 + 4.54 \sin \frac{2\pi}{24}t + 0.100 \cos \frac{2\pi}{24}t$$

$$y_p = 1.18 + 1.10 \sin \frac{2\pi}{24}t - 0.341 \cos \frac{2\pi}{24}t.$$

This particular solution, when added to the general solution (1.4) of the associated homogeneous system, yields the general solution

$$x = 4.34c_1e^{-2.34t} + 4.34c_2e^{-1.36t}$$
$$+ 4.41 + 4.54 \sin \frac{2\pi}{24}t + 0.100 \cos \frac{2\pi}{24}t$$

$$(1.9)$$

$$y = -0.58c_1e^{-2.34t} - 1.56c_2e^{-1.36t}$$
$$+ 1.18 + 1.10 \sin \frac{2\pi}{24}t - 0.341 \cos \frac{2\pi}{24}t$$

of the nonhomogeneous system (1.7).

The arbitrary constants c_1, c_2 are determined by initial glucose and insulin concentrations. Note that independent of their initial concentrations (i.e., for all values of c_1 and c_2) the amounts of glucose and insulin x and y are, in the long run, nearly equal to the particular solution we found; that is, for large t

$$x \approx 4.41 + 4.54 \sin \frac{2\pi}{24}t + 0.100 \cos \frac{2\pi}{24}t \qquad (1.10)$$

$$y \approx 1.18 + 1.10 \sin \frac{2\pi}{24}t - 0.341 \cos \frac{2\pi}{24}t.$$

This is because both exponential functions $e^{-2.34t}$ and $e^{-1.36t}$ tend to 0 as $t \to +\infty$. From this result we see that both the glucose and insulin concentrations in the bloodstream oscillate with a period equal to 24 hours (since this is true of both sine and cosine functions), as we conjectured. Another consequence of (1.10) is that the glucose and insulin concentrations oscillate with averages equal to the equilibrium levels 4.41 and 1.18 obtained when the ingestion rate is held constant at its average 18 [as in (1.3)]. Also see Exercise 24.

6.1.1 Second-Order Equations

The details of the method of undetermined coefficients become formidable as the forcing functions become more complicated and the method loses its efficiency as a shortcut. For systems arising from second-order equations the details are minimized by applying the procedure to the second-order equation directly, rather than its equivalent first-order system.

For example, the nonhomogeneous second-order equation

$$mx'' + kx = \alpha \sin \beta t \tag{1.11}$$

describes a simple harmonic oscillator subjected to a periodic external force $\alpha \sin \beta t$. The general solution has the form

$$x = x_h + x_p$$

where

$$x_h = c_1 \sin \omega t + c_2 \cos \omega t, \qquad \omega = \sqrt{\frac{k}{m}}$$

and x_p is a particular solution. Using the method of undetermined coefficients, we look for a particular solution in the form

$$x_p = k_1 \sin \beta t + k_2 \cos \beta t.$$

To determine the coefficients k_1 and k_2, we substitute this guess into the equation (1.11) and obtain, after some rearrangement of terms,

$$\left[(k - m\beta^2) k_1 - \alpha \right] \sin \beta t + \left[(k - m\beta^2) k_2 \right] \cos \beta t = 0.$$

This equation holds for all t if and only if the coefficients of $\sin \beta t$ and $\cos \beta t$ vanish.[1] The coefficient on $\sin \beta t$ cannot be made to vanish if $k - m\beta^2 = 0$, a case we put aside for the moment. If, on the other hand, $k - m\beta^2 \neq 0$, then we take

$$k_1 = \frac{1}{k - m\beta^2}\alpha = \frac{1}{\omega^2 - \beta^2}\frac{\alpha}{m}.$$

The coefficient on $\cos \beta t$ vanishes provided $k_2 = 0$. These choices for k_1 and k_2 lead to the particular solution

$$x_p = \frac{1}{\omega^2 - \beta^2}\frac{\alpha}{m} \sin \beta t.$$

To clear up the case when $k - m\beta^2 = 0$, we notice in this case that $\beta = \omega$, which means that the forcing function $\alpha \sin \beta t$ is a solution of the associated homogeneous equation $mx'' + kx = 0$. The method of undetermined coefficients requires that we modify our guess by a factor of t, that is,

$$x_p = k_1 t \sin \omega t + k_2 t \cos \omega t.$$

---

[1]To see that the coefficients in parentheses must equal 0, let $t = 0$ and $t = \pi/2\beta$.

To determine the coefficients k_1 and k_2, we substitute this modified expression into the equation (1.11) and obtain, after some rearrangement of terms,

$$- [2m\omega k_2 + \alpha] \sin \omega t + [2m\omega k_1] \cos \omega t = 0.$$

In that this equation holds for all t, it is necessary that both coefficients of $\sin \omega t$ and $\cos \omega t$ vanish. This leads to $k_1 = 0$ and $k_2 = -\alpha/2m\omega$ and the particular solution

$$x_p = -\frac{1}{2\omega} \frac{\alpha}{m} t \cos \omega t.$$

In summary, the general solution $x = x_h + x_p$ of the forced harmonic oscillator (1.11) is

$$x = \begin{cases} c_1 \sin \omega t + c_2 \cos \omega t + \frac{1}{\omega^2 - \beta^2} \frac{\alpha}{m} \sin \beta t & \text{if } \beta \neq \omega = \sqrt{\frac{k}{m}} \\ c_1 \sin \omega t + c_2 \cos \omega t - \frac{1}{2\omega} \frac{\alpha}{m} t \cos \omega t & \text{if } \beta = \omega. \end{cases}$$

The arbitrary constants c_1 and c_2 are determined by initial conditions. An interesting observation is that when $\beta \neq \omega$ all solutions (regardless of initial conditions) are bounded, since all three trigonometric functions in the general solution are bounded. However, if $\beta = \omega$ all solutions are unbounded because of the factor t appearing in the particular solution x_p. This phenomenon is called *resonance*. The frequency ω is called the *natural frequency* of the oscillator, since it will have this frequency when forcing is absent ($\alpha = 0$). *Resonance occurs when the forcing frequency β equals the natural frequency α of the oscillator.* Figures 6.3 and 6.4 show specific examples.

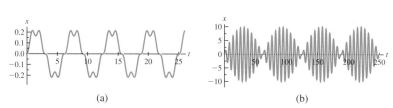

(a)   (b)

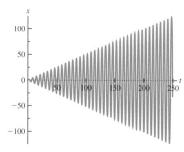

Figure 6.3 Solution graphs for the forced simple harmonic oscillator equation (1.11) are shown in two nonresonance cases. The coefficients are $m = k = \alpha = 1$ and the initial conditions are $x(0) = 0$, $x'(0) = 0$. In (a) the $\beta = 5 \neq 1 = \omega$ so that the forcing frequency $\beta = 5$ is greater than the natural frequency $\omega = 1$. In (b) the forcing frequency $\beta = 1.1$ is close to the natural frequency $\omega = 1$. The oscillatory pattern in case (b) is called a "beat."

Figure 6.4 The solution graph for the forced simple harmonic oscillator equation (1.11) is shown in a resonance case. The coefficients are $m = k = \alpha = 1$ and the initial conditions are $x(0) = 0$, $x'(0) = 0$. The forcing frequency is $\beta = 1$, which equals the natural frequency $\omega = 1$.

Exercises

Find a particular solution of the following systems using the method of undetermined coefficients.

1. $\begin{cases} x' = x + y - 2 \\ y' = x - y + 4 \end{cases}$

2. $\begin{cases} x' = -2x - y + 3 \\ y' = x - y - 6 \end{cases}$

3. $\begin{cases} x' = -x + y - 2e^{-t} \\ y' = 2x - y - e^{-t} \end{cases}$

4. $\begin{cases} x' = -x + y + 7e^t \\ y' = 2x - y - 8e^t \end{cases}$

5. $\begin{cases} x' = y \\ y' = -x - 3 \cos 2t \end{cases}$

6. $\begin{cases} x' = y \\ y' = -x - 8\sin 3t - 8\cos 3t \end{cases}$

7. $\begin{cases} x' = x + 3y - 5 - 4t^2 \\ y' = x - y + 1 \end{cases}$

8. $\begin{cases} x' = x + 3y + 2e^{-t}\sin t \\ y' = x - y \end{cases}$

9. For the systems in Exercises 1–8 solve the initial value problem $x(0) = 0$, $y(0) = 0$.

Solve the initial value problems below.

10. $\begin{cases} x' = -x + y \\ y' = 3x + y - e^{-t} \\ x(1) = 1, \ y(1) = 0 \end{cases}$ **11.** $\begin{cases} x' = -x + y + \sin t \\ y' = 3x + y \\ x(\pi) = 1, \ y(\pi) = 0 \end{cases}$

Find a particular solution of the following second-order equations using the method of undetermined coefficients.

12. $x'' + x = e^{-t}\sin t$ **13.** $x'' + x = te^{-t}$

14. $x'' - x = e^{t}$ **15.** $x'' - x = e^{-t}$

16. For the second-order equations in Exercises 12–15 solve the initial value problem $x(0) = 0$, $x'(0) = 0$.

Consider the initial value problem

$$x' = -3x + 2y + ae^{-3t}$$
$$y' = -x$$
$$x(0) = 0, \qquad y(0) = 0,$$

where $a > 0$ is a positive constant.

17. Use a computer program to obtain graphs of the solution pair components $x(t)$ and $y(t)$ for $0 \le t \le 5$ for a selection of a values ranging from $a = 5$ to 20. Observe that $x(t)$ has a root, but $y(t)$ does not. How does the root of $x(t)$ depend on a? What is the sign of $y(t)$? What happens to $x(t)$ and $y(t)$ as $t \to +\infty$?

18. Solve the initial value problem using the method of undetermined coefficients.

19. Use you answer in Exercise 18 to prove your answers in Exercise 17.

Consider the initial value problem

$$x'' + 2x' + 2x = e^{-t}$$
$$x(0) = 0, \qquad x'(0) = 0.$$

20. Use a computer program to plot the solution for $0 < t < 8$. How many roots does $x(t)$ have in this interval?

21. Solve the initial value problem using the method of undetermined coefficients.

22. Use your answer in Exercise 21 to validate or correct your answers in Exercise 20.

23. **(a)** Use the method of undetermined coefficients to find a particular solution of the glucose/insulin system (0.2) when the glucose ingestion rate g is a constant.

   (b) Find the general solution.

24. **(a)** Use the identity

$$k_1 \sin \beta t + k_2 \cos \beta t = \sqrt{k_1^2 + k_2^2}\ \sin(\beta(t + \theta)),$$

   in which the phase angle θ satisfies $\tan \beta\theta = k_2/k_1$, to rewrite the "asymptotic" solution pair (1.10).

   (b) Use your answer in (a) to find the (eventual) phase difference between the oscillation in glucose concentration x and the oscillation in the ingestion rate g. Also find the (eventual) phase difference between the oscillation in insulin concentration y and that of g.

25. Suppose the forcing functions h_1 and h_2 in the nonhomogeneous system (1.1) are both constant. If $ab - bc \ne 0$, show that there exists a unique equilibrium solution pair.

Consider the initial value problem

$$Lx'' + Rx' + \tfrac{1}{C}x = E(t)$$
$$x(0) = 0, \qquad x'(0) = 0$$

for the charge $x = x(t)$ on an electric circuit with a resistor (of R Ohms), inductor (of L Henrys), capacitor (of C Farads), and impressed voltage (e.g., from a battery). The circuit has no initial charge $x(0)$ and no initial current $x'(0)$. Suppose $E(t) = E_0$ is constant. Solve the initial value problem for the circuits below.

26. $L = 0.1$, $R = 250$, $C = 10^{-5}$

27. $L = 0.2$, $R = 200$, $C = 10^{-5}$

28. $L = 0.1$, $R = 0$, $C = 10^{-5}$

29. $L = 0.1$, $R = 200$, $C = 10^{-5}$

Solve the initial value problem

$$Lx'' + Rx' + \tfrac{1}{C}x = E(t)$$
$$x(0) = 0, \qquad x'(0) = 0$$

with $L = 1$, $R = 2$, $C = 1$ and the impressed voltages below.

30. $E(t) = \sin t$ **31.** $E(t) = \cos t$

32. $E(t) = e^{-t}$ **33.** $E(t) = te^{-t}$

6.2 The Variation of Constants Formula

In this section we study a method for calculating a formula for the general solution of a nonhomogeneous system

$$x' = a(t)x + b(t)y + h_1(t) \qquad (2.1)$$
$$y' = c(t)x + d(t)y + h_2(t).$$

The method requires the general solution of the associated homogeneous solution

$$x' = a(t)x + b(t)y \qquad (2.2)$$
$$y' = c(t)x + d(t)y$$

be available. That is, two independent solution pairs, (x_1, y_1) and (x_2, y_2), of this homogeneous system are known and therefore the general solution is given by the linear combination

$$x_h = c_1 x_1(t) + c_2 x_2(t) \qquad (2.3)$$
$$y_h = c_1 y_1(t) + c_2 y_2(t).$$

What is required for the general solution of (2.1)

$$x = x_h(t) + x_p(t)$$
$$y = y_h(t) + y_p(t)$$

is a particular solution pair $x_p(t)$ and $y_p(t)$.

To motivate the method recall, for a moment, the variation of constants formula for the general solution $x = x_h + x_p$ of the single linear equation

$$x' = p(t)x + q(t)$$

(Sec. 2.1 of Chapter 2). In that formula

$$x = ce^{P(t)} + e^{P(t)}Q(t), \qquad P(t) = e^{\int p(t)\,dt}$$

the particular solution

$$x_p = e^{P(t)}Q(t), \qquad Q(t) = \int e^{-\int p(t)dt} q(t)\,dt$$

is not a constant multiple of the homogeneous solution $e^{P(t)}$ but is a *function* $Q(t)$ times $e^{P(t)}$. Thus, for a single equation it is possible to find a particular solution x_p by allowing the arbitrary constant c in the general solution $x_h = ce^{P(t)}$ of the associated homogeneous equation to be replaced by a function $c = c(t)$.

Suppose we try a similar procedure for the nonhomogeneous system (2.1). That is, suppose we look for a particular solution in which the arbitrary constants c_1 and c_2 in the general solution (2.3) of the associated homogeneous system are replaced by functions of t. (This is the explanation for the odd and self-contradictory phrase *variation of constants*.) If we denote these functions by $k_1(t)$ and $k_2(t)$, then the particular solution we seek has the form

$$x_p = k_1(t)x_1(t) + k_2(t)x_2(t) \qquad (2.4)$$
$$y_p = k_1(t)y_1(t) + k_2(t)y_2(t).$$

The variation of constants formula results from determining the functions $k_1(t)$ and $k_2(t)$ so that this pair is a solution of the nonhomogeneous system (2.1).

First we illustrate the method with an example. The homogeneous system

$$x' = -2x + 2y \qquad (2.5)$$
$$y' = 2x - 5y$$

is an example of the equations (0.2) arising in the pesticide application discussed in Example 2 of the Introduction and in Sec. 5.1 of Chapter 5. The general solution of this homogeneous system is [see (1.9)]

$$x_h = 2c_1 e^{-t} + c_2 e^{-6t} \qquad (2.6)$$
$$y_h = c_1 e^{-t} - 2c_2 e^{-6t},$$

If pesticide is added to the trees at a rate that is steadily decreasing over time then the system becomes nonhomogeneous. Specifically, suppose pesticide is added at the exponentially decreasing rate e^{-2t}. Then the equations for the pesticide amounts in the trees x and soil y become

$$x' = -2x + 2y + e^{-2t} \qquad (2.7)$$
$$y' = 2x - 5y.$$

Figure 6.5 shows graphs of two solution pairs of the nonhomogeneous system (2.7) with initial conditions $x(0) \geq 0$, $y(0) = 0$ for which there is initially no pesticide in the soil. Notice that for both solution pairs the pesticide amounts x and y (in the trees and soil, respectively) tend to 0 as $t \to +\infty$. Nonetheless, the ratio y/x of pesticide in the soil to that in the trees approaches 0.5 (i.e., in the long run there is twice the amount in the trees as there is in the soil). Are these facts true for all solutions of (2.7) with initial condition $y(0) = 0$?

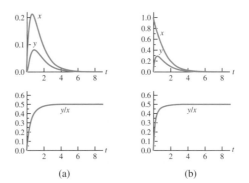

(a)   (b)

Figure 6.5 Solution pairs of the nonhomogeneous system (2.7) are shown for initial conditions (a) $x(0) = 0$, $y(0) = 0$ and (b) $x(0) = 1$, $y(0) = 0$. Also shown are the ratios y/x for both solution pairs.

To obtain the general solution of the nonhomogeneous system (2.7) we require, in addition to (2.6), a particular solution pair. By (2.4) we look for a particular solution pair in the form

$$x_p = 2k_1(t)e^{-t} + k_2(t)e^{-6t} \tag{2.8}$$

$$y_p = k_1(t)e^{-t} - 2k_2(t)e^{-6t}.$$

Our goal is to determine the unknown functions $k_1(t)$ and $k_2(t)$ so that this pair solves (2.7). We do this by substituting this pair into both equations of the system (2.7), which results in the two equations

$$2k_1'e^{-t} + k_2'e^{-6t} = e^{-2t}$$

$$k_1'e^{-t} - 2k_2'e^{-6t} = 0$$

for the derivatives k_1' and k_2'. After solving these two equations algebraically for

$$k_1' = \frac{2}{5}e^{-t}, \qquad k_2' = \frac{1}{5}e^{4t},$$

we integrate these answers to obtain

$$k_1(t) = -\frac{2}{5}e^{-t}, \qquad k_2(t) = \frac{1}{20}e^{4t}.$$

With these choices for $k_1(t)$ and $k_2(t)$ in (2.8) we get a particular solution pair

$$x_p = -\frac{3}{4}e^{-2t} \tag{2.9}$$

$$y_p = -\frac{1}{2}e^{-2t}.$$

We obtain the general solution by adding to this particular solution pair to the general solution pair (2.6) of the associated homogeneous system. The result is

$$x = 2c_1e^{-t} + c_2e^{-6t} - \frac{3}{4}e^{-2t} \tag{2.10}$$

$$y = c_1e^{-t} - 2c_2e^{-6t} - \frac{1}{2}e^{-2t}.$$

The arbitrary constants in the general solution (2.10) are determined by initial conditions. For example, the initial conditions

$$x(0) = x_0 \geq 0, \qquad y(0) = 0$$

are appropriate when there is initially no pesticide present in the soil, but there may be an amount x_0 present in the trees. The general solution (2.10), together with these initial conditions, leads to the equations

$$2c_1 + c_2 - \tfrac{3}{4} = x_0$$

$$c_1 - 2c_2 - \tfrac{1}{2} = 0$$

for c_1 and c_2. Their solution

$$c_1 = \frac{2}{5} + \frac{2}{5}x_0, \qquad c_2 = -\frac{1}{20} + \frac{1}{5}x_0$$

substituted into the general solution (2.10) yields the formulas

$$x = \frac{4}{5}(x_0 + 1)e^{-t} + \frac{1}{5}\left(x_0 - \frac{1}{4}\right)e^{-6t} - \frac{3}{4}e^{-2t}$$

$$y = \frac{2}{5}(x_0 + 1)e^{-t} - \frac{2}{5}\left(x_0 - \frac{1}{4}\right)e^{-6t} - \frac{1}{2}e^{-2t}$$

for the solution of the initial value problem.

In answer to the questions raised above about the solutions of the pesticide system (2.7), note that the solution formulas imply

$$\lim_{t \to +\infty} x = 0, \qquad \lim_{t \to +\infty} y = 0$$

(since all the exponential functions e^{-t}, e^{-6t} and e^{-2t} tend to 0 as $t \to +\infty$). Furthermore,

$$\lim_{t \to +\infty} \frac{y}{x} = \lim_{t \to +\infty} \frac{\frac{2}{5}(x_0 + 1)e^{-t} - \frac{2}{5}\left(x_0 - \frac{1}{4}\right)e^{-6t} - \frac{1}{2}e^{-2t}}{\frac{4}{5}(x_0 + 1)e^{-t} + \frac{1}{5}\left(x_0 - \frac{1}{4}\right)e^{-6t} - \frac{3}{4}e^{-2t}}$$

$$= \lim_{t \to +\infty} \frac{2(x_0 + 1) - 2\left(x_0 - \frac{1}{4}\right)e^{-5t} - \frac{5}{2}e^{-t}}{4(x_0 + 1) + \left(x_0 - \frac{1}{4}\right)e^{-5t} - \frac{15}{4}e^{-t}}$$

$$= \frac{2(x_0 + 1)}{4(x_0 + 1)} = \frac{1}{2}$$

provided $x_0 \neq -1$. This validates our observation obtained from the sample solution graphs in Fig. 6.5 and does so for all initial conditions $x_0 \geq 0$, $y_0 = 0$.

The procedure we used to obtain the particular solution (2.9) of the nonhomogeneous pesticide system (2.7) is applicable to other systems. The general method involves determining the unknown functions $k_1(t)$ and $k_2(t)$ in (2.4) so that the resulting pair solves the nonhomogeneous system (2.1). A substitution of (2.4) into the system (2.1) results in the equations

$$\begin{aligned} k_1'x_1 + k_2'x_2 + k_1x_1' + k_2x_2' &= a\,(k_1x_1 + k_2x_2) + b\,(k_1y_1 + k_2y_2) + h_1 \\ k_1'y_1 + k_2'y_2 + k_1y_1' + k_2y_2' &= c\,(k_1x_1 + k_2x_2) + d\,(k_1y_1 + k_2y_2) + h_2 \end{aligned} \qquad (2.11)$$

or, after some rearrangement of terms,

$$k_1'x_1 + k_2'x_2 + k_1\left[x_1' - (ax_1 + by_1)\right] + k_2\left[x_2' - (ax_2 + by_2)\right] = h_1$$

$$k_1'y_1 + k_2'y_2 + k_1\left[y_1' - (cx_1 + dy_1)\right] + k_2\left[y_2' - (cx_2 + dy_2)\right] = h_2.$$

However, because both pairs x_1, y_1 and x_2, y_2 are solution pairs of the associated homogeneous system (2.2), all four terms appearing in the square brackets are equal to 0. This leaves us with the two (algebraic) equations

$$k_1'x_1 + k_2'x_2 = h_1 \qquad (2.12)$$

$$k_1'y_1 + k_2'y_2 = h_2$$

for the two derivatives k_1' and k_2'. We obtain the sought-after functions k_1 and k_2 by solving these equations for

$$k_1' = \frac{h_1 y_2 - h_2 x_2}{x_1 y_2 - y_1 x_2}$$

$$k_2' = \frac{h_2 x_1 - h_1 y_1}{x_1 y_2 - y_1 x_2}$$

and integrating

$$k_1 = \int \frac{h_1 y_2 - h_2 x_2}{x_1 y_2 - y_1 x_2} \, dt$$

$$k_2 = \int \frac{h_2 x_1 - h_1 y_1}{x_1 y_2 - y_1 x_2} \, dt.$$

$$(2.13)$$

Notice that the denominators are not equal to 0 since the solution pairs x_1, y_1 and x_2, y_2 are independent (Theorem 5 in Chapter 4). With these formulas for k_1 and k_2 in hand, we have arrived at the rather formidable formula

$$x_p = x_1(t) \int \frac{h_1 y_2 - h_2 x_2}{x_1 y_2 - y_1 x_2} \, dt + x_2(t) \int \frac{h_2 x_1 - h_1 y_1}{x_1 y_2 - y_1 x_2} \, dt$$

$$y_p = y_1(t) \int \frac{h_1 y_2 - h_2 x_2}{x_1 y_2 - y_1 x_2} \, dt + y_2(t) \int \frac{h_2 x_1 - h_1 y_1}{x_1 y_2 - y_1 x_2} \, dt$$

$$(2.14)$$

for a particular solution pair of the nonhomogeneous system (2.1).

It follows that the general solution of the nonhomogeneous system is

$$x = c_1 x_1(t) + c_2 x_2(t) + x_1(t) \int \frac{h_1 y_2 - h_2 x_2}{x_1 y_2 - y_1 x_2} \, dt + x_2(t) \int \frac{h_2 x_1 - h_1 y_1}{x_1 y_2 - y_1 x_2} \, dt$$

$$y = c_1 y_1(t) + c_2 y_2(t) + y_1(t) \int \frac{h_1 y_2 - h_2 x_2}{x_1 y_2 - y_1 x_2} \, dt + y_2(t) \int \frac{h_2 x_1 - h_1 y_1}{x_1 y_2 - y_1 x_2} \, dt.$$

$$(2.15)$$

These are the *variation of constants formulas* for the general nonhomogeneous system (2.1).

EXAMPLE 1

The homogeneous system associated with the nonhomogeneous glucose/insulin system (1.3) has general solution (1.4) constructed from the independent solution pairs

$$x_1 = 4.34e^{-2.34t}, \qquad y_1 = -0.58e^{-2.34t}$$
$$x_2 = 4.34e^{-1.36t}, \qquad y_2 = -1.56e^{-1.36t}.$$

For these pairs

$$x_1 y_2 - y_1 x_2 = -4.2532e^{-3.7t}.$$

From (1.3), we have $h_1 = 18$ and $h_2 = 0$. In the variation of constants formula (2.15) we need the integrals

$$\int \frac{h_1 y_2}{x_1 y_2 - y_1 x_2} \, dt = \int 6.60 e^{2.34t} \, dt = 2.82 e^{2.34t}$$

$$\int \frac{-h_1 y_1}{x_1 y_2 - y_1 x_2} \, dt = -1.81 e^{1.36t}.$$

These results yield the general solution (rounded to three significant digits)

$$x = 4.34 c_1 e^{-2.34t} + 4.34 c_2 e^{-1.36t} + 4.41$$
$$y = -0.58 c_1 e^{-2.34t} - 1.56 c_2 e^{-1.36t} + 1.18.$$

This is the same general solution formula we found using the method of undetermined coefficients; see (1.5). ∎

Notice that unlike the method of undetermined coefficients, the variation of constants formulas apply to linear systems whose coefficients are not necessarily constants and whose forcing functions need not be of a special type. Here is an example.

EXAMPLE 2

In the pesticide application discussed in Example 2 of the Introduction (also see Chapter 4), an initial dose of pesticide is added to a stand of trees and the amounts of pesticide in the trees and in the soil are monitored over time. In this example we consider instead the case when pesticide is continuously added the trees at a rate $r(t)$. See Fig. 6.6. From the compartmental diagram in Fig. 6.6 we obtain the nonhomogeneous linear system

$$x' = -2x + 2y + r(t) \tag{2.16}$$
$$y' = 2x - 5y$$

for the amounts x and y of pesticide in the trees and soil, respectively. Suppose we want to add pesticide at a constant rate $\alpha > 0$ and begin the treatment at an initial time $t_0 = 0$. One way to model the pesticide treatment rate $r(t)$, which initially is 0 and increases to α, is to use a function that satisfies $r(0) = 0$, $r'(t) > 0$, and $\lim_{t \to +\infty} r(t) = \alpha$. A specific example of such a function is

$$\tanh t = \alpha \frac{e^t - e^{-t}}{e^t + e^{-t}};$$

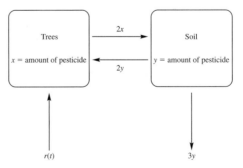

Figure 6.6

see Fig. 6.7. With $r(t) = \alpha \tanh t$, the system (2.16) becomes

$$x' = -2x + 2y + \alpha \tanh t \qquad (2.17)$$
$$y' = 2x - 5y.$$

Assuming there is initially no pesticide in either the trees or the soil, we have the initial conditions

$$x(0) = 0, \qquad y(0) = 0. \qquad (2.18)$$

A computer-drawn graph of the solution pair for this initial value problem with $\alpha = 1$ appears in Fig. 6.7. Notice that the amount of pesticide in both the trees and the soil increases as time increases, but not without bound. From the graphs in Fig. 6.7, it appears that the amount of pesticide in the trees levels off, in the long run, at approximately 0.8, while that in the soil levels off at approximately 0.3.

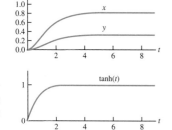

Figure   6.7 The solution graphs for (2.17) with $x(0) = 0$, $y(0) = 0$ with $\alpha = 1$.

The general solution of the homogeneous system associated with (2.17), given by (1.9), is constructed from the two independent solution pairs

$$x_1 = 2e^{-t}, \qquad x_2 = e^{-6t}$$
$$y_1 = e^{-t}, \qquad y_2 = -2e^{-6t}.$$

For these pairs, we have $x_1 y_2 - y_1 x_2 = -5e^{-7t}$ in the variation of constants formulas (2.15) for the general solution. Using $h_1 = \alpha \tanh t$ and $h_2 = 0$, we calculate the integrals (an algebraic computer program is helpful here)

$$\int \frac{h_1 y_2 - h_2 x_2}{x_1 y_2 - y_1 x_2} dt = \alpha \left(\frac{2}{5} e^t - \frac{4}{5} \tan^{-1} \left(e^t \right) \right)$$

$$\int \frac{h_2 x_1 - h_1 y_1}{x_1 y_2 - y_1 x_2} dt = \alpha \left(\frac{1}{5} e^{2t} - \frac{1}{10} e^{4t} + \frac{1}{30} e^{6t} - \frac{1}{5} \ln \left(1 + e^{2t} \right) \right).$$

The formulas (2.15) now yield the general solution

$$x = 2c_1 e^{-t} + c_2 e^{-6t}$$
$$+ \alpha \left(\frac{5}{6} - \frac{1}{10} e^{-2t} + \frac{1}{5} e^{-4t} - \frac{1}{5} e^{-6t} \ln \left(1 + e^{2t} \right) - \frac{8}{5} e^{-t} \tan^{-1} \left(e^t \right) \right)$$
$$\qquad (2.19)$$

$$y = c_1 e^{-t} - 2c_2 e^{-6t}$$
$$+ \alpha \left(\frac{1}{3} + \frac{1}{5} e^{-2t} - \frac{2}{5} e^{-4t} + \frac{2}{5} e^{-6t} \ln \left(1 + e^{2t} \right) - \frac{4}{5} e^{-t} \tan^{-1} \left(e^t \right) \right).$$

The arbitrary constants c_1 and c_2 are determined by initial conditions.

For example, to solve the initial value problem (2.18) we must choose c_1 and c_2 so that

$$x(0) = 2c_1 + c_2 + \alpha \left(\frac{14}{15} - \frac{1}{5} \ln 2 - \frac{2}{5}\pi \right) = 0$$

$$y(0) = c_1 - 2c_2 + \alpha \left(\frac{2}{15} + \frac{2}{5} \ln 2 - \frac{1}{5}\pi \right) = 0,$$

that is,

$$c_1 = \alpha \left(\frac{1}{5}\pi - \frac{2}{5} \right), \qquad c_2 = \alpha \left(\frac{1}{5} \ln 2 - \frac{2}{15} \right).$$

All of the exponential functions appearing in the solution (2.19) tend to 0 as $t \to +\infty$. Using l'Hôpital's rule, we can show that $e^{-6t} \ln \left(1 + e^{2t} \right)$ and $e^{-t} \tan^{-1} \left(e^t \right)$ also tend to 0 as $t \to +\infty$. It follows from the solution formula that

$$\lim_{t \to +\infty} x = \frac{5}{6}\alpha, \qquad \lim_{t \to +\infty} y = \frac{1}{3}\alpha.$$

Figure 6.7 shows graphs of the solution pair for the case $\alpha = 1$ when $\lim_{t \to +\infty} x = 5/6 \approx 0.8333$ and $\lim_{t \to +\infty} y = 1/3 \approx 0.3333$. ∎

The variation of constants formulas (2.15) are complicated. Nonetheless, they are important formulas in the theory and application of differential equations. They can be simplified notationally by use of matrix notation (Sec. 6.3).

6.2.1 Second-Order Equations

The second-order equation

$$x'' + x = \sin 2t$$

is an example of the forced harmonic oscillator equation (1.11). We can calculate its general solution using the variation of constants formulas (2.15) by considering the equivalent first-order system

$$x' = y$$
$$y' = -x + \sin 2t.$$

The general solution

$$x_h = c_1 \sin t + c_2 \cos t$$
$$y_h = c_1 \cos t - c_2 \sin t$$

of the associated homogeneous system

$$x' = y$$
$$y' = -x$$

is a linear combination of the two independent solution pairs

$$x_1 = \sin t, \qquad x_2 = \cos t$$
$$y_1 = \cos t, \qquad y_2 = -\sin t.$$

Using

$$x_1 y_2 - y_1 x_2 = -\sin^2 t - \cos^2 t = -1$$

and $h_1 = 0$, $h_2 = \sin 2t$, we calculate

$$x_1(t) \int \frac{h_1 y_2 - h_2 x_2}{x_1 y_2 - y_1 x_2} \, dt = \sin t \int \sin 2t \cos t \, dt$$

$$= -\frac{2}{3} \sin t \cos^3 t$$

$$x_2(t) \int \frac{h_2 x_1 - h_1 y_1}{x_1 y_2 - y_1 x_2} \, dt = -\cos t \int \sin 2t \sin t \, dt$$

$$= -\frac{2}{3} \sin t \cos t + \frac{2}{3} \sin t \cos^3 t.$$

From the first formula in (2.15) we obtain the general solution

$$x = c_1 \sin t + c_2 \cos t - \frac{2}{3} \sin t \cos t.$$

We can shorten the application of the variation of constants formulas (2.15) to second-order equations by skipping the conversion of the equation to an equivalent first order system and utilizing just the first formula in (2.15). In doing this we have $h_1 = 0$ and $y_1 = x_1'$, $y_2 = x_2'$. Thus, if $x_1(t)$ and $x_2(t)$ are two independent solutions of the associated homogeneous equation

$$x'' + \beta(t)x' + \gamma(t)x = 0,$$

then from (2.15) we obtain the formula

$$x = c_1 x_1(t) + c_2 x_2(t) - x_1(t) \int \frac{h x_2}{x_1 x_2' - x_1' x_2} \, dt + x_2(t) \int \frac{h x_1}{x_1 x_2' - x_1' x_2} \, dt$$

for the general solution of the nonhomogeneous, second-order equation

$$x'' + \beta(t)x' + \gamma(t)x = h(t). \tag{2.20}$$

The expression

$$W(x_1, x_2) = x_1 x_2' - x_1' x_2$$

is called the *Wronskian* of the solutions x_1 and x_2, and the variation of constants formula for the second-order equation (2.20) is

$$x = c_1 x_1(t) + c_2 x_2(t) - x_1(t) \int \frac{h x_2}{W(x_1, x_2)} \, dt + x_2(t) \int \frac{h x_1}{W(x_1, x_2)} \, dt. \tag{2.21}$$

Note: The coefficient of x'' in (2.20) is 1. For a second-order equation with another coefficient on x'' we divide both sides of the equation by this coefficient so that the coefficient on x'' for the resulting equation equals 1. Thus, we write

$$a(t)x'' + b(t)x' + c(t)x = g(t)$$

as

$$x'' + \frac{b(t)}{a(t)}x' + \frac{c(t)}{a(t)}x = \frac{g(t)}{a(t)}$$

so that

$$\beta(t) = \frac{b(t)}{a(t)}, \qquad \gamma(t) = \frac{c(t)}{a(t)}, \qquad h(t) = \frac{g(t)}{a(t)}$$

in the variation of constants formula.

EXAMPLE 3   The forced harmonic oscillator equation

$$mx'' + kx = \alpha \sin \beta t$$

is a second-order nonhomogeneous equation. The associated homogeneous (simple harmonic oscillator) equation $mx'' + kx = 0$ has the two independent solutions

$$x_1 = \sin \omega t, \qquad x_2 = \cos \omega t, \qquad \omega = \sqrt{\frac{k}{m}}.$$

The Wronskian of these solutions is

$$W(\sin \omega t, \cos \omega t) = -\omega \sin^2 \omega t - \omega \cos^2 \omega t = -\omega.$$

Using these terms in the variation of constants formula (2.21), together with

$$h(t) = \frac{\alpha}{m} \sin \beta t,$$

we obtain (a computer is useful here)

$$-x_1(t) \int h \frac{x_2}{W(x_1, x_2)} dt = \frac{\alpha \sin \omega t}{m\omega \left(\omega^2 - \beta^2\right)} (\beta \cos \beta t \cos \omega t + \omega \sin \beta t \sin \omega t)$$

and

$$x_2(t) \int h \frac{x_1}{W(x_1, x_2)} dt = \frac{\alpha \cos \omega t}{m\omega \left(\omega^2 - \beta^2\right)} (\omega \sin \beta t \cos \omega t - \beta \cos \beta t \sin \omega t).$$

These results are valid in the nonresonance case $\beta \neq \omega$. According to the formula (2.21), these terms are to be added, which results in

$$\frac{\alpha \sin \beta t}{m \left(\omega^2 - \beta^2\right)} \left(\sin^2 \omega t + \cos^2 \omega t\right) = \frac{\alpha}{m \left(\omega^2 - \beta^2\right)} \sin \beta t.$$

Thus, we arrive at the general solution

$$x = c_1 \sin \omega t + c_2 \cos \omega t + \frac{\alpha}{m \left(\omega^2 - \beta^2\right)} \sin \beta t. \qquad \blacksquare$$

Exercises

Consider the nonhomogeneous system

$$x' = -3x + 4y + e^{-t}$$
$$y' = -6x + 7y + 1.$$

1. Use the variation of constants formula (2.15) to find the general solution.

2. Solve the initial value problem $x(0) = 0$, $y(0) = 0$.

3. Solve the initial value problem $x(0) = 2$, $y(0) = 1$.

Consider the nonhomogeneous system

$$x' = -3x + 5y - 1$$
$$y' = -2x + 3y + 1.$$

4. Use the variation of constants formula (2.15) to find the general solution.

5. Solve the initial value problem $x(0) = 0$, $y(0) = 0$.

6. Solve the initial value problem $x(0) = -2$, $y(0) = 1$.

Consider the second-order equation $x'' + x = \cos t$.

7. Use the variation of constants formula (2.21) to find the general solution.

8. Solve the initial value problem $x(0) = 0$, $x'(0) = 0$.

9. Solve the initial value problem $x(0) = 1$, $x'(0) = -1$.

Consider the second-order equation $x'' - x = e^t$.

10. Use the variation of constants formula (2.21) to find the general solution.

11. Solve the initial value problem $x(0) = 0$, $x'(0) = 0$.

12. Solve the initial value problem $x(0) = 2$, $x'(0) = -2$.

Consider the nonhomogeneous equation $mx'' + cx' + kx = \alpha \sin \omega t$, where all five parameters are positive and $c > 2\sqrt{km}$.

13. Find a particular solution x_p using the variation of constants formula.

14. Show that the general solution tends to x_p as $t \to +\infty$. (*Hint*: $x_h = x - x_p$ tends to 0 as $t \to +\infty$.) Thus, x_p is called the steady state and x_h is called the transient part of the general solution.

15. Use the variation of constants formulas to obtain the general solution (1.9) of the glucose/insulin system (1.7).

If pesticide is removed from the soil at a constant rate $r > 0$ in the pesticide system

$$x' = -2x + 2y$$
$$y' = 2x - 5y,$$

the resulting modified system

$$x' = -2x + 2y$$
$$y' = 2x - 5y - r$$

is nonhomogeneous.

16. Use a computer to graph both x and y (versus t) for the initial value problem $x(0) = 1$, $y(0) = 0$. Do this for removal rate $r = 1$. What can you say about the amount of pesticide in the trees and the soil as time increases?

17. Solve the initial value problem using the variation of constants formulas (2.15).

18. Use your answer in Exercise 17 to justify your answer in Exercise 16.

An electric circuit has a capacitor ($0 < C$ farads), a resistor ($0 \le R$ ohms) and inductor ($0 \le L$ Henrys). The charge $q = q(t)$ on the capacitor satisfies the second-order equation

$$Lq'' + Rq' + \frac{1}{C}q = E(t),$$

where $E(t)$ is the electromotive force (volts) applied to the circuit. If the initial charge on the capacitor is q_0 and the initial amount of current in the circuit is i_0, then we have the initial conditions

$$q(0) = q_0, \qquad q'(0) = i_0.$$

19. To a circuit with $L = 1$, $R = 3$, and $C = \frac{1}{2}$, an electromotive force is turned on at $t = 0$ that increases to 1 (volt) as $t \to +\infty$. Specifically, $E(t) = 1 - e^{-2t}$. Use a computer program to graph the solution of this initial value problem when there is an initial charge of $\frac{1}{2}$ and no initial current in the circuit. What happens in the long run to the charge $q(t)$? What happens in the short run?

20. Use the variation of constants formula (2.21) to solve the initial value problem.

21. Use your answer in Exercise 20 to verify your answers in Exercise 19.

6.3 Matrix Notation

Our goal in this section is to reformulate the variation of constants formula derived in Sec. 6.2 into matrix notation. This will make the formula notationally and conceptually simpler and provide a straightforward way to extend the formula from systems of two equations to systems of any number of equations.

In Sec. 4.3 of Chapter 4 we used matrix notation to write the nonhomogeneous system

$$x' = a(t)x + b(t)y + h_1(t)$$
$$y' = c(t)x + d(t)y + h_2(t)$$

as

$$\begin{pmatrix} x \\ y \end{pmatrix}' = \begin{pmatrix} a(t) & b(t) \\ c(t) & d(t) \end{pmatrix} \begin{pmatrix} x \\ y \end{pmatrix} + \begin{pmatrix} h_1(t) \\ h_2(t) \end{pmatrix}$$

or, in more compact form, as

$$\tilde{x}' = A(t)\tilde{x} + \tilde{h}(t), \tag{3.1}$$

where

$$\tilde{x}(t) = \begin{pmatrix} x(t) \\ y(t) \end{pmatrix}, \qquad A(t) = \begin{pmatrix} a(t) & b(t) \\ c(t) & d(t) \end{pmatrix}, \qquad \tilde{h}(t) = \begin{pmatrix} h_1(t) \\ h_2(t) \end{pmatrix}.$$

The general solution

$$\tilde{x}_h = \begin{pmatrix} x_h(t) \\ y_h(t) \end{pmatrix}$$

of the associated homogeneous system

$$\tilde{x}' = A(t)\tilde{x} \tag{3.2}$$

is a linear combination

$$\tilde{x}_h = c_1 \begin{pmatrix} x_1(t) \\ y_1(t) \end{pmatrix} + c_2 \begin{pmatrix} x_2(t) \\ y_2(t) \end{pmatrix}$$

$$= \begin{pmatrix} c_1 x_1(t) + c_2 x_2(t) \\ c_1 y_1(t) + c_2 y_2(t) \end{pmatrix}$$

of two independent solutions pairs

$$\begin{pmatrix} x_1(t) \\ y_1(t) \end{pmatrix}, \qquad \begin{pmatrix} x_2(t) \\ y_2(t) \end{pmatrix}.$$

We wrote the general solution in the matrix form

$$\tilde{x}_h = \Phi(t)\tilde{c},$$

where

$$\tilde{c} = \begin{pmatrix} c_1 \\ c_2 \end{pmatrix}$$

is a vector of arbitrary constants c_1 and c_2 and

$$\Phi(t) = \begin{pmatrix} x_1(t) & x_2(t) \\ y_1(t) & y_2(t) \end{pmatrix}$$

is a fundamental solution matrix.

The fact that both columns of $\Phi(t)$ are solutions of the homogeneous system (3.2) is succinctly summarized by the matrix differential equation

$$\Phi'(t) = A(t)\Phi(t). \tag{3.3}$$

The general solution of the nonhomogeneous system (3.1) has the form $\tilde{x} = \tilde{x}_h + \tilde{x}_p$, that is,

$$\tilde{x} = \Phi(t)\tilde{c} + \tilde{x}_p,$$

where $\tilde{x}_p$ is any particular solution of (3.1). The variation of constants method described in Sec. 6.2 produces a particular solution $\tilde{x}_p$. In matrix notation this method and its resulting formula are as follows.

A particular solution is sought in the form

$$\tilde{x}_p = \Phi(t)\tilde{k}(t) \tag{3.4}$$

[which is (2.4) written in matrix form].[2] Our task is choose components $k_1(t)$ and $k_2(t)$ in

$$\tilde{k}(t) = \begin{pmatrix} k_1(t) \\ k_2(t) \end{pmatrix}$$

so that $\tilde{x}_p$ solves the nonhomogeneous system (3.1).

Substituting (3.4) into the left-hand side and the right-hand side of system (3.1), we obtain

$$\tilde{x}'_p = \Phi(t)\tilde{k}'(t) + \Phi'(t)\tilde{k}(t)$$
$$A(t)\tilde{x}_p + \tilde{h}(t) = A(t)\Phi(t)\tilde{k}(t) + \tilde{h}(t),$$

respectively. The function (3.4) solves the nonhomogeneous system (3.1) if and only if we can choose $\tilde{k}(t)$ so that these two terms are equal, that is,

$$\Phi(t)\tilde{k}'(t) + \Phi'(t)\tilde{k}(t) = A(t)\Phi(t)\tilde{k}(t) + \tilde{h}(t).$$

[This is the matrix form of equations (2.11)]. Using (3.3), we rewrite the second term on the left-hand side to obtain

$$\Phi(t)\tilde{k}'(t) + A(t)\Phi(t)\tilde{k}(t) = A(t)\Phi(t)\tilde{k}(t) + \tilde{h}(t),$$

which, after a cancellation of $A(t)\Phi(t)\tilde{k}(t)$ from both sides of this equation, yields the equation

$$\Phi(t)\tilde{k}'(t) = \tilde{h}(t)$$

for the derivative $\tilde{k}'$. This is the matrix formulation of the equations (2.12) for k'_1 and k'_2. The solution of this equation is

$$\tilde{k}'(t) = \Phi^{-1}(t)\tilde{h}(t). \tag{3.5}$$

Recall that we defined the derivative of a vector to be the vector of the derivatives of its components (Sec. 4.3 of Chapter 4):

$$\tilde{k}'(t) = \begin{pmatrix} k'_1(t) \\ k'_2(t) \end{pmatrix}.$$

Similarly, we define the integral (antiderivative) of a vector to be the vector of the integrals of its components. For example,

$$\int \tilde{k}'(t)\,dt = \begin{pmatrix} \int k'_1(t)\,dt \\ \int k'_2(t)\,dt \end{pmatrix} = \begin{pmatrix} k_1(t) \\ k_2(t) \end{pmatrix} = \tilde{k}(t).$$

With this notation we obtain from (3.5) the formula

$$\tilde{k}(t) = \int \Phi^{-1}(t)\tilde{h}(t)\,dt$$

---

[2]This expression comes from the general solution of the homogeneous system $\tilde{x}_h$ with the vector of constants $\tilde{c}$ replaced by a vector of functions $\tilde{k}$. Hence the name *variation of constants*.

for $\tilde{k}(t)$. This is (2.13) in matrix form. [The fundamental solution matrix has an inverse $\Phi^{-1}(t)$ because its determinant is nonzero by Theorem 5, Sec. 4.3 in Chapter 4.]

We have obtained the particular solution [see (2.14)]

$$\tilde{x}_p = \Phi(t) \int \Phi^{-1}(t)\tilde{h}(t)\, dt$$

and therefore the general solution

$$\tilde{x} = \Phi(t)\tilde{c} + \Phi(t) \int \Phi^{-1}(t)\tilde{h}(t)\, dt.$$

This formula is the matrix form of the *variation of constants formula* (2.15).

THEOREM 1

Variation of Constants Formula If $\Phi(t)$ is a fundamental solution matrix of the homogeneous system $\tilde{x}' = A(t)\tilde{x}$, then the general solution of the nonhomogeneous system

$$\tilde{x}' = A(t)\tilde{x} + \tilde{h}(t)$$

is given by the formula

$$\tilde{x} = \Phi(t)\tilde{c} + \Phi(t) \int \Phi^{-1}(t)\tilde{h}(t)\, dt, \tag{3.6}$$

where $\tilde{c}$ is a vector of arbitrary constants.

From the variation of constants formula we see that to find the general solution of a nonhomogeneous linear first-order system all we need are two independent solutions of the associated homogeneous system (3.2)! This is because two independent solutions of the homogeneous system produce a fundamental solution matrix $\Phi(t)$. However, in order to obtain an explicit formula for the general solution the integration in (3.6) must be carried out and in practice this might be very difficult (if not impossible).

EXAMPLE 1

In matrix notation the system [see (2.7)]

$$x' = -2x + 2y + e^{-2t}$$
$$y' = 2x - 5y$$

has the form $\tilde{x}' = A\tilde{x} + \tilde{h}(t)$ with

$$A = \begin{pmatrix} -2 & 2 \\ 2 & -5 \end{pmatrix}, \qquad \tilde{h}(t) = \begin{pmatrix} e^{-2t} \\ 0 \end{pmatrix}.$$

In Sec. 5.1 of Chapter 5 we found the two independent solution pairs

$$\tilde{x} = \begin{pmatrix} 2e^{-t} \\ e^{-t} \end{pmatrix}, \qquad \tilde{x} = \begin{pmatrix} e^{-6t} \\ -2e^{-6t} \end{pmatrix}$$

of the associated homogeneous system $\tilde{x}' = A\tilde{x}$. These give us the fundamental solution matrix

$$\Phi(t) = \begin{pmatrix} 2e^{-t} & e^{-6t} \\ e^{-t} & -2e^{-6t} \end{pmatrix}.$$

To use the variation of constants formula (3.6) we calculate

$$\Phi^{-1}(t) = \begin{pmatrix} \frac{2}{5}e^t & \frac{1}{5}e^t \\ \frac{1}{5}e^{6t} & -\frac{2}{5}e^{6t} \end{pmatrix}, \qquad \Phi^{-1}(t)\tilde{h}(t) = \begin{pmatrix} \frac{2}{5}e^{-t} \\ \frac{1}{5}e^{4t} \end{pmatrix}.$$

Thus

$$\int \Phi^{-1}(t)\tilde{h}(t)\,dt = \begin{pmatrix} -\frac{2}{5}e^{-t} \\ \frac{1}{20}e^{4t} \end{pmatrix}$$

and

$$\tilde{x}_p = \Phi(t)\int^t \Phi^{-1}(s)\tilde{h}(s)\,ds = \begin{pmatrix} -\frac{3}{4}e^{-2t} \\ -\frac{1}{2}e^{-2t} \end{pmatrix}.$$

The resulting general solution $\tilde{x} = \Phi(t)\tilde{c} + \tilde{x}_p$, that is,

$$\begin{pmatrix} x \\ y \end{pmatrix} = \begin{pmatrix} 2e^{-t} & e^{-6t} \\ e^{-t} & -2e^{-6t} \end{pmatrix}\begin{pmatrix} c_1 \\ c_2 \end{pmatrix} + \begin{pmatrix} -\frac{3}{4}e^{-2t} \\ -\frac{1}{2}e^{-2t} \end{pmatrix}$$

$$= \begin{pmatrix} 2c_1e^{-t} + c_2e^{-6t} - \frac{3}{4}e^{-2t} \\ c_1e^{-t} - 2c_2e^{-6t} - \frac{1}{2}e^{-2t} \end{pmatrix}$$

agrees with that we found using the method of undetermined coefficients above [see (2.10)]. ∎

In this section we have discussed the matrix form of the variation of constants formula in the context of systems of two linear equations. The matrix derivation of (3.6) shows, however, that the method applies to systems of any number of linear equations. For example, for systems of three linear equations the three columns in the 3×3 fundamental solution matrix $\Phi(t)$ are independent solution triples.

Computer programs that perform integrations and matrix algebra are useful in using the variation of constants formula (3.6), especially for higher-order systems.

EXAMPLE 2

In Example 1 (Sec. 5.6 of Chapter 5) the amounts of a radioactive tracer located in an aquatic food chain are modeled by the homogeneous system

$$\begin{aligned} x_1' &= -0.03x_1 + 0.06x_2 + 0.05x_3 \\ x_2' &= 0.02x_1 - 0.12x_2 \\ x_3' &= 0.01x_1 + 0.06x_2 - 0.05x_3. \end{aligned} \tag{3.7}$$

Tracer amounts in the water, zooplankton, and phytoplankton are denoted by x_1, x_2, and x_3, respectively. In that example we calculated the fundamental solution matrix

$$\Phi(t) \approx \begin{pmatrix} 0.5169e^{-0.07551t} & 0.2677e^{-0.1245t} & 0.9415 \\ 0.2324e^{-0.07551t} & -1.191e^{-0.1245t} & 0.1569 \\ -0.7493e^{-0.07551t} & 0.9234e^{-0.1245t} & 0.3766 \end{pmatrix}.$$

If the radioactive tracer is removed from the water at a constant rate $r > 0$, then the model equations become nonhomogeneous, namely

$$
\begin{aligned}
x_1' &= -0.03x_1 + 0.06x_2 + 0.05x_3 - r \\
x_2' &= 0.02x_1 - 0.12x_2 \\
x_3' &= 0.01x_1 + 0.06x_2 - 0.05x_3.
\end{aligned}
\tag{3.8}
$$

If initially 100 units of tracer are added to the water (and none is initially present in the phytoplankton and zooplankton), then we have the initial conditions

$$
x_1(0) = 100, \qquad x_2(0) = 0, \qquad x_3(0) = 0.
\tag{3.9}
$$

Figure 6.8 shows graphs of each component of the solution triple for this initial value problem when the removal rate is $r = 1$ (microcuries per hour). Negative values of x_1, x_2, and x_3 are not meaningful in this application, so we interpret the fact that all three components become negative in a finite amount of time as meaning that the tracer is eventually removed from each subsystem in the food chain. The graphs in Fig. 6.8 imply that the tracer is removed first from the water (after about 95 hours), then from the phytoplankton (after about 100 hours), and finally from the zooplankton (after about 115 hours).

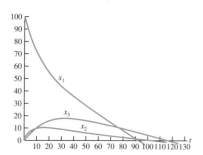

Figure 6.8 The solution components of the aquatic food chain system (3.8) with initial conditions $x_1(0) = 100$, $x_2(0) = 0$, $x_3(0) = 0$ and tracer removal rate $r = 1$.

We can calculate the general solution of the nonhomogeneous linear system (3.8) with $r = 1$ using the variation of constants formula (3.6) and the fundamental solution matrix $\Phi(t)$ above. For the result we can then calculate the solution of the initial value problem (3.9). In this calculation

$$
\tilde{h}(t) = \begin{pmatrix} -1 \\ 0 \\ 0 \end{pmatrix}.
$$

A computer program is very useful in carrying out the necessary integrations and matrix algebra in (3.6). The resulting general solution is

$$
\begin{pmatrix} x_1 \\ x_2 \\ x_3 \end{pmatrix} \approx
$$

$$
\begin{pmatrix}
0.5169c_1e^{-0.076t} + 0.2670c_2e^{-0.12t} + 0.9415c_3 - 4.504 - 0.6383t \\
0.2324c_1e^{-0.076t} - 1.191c_2e^{-0.12t} + 0.1569c_3 + 0.1356 - 0.1064t \\
-0.7493c_1e^{-0.076t} + 0.9234c_2e^{-0.12t} + 0.3766c_3 + 4.368 - 0.2553t
\end{pmatrix}.
\tag{3.10}
$$

The three arbitrary constants c_1, c_2, and c_3 are determined by initial conditions. For example, the initial conditions (3.9), together with the general solution, require the constants c_1, c_2, and c_3 to satisfy the equations

$$0.5169c_1 + 0.2670c_2 + 0.9415c_3 - 4.504 = 100$$
$$0.2324c_1 - 1.191c_2 + 0.1569c_3 + 0.1356 = 0$$
$$-0.7493c_1 + 0.9234c_2 + 0.3766c_3 + 4.368 = 0.$$

Using a computer to solve these equations, we obtain

$$c_1 \approx 67.2221, \qquad c_2 \approx 22.1635, \qquad c_3 \approx 67.8059.$$

With these values for the arbitrary constants in the general solution (3.10), we arrive at the solution

$$\widetilde{x}(t) = \begin{pmatrix} x_1(t) \\ x_2(t) \\ x_3(t) \end{pmatrix} \approx \begin{pmatrix} 35e^{-0.076t} + 5.9e^{-0.12t} + 59 - 0.64t \\ 16e^{-0.076t} - 26e^{-0.12t} + 11 - 0.11t \\ -50e^{-0.076t} + 20e^{-0.12t} + 30 - 0.26t \end{pmatrix}$$

of the initial value problem (3.9) (constants are rounded to two significant digits).

Returning to the observations concerning the removal of the radioactive tracer from each subsystem of the aquatic food chain, we can calculate the time the tracer is gone from the water by solving the equation $x_1(t) = 0$ for t, that is, we solve (using a computer)

$$35e^{-0.076t} + 5.9e^{-0.12t} + 59 - 0.64t = 0$$

for $t \approx 92$ (hours). Similarly, the (positive) roots of the equations $x_2(t) = 0$ and $x_3(t) = 0$ determine the times $t \approx 104$ (hours) and $t \approx 115$ (hours) at which the tracer is gone from the phytoplankton and zooplankton. ∎

We can derive a formula for the solution of an initial value problem from the variation of constants formula (3.6) by choosing a definite integral. For the initial value problem

$$\widetilde{x}' = A(t)\widetilde{x} + \widetilde{h}(t) \tag{3.11}$$
$$\widetilde{x}(t_0) = \widetilde{x}_0$$

we choose the definite integral

$$\widetilde{x}_p = \Phi(t) \int_{t_0}^{t} \Phi^{-1}(s)\widetilde{h}(s)\, ds$$

in (3.6), from which we obtain the general solution

$$\widetilde{x}(t) = \Phi(t)\widetilde{c} + \Phi(t) \int_{t_0}^{t} \Phi^{-1}(s)\widetilde{h}(s)\, ds.$$

To satisfy the initial condition $\widetilde{x}(t_0) = \widetilde{x}_0$ we must choose the constant $\widetilde{c}$ so that

$$\Phi(t_0)\widetilde{c} = \widetilde{x}_0,$$

that is,

$$\widetilde{c} = \Phi^{-1}(t_0)\widetilde{x}_0.$$

This choice for the arbitrary constants in the general solution yields the formula

$$\tilde{x}(t) = \Phi(t)\Phi^{-1}(t_0)\tilde{x}_0 + \Phi(t)\int_{t_0}^{t}\Phi^{-1}(s)\tilde{h}(s)\,ds \qquad (3.12)$$

for the solution of the initial value problem (3.11).

We can solve the initial value problem

$$x' = -2x + y + 1$$
$$y' = 2x - 3y$$
$$x(0) = -\frac{1}{4}, \qquad y(0) = \frac{1}{2}$$

by using the variation of constants formula (3.12). The coefficient matrix

$$A = \begin{pmatrix} -2 & 1 \\ 2 & -3 \end{pmatrix}$$

of the associated homogeneous system has eigenvalues

$$\lambda_1 = -1, \qquad \lambda_2 = -4$$

and corresponding eigenvectors

$$\tilde{v}_1 = \begin{pmatrix} 1 \\ 1 \end{pmatrix}, \qquad \tilde{v}_2 = \begin{pmatrix} 1 \\ -2 \end{pmatrix},$$

which yield two independent eigensolutions $\tilde{v}_1 e^{\lambda_1 t}$ and $\tilde{v}_2 e^{\lambda_2 t}$. These eigensolutions produce a fundamental solution matrix

$$\Phi(t) = \begin{pmatrix} e^{-t} & e^{-4t} \\ e^{-t} & -2e^{-4t} \end{pmatrix}$$

whose inverse is

$$\Phi^{-1}(t) = \begin{pmatrix} \frac{2}{3}e^{t} & \frac{1}{3}e^{t} \\ \frac{1}{3}e^{4t} & -\frac{1}{3}e^{4t} \end{pmatrix}.$$

With these matrices and

$$t_0 = 0, \qquad \tilde{x}_0 = \begin{pmatrix} -\frac{1}{4} \\ \frac{1}{2} \end{pmatrix}, \qquad \tilde{h}(s) = \begin{pmatrix} 1 \\ 0 \end{pmatrix}$$

we have all the ingredients for the variation of constants formula (3.12).
First we calculate

$$\Phi(t)\Phi^{-1}(0)\tilde{x}_0 = \begin{pmatrix} e^{-t} & e^{-4t} \\ e^{-t} & -2e^{-4t} \end{pmatrix}\begin{pmatrix} \frac{2}{3} & \frac{1}{3} \\ \frac{1}{3} & -\frac{1}{3} \end{pmatrix}\begin{pmatrix} -\frac{1}{4} \\ \frac{1}{2} \end{pmatrix} = -\frac{1}{4}\begin{pmatrix} e^{-4t} \\ -2e^{-4t} \end{pmatrix}$$

and then

$$\Phi(t)\int_{t_0}^{t}\Phi^{-1}(s)\widetilde{h}(s)\,ds = \begin{pmatrix} e^{-t} & e^{-4t} \\ e^{-t} & -2e^{-4t} \end{pmatrix}\int_{0}^{t}\begin{pmatrix} \frac{2}{3}e^{s} & \frac{1}{3}e^{s} \\ \frac{1}{3}e^{4s} & -\frac{1}{3}e^{4s} \end{pmatrix}\begin{pmatrix} 1 \\ 0 \end{pmatrix}ds$$

$$= \begin{pmatrix} e^{-t} & e^{-4t} \\ e^{-t} & -2e^{-4t} \end{pmatrix}\begin{pmatrix} \int_{0}^{t}\frac{2}{3}e^{s}\,ds \\ \int_{0}^{t}\frac{1}{3}e^{4s}\,ds \end{pmatrix}$$

$$= \begin{pmatrix} e^{-t} & e^{-4t} \\ e^{-t} & -2e^{-4t} \end{pmatrix}\begin{pmatrix} \frac{2}{3}e^{t} - \frac{2}{3} \\ \frac{1}{12}e^{4t} - \frac{1}{12} \end{pmatrix}$$

$$= \begin{pmatrix} \frac{3}{4} - \frac{2}{3}e^{-t} - \frac{1}{12}e^{-4t} \\ \frac{1}{2} - \frac{2}{3}e^{-t} + \frac{1}{6}e^{-4t} \end{pmatrix}.$$

From the sum of these two calculations we obtain the solution of the initial value problem,

$$\widetilde{x}(t) = \begin{pmatrix} -\frac{2}{3}e^{-t} - \frac{1}{3}e^{-4t} + \frac{3}{4} \\ -\frac{2}{3}e^{-t} + \frac{2}{3}e^{-4t} + \frac{1}{2} \end{pmatrix}. \qquad \blacksquare$$

Exercises

Use the variation of constants formula (3.6) to find the general solution of $\widetilde{x}' = A\widetilde{x} + \widetilde{h}(t)$.

1. $A = \begin{pmatrix} 0 & 1 \\ -1 & 0 \end{pmatrix}$, $\widetilde{h}(t) = \begin{pmatrix} 1 \\ -1 \end{pmatrix}$

2. $A = \begin{pmatrix} -2 & 1 \\ 2 & -3 \end{pmatrix}$, $\widetilde{h}(t) = \begin{pmatrix} 2e^{t} \\ 2e^{t} \end{pmatrix}$

3. Use the variation of constants formula (3.12) to solve the initial value problem

$$\widetilde{x}(0) = \begin{pmatrix} 0 \\ 0 \end{pmatrix}$$

associated with the system in Exercise 1.

4. Use the variation of constants formula (3.12) to solve the initial value problem

$$x(0) = \begin{pmatrix} 1 \\ -1 \end{pmatrix}$$

associated with the system in Exercise 2.

Consider the nonhomogeneous system $\widetilde{x}' = A(t)\widetilde{x} + \widetilde{h}(t)$ for $t > 0$ when

$$A(t) = \begin{pmatrix} \frac{3}{2t} & -\frac{1}{2} \\ -\frac{1}{2t^2} & \frac{1}{2t} \end{pmatrix}, \qquad \widetilde{h}(t) = \begin{pmatrix} t^2 \\ t \end{pmatrix}.$$

5. Verify that

$$\Phi(t) = \begin{pmatrix} t & t^2 \\ 1 & -t \end{pmatrix}$$

is a fundamental solution matrix for the associated homogeneous system $\widetilde{x}' = A(t)\widetilde{x}$.

6. Use the variation of constants formula (3.6) to find the general solution.

7. Use the variation of constants formula (3.12) to solve the initial value problem

$$\widetilde{x}(1) = \begin{pmatrix} 0 \\ 0 \end{pmatrix}.$$

Consider the second-order equation $x'' + x = \tan t$.

8. Use the variation of constants formula (3.6) to find the general solution.

9. Use the variation of constants formula (3.12) to solve the initial value problem $x(0) = 0$, $x'(0) = 0$.

10. Use the variation of constants formula (3.12) to solve the initial value problem $x(0) = 1$, $x'(0) = 0$.

Consider the second-order equation $t^2x'' - 2tx' + 2x = t^2$.

11. Find two independent solutions of the associated homogeneous equation $t^2x'' - 2tx' + 2x = 0$ of the form $x = t^m$ for a constant m.

12. Find the general solution of the nonhomogeneous equation by using the variation of constants formula (3.6).

13. Solve the initial value problem $x(1) = 0$, $x'(1) = 0$.

14. Solve the initial value problem $x(1) = 1$, $x'(1) = -1$.

Use the variation of constants formula (3.6) to find the general solution of the system $\tilde{x}' = A\tilde{x} + \tilde{h}(t)$.

15. $A = \begin{pmatrix} -5 & -10 & 14 \\ -4 & -5 & 8 \\ -5 & -8 & 12 \end{pmatrix}$, $\tilde{h}(t) = \begin{pmatrix} 1 \\ 0 \\ 1 \end{pmatrix}$

16. $A = \begin{pmatrix} 12 & 19 & -28 \\ 2 & 5 & -4 \\ 6 & 11 & -14 \end{pmatrix}$, $\tilde{h}(t) = \begin{pmatrix} 2e^{-t} \\ e^t \\ 1 - e^{-t} \end{pmatrix}$

6.4 Chapter Summary and Exercises

The general solution of a nonhomogeneous linear system

$$x' = a(t)x + b(t)y + h_1(t)$$
$$y' = c(t)x + d(t)y + h_2(t)$$

has the form

$$x = x_h(t) + x_p(t)$$
$$y = y_h(t) + y_p(t),$$

where $x = x_p(t)$, $y = y_p(t)$ is a particular solution pair of the system and $x = x_h(t)$, $y = y_h(t)$ is the general solution of the associated homogeneous system

$$x' = a(t)x + b(t)y$$
$$y' = c(t)x + d(t)y.$$

In Chapter 5 we learned that the general solution of a homogeneous system is the linear combination of two independent solution pairs. Moreover, we learned how to calculate two independent solution pairs in the case when the coefficients a, b, c and d are constants. In this chapter we focused on finding a particular solution, and hence the general solution, of a nonhomogeneous system. We found that if the general solution of the homogeneous system is known, then the variation of constants formula gives a particular solution of the nonhomogeneous system. We also saw that an adaptation of the method of undetermined coefficients from single linear equations provides a shortcut for calculating a particular solution when the coefficients are constants and the nonhomogeneous terms $h_1(t)$ and $h_2(t)$ are of special types. We also developed the variation of constants formula for the special case of linear second-order equations. The chapter concluded by formulating the variation of constants formula in matrix notation and extending the formula to systems of any number of linear equations.

Exercises

Use the variation of constants formula (3.6) to find the general solution of the following systems.

1. $\begin{cases} x' = -\frac{3}{2}x + y - \frac{1}{2} \\ y' = -2x + \frac{3}{2}y + 1 \end{cases}$

2. $\begin{cases} x' = -\frac{3}{2}x + y + 2 \\ y' = -2x + \frac{3}{2}y - \frac{1}{4} \end{cases}$

3. $\begin{cases} x' = 3x + 5y - 2 \\ y' = -2x - 3y + 1 \end{cases}$

4. $\begin{cases} x' = 3x + 5y - 13e^{-5t} \\ y' = -2x - 3y \end{cases}$

5. $\begin{cases} x' = 5x + 8y + r \\ y' = -3x - 5y, \end{cases}$ where r is a constant

6. $\begin{cases} x' = 5x + 8y + rt \\ y' = -3x - 5y + 2, \end{cases}$ where r is a constant

7. Solve the initial value problem $x(0) = 1$, $y(0) = -1$ for each system in Exercises 1–6.

The strengths of two opposing armies are $x = x(t)$ and $y = y(t)$ (as measured, for example, by the number of troops or armaments). In a battle an army's strength is reduced. Assume that the rate of reduction is proportional to the strength of the opposing army. During the battle reinforcements arrive at rates $h_1(t)$ and $h_2(t)$. Thus, x and y satisfy the nonhomogeneous linear system

$$x' = -ay + h_1(t)$$
$$y' = -bx + h_2(t).$$

Assume that the armies are evenly matched, by which we mean $a = b$. Let us assume $a = b = 1$. Finally, suppose reinforcements arrive at exponentially decreasing rates $h_1(t) = e^{-ct}$ and $h_2(t) = e^{-dt}$, where c and d are positive constants. Under these conditions we have the system

$$x' = -y + e^{-ct}$$
$$y' = -x + e^{-dt}$$
$$x(0) = x_0 > 0, \qquad y(0) = y_0.$$

The constants c and d measure how fast the reinforcement rates decrease.

8. Use a computer to study the solutions of the initial value problem when both armies are reinforced at the same (decreasing) rate (i.e., $d = c$). Organize your explorations in the following way. Choose a value for $c = d$ and have a computer draw graphs of x and y for a selection of initial conditions $x_0 \neq y_0$. Repeat this for several choices of $c = d$. Draw conclusions about the winner of the battle. (*Note:* An army loses if its strength equals 0 at some time t.)

9. Solve the initial value problem when $c = d$.

10. Use your answer in Exercise 9 to verify (or disprove) your conclusions in Exercise 8.

Suppose the rates at which two distinct groups move into and out of a city are proportional to the numbers $x = x(t)$ and $y = y(t)$ present. Specifically, members of group x are attracted to each other and so their numbers increase at a rate proportional to their numbers. However, members of group x do not like members of group y and they leave the city at a rate proportional to the number of y group members present.

The fundamental inflow/outflow (compartmental) rule implies $x' = ax - by$, where a and b are positive constants. Assume that group y feels the same way about its own members and those of group x so that $y' = -cx + dy$ for positive constants c and d. Suppose $a = d$ (i.e., both populations grow at the same exponential rate in the absence of the other). In fact, take $a = d = 1$.

11. Use a computer to study solutions of the initial value problem $x(0) = x_0 > 0$, $y(0) = y_0 > 0$. Do this for $b = c = 1$. In the long run can both groups live together in the city? Under what conditions is a group eventually gone from the city (i.e., is the city totally segregated)? Repeat for $b = 4$, $c = 1$.

12. Solve the initial value problem

$$x' = x - by$$
$$y' = -cx + y$$
$$x(0) = x_0 > 0, \qquad y(0) = y_0 > 0.$$

13. Use your answer in Exercise 12 to verify your conclusions in Exercise 11.

The system

$$x' = x - by$$
$$y' = -cx + y$$
$$x(0) = x_0 > 0, \qquad y(0) = y_0 > 0$$

models the numbers of two groups moving into and out of a city. Here b and c are positive constants. If $x_0 < y_0\sqrt{b/c}$, then group x disappears in the city in a finite amount of time (see Exercises 11–13). In particular, if initially there are no members of group x (so that $x_0 = 0$), then there will never be members of group x in the city (mathematically, $x < 0$ for all $t > 0$). Suppose individuals of group x are added to the city at a constant rate $r > 0$. Then we have the nonhomogeneous system

$$x' = x - by + r$$
$$y' = -cx + y$$
$$x(0) = 0, \qquad y(0) = y_0 > 0.$$

14. Use a computer to study solutions of this initial value problem. Organize your exploration as follows. Choose and fix an initial condition $y_0 > 0$ and graph x and y for an increasing sequence of immigration rates r. Repeat this for several choices of $y_0 > 0$. What do you conclude about the long-term group composition of the city?

15. Solve the initial value problem.

16. Use your answer in Exercise 15 to verify your conclusions in Exercise 14.

Find the general solution of the following second-order equations.

17. $x'' + 2x' + 2x = \cos t$

18. $x'' + 2x' + 2x = e^{-t} \sin t$

19. $x'' + 2x' + 2x = 2\cos t - e^{-t} \sin t$

20. $x'' + 6x' + 5x = e^{-2t}$

21. $x'' + 6x' + 5x = e^{at}$, where a is a constant

22. $x'' + 6x' + 5x = te^{-t}$

23. $x'' + 4x' + 4x = t$

24. $x'' + 4x' + 4x = e^{at}$, where a is a constant.

25. $x'' + 4x' + 4x = 3t - e^{-t} + 2e^{-2t}$

26. $x'' + k^2 x = e^{-t}$ 27. $x'' + k^2 x = \sin t$

To study blood flow through organs, tracer dyes are added intravenously and the concentrations in various organs monitored. The compartmental model below describes the tracer

concentrations (mg/liter) in arterial blood (x_1), venous blood in the liver (x_2), and venous blood in the right atrium (x_3).

$$x_1' = -2x_1 + 2x_3$$
$$x_2' = x_1 - x_2 - R$$
$$x_3' = x_1 + x_2 - 2x_3$$

R is the removal rate of the dye from the liver (mg/liter/min). Suppose an initial dose of $x_1(0) = 10$ is added to the arterial blood and no dye is initially present in the other compartments. Let $R = 1$.

28. Use a computer to graph each component of the solution. When is the dye gone from each component?

29. Solve the initial value problem.

30. Use your answer in Exercise 29 to determine when the dye is gone from each component.

6.5 APPLICATIONS

6.5.1 Drug Kinetics

In Sec. 5.8.1 of Chapter 5 we studied the homogeneous linear system

$$x' = -(r_1 + r_3) x + r_2 y \tag{5.1}$$
$$y' = r_3 x - r_2 y$$

as a model for the kinetics of a drug in the bloodstream and body tissues. The component x is the amount of drug in the bloodstream at time t and y is the amount in the body tissues. The application in Sec. 5.8.1 involved the injection of the drug into the bloodstream and the problem of determining the resulting time course of the drug amounts in the bloodstream. That application led to an initial value problem for the system (5.1).

In this section we consider a different problem. Instead of an injection of the drug into the body tissues, we consider a continuous, intravenous flow of the drug into the bloodstream. If we revisit the balance law derivation of the system (5.1) in Sec. 5.8.1, we see that this medical treatment adds another inflow of drug into the bloodstream compartment and hence a new inflow rate term to the right-hand side of the equation for x'. If the intravenous treatment adds the drug at a constant rate $n > 0$, then we have the nonhomogeneous, linear system

$$x' = -(r_1 + r_3) x + r_2 y + n \tag{5.2}$$
$$y' = r_3 x - r_2 y$$

for the dynamics of x and y. See Fig. 6.9.

Initial conditions are determined the amount of drug in the blood and tissues at initial time $t = 0$. One scenario of interest is when no drug is initially present in either the bloodstream or the body tissues. In this case, we have the initial conditions

$$x(0) = 0, \qquad y(0) = 0. \tag{5.3}$$

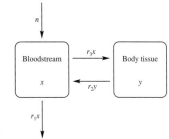

Figure 6.9

In other situations the drug may initially be present in one compartment or the other. See Exercise 2–4.

As an example, we recall the application studied Sec. 5.8.1 of Chapter 5 in which a patient was treated with lidocaine to treat an irregular heart rhythm. The flow rates for lidocaine are

$$r_1 = 2.40 \times 10^{-2} \text{ (per minute)}$$
$$r_2 = 3.80 \times 10^{-2} \text{ (per minute)}$$
$$r_3 = 6.60 \times 10^{-2} \text{ (per minute)}$$

and the model (5.2)–(5.3) for intravenous delivery of the drug becomes

$$x' = -9.00 \times 10^{-2}x + 3.80 \times 10^{-2}y + n$$
$$y' = 6.60 \times 10^{-2}x - 3.80 \times 10^{-2}y \qquad (5.4)$$
$$x(0) = 0, \qquad y(0) = 0.$$

Drugs must be present in sufficient quantity (or concentration) in the bloodstream in order to be effective. On the other hand, concentrations that are too high are often dangerous. One problem is, then, to determine the intravenous delivery rate n so that an effective concentration is reached without exceeding a dangerously high concentration.

In the case of lidocaine, the effective concentration in the bloodstream is 1.5 mg/liter and the lethal level is 6 mg/liter. As in Sec. 5.8.1 we consider the case of a patient whose blood volume is 5 liters, whose effective lidocaine amount (in the bloodstream) is

$$1.5 \text{ mg/liter} \times 5 \text{ liter} = 7.5 \text{ mg,}$$

and whose lethal lidocaine amount (in the bloodstream) is

$$6 \text{ mg/liter} \times 5 \text{ liter} = 30 \text{ mg.}$$

With these limits in mind, consider the results obtained from sample (computer calculated) solutions of (5.4) shown in Fig. 6.10 and Table 6.1 for a small selection of intravenous delivery rates n. A careful examination of these solutions suggests some important general properties of the solutions of (5.4). Until we analyze the initial value problem further, however, we must treat these observations as conjectures.

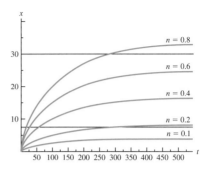

Figure 6.10 The x component of the initial value problem (5.4) is shown for selected values of the intravenous delivery rate n. The horizontal lines are located at the effective level of 7.5 mg and the lethal level of 30 mg.

Table 6.1   Some observations obtained from selected numerically computed solutions of (5.4). See Fig. 6.9. An "∞" means the computed solution x never attained the corresponding amount. These numbers were calculated for a patient with 5 liters of blood. Different results, and in particular different critical rates n_1 and n_2, are obtained for patients with different blood volumes. See Exercise 1.

n Intravenous Delivery Rate (mg/min)	t_e Time to Reach Effective Amount in Bloodstream (min)	t_l Time to Reach Lethal Amount in Bloodstream (min)	x_e Maximal Amount in Bloodstream (mg)
0.1	∞	∞	4.17
0.2	282	∞	8.34
0.4	58.0	∞	16.7
0.6	27.2	∞	25.0
0.8	16.3	282	33.4
1.0	11.4	147	41.7

Conjectures

1. The amount x of lidocaine in the bloodstream strictly increases and approaches a finite limit $x_e > 0$ as $t \to +\infty$.

2. The limit amount x_e of lidocaine in the bloodstream is proportional to the intravenous delivery rate n. (For example, x_e doubles if n doubles.)

3. There is a critical value n_1 of the intravenous delivery rate n below which the effective concentration is never attained.

4. There is a second critical value $n_2 > n_1$ of the intravenous delivery rate n below which the lethal concentration is never attained. However, if $n > n_2$, then the lethal concentration is attained in finite time.

5. The effective concentration is attained sooner at higher intravenous delivery rates n.

We can address the Conjectures (1)–(5) analytically by using formulas for the solution pair x and y of the initial value problem (5.4). As we learned in this chapter the general solution has the form

$$x = x_h + x_p$$
$$y = y_h + y_p,$$

where x_h, y_h is the general solution pair for the associated homogeneous system

$$x' = -9.00 \times 10^{-2}x + 3.80 \times 10^{-2}y \tag{5.5}$$
$$y' = 6.60 \times 10^{-2}x - 3.80 \times 10^{-2}y,$$

and x_p, y_p is any particular solution of the nonhomogeneous system

$$x' = -9.00 \times 10^{-2}x + 3.80 \times 10^{-2}y + n \tag{5.6}$$
$$y' = 6.60 \times 10^{-2}x - 3.80 \times 10^{-2}y.$$

Using the method of Chapter 5, we find the general solution of the homogeneous system (5.5) to be

$$x_h = -0.781c_1e^{-0.120t} + 0.431c_2e^{\left(-7.57\times10^{-3}\right)t}$$
$$y_h = 0.625c_1e^{-0.120t} + 0.936c_2e^{\left(-7.57\times10^{-3}\right)t},$$

where c_1 and c_2 are arbitrary constants.

One way to find a particular solution of the nonhomogeneous system (5.6) is to use the variation of constants formula (2.15) in Sec. 6.2 or (3.6) in Sec. 6.3. However, since the nonhomogeneous term n is a constant, it is easier to use the method of undetermined coefficients discussed in Sec. 6.1. Using this shortcut method, we look for a constant solution, that is, an equilibrium solution

$$x = x_e, \quad y = y_e.$$

Substituting these constants into the nonhomogeneous system (5.6), we obtain the two algebraic equations

$$0 = -\left(9.00 \times 10^{-2}\right)x_e + \left(3.80 \times 10^{-2}\right)y_e + n$$
$$0 = \left(6.60 \times 10^{-2}\right)x_e - \left(3.80 \times 10^{-2}\right)y_e$$

for x_e and y_e. Solving these simultaneous linear equations, we obtain the equilibrium solution pair

$$x_e = 41.7n, \quad y_e = 72.4n.$$

The general solution of the nonhomogeneous system (5.6) is

$$x(t) = -0.781c_1e^{-0.120t} + 0.431c_2e^{\left(-7.57\times10^{-3}\right)t} + 41.7n \tag{5.7}$$
$$y(t) = 0.625c_1e^{-0.120t} + 0.936c_2e^{\left(-7.57\times10^{-3}\right)t} + 72.4n.$$

To solve the initial value problem (5.4), we choose the arbitrary constants c_1 and c_2 so that

$$x(0) = -0.781c_1 + 0.431c_2 + 41.7n = 0$$
$$y(0) = 0.625c_1 + 0.936c_2 + 72.4n = 0.$$

The solutions of these two linear algebraic equations for c_1 and c_2 are

$$c_1 = 7.77n, \qquad c_2 = -82.5n.$$

After these choices are placed into the general solution (5.7), we have the formulas

$$x(t) = -6.07ne^{-0.120t} - 35.6ne^{\left(-7.57\times10^{-3}\right)t} + 41.7n \qquad (5.8)$$

$$y(t) = 4.86ne^{-0.120t} - 77.2ne^{\left(-7.57\times10^{-3}\right)t} + 72.4n$$

for the solution of the initial value problem (5.4).

We now turn our attention to the Conjectures (1)–(5). First, from (5.8) we calculate the derivative

$$x'(t) = 0.730ne^{-0.120t} + 0.270ne^{\left(-7.57\times10^{-3}\right)t},$$

which we observe is positive for all t. Therefore, x is a strictly increasing. Furthermore, from (5.8) we find

$$\lim_{t\to+\infty} x(t) = 41.7n.$$

This completes the justification of Conjecture (1) and provides a formula

$$x_e = 41.7n$$

for the maximal amount of the lidocaine in the bloodstream.

The formula for x_e shows that this limit amount of lidocaine in the bloodstream is a multiple of the intravenous delivery rate n. This establishes Conjecture (2).

The formula for x_e also implies that x never reaches the effective amount of 7.5 mg if

$$n < n_1 = \frac{7.5}{41.7} = 0.1799 \text{ mg/minute.}$$

This proves Conjecture (3).

If $x_e < 30$, then x never equals the lethal amount of 30 mg. This happens if

$$n < n_2 = \frac{30}{41.7} = 0.7194 \text{ mg/minute.}$$

On the other hand, if $n > n_2$, then the limit x_e is greater than 30 and x will equal 30 at some time t_l. This establishes Conjecture (4).

It is not possible to find a formula for the time t_l because this involves solving the equation $x(t_l) = 30$, that is, solving the equation

$$-6.07ne^{-0.120t_l} - 35.6ne^{\left(-7.57\times10^{-3}\right)t_l} + 41.7n = 30 \qquad (5.9)$$

for t_l. There is no algebraic formula for the solution of this equation. However, for any specified value of the intravenous delivery rate n we can solve (5.9) for t_l using a calculator or a computer.

For example, if $n = 0.8$, then the solution of the equation (5.9) is $t_l = 282$ (see Table 6.1). It is also possible, using a calculator or computer, to draw a graph of t_l as a function of n by using equation (5.9). One way to do this is to solve (5.9) for

$$n = \frac{30}{-6.07e^{-0.120t_l} - 35.6e^{\left(-7.57\times10^{-3}\right)t_l} + 41.7},$$

draw a graph of n as a function of t_l, and reflect the result through the $n = t_l$ (the 45 degree) line. The result appears in Fig. 6.10.

Finally, we consider Conjecture (5). If $n > n_1$, then we can calculate the time t_e at which the effective amount 7.5 is attained by solving the equation $x(t_e) = 7.5$ for t_e, that is, by solving the equation

$$-6.07ne^{-0.120t_e} - 35.6ne^{\left(-7.57 \times 10^{-3}\right)t_e} + 41.7n = 7.5 \tag{5.10}$$

for t_e. The solution of this equation depends on n. A graph of t_e as a function of n [obtained by solving (5.10) for n as a function of t and reflecting its graph through the line $n = t$] appears in Fig. 6.11. From this graph we see that t_e is a decreasing function of n.

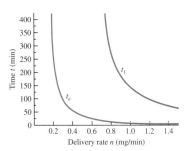

Figure 6.11 The critical times t_e and t_l, at which the effective level and lethal levels of lidocaine in the bloodstream are attained, respectively, are functions of the intravenous delivery rate n.

Another way to demonstrate this result is to calculate the derivative of t_e with respect to n by an implicit differentiation of equation (5.10). This results in

$$\frac{dt_e}{dn} = -\frac{1}{n} \frac{-6.07e^{-0.120t} - 35.6e^{\left(-7.57 \times 10^{-3}\right)t} + 41.7}{0.730e^{-0.120t} + 0.270e^{\left(-7.57 \times 10^{-3}\right)t}}$$

$$= -\frac{1}{n^2} \frac{37.5}{0.730e^{-0.120t} + 0.270e^{\left(-7.57 \times 10^{-3}\right)t}}$$

$$< 0$$

In the application of the (compartmental) drug kinetics model (5.2)–(5.3) studied above, we used specific values of the transfer rates r_i appropriate for the drug lidocaine. It is natural to speculate whether the conclusions drawn above in Conjectures (1)–(5) remain valid for other values of the model parameters r_i (i.e., for other drugs). The reader can study this question in Exercise 5. Also, in the application above, we considered only the case when no drug is initially present in the patient's bloodstream and body tissues. Another interesting problem arises when the drug is initially present in the bloodstream and the purpose of the intravenous administration of the drug is to maintain an effective (but not lethal) concentration. See Exercises 2–4.

6.5.2 Forced Mass-Spring Oscillations

Consider the mass-spring problem studied in Sec. 5.8.2 of Chapter 5. In that problem three forces act on the mass: gravity, friction, and that exerted by the spring. An interesting and important problem arises when, in addition, an external periodic force acts on the mass. By an external force we mean a force whose magnitude and direction do not depend on either the position or the velocity of the mass. By a periodic force we mean one whose magnitude and direction vary in time in a repetitious fashion. This could arise, for example, from an attached rod that physically pulls and pushes the mass or a fluctuating magnetic field that acts on the mass.

Let $F_e = F_e(t)$ denote an external force that is a periodic function of t. An example is

$$F_e(t) = \gamma_e \cos \beta_e t.$$

This represents a cosinusoidally oscillating external force with amplitude γ_e and frequency β_e (i.e., period $2\pi/\beta_e$). That $F_e(t)$ alternates between positive and negative values means the force alternately pushes and pulls the mass.

If the forces of gravity F_g, friction F_f, and the spring F_s are included with the external force, then the total force F acting on the mass is

$$F = F_g + F_s + F_f + F_e.$$

Following the derivation for the spring's displacement $x = x(t)$ from its rest position as given in Sec. 5.8.2, we have (from Newton's law $F = ma$)

$$mg - k\,(l + x(t)) - cx' + F_e(t) = mx''$$

or (remembering that $kl = mg$)

$$mx'' + cx' + kx = F_e(t). \tag{5.11}$$

We consider the case when friction is absent ($c = 0$) and the mass starts at rest ($x_0 = 0$, $v_0 = 0$). This leads to the the initial value problem

$$mx'' + kx = \gamma_e \cos \beta_e t \tag{5.12}$$
$$x(0) = 0, \quad x'(0) = 0.$$

Our goal is to use this model determine the motion of the mass. (For the case when friction is present, $c > 0$, see Exercises 14–17. See Exercise 10 for the case when the mass does not start at rest.)

Figure 6.12 shows solution graphs of selected the initial value problems (5.12). In these graphs the frequency β_e of the external forcing is varied. Notice in all but one case the solutions are oscillatory and bounded (and appear nearly periodic). An exception is the case $\beta_e = 1$ when the solution oscillates but appears unbounded. In Fig. 6.11 this case corresponds to the case when the external frequency $\beta_e = 1$ equals the frequency of oscillation when no external forcing is applied, namely,

$$\beta = \sqrt{\frac{k}{m}}. \tag{5.13}$$

(See Sec. 5.8.2.) This frequency is called the "natural frequency" of the mass-spring system. See Exercises 6–9 for further examples.

Based on these observations, we make the following conjectures. The solution of the forced, frictionless mass-spring initial value problem (5.12) is

(a) bounded and oscillatory when the external frequency β_e does not equal the natural frequency (5.13)

(b) unbounded and oscillatory when external frequency equals the natural frequency

$$\tag{5.14}$$

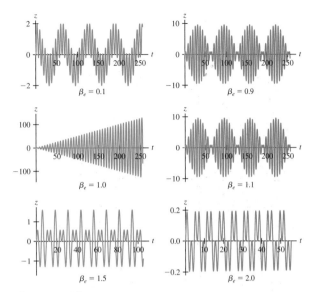

Figure 6.12 Solutions of (5.12) with $m = k = \gamma_e = 1$ for selected values of the forcing frequency β_e.

To address these conjectures, we calculate a formula for the solution of the initial value problem (5.12). The general solution of the nonhomogeneous second-order equation in (5.12) has the form $x = x_h + x_p$, where

$$x_h = c_1 \cos \beta t + c_2 \sin \beta t$$

is the general solution of the associated homogeneous equation $mx'' + kx = 0$ (Sec. 8.2 in Chapter 5) and x_p is a particular solution of (5.12).

We can find a particular solution x_p using the method of undetermined coefficients (Secs. 6.1 and 2.3). Since the nonhomogeneous term is a multiple of $\sin \beta_e t$, we consider a particular solution in the form

$$x_p = k_1 \cos \beta_e t + k_2 \sin \beta_e t, \tag{5.15}$$

where k_1 and k_2 are coefficients yet to be determined. However, this form is valid only if it contains no solutions of the associated homogeneous equation (i.e., only if $\beta_e \neq \beta$). When $\beta_e = \beta$ the appropriate form is

$$x_p = k_1 t \cos \beta_e t + k_2 t \sin \beta_e t. \tag{5.16}$$

(Because of the factor t in the latter case, we begin to see the reason why the case $\beta_e = \beta$ is exceptional and leads to unbounded solutions.)

We find the undetermined coefficients k_1 and k_2 by substituting the expression for x_p into the equation (5.12). If $\beta_e \neq \beta$, then (5.15) yields

$$mx_p'' + kx_p = \left(-m\beta_e^2 + k\right) k_1 \cos \beta_e t + \left(-m\beta_e^2 + k\right) k_2 \sin \beta_e t$$

and equation (5.12) requires

$$\left(-m\beta_e^2 + k\right) k_1 = \gamma_e, \qquad k_2 = 0.$$

Solving for

$$k_1 = \frac{\gamma_e}{-m\beta_e^2 + k} = \frac{\gamma_e}{m\left(\beta^2 - \beta_e^2\right)},$$

we have

$$x_p = \frac{\gamma_e}{m\left(\beta^2 - \beta_e^2\right)} \cos \beta_e t.$$

Therefore, in the case $\beta_e \neq \beta$ the general solution is

$$x(t) = c_1 \cos \beta t + c_2 \sin \beta t + \frac{\gamma_e}{m\left(\beta^2 - \beta_e^2\right)} \cos \beta_e t. \qquad (5.17)$$

To solve the initial value problem, we must determine the arbitrary constants c_1 and c_2 so that the initial conditions $x(0) = 0$, $x'(0) = 0$ are satisfied. The general solution (5.17) implies

$$x(0) = c_1 + \frac{\gamma_e}{m\left(\beta^2 - \beta_e^2\right)}$$

and hence

$$c_1 = -\frac{\gamma_e}{m\left(\beta^2 - \beta_e^2\right)}.$$

Also, the general solution (5.17) implies $x'(0) = c_2\beta$ and hence $c_2 = 0$. From these calculations we have the solution formula

$$x(t) = \frac{\gamma_e}{m\left(\beta^2 - \beta_e^2\right)} \left(-\cos \beta t + \cos \beta_e t\right) \qquad (5.18)$$

$$\beta_e \neq \beta = \sqrt{\frac{k}{m}}$$

for the initial value problem (5.12).

For the case $\beta_e = \beta$ we use the particular solution (5.16) and find the solution formula

$$x(t) = \frac{\gamma_e}{2m\beta} t \sin \beta t, \qquad \beta_e = \beta = \sqrt{\frac{k}{m}}. \qquad (5.19)$$

The details of this case are left as Exercise 11.

We can use formulas (5.18) and (5.19) to address the conjectures (a) and (b) made above about the solutions.

(a) When $\beta_e \neq \beta$, the solution is bounded for all t since both cosine terms in the formula (5.18) are bounded. The solution is also oscillatory. Specifically, we mean $x(t)$ has infinitely many, arbitrarily large roots. In other words, that the mass will pass through the rest point infinitely often. To see this, we use the trigonometric identity

$$-\cos A + \cos B = 2 \sin \left(\frac{A - B}{2}\right) \sin \left(\frac{A + B}{2}\right)$$

to rewrite the solution formula (5.18) as

$$x(t) = \frac{2\gamma_e}{m\left(\beta^2 - \beta_e^2\right)} \sin\left(\frac{\beta - \beta_e}{2}t\right) \sin\left(\frac{\beta + \beta_e}{2}t\right), \qquad \beta_e \neq \beta. \qquad (5.20)$$

Thus, $x(t) = 0$ when t satisfies

$$\frac{\beta - \beta_e}{2}t = n\pi \quad \text{or} \quad \frac{\beta + \beta_e}{2}t = n\pi$$

for $n = 0, \pm1, \pm2, \ldots$. These formulas give infinitely many times at which the mass is at the rest position, namely

$$t = \frac{2\pi n}{\beta - \beta_e} \quad \text{and} \quad \frac{2\pi n}{\beta + \beta_e} \quad \text{for } n = 0, \pm1, \pm2, \ldots.$$

(b) When $\beta_e = \beta$, the solution (5.19) is unbounded. This is because the amplitude of the sine term grows linearly with t. Nonetheless, the mass still crosses the rest point infinitely often, at those times at which $\beta t = n\pi$ or

$$t = \frac{2\pi n}{\beta} = 2\pi n\sqrt{\frac{m}{k}}, \quad n = 0, \pm1, \pm2, \ldots.$$

The solution formulas (5.18) and (5.19) explain other interesting and important features of the mass-spring oscillation, as seen in Fig. 6.12. See Exercises 12 and 13.

The fact that the solution is unbounded in the exceptional case $\beta_e = \beta$ is called *resonance*. Of course, in this case the model for the mass-spring system ultimately breaks down, when the spring becomes distorted or the mass hits the ceiling. Nonetheless, this case is important because it describes the conditions under which the system will collapse. The mass-spring system is a prototype for oscillations that occur in a wide variety of applications in many different fields. Thus, resonance is an important phenomena in a great many situations (electric circuits, structures such as buildings and bridges, moving parts in automobiles and aircraft, musical instruments and sound equipment, and so on). Resonance is not always a destructive phenomenon, however. For example, it is the means by which radios, televisions, and radio telescopes are tuned.

6.5.3 Electric Circuits

In Chapter 5, Sec. 8.2 we derived the second-order equation

$$Lq'' + Rq' + \frac{1}{C}q = V(t)$$

for the charge q present in an electric circuit containing an inductor, a resistor, and a capacitor. The nonhomogeneous term $V(t)$ in this linear equation denotes the voltage impressed on a simple series circuit (for example, by a battery). If no charge or current is initial present, we have the initial conditions

$$q(0) = 0, \quad q'(0) = 0. \qquad (5.21)$$

We can calculate a formula for the solution of this initial value problem from the general solution that has the form $q = q_h + q_p$. Formulas for the general solution q_h of the associated homogeneous equation appear in Tables 5.3 and 5.4 of Chapter 5. Recall that different formulas apply in the overdamped, underdamped, and critically damped cases. We can calculate a formula for q_p using the variations of constants formula or the method of undetermined coefficients.

As an example, consider a circuit with an inductor of $L = 2$ (Henrys), a resistor of $R = 100$ (Ohms), and capacitor of $C = 10^{-3}$ (Farads). Suppose a 10-volt battery is hooked to the circuit whose output decays exponentially according to the formula

$$V(t) = 10e^{-t/100}.$$

We would like to a formula that gives the charge q at future times $t > 0$ and, in particular, what the maximum charge is and when it is attained. The initial value problem for q is

$$2q'' + 100q' + 10^3 q = 10e^{-t/100}$$
$$q(0) = 0, \quad q'(0) = 0.$$

(5.22)

The characteristic polynomial

$$2\lambda^2 + 100\lambda + 10^3$$

of the associated homogeneous equation has roots

$$\lambda = -25 + 5\sqrt{5} \approx -13.82, \quad \lambda = -25 - 5\sqrt{5} \approx -36.18$$

and therefore the associated homogeneous equation has general solution

$$q_h = c_1 e^{-13.82t} + c_2 e^{-36.18t}.$$

We can find a particular solution of the nonhomogeneous equation, using the method of undetermined coefficients, by substituting the expression

$$q_p = ke^{-t/100}$$

into the differential equation to obtain

$$2\frac{1}{100^2} ke^{-t/100} - ke^{-t/100} + 10^3 ke^{-t/100} = 10e^{-t/100}$$

$$2\frac{1}{100^2} k - k + 10^3 k = 10$$

or $k = 1.111 \times 10^{-2}$. The general solution

$$q(t) = c_1 e^{-13.82t} + c_2 e^{-36.18t} + \left(1.111 \times 10^{-2}\right) e^{-t/100}$$

and the initial conditions (5.21) imply

$$c_1 + c_2 + 1.111 \times 10^{-2} = 0$$
$$-13.82c_1 - 36.18c_2 - \left(1.111 \times 10^{-4}\right) = 0$$

or

$$c_1 = -1.797 \times 10^{-2}, \quad c_2 = 6.862 \times 10^{-3}.$$

These calculations give the formula

$$q(t) = \left(-1.797 \times 10^{-2}\right) e^{-13.82t} + \left(6.862 \times 10^{-3}\right) e^{-36.18t}$$
$$+ \left(1.111 \times 10^{-2}\right) e^{-t/100}$$

(5.23)

for the solution of the initial value problem (5.22). The graph of this solution appearing in Fig. 6.13 shows that the charge on the circuit quickly rises to a maximum before slowing decreasing. The solution formula shows that $q(t)$ tends to 0 as $t \to +\infty$. To find the maximum of q, and the time at which it occurs, we calculate the derivative

$$\frac{dq(t)}{dt} = 0.2483e^{-13.82t} - 0.2483e^{-36.18t} - 1.111 \times 10^{-4}e^{-t/100}.$$

The maximum occurs at time t when this derivative equals 0. With aid of a computer, we find that the root of the derivative is $t = 0.5584$. Thus, the charge on the circuit attains a maximum of $q(0.5584) = 0.0110$ (coulombs) in $t = 0.5584$ (seconds).

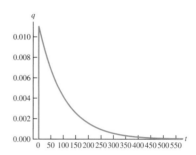

Figure 6.13 The graph of the solution (5.23) of the initial value problem (5.22) quickly reaches its maximum at before slowly decreasing to 0 as $t \to +\infty$.

Exercises

1. Recalculate the data in Table 6.1 for patients with blood capacities of 4 liters and 6 liters. What conclusions can you draw from your results concerning the relationship between patient blood capacity and the critical intravenous rates n_1 and n_2? Prove your conclusions using the solution formula for the initial value problem.

The compartmental model equations for the intravenous application of the drug lidocaine are (see 5.4)

$$x' = -0.09x + 0.038y + n$$
$$y' = 0.066x - 0.038y.$$

In the text we considered the case when no lidocaine is initially present in the patient (with blood capacity of 5 liters). Consider now the case when the initial amount $x(0) = x_0$ of lidocaine in the patient's bloodstream lies in the desired range $7.5 < x_0 < 30$ mg. Assume there is no lidocaine present in the body tissues, so that $y(0) = 0$. If not replenished, the amount $x = x(t)$ of lidocaine in the bloodstream of the patient will decrease and fall below the effective amount of 7.5 mg. To prevent this, an intravenous application of the drug is given to the patient at a rate of n mg/min.

2. Using a computer, study the component $x = x(t)$ of the solution of the initial value problem with $x_0 = 20$ for a selection of intravenous rates n and initial conditions x_0 in the range $7.5 < x_0 < 30$. Formulate a conjecture about the rate n necessary to keep the bloodstream amount within the desired range $7.5 < x(t) < 30$ for all time $t > 0$.

3. Find an analytical formula for $x = x(t)$ by solving the initial value problem.

4. Use the answer in Exercise 3 to address your conjectures in Exercise 2.

5. Assume that there is an effective amount x_e and a lethal amount $x_l > x_e$ of the drug for a given patient. Prove Conjectures (1)–(5) for the general two compartment drug model (5.2) under the initial conditions $x(0) = y(0) = 0$.

Use a computer to explore graphically the solutions of the initial value problem (5.12) for selected values of β_e in each of the cases below. Discuss whether your results support the conjectures (5.14).

6. $m = 2, k = 1, \gamma_e = 1$ 7. $m = 1, k = 2, \gamma_e = 1$

8. $m = 2, k = 1, \gamma_e = 2$ 9. $m = 1, k = 2, \gamma_e = 2$

10. Find a formula for the solution of the initial value problem $mx'' + kx = \gamma_e \cos \beta_e t, x(0) = x_0, x'(0) = v_0$.

11. Derive the solution formula (5.19).

12. In Fig. 6.12 the oscillations for β_e near, but unequal to, $\beta = 1$ have a peculiar shape. These oscillations, called beats, have the form of a rapid oscillation trapped within a slower oscillation. Explain this observation using the solution formula for the initial value problem (5.12).

13. In Fig. 6.12 the oscillations for β_e not near $\beta = 1$ have the form of a rapid oscillations around a slower oscillations of larger amplitude. For large β_e the amplitude of the oscillations decreases. Explain these observations using the solution formula for the initial value problem (5.12).

A forced mass-spring with friction is described by the second-order equation $mx'' + cx' + kx = a \cos \beta_e t$ with $c > 0$.

14. Use a computer program to explore the solutions this equations with initial conditions $x(0) = x'(0) = 0$. In particular, take $m = k = a = 1$ and $c = 1$. Describe the solutions for a selection of $\beta_e > 0$ values. Find a value $\beta_e = \beta_e^0$ at which the amplitude of the solution is maximum. Repeat this experiment for other values of the coefficient of friction $c < \sqrt{2}$. Formulate a conjecture about the relationship between β_e^0 and c (i.e., do they increase or decrease together, or are the inversely related?).

15. Find a formula for the solution when $c < 2\sqrt{mk}$.

16. Use your answer in Exercise 15 to show that as $t \to +\infty$ the solution becomes arbitrarily close to a cosine function. Find the amplitude of this cosine function and find its maximum as a function of β_e.

17. Compare yours results from Exercise 14 and Exercise 16.

An object of mass $m = 1$ is attached to the ceiling by a spring and an identical object is attached to the first by an identical spring. Suppose this coupled mass-spring system at rest until the second mass is subjected to a force $\sin t$. The displacements x_1, x_2 of the masses from their equilibrium positions

$$x_1'' + cx_1' + kx_1 - k(x_2 - x_1) = 0$$
$$x_2'' + cx_2' + k(x_2 - x_1) = \sin t,$$

where c is a coefficient of friction. In this exercise take both $c = 1$ and $k = 1$.

18. Use a computer to describe the behavior of the solutions for large t.

19. Prove mathematically that all solutions of the associated homogeneous system tend to 0 as $t \to +\infty$.

20. Use the method of undetermined coefficients to find a particular solution of the nonhomogeneous system.

21. Use your answers in Exercises 19 and 20 to corroborate your observations in Exercise 18.

22. Given L, R, and C and the impressed voltage $V(t) = \gamma_e \cos \beta_e t$, find the asymptotic amplitude (i.e., the amplitude for large t) of the oscillating charge q in an LCR circuit.

23. For what value of β_e is the asymptotic amplitude maximal?

CHAPTER

7

Approximations and Series Solutions

We have learned in previous chapters about special kinds of differential equations for which solution methods are available (i.e., methods for the calculation of solution formulas). Examples include single first-order linear equations (Chapter 2), separable first-order equations (Chapter 3), and linear systems with constant coefficients (Chapters 5 and 6). Even in these cases, however, the solution methods often involve the calculation of integrals and, as a result, may not always be successful or lead to useful solution formulas. Furthermore, many differential equations do not fall into a special class for which a solution method is available. In the absence of a method for calculating useful solution formulas, it is often fruitful to obtain formulas for approximations to solutions. In Chapter 3 (Sec. 3.4) we studied several such approximation methods for single first-order equations. In this chapter we show how to apply those methods to systems of differential equations and to higher-order equations.

7.1 Taylor Polynomials and Picard Iterates

In Chapter 3, Sec. 3.4.1, we calculated Taylor polynomial approximations for solutions of single differential equations. In this section we apply the same method to systems of equations.

The Taylor polynomial approximations of order n of a pair $(x(t), y(t))$ of functions are given by the formulas

$$x_n(t) = x(t_0) + x'(t_0)(t - t_0) + \frac{1}{2}x''(t_0)(t - t_0)^2 + \cdots$$
$$+ \frac{1}{n!}x^{(n)}(t_0)(t - t_0)^n$$

$$(1.1)$$

$$y_n(t) = y(t_0) + y'(t_0)(t - t_0) + \frac{1}{2}y''(t_0)(t - t_0)^2 + \cdots$$
$$+ \frac{1}{n!}y^{(n)}(t_0)(t - t_0)^n.$$

For the solution pair of the initial value problem

$$x\,(t_0) = x_0, \quad y\,(t_0) = y_0$$

for the system of differential equations

$$x' = f(t, x, y) \tag{1.2}$$
$$y' = g(t, x, y)$$

we can calculate the coefficients in the Taylor polynomials (1.1) by repeated differentiations of the differential equations.

The lowest-order coefficients $x(t_0)$, $y(t_0)$ are simply the initial conditions x_0, y_0. We calculate the first-order coefficients $x'(t_0)$, $y'(t_0)$ in (1.1) by letting $t = t_0$ in the differential equations. Thus,

$$x'(t_0) = f(t_0, x_0, y_0)$$
$$y'(t_0) = g(t_0, x_0, y_0).$$

To find the second-order coefficients in (1.1) we calculate the second derivatives x'', y'', by differentiating the equations (1.2), and evaluate the results at $t = t_0$. This differentiation and evaluation procedure is repeated until the desired order of the Taylor polynomial approximation is reached.

| EXAMPLE 1 |

The initial value problem

$$x' = -x + \frac{1}{2}e^{-t}y$$
$$y' = 6x - y \tag{1.3}$$
$$x(0) = 1, \quad y(0) = 0$$

involves a linear system of differential equations. However, we cannot apply the solution method in Chapter 5 because not all coefficients in the system of equations are constant. (The coefficient of y in the first equation is $e^{-t}/2$.) To find the coefficients in the Taylor polynomial approximations (1.1) of the solution pair x, y (with $t_0 = 0$), we begin by letting $t = 0$ in both differential equations in (1.3) and calculating $x'(0) = -1$ and $y'(0) = 6$. This calculation, together with in initial conditions, gives us the first-degree Taylor polynomial approximations to the solution pair:

$$x_1(t) = 1 - t$$
$$y_1(t) = 6t.$$

To calculate the second-degree Taylor polynomial approximations, we differentiate both differential equations in (1.3) to obtain

$$x'' = -x' + \frac{1}{2}e^{-t}y' - \frac{1}{2}e^{-t}y$$
$$y'' = 6x' - y'$$

and then let $t = 0$. This calculation yields $x''(0) = 4$ and $y''(0) = -12$ and results in the second-degree Taylor approximations [see (1.1)]

$$x_2(t) = 1 - t + 2t^2$$
$$y_2(t) = 6t - 6t^2.$$

A further differentiation yields

$$x''' = -x'' + \frac{1}{2}e^{-t}y'' - e^{-t}y' + \frac{1}{2}e^{-t}y$$

$$y''' = 6x'' - y''$$

and as a result $x'''(0) = -16$, $y'''(0) = 36$, from which we obtain the third-degree approximations

$$x_3(t) = 1 - t + 2t^2 - \frac{8}{3}t^3$$

$$y_3(t) = 6t - 6t^2 + 6t^3.$$

Continuing in this fashion, we can calculate Taylor polynomial approximations of any order. For example, we leave it to the reader to obtain the fourth-degree Taylor polynomial approximations

$$x_4(t) = 1 - t + 2t^2 - \frac{8}{3}t^3 + \frac{61}{24}t^4$$

$$y_4(t) = 6t - 6t^2 + 6t^3 - \frac{11}{2}t^4.$$ ■

Plots of the Taylor polynomial approximations calculated in Example 1 appear in Fig. 7.1. Notice that as the order of the Taylor polynomial increases, so does the accuracy of the polynomial as an approximation to the solution of the initial value problem. Also note, however, that the polynomials are good approximations to the solution only for t near $t_0 = 0$. This is, as we also saw in Chapter 3, Sec. 3.4, a general feature of Taylor polynomials. Namely, they are good approximations near t_0 (by design) and often are poor approximations for t far from t_0.

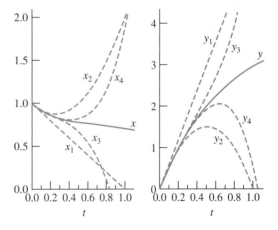

Figure 7.1 Four Taylor polynomial approximations to the solution pair of the initial value problem (1.3). The plots of the (exact) solution pair x, y were computed numerically using a Runge-Kutta method.

We can find Taylor polynomial approximations to the general solution of a system by finding the Taylor approximations to two independent solutions. For example, if we change the initial condition in (1.3) to

$$x(0) = 0, \quad y(0) = 1, \tag{1.4}$$

then the solution of this new problem will be independent from the one in Example 1. (See Theorem 5 in Chapter 4.) If we repeat the procedure in Example 1 for this new initial value problem, we will obtain Taylor approximations to this second, independent solution. The general solution of the system

$$x' = -x + \frac{1}{2}e^{-t}y \tag{1.5}$$
$$y' = 6x - y$$

is a linear combination of the Taylor approximations to these two independent solutions.

Leaving the details to the reader, we find the fourth-degree Taylor polynomial approximation for the solution pair of the initial value problem (1.4) is

$$x_4(t) = \frac{1}{2}t - \frac{3}{4}t^2 + \frac{5}{6}t^3 - \frac{39}{48}t^4$$
$$y_4(t) = 1 - t + 2t^2 - \frac{13}{6}t^3 + \frac{43}{24}t^4.$$

Therefore, the fourth-degree Taylor polynomial approximation to the general solution of the system (1.5) is the linear combination

$$x_4(t) = c_1\left(1 - t + 2t^2 - \frac{8}{3}t^3 + \frac{61}{24}t^4\right) + c_2\left(\frac{1}{2}t - \frac{3}{4}t^2 + \frac{5}{6}t^3 - \frac{39}{48}t^4\right)$$
$$y_4(t) = c_1\left(6t - 6t^2 + 6t^3 - \frac{11}{2}t^4\right) + c_2\left(1 - t + 2t^2 - \frac{13}{6}t^3 + \frac{43}{24}t^4\right),$$

where c_1 and c_2 are arbitrary constants.

Using the procedure of differentiation and evaluation, we can also calculate the Taylor polynomial approximations to solutions of higher-order differential equations.

| EXAMPLE 2 | Consider the linear nonautonomous second-order equation

$$x'' + (1 + \cos t)x = 0. \tag{1.6}$$

This is an example of Hill's equation $x'' + p(t)x = 0$, which, with periodic coefficients $p(t)$, arises in applications involving the motion of objects subject to Newton's laws of gravity (e.g., planetary motion, pendulum oscillations, etc.). We will calculate the fourth-degree Taylor polynomial approximation

$$x_4(t) = x(0) + x'(0)t + \frac{1}{2}x''(0)t^2 + \frac{1}{6}x'''(0)t^3 + \frac{1}{24}x^{(4)}(0)t^4$$

to the solution of the initial value problem

$$x(0) = 1, \quad x'(0) = 0. \tag{1.7}$$

From equation (1.6) we find by repeated differentiations that

$$x'' = -(1 + \cos t)x$$
$$x''' = -(1 + \cos t)x' + (\sin t)x$$
$$x^{(4)} = -(1 + \cos t)x'' + 2(\sin t)x' + (\cos t)x.$$

Letting $t = 0$ in these equations, we obtain

$$x''(0) = -2$$
$$x'''(0) = 0$$
$$x^{(4)}(0) = -3.$$

These calculations yield the Taylor polynomial approximations

$$x_1(t) = 1$$
$$x_2(t) = 1 - t^2$$
$$x_3(t) = 1 - t^2$$
$$x_4(t) = 1 - t^2 + \frac{5}{24}t^4$$

to this initial value problem.   ∎

In Fig. 7.2 we see, once again, how the Taylor polynomial is a good approximation of the solution only near $t_0 = 0$.

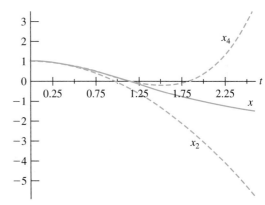

Figure 7.2 The two Taylor polynomial approximations x_2 and x_4 to the solution of the initial value problem (1.7) for equation (1.6).

Here is an example of the Taylor polynomial approximation method applied to a nonlinear system.

EXAMPLE 3   Consider the initial value problem

$$u' = -u\sqrt{u^2 + v^2}$$
$$v' = -32 - v\sqrt{u^2 + v^2} \qquad\qquad (1.8)$$
$$u(0) = \frac{\sqrt{2}}{2}, \quad v(0) = \frac{\sqrt{2}}{2}.$$

This problem arises in a model for the flight of a baseball (u and v are the horizontal and vertical components of the ball's velocity; see Sec. 4.6 of Chapter 4). To approximate the solution for t small, we calculate the first three Taylor polynomial approximations to the solution pair. From the initial conditions and the differential equations we have

$$u'(0) = -\frac{\sqrt{2}}{2}$$

$$v'(0) = -32 - \frac{\sqrt{2}}{2}.$$

Differentiating the equations for u and v, we get

$$u'' = -u\frac{uu' + vv'}{\sqrt{u^2 + v^2}} - u'\sqrt{u^2 + v^2}$$

$$v' = -v\frac{uu' + vv'}{\sqrt{u^2 + v^2}} - v'\sqrt{u^2 + v^2}$$

and, by letting $t = 0$,

$$u''(0) = \sqrt{2} + 16$$

$$v''(0) = \sqrt{2} + 48.$$

From these calculations we obtain the second-order Taylor polynomial approximations

$$u(t) = \frac{\sqrt{2}}{2} + \left(-\frac{\sqrt{2}}{2}\right)t + \frac{1}{2}\left(\sqrt{2} + 16\right)t^2$$

$$v(t) = \frac{\sqrt{2}}{2} + \left(-32 - \frac{\sqrt{2}}{2}\right)t + \frac{1}{2}\left(\sqrt{2} + 48\right)t^2$$

to the solution of the initial value problem.   ■

In Chapter 3 (Sec. 3.4) we studied another procedure for approximating solutions of initial value problems, namely, the Picard iteration method. This procedure extends, in a straightforward way, to initial value problems for systems of equations. For an initial value problem

$$x' = f(t, x, y)$$
$$y' = g(t, x, y)$$
$$x(t_0) = x_0, \quad y(t_0) = y_0$$

the sequence of Picard iterates $x_n(t)$, $y_n(t)$ begins with the initial conditions as an approximation to the solution pair for t near t_0

$$x_0(t) = x_0, \quad y_0(t) = y_0.$$

We calculate the next Picard iterate $x_1(t)$, $y_1(t)$ by substituting the current approximation pair $x_0(t)$, $y_0(t)$ into the right-hand sides of the differential equations. We obtain the next iterate pair $x_1(t)$, $y_1(t)$ by integrating the resulting equations. Specifically, from

$$x' = f(t, x_0(t), y_0(t))$$
$$y' = g(t, x_0(t), y_0(t))$$
$$x(t_0) = x_0, \quad y(t_0) = y_0$$

we obtain, by the fundamental theorem of calculus, the formulas

$$x_2(t) = x_0 + \int_{t_0}^{t} f(s, x_0(s), y_0(s)) \, ds$$

$$y_2(t) = y_0 + \int_{t_0}^{t} g(s, x_0(s), y_0(s)) \, ds.$$

To obtain the next Picard iterate we repeat these steps; that is, we substitute the current iterate pair $x_1(t)$, $y_1(t)$ into the right-hand sides of the differential equations and integrate the resulting equations. If this procedure is repeated we get the formulas

$$x_{i+1}(t) = x_0 + \int_{t_0}^{t} f(s, x_i(s), y_i(s)) \, ds$$

$$y_{i+1}(t) = y_0 + \int_{t_0}^{t} g(s, x_i(s), y_i(s)) \, ds$$

that define the sequence of Picard iterates. [Compare to the formula (4.24) in Sec. 3.4 of Chapter 3.]

Under the conditions of the fundamental existence and uniqueness theorem the Picard iterates will converge to the solution of the initial value problem for t near t_0 (on an interval containing t_0).

EXAMPLE 4

In this example we compute several Picard iterates for the initial value problem in Example 1, that is,

$$x' = -x + \frac{1}{2} e^{-t} y$$

$$y' = 6x - y \tag{1.9}$$

$$x(0) = 1, \quad y(0) = 0.$$

We begin with

$$x_0(t) = 1, \quad y_0(t) = 0.$$

From a substitution of these initial Picard iterates into the right-hand sides of the equations in (1.9), we obtain the initial value problem

$$x' = -1$$
$$y' = 6$$
$$x(0) = 1, \quad y(0) = 0.$$

An integration yields the first Picard iterates:

$$x_1(t) = 1 - t$$
$$y_1(t) = 6t.$$

A substitution of these into the right-hand sides of the equations in (1.9) yields the initial value problem

$$x' = -(1 - t) + 3te^{-t}$$
$$y' = 6(1 - t) - 6t$$
$$x(0) = 1, \quad y(0) = 0$$

whose solution by integration gives the second Picard iterates:

$$x_2(t) = 4 - t + \frac{1}{2}t^2 - (3 + 3t)\,e^{-t}$$

$$y_2(t) = 6t - 6t^2.$$

A substitution of these into the right-hand sides of the equations in (1.9) yields the initial value problem

$$x' = -4 + t - \frac{1}{2}t^2 + \left(3 + 6t - 3t^2\right)e^{-t}$$

$$y' = 24 - 12t + 9t^2 - 6\,(3 + 3t)\,e^{-t}$$

$$x(0) = 1, \quad y(0) = 0$$

whose solution gives the third Picard iterates;

$$x_3(t) = 4 - 4t + \frac{1}{2}t^2 - \frac{1}{6}t^3 - \left(3 - 3t^2\right)e^{-t}$$

$$y_3(t) = -36 + 24t - 6t^2 + 3t^3 + (36 + 18t)\,e^{-t}.$$

If this procedure is repeated one more time, we obtain the fourth Picard iterates

$$x_4(t) = \frac{25}{4} - 4t + 2t^2 - \frac{1}{6}t^3 + \frac{1}{24}t^4$$

$$+ \left(6 - 9t + \frac{3}{2}t^2 - \frac{3}{2}t^3\right)e^{-t} - \left(\frac{45}{4} + \frac{9}{2}t\right)e^{-2t}$$

$$y_4(t) = -36 + 60t - 24t^2 + 3t^3 - t^4 + \left(36 - 18t - 18t^2\right)e^{-t}. \qquad \blacksquare$$

Figure 7.3 shows plots of the Picard iterates calculated in Example 4. As the number of iterations increases, the Picard iterates show increasing accuracy. Similar to Taylor polynomials, the Picard iterates show greater accuracy for t near $t_0 = 0$.

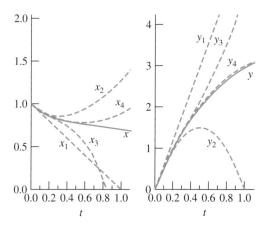

Figure 7.3 Four Picard iterate approximations to the solution pair of the initial value problem (1.9) in Example 4.

Picard iteration is also applicable to higher-order equations.

EXAMPLE 5

Write the second-order equation, initial value problem in Example 2 as

$$x'' = -(1 + \cos t)\, x$$
$$x(0) = 1, \quad x'(0) = 0.$$

Beginning with $x_0(t) = 1$, we compute the first Picard iterate by solving the initial value problem obtained by substituting $x_0(t)$ for x in the right-hand side of the differential equation. The solution of the resulting problem

$$x'' = -(1 + \cos t)$$
$$x(0) = 1, \quad x'(0) = 0$$

is

$$x_1(t) = -\frac{1}{2}t^2 + \cos t.$$

Substituting x_1 for x into the right-hand side of the differential equation, we obtain the initial value problem

$$x'' = -(1 + \cos t)\left(-\frac{1}{2}t^2 + \cos t\right)$$
$$x(0) = 1, \quad x'(0) = 0.$$

The solution

$$x_2(t) = -\frac{25}{8} - \frac{1}{4}t^2 + \frac{1}{24}t^4 + \left(4 - \frac{1}{2}t^2\right)\cos t + 2t \sin t + \frac{1}{8}\cos 2t$$

is the second Picard iterate. ∎

Figure 7.4 illustrates the accuracy of the two Picard iterates calculated in Example 5.

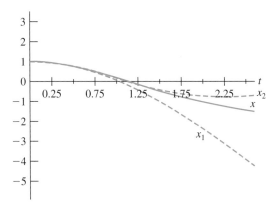

Figure 7.4 Two Picard iterate approximations to the solution of the initial value problem (1.7) for equation (1.6).

The next example illustrates both the Taylor polynomial approximation method and the Picard iteration method applied to a nonlinear system with an unspecified parameter.

EXAMPLE 6

Consider the initial value problem

$$\begin{aligned} x' &= y \\ y' &= -\alpha \left(x^2 - 1\right) y - x \\ x(0) &= 0, \quad y(0) = 1 \end{aligned} \tag{1.10}$$

(which is a system equivalent of the van der Pol equation). Setting $t = 0$ in the equations and using the initial conditions, we find

$$\begin{aligned} x'(0) &= 1 \\ y'(0) &= \alpha. \end{aligned}$$

Differentiating the equations to obtain

$$\begin{aligned} x'' &= y' \\ y'' &- \alpha \left(x^2 - 1\right) y' - 2\alpha x x' y - x' \end{aligned}$$

and then letting $t = 0$, we find

$$\begin{aligned} x''(0) &= \alpha \\ y''(0) &= \alpha^2 - 1. \end{aligned}$$

Another differentiation of the equations yields

$$\begin{aligned} x''' &= y'' \\ y''' &= -\alpha \left(x^2 - 1\right) y'' - 4\alpha x x' y' - 2\alpha x x'' y - 2\alpha (x')^2 y - x'' \end{aligned}$$

from which we obtain (upon setting $t = 0$)

$$\begin{aligned} x'''(0) &= \alpha^2 - 1 \\ y'''(0) &= \alpha \left(\alpha^2 - 4\right). \end{aligned}$$

From these calculations we obtain the third-order Taylor polynomial approximations

$$x_3(t) = t + \frac{1}{2}\alpha t^2 + \frac{1}{6}\left(\alpha^2 - 1\right) t^3$$

$$y_3(t) = 1 + \alpha t + \frac{1}{2}\left(\alpha^2 - 1\right) t^2 + \frac{1}{6}\alpha\left(\alpha^2 - 4\right) t^3$$

to the solution of the initial value problem. See Fig. 7.5.

To obtain another approximation to the solutions, we calculate several Picard iterates. Starting with the initial conditions

$$\begin{aligned} x_0(t) &= 0 \\ y_0(t) &= 1 \end{aligned}$$

we obtain, upon substitution for x and y into the right-hand sides of the differential equations, the initial value problem

$$\begin{aligned} x' &= 1 \\ y' &= \alpha \\ x(0) &= 0, \quad y(0) = 1 \end{aligned}$$

whose solutions

$$\begin{aligned} x_1(t) &= t \\ y_1(t) &= 1 + \alpha t \end{aligned}$$

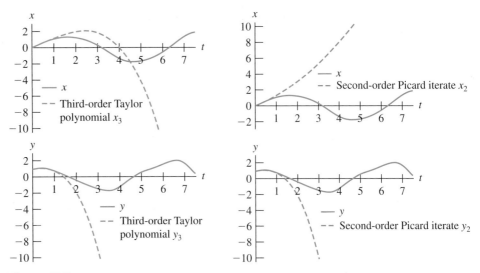

Figure 7.5 Taylor polynomial and Picard iterate approximations to the solution of the initial value problem (1.10) with $\alpha = 1/2$.

are the first Picard iterate approximations. Upon substitution of these terms for x and y into the right-hand sides of the differential equations, we obtain the initial value problem

$$x' = 1 + \alpha t$$
$$y' = -\alpha \left(t^2 - 1\right)(1 + \alpha t) - t$$
$$x(0) = 0, \quad y(0) = 1$$

whose solutions

$$x_2(t) = t + \tfrac{1}{2}\alpha t^2$$
$$y_2(t) = 1 + \alpha t + \tfrac{1}{2}\left(\alpha^2 - 1\right)t^2 - \tfrac{1}{3}\alpha t^3 - \tfrac{1}{4}\alpha^2 t^4$$

are the second Picard iterate approximations to the initial value problem. See Fig. 7.5.

■

Exercises

Find the Taylor polynomial approximations of orders $n = 1, 2,$ and 3 to the solutions of the initial value problems below.

1. $\begin{cases} x' = tx - y \\ y' = x + t^2 y \\ x(0) = 1, \quad y(0) = 1 \end{cases}$

2. $\begin{cases} x' = 2x - ty + t^2 \\ y' = tx + y \\ x(0) = 0, \quad y(0) = -1 \end{cases}$

3. $\begin{cases} x' = x + e^{-t}y \\ y' = -x + y + \cos t \\ x(0) = -2, \quad y(0) = 0 \end{cases}$

4. $\begin{cases} x' = (1 + t^2)x + y\sin t \\ y' = x - e^{2t}y - 2t \\ x(0) = 1, \quad y(0) = 1 \end{cases}$

5. $\begin{cases} x'' + e^{-t}x' + x = 0 \\ x(0) = 1, \quad x'(0) = 0 \end{cases}$

6. $\begin{cases} x'' + \dfrac{1}{1 + 2t}x' + x = 0 \\ x(0) = 0, \quad x'(0) = 1 \end{cases}$

7. $\begin{cases} x'' + x' + (1 + \cos t)x = 0 \\ x(0) = 1, \quad x'(0) = -1 \end{cases}$

8. $\begin{cases} x'' + x' + tx = 0 \\ x(0) = -1, \quad x'(0) = 1 \end{cases}$

Find the Taylor polynomial approximations (at the indicated t_0) of orders $n = 1, 2,$ and 3 to the solutions of the initial value problems below.

9. $\begin{cases} x' = tx - 2y \\ y' = 3x - y + \ln t \\ x(1) = 0, \quad y(1) = 0 \end{cases}$

10. $\begin{cases} x' = x + y - \dfrac{1}{t} \\ y' = -x + 2e^t y \\ x(-1) = 0, \quad y(-1) = 0 \end{cases}$

11. $\begin{cases} x'' + \left(1 - t^2\right) x' + x = \cos \pi t \\ x(-1) = 1, \quad x'(-1) = 0 \end{cases}$

12. $\begin{cases} x'' + \dfrac{1}{1+t} x' + x \sin t = 2 \\ x(\pi) = 0, \quad x'(\pi) = 1 \end{cases}$

Find Taylor polynomial approximations of order $n = 3$ (at $t_0 = 0$) to the general solution of the homogeneous systems below. Do this by finding the Taylor polynomial approximation of order $n = 3$ to the solutions of the two initial value problems $x(0) = 1, \ y(0) = 0$ and $x(0) = 0, \ y(0) = 1$.

13. $\begin{cases} x' = x \cos t - y \\ y' = x - y \sin t \end{cases}$ 14. $\begin{cases} x' = x - 2y \\ y' = -x + e^{-t} y \end{cases}$

Find Taylor polynomial approximations of order $n = 3$ to the general solution of the homogeneous, second-order equations below below. Do this by finding the Taylor polynomial approximation of order $n = 3$ to the solutions of the two initial value problems $x(0) = 1, \ x'(0) = 0$ and $x(0) = 0, \ x'(0) = 1$.

15. $x'' + (1 + \sin t) x = 0$

16. $x'' + e^{-t} x' + 2x = 0$

Find the first and second Picard iterate approximations to the solutions of the following initial value problems.

17. $\begin{cases} x' = t^2 x - y + t \\ y' = 2x + y - t \\ x(0) = 1, \quad y(0) = -1 \end{cases}$

18. $\begin{cases} x' = x - ty \\ y' = 2x + y + e^t \\ x(0) = 0, \quad y(0) = 1 \end{cases}$

19. $\begin{cases} x' = e^{2t} x - y + \sin t \\ y' = -x + y + \cos t \\ x(\pi) = 0, \quad y(\pi) = 0 \end{cases}$

20. $\begin{cases} x' = x + y - \sin t \\ y' = tx - y \\ x(0) = 1, \quad y(0) = 0 \end{cases}$

21. $\begin{cases} x'' + (1 + \sin t) x = 0 \\ x(0) = 1, \quad x'(0) = 0 \end{cases}$

22. $\begin{cases} x'' + e^{-t} x' + 2x = 0 \\ x(0) = 1, \quad x'(0) = 1 \end{cases}$

23. $\begin{cases} x' = y \\ y' = -\alpha \left(x^2 - 1 \right) y - x \\ x(0) = 0, \quad y(0) = 1 \end{cases}$

24. $\begin{cases} x' = \left(\frac{3}{2} - x - 2y \right) x \\ y' = \left(-\frac{1}{4} + x \right) y \\ x(0) = 1, \quad y(0) = 1 \end{cases}$

25. $\begin{cases} x'' = a \sin \pi x \\ x(0) = 0, \quad x'(0) = 1 \end{cases}$

26. $\begin{cases} x'' = -\alpha \left((x')^2 - 1 \right) x' - x \\ x(0) = 1, \quad x'(0) = -1 \end{cases}$

7.2 A Perturbation Method

In Chapter 3 (Sec. 3.4) we studied a method based on Taylor polynomials that constructs approximations to the solution of a single first-order differential equation in which there is a small parameter. In this section we extend this method to systems of equations and to higher-order equations.

The second-order equation (1.6) in Examples 2 and 5 of Sec. 7.1 is a specific case of the second-order, nonautonomous differential equation

$$x'' + (1 + \varepsilon \cos t) x = 0 \tag{2.1}$$

obtained by choosing $\varepsilon = 1$. The coefficient $1 + \varepsilon \cos t$ of x oscillates (cosinusoidally) around the average 1 with amplitude ε. If the amplitude ε is restricted to a small number, we can view this coefficient as a small perturbation of the average 1, and we can view the equation as a perturbation of the equation $x'' + x = 0$ (the harmonic oscillator equation). Solutions of the equation (2.1) depend, of course, on the numerical value assigned to the parameter ε. That is, solutions are functions not only of t, but also of ε. In Chapter 3 we saw, for first-order equations with small coefficients ε, how Taylor polynomials *in* ε (not polynomials in t, as in the previous Sec. 7.1) can provide good approximation formulas to solutions. This perturbation method extends in a natural way to higher-order equations such as (2.1).

For example, consider the initial value problem

$$x'' + (1 + \varepsilon \cos t)\, x = 0$$
$$x(0) = 1, \quad x'(0) = 0. \tag{2.2}$$

The perturbation method approximates the solution by the Taylor polynomials in ε:

$$x_n(t) = k_0(t) + k_1(t)\varepsilon + \cdots + k_n(t)\varepsilon^n, \quad n = 0, 1, 2, \dots .$$

The undetermined coefficients $k_i(t)$ are calculated by substituting $x_n(t)$ into the differential equation and equating coefficients of like powers of ε (up to power n) from both sides of the resulting equation. [In the case of equation (2.2) coefficients of the powers of ε on the left-hand side are all set equal to 0, since the right-hand side of this equation is 0.] The result of this step will be a differential equation for each of the coefficients $k_i(t)$. We obtain initial conditions for this differential equation by applying a similar substitution procedure to the initial conditions in (2.2).

EXAMPLE 1

To calculate the first order ($n = 1$) perturbation approximation

$$x_1(t) = k_0(t) + k_1(t)\varepsilon \tag{2.3}$$

to the initial value problem (2.2) we begin by substituting x_1 into the differential equation:

$$\left(k_0'' + k_0\right) + \left(k_1'' + k_1 + k_0 \cos t\right)\varepsilon + \cdots = 0 + 0 \cdot \varepsilon + \cdots ,$$

where the dots $\cdots$ indicate terms in ε of degree higher than $n = 1$. If we equate the coefficients of like powers of ε from both sides, we obtain

$$k_0'' + k_0 = 0$$
$$k_1'' + k_1 + k_0 \cos t = 0.$$

To supplement these differential equations for k_0 and k_1 with initial conditions, we substitute x_1 into the initial conditions in (2.2):

$$k_0(0) + k_1(0)\varepsilon = 1 + 0 \cdot \varepsilon$$
$$k_0'(0) + k_1'(0)\varepsilon = 0 + 0 \cdot \varepsilon.$$

Equating the coefficients of like powers of ε from both sides of these equations, we obtain initial conditions for k_0 and k_1. In summary, we have the initial value problems

$$k_0'' + k_0 = 0$$
$$k_0(0) = 1, \quad k_0'(0) = 0 \tag{2.4}$$

and

$$k_1'' + k_1 = -k_0 \cos t$$
$$k_1(0) = 0, \quad k_1'(0) = 0 \tag{2.5}$$

for the coefficients k_0 and k_1 in (2.3).

The differential equation for k_0 in (2.4) is a second-order, linear homogeneous equation with constant coefficients. We can solve this initial value problem (2.4) using the method in Chapter 5 (Sec. 5.3). After k_0 is known and substituted into the equation (2.5), this differential equation for k_1 becomes a second-order, linear nonhomogeneous equation with constant coefficients. We can solve the initial value problem (2.5) for k_1 by using the methods in Chapter 6 (Sec. 6.2.1).

The details are as follows. The solution of the differential equation in (2.4) is $k_0(t) = c_1 \cos t + c_2 \sin t$ (see Table 5.1 in Chapter 5). The initial conditions require $c_1 = 1, c_2 = 0$. Thus,

$$k_0(t) = \cos t$$

and the initial value problem (2.5) for $k_1(t)$ becomes

$$k_1'' + k_1 = -\cos^2 t$$
$$k_1(0) = 0, \quad k_1'(0) = 0.$$

We can solve this initial value problem either by the variation of constants formula [see (2.21) in Chapter 6, Sec. 6.2] or, perhaps more easily, by the method of undetermined coefficients (Chapter 6, Sec. 6.1). (In the latter case, the trigonometric identity

$$\cos^2 t = \frac{1}{2} + \frac{1}{2}\cos 2t$$

is useful.) The solution turns out to be

$$k_1(t) = -\frac{2}{3} + \frac{1}{3}\cos t + \frac{1}{3}\cos^2 t. \tag{2.6}$$

With the two coefficients $k_0(t)$ and $k_1(t)$ in hand, we have the first-order perturbation approximation (2.3)

$$x_1(t) = \cos t + \left(-\frac{2}{3} + \frac{1}{3}\cos t + \frac{1}{3}\cos^2 t\right)\varepsilon.$$

to the initial value problem (2.2). ∎

If we take $\varepsilon = 1$, then the initial value problem (2.2) is the same initial value problem studied in Sec. 7.1, where we approximated the solution by Taylor polynomials in t and also by Picard iterates. Figure 7.6 shows graphs of the zeroth and first-order perturbation approximations $x_0(t) = \cos t$ and $x_1(t)$ to this initial value problem that we calculated in Example 1.

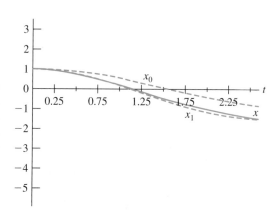

Figure 7.6 The zeroth- and first-order perturbation approximations to the solution x of the initial value problem (2.2) with $\varepsilon = 1$. Compare to the Taylor polynomial approximations in Fig. 7.2 and the Picard iterates in Fig. 7.4.

In principle we can calculate perturbation approximations $x_n(t)$ to any order n by continuing the procedure in Example 1, although the calculations become more and more tedious as n increases. For example, if we substitute

$$x_2(t) = k_0(t) + k_1(t)\varepsilon + k_2(t)\varepsilon^2$$

into the initial value problem (2.2) and calculate the coefficients of ε^2, we obtain the initial value problem

$$k_2'' + k_2 + k_1 \cos t = 0$$
$$k_2(0) = 0, \quad k_2'(0) = 0$$

for $k_2(t)$. Since we previously calculated $k_1(t)$ [see (2.6)], this initial value problem becomes

$$k_2'' + k_2 = \tfrac{2}{3} \cos t - \tfrac{1}{3} \cos^2 t - \tfrac{1}{3} \cos^3 t \qquad (2.7)$$
$$k_2(0) = 0, \quad k_2'(0) = 0.$$

The differential equation in this initial value problem is a second-order, nonhomogeneous linear equation with constant coefficients, and as a result we can solve the initial value problem using the methods in Chapter 6.

If we continue this procedure to even higher orders n, we find that each coefficient $k_n(t)$ satisfies an initial value problem of the same type:

$$k_n'' + k_n = -k_{n-1}(t) \cos t$$
$$k_n(0) = 0, \quad k_n'(0) = 0.$$

However, as the order n increases, the nonhomogeneous term $-k_{n-1}(t) \cos t$, which at each step involves the previous coefficient $k_{n-1}(t)$, becomes more and more complicated, and the solution of the initial value problem becomes more difficult and impractical.

The solution of (2.7) is

$$k_2(t) = -\frac{2}{9} + \frac{5}{72} \cos t + \frac{1}{9} \cos^2 t + \frac{1}{24} \cos^3 t + \frac{5}{24} t \sin t,$$

which gives the rather formidable formula

$$x_2(t) = \cos t + \left(-\frac{2}{3} + \frac{1}{3} \cos t + \frac{1}{3} \cos^2 t \right) \varepsilon$$
$$+ \left(-\frac{2}{9} + \frac{5}{72} \cos t + \frac{1}{9} \cos^2 t + \frac{1}{24} \cos^3 t + \frac{5}{24} t \sin t \right) \varepsilon^2$$

for the second-order perturbation approximation.

Setting $\varepsilon = 1$, we obtain the approximation

$$x_2(t) = -\frac{8}{9} + \frac{101}{72} \cos t + \frac{4}{9} \cos^2 t + \frac{1}{24} \cos^3 t + \frac{5}{24} t \sin t \qquad (2.8)$$

to the initial value problem (2.2). See Fig. 7.7.

The perturbation method also applies to systems of equations containing a small parameter ε. The fundamental steps are the same: Substitute a Taylor polynomial in the parameter ε for each unknown into the differential equations, and into the initial conditions, and then equate coefficients from like powers of ε from both sides of the equations. The results of these steps will produce a sequence of initial value problems to be solved for the undetermined coefficients.

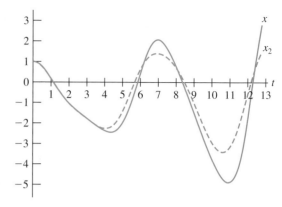

Figure 7.7 The second-order perturbation approximation $x_2(t)$, given by (2.8), to the solution of the initial value problem (2.2) with $\varepsilon = 1$.

EXAMPLE 2

Consider the initial value problem

$$
\begin{aligned}
x' &= -x + \varepsilon e^{-t} y \\
y' &= 6x - y \\
x(0) &= 1, \quad y(0) = 0,
\end{aligned}
\tag{2.9}
$$

which, when $\varepsilon = 1/2$, reduces to the initial value problem (1.3) studied in Sec. 7.1. To calculate the first-order perturbation approximations (in ε)

$$
\begin{aligned}
x_1(t) &= k_0(t) + k_1(t)\varepsilon \\
y_1(t) &= m_0(t) + m_1(t)\varepsilon
\end{aligned}
\tag{2.10}
$$

we substitute these expressions for x and y into the two differential equations and into the two initial conditions. We then equate the coefficients of like powers of ε from both sides of the resulting equations. From (2.9) we calculate

$$
\begin{aligned}
k_0' + k_1'\varepsilon &= -k_0 + (-k_1 + e^{-t}m_0)\varepsilon + \cdots \\
m_0' + m_1'\varepsilon &= (6k_0 - m_0) + (6k_1 - m_1)\,\varepsilon \\
k_0(0) + k_1(0)\varepsilon &= 1 + 0 \cdot \varepsilon \\
m_0(0) + m_1(0)\varepsilon &= 0 + 0 \cdot \varepsilon
\end{aligned}
$$

(where the dots $\cdots$ indicate terms in ε of degree higher than 1). By equating the coefficients of like powers of ε from both sides of these equations, we obtain the initial value problems

$$
\begin{aligned}
k_0' &= -k_0 \\
m_0' &= 6k_0 - m_0 \\
k_0(0) &= 1, \quad m_0(0) = 0
\end{aligned}
\tag{2.11}
$$

and

$$
\begin{aligned}
k_1' &= -k_1 + e^{-t}m_0 \\
m_1' &= 6k_1 - m_1 \\
k_1(0) &= 0, \quad m_1(0) = 0.
\end{aligned}
\tag{2.12}
$$

The first problem (2.11) for the coefficients $k_0(t)$ and $m_0(t)$ involves a homogeneous linear system. We can find formulas for the solution using the methods in Chapter 5.

[Alternatively, we can solve the first equation for $k_0(t)$ and use the result in the second equation.] The solution pair for the initial value problem (2.11) is

$$k_0(t) = e^{-t}$$
$$m_0(t) = 6te^{-t}.$$

This pair provides the zeroth-order perturbation approximation

$$x_0(t) = e^{-t}$$
$$y_0(t) = 6te^{-t}$$

to the initial value problem (2.9).

With the zeroth-order approximations in hand, we find that the initial value problem (2.12) for the coefficients k_1 and m_1 becomes

$$k_1' = -k_1 + 6te^{-2t}$$
$$m_1' = 6k_1 - m_1$$
$$k_1(0) = 0, \quad m_1(0) = 0.$$

This problem involves a nonhomogeneous linear system with constant coefficients. We can calculate a formula for its solution using the methods of Chapter 6 (e.g., using the method of undetermined coefficients). [Alternatively, we can solve the first equation for $k_1(t)$ and use the result in the second equation.] The solution is

$$k_1(t) = 6e^{-t} - 6e^{-2t} - 6te^{-2t}$$
$$m_1(t) = 36te^{-t} - 72e^{-t} + 72e^{-2t} + 36te^{-2t}.$$

With these calculations complete, the first-order perturbation approximations (2.10) to the initial value problem (2.9) become

$$x_1(t) = e^{-t} + \left(6e^{-t} - 6e^{-2t} - 6te^{-2t}\right)\varepsilon$$
$$y_1(t) = 6te^{-t} + \left(36te^{-t} - 72e^{-t} + 72e^{-2t} + 36te^{-2t}\right)\varepsilon. \tag{2.13}$$

∎

Setting $\varepsilon = 1/2$ in (2.13), we obtain from (2.13) the approximations

$$x_1(t) = 4e^{-t} - 3e^{-2t} - 3te^{-2t}$$
$$y_1(t) = 24te^{-t} - 36e^{-t} + 36e^{-2t} + 18te^{-2t}$$

to the solution of the initial value problem (1.3) studied in Sec. 7.1. See Fig. 7.8.

We conclude with an example of the perturbation method applied to a nonlinear second-order equation with an unspecified parameter.

EXAMPLE 3

Consider the initial value problem

$$x' = y$$
$$y' = -\alpha\left(x^2 - 1\right)y - x \tag{2.14}$$
$$x(0) = 0, \quad y(0) = 1$$

(which is a system equivalent of the van der Pol equation). In this example we study this problem when $\alpha = \varepsilon$ is a small number and calculate the first-order perturbation approximation

$$x_1(t) = k_0(t) + k_1(t)\varepsilon$$
$$y_1(t) = m_0(t) + m_1(t)\varepsilon$$

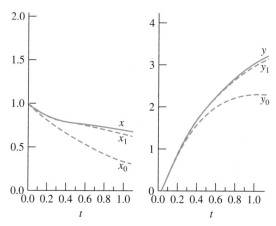

Figure 7.8 The zeroth- and first-order perturbation approximations to the solution of the initial value problem (2.9) with $\varepsilon = 1/2$. Compare to the Taylor polynomial approximations in Fig. 7.1 and the Picard iterates in Fig. 7.3.

to the solution of this initial value problem. Substituting x_1 and y_1 into the initial value problem, we obtain

$$k_0' + k_1'\varepsilon = m_0 + m_1\varepsilon$$
$$m_0' + m_1'\varepsilon = -k_0 + \left(-\left(k_0^2 - 1\right)m_0 - k_1\right)\varepsilon + \cdots$$
$$k_0(0) + k_1(0)\varepsilon = 0 + 0 \cdot \varepsilon$$
$$m_0(0) + m_1(0)\varepsilon = 1 + 0 \cdot \varepsilon$$

where the dots $\cdots$ denote terms of degree 2 or higher in ε. Equating the coefficients of like powers of ε from both sides of these equations, we obtain the initial value problems

$$k_0' = m_0$$
$$m_0' = -k_0$$
$$k_0(0) = 0, \quad m_0(0) = 1$$

for k_0 and m_0 and

$$k_1' = m_1$$
$$m_1' = -k_1 - \left(k_0^2 - 1\right)m_0$$
$$k_1(0) = 0, \quad m_1(0) = 0$$

for k_1 and m_1. The initial value problem for k_0 and m_0 involves a linear homogeneous system and its solution is

$$k_0(t) = \sin t$$
$$m_0(t) = \cos t.$$

With these formulas for k_0 and m_0, the initial problem for k_1 and m_1 becomes the initial value problem

$$k_1' = m_1$$
$$m_1' = -k_1 - \left(\sin^2 t - 1\right)\cos t$$
$$k_1(0) = 0, \quad m_1(0) = 0.$$

The differential equations form a linear, nonhomogeneous system. Using the trigono-metric identity

$$\left(\sin^2 t - 1\right) \cos t = -\frac{3}{4} \cos t - \frac{1}{4} \cos 3t$$

and the method of undetermined coefficients, we find the solution to be

$$k_1(t) = \frac{1}{32} \cos t - \frac{1}{32} \cos 3t + \frac{3}{8} t \sin t$$

$$m_1(t) = \frac{11}{32} \sin t + \frac{3}{32} \sin 3t + \frac{3}{8} t \cos t.$$

Thus, the first-order perturbation approximations to the solution of the initial value problem (2.14) are (recall $\alpha = \varepsilon$)

$$x_1(t) = \sin t + \frac{1}{32} \left(\cos t - \cos 3t + 12t \sin t\right) \alpha \tag{2.15}$$

$$y_1(t) = \cos t + \frac{1}{32} \left(11 \sin t + 3 \sin 3t + 12t \cos t\right) \alpha.$$

See Fig. 7.9.

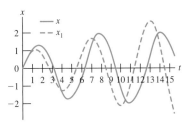

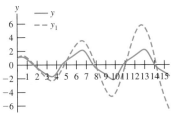

Figure 7.9 The first-order perturbation approximations (2.15) to the solution of the initial value problem (2.14) with $\varepsilon = \alpha = 1/2$. Compare to the Taylor polynomial and the Picard iterate approximations in Fig. 7.5.

Exercises

Calculate the first-order perturbation approximations to the so-lutions of the following initial value problems.

1. $\begin{cases} x' = x - \varepsilon y e^{-t} \\ y' = -x + y \\ x(0) = 1, \quad y(0) = 0 \end{cases}$

2. $\begin{cases} x' = x - \varepsilon y e^{-t} \\ y' = -x + y \\ x(0) = 0, \quad y(0) = 1 \end{cases}$

3. $\begin{cases} x' = -2x + y \\ y' = -\varepsilon e^{-t} x + y \\ x(0) = 0, \quad y(0) = 1 \end{cases}$

4. $\begin{cases} x' = -2x + y \\ y' = -\varepsilon e^{-t} x + y \\ x(0) = 1, \quad y(0) = 0 \end{cases}$

5. $\begin{cases} x' = (-1 + \varepsilon \sin t) x + y \\ y' = x - y \\ x(0) = 1, \quad y(0) = -1 \end{cases}$

6. $\begin{cases} x' = (-1 + \varepsilon \sin t) x + y \\ y' = x - y \\ x(0) = -2, \quad y(0) = 1 \end{cases}$

7. $\begin{cases} x' = -2y + t \\ y' = \varepsilon e^{-t} x - y \\ x(1) = 0, \quad y(1) = 0 \end{cases}$

8. $\begin{cases} x' = -2x + \varepsilon y \cos t + e^{-t} \\ y' = \varepsilon x \sin t - 2y \\ x(0) = 0, \quad y(0) = 0 \end{cases}$

9. $\begin{cases} x'' + (3 + \varepsilon e^{-t})x' + 2x = 0 \\ x(0) = 1, \quad x'(0) = 0 \end{cases}$

10. $\begin{cases} x'' + (3 + \varepsilon e^{-t})x' + 2x = 0 \\ x(0) = 0, \quad x'(0) = 1 \end{cases}$

11. $\begin{cases} x'' + x' + 6\left(-1 + 10\varepsilon e^{3t}\right)x = 0 \\ x(0) = 1, \quad x'(0) = -1 \end{cases}$

12. $\begin{cases} x'' + x' + 6\left(-1 + 10\varepsilon e^{3t}\right)x = 0 \\ x(0) = 0, \quad x'(0) = -5 \end{cases}$

13. $\begin{cases} x'' + (1 + \sin \varepsilon t)x = 0 \\ x(0) = 1, \quad x'(0) = 0 \end{cases}$

14. $\begin{cases} x'' + x \cos \varepsilon t = 0 \\ x(0) = 1, \quad x'(0) = -1 \end{cases}$

7.3 Power Series Solutions

In Chapter 3 (Sec. 3.4.1) we saw how to obtain power series representations of solutions of single differential equations. We can apply the same method to systems of equations and to higher-order equations. The first step of the power series method involves the calculation of a recursive formula for the coefficients of the power series. Subsequent steps, although not always easy (or even possible) to perform, involve solving the recursive formula and summing the series.

EXAMPLE 1   Consider the initial value problem

$$x'' - x = 0 \tag{3.1}$$
$$x(0) = 1, \quad x'(0) = 1.$$

To calculate the recursive formula for the coefficients k_i of a power series solution

$$x(t) = k_0 + k_1 t + k_2 t^2 + \cdots = \sum_{i=0}^{\infty} k_i t^i, \tag{3.2}$$

we begin by substituting this series into the (second-order) differential equation. From the calculations

$$x'(t) = k_1 + 2k_2 t + 3k_3 t^2 + \cdots = \sum_{i=0}^{\infty} (i + 1) k_{i+1} t^i \tag{3.3}$$

$$x''(t) = 2k_2 + (3)(2) k_3 t + (4)(3) k_4 t^2 + \cdots = \sum_{i=0}^{\infty} (i + 2)(i + 1) k_{i+2} t^i \tag{3.4}$$

we obtain for the left-hand side of the differential equation the power series

$$x'' - x = \sum_{i=0}^{\infty} (i + 2)(i + 1) k_{i+2} t^i - \sum_{i=0}^{\infty} k_i t^i$$

$$= \sum_{i=0}^{\infty} \left[(i + 2)(i + 1) k_{i+2} - k_i \right] t^i.$$

Since the right-hand side of the differential equation in (3.1) is 0, in order for the series (3.2) to be a solution, all of the coefficients in the power series for $x'' - x$ must equal 0. That is,

$$(i + 2)(i + 1) k_{i+2} - k_i = 0, \quad i = 0, 1, 2, 3, \ldots.$$

This leads to the recursive formula

$$k_{i+2} = \frac{1}{i+2}\frac{1}{i+1}k_i, \quad i = 0, 1, 2, 3, \ldots \tag{3.5}$$

for the coefficients of power series solutions.

The initial conditions in (3.1) give the first two coefficients

$$k_0 = 1, \quad k_1 = 1.$$

We use the recursive formula (3.5) to calculate the remaining coefficients $k_2, k_3, \ldots$.
For example, from the recursive formula we find

$$k_2 = \frac{1}{2}k_0 = \frac{1}{2}$$
$$k_3 = \frac{1}{3}\frac{1}{2}k_1 = \frac{1}{3}\frac{1}{2} = \frac{1}{3!}$$
$$k_4 = \frac{1}{4}\frac{1}{3}k_2 = \frac{1}{4}\frac{1}{3}\frac{1}{2} = \frac{1}{4!}$$
$$k_5 = \frac{1}{5}\frac{1}{4}k_3 = \frac{1}{5}\frac{1}{4}\frac{1}{3}\frac{1}{2} = \frac{1}{5!}.$$

From these results we recognize the pattern[1]

$$k_i = \frac{1}{i!} \quad i = 0, 1, 2, 3, \ldots \tag{3.6}$$

and consequently we obtain the power series solution

$$x(t) = \sum_{i=0}^{\infty} \frac{1}{i!}t^i$$

of the initial value problem. We recognize this power series as that of e^t (which is the solution formula we would obtain using the solution method in Chapter 5, Sec.5.3). ∎

In the example above, we were able to derive the recursive formula (3.5) for the power series coefficients, solve it [i.e., obtain the formulas (3.6) for the coefficients], and sum the resulting series (i.e., obtain the simpler formula e^t for the solution x). For most differential equations the successful completion of all these steps is usually not possible. In particular, it is not normally possible to carry out the last step and, as a result, the power series is the only available formula for the solution.

The next example applies the power series method to a nonautonomous equation.

EXAMPLE 2

Consider the nonautonomous, second-order differential equation

$$x'' + tx' + x = 0. \tag{3.7}$$

To find the recursive formula for the power series representation of the general solution we substitute the series (3.2) and its derivatives (3.3) and (3.4) into the equation. This

---

[1]Rigorously, we need to perform an induction proof to establish this formula for all i.

involves a calculation of the term

$$tx' = t\left(k_1 + 2k_2 t + 3k_3 t^2 + \cdots\right)$$
$$= k_1 t + 2k_2 t^2 + 3k_3 t^3 + \cdots \qquad (3.8)$$
$$= \sum_{i=0}^{\infty} i k_i t^i.$$

A substitution results in

$$x'' + tx' + x = \sum_{i=0}^{\infty} (i+2)(i+1)k_{i+2} t^i + \sum_{i=0}^{\infty} i k_i t^i + \sum_{i=0}^{\infty} k_i t^i$$
$$= \sum_{i=0}^{\infty} \left[(i+2)(i+1)k_{i+2} + (i+1)k_i\right] t^i.$$

Since the differential equation requires this expression equal 0, we must set the coefficients of all powers of t equal to 0. This leads to

$$(i+2)(i+1)k_{i+2} + (i+1)k_i = 0, \qquad i = 0, 1, 2, 3, \ldots$$

or

$$k_{i+2} = -\frac{1}{i+2}k_i, \qquad i = 0, 1, 2, 3, \ldots .$$

From this recursive formula we calculate

$$k_2 = -\frac{1}{2}k_0$$
$$k_3 = -\frac{1}{3}k_1$$
$$k_4 = -\frac{1}{4}k_2 = \frac{1}{4}\frac{1}{2}k_0$$
$$k_5 = -\frac{1}{5}k_3 = \frac{1}{5}\frac{1}{3}k_1$$
$$k_6 = -\frac{1}{6}k_4 = -\frac{1}{6}\frac{1}{4}\frac{1}{2}k_0$$
$$k_7 = -\frac{1}{7}k_5 = -\frac{1}{7}\frac{1}{5}\frac{1}{3}k_1$$

and so on. We are free to choose any values for k_0 and k_1. The recursive formula then allows us to calculate all the remaining coefficients $k_2, k_3, k_4, \ldots$ in the series solution. For example, if we take $k_0 = 1$ and $k_1 = 0$ we get the series solution

$$x_1(t) = 1 - \frac{1}{2}t^2 + \frac{1}{4}\frac{1}{2}t^4 - \frac{1}{6}\frac{1}{4}\frac{1}{2}t^6 + \cdots$$

whereas the choices $k_0 = 0$ and $k_1 = 1$ give the series solution

$$x_2(t) = t - \frac{1}{3}t^3 + \frac{1}{5}\frac{1}{3}t^5 - \frac{1}{7}\frac{1}{5}\frac{1}{3}t^7 + \cdots .$$

Using the recursive formulas for the coefficients, we can rewrite the series as

$$x(t) = k_0 + k_1 t - \frac{1}{2} k_0 t^2 - \frac{1}{3} k_1 t^3 + \frac{1}{4} \frac{1}{2} k_0 t^4 + \frac{1}{5} \frac{1}{3} k_1 t^5$$
$$- \frac{1}{6} \frac{1}{4} \frac{1}{2} k_0 t^6 - \frac{1}{7} \frac{1}{5} \frac{1}{3} k_1 t^7 + \cdots$$

or, after gathering together all terms that involve k_0 and those that involve k_1,

$$x(t) = k_0 \left[1 - \frac{1}{2} t^2 + \frac{1}{4} \frac{1}{2} t^4 - \frac{1}{6} \frac{1}{4} \frac{1}{2} t^6 + \cdots \right]$$
$$+ k_1 \left[t - \frac{1}{3} t^3 + \frac{1}{5} \frac{1}{3} t^5 - \frac{1}{7} \frac{1}{5} \frac{1}{3} t^7 + \cdots \right].$$

We usually use the letters c_1 and c_2 for the arbitrary constants in the general solution. If we let $c_1 = k_0$ and $c_2 = k_1$, then we find that the series solution is the linear combination

$$x(t) = c_1 \left[1 - \frac{1}{2} t^2 + \frac{1}{4} \frac{1}{2} t^4 - \frac{1}{6} \frac{1}{4} \frac{1}{2} t^6 + \cdots \right] \tag{3.9}$$
$$+ c_2 \left[t - \frac{1}{3} t^3 + \frac{1}{5} \frac{1}{3} t^5 - \frac{1}{7} \frac{1}{5} \frac{1}{3} t^7 + \cdots \right]$$

of the two independent solutions $x_1(t)$ and $x_2(t)$.

By carefully examining the series for $x_1(t)$ and $x_2(t)$, we can identify a pattern in their coefficients (i.e., we can solve the recursive formula) and write these solutions using summation notation as

$$x_1(t) = \sum_{m=0}^{\infty} (-1)^m \frac{1}{2^m} \frac{1}{m!} t^{2m}$$

$$x_2(t) = \sum_{m=0}^{\infty} (-1)^m \frac{1}{(2m+1)(2m-1)(2m-3) \cdots 3} t^{2m+1}. \qquad \blacksquare$$

An important issue regarding a power series

$$\sum_{i=0}^{\infty} k_i (t - t_0)^i$$

is its convergence. Power series always converge for $t = t_0$ (since all terms, except the first term, are equal to 0). If the series converges for any other values of t, then it converges for an interval of t values centered around t_0 (i.e., for all t satisfying $|t - t_0| < R$ for a number R called the *radius of convergence*). If the series converges for all values of t, we say the radius of convergence is infinite ($R = \infty$). If the series converges only for $t = t_0$, we say the radius of convergences is zero ($R = 0$).

If a function $x = x(t)$ has derivatives of all orders, $x^{(1)} = x'$, $x^{(2)} = x''$, ..., at $t = t_0$, then we can associate it with the power series

$$x(t) = \sum_{i=0}^{\infty} \frac{1}{i!} x^{(i)}(t_0) (t - t_0)^i.$$

This power series representation is useful as a formula for $x(t)$ only if it has a positive radius of convergent R. When calculating a power series formula for a solution of a differential equation, a important question that arises is whether or not the resulting series has a positive radius of convergence. For example, do the power series calculated in Example 2 for the solutions of the second-order equation (3.7) have positive radii of convergence? If not, these series do not define a solution on any interval for t.

In calculus, we study several methods for calculating the radius of convergence of a power series. Any of these tests can be applied to power series solutions of differential equations, such as those calculated in Example 2. However, the following theorem is usually more convenient (and simpler) to use for proving the convergence of power series solutions of a second-order differential equations.

THEOREM 1

Suppose the coefficients $p(t)$ and $q(t)$ and the nonhomogeneous term $h(t)$ of the second-order equation

$$x'' + p(t)x' + q(t)x = h(t)$$

have power series representations at $t = t_0$ with positive radii of convergence R_1, R_2, and R_3. Then the general solution[2] has a power series representation whose radius of convergence R is no smaller than the smallest R_i (i.e., $R \geq \min\{R_1, R_2, R_3\}$).

Any polynomial has a power series at any point $t = t_0$ with infinite radius of convergent $R = \infty$ (the series has only a finite number of terms). In Example 2 the coefficients $p(t) = t$ and $q(t) = 1$ and the nonhomogeneous term $h(t) = 0$ are all polynomials. Therefore, by Theorem 1 the power series solutions calculated in that example converge for all t.

The ratio $n(t)/d(t)$ of two polynomials $n(t)$ and $d(t)$ is called a rational function. Rational functions often occur as coefficients in second-order equations that arise in applications. A rational function has a power series representation with a positive radius of convergence at any point $t = t_0$, where the denominator $d(t)$ does not vanish. The radius of convergence is the distance from t_0 to the nearest (possibly complex) root of the denominator $d(t)$.

EXAMPLE 3

The equation

$$\left(1 - t^2\right) x'' - 2tx' + \alpha\left(\alpha + 1\right) x = 0 \tag{3.10}$$

is called the Legendre equation. In this nonautonomous second-order equation, α is a real number. The Legendre equation arises in a variety of applications, including studies of gravitational force fields resulting from matter distributed on a spherical surface, temperature distributions within solid spheres, and vibrations of hanging chains.

We can rewrite the Legendre equation (3.10) as

$$x'' - \frac{2t}{1 - t^2}x' + \frac{\alpha\left(\alpha + 1\right)}{1 - t^2}x = 0$$

and observe that the coefficients

$$p(t) = -\frac{2t}{1 - t^2}, \quad q(t) = \frac{\alpha(\alpha + 1)}{1 - t^2}$$

---

[2]By assumption the coefficients $p(t)$, $q(t)$, and $h(t)$ have derivatives of all orders. Since a differentiable function is necessarily continuous, Theorem 2 in Sec. 4.3 of Chapter 4 guarantees the existence of solutions.

are rational functions. The power series of each of these rational functions at $t_0 = 0$ has radius of convergence $R = 1$. (This is because the distance from $t_0 = 0$ to the roots ± 1 of the denominator $1 - t^2$ equals 1.) By Theorem 1 we conclude that the general solution has a power series representation with a radius of convergence $R \geq 1$. ■

To find a recursive formula for the coefficients of power series solutions of the Legendre equation (3.10) we substitute the series

$$x = \sum_{i=0}^{\infty} k_i t^i, \quad x' = \sum_{i=0}^{\infty} (i+1) k_{i+1} t^i, \quad x'' = \sum_{i=0}^{\infty} (i+2)(i+1) k_{i+2} t^i$$

into the left-hand side of the equation [recall (3.2), (3.3), and (3.4)]. To do this we need

$$tx' = \sum_{i=0}^{\infty} i k_i t^i$$

and

$$t^2 x'' = t^2 \left(2k_2 + (3)(2) k_3 t + (4)(3) k_4 t^2 + \cdots \right)$$

$$= 2k_2 t^2 + (3)(2) k_3 t^3 + (4)(3) k_4 t^4 + \cdots = \sum_{i=0}^{\infty} i(i-1) k_i t^i.$$

For the left-hand side of the Legendre equation we have

$$x'' - t^2 x'' - 2tx' + \alpha(\alpha+1)x = \sum_{i=0}^{\infty} (i+2)(i+1) k_{i+2} t^i - \sum_{i=0}^{\infty} i(i-1) k_i t^i$$

$$- 2 \sum_{i=0}^{\infty} i k_i t^i + \alpha(\alpha+1) \sum_{i=0}^{\infty} k_i t^i.$$

This expression requires that we perform some algebra on four power series. We can do this coefficient-wise since the power series have the same ranges, $i = 0$ to ∞, and all involve the same powers of t (namely, t^i). The calculation results in a single power series for the left hand side of the Legendre equation:

$$x'' - t^2 x'' - 2tx' + \alpha(\alpha+1)x$$

$$= \sum_{i=0}^{\infty} \left[(i+2)(i+1) k_{i+2} - i(i-1) k_i - 2i k_i + \alpha(\alpha+1) k_i \right] t^i.$$

After some algebraic manipulation of the coefficient of t^i we finally arrive at

$$x'' - t^2 x'' - 2tx' + \alpha(\alpha+1)x$$

$$= \sum_{i=0}^{\infty} \left[(i+2)(i+1) k_{i+2} + (\alpha-i)(\alpha+i+1) k_i \right] t^i.$$

For the power series for x to solve the Legendre equation, this expression must equal 0. By setting all of the coefficients of t^i equal to 0 and solving the resulting equation for k_{i+2}, we obtain the recursive formula

$$k_{i+2} = -\frac{(\alpha-i)(\alpha+i+1)}{(i+2)(i+1)} k_i, \qquad i = 0, 1, 2, 3, \ldots \tag{3.11}$$

for the coefficients k_i for the power series solution of the Legendre equation.

Notice that the coefficients produced by the formula (3.11), and hence the power series solution, depend on the number α appearing in the Legendre equation (3.10).

In some cases the recursive formula (3.11) for the Legendre equation treated in Example 3 produces some exceptional solutions.

For example, suppose we look for solutions satisfying $x(0) = 0$, in which case $k_0 = 0$. When $\alpha = 1$ the recursive formula gives

$$k_2 = 0, \quad k_3 = 0.$$

Because these two consecutive coefficients equal to 0, you can see from the recursive formula that all subsequent coefficients are also equal to 0 (i.e., $k_i = 0$ for all $i \geq 2$). This means that the power series solution is a polynomial in this case. Specifically, we have

$$x(t) = k_1 t,$$

where k_1 is an arbitrary constant [which would be specified by an initial condition $x'(0) = k_1$].

Similarly, when $k_0 = 0$ and $\alpha = 3$ the recursive formula (3.11) also produces two consecutive coefficients equal to 0:

$$k_2 = 0, \quad k_3 = -\frac{5}{3}k_1, \quad k_4 = 0, \quad k_5 = 0.$$

Therefore, once again, all subsequent coefficients equal 0 and the resulting solution is a polynomial. Specifically,

$$x(t) = k_1 t - \frac{5}{3}k_1 t^3,$$

where k_1 is an arbitrary constant.

In fact, when α is an odd integer [and $x(0) = 0$] the recursive formula always eventually produces two consecutive coefficients equal to 0 and hence a polynomial solution. This is because of two facts observable from the recursive formula (3.11). First, when $i = \alpha$, we have $k_{\alpha+2} = 0$ and second when $k_0 = 0$ all coefficients with even subscripts equal 0. Therefore, if α is an odd integer, then $\alpha + 1$ is an even number and the two consecutive coefficients $k_{\alpha+1}$ and $k_{\alpha+2}$ equal 0. As a result of this, all subsequent coefficients equal 0 and the solution is a polynomial of degree α. Again k_1 is an arbitrary constant appearing in this polynomial.

Another case also produces polynomial solutions of the Legendre equation. In Exercise 13 the reader is asked to show that when $x'(0) = 0$ and α is an even integer, then the recursive formula (3.11) eventually produces two consecutive coefficients equal to 0 and consequently a polynomial solution (of degree α and an arbitrary constant k_0).

The polynomial solutions of the Legendre equation are called *Legendre polynomials*. They are usually denoted by $P_0(t)$, $P_1(t)$, $P_2(t)$, ... The arbitrary constant that appears in these polynomial solutions of Legendre's equation are usually chosen so as to standardize or normalize the polynomials in way convenient in applications. For example, a common normalization requires that $P_n(1) = 1$. The first six Legendre polynomials standardized in this manner are

$$P_0(t) = 1 \qquad\qquad P_1(t) = t$$

$$P_2(t) = -\tfrac{1}{2} + \tfrac{3}{2}t^2 \qquad\qquad P_3(t) = -\tfrac{3}{2}t + \tfrac{5}{2}t^3 \qquad\qquad (3.12)$$

$$P_4(t) = \tfrac{3}{8} - \tfrac{15}{4}t^2 + \tfrac{35}{8}t^4 \qquad\qquad P_5(t) = \tfrac{15}{8}t - \tfrac{35}{4}t^3 + \tfrac{63}{8}t^5.$$

Legendre polynomials have been extensively studied and many facts about them are known. Alternative formulas for their calculation are also available. For example, it is known that

$$P_n(t) = \frac{1}{2^n}\frac{1}{n!}\frac{d^n}{dt^n}\left(t^2 - 1\right)^n.$$

Exercises

For the initial value problems below,

(a)   Find the recursive formula for the coefficients of the power series solution.

(b)   Use the recursive formula to obtain the sixth-degree Taylor polynomial approximation $x_6(t)$.

(c)   Find a formula for the solution using the method in Chapter 5, Sec. 5.3.

1. $x'' - 2x' + 2x = 0, x(0) = 0, x'(0) = 1$

2. $x'' + 2x' + 5x = 0, x(0) = 3, x'(0) = -5$

3. $x'' - 3x' + 2x = 0, x(0) = 1, x'(0) = 0$

4. $x'' - x = 0, x(0) = -1, x'(0) = 2$

For the initial value problems below,

(a)   Find the recursive formula for the coefficients of the power series solution.

(b)   Use the recursive formula to obtain the fourth-degree Taylor polynomial approximation $x_4(t)$.

5. $x'' - 2tx + 6x = 0, x(0) = 0, x'(0) = 1$

6. $x'' + 2tx = 0, \quad x(0) = 1, x'(0) = -1$

7. $(1 + t)x'' + tx = e^t, x(0) = 0, x'(0) = 0$

8. $x'' + (1 + t)x' - x = \dfrac{1}{1 - t}, x(0) = 0, x'(0) = 1$

Find the recursive formula for coefficients of the power series solutions of the equations below. Use the formula to write the

general solution as a linear combination of two power series solutions. (In your answer, give the first three nonzero terms in each power series.)

9. $x'' - x' + 2tx = 0$ **10.** $x'' - (1 + t)x = 0$

11. $\left(1 + t^2\right)x'' + x = 0$ **12.** $\left(1 - t^2\right)x'' + x = 0$

13. Suppose α is an even integer in the Legendre equation (3.10). Use the recursive formula (3.11) to show that solutions satisfying $x'(0) = 0$ are polynomials of degree α. Use the recursive formula to calculate the Legendre polynomials P_n in (3.12) for $n = 0, 2$, and 4.

The second-order equation $\left(1 - t^2\right)x'' - tx' + \alpha^2 x = 0$, in which α is a constant, is called the Tchebycheff equation.

14. Show that all power series solutions at $t_0 = 0$ have a positive radius of convergence.

15. Find the recursive formula for the coefficients of power series solutions at $t_0 = 0$.

16. If α is an (nonnegative) integer, show that there are polynomial solutions of degree $n = \alpha$. These polynomial solutions are called Tchebycheff polynomials.

17. If the polynomial solutions in Exercise 16 are normalized by requiring $x(1) = 1$, then they are called Tchebycheff polynomials of the first kind and denoted by $T_n(t)$. Find the first five Tchebycheff polynomials $T_n(t)$ of degrees $n = 0, 1, 2, 3$, and 4.

7.4   Chapter Summary and Exercises

In this chapter we learned that the solution approximation methods used for first-order equations in Sec. 3.4 of Chapter 3 are applicable to systems of first-order equations and to higher-order equations. We constructed polynomial approximations to solutions of initial value problems (using Taylor polynomials in the variable t centered at the initial point $t = t_0$). In general, these approximation are accurate only for t near t_0. The accuracy improves, however, as the order of the Taylor polynomial increases. The same is true for the Picard iteration procedure for calculating approximation formulas. The Picard iteration procedure for systems and higher order equations is a straightforward extension of the procedure for first order equations in Chapter 3 (Section 3.4.1).

We also studied another technique—the perturbation method—for calculating formulas for approximations by means of Taylor polynomials. The perturbation method is applicable to systems of equations or higher-order equations in which a small number ε appears as a coefficient. The procedure involves approximating the solution by a Taylor polynomial in ε. When the perturbation method is applicable, it often leads to approximations that are accurate on relatively large intervals.

Exercises

For the following initial value problems, find the fourth-order Taylor polynomial approximation and the first three Picard iterate approximations.

1. $x'' + (1 + \sin 2t)\, x = 0$, $x(0) = 1$, $x'(0) = 0$

2. $x'' + e^{-t} x' + x = 0$, $x(0) = 1$, $x'(0) = 0$

For the following initial value problems, find the third-order Taylor polynomial approximation and the first two Picard iterate approximations.

3. $\begin{cases} x' = x\,(1 + \cos t - x) - xy \\ y' = -y + xy \\ x(0) = 1,\ y(0) = 1 \end{cases}$

4. $\begin{cases} x' = x - y - x\left(x^2 + y^2\right) \\ y' = 2x - y\left(x^2 + y^2\right) \\ x(0) = 1,\ y(0) = 0 \end{cases}$

Find the first-order perturbation expansion approximation of the solutions of the following initial value problems.

5. $x'' + (1 + \varepsilon \sin 2t)\, x = 0$, $x(0) = 1$, $x'(0) = 0$

6. $x'' + \varepsilon e^{-t} x' + x = 0$, $x(0) = 1$, $x'(0) = 0$

7. $\begin{cases} x' = \varepsilon x\,(1 + \cos t - x) - \varepsilon xy \\ y' = -y + \varepsilon xy \\ x(0) = 1,\ y(0) = 1 \end{cases}$

8. $\begin{cases} x' = -5x + 9y \\ y' = x - 5y + \varepsilon y^2 \\ x(0) = 3,\ y(0) = 1 \end{cases}$

9. $\begin{cases} x' = x - y - \varepsilon y^2 \\ y' = 2x \\ x(0) = 1,\ y(0) = 0 \end{cases}$

Find the recursive formulas for coefficients of the power series solutions $x = \sum_{i=0}^{\infty} k_i t^i$, $y = \sum_{i=0}^{\infty} m_i t^i$ of the systems below. Use the formula to write the general solution as a linear combination of two power series solutions. (In your answer give at least the first two nonzero terms in each power series.)

10. $\begin{cases} x' = -tx + y \\ y' = x - ty \end{cases}$

11. $\begin{cases} x' = x + (1 + t)y \\ y' = x - 2y \end{cases}$

The second-order equation $x'' - 2tx' + \alpha x = 0$, in which α is a constant, is called the Hermite equation.

12. Show that all solutions have radius of convergence $R = \infty$.

13. Find the recursive formula for the coefficients of power series solutions at $t_0 = 0$.

14. Suppose α is an even integer, that is, suppose $\alpha = 2n$ for some $n = 0, 1, 2, 3, \ldots$. Show that there are polynomial solutions of degree n. These solutions are called Hermite polynomials and are denoted by $H_n(t)$.

15. Find the Hermite polynomials of degrees $n = 0, 1, 2, 3$, and 4.

16. Consider Hill's equation $x'' + (1 + \cos t)\, x = 0$ subject to the initial conditions $x(0) = 1$, $x'(0) = 0$.

    (a) Find the roots of the fourth-degree Taylor polynomial $x_4(t)$ of Hill's equation. Find the smallest positive root of $x_4(t)$.

    (b) Use a computer to obtain find the smallest positive root of the solution $x(t)$ of the initial value problem. Approximately what percent error does the root of $x_4(t)$ make as an approximation to the root of $x(t)$?

7.5 APPLICATIONS

7.5.1 Pesticide Example

In Chapter 4 we considered a model describing the application of a chemical pesticide to a stand of trees (see Sec. 4.1). In this model the amounts of pesticide in the trees and the soil, denoted by x and y, respectively, are governed by a linear homogeneous system of equations and the initial value problem

$$x' = -r_1 x + r_2 y$$
$$y' = r_1 x - (r_2 + r_3) y \qquad (5.1)$$
$$x(0) = d, \quad y(0) = 0.$$

The initial pesticide dose on the trees is $d > 0$ (pesticide units). There is no pesticide in the soil initially. These equations result from the inflow and outflow rates described in the compartmental diagram appearing in Fig. 7.10.

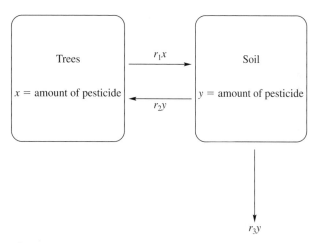

Figure 7.10 The compartmental diagram for the model system (5.1).

We studied a specific example in Chapter 4 (Sec. 4.3) for which $r_1 = r_2 = 2$ and $r_3 = 3$ (per year). Suppose in this example, however, that the ability of the soil to break down the pesticide decreases exponentially over time, so that $r_3 = 3e^{-t}$ (per year). Then, assuming an initial application of $d = 1$ unit of pesticide, the model equations become

$$x' = -2x + 2y$$
$$y' = 2x - \left(2 + 3e^{-t}\right) y \qquad (5.2)$$
$$x(0) = 1, \quad y(0) = 0.$$

Our goal in this application is to obtain approximation formulas for the pesticide amounts $x(t)$ and $y(t)$ that are valid at least for short periods of time after the initial application at time $t = 0$. To do this we will use the analytic approximation methods studied in this chapter. [Note that the coefficients of the linear system (5.2) are not all constants and as a result the solution method in Chapter 5 is not applicable.]

We begin by calculating Taylor polynomial approximations. Letting $t = 0$ in the differential equations (5.2) and using the initial conditions, we obtain

$$x'(0) = -2, \quad y'(0) = 2.$$

Two differentiations of the differential equations (5.2) yield

$$x'' = -2x' + 2y'$$
$$y'' = 2x' - \left(2 + 3e^{-t}\right) y' + 3e^{-t} y$$

and

$$x''' = -2x'' + 2y''$$
$$y''' = 2x'' - \left(2 + 3e^{-t}\right) y'' + 6e^{-t} y' - 3e^{-t} y.$$

Letting $t = 0$, we find

$$x''(0) = 8, \quad y''(0) = -14$$

$$x'''(0) = -44, \quad y'''(0) = 98.$$

These numbers yield the third-degree Taylor polynomial approximations

$$x_3(t) = 1 - 2t + 4t^2 - \frac{22}{3}t^3 \tag{5.3}$$

$$y_3(t) = 2t - 7t^2 + \frac{49}{3}t^3$$

to the solution pair of the initial value problem. These formulas approximate the amounts of pesticide in the trees and the soil for t small. We can use them to answer questions about these pesticide amounts during short time intervals after the initial application of pesticide at $t = 0$.

For example, we can estimate how long it takes for the amount of pesticide in the trees to decrease by 25%. Using a computer or calculator, we find the solution of the polynomial equation

$$1 - 2t + 4t^2 - \frac{22}{3}t^3 = \frac{3}{4}$$

to be approximately $t = 0.16$ year (or approximately 59 days).

We can obtain another approximation using the Picard iteration procedure. Starting with the initial conditions

$$x_0(t) = 1, \quad y_0(t) = 0$$

as the first approximation, we obtain an initial value problem for the next Picard iterate by substituting x_0 and y_0 for x and y in the right-hand sides of the two differential equations in (5.2). The solution pair of the resulting initial value problem

$$x' = -2$$
$$y' = 2$$
$$x(0) = 1, \quad y(0) = 0$$

is

$$x_1(t) = 1 - 2t, \quad y_1(t) = 2t.$$

Placing these Picard iterates into the right-hand sides of the differential equations in (5.2), we obtain the initial value problem

$$x' = -2(1 - 2t) + 2(2t) = -2 + 8t$$
$$y' = 2(1 - 2t) - \left(2 + 3e^{-t}\right)(2t) = 2 - 8t - 6te^{-t}$$
$$x(0) = 1, \quad y(0) = 0$$

whose solution gives the second Picard iterates

$$x_2(t) = 1 - 2t + 4t^2$$
$$y_2(t) = 6e^{-t} - 6 + 2t - 4t^2 + 6te^{-t}.$$

Finally, placing x_2 and y_2 into the right-hand sides of the differential equations in (5.2) and solving the resulting initial value problem, we obtain the third Picard iterates

$$x_3(t) = 25 - 14t + 4t^2 - \frac{16}{3}t^3 - 24e^{-t} - 12te^{-t}$$
$$y_3(t) = -\frac{3}{2} + 14t - 4t^2 + \frac{16}{3}t^3 - 12e^{-t} - 6te^{-t} - 12t^2e^{-t} + \frac{27}{2}e^{-2t} + 9te^{-2t}.$$

Using the third Picard iterate $x_3(t)$, we can approximate the time it takes the amount of pesticide in the trees to decrease 25% from its initial application amount. Using a computer or calculator, we find that the solution of the equation

$$25 - 14t + 4t^2 - \frac{16}{3}t^3 - 24e^{-t} - 12te^{-t} = \frac{3}{4}$$

is approximately $t = 0.1624$ year (or 59.28 days). This approximation compares favorably with that we obtained from the third-order Taylor polynomial, which provides us some confidence in their accuracy. (See Exercise 1.) However, neither of these Taylor nor Picard iterate approximations provide an accurate, long-term prediction for the pesticide amounts in the soil or the trees. See Exercise 2.

7.5.2 Objects in Motion

Consider a projectile shot upward from the surface of the earth and assume it is subject only to gravitational forces (e.g., air resistance is ignored). Let $x(t)$ denote the height above the surface of the earth. Then Newton's inverse square law of gravity yields, after some changes of variables (or "scaling"), the initial value problem[3]

$$x'' = -\frac{1}{(1 + \varepsilon x)^2} \tag{5.4}$$
$$x(0) = 0, \quad x'(0) = 1,$$

where $\varepsilon = v_0^2/gr$ is a (dimensionless) parameter. Here g is the acceleration due to gravity (approximately 9.8 meters/s^2), r is the radius of the earth (approximately 6.350×10^6 meters), and v_0 is the initial velocity of the projectile (meters/s). It turns out that the units for distance x and time t have been scaled in this model equation. Specifically, one unit of time in the model equals $v_0/g = v_0/9.8$ seconds and one unit of distance equals $v_0^2/g = v_0^2/9.8$ meters.

The differential equation in (5.4) is nonlinear and second-order. As a result, we have no method for calculating a solution formula. In this application we will approximate the solution using the perturbation method and use the approximation to estimate the time it takes the projectile to reach its maximum altitude.

---

[3]C. C. Lin and L. A. Segel, *Mathematics Applied to Deterministic Problems in the Natural Sciences*, Macmillan Publishing Co., Inc., New York, 1974.

For initial velocities v_0 small compared to $\sqrt{gr} \approx 7889$ (m/s) (i.e., about 1765 miles per hour), the parameter ε is small. We will calculate the second-order perturbation expansion approximation

$$x_1(t) = k_0(t) + k_1(t)\varepsilon + k_2(t)\varepsilon^2 \qquad (5.5)$$

to the initial value problem (5.4).

Substituting $x_1(t)$ in (5.5) for x in the left-hand side of the differential equation in (5.4), we obtain

$$k_0''(t) + k_1''(t)\varepsilon + k_2''(t)\varepsilon^2.$$

Substituting $x_1(t)$ into the right-hand side of the equation, we get

$$-\frac{1}{\left(1 + k_0(t)\varepsilon + k_1(t)\varepsilon^2 + \cdots\right)^2}, \qquad (5.6)$$

where the dots denote terms containing powers of ε greater than 2. Our goal is to equate the coefficients of like powers of ε (up to power 2) from these two expressions. To do this we need to determine the second-degree Taylor polynomial in ε of (5.6). It is helpful to recall the binomial formula

$$(1 + z)^n = 1 + nz + \frac{n(n-1)}{2}z^2 + \frac{n(n-1)(n-2)}{6}z^3 + \cdots.$$

With $n = -2$ and

$$z = k_0\varepsilon + k_1\varepsilon^2 + k_2\varepsilon^3$$

in (5.6) we obtain from the binomial formula the expression

$$-\frac{1}{\left(1 + k_0(t)\varepsilon + k_1(t)\varepsilon^2 + \cdots\right)^2} = -\Big[1 - 2\left(k_0\varepsilon + k_1\varepsilon^2 + k_2\varepsilon^3\right)$$

$$+ 3\left(k_0\varepsilon + k_1\varepsilon^2 + k_2\varepsilon^3\right)^2 + \cdots\Big]$$

$$= -1 + 2k_0\varepsilon + \left(2k_1 - 3k_0^2\right)\varepsilon^2 + \cdots,$$

where the dots indicate terms containing powers of ε higher than 2.

In summary, a substitution of (5.5) into the initial value problem (5.4) yields

$$k_0'' + k_1''\varepsilon + k_2''\varepsilon^2 = -1 + 2k_0\varepsilon + \left(2k_1 - 3k_0^2\right)\varepsilon^2 + \cdots$$

$$k_0(0) + k_1(0)\varepsilon + k_2(0)\varepsilon^2 = 0 + 0 \cdot \varepsilon + 0 \cdot \varepsilon^2$$

$$k_0'(0) + k_1'(0)\varepsilon + k_2'(0)\varepsilon^2 = 1 + 0 \cdot \varepsilon + 0 \cdot \varepsilon^2.$$

Equating coefficients of like powers of ε from both sides of these equations, we obtain three sets of equations, one from each power of ε ranging from power 0 to 2.

From the lowest-order terms in ε we get the initial value problem

$$k_0'' = -1$$

$$k_0(0) = 0, \quad k_0'(0) = 1$$

for $k_0(t)$, the solution of which is

$$k_0(t) = t - \frac{1}{2}t^2. \qquad (5.7)$$

From the first-order terms in ε we get the initial value problem

$$k_1'' = 2k_0$$
$$k_1(0) = 0, \quad k_1'(0) = 0$$

for $k_1(t)$. When the formula (5.7) for k_0 is used, this problem reduces to

$$k_1'' = 2t - t^2$$
$$k_1(0) = 0, \quad k_1'(0) = 0$$

whose solution is

$$k_1(t) = \frac{1}{3}t^3 - \frac{1}{12}t^4.$$

Finally, from the second-order terms in ε we get the initial value problem

$$k_2'' = 2k_1 - 3k_0^2$$
$$k_2(0) = 0, \quad k_2'(0) = 0,$$

which, given the formulas for k_0 and k_1 above, reduces to

$$k_2'' = -3t^2 + \frac{11}{3}t^3 - \frac{11}{12}t^4$$
$$k_2(0) = 0, \quad k_2'(0) = 0.$$

The solution of this problem is

$$k_2(t) = -\frac{1}{4}t^4 + \frac{11}{60}t^5 - \frac{11}{360}t^6.$$

With these calculations complete, we can substitute our answers for k_0, k_1, and k_2 into (5.5) and obtain the second-order perturbation approximation

$$x_2(t) = t - \frac{1}{2}t^2 + \left(\frac{1}{3}t^3 - \frac{1}{12}t^4 \right) \varepsilon + \left(-\frac{1}{4}t^4 + \frac{11}{60}t^5 - \frac{11}{360}t^6 \right) \varepsilon^2 \qquad (5.8)$$

to the solution of the initial value problem (5.4).

To calculate the time t_m at which the projectile reaches its highest point, we need to calculate the time at which $x'(t_m) = 0$. We can approximate this time by instead solving $x_2'(t_m) = 0$ for t_m, that is, solving equation

$$1 - t_m + \left(t_m^2 - \frac{1}{3}t_m^3 \right) \varepsilon + \left(-t_m^3 + \frac{11}{12}t_m^4 - \frac{11}{60}t_m^5 \right) \varepsilon^2 = 0 \qquad (5.9)$$

for t_m. The left-hand side of this equation is a fifth-degree polynomial for t_m, and therefore we have no general formula available to us for finding this root. By the fundamental theorem of algebra this polynomial can have as many as five real roots. We are only interested in the smallest positive root, which will estimate the time when the projectile first reaches its maximum height. Note that when $\varepsilon = 0$, the root of the equation is $t_m = 1$. Since ε is small, the root we seek is presumably near 1.

If we are interested in a specific value of $\varepsilon = v_0^2/gr$ (i.e., in a specific initial velocity v_0 for the projectile), then we can use a calculator or computer to solve equation (5.9) for t_m. For example, suppose $v_0 = 250$ m/s. Then

$$\varepsilon = \frac{(250)^2}{(9.8)(6.350 \times 10^6)} = 1.004 \times 10^{-3}.$$

We can use a calculator or computer to solve the equation (5.9) for t_m with this value of ε. The result is approximately $t_m \approx 1.00067$ model time units. Since each model time unit is

$$\frac{v_0}{9.8} = \frac{250}{9.8}$$

seconds, we find that the projectile will reach its maximum height in approximately 25.5 seconds.

We can also use the formula (5.8) to estimate the maximum height attained by the projectile. A calculation yields a maximum height of $x\,(t_m) \approx 0.500251$ model distance units, each of which equals

$$\frac{v_0^2}{9.8} = \frac{(250)^2}{9.8}$$

meters. Thus, the maximum height of the projectile is estimated to be about 3190 meters (10,466 feet, or nearly two miles).

Estimates for these quantities can also be obtained from Taylor and Picard iterate approximations to the solution of (5.4). See Exercises 8 and 9.

Exercises

1. Use a computer program to obtain a numerical approximation to the solution of the initial value problem (5.2) on the interval $0 \le t \le 0.25$. Use your results to approximate the time at which x equals 3/4. Compare your answer with the estimates obtained from the Taylor polynomial and Picard iterates calculated in the text.

2. **(a)** Use a computer program to graph the solutions x and y of the initial value problem (5.2). What do your graphs predict x and y do as $t \to +\infty$? What implications does this prediction have with respect to the long-term presence of the chemical pesticide in the trees and soil? How does the prediction compare with that when the soil's ability to degrade the chemical does not decrease [i.e., with that obtained from the solutions of the initial value problem (5.1) with $d = 1$]?

   (b) How do the behaviors of x and y as $t \to +\infty$ compare with those of the third order Taylor polynomial and the third Picard iteration approximations calculated in Exercise 3?

3. Consider the initial value problem

$$x' = -2x + 2y$$
$$y' = 2x - \left(2 + \varepsilon e^{-t}\right) y \qquad (5.10)$$
$$x(0) = 1, \quad y(0) = 0,$$

   where ε is a constant.

   (a) Calculate the first-order perturbation approximation $x_1(t)$ and $y_1(t)$ to the solution.

(b) If $\varepsilon = 3$, we obtain the pesticide model (5.2). Use your answer in (a) to approximate the time it takes the amount of pesticide in the trees to decrease 25%. Compare your answers with those obtained in the text from the Taylor and Picard iterate approximations.

(c) What do $x_1(t)$ and $y_1(t)$ do as $t \to +\infty$? How do you interpret the result?

Let $r_1 = r_2 = 2$ and $r_3 = 3$ (per year) in the pesticide model (5.1). Suppose the pesticide is added to the trees (but not the soil) at a rate that decreases exponential over time, specifically at the rate e^{-t} (per year). Assuming that no pesticide is initially present in either the trees or the soil, we have the initial value problem

$$x' = -2x + 2y + e^{-t}$$
$$y' = 2x - 5y \qquad (5.11)$$
$$x(0) = y(0) = 0.$$

4. **(a)** Find the third-order Taylor polynomial approximation to the solution of the initial value problem (5.11).

   (b) Use your answer in (a) to estimate the time at which the amount of pesticide in the soil first equals 0.01 (units).

   (c) Use your answer to estimate the time at which the amount of pesticide in the soil reaches its maximum.

5. **(a)** Find the third Picard iterate approximation to the solution of the initial value problem (5.11).

(b) Use your answer in (a) to estimate the time at which the amount of pesticide in the soil first equals 0.05 (units).

(c) Use your answer to estimate the time at which the amount of pesticide in the soil peaks.

6. (a) Use a computer program to estimate the solution of the initial value problem (5.11) on the interval $0 \leq t \leq 2$ and determine approximately the time at which the amount of pesticide in the soil first equals 0.01 (units).

(b) Use your results to estimate the time at which the amount of pesticide in the soil peaks.

(c) Compare your answers in (a) and (b) with those you obtained in Exercises 4 and 5.

7. Use a computer to plot the solutions x and y of the initial value problem (5.11) over the interval $0 \leq t \leq 20$. What do the plots suggest happens to x and y as $t \to +\infty$? How does this compare to the Taylor polynomial and Picard iterate approximations obtained in Exercises 4 and 5?

8. Consider the initial value problem (5.4) for the motion of a projectile under the influence of gravity. Let t_m denote the time at which the projectile reaches its maximum height.

(a) Starting with the initial Picard iterate $x_0(t) = 1$, calculate the next Picard iterate $x_1(t)$. (Your answer will depend on ε.)

(b) Use your answer in (a) to estimate the time t_m.

(c) Use your answer in (a) and (b) to estimate the maximum height reached by the projectile.

(d) Estimate the maximum height (in meters) and the time it is attained (in seconds) when the initial velocity is 250 m/s.

9. Consider the initial value problem (5.4) for the motion of a projectile under the influence of gravity. Let t_m denote the time at which the projectile reaches its maximum height.

(a) Calculate the third-order Taylor polynomial $x_3(t)$ approximation to the solution. (Your answer will depend on ε.)

(b) Use your answer in (a) to estimate the time t_m.

(c) Use your answer in (a) and (b) to estimate the maximum height reached by the projectile. (Your answer will depend on ε.)

(d) Estimate the maximum height (in meters) and the time it is attained (in seconds) when the initial velocity is 250 m/s.

C H A P T E R

8

Nonlinear Systems

In Chapters 5 and 6 we studied systems of linear differential equations. In this chapter we turn our attention to systems of nonlinear equations. Since any higher-order equation is equivalent to a first-order system, our study includes nonlinear higher-order equations. Unlike the case of single nonlinear equations, there are virtually no methods available for calculating solution formulas for nonlinear systems. Therefore, we must use other methods to analyze nonlinear systems. In this chapter we focus on methods for analyzing autonomous systems and their phase portraits.

8.1 Introduction

Consider the autonomous system

$$x' = f(x, y) \tag{1.1}$$
$$y' = g(x, y)$$

of two differential equations. As with linear systems consisting of two equations, the phase portrait of the nonlinear systems (1.1) consists of the orbits drawn in the (x, y)-plane. For this reason (1.1) is called a *planar autonomous system.* Our basic goal in this chapter is to develop methods for analyzing and drawing the phase plane portraits of planar autonomous systems.

We learned in Chapter 3 how to construct phase line portraits for single first-order autonomous equations. We found that equilibria play a key role and that all nonequilibrium solutions are monotonic. Equilibria also play a key role in phase plane portraits of systems. However, drawing a phase plane portrait for a system of equations is generally more difficult than drawing a phase line portrait for a single equation. Solutions of autonomous systems are not necessarily monotonic, and the possibility of solution oscillations (and even periodic solutions) introduces a new features to phase portraits. We have already seen this in the case of spirals and centers for linear, planar autonomous systems (Chapter 5). Despite these added complications, mathematicians have developed a nearly complete theory of phase portraits for planar autonomous systems. A full exposition of this theory is, however, beyond the scope of this introductory book. Nonetheless, we will study the basic ingredients of the theory and learn how to use them to obtain accurate sketches of phase plane portraits. Of course, numerical approximations and computer graphics can also play an important and helpful role in this endeavor.

Systems of three or more autonomous equations also arise in applications. Solutions and orbits of three- (or higher-) dimensional systems can exhibit extremely complicated dynamic behavior that is not possible for planar autonomous systems. So-called strange attractors and chaos are examples. For this reason (and because of the difficulty of drawing pictures in more than three dimensions), it is considerably more difficult to study phase portraits for higher-dimensional systems. We will take only a brief look at higher dimensional systems in Sec. 8.7.

Here are some examples of nonlinear planar autonomous systems that arise in applications. The planar autonomous system

$$x' = x - xy \tag{1.2}$$
$$y' = 2y - 2xy$$

is a model of competition between two populations or groups. This system is nonlinear because of the term xy. Planar autonomous systems often have coefficients (or parameters). For example, the system

$$x' = r_1 x - axy \tag{1.3}$$
$$y' = r_2 y - bxy$$

is a more general competition system that includes (1.2) as a special case. The system

$$x' = b - dx - c\frac{y}{x+y}x \tag{1.4}$$
$$y' = c\frac{y}{x+y}x - (d+a)y$$

is another example of a planar autonomous system, one that arises in an application to AIDS epidemics. This system is nonlinear because of the term $xy/(x+y)$. A third example of a nonlinear planar autonomous system is

$$x' = r\left(1 - \frac{x}{K}\right)x - m\frac{xy}{a+x} \tag{1.5}$$
$$y' = -dy + cm\frac{xy}{a+x},$$

a system that arises in theoretical ecology as a model of the interaction between a predator y and its prey x.

Planar autonomous systems also arise from autonomous second-order equations. For example, the system

$$x' = y \tag{1.6}$$
$$y' = -x - \alpha(x^2 - 1)y$$

is equivalent to the van der Pol equation

$$x'' + \alpha\left(x^2 - 1\right)x' + x = 0, \tag{1.7}$$

which arises in applications involving electric circuits. This equation is nonlinear because of the term $x^2 x'$.

An example of a nonlinear system of more than two equations is the Lorenz system

$$
\begin{aligned}
x' &= \sigma\,(y - x) \\
y' &= \rho x - y - xz \\
z' &= -\beta z + xy,
\end{aligned}
\tag{1.8}
$$

which arises in meteorological studies. It is nonlinear because of the terms xz and xy.

8.2 Phase Plane Portraits and Equilibria

In Chapter 3 we learned to draw phase line portraits for a single autonomous differential equation $x' = f(x)$. The equilibria [i.e., the roots of $f(x)$] and the signs of $f(x)$ between equilibria are enough to determine the entire phase line portrait. Equilibria also play an important role in the phase portraits of planar autonomous systems (1.1). Unfortunately, it is more difficult to determine the entire (global) phase portrait for systems—even systems of only two equations—than it is for a single equation.

We consider planar autonomous systems

$$
\begin{aligned}
x' &= f(x, y) \\
y' &= g(x, y)
\end{aligned}
\tag{2.1}
$$

for which initial value problems

$$
x(t_0) = x_0, \quad y(t_0) = y_0
\tag{2.2}
$$

have unique solutions. To guarantee this we require that f and g and their first-order derivatives with respect to x and y to be continuous on some domain of points D in the plane (Theorem 1 in Chapter 4).[1] In some cases D may not be the entire plane. For example, in the AIDS system (1.4) f and g both are undefined when $y = -x$. In this system D can be one of half planes determined by this line.

We begin with some remarks about the basic features of phase plane portraits. Recall that the *orbit* associated with a solution pair $x = x(t)$, $y = y(t)$ is the set of points $\{(x(t), y(t))\}$ in the (x, y)-plane (Definition 2 in Chapter 4). In general, an orbit is a curve in the plane. An orbit has a direction (or orientation) determined by the motion along the curve as t increases. Since initial value problems have unique solutions, an orbit passes through each point (x_0, y_0) in the domain D. To see this, simply use x_0 and y_0 as the initial condition in (2.2) and apply the fundamental existence theorem (Theorem 1 in Chapter 4); the orbit of the unique solution of this initial value problem passes through the point (x_0, y_0) at time $t = t_0$. Different initial times t_0 yield different initial value problems whose solutions produce the same orbit, except that they arrive at the point (x_0, y_0) at different times. See Exercise 1. This means there are infinitely many solutions associated with an orbit. (Recall that this is true for single nonlinear equations and their phase line orbits as well.) Thus, orbits *"fill up" the domain D* in the sense that there is an orbit through every point in D. Furthermore, *different orbits cannot cross each other* (Exercise 2).

---

[1] A domain is an "open" set. This means each point in the domain can be surrounded by a small circular disk all of which lies in the domain. The inside of a circle or a rectangle are examples of domains.

We now turn to the problem of determining the geometry of nonlinear phase portraits. Equilibria are of fundamental importance in solving this problem and we begin with their study. Recall that equilibria are constant solutions:

$$x(t) = x_e = \text{constant}$$
$$y(t) = y_e = \text{constant.}$$

The orbit associated with an equilibrium is just the single point (x_e, y_e) in the (x, y)-phase plane. Since the derivative of a constant is zero, it follows that the equilibrium (x_e, y_e) points of (2.1) satisfy the two equations

$$0 = f(x_e, y_e)$$
$$0 = g(x_e, y_e).$$

Thus, the equilibria of the planar autonomous system (2.1) are the roots of the equilibrium equations

$$f(x, y) = 0$$
$$g(x, y) = 0.$$

For nonlinear systems, these algebraic equations are nonlinear. As a result, their solution is likely to be a difficult algebraic problem. In some applications, we can solve the equilibrium equations explicitly. In other applications, we can use computers to approximate solutions.

EXAMPLE 1

The equilibrium equations for the competition system (1.2) are

$$x - xy = 0$$
$$2y - 2xy = 0$$

or

$$(1 - y)x = 0$$
$$2(1 - x)y = 0.$$

There are two ways to solve the first equation: either $x = 0$ or $y = 1$. In the first case, the second equation implies $y = 0$. This gives the equilibrium point $(x_e, y_e) = (0, 0)$. In the second case $y = 1$, the second equation implies $x = 1$. This gives the equilibrium point $(x_e, y_e) = (1, 1)$.

The equilibrium equations for the system (1.3) are

$$r_1 x - axy = 0$$
$$r_2 y - bxy = 0.$$

Similar reasoning produces the two equilibrium points

$$(x_e, y_e) = (0, 0) \quad \text{and} \quad \left(\frac{r_2}{b}, \frac{r_1}{a}\right). \qquad \blacksquare$$

Notice in Example 1 that the system (1.2) has more than one equilibrium (as does (1.3). This is not an uncommon feature of nonlinear systems.

EXAMPLE 2

The equilibrium equations for the planar autonomous system

$$x' = \sin x - y \qquad (2.3)$$
$$y' = x - 2y$$

are

$$\sin x - y = 0$$
$$x - 2y = 0.$$

Solving the second equation for

$$y = \frac{1}{2}x \qquad (2.4)$$

and substituting this result into the first equation, we obtain the equation

$$\sin x = \frac{1}{2}x \qquad (2.5)$$

for x. Any solution x of this equation, together with (2.4), yields an equilibrium of the system (2.3).

One solution of equation (2.5) is rather easy to observe: $x = 0$. This yields the equilibrium point $(x_e, y_e) = (0, 0)$. Are there any other equilibria? That is, are there any other solutions of the equation (2.5)?

It is not possible to solve equation (2.5) algebraically for x. However, from the graph in Fig. 8.1 we see there are two other solutions, one positive and one negative. Using a computer or calculator to approximate these solutions, we obtain $x \approx 1.8955$ and -1.8955. Thus, $(x_e, y_e) \approx (1.8955, 0.9478)$ and $(-1.895, -0.9478)$ are the remaining two equilibria of the system (2.3). ∎

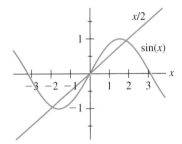

Figure 8.1 Plots of $\sin x$ and the straight line $x/2$ reveal three intersection points.

We can find the equilibria of a higher-order differential equation by finding the equilibrium points of an equivalent first-order system. Or, we can proceed more directly and derive an equilibrium equation by setting derivatives of all orders equal to zero in the differential equation.

For example, $x' = 0$ and $x'' = 0$ in the van der Pol equation (1.7) yields the equilibrium equation $x = 0$. Thus, the only equilibrium of this second-order equation is $x_e = 0$. In the phase plane, for its equivalent system (1.6), this equilibrium corresponds to the point $(x_e, y_e) = (0, 0)$.

Systems of $n \geq 3$ equations can also have equilibrium solutions. The equilibrium equations are obtained from the differential system by setting all derivatives equal to zero. The result is a system of n algebraic equations whose solutions are the equilibria of the differential system.

EXAMPLE 3

The equilibrium equations for the Lorenz system

$$x' = \sigma (y - x)$$
$$y' = \rho x - y - xz$$
$$z' = -\beta z + xy$$

are

$$\sigma (y - x) = 0$$
$$\rho x - y - xz = 0$$
$$-\beta z + xy = 0.$$

To find the equilibria of the Lorenz system, we must solve these three algebraic equations for x, y, and z. One way to do this is as follows. The first equilibrium equation implies $x = y$. Letting $x = y$ in remaining equations yields the two equations

$$\rho y - y - yz = 0$$
$$-\beta z + y^2 = 0$$

or

$$(\rho - 1 - z) y = 0$$
$$-\beta z + y^2 = 0$$

for y and z. The first of these equations leads to two alternatives: either $y = 0$ or $z = \rho - 1$.

Consider the first alternative $y = 0$. The second equation implies $z = 0$, and we obtain the equilibrium point $(x_e, y_e, z_e) = (0, 0, 0)$.

Now consider the second alternative $z = \rho - 1$. In this case the second equation implies $y^2 = \beta(\rho - 1)$. If $\rho \geq 1$ this equation has solutions $y = \pm\sqrt{\beta(\rho - 1)}$. Consequently, this alternative yields the equilibrium points

$$(x_e, y_e, z_e) = (\sqrt{\beta(\rho - 1)}, \pm\sqrt{\beta(\rho - 1)}, \rho - 1)$$

when $\rho \geq 1$. Notice that these two points coincide with each other (and equal the origin $(0, 0, 0)$) when $\rho = 1$.

To summarize, the Lorenz system (1.8) has one equilibrium if $\rho \leq 1$ and three equilibria if $\rho > 1$. ∎

Once we have found the equilibria our next step in the construction of a phase portrait is to determine the properties of nonequilibrium orbits. In general, this is a difficult task. The monotonicity property of solutions of single autonomous equations, which is so fundamental in the construction of phase line portraits for single autonomous equations, has no general counterpart in the two-dimensional case of planar systems. One of the techniques we used for analyzing single equations in Chapter 3, however, does carry over to planar systems (and also to higher-order systems). In the next section we will use the linearization method to determine the behavior of orbits, at least when they are near equilibrium points.

Exercises

1. (a) If $x = x(t)$, $y = y(t)$ is a solution pair of the planar autonomous system (1.1) and if τ is any real number, show the translations $x = x(t + \tau)$, $y = y(t + \tau)$ also form a solution pair.

(b) Show a solution pair and any of its translates give the same orbit in the phase plane.

(c) Prove the following statement: If two solution pairs $(x_1(t), y_1(t))$ and $(x_2(t), y_2(t))$ of a planar autonomous system have the same orbit, then each is a translate of the other. [*Hint*: Pick any point (x_0, y_0) on the common orbit. Then for some t_1 and t_2 we have $(x_1(t_1), y_1(t_1)) = (x_0, y_0)$ and $(x_2(t_2), y_2(t_2)) = (x_0, y_0)$. Take $\tau = t_1 - t_2$.]

2. Suppose two orbits have a point (x_0, y_0) in common. Prove that the orbits are identical. This fact shows different orbits cannot cross each other in the phase plane. [*Hint*: For some t_1 and t_2 we have $(x_1(t_1), y_1(t_1)) = (x_0, y_0)$ and $(x_2(t_2), y_2(t_2)) = (x_0, y_0)$. Take $\tau = t_1 - t_2$.]

Find all equilibria of the following systems and higher-order equations.

3. $\begin{cases} x' = x - y^2 \\ y' = x - y \end{cases}$

4. $\begin{cases} x' = x\,(2x - y) \\ y' = x^2 + y - 8 \end{cases}$

5. $\begin{cases} x' = x^2 + y^2 - 4 \\ y' = (x - 3)^2 + y^2 - 4 \end{cases}$

6. $\begin{cases} x' = 18 - x^2 - y^2 \\ y' = x - y \end{cases}$

7. $\begin{cases} x' = y^2 - (x^2 - 1)(x - 2) \\ y' = x - y - 2 \end{cases}$

8. $\begin{cases} x' = x^2 - y \\ y' = xy + y^2 \end{cases}$

9. $x'' + x' + \sin x = 0$

10. $x'' + xx' + x - x^3 = 0$

11. $x'' + px' + qx^3 = 0$, where p and q are constants (Duffing's equation)

12. $mx'' + k \sin x = 0$, where $m \neq 0$ and k are constants (the frictionless pendulum equation)

Use a computer or calculator to find all equilibria of the following systems and higher-order equations.

13. $\begin{cases} x' = x - e^{-y} \\ y' = x - y \end{cases}$

14. $\begin{cases} x' = y - \dfrac{1}{2 - x} \\ y' = 1 - 3ye^{-x^2} \end{cases}$

15. $\begin{cases} x' = \ln\left(\dfrac{1}{1 + 2x^2}\right) - y \\ y' = -3x - 4y \end{cases}$

16. $\begin{cases} x' = \dfrac{3x}{1 + y^2} - 1 \\ y' = y - x^2 \end{cases}$

17. $x'' + 2x' + xe^{-x} = \frac{1}{4}$ **18.** $x'' + x - \cos x = 0$

In the systems and equations below, $r > 0$ is a positive constant. Use geometric methods to study the equilibria. Without solving the equilibrium equations algebraically, determine those values of r for which there are no equilibria and those for which there are equilibria. In the latter case, determine how many equilibria there are.

19. $\begin{cases} x' = x^2 + y^2 - r^2 \\ y' = (x - 3)^2 + y^2 - 4 \end{cases}$

20. $\begin{cases} x' = x^2 + y^2 - r^2 \\ y' = x + y - 1 \end{cases}$

21. $\begin{cases} x' = rx - y \\ y' = 6x + y - 8x^2 + 2x^3 \end{cases}$

22. $\begin{cases} x' = y + x^2y - x^2 \\ y' = -x + \dfrac{1}{r}y \end{cases}$

23. $x'' + 2x' + xe^{-x} - r = 0$

24. $x'' + x + \dfrac{r}{x - 1} = 0$

25. Find all equilibria for the chemostat equations

$$x' = (x_{in} - x)d - \frac{1}{\gamma}\frac{mx}{a + x}y,$$

$$y' = \frac{mx}{a + x}y - dy.$$

In these equations all coefficients are positive constants.

8.3   The Linearization Principle

In Chapter 3 we learned that equilibria of single equations $x' = f(x)$ are of three types: sinks, sources, or shunts. The linearization principle (Theorem 5 in Chapter 3) tells us that an equilibrium has the same type as that of the linearization of the equation at the equilibrium

$$u' = \lambda u, \qquad \lambda = \left. \frac{df}{dx} \right|_{x_e} \tag{3.1}$$

provided that the equilibrium x_e is hyperbolic (i.e., provided $\lambda \neq 0$). Specifically, the equilibrium is a sink if $\lambda < 0$ and a source if $\lambda > 0$. Our goal in this section is to extend the linearization principle to systems of equations.

In Chapter 3 (Sec. 3.1.3) we derived the linearization (3.1) of the scalar equation $x' = f(x)$ at an equilibrium $x = x_e$ from the linear Taylor polynomial approximation $f(x_e) + \lambda(x - x_e)$ to $f(x)$. We did this as follows. Using $f(x_e) = 0$ and the resulting Taylor approximation

$$f(x) \approx \lambda(x - x_e),$$

together with the notation $u = x - x_e$, we obtain the linearization (3.1) of $x' = f(x)$ at the equilibrium x_e.

We proceed in a similar manner for a planar autonomous system

$$\begin{aligned} x' &= f(x, y) \\ y' &= g(x, y). \end{aligned} \tag{3.2}$$

We first approximate $f(x, y)$ and $g(x, y)$ by linear Taylor series polynomials centered at an equilibrium point (x_e, y_e). These polynomials are

$$\begin{aligned} f(x_e, y_e) + a\,(x - x_e) + b\,(y - y_e) \\ g(x_e, y_e) + c\,(x - x_e) + d\,(y - y_e), \end{aligned}$$

where coefficients are calculated from the derivatives of f and g by the formulas[2]

$$a = \left. \frac{\partial f}{\partial x} \right|_{(x_e, y_e)}, \qquad b = \left. \frac{\partial f}{\partial y} \right|_{(x_e, y_e)} \tag{3.3}$$

$$c = \left. \frac{\partial g}{\partial x} \right|_{(x_e, y_e)}, \qquad d = \left. \frac{\partial g}{\partial y} \right|_{(x_e, y_e)}.$$

Since $f(x_e, y_e) = 0$, $g(x_e, y_e) = 0$, the Taylor polynomial approximations become

$$\begin{aligned} f(x, y) &\approx a\,(x - x_e) + b\,(y - y_e) \\ g(x, y) &\approx c\,(x - x_e) + d\,(y - y_e). \end{aligned}$$

Using the notation

$$u = x - x_e, \qquad v = y - y_e,$$

---

[2] This notation means the first the derivative is calculated and then the answer is evaluated at $(x, y) = (x_e, y_e)$.

we obtain the system

$$u' = au + bv \tag{3.4}$$
$$v' = cv + dv$$

as an approximation to the system (3.2).

The linear homogeneous system (3.4), with coefficients given by the formulas (3.3), is called the *linearization of (3.2) at the equilibrium point* (x_e, y_e). The coefficient matrix of the linearization

$$J = \begin{pmatrix} a & b \\ c & d \end{pmatrix}$$

is called the *Jacobian* of the nonlinear system (3.2) at the equilibrium point. This matrix depends on both the derivatives of f and g and on the equilibrium point (x_e, y_e). Specifically, $J = J(x_e, y_e)$, where

$$J(x, y) = \begin{pmatrix} \frac{\partial f}{\partial x} & \frac{\partial f}{\partial y} \\ \frac{\partial g}{\partial x} & \frac{\partial g}{\partial y} \end{pmatrix}.$$

A nonlinear planar autonomous system may have more than one equilibrium (e.g., see Example 1). We cannot refer to "the" linearization of a planar autonomous system but instead must refer to the linearization *at an equilibrium point*.

EXAMPLE 1

In Example 1 we found that the competition system

$$x' = x - xy$$
$$y' = 2y - 2xy$$

has two equilibrium points: $(x_e, y_e) = (0, 0)$ and $(1, 1)$. The Jacobian matrix for this system is

$$J(x, y) = \begin{pmatrix} 1 - y & -x \\ -2y & 2 - 2x \end{pmatrix}.$$

At the equilibrium point $(0, 0)$

$$J(0, 0) = \begin{pmatrix} 1 & 0 \\ 0 & 2 \end{pmatrix},$$

which is the coefficient matrix of the linearization

$$u' = u$$
$$v' = 2v$$

at $(0, 0)$. The Jacobian at the equilibrium point $(1, 1)$ is

$$J(1, 1) = \begin{pmatrix} 0 & -1 \\ -2 & 0 \end{pmatrix},$$

which is the coefficient matrix of the linearization

$$u' = -v$$
$$v' = -2u$$

at $(1, 1)$. ■

We can also linearize second-order equations at an equilibrium. The next example illustrates this.

EXAMPLE 2

The van der Pol equation

$$x'' + \alpha\left(x^2 - 1\right)x' + x = 0$$

has equilibrium $x_e = 0$. The equivalent system

$$x' = y \tag{3.5}$$
$$y' = -x - \alpha(x^2 - 1)y$$

has equilibrium point $(x_e, y_e) = (0, 0)$. The Jacobian of this system is

$$J(x, y) = \begin{pmatrix} 0 & 1 \\ -1 - 2\alpha xy & -\alpha(x^2 - 1) \end{pmatrix}.$$

At the equilibrium point

$$J(0, 0) = \begin{pmatrix} 0 & 1 \\ -1 & \alpha \end{pmatrix},$$

which is the coefficient matrix of the linearization

$$u' = v \tag{3.6}$$
$$v' = -u + \alpha v$$

at $(0, 0)$. This first-order system is equivalent to the linear second-order equation

$$u'' - \alpha u' + u = 0,$$

which is the linearization of the van der Pol equation at $x_e = 0$. ∎

Now that we have learned how to linearize a system of differential equations at an equilibrium, we ask, What can we learn about the solutions of a planar autonomous system from its linearization? Can we learn anything about the system's orbits and its phase portrait? We studied phase portraits for linear homogeneous systems with constant coefficients in Chapter 5. Using the classification scheme developed in that chapter, we can identify the phase portrait type of the linearization (3.4). Since the linearization is an approximation to the nonlinear system for u and v small, we anticipate that the phase portrait of the nonlinear system will resemble that of its linearization—at least in a neighborhood of the equilibrium point. Before examining some fundamental theorems that support this conclusion is valid, we look at some examples.

Figure 8.2 shows graphs of several orbits of the competition system

$$x' = x - xy \tag{3.7}$$
$$y' = 2y - 2xy$$

in a magnified neighborhood of the equilibrium point $(0, 0)$. These orbits show that near $(0, 0)$ the phase plane portrait appears very much like an unstable node for a linear system. In fact, from Example 1 we see that the phase portrait of the linearization at the origin is indeed an unstable node! This is because the roots $\lambda = 1, 2$ of the characteristic polynomial $\lambda^2 - 3\lambda + 3$ [i.e., the eigenvalues of $J(0, 0)$] are real and positive. (See Table 5.2 in Chapter 5.)

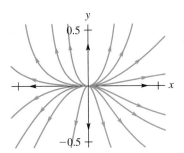

Figure 8.2 The phase portrait of system (3.7) near the equilibrium point $(0, 0)$.

Figure 8.3 shows graphs of orbits near the equilibrium point $(0, 0)$ of the van der Pol equation for $\alpha = -1$ and 1. The phase portrait resembles a stable spiral for $\alpha = -1$ and an unstable spiral for $\alpha = 1$. What about the phase portrait of the linearization (3.6) at $(0, 0)$? The roots

$$\lambda = \frac{1}{2}\alpha \pm \frac{1}{2}\sqrt{\alpha^2 - 4}$$

of the characteristic polynomial $\lambda^2 - \alpha\lambda + 1$ are complex conjugates when $|\alpha| < 2$ with a real part equal to α. The phase portrait in this case is a spiral (stable if $\alpha < 0$ and unstable for $\alpha > 0$). For $\alpha = -1$ the linearization has a stable spiral and for $\alpha = 1$ it has an unstable spiral.

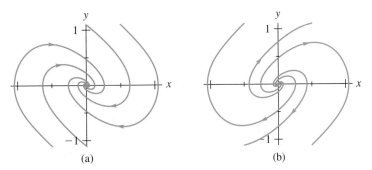

(a)   (b)

Figure 8.3 The phase plane portraits of the van der Pol equation near the equilibrium point $(0, 0)$ for (a) $\alpha = -1$ and (b) $\alpha = 1$.

In Chapter 5 we identified stable and unstable phase portraits for linear homogeneous systems. (See Table 5.2 of Sec. 5.4.) We will not give a technical definition of stability for an equilibrium of a nonlinear system. Roughly speaking, an equilibrium is *(locally) stable* means that all orbits will remain (arbitrarily) close to the equilibrium point for all $t > t_0$ if they start sufficiently close at $t = t_0$. Otherwise the equilibrium is *unstable*. If, in addition, all orbits starting sufficiently close to the equilibrium tend to the equilibrium as $t \to +\infty$, then the equilibrium is *(locally) asymptotically stable*. The word *local* is used in these definitions because they deal with the behavior of orbits nearby the equilibrium point. It is common to refer to locally asymptotically stable equilibrium points as simply "stable" equilibrium points.

In the examples of Figs. 8.2 and 8.3 the stability properties of the equilibrium points are the same as those of the linearization at the equilibrium point. These are examples of the following general *linearization principle* for planar autonomous systems.

THEOREM 1

The Linearization Principle for Planar Systems   Suppose (x_e, y_e) is an equilibrium point of the autonomous system

$$x' = f(x, y)$$
$$y' = g(x, y)$$

and let λ_1, λ_2 denote the roots of the characteristic polynomial $\lambda^2 - (a+d)\lambda + ad - bc$, where

$$a = \frac{\partial f}{\partial x}\bigg|_{(x_e, y_e)}, \qquad b = \frac{\partial f}{\partial y}\bigg|_{(x_e, y_e)}$$
$$c = \frac{\partial g}{\partial x}\bigg|_{(x_e, y_e)}, \qquad d = \frac{\partial g}{\partial y}\bigg|_{(x_e, y_e)}.$$

If λ_1 and λ_2 both have negative real parts, then the equilibrium point (x_e, y_e) is (locally asymptotically) stable.[3] If at least one root has a positive real part, then the equilibrium point is unstable.

A number λ with a negative real part is said to lie in the left half of the complex number plane. Thus, an equilibrium is stable if both roots of the characteristic polynomial at the equilibrium lie in the left half-plane. It is unstable if at least one root lies in the right half-plane. [In matrix terminology, the equilibrium is stable if the eigenvalues of the Jacobian $J(x_e, y_e)$ matrix evaluated at the equilibrium lie in the left half-plane and is unstable if at least one eigenvalue lies in the right half plane.]

Here are some facts about quadratic polynomials $\lambda^2 + \beta\lambda + \gamma$. Both roots have negative real parts if and only if $\beta > 0$ and $\gamma > 0$, and if either $\beta < 0$ or $\gamma < 0$, then at least one root has a positive real part. You can convince yourself of these assertions by using the quadratic formula for the roots (Exercises 15–16). (You can also refer to the trace-determinant criterion in Sec. 5.4.5 of Chapter 5.) From Theorem 1 we obtain the following theorem.

THEOREM 2

An equilibrium point (x_e, y_e) of the autonomous system

$$x' = f(x, y)$$
$$y' = g(x, y)$$

is stable if both inequalities

$$\frac{\partial f}{\partial x} + \frac{\partial g}{\partial y} < 0 \quad \text{and} \quad \frac{\partial f}{\partial x}\frac{\partial g}{\partial y} - \frac{\partial f}{\partial y}\frac{\partial g}{\partial x} > 0$$

hold at the point $(x, y) = (x_e, y_e)$.[4]

---

[3]*Note*: Real roots are equal to their own real part.
[4]The sum $\frac{\partial f}{\partial x} + \frac{\partial g}{\partial y}$ is the *trace* of the Jacobian matrix J and is denoted by $\mathrm{tr}J$. The expressions $\frac{\partial f}{\partial x}\frac{\partial g}{\partial y} - \frac{\partial f}{\partial y}\frac{\partial g}{\partial x}$ is the *determinant* of the Jacobian matrix J and is denoted by $\det J$.

**THEOREM 2
(Continued)**

If

$$\frac{\partial f}{\partial x} + \frac{\partial g}{\partial y} > 0 \quad \text{or} \quad \frac{\partial f}{\partial x}\frac{\partial g}{\partial y} - \frac{\partial f}{\partial y}\frac{\partial g}{\partial x} < 0$$

at $(x, y) = (x_e, y_e)$, then the equilibrium point is unstable.

As an example of the application of Theorem 2 we calculate, for the system (3.7), the expressions

$$\frac{\partial f}{\partial x} + \frac{\partial g}{\partial y} = 3 - y - 2x$$

$$\frac{\partial f}{\partial x}\frac{\partial g}{\partial y} - \frac{\partial f}{\partial y}\frac{\partial g}{\partial x} = 2 - 2x - 2y.$$

For the equilibrium $x = 0$, $y = 0$ we find that

$$\frac{\partial f}{\partial x} + \frac{\partial g}{\partial y} = 3 > 0$$

and conclude by Theorem 2 that the equilibrium point $(x_e, y_e) = (0, 0)$ is unstable. For the equilibrium $x = 1$, $y = 1$ we find that

$$\frac{\partial f}{\partial x}\frac{\partial g}{\partial y} - \frac{\partial f}{\partial y}\frac{\partial g}{\partial x} = -2 < 0$$

and conclude by Theorem 2 that the equilibrium point $(x_e, y_e) = (1, 1)$ is unstable.

As a second example, consider the van der Pol system (1.6)

$$x' = y$$

$$y' = -x - \alpha(x^2 - 1)y.$$

For this system we calculate

$$\frac{\partial f}{\partial x} + \frac{\partial g}{\partial y} = -\alpha(x^2 - 1)$$

$$\frac{\partial f}{\partial x}\frac{\partial g}{\partial y} - \frac{\partial f}{\partial y}\frac{\partial g}{\partial x} = 1 + 2\alpha xy.$$

At the equilibrium point $(x_e, y_e) = (0, 0)$

$$\frac{\partial f}{\partial x} + \frac{\partial g}{\partial y} = \alpha$$

$$\frac{\partial f}{\partial x}\frac{\partial g}{\partial y} - \frac{\partial f}{\partial y}\frac{\partial g}{\partial x} = 1 > 0.$$

By Theorem 2 the equilibrium point $(x_e, y_e) = (0, 0)$ is stable if $\alpha < 0$ and unstable if $\alpha > 0$.

In the examples of Figs. 8.2 and 8.3, not only do the stability properties of the equilibrium and the linearization agree, but the geometry of the phase portraits is similar. That is, the graph in Fig. 8.2 looks like an unstable node for a linear system and the graphs in Fig. 8.3 look like spirals for linear systems.

Here is another example involving system (3.7). In Fig. 8.4 appear several orbits in a magnified neighborhood of the equilibrium point $(1, 1)$. Near this equilibrium the phase portrait is very much like a saddle for a linear system. Indeed, from Example 1 we find that the phase portrait of the linearization at $(1, 1)$ is a saddle. This is because the roots $\lambda = \pm\sqrt{2}$ of the characteristic polynomial $\lambda^2 - 2$ have opposite signs. Note that one root, $\lambda = \sqrt{2}$, is positive and Theorem 1 implies that $(1, 1)$ is unstable. It appears that not only can we learn about stability and instability of hyperbolic equilibria from the linearization, but that we can also learn about the geometry of the phase portrait. This is the subject of the next section.

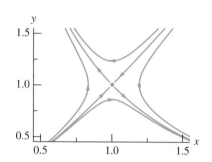

Figure 8.4 The phase portrait of system (3.7) near the equilibrium point $(1, 1)$.

Exercises

Find the equilibria of the following systems and calculate Jacobian at each of them.

1. $\begin{cases} x' = x - y^2 \\ y' = x - y \end{cases}$

2. $\begin{cases} x' = x\,(2x - y) \\ y' = x^2 + y - 8 \end{cases}$

3. $\begin{cases} x' = x\,(1 - x - y) \\ y' = y\,(2 - x - 4y) \end{cases}$

4. $\begin{cases} x' = y - xy \\ y' = -x + xy \end{cases}$

5. $\begin{cases} x' = 1 - x^2 - y^2 \\ y' = x - y \end{cases}$

6. $\begin{cases} x' = x - x^2 y \\ y' = 1 - x^2 + xy \end{cases}$

7. Find the linearization at each equilibrium of the systems in Exercises 1–6.

8. Apply Theorem 1 and/or 2, if possible, to determine the stability of each equilibrium of the systems in Exercises 1–6.

Consider the nonlinear system

$$x' = \left(\frac{3}{2} - x - 2y\right)x$$

$$y' = \left(-\frac{1}{4} + x\right)y.$$

This is an example of a predator-prey system in which x is the density of prey and y is the density of predator.

9. Find all equilibrium points.

10. Find the linearization at each equilibrium point.

11. Determine the stability of each equilibrium point.

Consider the nonlinear system

$$x' = x - e^{-y}$$

$$y' = x - y.$$

12. Find numerical approximations to all equilibrium points.

13. Find the linearization at each equilibrium point.

14. Determine the stability of each equilibrium point.

15. Show that both roots of the quadratic $\lambda^2 + \beta\lambda + \gamma$ have negative real parts if and only if $\beta > 0$ and $\gamma > 0$.

16. Show that if either $\beta < 0$ or $\gamma < 0$, then at least one root of the quadratic $\lambda^2 + \beta\lambda + \gamma$ has a positive real part.

8.4 Local Phase Plane Portraits

We classified phase portraits of *linear* systems in Chapter 5, Sec. 5.4. The examples in Figs. 8.2, 8.3, and 8.4 suggest that the phase portrait of a nonlinear system near an equilibrium point is similar to that of the linearization (at that equilibrium point). For example, the phase portrait of competition system

$$x' = x - xy$$
$$y' = 2y - 2xy$$

near the equilibrium point $(0, 0)$ shown in Fig. 8.2 is very similar to the unstable node of its linearization

$$x' = x$$
$$y' = 2y$$

$$(4.1)$$

shown in Fig. 8.5.

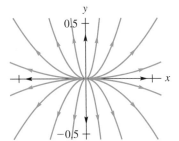

Figure 8.5 The phase portrait of system (4.1).

Will it always be the case for a nonlinear system of equations that the phase portrait near an equilibrium is a node if the phase portrait of the linearization at the equilibrium is a node? Will the phase portrait near an equilibrium be a spiral if that of its linearization at the equilibrium is a spiral (for example, as in Fig. 8.3)? To give an answer we must first consider what we mean by a node and a spiral for a nonlinear system.

The characteristic feature of a node is that all orbits approach the origin from a definite direction. That is, for a stable node the polar angle θ determined by a point on an orbit, using the origin as a reference point, approaches a limit θ_0 as $t \to +\infty$. See Fig. 8.6. For an unstable node the angle θ approaches a limit as $t \to -\infty$. (The limit angle is not necessarily the same for all orbits.) For example, in Fig. 8.5 as $t \to -\infty$ all orbits approach the origin tangential to the x-axis (i.e., θ approaches 0), except for the two orbits lying on the y-axis that approach the origin vertically (i.e., θ approaches $\pi/2$). This characteristic of a node is in distinction to that of a spiral for which the angle θ does not approach a limit. For a spiral the angle θ instead increases without bound as orbits spiral around the equilibrium.

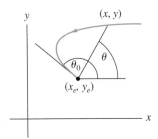

Figure 8.6

For stable phase portraits (nodes or spirals) the distance r from points on the orbit to the equilibrium point approaches 0 as $t \to +\infty$, whereas for unstable phase portraits the distance r approaches 0 as $t \to -\infty$.

We can use these characteristics of linear phase portraits to define nodes and spirals for nonlinear phase portraits.

DEFINITION 1. The "local" phase portrait near an equilibrium of a nonlinear system is a **stable node** if $r \to 0$ and for each orbit the angle θ approaches a limit as $t \to +\infty$. It is an **unstable node** if $r \to 0$ and for each orbit the angle θ approaches a limit as $t \to -\infty$.

The "local" phase portrait near an equilibrium of a nonlinear system is a **stable spiral** if $r \to 0$ and for each orbit the angle θ increases without bound as $t \to +\infty$. It is an **unstable spiral** if $r \to 0$ and for each orbit the angle θ increases without bound as $t \to -\infty$.

In Sec. 5.4 (Chapter 5) we studied one other fundamental type of phase portrait for linear systems, namely, the saddle point. The distinguishing characteristics of a saddle are the existence of two half-line orbits that tend to the equilibrium point $(0, 0)$ as $t \to +\infty$ (forming the "stable manifold") and two half-line orbits that tend to the equilibrium point $(0, 0)$ as $t \to -\infty$ (forming the "unstable manifold"). No other orbits approach $(0, 0)$ for $t \to +\infty$ or $t \to -\infty$. For nonlinear systems we define a saddle as follows.

DEFINITION 2. An equilibrium is a **saddle** if

1. There are two orbits that tend to the equilibrium, each of whose angle θ approaches a limit as $t \to +\infty$.
2. There are two orbits that tend to the equilibrium each of whose angle θ approaches a limit as $t \to -\infty$.
3. No other orbits approach the equilibrium as $t \to +\infty$ or $t \to -\infty$.

The two orbits that tend to the equilibrium as $t \to +\infty$ form the *stable manifold.* However, for a nonlinear system these orbits are not necessarily straight lines, as they are for a linear system. Similarly, the two orbits that tend to the equilibrium as $t \to -\infty$ and that together form the *unstable manifold* are not necessarily straight lines.

Figure 8.4 shows an example of a saddle for a nonlinear system.

For single nonlinear equations the linearization principle (Theorem 5, Sec. 3.1.3 of Chapter 3) tells us that the equilibrium type (either sink or source) is the same as that of the linearization at the equilibrium provided the equilibrium is hyperbolic; that is, provided

$$\lambda = \left.\frac{df}{dx}\right|_{x_e} \neq 0.$$

Similarly, for a nonlinear system the type of an equilibrium is the same as that of its linearization provided the equilibrium is hyperbolic, as defined next.

DEFINITION 3. An equilibrium point is **hyperbolic** if the roots λ of the characteristic polynomial obtained from the linearization at the equilibrium all have nonzero real parts.

THEOREM 3

A hyperbolic equilibrium of the planar autonomous system

$$x' = f(x, y) \qquad (4.2)$$
$$y' = g(x, y)$$

is a

stable (unstable) spiral if the linearization is a stable (unstable) spiral
stable (unstable) node if the linearization is a stable (unstable) node
saddle if the linearization has a saddle

In the latter case, at the equilibrium point the stable and unstable manifolds of the saddle are tangent to those of the linearization.

The phase portrait of the linearization system is determined by the roots of its characteristic equation (i.e., the eigenvalues of its coefficient matrix). See Table 5.2 in Sec. 5.4 (Chapter 5).

◆ **Corollary 1**

Suppose (x_e, y_e) is a hyperbolic equilibrium of a planar autonomous system (4.2).

Suppose the roots of the characteristic polynomial obtained from the linearization at (x_e, y_e) are complex. Then the equilibrium point (x_e, y_e) is a stable spiral if the real parts are negative. The point (x_e, y_e) is an unstable spiral if the real parts are positive.

Suppose the roots of the characteristic polynomial are real. If they are both negative, then the equilibrium point (x_e, y_e) is a stable node. If they are both positive (x_e, y_e) is an unstable node. If they have different signs, (x_e, y_e) is a saddle. In the latter case the stable and unstable manifolds are tangent to those of the linearization.◆

Here are some examples. We see from Example 1 in Sec. 8.3 that roots of the characteristic polynomial obtained from the linearization at the equilibrium point $(0, 0)$ of the nonlinear system (3.7) are $\lambda = 1, 2$. By Corollary 1, $(0, 0)$ is an unstable node for this nonlinear system (as seen in Fig. 8.2). We also see from Example 1 that the roots of the characteristic polynomial obtained from the linearization of system (3.7) at the equilibrium point $(1, 1)$ are $\lambda = \pm\sqrt{2}$. By Corollary 1, $(1, 1)$ is a saddle (as seen in Fig. 8.4).

In the previous section we saw that the roots of the characteristic polynomial of the linearization (3.6) of the van der Pol equation with $\alpha = -1$ at the equilibrium point $(0, 0)$ are the complex conjugates

$$\lambda = -\frac{1}{2} \pm i\frac{1}{2}\sqrt{3}.$$

Since the real part $-1/2$ is negative, Corollary 1 implies $(0, 0)$ is a stable spiral. See Fig. 8.3(a). When $\alpha = 1$ in the van der Pol equation, the roots

$$\lambda = \frac{1}{2} \pm i\frac{1}{2}\sqrt{3}$$

of the characteristic polynomial are complex with positive real part $1/2$. By Corollary 1, the equilibrium points $(0, 0)$ is an unstable spiral. See Fig. 8.3(b).

EXAMPLE 1   The nonlinear system

$$x' = x_{in} - x - \frac{2x}{1+x}y$$

$$y' = \frac{2x}{1+x}y - y$$

is a particular case of the chemostat model. A chemostat is a container used to contain chemical and biological reactions. Chemostats are used in scientific studies and diverse applications ranging from gene splicing to brewing. The quantity x is the concentration of a substrate that is continuously pumped into the chemostat at a fixed rate (with concentration $x_{in} > 0$) and reacts with (or is consumed by) the quantity y. The mixture is well stirred and continuously pumped out at the same rate so as to maintain a fixed volume. (For more details see Sec. 4.6.1 in Chapter 4.)

There are two equilibria (Exercise 13):

$$(x_e, y_e) = (x_{in}, 0) \qquad \text{and} \qquad (1, x_{in} - 1).$$

The Jacobian matrix is

$$J(x, y) = \begin{pmatrix} -1 - \frac{2y}{(1+x)^2} & -\frac{2x}{1+x} \\ \frac{2}{(1+x)^2}y & \frac{x-1}{1+x} \end{pmatrix}.$$

For the first equilibrium

$$J(x_{in}, 0) = \begin{pmatrix} -1 & -\frac{2x}{1+x} \\ 0 & \frac{x_{in}-1}{1+x_{in}} \end{pmatrix}.$$

The roots of the characteristic polynomial

$$\lambda^2 + \frac{2}{1+x_{in}}\lambda + \frac{1-x_{in}}{1+x_{in}}$$

are

$$\lambda_1 = -1, \qquad \lambda_2 = \frac{x_{in}-1}{1+x_{in}}.$$

If $x_{in} > 1$, then $\lambda_2 > 0$ and the roots have opposite signs. By Corollary 1 the equilibrium point $(x_{in}, 0)$ is a saddle.

If $x_{in} < 1$, then $\lambda_2 < 0$ and both roots are negative. By Corollary 1 the equilibrium point $(x_{in}, 0)$ is a stable node.

For the second equilibrium

$$J(1, x_{in} - 1) = \begin{pmatrix} -\frac{1}{2}(1+x_{in}) & -1 \\ \frac{1}{2}(x_{in}-1) & 0 \end{pmatrix}.$$

The roots of the characteristic polynomial

$$\lambda^2 + \frac{1}{2}(1+x_{in})\lambda + \frac{1}{2}(x_{in}-1)$$

are

$$\lambda_1 = -1, \qquad \lambda_2 = \frac{1}{2}(1 - x_{in}).$$

If $x_{in} > 1$, then $\lambda_2 < 0$ and both roots are negative. By Corollary 1 the equilibrium point $(1, x_{in} - 1)$ is a stable node.

If $x_{in} < 1$, then $\lambda_2 > 0$ and the roots have opposite signs. By Corollary 1 the equilibrium point $(1, x_{in} - 1)$ is a saddle.   ∎

For nonlinear systems we have not defined different types of nodes (improper, star, etc.) as we did for linear phase portraits in Chapter 5; nor have we defined a center for nonlinear systems. For these cases the relationship between the phase portrait of a nonlinear system and that of its linearization is complicated. Also, in the neighborhood of a *nonhyperbolic* equilibrium the phase portraits of a planar autonomous system and that of its linearization need not be, and often are not, of the same type.

We emphasize that Theorem 3 and its Corollary 1 describe phase portraits only in a neighborhood of an equilibrium point. The *global* phase portrait of a planar autonomous system may look considerably different from the phase portrait of its linearization, as we will see in the next section.

Exercises

Find the equilibria of the following systems and determine whether they are hyperbolic or nonhyperbolic. Determine the phase portrait in the neighborhood of all hyperbolic equilibria.

1. $\begin{cases} x' = x - y^2 \\ y' = x - y \end{cases}$

2. $\begin{cases} x' = x\,(2x - y) \\ y' = x^2 + y - 8 \end{cases}$

3. $\begin{cases} x' = x\,(1 - x - y) \\ y' = y\,(2 - x - 4y) \end{cases}$

4. $\begin{cases} x' = y - xy \\ y' = -x + xy \end{cases}$

5. $\begin{cases} x' = 1 - x^2 - y^2 \\ y' = x - y \end{cases}$

6. $\begin{cases} x' = x - x^2 y \\ y' = 1 - x^2 + xy \end{cases}$

Consider the predator-prey system

$$x' = \left(\frac{3}{2} - x - 2y\right) x$$

$$y' = \left(-\frac{1}{4} + x\right) y.$$

7. Find all equilibria.

8. Calculate the Jacobian matrix $J(x, y)$.

9. Which equilibria are hyperbolic?

10. Determine the phase portrait in the neighborhood of each equilibrium.

The system

$$x' = y$$

$$y' = -x - \alpha\,(x^2 - 1)\,y$$

is equivalent to the van der Pol equation. The only equilibrium point is $(x_e, y_e) = (0, 0)$.

11. Calculate the Jacobian matrix at $(0, 0)$.

12. Determine the phase portrait in the neighborhood of $(0, 0)$. Identify any nonhyperbolic cases. Your answer will depend on the coefficient α.

13. Find the equilibria of the nonlinear system in Example 1.

14. The general chemostat model equations are

$$x' = (x_{in} - x)\,d - \frac{1}{\gamma}\frac{mx}{a + x}\,y$$

$$y' = \frac{mx}{a + x}\,y - dy.$$

All coefficients are positive constants. When $m \neq d$ there are two equilibria

$$(x_e, y_e) = \begin{cases} (x_{in}, 0) \\ \left(\frac{ad}{m-d},\ \left(x_{in} - \frac{ad}{m-d}\right)\gamma\right) \end{cases}.$$

In applications only nonnegative solutions are of interest. Therefore, assume

$$m > d \quad \text{and} \quad x_{in} > \frac{ad}{m - d}.$$

[If $x_{in} = ad/(m - d)$, then the second equilibrium point coincides with the first.]

(a) Calculate the Jacobian matrix $J(x_e, y_e)$ for each equilibrium.

(b) Determine the phase portrait near each equilibrium. Identify any nonhyperbolic cases. (*Hint:* Use the trace/determinant criteria.)

8.5 Global Phase Plane Portraits

The *global* phase portrait of a planar autonomous system may look considerably different from the phase portrait of its linearization. For example, the phase portrait of the competition system

$$x' = x - xy \tag{5.1}$$
$$y' = 2y - 2xy$$

appears in Fig. 8.7. The phase portraits shown in Figs. 8.2 and 8.4 are magnifications of this phase portrait near the two equilibrium points $(0, 0)$ and $(1, 1)$.

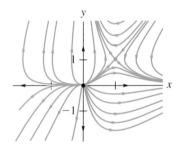

Figure 8.7 The phase portrait of (5.1). The phase portraits in Figs. 8.2 and 8.3 are magnifications near the equilibrium points $(0, 0)$ and $(1, 1)$.

Figure 8.8 shows the phase portrait of the system

$$x' = y \tag{5.2}$$
$$y' = -x - \alpha(x^2 - 1)y$$

associated with the van der Pol equation

$$x'' + \alpha\left(x^2 - 1\right)x' + x = 0$$

for $\alpha = 1$. The unstable spiral shown in Fig. 8.3(b) is a magnification of this phase portrait near the equilibrium point $(0, 0)$. It shows orbits near this unstable spiral equilibrium spiraling outward. In Fig. 8.8, however, we see that orbits far enough away from the equilibrium do not spiral outward, but instead spiral *inward*.

A notable feature of the phase portrait shown in Fig. 8.8 is what appears to be a closed loop that separates the outward spiraling orbits from the inward spiraling orbits. This closed loop is apparently itself an orbit of the van der Pol system. This can be seen by choosing a point on the loop [e.g., $(x_0, y_0) = (-1.14, -2.57)$] and calculating the orbit starting at that initial point. The orbit will trace out the loop, as shown in Fig 8.8. Graphs of both components $x(t)$ and $y(t)$ of the solution pair corresponding to this loop orbit appear in Fig. 8.9. Notice both $x(t)$ and $y(t)$ appear to be periodic functions of t (repeating with a period of between 6 and 7 time units).

A periodic function $x = x(t)$ of period p satisfies $x(t + p) = x(t)$ for all t. The graph of a periodic function on the interval $0 \leq t \leq p$ repeats itself identically on intervals of length p (i.e., on the intervals $p \leq t \leq 2p$, $2p \leq t \leq 3p$ and so on) (and also on the intervals $-2p \leq t \leq -p$, $-3p \leq t \leq -2p, \ldots$). Recall that nonequilibrium solutions of a *single* autonomous equation are either monotonically increasing

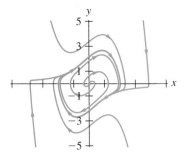 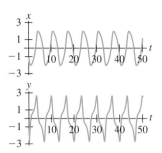

Figure 8.8 The phase portrait of (5.2) with $\alpha = 1$. The phase portrait in Fig. 8.3(b) is a magnification near the equilibrium point $(0, 0)$.

Figure 8.9 A solution pair of (5.2) with $\alpha = 1$ whose orbit is the closed loop seen in Fig. 8.8.

or decreasing (Chapter 3) and therefore cannot be periodic. That nonequilibrium solutions of systems can be periodic is an important feature of systems of two (or more) equations.

A (nonequilibrium) periodic solution pair $x(t)$ and $y(t)$ satisfies $x(t + p) = x(t)$ and $y(t + p) = y(t)$ for all t. As a result, *the orbit of periodic solution pair is a closed loop in the phase plane.* Moreover, the converse is true: *A closed loop orbit is associated with periodic solution pairs* (see Exercise 20). That is, closed loop orbits arise from and only from periodic solutions.

The closed loop orbit of a periodic solution is called a *cycle.* If other orbits approach a cycle as $t \to +\infty$ (or $-\infty$), then it is called a *limit cycle.* Figure 8.8 shows that the van der Pol equation (5.2) with $\alpha = 1$ has a limit cycle.

In Chapter 3 (Sec. 3.1) we learned that the only points that can be approached by orbits of single autonomous equations (as $t \to +\infty$ or $-\infty$) are equilibrium points. From the van der Pol equation example we see that for systems of equations, on the other hand, orbits do not necessarily approach equilibrium points. A new type of limit set is possible for systems, namely a limit cycle. In constructing the phase plane portrait of a system, an important problem is then to determine the sets of points approached by orbits as $t \to +\infty$ and $-\infty$.

Consider orbits bounded as $t \to +\infty$.[5] We call the set of points approached by the orbit as $t \to +\infty$ the *forward (or omega) limit set* of the orbit and denote the set by S^+. More precisely, a point (x^*, y^*) is in the set S^+ if there is a sequence $t_n \to +\infty$ such that the corresponding points on the orbit $(x(t_n), y(t_n))$ approach (x^*, y^*) :

$$S^+ = \left\{ (x^*, y^*) \mid \text{there exist } t_n \to +\infty \text{ such that } (x(t_n), y(t_n)) \to (x^*, y^*) \right\}.$$

[Note that this does not necessarily mean $(x(t), y(t)) \to (x^*, y^*)$ as $t \to +\infty$. For other sequences $t_i \to +\infty$ the orbit points $(x(t), y(t))$ might approach a different point.]

---

[5] An orbit is *bounded* as $t \to +\infty$ if it does not get arbitrarily far from the origin, or, in other words, remains inside a circle of sufficiently large radius for all $t \geq 0$. Similarly, an orbit is bounded as $t \to -\infty$ if it remains inside a circle of sufficiently large raduis for all $t \leq 0$.

The *backward (or alpha) limit* set S^- is similarly defined as the set of points approached by an orbit bounded as $t \to -\infty$, or

$$S^- = \left\{ (x^*, y^*) \mid \text{there exist } t_n \to -\infty \text{ such that } (x(t_n), y(t_n)) \to (x^*, y^*) \right\}.$$

A fundamental fact about the limit sets $S^\pm$ is that they are *invariant sets*. This means that if a point (x^*, y^*) lies in a limit set, then the entire orbit passing throughout that point lies in the limit set (for *all* t, positive and negative). Thus, forward and backward limit sets are made up of orbits.

To illustrate these ideas, consider the van der Pol system (5.2) with $\alpha = 1$. The phase plane portrait in Fig. 8.8 indicates that all orbits are bounded as $t \to +\infty$. The limit cycle is the forward limit set of all orbits [except for the equilibrium point at $(0, 0)$]. The equilibrium $(0, 0)$ is the backward limit set of all orbits inside the the limit cycle. The orbits outside the limit cycle are unbounded as $t \to -\infty$ and have no backward limit set. In this example, the forward and backward limit sets consist of single orbits (either the equilibrium point or the limit cycle). As we will see (Example 2), it is possible for limit sets to consist of more than one orbit.

The following famous theorem provides information about the limit sets of orbits. As always, in the planar autonomous system

$$x' = f(x, y)$$
$$y' = g(x, y)$$

we assume $f(x, y)$, $g(x, y)$ and their partial derivatives with respect to x and y are continuous.

THEOREM 4

Poincaré-Bendixson   Let S^+ be the forward limit set of an orbit bounded as $t \to +\infty$. Either S^+ contains an equilibrium point or S^+ is a limit cycle.

These two alternatives also hold for the backward limit set S^- of an orbit bounded as $t \to -\infty$.

Theorem 4 provides two alternatives for the limit set of a bounded orbit. If, for a particular system, we can rule out one alternative, then the remaining alternative must hold. For example, if by some means or other we can be show that an orbit cannot have an equilibrium in its limit set, then it follows that the orbit approaches a limit cycle. This is the case in the van der Pol equation (5.2) with $\alpha = 1$, as seen in Fig. 8.9. The only equilibrium of the system, namely the origin $(0, 0)$, is an unstable spiral (as shown by an application of the linearization principle) and therefore cannot be in the forward limit set of an orbit. By so ruling out the first alternative in Theorem 4, we conclude that any bounded orbit of the van der Pol system approaches a limit cycle. The phase plane portrait shown in Fig. 8.8 indicates (but does not rigorously prove) that in fact all orbits are bounded as $t \to +\infty$.

It is in general a difficult mathematical problem to establish the existence of a limit cycle (i.e., periodic solutions) of a planar autonomous system. The second alternative in Theorem 4 is a powerful tool that can often provide the existence of limit cycles in applications.

In addition to the linearization principle and the Poincaré-Bendixson theorem (Theorem 4) many other facts and techniques are known that help to sketch the phase plane portrait of autonomous systems. The following list contains a few. Also see Exercises 12–25.

▶ A cycle must surround at least one equilibrium point.
▶ If the forward limit set S^+ of an orbit contains a stable node or a stable spiral, then it consists only of that equilibrium point (that is, the orbit approaches the equilibrium as $t \to +\infty$).
▶ A forward limit set S^+ cannot contain an unstable node or unstable spiral point.

EXAMPLE 1

Consider the planar autonomous system

$$x' = -x - xy^2 \tag{5.3}$$
$$y' = -2y - x^2 y.$$

The equilibrium equations

$$-x\left(1 + y^2\right) = 0$$
$$-y\left(2 + x^2\right) = 0$$

have only one solution

$$(x_e, y_e) = (0, 0).$$

The Jacobian

$$J(x, y) = \begin{pmatrix} -1 - y^2 & -2xy \\ -2xy & -2 - x^2 \end{pmatrix}$$

evaluated at the equilibrium

$$J(x, y) = \begin{pmatrix} -1 & 0 \\ 0 & -2 \end{pmatrix}$$

has characteristic polynomial $\lambda^2 + 3\lambda + 2$ whose roots are $\lambda = -1, -2$. The origin is a stable node.

Our next step is to show there exists no cycle (i.e., periodic solutions). One way to do this is to examine the sketch of the direction field shown in Fig. 8.10. A cycle would have to encircle the origin. However, such a cycle would have to pass sequentially from each quadrant to the next, a path that is simply not possible, as the sketch in Fig. 8.10 shows. This direction field also indicates that orbits (at least those in the region shown) are bounded (and move toward the origin) as $t \to +\infty$. For a more rigorous validation of these conclusions, see Exercises 12–15 and 20–25.

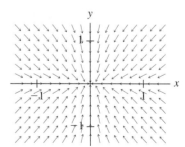

Figure 8.10 The direction field of the system (5.3).

An application of the Poincaré-Bendixson theorem (Theorem 4) implies the forward limit set S^+ of any orbit must contain the (one and only one) equilibrium $(0, 0)$.

Since this equilibrium is a stable node, S^+ must in fact consist solely of this equilibrium. As a result, we conclude that all orbits approach the stable node at the origin $(0, 0)$ as $t \to +\infty$. ■

The limit set of an orbit does not necessarily consists of a single equilibrium point or a limit cycle. Limit sets can consist of more than one orbit. In particular, a limit set might contain several equilibria. The following example illustrates this possibility.

EXAMPLE 2

Consider the planar autonomous system

$$x' = (x^2 - 1) y \tag{5.4}$$

$$y' = (1 - y^2) \left(x + \frac{3}{10} y \right).$$

We wish to determine the limit set of orbits starting near the origin $(0, 0)$, which is an equilibrium. The other equilibria are

$$(1, 1), \quad (-1, 1), \quad (-1, -1), \quad (1, -1) \tag{5.5}$$

and

$$\left(1, -\frac{10}{3} \right), \quad \left(-1, \frac{10}{3} \right).$$

The direction field sketch in Fig. 8.11 shows an orbit spirally counterclockwise out from the origin and seemingly approaching the square whose corners are the equilibria (5.5).

The sides of the square are each themselves orbits. This can be seen by noting that choosing an initial condition on a side. For example, let $(x_0, y_0) = (1, 0)$. The first equation in (5.4) implies $x(t) = 1$ for all t. To determine the solution pair of the system (5.4) we consider the second equation with $x = 1$. The result is the single autonomous equation

$$y' = (1 - y^2) \left(1 + \frac{3}{10} y \right)$$

for $y(t)$. The phase line portrait of this equation is

$$\longrightarrow -\frac{10}{3} \longleftarrow -1 \longrightarrow 1 \longleftarrow .$$

The initial condition $y_0 = 0$ yields a solution $y_1(t)$ associated with the orbit from -1 to 1. Putting this all together, we get the solution pair $(x, y) = (1, y_1(t))$ of (5.4). The orbit of this pair is the right-hand side of the square with corners (5.5). Similar considerations show the remaining three sides of the square are also orbits of (5.4).

No orbit starting inside the square can leave the square, since the square (sides and corners) consists entirely of orbits and orbits cannot cross. Therefore, the nonequilibrium orbits starting inside the square are bounded. Figure 8.11 indicates that the square is the forward limit set of these orbits. ■

The forward limit set S^+ described in Example 2 is the square with corners (5.5). In this example S^+ consists of eight different orbits, four equilibria and four orbits that "connect" the equilibria. These connecting orbits are examples of what are called *heteroclinic orbits* (i.e., orbits that connect two different equilibria as $t \to +\infty$ and $t \to -\infty$) and the limit set is an example of a *cycle chain*.

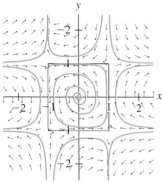

Figure 8.11 The direction field of the system (5.4). Also shown is an orbit spirally counterclockwise out from the origin and approaching the square whose corners are the equilibria (5.5).

   In this section we have seen, for planar autonomous systems, that limit sets of orbits are not necessarily equilibria, as they are for single autonomous equations. Limit cycles or cycle chains are also possible limit sets. While the possibilities have become more complicated as we moved from one-dimensional phase line portraits to two-dimensional phase plane portraits, nonetheless we have at least managed to describe the fundamental types of limit sets available for planar autonomous systems. In Sec. 8.7 we will see that further increases in dimension (i.e., in the number of equations in the system) can increase the complexity of limit sets even further.

Exercises

Use a computer to sketch the direction field of each planar autonomous systems below. Use your sketch and the tools in this section to determine the limit set S^+ of all orbits.

1. $\begin{cases} x' = -x + y - xy^2 \\ y' = x - 2y \end{cases}$ 2. $\begin{cases} x' = -x - xy^2 \\ y' = -2y - x^2y \end{cases}$

3. $\begin{cases} x' = y - x^3 \\ y' = -x + y - y^3 \end{cases}$

4. $\begin{cases} x' = x - y - x\left(x^2 + y^2\right) \\ y' = 2x - y\left(x^2 + y^2\right) \end{cases}$

5. $\begin{cases} x' = 1 - x - xy^2 \\ y' = -2y - y^3 \end{cases}$ 6. $\begin{cases} x' = -2x \\ y' = 2 - y - x^2y \end{cases}$

7. Consider the system
$$x' = x + y - x\left(x^2 + y^2\right)$$
$$y' = -x + y - y\left(x^2 + y^2\right).$$

   (a) Show that $(x, y) = (0, 0)$ is the only equilibrium.

   (b) Use a computer to explore the phase portrait of the system. Formulate a conjecture about nonequilibrium orbits. Does there appear to be a limit cycle?

   (c) Calculate the Jacobian and determine the phase portrait in the neighborhood of the equilibrium $(0, 0)$. Does this agree with your observations in (b)?

   (d) The distance from the equilibrium $(0, 0)$ to a point $(x(t), y(t))$ on an orbit is $r(t) = \sqrt{x^2(t) + y^2(t)}$. Show that $r(t)$ satisfies the first-order differential equation $r' = \left(1 - r^2\right)r$. Draw the phase line portrait of this equation for $r \geq 0$.

   (e) Show that the polar angle $\theta(t) = \tan^{-1}\left(y(t)/x(t)\right)$ satisfies $\theta' = -1$.

   (f) Use (d) and (e) to draw the phase plane portrait of the system. Compare the result with your observations in (b).

8. Consider the system
$$x' = px + y - x\left(x^2 + y^2\right)$$
$$y' = -x + py - y\left(x^2 + y^2\right),$$

   where α is a constant.

   (a) Show that $(x, y) = (0, 0)$ is the only equilibrium of the system.

   (b) Use a computer to explore the phase portrait of the system for selected values of p, both positive and negative. Formulate a conjecture about nonequilibrium orbits and limit cycles.

(c) Calculate the Jacobian and determine the phase portrait in the neighborhood of the equilibrium $(0, 0)$. Does this agree with your observations in (b)?

(d) The distance from the equilibrium $(0, 0)$ to a point $(x(t), y(t))$ on an orbit is $r(t) = \sqrt{x^2(t) + y^2(t)}$. Show that $r(t)$ satisfies the first-order differential equation $r' = (p - r^2)r$. Draw the phase line portrait of this equation for $r \geq 0$.

(e) Show that the polar angle $\theta(t) = \tan^{-1}(y(t)/x(t))$ satisfies $\theta' = -1$.

(f) Use (d) and (e) to draw the phase plane portrait of the system. How does the portrait depend on p? Compare your result with your observations in (b).

9. Consider the planar autonomous system

$$x' = (1 - x - y)x$$
$$y' = (2 - x - y)y.$$

(a) Show that the x-axis and y-axis consist of orbits.

(b) Show that solutions starting in the first quadrant remain in the first quadrant for all t.

(c) Show that there are no cycles in the first quadrant.

(d) What happens to a bounded orbit as $t \to +\infty$? Explain your answer.

10. Consider the planar autonomous system

$$S' = -SI$$
$$I' = SI - I.$$

(a) Show that the S- and I-axes consist of orbits.

(b) Show that solutions starting in the first quadrant remain in the first quadrant for all t.

(c) Show that there are no cycles in the first quadrant.

(d) What happens to a bounded orbit as $t \to +\infty$? Explain your answer.

11. Apply the linearization principle to each of the six equilibria of the system (5.4). Classify the hyperbolic equilibria and identify any nonhyperbolic equilibria.

The distance from the origin $(0, 0)$ to a point $(x(t), y(t))$ on an orbit is $r(t) = \sqrt{x^2(t) + y^2(t)}$. If $r(t)$ is decreasing at t, then the orbit is moving toward the origin at the point $(x(t), y(t))$. A calculation shows that

$$\frac{dr(t)}{dt} = \frac{xx' + yy'}{r} = \frac{xf(x, y) + yg(x, y)}{r}.$$

One way to show that orbits are bounded as $t \to +\infty$ is to show they cannot move away from the origin, at least at points

far away from the origin. Thus, if $xf(x, y) + yg(x, y) \leq 0$ for r sufficiently large (say for r greater than some positive number r_0), then orbits are bounded. Use this test to show that the orbits of the following systems are bounded as $t \to +\infty$.

12. $\begin{cases} x' = -x - xy^2 \\ y' = -2y - x^2y \end{cases}$

13. $\begin{cases} x' = -x + y - x\left(x^2 + y^2\right) \\ y' = -x - 2y - y\left(x^2 + y^2\right) \end{cases}$

14. $\begin{cases} x' = x + y - x\left(x^2 + y^2\right) \\ y' = -x + y - y\left(x^2 + y^2\right) \end{cases}$

15. $\begin{cases} x' = px + y - x\left(x^2 + y^2\right) \\ y' = -x + py - y\left(x^2 + y^2\right), \end{cases}$
where p is a real number

The following is called the Dulac criterion. Suppose there exists a function $\mu = \mu(x, y)$ such that

$$\frac{\partial}{\partial x}(\mu f) + \frac{\partial}{\partial y}(\mu g) \neq 0$$

for all (x, y) in a simply connected region D of the plane.[6] Then the planar autonomous system

$$x' = f(x, y), \quad y' = g(x, y)$$

has no cycle in D.

16. Use the Dulac criterion with $\mu = 1/xy$ to prove part (c) in Exercise 9.

17. Use the Dulac criterion with $\mu = 1/I$ to prove part (c) in Exercise 10.

18. Use the Dulac criterion with $\mu = 1$ to show the system

$$x' = x - xy^2 + y^3$$
$$y' = 3y - yx^2 + x^3$$

has no cycle in the circle of radius 2 centered at the origin.

19. Use the Dulac criterion with $\mu = 1$ to show the system

$$x' = x + y - xy^2 - x^2y + y^2$$
$$y' = y - 2yx^2 + x^2$$

has no cycle inside the ellipse $2x^2 + 2xy + y^2 = 2$.

_____
[6] A simply connected region in the plane is one with no holes in it. In a simply connected domain a closed loop can encircle no point lying outside of the domain.

An extended Dulac criterion states that if

$$\frac{\partial}{\partial x}(\mu f) + \frac{\partial}{\partial y}(\mu g) \neq 0$$

for some function $\mu = \mu(x, y)$ and all x and y, then an orbit of $x' = f(x, y)$, $y' = g(x, y)$ is either unbounded or approaches an equilibrium as $t \to +\infty$.[7] Apply this criterion to the following systems. Hint: First try $\mu = 1$; then, if that fails to work, try $\mu = x^p y^q$ for some appropriate numbers p and q.

20. $\begin{cases} x' = -x - xy^2 \\ y' = -2y - x^2 y \end{cases}$

21. $\begin{cases} x' = -x + y - xy \\ y' = 2x - y + \frac{1}{2}y^2 \end{cases}$

22. $\begin{cases} x' = x + y + 2x^3 y \\ y' = 1 - \frac{1}{2}y + x^2 \end{cases}$

23. $\begin{cases} x' = 1 + x^2 - xy^2 \\ y' = -3xy + \frac{2}{3}y^3 \end{cases}$

24. $\begin{cases} x' = x + y^2 + x^3 \\ y' = -x + y + yx^2 \end{cases}$

25. $\begin{cases} x' = -x - y^2 - x^3 \\ y' = x - y - yx^2 \end{cases}$

26. (For readers who have studied multivariable calculus.) Use Green's theorem to prove the Dulac criterion in Exercises 16–19.

8.6 Bifurcations

The phase plane portrait of a system might change in significant ways if the numerical values of parameters appearing in the equations are changed. For example, the phase portrait of the linear system

$$x' = px + y$$
$$y' = -x + py$$

changes from a stable spiral to an unstable spiral as the value assigned to the parameter p changes from negative to positive. The reason for this is that the roots of the characteristic polynomial of the Jacobian

$$\begin{pmatrix} p & 1 \\ -1 & p \end{pmatrix}$$

are the complex numbers $\lambda = p \pm i$, whose real parts are p. Such a fundamental change in the phase portrait of a system is called a bifurcation.

We studied bifurcations in the phase line portraits of single autonomous equations in Chapter 3 (Sec. 3.1.4). We classified a few fundamental types of bifurcations, which for single equations necessarily involve only equilibrium configurations. In this section we will see that each of those types of bifurcations—saddle-node, pitchfork, and transcritical bifurcations—can also occur in systems of autonomous equations. For systems, however, we will also see that bifurcations can involve limit sets other than equilibria (e.g., limit cycles).

8.6.1 Local Bifurcations of Equilibria

One way to see that the bifurcation types we classified for single equations in Chapter 3 can also occur in systems of equations is to consider a system of the form

$$x' = f(x, p) \tag{6.1}$$
$$y' = -y.$$

In this kind of a system, the equations are uncoupled from one another and can be treated separately. The second equation implies $y(t) \to 0$ as $t \to +\infty$ for all solutions and, as a result, all orbits in the phase plane approach the x-axis. If we take the initial condition $y_0 = 0$, then $y(t) = 0$ for all t and, consequently from solutions of the first equation, we obtain orbits that lie on the x-axis. The first equation is, however, a

---

[7]C. C. McCluskey and J. S. Muldowney, *SIAM Review* 40, No. 4 (1998), 931–934.

single autonomous equation for x of the type studied in Chapter 3 and any bifurcations that its phase line portrait undergoes will correspond to a bifurcation for the planar autonomous system (6.1).

For example, consider the system

$$x' = x^2 - p \qquad (6.2)$$
$$y' = -y.$$

In Example 16 of Chapter 3 we saw that the equation $x' = x^2 - p$ undergoes a saddle-node bifurcation at the critical value $p_0 = 0$. For $p < 0$ this equation has no equilibria and for $p > 0$ it has two equilibria $x_e = \pm\sqrt{p}$. As a result, the phase plane portrait of the system (6.2) for $p < 0$ and for $p > 0$ appear similar to those shown in Fig. 8.12. For $p < 0$ orbits are unbounded. For $p > 0$ there are two equilibrium points

$$(x_e, y_e) = \left(\sqrt{p}, 0\right) \quad \text{and} \quad \left(-\sqrt{p}, 0\right)$$

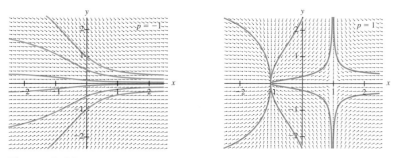

Figure 8.12 The direction fields for system (6.2) with $p = -1$ and $p = 1$.

The Jacobian

$$J(x, y) = \begin{pmatrix} 2x & 0 \\ 0 & -1 \end{pmatrix}$$

evaluated at the equilibrium $\left(\sqrt{p}, 0\right)$ is

$$J\left(\sqrt{p}, 0\right) = \begin{pmatrix} 2\sqrt{p} & 0 \\ 0 & -1 \end{pmatrix}.$$

The roots of the characteristic equation are $\lambda_1 = 2\sqrt{p} > 0$ and $\lambda_2 = -1 < 0$ and therefore this equilibrium is a saddle.

The Jacobian evaluated at the equilibrium $\left(-\sqrt{p}, 0\right)$ is

$$J\left(-\sqrt{p}, 0\right) = \begin{pmatrix} -2\sqrt{p} & 0 \\ 0 & -1 \end{pmatrix}.$$

The roots of the characteristic equation are $\lambda_1 = -2\sqrt{p} < 0$ and $\lambda_2 = -1 < 0$ and therefore this equilibrium is a stable node.

In summary, for p less than the critical (bifurcation) point $p_0 = 0$ there are no equilibria and for p greater than $p_0 = 0$ there is a saddle and a node.

For a single equation, the phase portrait scenario described above—namely, no equilibria for $p < p_0$ and two equilibria for $p > p_0$—is what we called a saddle-node bifurcation in Chapter 3. It is the phase plane case for two-dimensional systems, when the bifurcation involves a saddle and a node (as in the preceding example), that gives this type of bifurcation its name.

In Chapter 3 we drew bifurcation diagrams as a convenient graphical way to summarize the bifurcations that occur for single equations. These diagrams show the equilibria plotted against the parameter p. We can also draw bifurcation diagrams for planar systems of equations. If, however, we plot equilibria against the parameter p, we will have to draw in three dimensions, since equilibria are now ordered pairs. While this certainly can be done, three-dimensional pictures are often difficult to draw and understand. Therefore, the usual practice to plot a single representative of the equilibria against the parameter p. For example, we might plot just the x-coordinate or just the y-coordinate of the equilibria against p.

A bifurcation diagram for system (6.2) appears in Fig. 8.13.

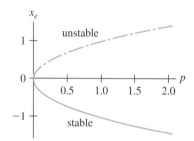

Figure 8.13 A plot of the x component of the equilibria against the parameter p results in this bifurcation diagram for system system (6.2).

In a similar fashion, we can provide illustrative examples of pitchfork and transcritical bifurcations in planar autonomous systems by choosing appropriate expressions for $f(x, p)$ in (6.1). Referring to Examples 17 and 18 in Chapter 3, the choices $f(x, p) = px - x^3$ and $f(x, p) = px - x^2$ in (6.2) produce systems with a pitchfork and a transcritical bifurcation at $p_0 = 0$, respectively. See Exercises 1 and 2.

EXAMPLE 1

The system

$$x' = (p - x)x - xy \qquad (6.3)$$
$$y' = (1 - 2y)y - xy$$

arises in competition theory. This system describes the dynamics of two populations which, when isolated from one another, grow according to the logistic equations

$$x' = (p - x)x$$
$$y' = (1 - 2y)y,$$

but when interacting have negative effects on each other's per capita growth rates. For example, when the two populations are interacting, population y negatively effects the per capita growth rate x'/x of population x by an amount proportional to its population size y. This gives rise to the term $-xy$ in the first equation of system (6.3). A similar effect of population x on the per capita growth rate of population y gives rise to the same term in the second equation of the system. The parameter $p > 0$ is positive.

The equilibrium equations

$$(p - x)x - xy = 0$$
$$(1 - 2y)y - xy = 0$$

yield four equilibria:

$$(x_e, y_e) = (0, 0), \quad (p, 0), \quad \left(0, \frac{1}{2}\right) \quad \text{and} \quad \left(2\left(p - \frac{1}{2}\right), 1 - p\right). \qquad (6.4)$$

The first equilibrium represents the absence of both species. The second and third equilibria represent the absence of one population and the presence of the other. The fourth equilibrium allows for the coexistence of the two competing populations (provided its components x_e and y_e are both positive), and it is to this equilibrium that we turn our attention.

In this application only equilibria and orbits with nonnegative values of x and y are relevant (since they denote population numbers or densities). The first three equilibria in (6.4) are therefore relevant in applications for all $p > 0$ values, while the fourth equilibrium is relevant if and only if $1/2 \le p \le 1$.

Note that when $p = 1/2$, the third and fourth equilibria in (6.4) coincide; otherwise they are distinct equilibria. This intersection of the two equilibria is a transcritical bifurcation (see Chapter 3). Another transcritical bifurcation occurs at $p = 1$, where the second and fourth equilibrium in (6.4) coincide. See Fig. 8.14.   ∎

An investigation of the stability properties of the equilibria in (6.4), using the linearization principle, shows a characteristic of transcritical bifurcations pointed out in Chapter 3, namely, an exchange of stability between the two intersecting equilibria. We will show this for the transcritical bifurcation at $p = 1/2$ that occurs between the third and fourth equilibria in (6.4). The reader is asked to investigate the bifurcation at $p = 1$ in Exercise 12.

We wish to apply the linearization principle to the third and fourth equilibria. The Jacobian of the system is

$$J(x, y) = \begin{pmatrix} p - 2x - y & -x \\ -y & 1 - 4y - x \end{pmatrix}.$$

When the Jacobian is evaluated at the equilibrium $(x_e, y_e) = (0, 1/2)$ we obtain the matrix

$$J\left(0, \frac{1}{2}\right) = \begin{pmatrix} p - \frac{1}{2} & 0 \\ -\frac{1}{2} & -1 \end{pmatrix}$$

whose characteristic polynomial has roots

$$\lambda = -1 \quad \text{and} \quad p - \frac{1}{2}.$$

If follows that

$$(x_e, y_e) = \left(0, \frac{1}{2}\right) \quad \text{is a} \quad \begin{cases} \text{stable node} & \text{if } p < \dfrac{1}{2} \\ \text{saddle} & \text{if } \dfrac{1}{2} < p. \end{cases}$$

An analysis of the equilibrium $(x_e, y_e) = \left(2\left(p - \frac{1}{2}\right), 1 - p\right)$ is a little more complicated but still algebraically manageable. The Jacobian at this equilibrium

$$J\left(2\left(p - \frac{1}{2}\right), 1 - p\right) = \begin{pmatrix} 1 - 2p & 1 - 2p \\ p - 1 & -2 + 2p \end{pmatrix}$$

whose characteristic polynomial has roots

$$\lambda_1 = \frac{1}{2}\left(-1 + \sqrt{5 - 12p + 8p^2}\right)$$

$$\lambda_2 = \frac{1}{2}\left(-1 - \sqrt{5 - 12p + 8p^2}\right).$$

A little bit of algebra shows

$$0 < 5 - 12p + 8p^2 < 1 \quad \text{if } \tfrac{1}{2} < p < 1$$

$$1 < 5 - 12p + 8p^2 \quad \text{otherwise.}$$

As a result, $\lambda_2 < 0$ for all $p > 0$ and

$$\lambda_1 < 0 \quad \text{if } \tfrac{1}{2} < p < 1$$
$$\lambda_1 > 0 \quad \text{otherwise.}$$

It follows that

$$(x_e, y_e) = \left(2\left(p - \tfrac{1}{2}\right), 1 - p\right) \quad \text{is a} \quad \begin{cases} \text{saddle} & \text{if } p < \tfrac{1}{2} \\ \text{stable node} & \text{if } \tfrac{1}{2} < p < 1 \\ \text{saddle} & \text{if } 1 < p. \end{cases}$$

One result of this analysis is that as p increases through the critical value $p_0 = 1/2$ and the third and fourth equilibria intersect, an exchange of stability occurs between these two equilibria. See Fig. 8.14.

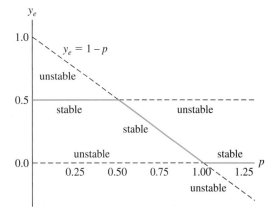

Figure 8.14 A plot of the y_e component of the equilibria (x_e, y_e) against the parameter p results in this bifurcation diagram for system system (6.3).

In terms of the application to competition theory, when $p < p_0$ the extinction equilibrium $(0, 1/2)$ is stable, a fact indicating that population x goes extinct as $t \to +\infty$. Only when p is increased greater than p_0 can population x survive, as indicated by the stability of the fourth equilibrium in which the x component is positive. (To strengthen these conclusions obtained from *local* stability results supplied by the linearization principle, we would need to investigate the global phase portraits.)

In this section we have looked at bifurcations involving equilibria. In the next section we consider a bifurcation that involves cycles.

8.6.2 Hopf Bifurcation of Limit Cycles

An investigation of the bifurcation examples in the previous section reveals that at the bifurcation value $p = p_0$ the equilibria are nonhyperbolic. Moreover, they are non-hyperbolic because one of the roots of the characteristic polynomial of the Jacobian equals 0. This is characteristic of equilibrium bifurcations. Another important type of bifurcation occurs when an equilibrium is nonhyperbolic because a root of the characteristic polynomial of the Jacobian has zero real part but is *not* equal to 0; that is, a root of the form $\lambda = \beta i$, $\beta \neq 0$. (Actually, since complex roots appear in conjugate pairs, at such a nonhyperbolic equilibrium the two roots of the form $\lambda = \pm\beta i$, $\beta \neq 0$.)

Here is an example. The origin $(x_e, y_e) = (0, 0)$ is an equilibrium of the planar autonomous system

$$x' = px + y - x^3 \tag{6.5}$$
$$y' = -x + py - y^3.$$

The Jacobian evaluated at $(0, 0)$ is

$$J(0, 0) = \begin{pmatrix} p & 1 \\ -1 & p \end{pmatrix}.$$

The roots of the characteristic polynomial are

$$\lambda = p \pm i.$$

By the linearization principle $(0, 0)$ is a stable spiral for $p < 0$ and an unstable spiral for $p > 0$. The value $p_0 = 0$ is therefore a bifurcation point. Furthermore, for $p = 0$ the origin is nonhyperbolic (because Re $\lambda = 0$).

For $p < 0$, orbits near the origin spiral into the origin as $t \to +\infty$. For $p > 0$, orbits near the origin spiral away. Where do these orbits go? Are the unbounded as $t \to +\infty$? Figure 8.15 shows an example that indicates that these orbits are bounded as $t \to +\infty$ and that the limit set of these orbits is a cycle!

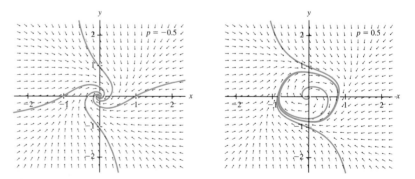

Figure 8.15 Sample direction fields and orbits of the system (6.5) for a negative and a positive value of p. In the latter case, orbits approach a limit cycle.

In summary, as p increases through $p_0 = 0$ in the system (6.5), the equilibrium $(0, 0)$ loses stability and a limit cycle is created. This is an example of a famous theorem concerning bifurcation of limit cycles. This theorem—called the Hopf bifurcation theorem—guarantees that a limit cycle is created under the circumstances described previously (provided certain other conditions are met). A full detailed statement of this theorem is too technical for the level of this book. Roughly speaking, however, the Hopf bifurcation theorem says the following:

Suppose an equilibrium of the system

$$x' = f(x, y, p)$$
$$y' = g(x, y, p)$$

loses stability as the parameter p passes through a critical value p_0 (increasing or decreasing) and does so because a pair of complex conjugate characteristic roots

$$\lambda(p) = \alpha(p) \pm \beta(p)i$$

of the Jacobian evaluated at the equilibrium crosses transversely from the left to the right half complex plane (or vice versa). By this is meant

$$\alpha(p_0) = 0, \quad \beta(p_0) \neq 0, \quad \left.\frac{d\alpha}{dp}\right|_{p_0} \neq 0. \tag{6.6}$$

Then (if a certain technical condition is met) limit cycles bifurcate from the equilibrium at $p = p_0$. These limit cycles exist for p (near and) either greater than or less than p_0. The cycles encircle the equilibrium and have amplitudes that shrink to zero and periods that approach $2\pi/\beta(p_0)$ as $p \to p_0$.

The technical condition mentioned in this statement requires that a quantity, calculated from higher-order partial derivatives of f and g, be nonzero. While general formulas for this quantity are available, they are complex and often difficult to apply in applications. It is also possible to determine, by means of complicated formulas, whether the bifurcating limit cycles are stable (i.e., attract all nearby orbits as $t \to +\infty$) or unstable (not attract all nearby orbits) and whether they exist for $p < p_0$ or for $p > p_0$. In practice, we use the criterion (6.6) to locate possible Hopf bifurcation points and rely on computer examples to determine whether limit cycles exist and are stable or unstable. For example, we can use a computer to sketch direction fields and draw some sample orbits for selected values of $p < p_0$ and $p > p_0$.

For the system (6.5) we have $\alpha(p) = p$ and $\beta(p) = 1$ and the Hopf criteria (6.6) are satisfied with $p_0 = 0$. The computer-generated examples in Fig. 8.15 show that a bifurcation to a stable limit cycle occurs at this value of p.

The technical condition mentioned in the Hopf theorem cannot be ignored. That is, if it is eliminated, then the assertion of the Hopf theorem might not be true. An example is the van der Pol system

$$x' = y \tag{6.7}$$
$$y' = -x - p(x^2 - 1)y.$$

The roots

$$\lambda = \frac{1}{2}p \pm \frac{1}{2}i\sqrt{4 - p^2}$$

of the characteristic polynomial of the Jacobian evaluated at the equilibrium $(x_e, y_e) = (0, 0)$ satisfy the criteria (6.6) at $p = p_0 = 0$. Yet a Hopf bifurcation does not occur; see Exercise 20. It turns out that the technical condition is not met.

EXAMPLE 2

The system

$$x' = 1 - (p+1)x + x^2 y \qquad (6.8)$$
$$y' = px - x^2 y$$

is an example of a model for a idealized chemical reaction called the Brusselator reaction. In this system x and y are chemical concentrations and p is a positive constant. We can algebraically solve the equilibrium equations

$$1 - (p+1)x + x^2 y = 0$$
$$px - x^2 y = 0$$

to find the equilibrium $(x_e, y_e) = (1, p)$. (*Hint*: Add the two equations.) The Jacobian

$$J(x, y) = \begin{pmatrix} -p - 1 + 2xy & x^2 \\ p - 2xy & -x^2 \end{pmatrix}$$

evaluated at the equilibrium is

$$J(1, p) = \begin{pmatrix} p - 1 & 1 \\ -p & -1 \end{pmatrix}.$$

The roots

$$\lambda(p) = \frac{1}{2}(p - 2) \pm \frac{1}{2} i \sqrt{p(4 - p)}$$

of the characteristic polynomial are complex for $p < 4$. The real part $\alpha(p) = (p - 2)/2$ and imaginary part $\beta(p) = \sqrt{p(4 - p)}/2$ satisfy the criteria (6.6) with $p_0 = 2$. The graphs in Fig. 8.16 indicate that a Hopf bifurcation of stable limit cycles occurs at $p_0 = 2$.

In terms of the application, we find that the chemical concentrations will equilibrate if $p < 2$ but will settle into sustained periodic oscillations if $p > 2$. ∎

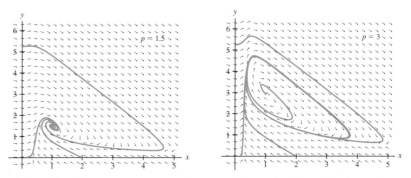

Figure 8.16 These two sample direction fields and orbits of the Brusselator system (6.8) show the a stable spiral equilibrium for $p < 2$ and an unstable spiral equilibrium encircled by a stable limit cycle for $p > 2$.

Bifurcation diagrams provide a useful way to summarize equilibrium bifurcations that occur in a system. To include Hopf bifurcations in such a graph we need to devise a way to "plot" a cycle in the diagram. One way to do this is to plot both the maximum and the minimum of one component of the cycles (say, the x component) against the parameter value p. Figure 8.17(a) shows an example of how a Hopf bifurcation would appear in such a plot. We must indicate on such a graph, or in its caption, that the plot represents a cycle, so that the plot is not mistaken for a pitchfork bifurcation of equilibria. Another way to plot a Hopf bifurcation is shown in Fig. 8.17(b). In this plot, all values of one component of the cycle are plotted above the corresponding value of p. Since these values will cover an entire interval of numbers ranging from the maximum to the minimum of the component, the plot appears as a vertical line segment lying above the corresponding value of p.

Figure 8.17 Two ways to plot a Hopf bifurcation in a bifurcation diagram are shown. In (a) only the maximum and the minimum of the x component of the cycle are plotted. In (b) all values of the x component of the cycle are plotted.

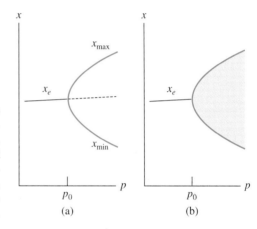

(a)   (b)

Exercises

1. Find all equilibria of the system

$$x' = px - x^3$$
$$y' = -y.$$

Apply the linearization principle to classify each equilibrium. Describe the phase plane portrait, and how it depends on the parameter p. Show that a pitchfork bifurcation occurs in this system.

2. Find all equilibria of the system

$$x' = px - x^2$$
$$y' = -y.$$

Describe the phase plane portrait and how it depends on the parameter p. Show that a transcritical bifurcation occurs in this system.

Classify the bifurcations that occur in the following systems with parameter p.

3. $\begin{cases} x' = p - x^2 - y^2 \\ y' = 1 - x - y \end{cases}$

4. $\begin{cases} x' = [(x-1)^2 + y^2 - p](x^2 + y^2) \\ y' = x(x^2 + y^2) \end{cases}$

5. $\begin{cases} x' = (p - 2x^2 - y^2)[(x-1)^2 + y^2] \\ y' = (x-1)[(x-1)^2 + y^2] \end{cases}$

6. $\begin{cases} x' = y + (1-x)(2-x) \\ y' = y - px^2 \end{cases}$

7. $\begin{cases} x' = px - xy \\ y' = py - xy \end{cases}$

8. $\begin{cases} x' = (x - 2p)(p - y) \\ y' = (y + p)(p - x) \end{cases}$

9. $\begin{cases} x' = y - \ln x \\ y' = px - y \end{cases}$

10. $\begin{cases} x' = y - e^{-x} \\ y' = y + x^2 - p \end{cases}$

11. The nonlinear, second-order equation $x'' + (p - \cos x) \sin x = 0$, $p > 0$, is called the rotating pendulum equation. It models the motion of a swinging pendulum whose pivot rotates in a circle; x is the angle made by the pendulum with the vertical. Note that $x_e = 0$ is an equilibrium for all p. Does this equilibrium undergo a bifurcation?

12. Show that a transcritical bifurcation of equilibria and an exchange of stability occur in the competition system (6.3) at the critical value $p_0 = 1$.

13. Apply the linearization principle to the equilibrium $(x_e, y_e) = (0, 0)$ of the competition system (6.3).

For each of the systems below show the equilibrium $(x_e, y_e) = (0, 0)$ undergoes a Hopf bifurcation by supplementing the criteria (6.6) with computer sketches of the phase plane portrait. Identify the bifurcation value p_0.

14.
$$\begin{cases} x' = 1 + (p - 1)x + (p - 2)y - e^{xy} \\ y' = x + y - y^3 \end{cases}$$

15.
$$\begin{cases} x' = 1 - e^{px+y} \\ y' = 1 - x^2 y - e^{-x+py} \end{cases}$$

16.
$$\begin{cases} x' = px + y - x\left(x^2 + y^2 - 1\right)\left(x^2 + y^2 - 4\right) \\ y' = -3x + py - y\left(x^2 + y^2 - 1\right)\left(x^2 + y^2 - 4\right) \end{cases}$$

17.
$$\begin{cases} x' = px + 2y + x\left(x^2 + y^2 - 1\right) \\ y' = -x + py + 3y\left(x^2 + y^2 - 1\right) \end{cases}$$

18.
$$\begin{cases} x' = y - x^3 \\ y' = -2x - py - x^3 - xy^2 \end{cases}$$

19.
$$\begin{cases} x' = 1 + x - e^x + y \\ y' = -x - y \ln\left(p + x^2 + y^2\right) \end{cases}$$

20. Use a computer to calculate the limit cycles of the van der Pol system (6.7) for $p > 0$. What happens to the limit cycle as $p \to 0$? Does it bifurcate from the equilibrium $(x_e, y_e) = (0, 0)$?

8.7 Higher-Dimensional Systems

As the number of equations in a system of differential equations increases, the dynamics of solutions and orbits can become more complicated. The solutions of a single autonomous equation are monotonic and bounded orbits approach an equilibrium as $t \to +\infty$. Bounded orbits of systems of two autonomous equations, however, do not have to approach equilibria; solutions can oscillate and orbits approach limit cycles or cycle chains. Bounded orbits from systems of three (or more) autonomous equations can also approach equilibria or cycles (or cycle chains). However, it is also possible for bounded orbits in three- (or higher-) dimensional phase space to approach limit sets that are none of these. The behavior of solutions and orbits of higher-dimensional systems can even be so irregular that they are called chaotic.

The analysis of equilibria, by means of the linearization principle, can be extended to higher-dimensional systems. The analytic study of more complex solutions and exotic limit sets, on the other hand, is difficult, and for their investigation we usually rely a great deal on computer explorations.

8.7.1 The Linearization Principle

In 1963 E. N. Lorenz, a meteorologist studying the dynamics of a layer of fluid heated from below, investigated the nonlinear system of three first-order equations

$$\begin{align} x' &= \sigma (y - x) \\ y' &= \rho x - y - xz \\ z' &= -\beta z + xy. \end{align} \tag{7.1}$$

The system contains three positive parameters: σ (the Prandtl number), ρ (the Rayleigh number), and β (the aspect ratio). Lorenz investigated the dynamics of solutions of (7.1) for values of ρ considerably larger than 1 and found that some very complicated orbits can result. Although it was later shown that this system is an accurate approximation to the original fluid dynamic problem only for Rayleigh numbers ρ near 1, the

Lorenz equations (7.1) have become a prototypical mathematical example of a system with a strange attractor. In this section we will look, however, at only the equilibrium solutions of the Lorenz system. We investigate a strange attractor for this system in Sec. 8.7.2.

A solution of the system (7.1) is a triple of functions $x = x(t)$, $y = y(t)$, $z = z(t)$ for which the points $(x(t), y(t), z(t))$ define an orbit. Because these orbits lie in three dimensional phase space, we say (7.3) is a three-dimensional system. More generally, a system

$$
\begin{aligned}
x_1' &= f_1(x_1, \ldots, x_n) \\
x_2' &= f_2(x_1, \ldots, x_n)
\end{aligned}
\tag{7.2}
$$

$$\vdots$$

$$
x_n' = f_n(x_1, \ldots, x_n)
$$

of n first-order (autonomous) equations for n unknown functions $x_1 = x_1(t)$, $x_2 = x_2(t), \ldots, x_n = x_n(t)$ is an *n-dimensional system*. An equilibrium solution consists of n constant functions $x_1 = e_1$, $x_2 = e_2, \ldots, x_n = e_n$, where the constants satisfy the n *equilibrium equations*

$$
\begin{aligned}
f_1(x_1, \ldots, x_n) &= 0 \\
f_2(x_1, \ldots, x_n) &= 0
\end{aligned}
$$

$$\vdots$$

$$
f_n(x_1, \ldots, x_n) = 0.
$$

EXAMPLE 1

A commonly studied special case of the Lorenz system (7.1) is

$$
\begin{aligned}
x' &= 10(y - x) \\
y' &= \rho x - y - xz \\
z' &= -\frac{8}{3}z + xy.
\end{aligned}
\tag{7.3}
$$

This system arises from (7.1) by setting $\sigma = 10$ and $\beta = 8/3$ and leaving only the one parameter ρ unspecified. The equilibrium equations for this system are

$$
\begin{aligned}
10(y - x) &= 0 \\
\rho x - y - xz &= 0 \\
-\frac{8}{3}z + xy &= 0.
\end{aligned}
$$

The first equation implies $y = x$, which, when substituted into the remaining equations, yields two equations

$$
\begin{aligned}
(\rho - 1 - z)\, y &= 0 \\
-\frac{8}{3}z + y^2 &= 0
\end{aligned}
$$

for two unknowns y and z. The first of these equations implies

$$
y = 0 \quad \text{or} \quad z = \rho - 1.
$$

The first choice $y = 0$ and the second equation imply $z = 0$ and we have the equilibrium

$$(x, y, z) = (0, 0, 0).$$

The second choice $z = \rho - 1$ and the second equation imply $y^2 = \frac{8}{3}(\rho - 1)$. If $0 < \rho \le 1$, the only equilibrium of the system (7.3) is the origin,

$$(x, y, z) = (0, 0, 0).$$

On the other hand, if $\rho > 1$, there are three equilibria:

$$(x, y, z) = \begin{cases} (0, 0, 0) \\ \left(\sqrt{\frac{8}{3}(\rho - 1)}, \sqrt{\frac{8}{3}(\rho - 1)}, \rho - 1\right) \\ \left(-\sqrt{\frac{8}{3}(\rho - 1)}, -\sqrt{\frac{8}{3}(\rho - 1)}, \rho - 1\right). \end{cases}$$

Thus, there is a pitchfork bifurcation of equilibria at the bifurcation value $\rho_0 = 1$. ∎

Suppose $(x_1, x_2, \ldots, x_n) = (e_1, e_2, \ldots, e_n)$ is an equilibrium point of the system (7.2). We obtain the linearization of the system at the equilibrium from the Taylor approximation of

$$f_i(x_1, \ldots, x_n) \approx \sum_{j=1}^{n} \left.\frac{\partial f_i}{\partial x_j}\right|_{(e_1, \ldots, e_n)} \left(x_j - e_j\right).$$

Namely, letting $y_i = x_i - e_i$, we obtain the system of n linear equations

$$y_i' = \sum_{j=1}^{n} \left.\frac{\partial f_i}{\partial x_j}\right|_{(e_1, \ldots, e_n)} y_j,$$

which is the *linearization of (7.2) at the equilibrium point* $(e_1, e_2, \ldots, e_n)$. The coefficient matrix of this linear system is the Jacobian

$$J(x_1, x_2, \ldots, x_n) = \begin{pmatrix} \frac{\partial f_1}{\partial x_1} & \frac{\partial f_1}{\partial x_2} & \cdots & \frac{\partial f_1}{\partial x_n} \\ \frac{\partial f_2}{\partial x_1} & \frac{\partial f_2}{\partial x_2} & \cdots & \frac{\partial f_2}{\partial x_n} \\ \vdots & \vdots & & \vdots \\ \frac{\partial f_n}{\partial x_1} & \frac{\partial f_n}{\partial x_2} & \cdots & \frac{\partial f_n}{\partial x_n} \end{pmatrix}$$

of the nonlinear system (7.2) in which all partial derivatives are evaluated at the equilibrium. That is, the coefficient matrix of the linearization is $J(e_1, e_2, \ldots, e_n)$. The *characteristic polynomial*

$$\det(J(e_1, e_2, \ldots, e_n) - \lambda I)$$

of the coefficient matrix has degree n. Here

$$I = \begin{pmatrix} 1 & 0 & \cdots & 0 \\ 0 & 1 & \cdots & 0 \\ \vdots & \vdots & & \vdots \\ 0 & 0 & \cdots & 1 \end{pmatrix}$$

is called the $n \times n$ identity matrix. By the fundamental theorem of algebra, the characteristic polynomial has n roots (allowing for complex roots and counting multiplicities); these roots are the *characteristic roots* (or the *eigenvalues*) of the coefficient matrix.

The following theorem is the n-dimensional generalization of Theorem 1 and embodies the *linearization principle* for systems of arbitrary dimension. The stability definitions given in Sec. 8.3 for planar autonomous systems extend straightforwardly to systems of any dimension.

THEOREM 5   **The Linearization Principle**   Suppose $(x_1, \ldots, x_n) = (e_1, e_2, \ldots, e_n)$ is an equilibrium point of the of the n-dimensional system (7.2). If all characteristic roots (eigenvalues) of the Jacobian $J(e_1, e_2, \ldots, e_n)$ have negative real parts, then the equilibrium point is stable. If at least one characteristic root (eigenvalue) has positive real part, then the equilibrium is unstable.

EXAMPLE 2   The Jacobian of the Lorenz system (7.3) is

$$J(x, y, z) = \begin{pmatrix} -10 & 10 & 0 \\ \rho - z & -1 & -x \\ y & x & -\frac{8}{3} \end{pmatrix}.$$

For the equilibrium $(0, 0, 0)$ the Jacobian

$$J(0, 0, 0) = \begin{pmatrix} -10 & 10 & 0 \\ \rho & -1 & 0 \\ 0 & 0 & -\frac{8}{3} \end{pmatrix}$$

has the cubic characteristic polynomial

$$\det(\lambda I - J(0, 0, 0)) = \left(\lambda + \frac{8}{3} \right)\left(\lambda^2 + 11\lambda - 10\rho + 10 \right)$$

whose three roots are

$$\lambda = -\frac{8}{3}, \quad \frac{1}{2}\left(-11 \pm \sqrt{121 - 40(1 - \rho)} \right).$$

If $\rho < 1$, all three roots are real and negative. By Theorem 5, the origin $(x, y, z) = (0, 0, 0)$ is stable if $\rho < 1$.

If $\rho > 1$, then one root, namely

$$\lambda = \frac{1}{2}\left(-11 + \sqrt{121 - 40(1 - \rho)} \right),$$

is positive and the origin is unstable. See Fig. 8.18.   ■

When $\rho = 1$ in the preceding example, $\lambda = 0$ is a characteristic root (eigenvalue) and Theorem 5 does not apply. (This does not mean the equilibrium is unstable. It simply means we cannot draw any conclusion from the Theorem 5.) In general, if the Jacobian at an equilibrium has a characteristic root (eigenvalue) with real part equal to 0, then the equilibrium is called *nonhyperbolic*; otherwise (when no eigenvalue has real part equal to 0) the equilibrium is *hyperbolic*. Theorem 5 applies to hyperbolic equilibria.

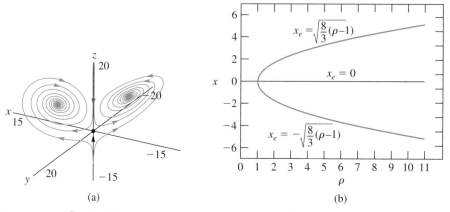

Figure 8.18 (a) Orbits of the Lorenz system (7.3) with $\rho = 10$. The equilibrium $(x, y, z) = (0, 0, 0)$ is unstable. The two equilibria $(x, y, z) = (2\sqrt{6}, 2\sqrt{6}, 9)$ and $(-2\sqrt{6}, -2\sqrt{6}, 9)$ are stable spirals. (b) A bifurcation diagram showing the pitchfork bifurcation that occurs in the Lorenz system (7.3) at $\rho = 1$.

In Example 2 the equilibrium $(x, y, z) = (0, 0, 0)$ of the Lorenz system (7.3) loses stability as ρ increases through 1. The origin $(0, 0, 0)$ is the only equilibrium for $\rho < 1$ (Example 1), but for $\rho > 1$ there are two additional equilibria (i.e., a pitchfork bifurcation occurs at $\rho = 1$) (Sec. 3.1.4 in Chapter 3). It turns out that these two additional equilibria are stable for $\rho > 1$ close to 1 (Exercise 9). For ρ sufficiently large, however, each of these stable equilibria loses stability and all three equilibria are unstable! We see in the next section that solutions of the Lorenz equation exhibit unusual behavior for ρ large.

At the bifurcation point $\rho = 1$ in the Lorenz system (7.3), $\lambda = 0$ is a characteristic root of the Jacobian at the equilibrium $(0, 0, 0)$. We have seen that this kind of nonhyperbolicity is generally associated with the bifurcation of equilibria (such as, in this case, a pitchfork bifurcation). For nonhyperbolic cases associated with eigenvalues $\lambda = \pm \beta i$, $\beta \neq 0$, the Hopf bifurcation theorem for planar systems implies a limit cycle bifurcation can occur. It turns out that the Hopf theorem also holds for systems of three or more equations. The only additional requirement is that at the bifurcation point all other characteristic roots have nonzero real part.

EXAMPLE 3

In Exercise 10 the reader is asked to show, by means of the linearization principle, that the equilibrium $(x_e, y_e, z_e) = (0, 0, 0)$ of the system

$$x' = 2x + 2y - 2xy - \tfrac{1}{10}x^2$$
$$y' = pz - y - xy \tag{7.4}$$
$$z' = \tfrac{1}{5}(x - z)$$

is unstable for all p. In this example we are interested in other equilibria of this system. We consider two cases: $p = 1/2$ and 1.

For $p = 1/2$ the system has two other equilibria:

$$E_1 : (x_e, y_e, z_e) = (11.59, \ -0.2897, \ 11.59)$$
$$E_2 : (x_e, y_e, z_e) = \left(-2.589, \ 6.472 \times 10^{-2}, \ -2.589\right).$$

The characteristic roots of the Jacobian

$$J(x, y, z) = \begin{pmatrix} 2 - 2y - \frac{1}{5}x & 2 - 2x & 0 \\ -y & -1 - x & p \\ \frac{1}{5} & 0 & -\frac{1}{5} \end{pmatrix}$$

evaluated at these equilibria are

$$E_1 \colon \lambda \approx -12.11 \text{ and } -0.2100 \pm 0.4216i$$
$$E_2 \colon \lambda \approx 1.729, 2.085 \text{ and } -3.717 \times 10^{-2}.$$

Therefore, by the linearization principle,

$$\text{for } p = \tfrac{1}{2}, E_1 \text{ is stable and } E_2 \text{ is unstable.}$$

For $p = 1$ the system has two equilibria:

$$E_1 \colon (x_e, y_e, z_e) = (5.844, \ 0.8540, \ 5.844)$$
$$E_2 \colon (x_e, y_e, z_e) = (-6.8443, \ 1.171, \ -6.844).$$

The characteristic roots of the Jacobian evaluated at these equilibria are

$$E_1 \colon \lambda \approx -8.035 \text{ and } 5.684 \times 10^{-2} \pm 0.4259i$$
$$E_2 \colon \lambda \approx, -7.392 \times 10^{-2} \text{ and } 3.372\,4 \pm 3.482i.$$

Therefore, by the linearization principle

$$\text{for } p = 1, E_1 \text{ is unstable and } E_2 \text{ is unstable.}$$

We see, as p is increased from $1/2$ to 1, that the equilibrium E_2 remains unstable. The equilibrium E_1, however, loses stability because the real part of a complex pair of characteristic roots changes from negative to positive. This suggests that a Hopf bifurcation might have occurred at some critical value of p between $1/2$ and 1. The existence of a limit cycle when $p = 1$ is corroborated by Fig. 8.19. ■

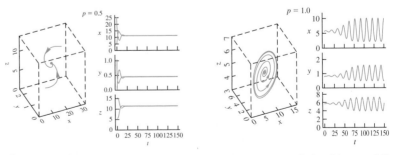

Figure 8.19 Two orbits of the three-dimensional system (7.4) with $p = 1/2$ are shown approaching the stable equilibrium E_1. For $p = 1$, there exists a limit cycle, which is shown approached by an orbit starting near the unstable equilibrium E_1.

8.7.2 Strange Attractors and Chaos

Orbits of systems consisting of three or more equations can be very complicated. Bounded orbits can approach unusual sets in phase space that are neither equilibrium points or cycles. Solutions can exhibit such irregular oscillations that they are called chaotic and limit sets can be so exotic that they are called strange attractors. A mathematical study of these kinds of solutions is in general very difficult to carry out. In this section we will look briefly at only one example of an exotic chaotic solution, using the computer as an aide. The example comes from the Lorenz system (7.3).

We saw in Example 2 that when $\rho < 1$ the point $(x, y, z) = (0, 0, 0)$ is the only equilibrium of the Lorenz system (7.3) and that this equilibrium is stable. For $\rho > 1$ there are two additional equilibria (see Example 1), both of which are stable for ρ near 1 (Exercise 9). For large ρ, however, these two equilibria are unstable and orbits of the Lorenz system (7.3) can become extremely complicated.

For example, Fig. 8.20 shows an orbit when $\rho = 28$. This orbit lies on a fascinating "double-winged" looking surface whose two branches are centered on two of the equilibria. The orbit moves on this surface in a complicated manner. It circles around one of the branches several times before embarking on an excursion to the other branch, around which it then circles before returning to the first branch, and so on indefinitely. The number consecutive circuits flown around each loop is irregular and unpredictable.

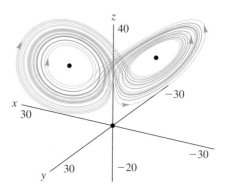

Figure 8.20 An orbit for the Lorenz system (7.3) with $\rho = 28$. All three equilibria $(x, y, z) = (0, 0, 0)$, $(6\sqrt{2}, 6\sqrt{2}, 27)$, and $(-6\sqrt{2}, -6\sqrt{2}, 27)$ are unstable.

The irregular nature of the oscillations in x, y and z for the orbit in Fig. 8.20 is seen in Fig. 8.21. This kind of solution is called *chaotic*. An important feature of chaos is that solutions whose initial conditions are very close together do not remain close together as t increases. This divergence of solutions (or orbits) that start arbitrarily close together is called *sensitivity to initial conditions*. See Fig. 8.22. This property is a hallmark of chaos and has important consequences in applications. It means that small errors or perturbations in initial conditions result in drastically different long-term predictions. Given that errors in measuring initial conditions and/or external disturbances are inevitable in applications, this property raises serious questions concerning the ability to make long-term predictions.

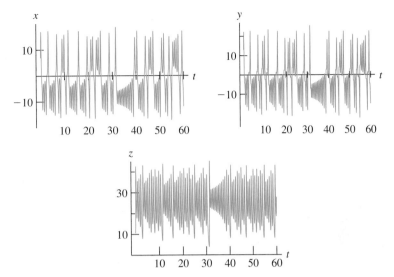

Figure 8.21 The x, y, and z components of the orbits in Fig. 8.20 show irregular oscillations.

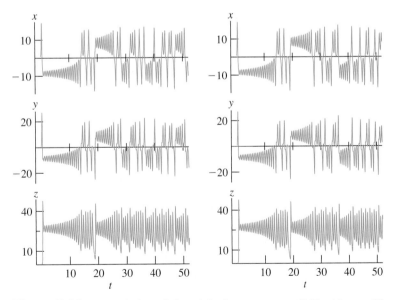

Figure 8.22 Two solution triples of the Lorenz system (7.3) with $\rho = 28$ with slightly different initial conditions. For $t > 30$ the solutions have little in common. (a) Initial condition $(x, y, z) = (1, 1, 1)$. (b) Initial condition $(x, y, z) = (1.0001, 1, 1)$.

Exercises

The origin $(x, y, z) = (0, 0, 0)$ is an equilibrium of each of the systems below. Find the Jacobian of the system, write down the linearization at the origin, and apply the linearization principle to determine the stability of the origin (if possible).

1. $\begin{cases} x' = x(1-x) - xy - xz \\ y' = -y + xy \\ z' = -z + xz \end{cases}$

2. $\begin{cases} x' = x - ye^{-z} \\ y' = y - ze^{-x} \\ z' = z - xe^{-y} \end{cases}$

3. $\begin{cases} x' = -x^2 - xy^2 - xz \\ y' = y - xy - y^2 - yz^2 \\ z' = -z - zx^2 - yz - z^2 \end{cases}$

4. $\begin{cases} x' = \sin(x + y + z) \\ y' = \ln(1 + 2x + y - z) \\ z' = -z + x^2 + y^2 \end{cases}$

5. $\begin{cases} x' = -1 + e^{y-2x} \\ y' = -z \\ z' = -1 + e^{y-2z} \end{cases}$

6. $\begin{cases} x' = x - \sin(5x - y) \\ y' = -2y + z - x^2 \\ z' = x - 5ze^{-y} \end{cases}$

Find the equilibria of the systems below and determine their stability properties by applying the linearization principle.

7. $\begin{cases} x' = x + y - xy - x^2 \\ y' = 4z - y - xy \\ z' = x - z \end{cases}$

8. $\begin{cases} x' = x + y - xy - x^2 \\ y' = -4z - y - xy \\ z' = x - z \end{cases}$

9. Show that the two equilibria

$$(x_e, y_e, z_e) = \left(\sqrt{8(\rho-1)/3}, \sqrt{8(\rho-1)/3}, \rho-1\right)$$
$$(x_e, y_e, z_e) = \left(-\sqrt{8(\rho-1)/3}, -\sqrt{8(\rho-1)/3}, \rho-1\right)$$

of the Lorenz equations (7.3) are asymptotically stable for $\rho > 1$ close to 1. [*Hint*: Let $\lambda = \lambda(\rho)$ be the eigenvalue that equals 0 at $\rho = 1$. Show that $d\lambda(1)/d\rho < 0$.]

10. Use the linearization principle to show the equilibrium $(x_e, y_e, z_e) = (0, 0, 0)$ of the system (7.4) is unstable for all $p > 0$. (*Hint*: Find the characteristic polynomial and argue that it always has a positive root. You need not calculate this root, nor the other two roots.)

11. This exercise is a computer investigation of some bifurcations and chaos that occurs in the three-dimensional (Roessler) system

$$x' = -y - z$$
$$y' = x + by$$
$$z' = c + z(x - a)$$

with $a = 5.7$ and $b = 0.2$.

(a) Use a computer program to determine what orbits do as $t \to +\infty$ for values of c *ranging* from 10.0 down to 0.2. Describe the bifurcations that occur. (A "bifurcation" here means a significant change in what orbits approach as $t \to +\infty$.)

(b) How would you describe the orbits when $b = 0.2$?

8.8   Chapter Summary and Exercises

In this chapter we studied nonlinear systems of autonomous equations. We studied techniques used to determine the stability properties of equilibrium points and learned methods that aid in the construction of phase portraits. The linearization principle (Theorems 1 and 5 and Corollary 1) relates the phase portrait near an equilibrium to that of the linearization at the equilibrium. The roots of the characteristic polynomial associated with the linearization (i.e., the eigenvalues of the Jacobian of the system evaluated at the equilibria) determine the nature of the phase portrait in a neighborhood of the equilibrium (provided it is hyperbolic). We saw how the bifurcation scenarios for equilibria that we classified for single autonomous equations in Chapter 3 (saddle-node, pitchfork, and transcritical) occur in planar autonomous systems as well. Unlike the phase line portrait of a single autonomous equation, however, the global phase portrait of a system is not easily determined from the local phase portraits near equilibrium points. For example, phase plane portraits can contain closed loop orbits or cycles, which correspond to periodic solutions. The Poincaré-Bendixson and the Hopf bifurcation theorem are important tools for the study of such cycles. In systems consisting of three or more equations even more complicated orbits can arise.

Exercises

Find the equilibria of the planar autonomous systems below. If it applies, use the linearization principle to determine their stability properties.

1. $\begin{cases} x' = x + y + x^2 + y^2 \\ y' = y + xy \end{cases}$

2. $\begin{cases} x' = y^2 - (x-1)^2 \\ y' = 1 + x - 2y \end{cases}$

3. $\begin{cases} x' = -y - x^2 \\ y' = -x + xy^2 \end{cases}$

4. $\begin{cases} x' = y - x^2 \\ y' = -x + xy^2 \end{cases}$

5. $\begin{cases} x' = -x - xy^2 \\ y' = x - 2y^2 \end{cases}$

6. $\begin{cases} x' = x - xy^2 \\ y' = x - 2y^2 \end{cases}$

Find the equilibria of the following planar autonomous systems. Using p as a parameter, identify all equilibrium bifurcations.

7. $\begin{cases} x' = x + py + x^2 + y^2 \\ y' = y + xy \end{cases}$

8. $\begin{cases} x' = y^2 - (x-1)^2 \\ y' = p + x - 2y \end{cases}$

9. $\begin{cases} x' = x - xy^2 \\ y' = p + x - 2y^2 \end{cases}$

10. $\begin{cases} x' = py - x^2 \\ y' = -x + xy^2 \end{cases}$

11. $\begin{cases} x' = x + y - y^2 \\ y' = x - y + p^2 y^2 \end{cases}$

12. $\begin{cases} x' = px - xy^2 \\ y' = x - 2y^2 \end{cases}$

13. $\begin{cases} x' = x - xy \\ y' = p + x - 2y \end{cases}$

14. $\begin{cases} x' = x - x^3 y - p \\ y' = -y + xy^2 \end{cases}$

The origin $(x, y) = (0, 0)$ is an equilibrium of the systems below. For which values of p do the Hopf bifurcation criteria (6.6) hold at $(0, 0)$? Use a computer to determine if stable limit cycles bifurcate or not. The coefficient p satisfies $-1 < p < 1$.

15. $\begin{cases} x' = px - 2y - x \sin\left(x^2 + y^2\right) \\ y' = x - y \sin\left(x^2 + y^2\right) \end{cases}$

16. $\begin{cases} x' = (p-2)x + (p+5)y - 10xy^2 \\ y' = -x + (p+2)y - 10yx^2 \end{cases}$

17. (a) Find all equilibria of the system

$$x' = (1 - x - 2y)x$$
$$y' = (1 - 2x - y)y.$$

(b) Find the Jacobian of this system.

(c) Find the linearization at each equilibrium.

18. (a) Find the equilibria of the system

$$x' = y + \alpha\left(x - \frac{1}{3}x^3\right), \quad \alpha > 0$$

$$y' = -x.$$

(b) Find the Jacobian of this system and evaluate it at the equilibrium. Show the equilibrium is hyperbolic for all $\alpha > 0$.

(c) Determine the local phase plane portrait near the equilibrium.

(d) Use a computer to study the phase plane portrait for selected values of α between 0 and 2. What do you conclude about the local phase portrait and stability of the equilibrium? What do orbits do as $t \to +\infty$?

(e) Use a computer to study the phase plane portrait in the case $\alpha = 0$. What do orbits do as $t \to +\infty$. (*Hint*: Be sure to use a sufficiently small step size.)

(f) If $x = x(t)$, $y = y(t)$ is a solution pair of, show that $x = x(t)$ solves the van der Pol equation (1.7).

19. (a) Show that $(x, y) = (b/d, 0)$ is an equilibrium of the AIDS equations (1.4). Find all other equilibria.

(b) Find the Jacobian of the system.

(c) Find the linearization at $(x, y) = (b/d, 0)$.

(d) Determine the phase portrait in the neighborhood of $(x, y) = (b/d, 0)$. (All coefficients are positive.)

20. The orbit of a (nonequilibrium) periodic solution pair of a planar autonomous system (1.1) is a closed loop. The object of this exercise is to prove the converse. That is, if the orbit of a solution is a closed loop, then the solution pair is periodic.

(a) Let $(x(t), y(t))$ be a solution pair of (1.1). Let τ be any real number. Show that $(x(t + \tau), y(t + \tau))$ is also a solution of (1.1).

(b) Suppose the orbit associated with a nonequilibrium solution $(x(t), y(t))$ is self-intersecting. That is, suppose there are two different values of t, say $t_1 > t_2$, for which $(x(t_1), y(t_1)) = (x(t_2), y(t_2))$. Show that $(x(t), y(t))$ is periodic. [*Hint*: Let $\tau = t_2 - t_1$ and use (a) and the fundamental existence and uniqueness theorem to prove

$$(x(t + \tau), y(t + \tau)) = (x(t), y(t))$$

for all t.]

Since a nonequilibrium loop orbit is self-intersecting, it follows that it is associated with a periodic orbit.

21. Use a computer program to investigate the nonlinear system

$$x' = (1 - y)x$$
$$y' = (-1 + x)y.$$

Describe the orbits lying in the positive quadrant (i.e., the first quadrant $x > 0$, $y > 0$). Are there any cycles? Any limit cycles? What does the linearization principle tell you about the equilibrium $(x_e, y_e) = (1, 1)$?

8.9 APPLICATIONS

We consider three applications of nonlinear systems. The first two applications, from biology and physics, use the Poincaré-Bendixson theory to analyze equilibria. Bifurcations and thresholds play an important role in these applications. The third application arises in ecology and involves limit cycles and chaos.

8.9.1 Epidemics

A large variety of diseases afflict biological populations and consequently scientists have formulated and applied many different kinds of mathematical models to study the spread of diseases through a population. We consider two epidemic models, the classic SIR (susceptibles, infected, recovered) model and a more recent model for HIV/AIDS. Both models are compartmental models. They are derived from balance laws for input and output rates.

An SIR Model   For many diseases, a susceptible individual becomes infected and either dies or recovers with permanent immunity to the disease. We consider the problem of determining the spread of such a disease through a population of fixed size. We divide all individuals in the population into three distinct classes: susceptibles, infected, and removed. Susceptible individuals are those who can catch the disease (but cannot spread it). Infected individuals are those who have the disease and can transmit it to susceptible individuals. The class of removed individuals includes all others (e.g., other individuals who had the disease but recovered and gained permanent immunity, diseased individuals who are in quarantine, and dead individuals). Let $S = S(t)$, $I = I(t)$, and $R = R(t)$ denote the number of individuals in these three respective classes. As the disease takes its course, these numbers change in time t.

The modeling step in the modeling cycle (see Fig. 1 in the Introduction) involves specifying the rates of transfer between the three specified classes in order to determine the rates S' and I' by means of the balance laws:

$$S' = \textit{inflow rate of susceptibles} - \textit{outflow rate of susceptibles} \qquad (9.1)$$
$$I' = \textit{inflow rate of infected} - \textit{outflow rate of infected.}$$

We need not model the rate R' since the total population size $S + I + R$ remains fixed (by assumption) and as a result $(S + I + R)' = 0$. This means $R' = -S' - I'$ and hence R can be determined (by an integration) from S' and I'.

The classic *SIR model* (also called the *Kermack-McKendrick Model*) is based on the compartmental diagram in Fig. 8.23. In words, this model assumes the per capita rate at which susceptibles become infected is proportional to the number of infected individuals present, no new susceptibles enter the population, and the per capita rate at which infected individuals move to the class of recovered individuals is constant. In

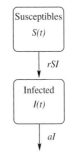

Figure 8.23 The compartmental diagram for the SIR model of Kermack and McKendrick.

mathematical terms, from (9.1) we have

$$S' = 0 - rSI$$
$$I' = rSI - aI$$

for a positive constants a and $r > 0$. This is a nonlinear planar autonomous system. The initial value problem for this model is

$$S' = -rSI$$
$$I' = rSI - aI \qquad (9.2)$$
$$S(0) = S_0 \geq 0, \quad I(0) = I_0 \geq 0.$$

Suppose at time $t_0 = 0$ a number $I_0 > 0$ of individuals become infected. Our goal is to determine how the class of infected individuals changes with time. Does $I(t)$ increase or decrease? If the number of infected individuals $I(t)$ initially increases, then we say that an epidemic occurs. We wish to determine under what conditions the SIR model (9.2) predicts that an epidemic will occur. We also want to determine the ultimate fate (as $t \to +\infty$) of each of the three classes of individuals.

We begin with an example that explores some sample solutions of the SIR model (9.2) with the help of a computer. Figure 8.24(a) shows selected solution graphs when $a = 0.6$ and $r = 0.003$. Figure 8.24(b) shows the corresponding orbits in the phase plane. A close examination of these graphs reveals the following. For an initial condition S_0 small enough (less than approximately 200), the number of infected individuals $I(t)$ decreases to 0 as $t \to +\infty$ (i.e., no epidemic occurs). On the other hand, for S_0 large (greater than approximately 200), the infected population $I(t)$ first increases from 20 to a maximum before it decreases to 0 as $t \to +\infty$. For these initial conditions an epidemic occurs.

Thus, there is a threshold value $S_{th} \approx 200$ for the initial susceptible population numbers S_0 above which an epidemic occurs and below which an epidemic does not occur. Notice in Fig. 8.24(b) how orbits tend (as $t \to +\infty$) to a point on the S-axis. This means the disease ultimately dies out [($I(t) \to 0$ as $t \to +\infty$)]. It also means that the class of susceptibles tends to a limit which is less than its initial size S_0.

Our goal in the solution step of the modeling cycle is to show that the features observed in the example in Fig. 8.24 are in fact general features of the SIR model (9.2) (i.e., that they occur for all numerical values of a and r and for all orbits [initial conditions]).

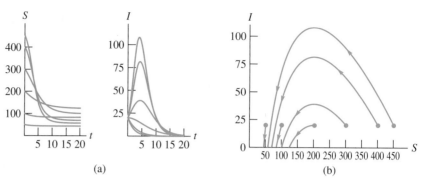

Figure 8.24 Selected solutions of the SIR model (9.2) when $a = 0.6$ and $r = 0.003$. The time series of both components S and I appear in (a). The corresponding orbits in the phase plane appear in (b). In these graphs the initial number of infected individuals is $I_0 = 20$ and the initial number of susceptibles S_0 ranges from 50 to 400.

We begin by investigating equilibria. Only the nonnegative equilibrium points are meaningful in this application [i.e., equilibrium points $(S, I) = (S_e, I_e)$ for which $S_e \geq 0$ and $I_e \geq 0$]. The equilibrium equations for the system (9.2) are

$$-rSI = 0$$
$$rSI - aI = 0.$$

The first equation implies either $S = 0$ or $I = 0$ (or both). If $S = 0$, then the second equation implies $I = 0$, which means the origin $(S, I) = (0, 0)$ is an equilibrium. If $I = 0$, then both equations are satisfied for any value of S. We conclude that *every point $(S, I) = (S_e, 0)$ lying on the S-axis in the (S, I)-phase plane is an equilibrium.* We are interested only in these equilibria for which $S_e \geq 0$.

To investigate the local stability properties of an equilibrium $(S_e, 0)$, we calculate the Jacobian matrix

$$J(S, I) = \begin{pmatrix} -rI & -rS \\ rI & rS - a \end{pmatrix}$$

and evaluate it at the equilibrium $(S_e, 0)$ to obtain

$$J(S_e, 0) = \begin{pmatrix} 0 & -rS_e \\ 0 & rS_e - a \end{pmatrix}.$$

The roots of the characteristic polynomial (eigenvalues) are

$$\lambda_1 = 0, \quad \lambda_2 = rS_e - a.$$

Note that one of the roots (namely, $\lambda_1 = 0$) has real part equal to 0. Therefore, *the equilibria $(S_e, 0)$ are nonhyperbolic.*

If $\lambda_2 = rS_e - a > 0$ (i.e., $S_e > a/r$), then the fundamental theorem of stability (Theorem 1) implies the equilibrium $(S_e, 0)$ is unstable. On the other hand, if $\lambda_2 = rS_e - a \leq 0$ (i.e., if $S_e \leq a/r$), then the linearization principle fails to provide us any information about the equilibrium $(S_e, 0)$. This is because the equilibrium is nonhyperbolic and Theorems 1 and 3 do not apply. The Poincaré-Bendixson theory, however, can help us with further analysis.

In order to apply Theorem 4 we must show that orbits are bounded. First we note that orbits with initial points in the positive quadrant must remain in the positive quadrant. This is because in order to leave the quadrant an orbit would have to cross one of the positive coordinate axis. It cannot cross the S-axis, because it consists of orbits (equilibria) and orbits cannot cross. Nor can it cross the positive I-axis because, it turns out (Exercise 2), the positive I-axis is itself an orbit.

To show an orbit in the positive quadrant is bounded for $t \geq 0$, consider the ($45°$ right) triangle containing the initial point (S_0, I_0) in Fig. 8.25. We claim that the orbit remains inside this triangle for $t \geq 0$ (and is therefore bounded for $t \geq 0$). Since we have already shown the orbit cannot cross the legs of the triangle (they lie on the coordinate axes), we need only know that the orbit cannot cross the hypotenuse of the triangle. This fact follows from the vector field along the hypotenuse, which points into the triangle (Exercise 3.) See Fig. 8.25.

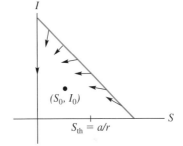

Figure 8.25 The direction field of the system (9.2) points inward along the hypotenuse of the $45°$ triangle shown.

With these preliminaries done, we can apply the Poincaré-Bendixson theorem to orbits in the positive quadrant. First, note that there can be no cycle in the positive quadrant because there is no equilibrium in the quadrant. (Recall that a cycle must encircle an equilibrium.) It follows from Theorem 4 that the limit set S^+ of an orbit must contain an equilibrium $(S_e, 0)$ on the positive S-axis. This theorem does not tell us, however, that the orbit actually approaches an equilibrium (as we suspect from the example in Fig. 8.23). We have to deduce this further fact from another consideration.

Notice from the first equation in the system (9.2) that $S' < 0$ for orbits in the first quadrant. It follows that $S(t) > 0$ is a monotonically decreasing, positive function and therefore must approach a limit as $t \rightarrow +\infty$. As a result, the orbit must approach the equilibrium $(S_e, 0)$ in its limit set as $t \rightarrow +\infty$.

We are now ready for the interpretation step in the modeling cycle in which we draw some conclusions about SIR epidemics from the mathematical results we obtained for the model system (9.2). We have found that $S_0 > a/r$ implies the number of infected individuals $I(t)$ increases before decreasing to 0 as $t \rightarrow +\infty$ (i.e., an epidemic occurs). On the other hand, $S_0 \leq a/r$ implies $I(t)$ strictly decreases to 0 as $t \rightarrow +\infty$ (i.e., an epidemic does not occur). See Fig. 8.26. Thus, there is a *threshold level* of susceptibles above which an epidemic will occur and below which an epidemic will not occur. (Note, by the way, that in the example shown in Fig. 8.24 this formula for the threshold level gives $a/r = 200$.)

Threshold phenomena are common in epidemic models. We will see another in the AIDS/HIV model in the next section.

A feature of an SIR epidemic implied by the direction field of (9.2) in Fig. 8.26 is that the maximum number I_{max} of infected individuals attained during the epidemic increases with an increase the initial number of susceptibles S_0. Furthermore, the number of susceptibles S_e remaining after the epidemic has run its course is smaller for those populations with a larger initial number of susceptibles. Thus, a larger number of susceptibles means an epidemic is more severe and that there are fewer susceptibles remaining after the epidemic is over.

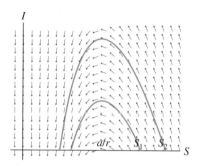

Figure 8.26 The direction field and two sample orbits for (9.2) are shown. If $S_0 > a/r$, orbits follow a parabolic-like trajectory. If $S_0 \leq a/r$, orbits decreases to the S-axis. An orbit of (9.2) starting with an initial condition $S_0 = S_1$ less than the initial condition $S_0 = S_2 > S_1$ of another orbit reaches an equilibrium $(S_e, 0)$ on the S-axis with a larger S_e component and (if $S_0 > a/r$), the component $I(t)$ attains a lower maximum.

The Kermack-McKendrick SIR model (9.2) is based on modeling assumptions too simplistic to make it realistic in most real world applications. (For one example in which the simplifying assumptions are reasonable, see Exercise 5.) Despite its weaknesses, however, the Kermack-McKendrick model illustrates a fundamental threshold phenomenon found in many epidemic models, namely, that the number of susceptible individuals must exceed a critical value before an epidemic can occur.

An HIV/AIDS Model   The Kermack-McKendrick SIR system (9.2) models the rapid spread of a disease through a population. By "rapid" is meant that the course of the disease is considerably faster than the reproductive cycle of the population and that the probability of dying of causes other than the disease is negligible. Because birth and (nondisease) death rates are ignored, the total population in that model remains fixed in time. (See Exercise 8.) For a "slower moving" disease the assumption of a constant population is not a reasonable one to make and the SIR model is inappropriate.

One such disease is AIDS whose incubation period (from exposure to the HIV virus to the appearance of symptoms of AIDS) is many years. HIV infection and the development of AIDS is very complicated and not completely understood. It is therefore difficult to model the course of an HIV/AIDS epidemic. Nonetheless, in this section we consider one basic model for the spread of HIV infection that has been used as a starting point.

Consider a population of individuals susceptible to the HIV virus. This population might be the inhabitants of a certain country, a state, a city, or one of many high-risk subgroups of such populations. To begin the model derivation step of the modeling cycle, let

$$x = x(t) \text{ denote the number of susceptibles at time } t$$
$$y = y(t) \text{ denote the number of individuals with HIV (but not AIDS).}$$

The individuals in class y are infectious and therefore capable of passing the HIV virus to susceptibles. We assume all other individuals in the population are susceptible, so that the total population size at any point in time is $x(t) + y(t)$. We seek differential equations for x and y that result from the balance laws

> **(a)** $x' = x\text{-inflow rate} - x\text{-outflow rate}$
> **(b)** $y' = y\text{-inflow rate} - y\text{-outflow rate}$. (9.3)

We must obtain mathematical expressions for each of the four rates on the right-hand sides of these equations. In the model we will consider, these expressions are based on the modeling assumptions summarized in the compartmental diagram shown in Fig. 8.27.

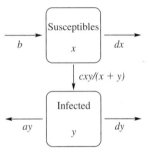

Figure 8.27 The compartmental diagram for an HIV/AIDS model.

Referring to Fig. 8.27, we see that in this model susceptibles arrive in the population at a fixed constant rate $b > 0$ so that in the balance equation (9.3a):

$$x\text{-inflow rate} = b.$$

In Fig. 8.27 there are only two outflows from the x compartment. Either a susceptible dies of causes unrelated to HIV or becomes infected with HIV and moves into the infected class. These two rates, according to Fig. 8.27, are, respectively, dx, where $d > 0$ constant per capita death rate, and

$$c\frac{y}{x+y}x.$$

The latter term expresses the assumption in this model that HIV infections occur at a (per capita) rate proportional to the fraction

$$\frac{y}{x+y}$$

of infectious individuals in the population. (This assumption is based on the assumption that individuals come into contact with one another with equal likelihood, i.e., there is a "uniform" mixing of individuals in the population. Thus, the likelihood that an individual will come into contact with an infectious individual depends on the fraction of the population that is infectious.)

In summary, from Fig. 8.27 we have

$$x\text{-outflow rate} = dx + c\frac{y}{x+y}x, \quad d, c > 0$$

and the differential equation (9.3a) for x becomes

$$x' = b - dx + c\frac{y}{x+y}x.$$

The expression

$$c\frac{y}{x+y}$$

measures the (per capita) rate of infection of the disease. The maximum of this rate is c and we will refer to this coefficient as the force of infection.

We turn now to the compartmental diagram in Fig. 8.24 and the balance equation (9.3b) for y'. First, the HIV-infected individuals arrive from the susceptible class, so that

$$y\text{-inflow rate} = c\frac{y}{x+y}x.$$

Second, with regard to the y-outflow rate, the model distinguishes between those infected individuals who develop AIDS (and are therefore removed from the susceptible population) and those individuals who die of causes unrelated to HIV/AIDS. Thus, as displayed in Fig. 8.24, the y-outflow rate is the sum of two removal rates: the HIV-infected individuals develop AIDS at a constant per capita rate $a > 0$ and the non-AIDS-related per capita death rate d (which is the same for the infected class as it is for the susceptible class) This means

$$y\text{-outflow rate} = dy + ay$$

and the differential equation (9.3b) for y becomes

$$y' = c\frac{y}{x+y}x - (d+a)y.$$

In summary, the compartmental diagram in Fig. 8.24 yields the initial value problem

$$x' = b - dx - c\frac{y}{x+y}x$$

$$y' = c\frac{y}{x+y}x - (d+a)y \tag{9.4}$$

$$x(0) = x_0, \quad y(0) = y_0$$

for x and y. In this planar autonomous system the four coefficients a, b, c, and d are positive and the initial conditions $x_0 \geq 0$, $y_0 \geq 0$ are nonnegative (but not both 0). Notice that the right-hand sides

$$f(x, y) = b - dx - c\frac{y}{x+y}x$$

$$g(x, y) = c\frac{y}{x+y}x - (d+a)y$$

of the equations in (9.4) are well defined and (continuously) differentiable with respect to both $x \geq 0$ and $y \geq 0$ (not both 0). Therefore, Theorem 1 tells us that the initial value problem (9.4) has a unique solution.

Our goal is to address the following question: After an introduction into the population, does the HIV virus persist (i.e., will there be an epidemic) or will it die out? That is, under what conditions will the solution component $y(t)$ tend to 0 as as $t \to +\infty$? Before we embark on the solution step of the modeling cycle—in which we will use the methods of this chapter analyze the initial value problem (9.4) and address this question—we look at an example with the help of a computer.

Consider a population into which 1000 susceptibles enter per year. If time is measured in years and population numbers are measured in units of 1000 individuals, then $b = 1$ in (9.4). Suppose an HIV population has a halving time of 8 years (i.e., one-half of an HIV population will develop AIDS in 8 years). Then $a = (\ln 2)/8 \approx 0.08664$. Suppose the annual non-AIDS-related per capita death rate d is 3% (i.e., $d = 0.03$). Figure 8.28(a) shows several typical orbits calculated by a computer from the model equations (9.4) with these parameter values and with the force of infection set at $c = 0.1$. These orbits appear to approach an equilibrium located on the positive x-axis. We conclude, in this case, that the class of infected individuals dies out in the long run and there is no epidemic. However, the situation changes significantly if the force of infection increases to $c = 0.13$, as shown by Fig. 8.28(b). In this case orbits tend to an equilibrium point lying in the positive quadrant. This means that the population of HIV-infected individuals persists indefinitely and the disease becomes an epidemic.

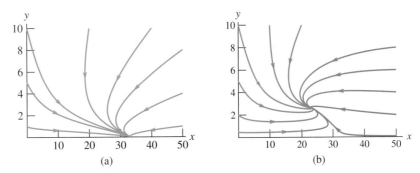

Figure 8.28 Orbits from the epidemic model (9.4) are shown for $a = (\ln 2)/8$, $b = 1$, and $d = 1$. In (a) the force of infection is $c = 0.1$, while in (b) it is $c = 0.13$.

Our goal is to show that the threshold phenomenon for the force of infection c observed in the example shown in Fig. 8.28 holds in general for the model (9.4). That is, we want to show that no matter what the numerical values of the parameters a, b, and d are, there is a threshold value c_{th} of the force of infection c above which an epidemic occurs (i.e., there is a stable equilibrium in the positive quadrant) and below which the HIV virus dies out of the population (i.e., there is a stable equilibrium on the x-axis).

The equilibrium equations associated with the system (9.4) are

$$b - dx - c\frac{y}{x+y}x = 0$$

$$\left(c\frac{1}{x+y}x - (d+a)\right)y = 0.$$

From the second equation there are two choices: either $y = 0$ or

$$c\frac{1}{x+y}x - (d+a) = 0. \tag{9.5}$$

In the first case, $y = 0$, the first equilibrium equation implies $x = b/d$ and we obtain the equilibrium

$$(x_e, y_e) = \left(\frac{b}{d}, 0\right)$$

on the positive x-axis. In the second case ($y \neq 0$) equation (9.5) implies

$$y = \frac{c - (d+a)}{d+a}x, \tag{9.6}$$

which, when substituted into the second equation, yields (after some algebraic details)

$$x = \frac{b}{c-a}$$

(provided $c \neq a$). Then from (9.6) we have

$$y = \frac{c-d-a}{d+a}\frac{b}{c-a}.$$

In summary, we have found the following equilibria for the system (9.4):

$$(x_e, y_e) = \begin{cases} \left(\frac{b}{d}, 0\right) \\ \left(\frac{b}{c-a}, \frac{b}{c-a}\frac{c-a-d}{a+d}\right) & \text{if } c \neq a. \end{cases} \tag{9.7}$$

The second equilibrium is positive if and only if $c > d + a$.

Note that the two equilibria in (9.7) coincide when $c = d + a$. Thus, there are two equilibrium branches that cross (i.e., undergo a transcritical bifurcation) as c is increased through the threshold (bifurcation) value

$$c_{th} \doteq d + a.$$

Our next step is to study the stability properties of the equilibria (9.7) using the linearization principle. The Jacobian of the system (9.4)

$$J(x, y) = \begin{pmatrix} -d + c\frac{y}{(x+y)^2}x - c\frac{y}{x+y} & -\frac{c}{x+y}x + c\frac{y}{(x+y)^2}x \\ -c\frac{y}{(x+y)^2}x + c\frac{y}{x+y} & \frac{c}{x+y}x - c\frac{y}{(x+y)^2}x - d - a \end{pmatrix}$$

evaluated at the x-axis equilibrium is

$$J\left(\frac{b}{d}, 0\right) = \begin{pmatrix} -d & -c \\ 0 & c-d-a \end{pmatrix}.$$

The roots of the characteristic polynomial (eigenvalues) of this matrix are

$$\lambda_1 = -d < 0, \qquad \lambda_2 = c - c_{\text{th}}.$$

Since λ_1 is negative, the stability properties of this equilibrium are determined by the sign of λ_2. If $c < c_{\text{th}}$, then $\lambda_2 < 0$ and the x-axis equilibrium is a stable node. However, if $c > c_{\text{th}}$ this equilibrium is unstable (a saddle). (If $c = c_{\text{th}}$, the x-axis equilibrium is nonhyperbolic and nothing can be deduced from the linearization principle.)

The Jacobian evaluated at the equilibrium

$$(x_e, y_e) = \left(\frac{b}{c-a}, \frac{b}{c-a} \frac{c-a-d}{a+d} \right)$$

is

$$J(x_e, y_e) = \frac{1}{c} \left(\begin{array}{cc} -(c - c_{\text{th}})^2 - cd & -c_{\text{th}}^2 \\ (c - c_{\text{th}})^2 & -c_{\text{th}}(c - c_{\text{th}}) \end{array} \right).$$

To apply Theorem 2 we calculate the trace and determinant:

$$\text{tr } J(x_e, y_e) = -\frac{1}{c}(c - c_{\text{th}})^2 - d - \frac{1}{c}c_{\text{th}}(c - c_{\text{th}})$$

$$\det J(x_e, y_e) = \frac{c_{\text{th}}}{c}(c - c_{\text{th}})(c - c_{\text{th}} + d).$$

If $c > c_{\text{th}}$, then $\text{tr } J(x_e, y_e) < 0$ and $\det J(x_e, y_e) > 0$. By Theorem 2 the equilibrium (x_e, y_e) is stable. If $a < c < c_{\text{th}}$, then $\det J(x_e, y_e) < 0$ and Theorem 2 implies that (x_e, y_e) is unstable.

Table 8.1 The equilibria of system (9.4) and their stability properties as they depend on the force of infection c. The threshold value is $c_{\text{th}} = a + d$.

Equilibrium	$a < c < c_{\text{th}}$	$c_{\text{th}} < c$
$\left(\frac{b}{d}, 0 \right)$	Stable	Unstable
$\left(\frac{b}{c-a}, \frac{b}{c-a}\frac{c-a-d}{a+d} \right)$	Unstable (and not positive)	Stable (and positive)

The local, linearization analysis of equilibria, summarized in Table 8.1, shows that a transcritical bifurcation of equilibria occurs in the HIV/AIDS model (9.4) as the force of infection c increases through the critical value $c_{\text{th}} = a + d$. In Exercises 10, 11, 12, and 13, you are asked to show (using Poincaré-Bendixson theory) that the stability assertions in Table 10.22 are in fact global. That is, all orbits in the positive quadrant approach the axis equilibrium $(b/d, 0)$ (and the HIV virus vanishes from the population) when $c < c_{\text{th}}$, whereas they all approach the positive equilibrium (and an epidemic occurs) when $c > c_{\text{th}}$.

8.9.2 Objects in Motion

In Chapter 4, Sec. 4.6, we studied systems of first-order equations that arise in applications to objects moving in a plane subject to Newton's laws of motion. When frictional forces due to air resistance are modeled by a quadratic law, we found the velocity components u and v of the object satisfy the nonlinear system

$$u' = -\frac{c}{m}u\sqrt{u^2 + v^2} \tag{9.8}$$

$$v' = -g - \frac{c}{m}v\sqrt{u^2 + v^2}.$$

Here m is the mass of the object, g is the (constant) acceleration due to gravity, and $c > 0$ is a coefficient of friction. We used this nonlinear system in Sec. 6.2 of Chapter 4 to study of the flight of a batted baseball. In that application, we relied only on computer calculated graphs of solutions. In this section we analyze the system using the methods studied in this chapter. Our goal is to determine the phase plane portrait of this system, and see what conclusions we can draw from it about the object's flight.

The equilibrium equations for (9.8) are

$$-\frac{c}{m}u\sqrt{u^2 + v^2} = 0$$

$$-g - \frac{c}{m}v\sqrt{u^2 + v^2} = 0.$$

Since the origin $(u, v,) = (0, 0)$ is not an equilibrium (the second equation is not satisfied), the first equation implies $u = 0$. From the second equation we find that $v = -\sqrt{mg/c}$ and the system (9.8) has only the one equilibrium point:

$$(u_e, v_e) = \left(0, -\sqrt{\frac{mg}{c}}\right). \tag{9.9}$$

To apply the linearization principle (Theorem 1 and Corollary 1) we evaluate the Jacobian of (9.8)

$$J(u, v) = \begin{pmatrix} -\frac{c}{m}\frac{2u^2 + v^2}{\sqrt{u^2 + v^2}} & -\frac{c}{m}\frac{uv}{\sqrt{u^2 + v^2}} \\ -\frac{c}{m}\frac{uv}{\sqrt{u^2 + v^2}} & -\frac{c}{m}\frac{u^2 + 2v^2}{\sqrt{u^2 + v^2}} \end{pmatrix}$$

at the equilibrium and obtain

$$J\left(0, -\sqrt{g\frac{m}{k}}\right) = \begin{pmatrix} -\frac{cg}{\sqrt{gmc}} & 0 \\ 0 & -2\frac{cg}{\sqrt{gmc}} \end{pmatrix}.$$

The roots of the characteristic polynomial (eigenvalues) of this matrix

$$\lambda = -\frac{cg}{\sqrt{gmc}} \quad \text{and} \quad -2\frac{cg}{\sqrt{gmc}}$$

are real, negative, and unequal. By the linearization principle we conclude that the equilibrium is a stable node.

Our next goal is to determine the global phase portrait and show all orbits tend to the equilibrium as $t \to +\infty$. We will do this using Poincaré-Bendixson theory.

First, we must show orbits are bounded for $t \geq 0$. We can do this by considering a rectangle in the (u, v)-plane with sides located at $u = \pm a$ and $v = \pm b$ as shown in Fig. 8.29. We choose the rectangle so large that it encloses the initial point $(u(0), v(0))$

of an orbit and the equilibrium (i.e., $b > \sqrt{mg/c}$). On each side of the rectangle the direction field of the system (9.8) points inward; see the caption of Fig. 8.29. Therefore, the orbit cannot leave the rectangle and, as a result, remains bounded for $t \geq 0$.

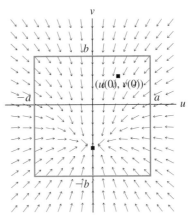

Figure 8.29 On the vertical side $u = a$ of the rectangle, the derivative $u' = -\frac{k}{m}a\sqrt{a^2 + v^2}$ is negative and the direction field of the system (9.8) points to the left. On the vertical side $u = -a$, the derivative $u' = \frac{k}{m}a\sqrt{a^2 + v^2}$ is positive and the direction field points to the right. On the top of the rectangle $v = b$, the derivative $v' = -g - \frac{k}{m}b\sqrt{u^2 + b^2}$ is negative and the direction field points downward. Finally, on the bottom of the rectangle $v = -b$ we have

$$v' = -g + \frac{k}{m}b\sqrt{u^2 + b^2} \geq -g + \frac{k}{m}b\sqrt{b^2} = -g + \frac{k}{m}b^2,$$

which shows v' is positive (since $b > \sqrt{mg/c}$) and the direction field points upward.

The Poincaré-Bendixson theorem (Theorem 4) gives two alternatives for an orbit of the system (9.8). Either its limit set S^+ contains an equilibrium or is a limit cycle. In the first case, since there is only one equilibrium and it is a stable node, the limit set would consists of exactly that point. That is, the first alternative implies orbits tend to the equilibrium as $t \to +\infty$. All that remains for us to do is to rule out the second alternative (i.e., the existence of a limit cycle).

We can show that the system (9.8) has no cycle by applying the method in Exercise 16–19 in Sec. 8.5. From the calculation

$$\frac{\partial}{\partial u}\left(-\frac{c}{m}u\sqrt{u^2 + v^2}\right) + \frac{\partial}{\partial v}\left(-g - \frac{c}{m}v\sqrt{u^2 + v^2}\right) = -3\frac{c}{m}\sqrt{u^2 + v^2} < 0$$

and the fact that this expression is negative for all u and v, the so-called Dulac criterion described in those exercises tells us there is no limit cycle. (Also see the extended Dulac criterion in Exercises 20–25 in Sec. 8.5.)

What conclusions can we draw from the fact that all solutions of the system (9.8) tend to the equilibrium (9.9)? Since this implies $u \to 0$ as $t \to +\infty$, we see that motion in the horizontal direction deceases to 0 as $t \to +\infty$. Since $v \to -\sqrt{mg/c}$ the velocity in the vertical direction becomes constant as $t \to +\infty$ (acceleration decreases to 0). That is, the object approaches a "terminal" velocity in an increasingly vertical fall.

8.9.3 Ecology

The dynamics of interacting biological species are often modeled by systems of nonlinear differential equations. We consider two examples of a particular type of interaction between different species, namely, that of a predator and its prey. In the first application, involving a single predator and its prey, a planar autonomous system arises as the model equations, and the methods of Secs. 8.3 and 8.6 are available for analysis. The example includes a transcritical bifurcation and a Hopf bifurcation. The second application involves two predators and a prey. We will find for this model that the ecological system can exhibit chaos.

Predator-Prey Interactions and Limit Cycles   Consider a predator population and its prey. We begin the modeling derivation step by assuming the dynamics of the prey population described by the logistic equation (see Chapter 3)

$$x' = r \left(1 - \frac{x}{K} \right) x, \quad r > 0, \quad K > 0.$$

From the phase line portrait of this equation

$$0 \longrightarrow K \longleftarrow$$

we see that all solutions with initial conditions $x_0 > 0$ tend to the equilibrium K (the prey's carrying capacity) as $t \to +\infty$. We wish to place this population in the presence of a predator.

Let y denote density of the predator population and assume, in the absence of the prey, that the dynamics of the predator are described by the linear equation $y' = -dy$, $d > 0$. This modeling assumption means that the predator goes extinct, exponentially as $t \to +\infty$, in the absence of prey.

To model the interaction between the predator and the prey, we let $p \geq 0$ denote the per capita rate at which the predator consumes prey. It is reasonable to suppose that p is not a constant but instead depends on the amount of prey available [i.e., $p = p(x)$]. A common situation is that an increase in prey abundance results in an increase in the per capita predator consumption. (For example, each predator eats more prey if more is available, prey might be easier to capture when abundant, and so on.) In this case, $p(x)$ is an increasing function of x and satisfies $p(0) = 0$. The precise mathematical expression chosen for $p(x)$ depends on the behavior and the characteristics of individual predators (searching strategies, level of hunger, etc.) and prey (evasion strategies, etc.).

When the predator is present, the logistic equation for the prey becomes

$$x' = r \left(1 - \frac{x}{K} \right) x - p(x) y. \tag{9.10}$$

In addition, if the (per predator) rate at which new predators are born is proportional to the prey consumption rate p, then the equation for the predator population becomes

$$y' = -dy + cp(x) y. \tag{9.11}$$

Here $c > 0$ is a constant of proportionality (a conversion factor).

One popular choice for the consumption rate $p(x)$ (called the Holling type II model) is the formula

$$p(x) = m \frac{x}{a+x}, \quad m > 0, \quad a > 0. \tag{9.12}$$

This expression has the following features: $p(0) = 0$ (if no prey is present, then no predation occurs); $p(x)$ is increasing for $x \geq 0$ (increasing prey populations causes an increase in per capita consumption by predators); and $p(x) \to m$ as $x \to +\infty$ (the per capita predation rate approaches to a maximal attainable rate m as the prey population increases without bound). With this model of the consumption rate, the predator-prey system (9.10)–(9.11) becomes

$$x' = r \left(1 - \frac{x}{K}\right) x - m \frac{xy}{a+x} \tag{9.13}$$

$$y' = -dy + cm \frac{xy}{a+x}.$$

Our goal is to determine under what circumstances both species will persist and, in this event, to describe the dynamics of the interaction (do their populations equilibrate? do they oscillate?). We will approach these questions by describing (in the solution step of the modeling cycle) the phase plane portrait associated with the planar autonomous system (9.13) in the positive quadrant $x \geq 0$, $y \geq 0$.

There are six parameters in the predator-prey model (9.13). We will investigate a special case by choosing numerical values for five of the parameters, leaving one—namely, K—as a bifurcation parameter. This example (it turns out) will typify the general case. The reader is invited to carry out some more general analysis in Exercises 28–34.

Let

$$r = m = 10, \quad a = 1, \quad d = 1, \quad c = \frac{1}{4}$$

in (9.13). The equilibrium equations of the resulting system

$$x' = 10 \left(1 - \frac{x}{K}\right) x - 10 \frac{xy}{1+x} \tag{9.14}$$

$$y' = -y + \frac{5}{2} \frac{xy}{1+x}$$

are (after some algebra)

$$\left(1 - \frac{x}{K} - \frac{y}{1+x}\right) x = 0$$

$$\left(-1 + \frac{5}{2} \frac{x}{1+x}\right) y = 0.$$

There are three solution pairs to these algebraic equations:

$$(x_e, y_e) = \begin{cases} (0, 0) \\ (K, 0) \\ \left(\frac{2}{3}, \frac{5}{3K}\left(K - \frac{2}{3}\right)\right) \end{cases}. \tag{9.15}$$

Notice that the third equilibrium lies in the positive quadrant if and only if $K > 2/3$.

Next we turn our attention to the stability properties of each equilibrium, as ascertained by the linearization principle. The Jacobian of the system (9.14) is

$$J(x, y) = \begin{pmatrix} 10\left(1 - 2\frac{x}{K}\right) + \frac{10}{(1+x)^2}y & -\frac{10x}{1+x} \\ \frac{1}{2}\frac{5}{(1+x)^2}y & \frac{1}{2}\frac{3x-2}{1+x} \end{pmatrix}.$$

At the equilibrium $(x, y) = (0, 0)$ the characteristic polynomial of the Jacobian

$$J(0, 0) = \begin{pmatrix} 10 & 0 \\ 0 & -1 \end{pmatrix}$$

has a positive and a negative root (eigenvalue), namely,

$$\lambda_1 = 10 > 0, \qquad \lambda_2 = -1 < 0.$$

It follows that the equilibrium $(0, 0)$ is a saddle.

At the equilibrium $(x, y) = (K, 0)$ the characteristic polynomial of the Jacobian

$$J(K, 0) = \begin{pmatrix} -10 & -10\frac{K}{1+K} \\ 0 & \frac{3}{2(1+K)}\left(K - \frac{2}{3}\right) \end{pmatrix}$$

has roots (eigenvalues)

$$\lambda_1 = -10 < 0, \quad \lambda_2 = \frac{3}{2(1+K)}\left(K - \frac{2}{3}\right).$$

Since λ_1 is negative, the stability properties of the equilibrium depend on the sign of the second root λ_2. Specifically, $(K, 0)$ is a stable node if λ_2 is negative and a saddle if λ_2 is positive. These cases occur if $K < 2/3$ and $K > 2/3$, respectively.

Finally, at the third equilibrium in (9.15)—the coexistence equilibrium—the Jacobian is

$$J(x_e, y_e) = \begin{pmatrix} \frac{4}{3}\frac{3K-7}{K} & -4 \\ \frac{1}{2}\frac{3K-2}{K} & 0 \end{pmatrix}. \tag{9.16}$$

We could determine the stability properties of this equilibrium by finding the roots of the characteristic polynomial (eigenvalues). Another way is to use Theorem 2 by calculating the determinant

$$\det J(x_e, y_e) = \frac{6}{K}\left(K - \frac{2}{3}\right)$$

and the trace

$$\operatorname{tr} J(x_e, y_e) = \frac{4}{K}\left(K - \frac{7}{3}\right)$$

of the Jacobian (9.16). The determinant is positive if $K > 2/3$ and negative if $K < 2/3$. The trace is positive if $K > 7/3$ and is negative if $K < 7/3$. It follows from the (tr, det)-criteria that the coexistence equilibrium is stable if $2/3 < K < 7/3$ and unstable if $K > 7/3$.

It is left as Exercises 24–25 to show the coexistence equilibrium is a stable node if $2/3 < K < K^*$ and a stable focus if $K^* < K < 7/3$, where

$$K^* = -\frac{11}{3} + \frac{1}{3}\sqrt{219} \approx 1.266.$$

These facts about the equilibria are summarized in Table 8.2.

Figure 8.30 illustrates the four cases in Table 8.2 by showing examples of the phase plane portraits of each case. Notice that two bifurcations occur. First, as K increases through $K = 1$, a transcritical bifurcation and an exchange of stability occurs between the second and third equilibria in (9.15). Second, as K increases through $K = 7/3$ a Hopf bifurcation to a limit cycle occurs.

Table 8.2   Summary of local phase portraits for (9.14).

	$(0, 0)$	$(K, 0)$	$\left(\frac{2}{3}, \frac{5}{3K}\left(K - \frac{2}{3}\right)\right)$
(a) $0 < K < \frac{2}{3}$	Saddle	Stable node	Saddle (unstable)
(b) $\frac{2}{3} < K < K^* \approx 1.266$	Saddle	Unstable node	Stable node
(c) $K^* < K < \frac{7}{3}$	Saddle	Unstable node	Stable spiral
(d) $\frac{7}{3} < K$	Saddle	Unstable node	Unstable spiral

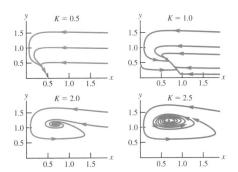

Figure 8.30 Shown are selected orbits of the predator-prey system (9.14) for four values of K.

Table 8.2 is the result of a linearization analysis at each equilibrium. The numerical examples in Fig. 8.30 suggest some global conjectures about orbits in the positive quadrant.

Case (a)	all orbits tend to the equilibrium $(K, 0)$	
Cases (b)–(c)	all orbits tend to the equilibrium $\left(\frac{2}{3}, \frac{5}{3K}\left(K - \frac{2}{3}\right)\right)$	(9.17)
Case (d)	all (nonequilibrium) orbits tend to a limit cycle	

To address these conjectures mathematically we need to know the global phase plane portrait for the predator-prey system (9.14). The technical details of such an analysis are beyond the scope of this book. However, the methods studied in Sec. 8.5 are keys tools for carrying out these analyses. For example, from Exercises 26–27 we find that orbits of (9.14) starting in the positive quadrant are bounded and remain in the positive quadrant for all $t \geq 0$. This means the Poincaré-Bendixson theorem (Theorem 4) applies to these orbits. Since in Case (a) there is no equilibrium in the positive quadrant, there can be no limit cycle in the positive quadrant. (Recall from Sec. 8.5 that a cycle must surround an equilibrium.) Therefore, the forward limit set of an orbit in the positive quadrant must contain an equilibrium. The equilibria in Case (a) are $(0, 0)$ and $(0, K)$. It turns out that $(0, 0)$ can be ruled out because it is a saddle.[8] This leaves $(0, K)$, which, because it is a stable node, is the only point in the limit set—that is, the orbit approaches $(0, K)$ as $t \to +\infty$ (see Sec. 8.5).

Another application of the Poincaré-Bendixson theorem arises in Case (d). Because all three equilibria are unstable in Case (d), they can be ruled out as members of the limit set of an orbit (we skip the technical details). Thus, the theorem tells us the limit set of an orbit is a cycle.

We turn briefly to the interpretation step of the modeling cycle. The biological interpretation of the results in Table 8.2 and Fig. 8.30 are as follows. The constant K is the carrying capacity of the prey in the absence of the predator. We can view K as a measure of the quality or richness of the environmental habitat (with regard to its ability to grow prey). From this point of view, an increase in K reflects an enrichment of the habitat. If the habitat is too poor in quality ($K < 1$), it cannot support both the predator and prey [because orbits tend to the equilibrium $(K, 0)$ in which the predator is absent]. By enriching the habitat we allow the survival of both the predator and the prey populations at a stable equilibrium level, provided we do not increase $K > 1$ too high, namely, provided $K < 7/3$. Too large an increase in habitat quality ($K > 7/3$) destabilizes the coexistence equilibrium and causes the predator-prey system to undergo crash and boom cycles. Equilibrium destabilization caused by habitat enrichment occurs in many predator-prey models. This unexpected phenomenon is called the paradox of enrichment.

A Chaotic Predator-Prey System One consequence of the Poincaré-Bendixson theory is that planar (two-dimensional) autonomous systems cannot have dynamics that are too complicated. On the other hand, autonomous systems of dimension three or higher can exhibit dynamics of considerable complexity. The Lorenz system (1.8) is an example. In this section we look at another application in which chaos occurs.

Ecological systems usually contain a large number of interacting species. In the previous section we considered a system of just two species, a predator and a prey, and found that the interaction exhibited either equilibrium or limit cycle dynamics. We now consider a system of three species obtained by adding another predator, namely, a second predator that preys on the first predator. Thus, in this interaction we have three trophic levels. At the lowest level is a prey species; at the intermediate level is a middle predator; and at the highest level is a top predator. We can think, for example, of a plant-herbivore-predator system (such as a grass-caribou-wolf system).

We model the two-predator, one-prey system by modifying the predator-prey model (9.13) studied in the previous section (following the model modification step of the modeling cycle). Denote the density of the top predator by $z = z(t)$. We make the two following modeling assumptions about the top predator: It dies out exponentially in the absence of its prey y (the middle predator) and its predation rate on y is described by a Holling type II expression (9.12).

---

[8]This is basically because its stable and unstable manifolds lie on the axes.

Introducing subscripts for the parameters associated with the two different predators, we have the autonomous system[9]

$$x' = r\left(1 - \frac{x}{K}\right)x - \frac{m_1 x}{a_1 + x}y$$

$$y' = -d_1 y + c_1\frac{m_1 x}{a_1 + x}y - \frac{m_2 y}{a_2 + y}z \tag{9.18}$$

$$z' = -d_2 z + c_2\frac{m_2 y}{a_2 + y}z.$$

This formidable system involves ten model parameters. Needless to say, we will not attempt a general analysis. We take it as our limited goal to show that very complicated (indeed, chaotic) dynamics can arise in this model. To do this, we consider an example obtained by assigning numerical values to all but one parameter (namely, K) in the system (9.18).

For the middle predator we use the same parameter values used in the previous section, that is,

$$r = 10, \quad m_1 = 10, \quad a_1 = 1, \quad d_1 = 1, \quad c_1 = \frac{1}{4}.$$

For the top predator we take

$$m_2 = 10, \quad a_2 = 1, \quad d_2 = 1, \quad c_2 = \frac{3}{4}.$$

These choices yield the nonlinear autonomous system

$$x' = 10\left(1 - \frac{x}{K}\right)x - \frac{10x}{1 + x}y$$

$$y' = -y + \frac{1}{4}\frac{10x}{1 + x}y - \frac{10y}{1 + y}z \tag{9.19}$$

$$z' = -z + \frac{3}{4}\frac{10y}{1 + y}z$$

for the two predators and the prey. This system has one unspecified parameter, namely, the prey carrying capacity $K > 0$.

As we did for the predator-prey model (9.13), we wish investigate the dynamics of the two predator, one prey model (9.19) as K is increased (i.e., as the habitat is enriched).

Note, first, that the system (9.19) has the equilibrium $(x_e, y_e, z_e) = (0, 0, 0)$. This equilibrium represents the absence of all three species. The Jacobian of (9.19)

$$J(x, y, z) = \begin{pmatrix} 10\frac{K-2x}{K} - 10\frac{y}{(1+x)^2} & -10\frac{x}{1+x} & 0 \\ \frac{5}{2}\frac{y}{(1+x)^2} & \frac{1}{2}\frac{3x-2}{1+x} - 10\frac{z}{(1+y)^2} & -10\frac{y}{1+y} \\ 0 & \frac{15}{2}\frac{z}{(1+y)^2} & \frac{1}{2}\frac{13y-2}{1+y} \end{pmatrix} \tag{9.20}$$

---

[9]A. Hastings and T. Powell, "Chaos in a Three Species Food Chain." *Ecology* 72 (1991), 896–903.

evaluated at $(0, 0, 0)$ gives

$$J(0,0,0) = \begin{pmatrix} 10 & 0 & 0 \\ 0 & -1 & 0 \\ 0 & 0 & -1 \end{pmatrix}.$$

The roots of the characteristic polynomial (eigenvalues) of this matrix are $\lambda = -1$ (a double root) and $\lambda = 10$. The positive root implies the equilibrium $(0, 0, 0)$ is unstable. This is true for all values of K.

   The system (9.19) also has several other equilibria. We leave their analysis, however, to the reader in Exercises 35, 36, and 37. We limit our investigation here to some computer explorations of solutions for selected values of K.

   The graphs in Fig. 8.31 illustrate a sequence of bifurcations that the dynamics of the model system (9.19) undergo as K increases.

   For $K = 0.5$ both predators go extinct while the prey tends to its carrying capacity $K = 0.5$.

   For the larger value $K = 0.7$, the top predator goes extinct while the middle predator and the prey survive in a stable equilibrium state.

   For $K = 0.9$ all three species coexist in a stable equilibrium. However, this coexistence equilibrium loses stability, and a Hopf bifurcation to a limit cycle occurs, as we increase K further. For example, see Fig. 8.31(b), where $K = 1.5$.

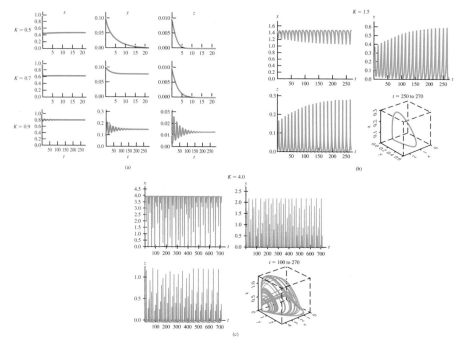

Figure 8.31 As the habitat in the two predator, one prey model (9.19) is enriched (K is increased) the behavior of the three species interaction changes dramatically. (a) As K ranges from 0.5 to 0.9, we see the long-term outcome of the interaction change radically—from one that eliminates both predators to one that allows the middle predator to survive, to one in which all three species survive. (b) A Hopf bifurcation to a limit cycle occurs as K is increased beyond 1.5. (c) For $K = 4$ a chaotic attractor exists.

For even larger values of K, the periodic coexistence state (limit cycle) "bifurcates" to a chaotic coexistence state. See Fig. 8.31(c).

From the one-predator system (9.13) we found a destabilization of equilibrium coexistence as K increases. In the two-predator system (9.19) we find an even more dramatic illustration of this paradox of enrichment, in which the dynamics of the three species system becomes chaotic (i.e., a very complicated oscillatory motion) if the habitat quality is sufficiently enriched.

In Sec. 8.7.2 we saw that an important feature of chaos is sensitivity to initial conditions. As illustrated in Fig. 8.32, this property is present in the chaotic dynamics of the predator-prey model (9.19). After a sufficient length of time, two solutions that initially differ very little become very different. This makes long-term prediction very difficult for the reason that very small differences (such as measurement errors) in initial population sizes lead in the long run to different oscillatory patterns. For example, the times at which outbreaks of the top predator density occur in Fig. 8.32(b) are ultimately very different for these two solutions.

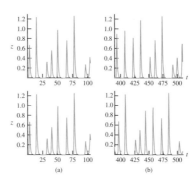

Figure 8.32 These two graphs are those of the z component of two solutions of the two-predator, one-prey model (9.19) with $K = 4$. The solution have nearly identical initial conditions, namely, $(x_0, y_0, z_0) = (1.0, 0.1, 0.01)$ in the top graph and $(x_0, y_0, z_0) = (1.01, 0.1, 0.01)$ in the lower graph. (a) For the time interval displayed ($0 < t < 100$), these two solution graphs appear nearly identical. The two plots in (b) show these solutions over the much later time interval $400 < t < 600$, during which the graphs are very different.

Exercises

1. The balance law applied to the removed class in the SIR model yields $R' = aI$. Show all solutions of the system $S' = -rSI$, $I' = rSI - aI$, $R' = aI$ satisfy $S(t) + I(t) + R(t) = $ constant for all t.

2. Show that the positive I-axis is an orbit of the system (9.2). (*Hint*: Solve the initial value problem with $S_0 = 0$.)

3. Show that the vector field associated with (9.2) points into the hypotenuse of the $45°$ right triangle in Fig. 8.25.

4. Let $(S(t), I(t))$ be the solution pair of the initial value problem (9.2) with initial conditions $S_0 > 0$ and $I_0 > 0$.

   (a) Show that $S = S(t)$ and $I = I(t)$ satisfy the equation

   $$I = \frac{a}{r} \ln S - S - \frac{a}{r} \ln S_0 + S_0 + I_0$$

   for all $t \geq 0$. This equation is called a first integral of the system (9.2). [*Hint*: Treat $I = I(S)$ as a function of S and then derive and solve an equation for $\frac{dI}{dS} = \frac{I'}{S'}$.]

(b) Assume $S_0 > S_{th} = a/r$. Derive a formula for the maximum I_{max} of $I(t)$.

(c) Use your answer in (b) prove that I_{max} is an increasing function of the initial number of susceptibles S_0.

(d) Use the equation in (a) to obtain an equation satisfied by S_e.

(e) Use your answer in (d) to prove that the number of susceptibles S_e left after the epidemic has run its course is a decreasing function of S_0.

5. During a two-week period in 1978 an influenza epidemic swept through a boarding school of 763 boys in northern England. The number of infected boys reported daily appears in the following table.[10]

Date	t (days)	Number infected
January 22	0	3
January 23	1	6
January 24	2	25
January 25	3	74
January 26	4	224
January 27	5	294
January 28	6	259
January 29	7	235
January 30	8	191
January 31	9	124
February 1	10	68
February 2	11	25
February 3	12	12
February 4	13	5

(a) Assume all of the boys were initially susceptible so that $S_0 = 763$. Use the equation

$$I = \frac{a}{r} \ln S - S - \frac{a}{r} \ln S_0 + S_0 + I_0$$

from Exercise 4 together with the initial conditions $I_0 = 3$ from the table to approximate the threshold $S_{th} = a/r$.

(b) Use your answer in (a) to eliminate a from the equations (9.2).

(c) Use a computer to plot the data in the table together with solutions of (9.2) for selected values of $r > 0$. Find a value of r that give a solution graph that visually fits the data well. What weaknesses do the model solutions have with regard to fitting the data?

(d) Use your approximate values of a and r to estimate the number of boys who did not become infected during the epidemic.

6. Prove that the planar autonomous system (9.2) has no cycles.

7. In the SIR model (9.2) of Kermack and McKendrick the parameter $a > 0$ is the rate at which infected individuals are removed. This "removal" includes not only fatalities and recovered individuals but individuals who are quarantined so that they can no longer transmit the disease. Thus, a quarantine policy would decrease the value of $a > 0$. Show that, when an epidemic occurs, the severity of the epidemic [defined to be the maximum I_{max} of the number of infected individuals $I(t)$ attained during the course of the epidemic] decreases with an increase in a.

8. In the SIR model (9.2) of Kermack and McKendrick deaths due to natural causes (i.e., deaths due to any cause other than the disease) are ignored. The assumption is that the epidemic is occurring so rapidly that other causes of death are insignificant. Similarly, births are ignored (and the population, counting dead individuals, remains constant over time). Suppose, however, we wish to take other causes of death into account in an SIR model, at least for susceptibles. Specifically, suppose the class of susceptibles suffers a per capita natural death rate of $d > 0$. Then, assuming the class of infected individuals does not have any additional mortality, we have the model equations

$$S' = -rSI - dS$$
$$I' = rSI - aI$$
$$R' = aI + dS.$$

The equation for R is uncoupled from the first two equations, so this model can be analyzed by studying the planar autonomous system for S and I.

(a) Explore the orbits of the planar autonomous system

$$S' = -rSI - dS$$
$$I' = rSI - aI$$

using a computer program. Formulate conjectures about the asymptotic dynamics of orbits starting in the first quadrant. For example, formulate answers to the questions: Is the first quadrant forward invariant? Are orbits bounded for $t \geq 0$? What are the possible limit sets S? What are the long-term fates of the classes of susceptibles and infected individuals? How are these fates different from those predicted by the Kermack-McKendrick model (9.2) in which $d = 0$? Is there a threshold effect for the occurrence of an epidemic? You may find other interesting conjectures to make as well.

(b) Sketch a vector field for the planar autonomous system in (a).

(c) Find all equilibria of the system in (a) satisfying $S \geq 0$, $I \geq 0$.

(d) Determine the local stability properties and phase portraits of each equilibrium by a linearization analysis. If a linearization analysis is not possible, explain why.

(e) Apply the Poincaré-Bendixson theory to determine the global phase portrait in the first quadrant.

(f) Address the conjectures made in (a).

(g) Prove that the total population $S + I + R$ remains constant. What do you conclude about the asymptotic dynamics of R?

(h) Show that I is maximum when $S = a/r$.

(i) Find a first integral. (See Exercise 4.)

(j) Find an equation for the maximum $I_{\max}$ of I. Show that $I_{\max}$ is an increasing function of S_0 and a decreasing function of a.

9. If individuals lose susceptibility to HIV infection, then $c = 0$ in the system (9.4). Analyze the initial value problem (9.4) in this case.

10. (a) Show that orbits of the system (9.4) starting at points $(x_0, 0)$, $x_0 \geq 0$, on the x-axis remain on the x-axis for all $t \geq 0$.

(b) Show that orbits of the system (9.4) starting at points $(x_0, 0)$, $x_0 \geq 0$, on the x-axis approach the equilibrium point $(x_e, y_e) = (b/d, 0)$ as $t \to +\infty$.

(c) Show that all orbits of (9.4) starting at a point (x_0, y_0) in the first quadrant (i.e., with $x_0 \geq 0$, $y_0 \geq 0$) remain in the first quadrant for all $t > 0$.

11. Show that all orbits of (9.4) starting at a point (x_0, y_0) in the first quadrant (i.e., with $x_0 \geq 0$, $y_0 \geq 0$) remain bounded for all $t \geq 0$. [*Hint*: Investigate the vector field on a rectangle in the first quadrant with sides on the axes and large enough to contain (x_0, y_0) inside.]

12. Use the Dulac criterion (Exercises 16–19 in Sec. 8.5) to show that the system (9.4) has no cycles in the first quadrant. (*Hint*: Try $\mu = \frac{1}{y}$.)

13. (a) If $a < c < c_{\text{th}} = a + d$, show that all orbits of (9.4) lying in the first quadrant tend to the equilibrium $(b/d, 0)$ as $t \to +\infty$.

(b) If $c > c_{\text{th}}$, show that all orbits in the first quadrant tend to the equilibrium $(x_e, y_e) = \left(\frac{b}{c-a}, \frac{b}{c-a} \frac{c-a-d}{a+d} \right)$ as $t \to +\infty$.

14. If $a < c < c_{\text{th}} = a + d$ in the HIV model (9.4), then it was shown that all orbits in the first quadrant tend to the x-axis equilibrium $(b/d, 0)$ as $t \to +\infty$. In particular, $y(t) \to 0$

as $t \to +\infty$ and the HIV virus dies out of the population. Prove that y is a monotonically decreasing function of t.

15. In the HIV/AIDS model (9.4) it is assumed that the input rate of new susceptibles into the population is a constant $b > 0$. Suppose instead that this input rate is proportional to susceptible population x. Then a modified HIV/AIDS model is

$$x' = bx - dx - c\frac{y}{x + y}x$$

$$y' = c\frac{y}{x + y}x - (d + a)\, y,$$

where $b > 0$.

(a) What is the implication of the assumption $d > b$? What is the implication of $d < b$? (*Hint*: Consider what happens to the susceptible population x if there are no HIV infected individuals, i.e., y is identically 0.)

(b) Explore the phase portrait of the modified HIV/AIDS model from (a) in the first quadrant using a computer program. Formulate a conjecture about the asymptotic behavior of all orbits when $d > b$ and when $d < b$.

(c) Find all equilibria of the modified HIV/AIDS model.

(d) Show that the first quadrant is invariant.

(e) If $d > b$, show that all orbits in the first quadrant are bounded for $t \geq 0$.

(f) Show that there are no cycles in the first quadrant.

Suppose a pendulum is subject to a frictional force that is proportional to its angular velocity. This could be due, for example, to air resistance, or friction in the pivot. If the constant of proportionality (the "coefficient of friction") is denoted by $c > 0$, then the model equation for the angular displacement of the pendulum from the vertical is the nonlinear, second-order equation

$$\theta'' + \frac{c}{m}\theta' + \frac{g}{l}\sin\theta = 0.$$

16. Write the equivalent first-order system (using $x = \theta$, $y = \theta'$).

17. Take $m = 1$ kg, $l = 1$ m, and $g = 9.8$ m/s^2. Use a computer to sketch the vector field and selected orbits when $c = 0.5$ and $x(0) = \theta_0$, $0 < \theta_0 < \pi$, $y(0) = 0$. Based on your sketch, what do orbits do as $t \to +\infty$? How does this differ from the pendulum without friction ($c = 0$)?

18. Consider the case of no initial displacement $x(0) = 0$ and an initial velocity $y(0) = \theta_1 > 0$. Use a computer to obtain the orbits for $\theta_1 = 7$ and 8. What happens to these orbits as $t \to +\infty$? Interpret these orbits in terms of the pendulum's motion.

19. Find all equilibria and determine their local stability and phase portraits.

20. Use the Dulac criterion (Exercises 16–19 in Sec. 8.5) to prove there are no limit cycles.

21. From the results of Exercises 19 and 20, what can you conclude about orbits bounded for $t \geq 0$?

22. Show that the energy $E(x, y) = \frac{1}{2}ml^2y^2 + mgl(1 - \cos x)$ does not increase along orbits.

23. Consider just those orbits starting from an initial displacement only [i.e., for which $x(0) = \theta_0$ lies in the range $-\pi < \theta_0 < \pi$ and $y(0) = 0$]. Use Exercise 22 to show that orbits are bounded for $t \geq 0$ (and hence tend asymptotically to an equilibrium).

Consider the Jacobian matrix (9.16).

24. Show that the roots of the characteristic polynomial (the eigenvalues) are real and negative if $2/3 < K < -\frac{11}{3} + \frac{1}{3}\sqrt{219}$.

25. Show that the roots of the characteristic polynomial (the eigenvalues) are complex conjugates with negative real parts if $-\frac{11}{3} + \frac{1}{3}\sqrt{219} < K < 7/3$.

26. The goal of this exercise is to show that any orbit of the predator-prey model (9.14) that starts in the first quadrant is bounded for $t \geq 0$. The idea is to construct a triangle surrounding the initial point $(x(0), y(0)) = (x_0, y_0)$, $x_0 > 0$, $y_0 > 0$, out of which the orbit cannot escape for all $t \geq 0$ because the vector field defined by these equations points inward along all sides of the triangle. Take a right triangle with legs lying on the x- and y-axes and a hypotenuse running across the positive quadrant from the y-axis to the x-axis. Specifically, consider the triangle described by the three sides

$$S_1: x = 0, \qquad 0 \leq y \leq \tfrac{1}{4}\beta$$
$$S_2: y = 0, \qquad 0 \leq x \leq \beta$$
$$S_3: x + 4y = \beta, \quad 0 \leq x \leq \beta,$$

where the x-intercept $\beta > 0$ is yet to be chosen.

(a) Sketch the triangle.

(b) The orbit cannot escape the first quadrant through the sides S_1 or the side S_2. Why not?

(c) Sketch the family of straight lines defined by the equation $x + 4y = \beta$ for $\beta > 0$. What happens to the line as β decreases?

(d) If $(x, y) = (x(t), y(t))$ is an orbit that crosses a line $x + 4y = \beta$ at a time t for which $\frac{d}{dt}(x(t) + ry(t)) < 0$, then in what direction is the orbit crossing the line? In other words, is the orbit moving into or out of the triangle as t increases?

(e) Show that for all sufficiently large β, the derivative $\frac{d}{dt}(x(t) + 4y(t))$ is negative at every point along the hypotenuse S_3.

(f) Use (b)–(e) to prove that all orbits in the first quadrant are bounded for $t \geq 0$.

27. Consider the general predator-prey model

$$x' = r\left(1 - \frac{x}{K}\right)x - p(x)y$$
$$y' = -dy + cp(x)y$$

under the assumptions that $p(0) = 0$, $p'(x) > 0$, and $\lim_{x \to +\infty} p(x) = m < +\infty$.

(a) Show that $(x, y) = (0, 0)$ and $(K, 0)$ are equilibria.

(b) Show that the positive y-axis is an orbit and the positive x-axis consists of three orbits.

(c) Prove that the first quadrant is forward invariant (i.e., all solutions with initial conditions in the first quadrant remain in the first quadrant for all t). (*Hint:* Show that the quadrant boundaries are made up of orbits.)

(d) Compute the Jacobian at the origin $(0, 0)$ and find its characteristic roots (eigenvalues). Determine the local phase portrait near $(0, 0)$. Draw a sketch of orbits near the origin.

(e) Compute the Jacobian at $(K, 0)$ and find its characteristic roots (eigenvalues).

(f) Find conditions on the model parameters under which the equilibrium $(K, 0)$ is (locally) asymptotically stable and determine the nature of its local phase portrait.

(g) Find conditions on the model parameters under which the equilibrium $(K, 0)$ is unstable and determine the nature of its local phase portrait.

(h) Imagine that r, d, and c are fixed. Using your answers for (f) and (g) interpret the fate of the predator as it depends on K.

28. Suppose $x(t)$ and $y(t)$ solve the predator-prey system (9.13). Define new scaled variables $X(\tau)$, $Y(\tau)$, and τ by $X(\tau) \doteq \frac{1}{K}x\left(\frac{t}{r}\right)$, $Y(\tau) \doteq \frac{m}{rK}y\left(\frac{t}{r}\right)$, and $\tau \doteq rt$. Show that these new functions X and Y of τ satisfy a system of the form

$$x' = (1 - x)x - \frac{xy}{a^* + x}$$
$$y' = -d^*y + m^*\frac{xy}{a^* + x}.$$

Thus, by rescaling the variables in (9.13) we have reduced the number of parameters in this predator-prey model from six to three.

Consider the predator-prey system (9.13) with $c = 1$.

29. Find all equilibria.

30. Show that there is an equilibrium with positive coordinates if and only if $m > 1$.

31. Calculate the Jacobian of the system.

32. Assume $m > 1$. Determine how the stability properties of each equilibrium depend on K.

33. Use your answer in Exercise 32 to find and classify the equilibrium bifurcations, using $K > 0$ as the bifurcation parameter.

34. Show that the positive quadrant is invariant. That is, if $x(0) > 0$ and $y(0) > 0$, then the solution satisfies $x(t) > 0$ and $y(t) > 0$ for all t.

35. **(a)** Find all equilibria of the two-predator and one-prey system (9.19) in which no predators are present.

    (b) Use the Jacobian (9.20) to analyze the stability of each equilibrium you found in (a).

36. **(a)** Find all equilibria of the two-predator and one-prey system (9.19) in which the middle predator y is present, but the top predators is absent.

    (b) Use the Jacobian (9.20) to analyze the stability of each equilibrium you found in (a).

37. **(a)** Show that there is an equilibrium of the two-predator and one-prey system (9.19) in which all three species are present (i.e., $x > 0$, $y > 0$, $z > 0$) if and only if $K > \frac{130}{177}$.

    (b) Numerically compute the roots of the characteristic equation (eigenvalues) of the Jacobian at this equilibrium for $K = 1$ and $K = 2$. Describe the roots. What happens to the stability of the equilibrium as K is increased from 1 to 2?

    (c) Why is a Hopf bifurcation to a limit cycle suggested?

38. Suppose $p = mx$ in the predator-prey model (9.10)–(9.11). The result is the so-called Lotka-Volterra predator-prey model. Let $r = K = c = 1$.

    (a) Find all equilibria. Only nonnegative equilibrium $x \geq 0$, $y \geq 0$ are of interest. Which equilibria are nonnegative?

    (b) Determine the local phase portraits and the stability properties of each nonnegative equilibrium.

    (c) Identify any bifurcations that occur, using d as a parameter (holding m fixed).

(d) Interpret your results in (b).

39. **(a)** Write down a predator-prey model in which the prey grows exponentially and the predator dies exponentially when the species are alone and the per predator predation rate is $p = mx$.

    (b) Find all nonnegative equilibria.

    (c) Determine the stability properties and local phase portraits for each nonnegative and hyperbolic equilibrium.

    (d) Study nonhyperbolic equilibria using a computer program. Sketch the phase portrait in the first quadrant $x > 0$, $y > 0$.

    (e) Interpret your results.

40. Consider the Lotka-Volterra competition model

$$x' = r_1 x - c_{12} x y$$
$$y' = r_2 y - c_{21} x y.$$

All coefficients are positive.

    (a) Use a computer program to explore the orbits with nonnegative initial conditions: $x(0) \geq 0$, $y(0) \geq 0$. Formulate conjectures about the asymptotic dynamics of an orbit that starts in the first quadrant [i.e., whose initial conditions satisfy $x(0) > 0$, $y(0) > 0$]. (Are the orbits bounded for $t \geq 0$? Are there any equilibria? Do all orbits tend to equilibria? Or are their limit cycles? etc.)

    (b) Find all equilibria of the equations.

    (c) Use the linearization procedure to classify each equilibrium and to determine the stability of each equilibrium in (a) (unstable saddle, stable node, etc.).

    (d) Show that both the positive x-axis and the positive y-axis consist of orbits. What does this imply about orbits that start in the first quadrant [i.e., with initial conditions $x(0) > 0$ and $y(0) > 0$]?

    (e) Give an argument that all orbits in the first quadrant, except two, are forward unbounded.

    (f) As best you can, use the results of (b)–(e) to address the conjectures you made in (a).

Laplace Transforms

In previous chapters we learned methods for solving linear differential equations with constant coefficients. In Chapter 2 we studied the variation of constants formula (the integrating factor method) and the method of undetermined coefficients for first-order equations of the form

$$x' = px + q(t),$$

where p is a constant. In Chapters 5 and 6 we extended these methods to linear systems

$$x' = ax + by + h_1(t)$$
$$y' = cx + dy + h_2(t)$$

with constant coefficients a, b, c, d and to second-order equations

$$\alpha x'' + \beta x' + \gamma x = g(t)$$

with constant coefficients α, β, and γ. In this section we study another method for obtaining solution formulas for linear equations with constant coefficients, namely, the *Laplace transform method*. Why learn yet another solution method for these equations? We have several reasons. First, the Laplace transform method provides a short-cut method for equations with types of nonhomogeneous terms $q(t)$ and $g(t)$ we have not considered before and that arise in applications (namely, discontinuous terms). Second, the Laplace transform method is a classical method that is commonly used in certain disciplines (such as engineering). To read and understand the literature in these disciplines it is necessary to have familiarity with Laplace transforms. Finally, the Laplace transform is one of several types of transforms used in solving differential equations (and many other kinds of equations as well), and therefore a study of Laplace transforms serves as an introduction to transform methods in general.

9.1 Introduction

In some applications a quantity is switched from one value to another during a very short period of time or, theoretically, at an instantaneous point of time. Examples include a switch that turns on an electrical current, a valve that suddenly releases a fluid flow, an intravenous injection of a drug into a patient's bloodstream, the force of a bat or racket on a struck ball, or a resource management decision that initiates harvesting of a biological population at a specified time. We can mathematically model such situations by using functions that discontinuously jump from one value to another at a certain value of its argument.

The equation

$$x' = rx - H, \qquad r > 0, \quad H > 0$$

arises in an application to harvested populations. Here $x = x(t)$ is the population size at time t and H is the (constant) rate at which the population harvested. In this model the population is continuously harvested for all time $t > 0$. Suppose, however, that harvesting is suddenly ceased at a certain point in time, say at time $t = 1$. Then the differential equation for population size x becomes

$$x' = rx + q(t), \tag{1.1}$$

where

$$q(t) = \begin{cases} 0 & \text{for } 0 < t < 1 \\ -H & \text{for } 1 \leq t \end{cases}. \tag{1.2}$$

This function $q(t)$ is not continuous for all $t > 0$. It has a point of discontinuity at $t = 1$.[1] For this reason the differential equation (1.1) does not fall under the purview of the general existence and uniqueness theorem 1 in Chapter 1. Nonetheless, we can show that the initial value problem $x(0) = x_0$ has a unique solution if we modify the definition of solution given in Definition 1 in Chapter 1.

We can break the initial value problem (1.1)–(1.2) into two parts. Observe that $q(t)$ is continuous on the separate intervals

$$0 < t < 1, \qquad 1 < t < +\infty.$$

Equation (1.1) has (infinitely many) solutions on each of these intervals as defined in Definition 1 in Chapter 1 [i.e., differentiable functions that reduce the differential equation (1.1) to an identity on these subintervals]. Perhaps we can choose a solution on each subinterval in such a way so as to define a solution over the whole interval $t > 0$. A key point in doing this is that in applications we are usually interested in solutions $x = x(t)$ that are continuous functions. If we impose the requirement that the solution we are trying to construct is continuous for all $t \geq 0$, then the solutions on the subintervals must be made to match at the endpoints of the two subintervals, namely at $t = 1$. The next example shows how to do this for the equation (1.1) with the discontinuous $q(t)$ defined by (1.2).

EXAMPLE 1

Consider the initial value problem

$$x' = rx + q(t), \qquad x(0) = x_0, \tag{1.3}$$

where $q(t)$ is the discontinuous function defined by (1.2). On the interval $0 < t < 1$ this problem reduces to

$$x' = rx, \qquad x(0) = x_0$$

---

[1] We have assigned the value of 0 to $q(t)$ at $t = 1$ rather than the value of $-H$. For our purposes we could just as well assign the value of $-H$ to $q(t)$ at $t = 1$. In other words, it is irrelevant for our purposes here whether $q(t)$ is continuous "from the right" or "from the left."

whose solution is

$$x_1(t) = x_0 e^{rt} \qquad \text{for } 0 \leq t < 1.$$

On the interval $1 < t < +\infty$ the differential equation reduces to the equation

$$x' = rx - H.$$

In order for the solution $x_2(t)$ on the interval $1 < t < +\infty$ to "match up" continuously with the solution $x_1(t)$ on the interval $0 < t < 1$ at the common endpoint point $t = 1$, we require $x_2(1) = x_1(1)$. This yields the initial value problem

$$x' = rx - H, \qquad x(1) = x_0 e^r$$

for $x_2(t)$ on the interval $t > 1$. The solution of this linear (autonomous, non-homogeneous) problem is (see Exercise 19)

$$x_1(t) = \left(x_0 e^r - \frac{H}{r}\right) e^{r(t-1)} + \frac{H}{r} \qquad \text{for } t \geq 1. \tag{1.4}$$

In summary, the continuous "solution" of (1.3) is

$$x(t) = \begin{cases} x_0 e^{rt} & \text{for } 0 \leq t < 1 \\ \left(x_0 e^r - \frac{H}{r}\right) e^{r(t-1)} + \frac{H}{r} & \text{for } 1 \leq t \end{cases}. \tag{1.5}$$

While the solution (1.5) in Example 1 was constructed so as to be continuous for all $t > 0$, it is *not differentiable* for all $t > 0$. To see this we calculate from (1.5) the derivatives on the subintervals

$$x'(t) = \begin{cases} rx_0 e^{rt} & \text{for } 0 < t < 1 \\ r\left(x_0 e^r - \frac{H}{r}\right) e^{r(t-1)} & \text{for } 1 < t \end{cases}$$

and notice that they do not match at $t = 1$; that is,

$$\lim_{t \to 1-} x'(t) = rx_0 e^r \neq r\left(x_0 e^r - \frac{H}{r}\right) = \lim_{t \to 1+} x'(t).$$

Consequently, $x(t)$ is *not* differentiable at $t = 1$. See Fig. 9.1.

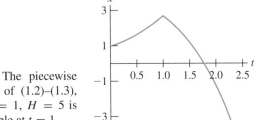

Figure 9.1 The piecewise solution (1.5) of (1.2)–(1.3), with $x_0 = r = 1$, $H = 5$ is not differentiable at $t = 1$.

Since $x(t)$ defined by (1.5) is not differentiable on the entire interval $0 < t < +\infty$, it is not a solution of the initial value problem (1.3) according to Definition 1 in Chapter 1. On the other hand, this $x(t)$ *is* a solution on each of the individual subintervals $0 < t < 1$ and $1 < t$. For these kinds of piecewise solutions we need a more general definition of solution. We will give such a definition, but first we look at a second-order example.

EXAMPLE 2

Consider the second-order, initial value problem

$$x'' + x = g(t), \qquad x(0) = 1, \qquad x'(0) = 0,$$

where

$$g(t) = \begin{cases} 0 & \text{for } t < \frac{\pi}{6} \\ 1 & \text{for } \frac{\pi}{6} \le t \end{cases}.$$

On the interval $0 < t < \frac{\pi}{6}$ the initial value problem reduces to

$$x'' + x = 0, \quad x(0) = 1, \quad x'(0) = 0$$

whose solution is

$$x_1(t) = \cos t \qquad \text{for } 0 \le t < \frac{\pi}{6}.$$

On the interval $\frac{\pi}{6} < t$ the equation is

$$x'' + x = 1$$

whose general solution is

$$x_2(t) = 1 + c_1 \sin t + c_2 \cos t \qquad \text{for } \frac{\pi}{6} < t.$$

To get a continuous solution at $t = \frac{\pi}{6}$, we require $x_2\left(\frac{\pi}{6}\right) = x_1\left(\frac{\pi}{6}\right)$, or

$$1 + \frac{1}{2}c_1 + \frac{\sqrt{3}}{2}c_2 = \frac{\sqrt{3}}{2}. \tag{1.6}$$

Any of the infinitely many choices for the two constants c_1, c_2 that satisfy this equation yields a solution $x_2(t)$ that "matches" $x_1(t)$ at $t = \pi/6$ and therefore form a continuous function for all $t \ge 0$. Having two constants c_1 and c_2 at our disposal, we can obtain a solution with more smoothness, namely, we can also match the derivatives (or graphically, the slopes) of $x_1(t)$ and $x_2(t)$ at $t = \pi/6$ and obtain a solution that is differentiable for all $t \ge 0$. The requirement that $x_2'(\pi/6) = x_1'(\pi/6)$ yields

$$\frac{\sqrt{3}}{2}c_1 - \frac{1}{2}c_2 = -\frac{1}{2}. \tag{1.7}$$

Thus, we obtain a unique, differentiable solution by solving the two equations (1.6)–(1.7) for c_1 and c_2. We obtain

$$c_1 = -\frac{1}{2}, \quad c_2 = 1 - \frac{\sqrt{3}}{2}$$

and therefore

$$x(t) = \begin{cases} \cos t & \text{for } 0 \le t < \frac{\pi}{6} \\ 1 - \frac{1}{2}\sin t + \left(1 - \frac{\sqrt{3}}{2}\right)\cos t & \text{for } \frac{\pi}{6} \le t \end{cases}. \tag{1.8}$$

The piecewise defined solution in the preceding example is not a solution by Definition 1 in Chapter 4, which requires that $x(t)$ be twice differentiable. The function (1.8) is once, but not twice, differentiable for all $t \geq 0$. The calculation

$$x''(t) = \begin{cases} -\cos t & \text{for } 0 < t < \frac{\pi}{6} \\ \frac{1}{2}\sin t - \left(1 - \frac{\sqrt{3}}{2}\right)\cos t & \text{for } \frac{\pi}{6} < t \end{cases}$$

shows

$$\lim_{t \to \frac{\pi}{6}-} x_1''(t) = -\frac{\sqrt{3}}{2} \neq -\frac{\sqrt{3}}{2} + 1 = \lim_{t \to \frac{\pi}{6}+} x_2''(t)$$

and therefore $x(t)$ does not have a second derivative, at $t = \pi/6$. See Fig. 9.2. Thus, as for first-order equation, we need a more general definition of solution for second-order equations. We begin with the following definition.

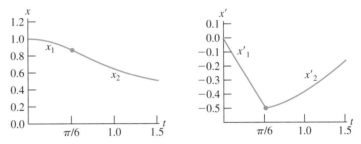

Figure 9.2 The piecewise solution (1.8) of the second-order initial value problem in Example 2 is continuous and differentiable, but not twice differentiable at $t = \frac{\pi}{6}$.

DEFINITION 1. A function $f(t)$ is **piecewise continuous on a finite interval** $a < t < b$ if there is a finite number of points t_i

$$a = t_1 < t_2 < t_3 < \cdots < t_{m-1} < t_m = b$$

such that $f(t)$ is continuous on each subinterval $t_i < t < t_{i+1}$ and (one-sided) limits at each endpoint t_i exist (and are finite). A function $f(t)$ is **piecewise continuous on an infinite interval** if it is piecewise continuous on every finite subinterval.

If the one-sided limits at a point $t = t_i$ are equal, that is, if

$$\lim_{t \to t_i-} f(t) = \lim_{t \to t_i+} f(t),$$

then $f(t)$ is continuous at $t = t_i$. On the other hand, if these one-sided limits are unequal, then $f(t)$ has a *jump discontinuity* at $t = t_i$.

The harvest rate (1.2) is an example of a piecewise continuous function on $0 < t < +\infty$.

DEFINITION 2. A function $x(t)$ is a **piecewise solution** of a first-order differential equation on a finite interval $a < t < b$ if $x(t)$ is continuous on the interval $a < t < b$ and if there is a finite number of endpoints t_i

$$a = t_1 < t_2 < t_3 < \cdots < t_{m-1} < t_m = b$$

such that $x(t)$ is a solution on each subinterval $t_i < t < t_{i+1}$ (by Definition 1 in Chapter 1). A function $x(t)$ is a piecewise solution on an infinite interval if it is a piecewise solution on every finite subinterval.

A function $x(t)$ is a **piecewise solution** of a second-order differential equation on a finite interval $a < t < b$ if $x(t)$ is differentiable (and therefore continuous) on the interval $a < t < b$ and if there is a finite number of endpoints t_i

$$a = t_1 < t_2 < t_3 < \cdots < t_{m-1} < t_m = b$$

such that $x(t)$ is a solution on each subinterval $t_i < t < t_{i+1}$ (by Definition 1 in Chapter 4). A function $x(t)$ is a piecewise solution on an infinite interval if it is a piecewise solution on every finite subinterval.

A piecewise solution of a first-order equation may not be differentiable at the points t_i. A piecewise solution of a second-order equation may not have a second derivative at the points t_i. Examples 1 and 2 illustrate these facts.

If $q(t)$ and $g(t)$ are continuous on $a < t < b$, then we know that initial value problems for the linear equations

$$x' = px + q(t)$$
$$\alpha x'' + \beta x' + \gamma x = g(t), \quad \alpha \neq 0$$

(with constant coefficients p and α, β, γ) have unique solutions defined on the whole interval $a < t < b$ (Theorem 2, Chapter 4). If $q(t)$ and $g(t)$ are piecewise continuous, a similar result holds. See Exercise 23.

THEOREM 1

(a) Suppose the coefficient $q(t)$ is piecewise continuous on an interval $a < t < b$. Then for any initial time t_0 in this interval and for any x_0 the initial value problem

$$x' = px + q(t), \quad x(t_0) = x_0$$

has a unique piecewise solution on the interval $a < t < b$.

(b) Similarly, suppose $g(t)$ is piecewise continuous on an interval $a < t < b$. Then for any initial time t_0 in this interval and for any x_0 and y_0 the initial value problem

$$\alpha x'' + \beta x' + \gamma x = g(t), \quad x(t_0) = x_0, \quad y(t_0) = y_0$$

($\alpha \neq 0$) has a unique piecewise pair on the interval $a < t < b$.

Although we can find formulas for piecewise solutions by the piecewise construction method used in Example 1, there is another useful solution method for these problems. Before turning our attention to the Laplace transform method, we consider some useful notation for piecewise defined functions.

We can obtain convenient formulas for piecewise functions using the *unit step function* at $t = 0$ defined by

$$u(t) = \begin{cases} 0 & \text{for } t < 0 \\ 1 & \text{for } 0 \leq t \end{cases}.$$

We obtain a unit step function at other values of $t = c$ by shifting this function

$$u(t - c) = \begin{cases} 0 & \text{for } t < c \\ 1 & \text{for } c \leq t \end{cases}.$$

The unit step function at $t = c$ mathematically describes a "switch" that "turns on" at $t = c$. It is a piecewise continuous function with a jump discontinuity at $t = c$. See Fig. 9.3.

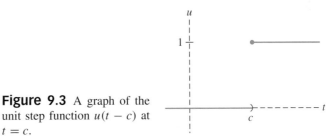

Figure 9.3 A graph of the unit step function $u(t - c)$ at $t = c$.

EXAMPLE 3

We can write the piecewise continuous harvest function (1.2) as

$$q(t) = -Hu(t - 1).$$

One way to understand this formula is to note that for $t < 1$ the switch $u(t - 1)$ is off, so that $q(t) = 0$ for $t < 1$, and as t increases through 1 the switch turns on, so that $q(t) = -H$ for $t > 1$. The differential equation (1.1)–(1.2) becomes

$$x' = rx - Hu(t - 1).$$ ∎

We can write a piecewise defined function as

$$f(t) = \begin{cases} f_1(t) & \text{for } a < t < t_1 \\ f_2(t) & \text{for } t_1 \leq t < t_2 \\ \vdots & \vdots \\ f_m(t) & \text{for } t_{m-1} \leq t < t_m \\ f_{m+1}(t) & \text{for } t_m \leq t < b. \end{cases}$$

It is shown in Exercise 25 that such a function can also be written in terms of unit step functions $u(t - t_i)$ as follows:

$$f(t) = f_1(t) + \sum_{n=1}^{m} \left[f_{n+1}(t) - f_n(t) \right] u(t - t_n). \tag{1.9}$$

For example, using this formula we can write the piecewise solution (1.5) as

$$x(t) = x_0 e^{rt} + \left[\left(x_0 e^r - \frac{H}{r} \right) e^{r(t-1)} + \frac{H}{r} - x_0 e^{rt} \right] u(t - 1)$$

or

$$x(t) = x_0 e^{rt} + \frac{H}{r} \left[1 - e^{r(t-1)}\right] u(t-1).$$

EXAMPLE 4   Consider the piecewise continuous harvesting function

$$q(t) = \begin{cases} 0 & \text{for } 0 < t < 1 \\ -H & \text{for } 1 \le t < 2 \\ 0 & \text{for } 2 \le t \end{cases}.$$

This represents a case when harvesting starts at time $t = 1$ and ceases at time $t = 2$. Using formula (1.9),

$$q(t) = 0 + [-H - 0] u(t-1) + [0 - (-H)] u(t-2),$$

we have

$$q(t) = -Hu(t-1) + Hu(t-2).$$

One way to understand this formula is to note that for $t < 1$ both switches $u(t-1)$ and $u(t-2)$ are off, so that $q(t) = 0$ for $t < 0$. As t increases through 1, only the first switch $u(t-1)$ turns on, so that $q(t) = -H$ until t increases through 2, at which time the second switch $u(t-2)$ also turns on causing the cancellation $q(t) = -H + H = 0$ for $t > 2$. In Exercise 18 you are asked to find a formula for the solution of the resulting differential equation

$$x' = rx - Hu(t-1) + Hu(t-2). \qquad \blacksquare$$

Exercises

Write the following piecewise functions in terms of unit step functions.

1. $\begin{cases} -2 & \text{for } t < 2 \\ 3 & \text{for } 2 \le t \end{cases}$

2. $\begin{cases} 5 & \text{for } t < -1 \\ -3 & \text{for } -1 \le t \end{cases}$

3. $\begin{cases} 1 & \text{for } t < 1 \\ 0 & \text{for } 1 \le t < 2 \\ -1 & \text{for } 2 \le t \end{cases}$

4. $\begin{cases} -5 & \text{for } t < 0 \\ 5 & \text{for } 0 \le t < 4 \\ 10 & \text{for } 4 \le t \end{cases}$

5. $\begin{cases} 1 & \text{for } t < 1 \\ 2 & \text{for } 1 \le t < 2 \\ 3 & \text{for } 2 \le t < 3 \\ 4 & \text{for } 3 \le t \end{cases}$

6. $\begin{cases} 1 & \text{for } t < -1 \\ -1 & \text{for } -1 \le t < 0.5 \\ 1 & \text{for } 0.5 \le t < \pi \\ -1 & \text{for } \pi \le t \end{cases}$

7. $\begin{cases} 1 & \text{for } t < 0 \\ 1-t & \text{for } 0 \le t < 1 \\ 0 & \text{for } 1 \le t \end{cases}$

8. $\begin{cases} 0 & \text{for } t < 0 \\ \sin t & \text{for } 0 \le t < \pi \\ 0 & \text{for } \pi \le t \end{cases}$

9. $\begin{cases} t^2 & \text{for } t < 1 \\ 2-t & \text{for } 1 \le t < 3 \\ -1 & \text{for } 3 \le t < 3\pi \\ \cos t & \text{for } 3\pi \le t \end{cases}$

10. $\begin{cases} \cos t & \text{for } t < 0 \\ 2t & \text{for } 0 \le t < 3.5 \\ 7 & \text{for } 3.5 \le t \end{cases}$

11. Sketch a graph of the functions in Exercises 1–10. Which functions are continuous?

Using step functions, write formulas for the piecewise functions whose graphs appear in the following exercises.

12.

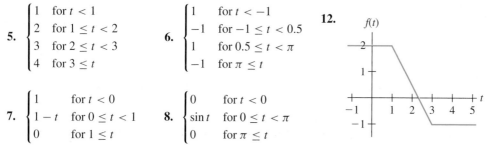

13.

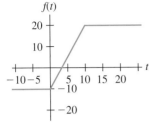

Using step functions, find formulas for the piecewise solutions of the following initial value problems. Sketch a graph of the solution.

14. $x' = x + 1 - 2u(t - 1), x(0) = 0$

15. $x' = -2x + t - tu(t - 2), x(0) = 0$

16. $x'' - x = u(t - 1), x(0) = 1, x'(0) = 0$

17. $x'' + x = (\sin 2t) u(t - \pi), x(0) = 1, x'(0) = 0$

18. Using unit step functions, find a formula for the piecewise solution of the equation $x' = rx - Hu(t-1) + Hu(t-2)$. Sketch a graph of the solution.

19. Solve the initial value problem $x' = rx - H, x(1) = x_0 e^r$.

In the exercises below, assume that the m functions $q_1(t)$, $q_2(t)$, ..., $q_m(t)$ are each piecewise continuous on a finite interval $a < t < b$.

20. Prove that the linear combination $f(t) = c_1 q_1(t) + \cdots + c_m q_m(t)$ is piecewise continuous on $a < t < b$. Here $c_i =$ constant.

21. Prove that the product $p(t) = q_1(t)q_2(t) \cdots q_m(t)$ is piecewise continuous on $a < t < b$.

22. If the functions $h_i(t)$ are continuous on $a < t < b$, prove that $g(t) = h_1(t)q_1(t) + \cdots + h_m(t)q_m(t)$ is piecewise continuous on $a < t < b$.

23. Prove Theorem 1.

24. Assume $q(t)$ is piecewise continuous on an interval $a < t < b$. Prove that a piecewise continuous solution of $x' = rx + q(t)$ has a piecewise continuous derivative.

25. Prove (1.9).

26. For $t \geq 0$ write the square wave function

$$f(t) = \begin{cases} 1 & \text{for } 0 \leq t < 1 \\ 0 & \text{for } 1 \leq t < 2 \\ 1 & \text{for } 2 \leq t < 3 \\ 0 & \text{for } 3 \leq t < 4 \\ \vdots & \quad \vdots \end{cases}$$

in terms of unit step functions.

27. For $t \geq 0$ write the sawtooth function, whose graph appears below, in terms of unit step functions. Assume that the graph continues with the same pattern indefinitely to the right.

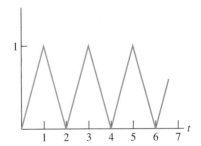

9.2   The Laplace Transform

A common strategy in mathematics when dealing with a difficult problem is to somehow transform the problem so as to obtain a simpler problem. After obtaining the solution of the simpler problem one reverses, or inverts, the transformation to find the solution of the original problem.

For example, we can transform the equation $e^{2x} - 5e^x + 6 = 0$ into the quadratic polynomial equation $y^2 - 5y + 6 = 0$ by means of the transformation (or change of variables) $y = e^x$. From the roots $y = 2$ and $y = 3$ of this transformed quadratic polynomial equation we obtain the roots of the original equation from the inverse transform $x = \ln y$. Thus, the roots of the original equation are $x = \ln 2$ and $\ln 3$. Other examples of this transform strategy include the use of logarithms to convert multiplication problems into addition problems and changes of variables that change complicated integrals into simpler integrals.

Let us write out the steps of the example in the preceding paragraph in order to expose fully all the steps involved. We will use this same sequence of steps in the

Laplace transform method for solving differential equations.

Steps to solve the equation $e^{2x} - 5e^x + 6 = 0$ for x.

STEP 1: $y = e^x$ Define the transformation.

STEP 2: $y^2 - 5y + 6 = 0$ Transform the equation.

STEP 3: $y = 2$ and 3 Solve the transformed equation.

STEP 4: $x = \ln 2$ and $\ln 3$ Inverse transformation of solutions.

Step 1 involves the selection of a suitable transformation of the unknown x to a new unknown y. Suitable means here that the transformation successfully transforms the problem into a simpler problem, namely, one that can be solved. Step 2 involves the actual transformation of the original equation for x into an equation for the new unknown y. Step 3 is the solution step of the transformed problem and Step 4 is the inverse transformation of the answer y in Step 3 performed to obtain the answer x of the original equation.

The Laplace transformation method follows the same steps to solve linear differential equations with constant coefficients. We will discuss the definition of this transform (Step 1) in this section. The Laplace transform will transform one function into another function, in our case the unknown function x into another unknown function denoted by $\mathcal{L}(x)$ or sometimes by X. The transformed equation for $\mathcal{L}(x)$ (Step 2) will turn out to be simpler in the sense that it is an easily solvable *algebraic* equation. Step 4 is the most difficult step in the Laplace transform method [i.e., once the solution $\mathcal{L}(x)$ of the transformed equation is calculated we must perform an inverse Laplace transform in order to obtain the solution x of the original differential equation]. A full treatment of this last inversion step is beyond the scope of this text; we will have to rely on tables of transforms and a collection of basic properties of Laplace transforms. (Some symbolic computer programs can calculate inverse Laplace transforms.)

The first step is to define the Laplace transform.

The basic idea underlying the Laplace transform method lies in the transformation of the calculus operation of differentiation into the algebraic operation of multiplication. If this seems like magic, recall that the derivative of some well-known functions can be calculated by means of multiplication (for example, the derivative of e^{ct} is ce^{ct}). Because exponential functions have this property, they play a prominent role in the definition of the Laplace transform.

Another important ingredient underlying the Laplace transform is the integration-by-parts formula

$$\int_0^a f(t)x'(t)\,dt = f(t)x(t)|_{t=0}^{t=a} - \int_0^a f'(t)x(t)\,dt,$$

where a is, for the moment, a fixed constant. If we choose $f(t)$ as an exponential $f(t) = e^{-st}$, where s is an arbitrary constant, then $f'(t) = -se^{-st}$ and we have

$$\int_0^a e^{-st}x'(t)\,dt = e^{-st}x(t)\Big|_{t=0}^{t=a} + s\int_0^a e^{-st}x(t)\,dt. \tag{2.1}$$

What does the formula (2.1) do for us with regard to the transformation strategy we described above? If we view the arbitrary constant s as a new independent variable, then the integral $\int_0^a e^{-st}x(t)\,dt$ transforms the function $x(t)$ to a new function of s (the variable t has been "integrated away"). The integral $\int_0^a e^{-st}x'(t)\,dt$ is the same transformation performed on the derivative $x'(t)$. Equation (2.1) says that the transformation of $x'(t)$ can be calculated (partly) by multiplying the transform of $x(t)$ by s. In other words, differentiation has been transformed into multiplication!

However, also involved in the transformation of $x'(t)$, on the right-hand side of (2.1) is the boundary term

$$e^{-st}x(t)\Big|_{t=0}^{t=a} = e^{-sa}x(a) - x(0),$$

which involves the evaluation of $x(t)$ at the 0 and a. If we are solving an initial value problem, then $x(0)$ is known and we lack only $x(a)$. Up to now we have left a as an unspecified, but fixed, constant. In order to circumvent the problem of unknown $x(a)$, what is done is the following. If we choose $a = +\infty$, then the boundary term becomes a limit

$$e^{-st}x(t)\Big|_{t=0}^{t=+\infty} = \lim_{t\to+\infty} e^{-st}x(t) - x(0).$$

For real $s > 0$ the exponential e^{-st} approaches 0 as $t \to +\infty$. If the product $e^{-st}x(t)$ also tends to 0, the boundary term reduces to $-x(0)$. For the moment we assume

$$\lim_{t\to+\infty} e^{-st}x(t) = 0. \tag{2.2}$$

We will say more about this condition below. Formula (2.1) becomes

$$\int_0^{+\infty} e^{-st}x'(t)\,dt = -x(0) + s\int_0^{+\infty} e^{-st}x(t)\,dt. \tag{2.3}$$

Before proceeding with Steps 2, 3, and 4 of our transformation strategy, we formalize what we have found so far in the following definition.

> **DEFINITION 3.** Let $x(t)$ be a piecewise continuous function defined for $t \geq 0$. The **Laplace transform** $\mathcal{L}(x)(s)$ of $x(t)$ is defined by the integral
>
> $$\mathcal{L}(x)(s) \doteq \int_0^{+\infty} e^{-st}x(t)\,dt,$$
>
> where s is a real number.[2] An alternative notation for the Laplace transform of $x(t)$ is
>
> $$X(s) = \int_0^{+\infty} e^{-st}x(t)\,dt.$$
>
> For notational simplification we will sometimes denote $\mathcal{L}(x)(s)$ by $\mathcal{L}(x)$.

Using Laplace transform notation, we write formula (2.3) as

$$\mathcal{L}(x')(s) = s\mathcal{L}(x)(s) - x(0). \tag{2.4}$$

In Definition 3 it is assumed, of course, that the improper integral exists (at least for some values of s). When can we be sure that the Laplace transform of a function $x(t)$ exists? Or, more precisely, how can we determine those values of s for which $\mathcal{L}(x)(s)$ exists? And when can we be sure (2.2) holds? One way, for a specific function $x(t)$, is by direct calculation of the improper integral in the Definition 3. Another way is

---

[2]For a full treatment of Laplace transforms it is necessary to allow s to be a complex number. For our purposes here, however, we will restrict our attention to real s (Exercises 5–10).

to use a theorem that describes conditions on $x(t)$ that are sufficient to guarantee the existence of $\mathcal{L}(x)(s)$. Such a theorem appears below (Theorem 2).

EXAMPLE 1

Consider the unit step function at $t = c > 0$

$$u(t - c) = \begin{cases} 0 & \text{for } t < c \\ 1 & \text{for } c \leq t \end{cases}.$$

Then for $s > 0$

$$\mathcal{L}(u(t - c))(s) = \int_0^{+\infty} e^{-st} u(t - c)\, dt$$

$$= \int_c^{+\infty} e^{-st}\, dt$$

$$= -\frac{1}{s} e^{-st} \Big|_{t=c}^{t=+\infty}$$

$$= \lim_{t \to +\infty} \left(-\frac{1}{s} e^{-st} \right) + \frac{1}{s} e^{-sc}.$$

For $s > 0$

$$\lim_{t \to +\infty} \left(-\frac{1}{s} e^{-st} \right) = 0,$$

that is, (2.2) holds. Thus,

$$\mathcal{L}(u(t - c))(s) = \frac{1}{s} e^{-cs} \quad \text{for} \quad s > 0.$$ ■

EXAMPLE 2

To calculate the Laplace transform of an exponential e^{ct}, where c is a real constant, we perform the integration

$$\mathcal{L}(e^{ct})(s) = \int_0^{+\infty} e^{-st} e^{ct}\, dt$$

$$= \int_0^{+\infty} e^{(c-s)t}\, dt$$

$$= \frac{1}{c - s} e^{(c-s)t} \Big|_{t=0}^{t=+\infty}$$

$$= \lim_{t \to +\infty} \frac{1}{c - s} e^{(c-s)t} - \frac{1}{c - s}.$$

If $s > c$, then

$$\lim_{t \to +\infty} \frac{1}{c - s} e^{(c-s)t} = 0,$$

that is, (2.2) holds. Thus, for $s > c$

$$\mathcal{L}(e^{ct})(s) = \frac{1}{s - c}. \tag{2.5}$$

Notice that this formula includes (by taking $c = 0$) the formula

$$\mathcal{L}(1)(s) = \frac{1}{s}.$$ ■

In the preceding examples we calculated the Laplace transform of exponential functions and unit step functions (also see Exercise 5–10). Formulas for the Laplace transforms of many other basic functions, such as polynomials and trigonometric functions, can also be calculated directly from the Definition 3. See Table 9.2 in Sec. 9.4.

Before studying further methods of calculating and using Laplace transforms, we consider the following question: Under what conditions does a function $x(t)$ have a Laplace transform and condition (2.2) hold? This is an important question when we apply the Laplace transform method to a differential equation. How can we be sure that the transform of the solution exists, so that Laplace transform techniques can be used to find it?

For $s > 0$, the exponential e^{-st} tends to 0 as $t \to +\infty$. For (2.2) to hold, $x(t)$ must not grow faster than e^{st}. Thus, by requiring that $x(t)$ grows no faster than exponential, we can expect the limit (2.2) to hold. Moreover, under this condition the improper integral in the definition of the Laplace transform will converge. More specifically, if $x(t)$ satisfies a bound of the form

$$|x(t)| \le me^{ct}, \quad t > 0$$

for some constants $m \ge 0$ and c, then

$$\left| e^{-st} x(t) \right| \le e^{-st} me^{ct} = me^{(c-s)t}.$$

which in turn implies

$$0 \le \lim_{t \to +\infty} \left| e^{-st} x(t) \right| \le \lim_{t \to +\infty} me^{(c-s)t} = 0 \quad \text{for } s > c$$

and the limit (2.2) holds. By the comparison theorem for improper integrals, the improper integral $\int_0^{+\infty} e^{-st} x(t) \, dt$ converges for $s > c$ because the improper integral $\int_0^{+\infty} me^{(c-s)t} \, dt$ converges (to $m/(s - c)$). We summarize these observations in the following definition and theorem.

> **DEFINITION 4.** A function $x(t)$ is **exponentially bounded** for $t \ge 0$ if there are constants $m \ge 0$ and c such that
>
> $$|x(t)| \le me^{ct} \quad \text{for all} \quad t \ge 0.$$

THEOREM 2

> The Laplace transform $\mathcal{L}(x)(s)$ of a piecewise continuous, exponentially bounded function $x(t)$ exists for $s > c$.

From this theorem we can deduce that many familiar functions have Laplace transforms. For example, any bounded (piecewise continuous) function is exponentially bounded (use $c = 0$ in the definition above). Thus, for example, the trigonometric $\sin t$ and $\cos t$ have Laplace transforms for all $s > c = 0$. Also, it can be shown by Theorem 2 that any polynomial, of any degree, has a Laplace transform for $s > 0$ (see Exercise 1). Moreover, sums and products of exponentially bounded functions are exponentially bounded (see Exercises 2–3). Therefore, sums of products of exponential functions, polynomials, and sine and cosine functions are all exponentially bounded and by Theorem 2 have Laplace transforms for all $s > 0$. We will learn some useful methods for calculating such Laplace transforms in the following section.

Theorem 2 and the following theorem answers our question about the Laplace transform of a solutions of linear differential equations.

THEOREM 3   (a) Suppose $q(t)$ is piecewise continuous and exponentially bounded for $t \geq 0$. The piecewise solution $x(t)$ of the initial value problem

$$x' = px + q(t), \quad x(0) = x_0 \tag{2.6}$$

is exponentially bounded for $t \geq 0$.

(b) Suppose $g(t)$ is piecewise continuous and exponentially bounded for $t \geq 0$. The piecewise solution $x(t)$ of the initial value problem

$$\alpha x'' + \beta x' + \gamma x = g(t), \quad x(t_0) = x_0, \quad x'(t_0) = y_0 \tag{2.7}$$

is exponentially bounded for $t \geq 0$.

From Theorem 2 we see that the piecewise solutions of the initial value problems (2.6) and (2.7) have Laplace transforms. We will give a proof for part (a) of the theorem. [Part (b) is proved in a similar manner.] Suppose me^{ct} is an exponential bound for $q(t)$ (Definition 4). Since e^{ct} is an increasing function of c for $t > 0$, the inequality $|q(t)| \leq me^{c^*t}$ is also valid for any constant $c^* \geq c$. Choose any c^* larger than both c and p (i.e., $c^* > \max\{c, p\}$). In Exercise 4 it is shown that the solution of (2.6) is given by the variation of constants formula

$$x(t) = x_0 e^{pt} + e^{pt} \int_0^t e^{-p\tau} q(\tau)\, d\tau. \tag{2.8}$$

Then for $t \geq 0$

$$|x(t)| \leq |x_0|\, e^{pt} + e^{pt} \int_0^t e^{-p\tau}\, |q(\tau)|\, d\tau$$

$$\leq |x_0|\, e^{pt} + e^{pt} \int_0^t e^{-p\tau} me^{c^*\tau}\, d\tau$$

$$= |x_0|\, e^{pt} + me^{pt}\, \frac{1}{c - p}\left(e^{(c^* - p)t} - 1\right)$$

$$\leq |x_0|\, e^{c^*t} + m\, \frac{1}{c^* - p} e^{c^*t} = m_1 e^{c^*t},$$

where $m_1 = |x_0| + m/(c^* - p)$. This shows that $x(t)$ is exponentially bounded for $t \geq 0$.

The formula (2.4) for the Laplace transform of the first derivative x' allows us to obtain a formula for the Laplace transform of the second-order derivative x''. Since $x'' = (x')'$ we have

$$\mathcal{L}\left(x''\right) = \mathcal{L}\left((x')'\right)$$

$$= s\mathcal{L}\left(x'\right) - x'(0)$$

$$= s\left(s\mathcal{L}\left(x\right) - x(0)\right) - x'(0)$$

or

$$\mathcal{L}\left(x''\right) = s^2 \mathcal{L}\left(x\right) - sx(0) - x'(0). \tag{2.9}$$

Similarly, we can obtain a formula for the Laplace transform of higher order derivatives. See Exercises 11 and 12.

Exercises

1. **(a)** Suppose $x(t)$ is piecewise continuous and for some constant c

$$\lim_{t \to +\infty} \left| e^{-ct} x(t) \right| = 0, \quad t \geq 0. \tag{2.10}$$

   Prove $x(t)$ is exponentially bounded.

   (b) Prove that $x(t) = t$ satisfies (2.10).

   (c) Prove that $x(t) = t^i$, where i = positive integer, satisfies (2.10).

   (d) Prove that any polynomial of any degree has a Laplace transform for all $s > 0$.

2. If $f(t)$ and $g(t)$ are exponentially bounded, show that the sum $f(t) + g(t)$ is exponentially bounded.

3. If $f(t)$ and $g(t)$ are exponentially bounded, show that the product $f(t)g(t)$ is exponentially bounded.

4. Show that the formula (2.8) gives the piecewise solution of the first-order initial value problem (2.6).

Write the complex number $s = \alpha + i\beta$, where $\alpha = \operatorname{Re} s$ is the real part of s and $\beta = \operatorname{Im} s$ is the imaginary part. Let s be complex in the definition of the Laplace transform.

5. Show that $\left| e^{-st} \right| = e^{-\alpha t}$ for real $t \geq 0$.

6. Show that $\lim_{b \to +\infty} e^{-sb} = 0$ if $\alpha > 0$.

7. For complex s show that by direct calculation

$$\mathcal{L}(e^{ct})(s) = \frac{1}{s - c}$$

   if $\alpha > c$.

8. For $c \geq 0$ and complex s show that by direct calculation

$$\mathcal{L}(u(t - c))(s) = \frac{1}{s} e^{-cs}$$

   if $\alpha > 0$.

9. Prove Theorem 2 for complex s provided $\alpha > c$.

10. Prove Theorem 3 for complex s.

11. Derive a formula for $\mathcal{L}\left(x^{(3)}\right)(s)$, where $x^{(3)}$ is the third-order derivative of x.

12. Use (2.4) and induction to derive the formula

$$\mathcal{L}\left(x^{(n)}\right)(s) = s^n \mathcal{L}(x)(s) - s^{n-1} x(0)$$
$$- s^{n-2} x'(0) - \cdots - x^{(n-1)}(0),$$

   where $x^{(n)}$ is the nth derivative of x.

9.3   Linearity and the Inverse Laplace Transform

By Theorem 3 we know that the solution of the initial value problem

$$x' = px + q(t), \quad x(0) = x_0 \tag{3.1}$$

has a Laplace transform $\mathcal{L}(x)$ when $q(t)$ is exponentially bounded. Step 2 in the transform method is to transform this problem for x into a problem for the Laplace transform $\mathcal{L}(x)$. We have learned how to transform the left-hand side of the equation, namely, $\mathcal{L}(x') = s\mathcal{L}(x) - x_0$. What about the transform of the right-hand side of the equation? How do we transform the sum $px + q(t)$?

One of the basic properties of the Laplace transform is its linearity. This property implies, for example, that the Laplace transform of a sum $x(t) + y(t)$ of two functions is the same as the sum of their Laplace transforms $\mathcal{L}(x) + \mathcal{L}(y)$. It also says that the Laplace transform of a multiple $kx(t)$ is the multiple of the Laplace transform $k\mathcal{L}(x)$. More generally, linearity of the Laplace transform means

$$\mathcal{L}(k_1 x + k_2 y) = k_1 \mathcal{L}(x) + k_2 \mathcal{L}(y), \tag{3.2}$$

where k_1 and k_2 are real numbers. See Exercise 11.

EXAMPLE 1

From Example 4 consider the function

$$q(t) = -Hu(t - 1) + Hu(t - 2).$$

Using the linearity property (3.2) we have

$$\mathcal{L}(q(t)) = -H\mathcal{L}(u(t-1)) + H\mathcal{L}(u(t-2))$$

and hence (see Example 1 in Sec. 9.2)

$$\mathcal{L}(q(t)) = -H\frac{1}{s}e^{-s} + H\frac{1}{s}e^{-2s}.$$ ■

EXAMPLE 2

Using the linearity property (3.2), together with the exponential transform formula (2.5), we calculate

$$\mathcal{L}(5e^{-2t} - 3e^{-3t}) = 5\mathcal{L}(e^{-2t}) - 3\mathcal{L}(e^{-3t})$$

$$= 5\frac{1}{s+2} - 3\frac{1}{s+3}$$

$$= \frac{2s+9}{(s+2)(s+3)}.$$ ■

So far we have seen three fundamental properties of Laplace transforms:

First derivative formula	$\mathcal{L}\left(x'\right) = s\mathcal{L}\left(x\right) - x(0)$
Second derivative formula	$\mathcal{L}\left(x''\right) = s^2\mathcal{L}\left(x\right) - sx(0) - x'(0)$
Linearity	$\mathcal{L}(k_1 x + k_2 y) = k_1\mathcal{L}(x) + k_2\mathcal{L}(y).$

We can be use these properties to transform the initial value problem (3.1) to a problem for the Laplace transform of x as follows:

$$\mathcal{L}(x') = \mathcal{L}(px + q(t))$$

$$s\mathcal{L}(x) - x_0 = p\mathcal{L}(x) + \mathcal{L}(q).$$

Step 3 of the transform method is to solve this transformed equation for

$$\mathcal{L}(x) = \frac{\mathcal{L}(q) + x_0}{s - p}. \tag{3.3}$$

EXAMPLE 3

Consider the initial value problem

$$x' = -2x + 3e^{-3t}, \quad x(0) = 2.$$

Transforming this problem, we perform the calculations

$$\mathcal{L}(x') = \mathcal{L}(-2x + 3e^{-3t})$$

$$s\mathcal{L}(x) - 2 = -2\mathcal{L}(x) + 3\mathcal{L}(e^{-3t}).$$

Since

$$\mathcal{L}(e^{-3t}) = \frac{1}{s+3}$$

we have

$$s\mathcal{L}(x) - 2 = -2\mathcal{L}(x) + 3\frac{1}{s+3},$$

an equation we can solve algebraically solved for

$$\mathcal{L}(x) = \frac{1}{s+2}\left(2 + 3\frac{1}{s+3}\right)$$

$$= \frac{2s+9}{(s+2)(s+3)}.$$ ■

To apply the transform method to the second-order initial value problems we use the second derivative formula. Here is an example.

EXAMPLE 4

To transform the second-order, initial value problem

$$x'' + x = e^{-3t}$$
$$x(0) = 2$$
$$x'(0) = -1$$

we calculate

$$\mathcal{L}\left(x'' + x\right) = \mathcal{L}\left(e^{-3t}\right)$$

$$\mathcal{L}\left(x''\right) + \mathcal{L}\left(x\right) = \frac{1}{s+3}$$

$$s^2 \mathcal{L}\left(x\right) - sx(0) - x'(0) + \mathcal{L}\left(x\right) = \frac{1}{s+3}$$

or finally

$$s^2 \mathcal{L}\left(x\right) - 2s + 1 + \mathcal{L}\left(x\right) = \frac{1}{s+3}.$$

We algebraically solve this (transformed) equation for

$$\mathcal{L}\left(x\right) = \left(\frac{1}{s+3} + 2s - 1\right)\frac{1}{s^2+1}$$

$$= \frac{2s^2 + 5s - 2}{(s+3)\left(s^2+1\right)}. \qquad \blacksquare$$

To transform a general second-order initial value problem second-order initial value problem

$$\alpha x'' + \beta x' + \gamma x = g(t), \quad x(0) = x_0, \quad x'(0) = x_1 \qquad (3.4)$$

we calculate

$$\mathcal{L}\left(\alpha x'' + \beta x' + \gamma x\right) = \mathcal{L}\left(g\right)$$
$$\alpha \mathcal{L}\left(x''\right) + \beta \mathcal{L}\left(x'\right) + \gamma \mathcal{L}\left(x\right) = \mathcal{L}\left(g\right)$$
$$\alpha \left(s^2 \mathcal{L}\left(x\right) - sx(0) - x'(0)\right) + \beta \left(s\mathcal{L}\left(x\right) - x(0)\right) + \gamma \mathcal{L}\left(x\right) = \mathcal{L}\left(g\right)$$

an equation we can algebraically solve for

$$\mathcal{L}\left(x\right) = \frac{\mathcal{L}\left(g\right) + (\alpha s + \beta)\, x(0) + \alpha x'(0)}{\alpha s^2 + \beta s + \gamma}. \qquad (3.5)$$

The formulas (3.3) and (3.5) give the Laplace transforms L(x) of the solutions x of the initial value problems (3.1) and (3.4), respectively. The final step (Step 4) of the transform method is to determine x from its Laplace transform $\mathcal{L}(x)$.

For example, we must determine the solutions x of the initial value problems in Examples 3 and 4 from their Laplace transforms $\mathcal{L}(x) = X(s)$

$$X(s) = \frac{2s+9}{(s+2)\,(s+3)}$$

and

$$X(s) = \frac{2s^2 + 5s - 2}{(s + 3)(s^2 + 1)}.$$

> **DEFINITION 5.** A function $x(t)$ whose Laplace transform is $X(s)$ is denoted by $x = \mathcal{L}^{-1}(X)$ and is called an **inverse Laplace transform** of $X(s)$.

Symbolically, the solutions of the initial value problems in Examples 3 and 4 are

$$x(t) = \mathcal{L}^{-1}\left(\frac{2s + 9}{(s + 2)(s + 3)}\right)$$

and

$$x(t) = \mathcal{L}^{-1}\left(\frac{2s^2 + 5s - 2}{(s + 3)(s^2 + 1)}\right).$$

To solve initial value problems using the Laplace transform method, we must be able to calculate transforms and inverse transforms. To do this by hand requires some further properties of the Laplace transform, a table containing the transforms of some basic functions, and considerable practice. (Alternatively, some modern computer programs can carry out these calculations.) In the next section we study certain useful properties of the Laplace transform and learn the transforms of a few more elementary functions. We also study methods for calculating inverse Laplace transforms (which is usually a more difficult problem). However, a full treatment of the inverse problem requires complex analytical function theory and is beyond the scope of this book.

Exercises

Find the Laplace transform of the solution of the following initial value problems.

1. $x' = -2x + 3e^t$, $x(0) = -5$
2. $x' = \frac{1}{2}x - 2e^{-t}$, $x(0) = 2$
3. $x' = x + 3e^t - e^{2t}$, $x(0) = 1$
4. $x' = -x + 4e^{-3t} + 2e^{-4t}$, $x(0) = -1$
5. $x' = 10x - 5u(t - 3) + 3e^{-t}$, $x(0) = -5$
6. $x' = 4x - 7e^{-2t} + 10u(t + 5)$, $x(0) = 0$

Find the Laplace transform of the solution of the following initial value problems.

7. $x'' + x = u(t - 1)$, $x(0) = 1$, $x'(0) = 0$
8. $x'' + x = 2u(t - 1) + u(t - 2)$, $x(0) = 0$, $x'(0) = 1$
9. $x'' + 4x' + x = e^t - e^{-t}$, $x(0) = 1$, $x'(0) = 2$
10. $x'' + 2x' + 3x = 2e^t - 3e^{-t}$, $x(0) = -2$, $x'(0) = -3$
11. Prove the linearity property of the Laplace transform.

9.4 Properties of the Laplace Transform

In this section we will study some important properties of Laplace transforms and derive the transforms of some basic functions. With some practice we can then calculate both the transforms and inverse transforms for a large class of useful functions. A list of properties appears in Table 9.1. Derivations of these formulas can be found in the Exercises.

Table 9.1   Properties of Laplace transforms (c, k_1, k_2 are constants).

Linearity property	$\mathcal{L}(k_1 x(t) + k_2 y(t)) = k_1 \mathcal{L}(x) + k_1 \mathcal{L}(y)$
First derivative formula	$\mathcal{L}(x') = s\mathcal{L}(x) - x(0)$
Second derivative formula	$\mathcal{L}(x'') = s^2 \mathcal{L}(x) - sx(0) - x'(0)$
First translation formula	$\mathcal{L}(e^{ct} x(t)) = X(s - c)$
Second translation formula	$\mathcal{L}(x(t - c)u(t - c)) = e^{-cs} X(s), c \geq 0$
Derivative formula	$\mathcal{L}(t^n x(t)) = (-1)^n \frac{d^n X(s)}{ds^n}, n = 1, 2, 3, \ldots$
Convolution formula	$\mathcal{L}\left(\int_0^t x(t - \tau) y(\tau)\, d\tau\right) = X(s)Y(s)$
Integral formula	$\mathcal{L}\left(\int_0^t x(\tau)\, d\tau\right) = \frac{1}{s} X(s)$

Any formula for a Laplace transform, when viewed backward, is also a formula for an inverse Laplace transform. Indeed some of the formulas in Table 9.1 are more useful for finding inverse Laplace transforms. In the remainder of this section, we give examples that illustrate both uses of these formulas.

By direct calculation the Laplace transform of the two basic trigonometric functions $\sin \beta t$ and $\cos \beta t$ are

$$\mathcal{L}(\sin \beta t) = \int_0^{+\infty} e^{-st} \sin \beta t\, dt = \frac{\beta}{s^2 + \beta^2}, \quad s > 0$$

and

$$\mathcal{L}(\cos \beta t) = \int_0^{+\infty} e^{-st} \cos \beta t\, dt = \frac{s}{s^2 + \beta^2}, \quad s > 0$$

(Exercise 5). An important application of the first translation property in Table 9.1 is to functions of the form $e^{\alpha t} \sin \beta t$ and $e^{\alpha t} \cos \beta t$, which often appear in applications. Since the factors $\sin \beta t$ and $\cos \beta t$ have known transforms, we obtain, from the first translation property,

$$\mathcal{L}(e^{\alpha t} \sin \beta t) = \frac{\beta}{(s - \alpha)^2 + \beta^2}, \quad s > 0$$

$$\mathcal{L}\left(e^{\alpha t} \cos \beta t\right) = \frac{s - \alpha}{(s - \alpha)^2 + \beta^2}, \quad s > 0.$$

EXAMPLE 1

$$\mathcal{L}\left(e^{3t} (\cos t - 4 \cos 2t)\right) = \mathcal{L}\left(e^{3t} \cos t\right) - 4\mathcal{L}\left(e^{3t} \cos 2t\right)$$

$$= \frac{s - 3}{(s - 3)^2 + 1} - 4\frac{s - 3}{(s - 3)^2 + 4}$$

$$= -3\frac{(s - 3)^3}{\left(s^2 - 6s + 10\right)\left(s^2 - 6s + 13\right)} \qquad \blacksquare$$

A polynomial in t is a linear combination of integer powers of t (i.e. of $1, t, t^2, t^3$, $\ldots, t^n, \ldots$). In order to calculate the Laplace transform of a polynomial, we need to know the Laplace transforms of the powers $1, t, t^2, t^3, \ldots, t^n, \ldots$. One way to do this is to use the derivative formula in Table 9.1.

Treating t^n as the product $t^n = t^n x(t)$, where $x(t) = 1$, and recalling the transform

$$F(s) = \mathcal{L}(1) = \frac{1}{s},$$

we find from the derivative formula that

$$\mathcal{L}(t) = -\frac{d}{ds}\left(\frac{1}{s}\right) = \frac{1}{s^2}.$$

Treating t^2 as the product $t^2 = tx(t)$, where $x(t) = t$ [and hence $X(s) = 1/s^2$], we have

$$\mathcal{L}(t^2) = -\frac{d}{ds}\left(\frac{1}{s^2}\right) = \frac{2}{s^3}.$$

An induction argument shows

$$\mathcal{L}(t^n) = \frac{n!}{s^{n+1}}.$$

See Exercise 6.

EXAMPLE 2

$$\mathcal{L}\left(t^2 - \frac{2}{3}t + 2\right) = \mathcal{L}(t^2) - \frac{2}{3}\mathcal{L}(t) + 2\mathcal{L}(1)$$

$$= \frac{2}{s^3} - \frac{2}{3}\frac{1}{s^2} + 2\frac{1}{s}$$

$$= \frac{2}{3}\frac{3s^2 - s + 3}{s^3}$$

Using the first translation formula, we can calculate

$$\mathcal{L}\left(\left(t^2 - \frac{2}{3}t + 2\right)e^{2t}\right) = \frac{2}{3}\frac{3(s-2)^2 - (s-2) + 3}{(s-2)^3}$$

$$= \frac{2}{3}\frac{3s^2 - 13s + 17}{(s-2)^3}. \qquad \blacksquare$$

The properties of the Laplace transform can be used in various combinations. Here are some other examples.

EXAMPLE 3

One way to calculate the transform $\mathcal{L}\left(te^{-3t}\sin\pi t\right)$ is to use both the first translation formula (because of the exponential factor) and the derivative formula (because of the factor t). The calculation begins with

$$\mathcal{L}(\sin\pi t) = \frac{\pi}{s^2 + \pi^2}$$

and then calculates

$$\mathcal{L}\left(e^{-3t}\sin\pi t\right) = \frac{\pi}{(s+3)^2 + \pi^2}$$

and finally

$$\mathcal{L}\left(te^{-3t}\sin \pi t\right) = -\frac{d}{ds}\left(\frac{\pi}{(s+3)^2 + \pi^2}\right)$$

$$= \frac{2\pi\,(s+3)}{\left(s^2 + 6s + 9 + \pi^2\right)^2}.$$ ■

In the preceding examples we used the first translation formula to calculate Laplace transforms. We can also use this formula to calculate inverse transforms. To do this, however, we have to be able to recognize the given function of s as a translation $X(s - c)$ of a known transform $X(s)$. Thus, if the inverse $x(t)$ of $X(s)$ is known, then the inverse of the original function $X(s - c)$ is $e^{ct}x(t)$. That is to say, an equivalent way to write the first translation formula is

$$\mathcal{L}^{-1}(X(t - c)) = e^{ct}x(t).$$

EXAMPLE 4   The formula

$$\mathcal{L}(\sin \beta t) = \frac{\beta}{s^2 + \beta^2}$$

implies

$$\sin \beta t = \mathcal{L}^{-1}\left(\frac{\beta}{s^2 + \beta^2}\right).$$

With this in mind, to calculate the inverse Laplace transform of

$$\frac{2}{(s-4)^2 + 4}$$

we recognize this function as a translation of

$$X(s) = \frac{2}{s^2 + 4}$$

and that $X(s)$ has the inverse $x(t) = \sin 2t$. Thus,

$$\mathcal{L}^{-1}(X(t - 4)) = e^{4t}x(t)$$

or

$$\mathcal{L}^{-1}\left(\frac{2}{(s-4)^2 + 4}\right) = e^{4t}\sin 2t.$$ ■

In the preceding example we immediately recognized the function

$$\frac{2}{(s-4)^2 + 4}$$

as a translation of

$$\frac{2}{s^2 + 4}.$$

If, instead, this function were written in the algebraically equivalent form

$$\frac{2}{s^2 - 8s + 20}$$

we probably would not have so easily recognized it as a translation. We would have first to notice that

$$\frac{2}{s^2 - 8s + 20} = \frac{2}{(s-4)^2 + 4}.$$

Here we have algebraically completed the square on the denominator. This is often a useful step to keep in mind when dealing with quadratic factors (which often arise in the denominator of Laplace transforms).

EXAMPLE 5

To calculate the inverse Laplace transform of

$$\frac{4}{3s^2 - 6s + 8}$$

we complete the square on the denominator

$$3s^2 - 6s + 8 = 3\left(s^2 - 2s + \frac{8}{3}\right)$$

$$= 3\left((s-1)^2 + \frac{5}{3}\right)$$

and write

$$\frac{4}{3s^2 - 6s + 8} = \frac{4}{3}\frac{1}{(s-1)^2 + \frac{5}{3}}.$$

The second factor is nearly a translation of an expression of the form

$$\frac{\beta}{s^2 + \beta^2}$$

with

$$\beta = \sqrt{\frac{5}{3}}.$$

The problem is that the numerator is 1, not $\beta = \sqrt{5/3}$. We can obtain an expression of this form if we put $\sqrt{5/3}$ in the numerator and, in order not to change the expression, also divide by β. That is, we write

$$\frac{4}{3s^2 - 6s + 8} = \frac{4}{3}\frac{1}{\sqrt{\frac{5}{3}}}\frac{\sqrt{\frac{5}{3}}}{(s-1)^2 + \frac{5}{3}}$$

$$= \frac{4\sqrt{15}}{15}\frac{\sqrt{\frac{5}{3}}}{(s-1)^2 + \frac{5}{3}}.$$

We have now written this expression as a translation (by $c = 1$). Namely,

$$\frac{4}{3s^2 - 6s + 8} = \frac{4\sqrt{15}}{15}X(s-1),$$

where

$$X(s) = \frac{\sqrt{\frac{5}{3}}}{s^2 + \frac{5}{3}}.$$

The function $X(s)$ has the inverse

$$x(t) = \mathcal{L}^{-1}(X(s)) = \sin\left(\sqrt{\frac{5}{3}}\, t\right).$$

We now apply the first translation formula to obtain the desired inverse transform:

$$\mathcal{L}^{-1}\left(\frac{4}{3s^2 - 6s + 8}\right) = \frac{4\sqrt{15}}{15} e^t \sin\left(\sqrt{\frac{5}{3}}\, t\right).$$ ∎

The linearity property also implies that the inverse Laplace transform is a linear operator, that is,

$$\mathcal{L}^{-1}\left(k_1 X(s) + k_2 Y(s)\right) = k_1 x(t) + k_2 y(t).$$

Here is an example that uses this fact, together with the first translation formula.

EXAMPLE 6

To calculate the inverse transform of

$$X(s) = \frac{3s + 2}{s^2 + 4s + 6}$$

we first complete the square on the denominator:

$$X(s) = \frac{3s + 2}{(s + 2)^2 + 2}.$$

Writing the numerator as a translation $3s + 2 = 3(s + 2) - 4$, we have

$$\begin{aligned}
X(s) &= \frac{3(s + 2) - 4}{(s + 2)^2 + 2} \\[2mm]
&= 3\frac{s + 2}{(s + 2)^2 + 2} - 4\frac{1}{(s + 2)^2 + 2} \\[2mm]
&= 3\frac{s + 2}{(s + 2)^2 + 2} - 4\frac{1}{\sqrt{2}}\frac{\sqrt{2}}{(s + 2)^2 + 2}.
\end{aligned}$$

Each step above is motivated by an attempt to introduce terms of the form

$$\frac{\beta}{s^2 + \beta^2} \quad \text{and} \quad \frac{s}{s^2 + \beta^2}$$

or translations of such functions, since these functions have known inverse Laplace transforms. Using the linearity of the inverse transform, we have

$$\begin{aligned}
\mathcal{L}^{-1}(X(s)) &= 3\mathcal{L}^{-1}\left(\frac{s + 2}{(s + 2)^2 + 2}\right) - 2\sqrt{2}\mathcal{L}^{-1}\left(\frac{\sqrt{2}}{(s + 2)^2 + 2}\right) \\[2mm]
&= 3e^{-2t}\cos\sqrt{2}t - 2\sqrt{2}e^{-2t}\sin 2t.
\end{aligned}$$ ∎

The second translation formula

$$\mathcal{L}(x(t - c)u(t - c)) = e^{-cs} X(s)$$

is useful as a means of calculating both transforms and inverse transforms. To use the formula for inverse transforms, we must recognize the given function of s as the product of an exponential e^{-cs} and a function $X(s)$ with a known inverse transform $x(t)$. In symbols,

$$\mathcal{L}^{-1}(e^{-cs}X(s)) = x(t - c)u(t - c).$$

Here is an example.

EXAMPLE 7   We recognize the expression

$$e^{-2s}\frac{1}{s + 3}$$

as the product of an exponential and a factor

$$X(s) = \frac{1}{s + 3}$$

with a known inverse $x(t) = e^{-3t}$. Thus, by the second translation formula

$$\mathcal{L}^{-1}(e^{-2s}X(s)) = x(t - 2)u(t - 2)$$
$$= e^{-3(t-2)}u(t - 2). \qquad \blacksquare$$

To use the second translation formula to calculate a Laplace transform, we must recognize the function (to be transformed) as the product of a step function $u(t - c)$ and a factor that is the translation of a function $x(t)$ with known transform $X(s)$.

EXAMPLE 8   To calculate the transform $\mathcal{L}(tu(t - 10))$, we first observe the step function factor $u(t-10)$, which suggests the use of the second translation formula. The other factor t is the translation (by 10) of the function $x(t) = t+10$; that is, $x(t-10) = (t - 10)+10 = t$. Thus,

$$tu(t - 10) = x(t - 10)u(t - 10),$$

where the function

$$x(t) = t + 10$$

has transform

$$X(s) = \mathcal{L}(t) + 10\mathcal{L}(1) = \frac{1}{s^2} + 10\frac{1}{s}.$$

By the second translation formula

$$\mathcal{L}(tu(t - 10)) = \mathcal{L}(x(t - 10)u(t - 10))$$
$$= e^{-10s}X(s)$$
$$= e^{-10s}\left(\frac{1}{s^2} + 10\frac{1}{s}\right). \qquad \blacksquare$$

The Laplace transform is an (improper) integral. Since the integral of a product is not (in general) equal to the product of integrals, the same is true of Laplace transforms (i.e., *the Laplace transform of a product is not [in general] equal to the product of the Laplace transforms*). *That is, the inverse transform of a product $X(s)Y(s)$ is not the product $x(t)y(t)$.*

The *convolution formula*

$$\mathcal{L}\left(\int_0^t x(t - \tau)y(\tau)\,d\tau\right) = X(s)Y(s)$$

gives the correct answer. The inverse transform of the product $X(s)Y(s)$ is the convolution of $x(t)$ and $y(t)$ denoted

$$x * y = \int_0^t x(t - \tau)y(\tau)\, d\tau.$$

The convolution formula is equivalently written

$$\mathcal{L}(f * g) = \mathcal{L}(f)\mathcal{L}(g)$$

or, in inverse form,

$$f * g = \mathcal{L}^{-1}\left(F(s)G(s)\right).$$

EXAMPLE 9

The function

$$\frac{1}{s(s+1)} = \frac{1}{s}\frac{1}{s+1} \tag{4.1}$$

is a product of the two factors $X(s) = \frac{1}{s}$ and $Y(s) = \frac{1}{s+1}$, both of which have known inverses, namely $x(t) = 1$ and $y(t) = e^{-t}$. Thus, the inverse is the convolution of these two functions:

$$\mathcal{L}^{-1}\left(\frac{1}{s}\frac{1}{s+1}\right) = 1 * e^{-t}$$
$$= \int_0^t 1 \cdot e^{-\tau}\, d\tau$$
$$= 1 - e^{-t}. \qquad \blacksquare$$

It is shown in Exercise 7 that $x * y = y * x$, that is, that

$$\int_0^t x(t - \tau)y(\tau)\, d\tau = \int_0^t y(t - \tau)x(\tau)\, d\tau.$$

In the previous example we can also calculate the inverse as the convolution of $g = e^{-t}$ and $f = 1$, that is,

$$\mathcal{L}^{-1}\left(\frac{1}{s+1}\frac{1}{s}\right) = e^{-t} * 1$$
$$= \int_0^t e^{-(t-\tau)} \cdot 1\, d\tau$$
$$= e^{-t} \int_0^t e^{\tau}\, d\tau$$
$$= 1 - e^{-t}.$$

The product (4.1) in the previous example has the factor $1/s$. Thus, we could also use the integral formula to calculate its inverse transform. The integral formula implies

$$\mathcal{L}^{-1}\left(\frac{1}{s}X(s)\right) = \int_0^t x(\tau)\, d\tau.$$

Writing

$$\frac{1}{s}\frac{1}{s+1} = \frac{1}{s}X(s),$$

where

$$X(s) = \frac{1}{s+1}$$

has the inverse transform $x(t) = e^{-t}$, we have

$$\mathcal{L}^{-1}\left(\frac{1}{s}\frac{1}{s+1}\right) = \mathcal{L}^{-1}\left(\frac{1}{s}X(s)\right)$$

$$= \int_0^t x(\tau)\,d\tau$$

$$= \int_0^t e^{-\tau}\,d\tau$$

$$= 1 - e^{-t}.$$

Yet another way to calculate the inverse transform of (4.1) is to notice that

$$\frac{1}{s}\frac{1}{s+1} = \frac{1}{s} - \frac{1}{s+1} \qquad (4.2)$$

and that the terms $1/s$ and $1/(s+1)$ both have known transforms (1 and e^{-t}, respectively). Thus, by the linearity property of the inverse Laplace transform we have

$$\mathcal{L}^{-1}\left(\frac{1}{s}\frac{1}{s+1}\right) = \mathcal{L}^{-1}\left(\frac{1}{s}\right) - \mathcal{L}^{-1}\left(\frac{1}{s+1}\right)$$

$$= 1 - e^{-t}.$$

The key to this calculation is the algebraic identity (4.2). This identity is called a *partial fraction decomposition*. The reader may have encountered partial fraction decompositions in other mathematical contexts, since this algebraic procedure has applications to many problems. It can be very useful for calculating inverse Laplace transforms because it decomposes a rational function (that is, a ratio of polynomials) into a linear combination of simpler fractions, where *simpler* here means fractions with easily calculated inverses. The decomposition is possible if the numerator has degree less than that of the denominator. Computer programs often are capable of calculating partial fraction decompositions. Simpler rational functions can be decomposed by hand in a rather straightforward manner, but we will not pursue this algebraic problem here (see Exercises 8–10).

EXAMPLE 10

To find the inverse Laplace transform of the function

$$\frac{s+4}{s^2+4s+3}$$

we can complete the square in the denominator and try to proceed as in Examples 4, 5, and 6. However, in this case the resulting expressions

$$\frac{s+4}{s^2+4s+3} = \frac{s+4}{(s+2)^2 - 1}$$

do not lead us to translated sine and cosine transforms (because of the minus sign in the denominator). In this case, the denominator is factorable

$$\frac{s+4}{s^2+4s+3} = \frac{s+4}{(s+3)(s+1)},$$

and this suggests that we should use a partial fraction decomposition. If available, an algebraic computer program will also find the partial fraction decomposition; otherwise, we must find the decomposition by hand. (See Exercises 8–10 for a general treatment of quadratic denominators and how to use partial fraction decomposition.) The partial fraction decomposition is a linear combination of the terms $1/(s + 3)$ and $1/(s + 1)$:

$$\frac{s + 4}{s^2 + 4s + 3} = A\frac{1}{s + 3} + B\frac{1}{s + 1}.$$

Our problem is to find the constants A and B so that this is an algebraic identity. From the numerators in the identity

$$\frac{s + 4}{s^2 + 4s + 3} = \frac{(A + B)s + (A + 3B)}{(s + 3)(s + 1)}$$

we obtain the two equations

$$A + B = 1$$
$$A + 3B = 4$$

whose solution is $A = -1/2$ and $B = 3/2$. Thus, the partial fraction decomposition is

$$\frac{s + 4}{s^2 + 4s + 3} = -\frac{1}{2}\frac{1}{s + 3} + \frac{3}{2}\frac{1}{s + 1}.$$

By the linearity property of the inverse transform, we have

$$\mathcal{L}^{-1}\left(\frac{s + 4}{s^2 + 4s + 3}\right) = -\frac{1}{2}\mathcal{L}^{-1}\left(\frac{1}{s + 3}\right) + \frac{3}{2}\mathcal{L}^{-1}\left(\frac{1}{s + 1}\right)$$

$$= -\frac{1}{2}e^{-3t} + \frac{3}{2}e^{-t}. \qquad \blacksquare$$

Table 9.2 contains a list of the Laplace transforms of the elementary functions we have encountered in this section. In reference books, we find tables containing many other entries.

Table 9.2   Laplace transforms of some elementary functions.

$$\mathcal{L}(u(t - c)) = \frac{1}{s}e^{-cs}$$

$$\mathcal{L}(e^{ct}) = \frac{1}{s - c}$$

$$\mathcal{L}(e^{\alpha t}\sin \beta t) = \frac{\beta}{(s - \alpha)^2 + \beta^2}$$

$$\mathcal{L}\left(e^{\alpha t}\cos \beta t\right) = \frac{s - \alpha}{(s - \alpha)^2 + \beta^2}$$

$$\mathcal{L}(t^n) = \frac{n!}{s^{n+1}}, \quad n = 0, 1, 2, \ldots$$

Exercises

1. Derive the first translation formula $\mathcal{L}(e^{ct}x(t)) = X(s-c)$, $c =$ any constant. (*Hint*: Use the definition of the Laplace transform.)

2. Derive the second translation formula $\mathcal{L}(x(t - c)u(t - c)) = e^{-cs}X(s)$, $0 < c =$ constant. (*Hint*: Use the definition of the Laplace transform and the unit step function, and then make a change of variable in the resulting integral.)

3. Derive the formula

$$\mathcal{L}(t^n x(t)) = (-1)^n \frac{d^n X(s)}{ds^n}, n = 1, 2, 3, \ldots .$$

4. Derive the integral formula

$$\mathcal{L}\left(\int_0^t x(u)\, du\right) = \frac{1}{s}X(s).$$

[*Hint*: In the definition of the Laplace transform of the integral, write $e^{-st} = \frac{1}{-s}\frac{d}{dt}\left(e^{-st}\right)$ and use the integration by parts formula from calculus.]

5. Calculate $\mathcal{L}(\sin \beta t)$ and $\mathcal{L}(\cos \beta t)$ from the definition of the Laplace transform.

6. Give an induction argument to prove

$$\mathcal{L}(t^n) = \frac{n!}{s^{n+1}}$$

for all $n = 0, 1, 2, \ldots$.

7. Show $x * y = y * x$.

Consider the ratio of a linear and a quadratic polynomial

$$X(s) = \frac{ms + k}{as^2 + bs + c}, \quad a \neq 0.$$

8. Suppose the denominator $as^2 + bs + c$ has two unequal real roots $r_1 \neq r_2$. Find the partial fraction decomposition of $X(s)$ and use it to find its inverse Laplace transform $x(t) = \mathcal{L}^{-1}(X(s))$. (*Hint*: $s - r_1$ and $s - r_2$ are factors of the denominator so find a decomposition that is a linear combination of $\frac{1}{s-r_1}$ and $\frac{1}{s-r_2}$.)

9. If the denominator $as^2 + bs + c$ has no real root, show that the inverse Laplace transform $x(t)$ is a linear combination of $e^{\alpha t}\sin \beta t$ and $e^{\alpha t}\cos \beta t$ for certain constants α and β that depend on a, b, c, m and k. (*Hint*: Complete the square on the denominator.)

10. Suppose the denominator $as^2 + bs + c$ has two real and equal roots, i.e., a double root r. Find the inverse Laplace transform $x(t)$.

Use the linearity property to find the Laplace transforms of the following functions of t. (a, b, c, k_1, k_2 and β are constants.)

11. $2 + 3e^{2t}$

12. $\frac{1}{2}e^{-2t} - \frac{1}{2}e^{2t}$

13. $2\sin \pi t - 4\cos 2\pi t$

14. $-2e^{3t} - \frac{\pi}{2}\sin \pi t$

15. $1 - 2u(t - 1)$

16. $-\frac{1}{3}u(t - 2\pi) - 2\sin \frac{1}{4}t$

17. $k_1 \sin \beta t + k_2 \cos \beta t$

18. $(at + b)e^{ct}$

Use the first translation formula $\mathcal{L}(e^{ct}x(t)) = X(s - c)$ to find the Laplace transforms of the following functions.

19. te^{-t}

20. $t^2 e^t$

21. $7e^{-5t}u(t - 6)$

22. $2e^{3t}u(t - 2\pi)$

23. $e^{2\pi t}\cos 2\pi t$

24. $e^{-t}\sin \frac{1}{4}t$

25. $e^{\pi t}\cos(t + 1)$

26. $e^{-t}\sin\left(t - \frac{\pi}{3}\right)$

27. $e^{rt}\cos(\beta t + \varphi)$

28. $e^{rt}\sin(\beta t + \varphi)$

Use the second translation formula $\mathcal{L}(x(t - c)u(t - c)) = e^{-cs}X(s)$ to find the Laplace transforms of the following functions. (a, b, c, and β are constants.)

29. $u(t - 2\pi)\sin t$

30. $3u(t - 1)\cos 2\pi t$

31. $e^{2t}u(t - 2)$

32. $e^{-3t}u\left(t - \frac{1}{2}\right)$

33. $tu(t - 1)$

34. $(t + 2)u(t - 2)$

35. $(t - \beta)u(t - c)$

36. $(at + b)u(t - 1)$

Use the derivative formula $\mathcal{L}(t^n x(t)) = (-1)^n \frac{d^n X(s)}{ds^n}$ to find the Laplace transforms of the following functions. (c, α, and β are constants.)

37. te^t

38. $t^2 e^{-t}$

39. $t \sin t$

40. $t \cos t$

41. $t^2 \sin 2\pi t$

42. $t^2 \cos 2\pi t$

43. $t^3 e^{ct}$

44. $t^2 e^{\alpha t}\sin \beta t$

Find the Laplace transform of the following functions. (α, β, and φ are constants.)

45. $te^{-t}\sin t$

46. $te^{-t}\cos t$

47. $2t^2 e^t \sin 4t$

48. $-\frac{1}{3}t^2 e^{-2t}\cos \pi t$

49. $1 + \sin 2\pi t + \cos 2\pi t$

50. $\frac{2}{3} - 5\sin t + 3\sin 2t$

51. $(1 + t^2)e^{-t}u(t - 10)$

52. $t^2 e^t u(t - 1)$

53. $u\left(t - \frac{\pi}{4}\right)\sin 2t$

54. $u\left(t - \frac{\pi}{6}\right)\cos t$

55. $e^{\alpha t}\cos(\beta t + \varphi)$

56. $e^{\alpha t}\sin(\beta t + \varphi)$

Use the first translation formula $\mathcal{L}(e^{ct}x(t)) = X(s - c)$ to find the inverse Laplace transforms of the following functions. (a, b, and r are constants.)

57. $\dfrac{1}{(s - 2)^3}$

58. $\dfrac{1}{(s + 1)^4}$

59. $\dfrac{2+s}{s^2-4s+13}$

60. $\dfrac{2s+3}{s^2+14s+54}$

61. $\dfrac{as+b}{s^2+2s+2}$

62. $\dfrac{a}{s-r}+\dfrac{bs}{(s-r)^2}$

Use the second translation formula $\mathcal{L}(x(t-c)u(t-c)) = e^{-cs}X(s)$ to find the inverse Laplace transforms of the following functions. (a and r are constants.)

63. $e^{-s}\dfrac{1}{s}\dfrac{1}{s-10}$

64. $e^{-2s}\dfrac{2}{s^2-1}$

65. $e^{-0.5s}\dfrac{3s-1}{2s^2-16s+40}$

66. $e^{-\pi s}\dfrac{1}{s(s^2+1)}$

67. $e^{-3s}\dfrac{1}{s-a}$

68. $e^{-\sqrt{2}s}\dfrac{1}{s^2-r^2}$

Use the convolution formula to find the inverse Laplace transforms of the following functions. (a, b, α, and β are constants.)

69. $\dfrac{1}{(s+1)(s+2)}$

70. $\dfrac{1}{s^2-1}$

71. $\dfrac{1}{s^2-a^2}$

72. $\dfrac{1}{(s^2+1)^2}$

73. $\dfrac{s^2}{(s^2+1)^2}$

74. $\dfrac{1}{(s+a)(s+b)}$

75. $\dfrac{1}{s^2(s^2+1)}$

76. $\dfrac{1}{s^3(s^2+4)}$

77. $\dfrac{\beta}{(s-a)(s^2+\beta^2)}$

78. $\dfrac{s}{(s-a)^2(s^2+\beta^2)}$

79. $\dfrac{1}{s+2}\dfrac{1}{s}e^{-s}$

80. $e^{-2s}\dfrac{1}{s(s^2+2)}$

81. $\dfrac{1}{s^2}e^{-s}$

82. $\dfrac{1}{s^3}e^{-\alpha s}$

Use the integral formula $\mathcal{L}\left(\int_0^t x(\tau)\,d\tau\right) = \frac{1}{s}X(s)$ to find the inverse Laplace transforms of the following functions. (a, b, and r are constants.)

83. $\dfrac{1}{s(s-1)}$

84. $\dfrac{1}{s(s+1)}$

85. $\dfrac{3}{s(s^2+4)}$

86. $\dfrac{2}{s(s-1)^2}$

87. $\dfrac{1}{s(s^2-a^2)}$

88. $\dfrac{1}{s(s+a)(s+b)}$

89. $\dfrac{1}{s((s-1)^2+1)}$

90. $\dfrac{1}{s(s-r)^4}$

Find the inverse Laplace transform of the following functions. (c, p, q, α, and β are constants.)

91. $-3\dfrac{s+5}{(s+3)s}$

92. $-\dfrac{1}{2}\dfrac{s^2-6s+12}{(s-2)s^2}$

93. $\dfrac{cqs-ps+p^2}{(s-p)qs}$

94. $\dfrac{cs^s+c\beta^2+\beta s-\beta p}{(s-p)(s^s+\beta^2)}$

95. $\dfrac{s-2}{s^2+1}$

96. $\dfrac{s+2\pi}{s^2+4\pi^2}$

97. $6\dfrac{s^4+8s^2+11}{(s^2+1)(s^2+4)(s^2+9)}$

98. $6s\dfrac{s^4+8s^2+11}{(s^2+1)(s^2+4)(s^2+9)}$

99. $\dfrac{3s-5}{(s-2)(s-1)}$

100. $-\dfrac{s-1}{(s+1)(s+3)}$

101. $-\dfrac{2s^3+s^2-3}{(s+1)(s+2)(s^2+1)}$

102. $2\dfrac{s^3-s^2-1}{(s-1)^2(s^2+1)}$

103. $\dfrac{1}{s}\left(e^{-as}-e^{-\beta s}\right)$

104. $\dfrac{1}{s}\left(e^{-s}+e^{-2s}+e^{-3s}\right)$

105. $\dfrac{1}{s^2}\left(e^{-s}-e^{-2s}\right)$

106. $\dfrac{1}{s-1}e^{-(s-1)}-\dfrac{1}{s-2}e^{-2(s-2)}$

9.5   Solution of Initial Value Problems

Applying the Laplace transform method to solve the first-order initial value problem

$$x' = px + q(t), \quad x(0) = x_0,$$

we arrive at formula (3.3), that is,

$$X(s) = \frac{Q(s)+x_0}{s-p}$$

for the transform of the solution $x(t)$. To use this formula to find $x(t)$ we first calculate the Laplace transform $Q(s)$ of $q(t)$ and then the inverse transform of $X(s)$. Similarly, we can find the solution of the second-order initial value problem

$$\alpha x'' + \beta x' + \gamma x = g(t), \quad x(0) = x_0, \quad x'(0) = x_1$$

by means of the formula (3.5) for the transform of the solution, that is, by inverting

$$X(s) = \frac{G(s) + (\alpha s + \beta) x_0 + \alpha x_1}{\alpha s^2 + \beta s + \gamma}.$$

While these formulas are often valuable for general theoretical studies, for specific initial value problems it is usually more straightforward to perform the individual steps of the transform method to find $X(s)$. In this section we give some examples of the use of Laplace transforms to solve first- and second-order initial value problems. We also show how to use Laplace transforms to solve first-order systems.

9.5.1 First- and Second-Order Equations

In Example 3 we transformed the initial value problem

$$x' = -2x + 3e^{-3t}, \quad x(0) = 2$$

by using the first derivative formula and the linearity property of Laplace transforms. Thus,

$$sX(s) - 2 = -2X(s) + 3\frac{1}{s+3}.$$

Solving for the transform $X(s)$, we have

$$X(s) = \frac{2s + 9}{(s+2)(s+3)}.$$

The solution x is the inverse Laplace transform of this expression. Using the partial fraction decomposition

$$\frac{2s + 9}{(s+2)(s+3)} = 5\frac{1}{s+2} - 3\frac{1}{s+3}$$

and the linearity of the inverse Laplace transform, we find the solution

$$x(t) = 5e^{-2t} - 3e^{-3t}.$$

In Example 1 we considered the initial value problem

$$x' = rx - Hu(t-1), \quad x(0) = x_0, \tag{5.1}$$

where $u(t-1)$ is the unit step function at $t = 1$. Here x is a population that is harvested at the constant rate H, starting at $t = 1$. The population grows exponentially at rate r in the absence of harvesting. Taking the Laplace transform of this equation, we obtain

$$sX(s) - x_0 = rX - H\frac{1}{s}e^{-s}$$

and hence

$$X(s) = x_0\frac{1}{s-r} - H\frac{1}{s-r}\frac{1}{s}e^{-s}.$$

The solution x is the inverse transform of this expression. To use the linearity of the inverse Laplace transform, we need the inverse transform of the two terms

$$\frac{1}{s-r} \quad \text{and} \quad \frac{1}{s-r}\frac{1}{s}e^{-s}.$$

The inverse of the first term is e^{rt}. The presence of the factor e^{-s} in the second term suggests we use the second translation formula, which requires the inverse of the other factor

$$\frac{1}{s-r}\frac{1}{s}.$$

The partial fraction decomposition

$$\frac{1}{s-r}\frac{1}{s} = \frac{1}{r}\frac{1}{s-r} - \frac{1}{r}\frac{1}{s}$$

yields the inverse

$$\frac{1}{r}e^{rt} - \frac{1}{r} = \frac{1}{r}\left(e^{rt} - 1\right).$$

Applying the second translation formula, we find the inverse of

$$\frac{1}{s-r}\frac{1}{s}e^{-s}$$

to be

$$\frac{1}{r}\left(e^{r(t-1)} - 1\right)u(t-1).$$

Finally, returning to the formula for $X(s)$, we obtain the inverse

$$x(t) = x_0 e^{rt} - H\frac{1}{r}\left(e^{r(t-1)} - 1\right)u(t-1). \tag{5.2}$$

This is identical to the solution (1.5) found in Example 1 of Sec. 9.1.

EXAMPLE 1

Suppose an insect pest population grows exponentially with rate $r = 1/10$ (per day). In an attempt to eliminate the pest, a farmer applies a pesticide. The pesticide kills the insects at a rate that, in one day, decreases linearly with time from the maximum rate $H > 0$ to 0 (in units of 100,000 per day). Regulations allow the farmer to apply the pesticide only once. If the initial pest population is approximately 500,000, what dosage H must the farmer use to eliminate the pest population?

If the insect population is measured in units of 100,000, we have the equation

$$x' = \frac{1}{10}x + q(t)$$

for the pest population size $x(t)$, where

$$q(t) = \begin{cases} -(1-t)H & \text{if } 0 \le t < 1 \\ 0 & \text{if } 1 \le t \end{cases}$$

is the death rate caused by the pesticide. We can write this death rate in terms of unit step functions in the following way:

$$q(t) = -(1 - t)H + (1 - t)\,Hu(t - 1)$$
$$= (t - 1)H - (t - 1)\,Hu(t - 1).$$

Thus, we arrive at the initial value problem

$$x' = \frac{1}{10}x + (t - 1)H - (t - 1)\,Hu(t - 1), \quad x(0) = 5. \tag{5.3}$$

Taking the Laplace transform of this equation, we obtain (using the second translation formula)

$$sX(s) - 5 = \frac{1}{10}X(s) + \left(\frac{1}{s^2} - \frac{1}{s}\right)H - \frac{1}{s^2}He^{-s}$$

or

$$X(s) = 5\frac{1}{s - \frac{1}{10}} + \frac{1}{s - \frac{1}{10}}\left(\frac{1}{s^2} - \frac{1}{s}\right)H - \frac{1}{s - \frac{1}{10}}\frac{1}{s^2}e^{-s}H$$

$$= 5\frac{1}{s - \frac{1}{10}} + \frac{1 - s}{s^2\left(s - \frac{1}{10}\right)}H - \frac{1}{s - \frac{1}{10}}\frac{1}{s^2}e^{-s}H.$$

We can invert the first two terms using partial fraction decomposition and the last term using the second translation formula. The details are left as Exercise 18–20. The results are

$$\mathcal{L}^{-1}\left(\frac{1}{s - \frac{1}{10}}\right) = e^{\frac{1}{10}t}$$

$$\mathcal{L}^{-1}\left(\frac{1 - s}{s^2\left(s - \frac{1}{10}\right)}\right) = -90 - 10t + 90e^{\frac{1}{10}t}$$

$$\mathcal{L}^{-1}\left(\frac{1}{s - \frac{1}{10}}\frac{1}{s^2}e^{-s}\right) = \left(-100 - 10(t - 1) + 100e^{\frac{1}{10}(t-1)}\right)u(t - 1),$$

which yield the solution

$$x(t) = 5e^{\frac{1}{10}t} + \left(-90 - 10t + 90e^{\frac{1}{10}t}\right)H$$
$$- \left(-100 - 10(t - 1) + 100e^{\frac{1}{10}(t-1)}\right)Hu(t - 1). \tag{5.4}$$

The insect population is eliminated if $x(t)$ equals 0 at some time $t > 0$. Figure 9.4 shows some typical graphs of $x(t)$ for selected values of H. Using a computer to explore the graphs of solution we find that in order to eliminate the pest, the farmer must use a dosage H greater than a critical value H_0 that is approximately equal to 10.34. In Exercise 21 you are asked to show that

$$H_0 = \frac{1}{2}\frac{e^{0.1}}{10 - 9e^{0.1}} \approx 10.3361.$$

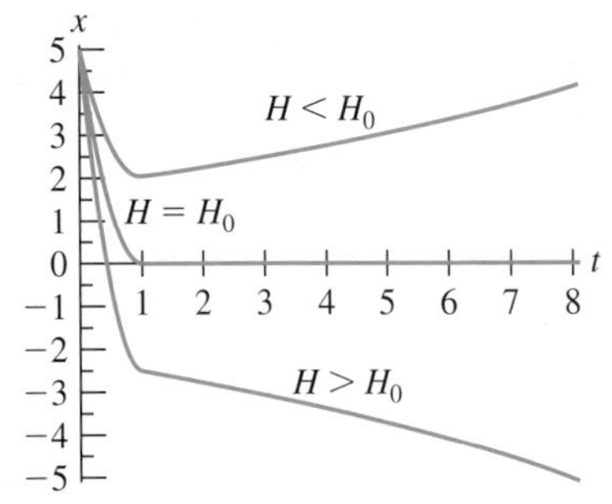

Figure 9.4 Solutions of (5.3) for $H > H_0 \approx 10.3361$ reach 0 while those for $H < H_0$ do not.

In Example 4 of Sec. 9.3 we transformed the initial value problem

$$x'' + x = e^{-3t}, \quad x(0) = 2, \quad x'(0) = -1$$

by using the first and second derivative formulas and the linearity property of Laplace transforms. Thus,

$$s^2 X(s) - 2s + 1 + X(s) = \frac{1}{s+3}.$$

Solving for the transform $X(s)$, we have

$$X(s) = \frac{2s^2 + 5s - 2}{(s+3)\left(s^2+1\right)}.$$

The solution x is the inverse Laplace transform of this expression. Using the partial fraction decomposition

$$\frac{2s^2 + 5s - 2}{(s+3)\left(s^2+1\right)} = \frac{1}{10}\frac{1}{s+3} - \frac{1}{10}\frac{7-19s}{s^2+1}$$

$$= \frac{1}{10}\frac{1}{s+3} - \frac{7}{10}\frac{1}{s^2+1} + \frac{19}{10}\frac{s}{s^2+1}$$

and the linearity of the inverse Laplace transform, we find the solution

$$x(t) = \frac{1}{10}e^{-3t} - \frac{7}{10}\sin t + \frac{19}{10}\cos t.$$

In Example 2 of Sec. 9.1 we considered the initial value problem

$$x'' + x = u\left(t - \frac{\pi}{6}\right), \quad x(0) = 1, \quad x'(0) = 0,$$

where $u\left(t - \pi/6\right)$ is the unit step function at $t = \pi/6$. Taking the Laplace transform of this equation, we obtain

$$s^2 X(s) - s + X(s) = \frac{1}{s}e^{-\frac{\pi}{6}s}$$

and hence

$$X(s) = \frac{s}{s^2+1} + \frac{1}{s}\frac{1}{s^2+1}e^{-\frac{\pi}{6}s}.$$

The solution $x(t)$ is the inverse transform of this expression of $X(s)$. Using the linearity of the inverse Laplace transform, we need the inverse transform of the two terms

$$\frac{s}{s^2+1} \quad \text{and} \quad \frac{1}{s}\frac{1}{s^2+1}e^{-\frac{\pi}{6}s}.$$

The inverse of the first term is $\cos t$. The presence of the factor $e^{-\pi s/6}$ in the second term suggests we use the second translation formula, which requires the inverse of the other factor

$$\frac{1}{s}\frac{1}{s^2+1}.$$

Observing the factor $1/s$, we use the integral formula to find that the inverse of this expression is

$$\int_0^t \sin\tau\, d\tau = 1 - \cos t.$$

Applying the second translation formula, we find the inverse of

$$\frac{1}{s}\frac{1}{s^2+1}e^{-\frac{\pi}{6}s}$$

to be

$$\left(1 - \cos\left(t - \frac{\pi}{6}\right)\right)u\left(t - \frac{\pi}{6}\right).$$

Finally, returning to the formula for $X(s)$, we obtain the inverse

$$x(t) = \cos t + \left(1 - \cos\left(t - \frac{\pi}{6}\right)\right)u\left(t - \frac{\pi}{6}\right).$$

The trigonometric identity

$$\cos\left(t - \frac{\pi}{6}\right) = \frac{1}{2}\sin t + \frac{\sqrt{3}}{2}\cos t$$

shows this solution is identical to the solution (1.8) found in Example 2 of Sec. 9.1.

EXAMPLE 2   The harmonic oscillator equation

$$x'' + x = 0$$

has general solution

$$x(t) = c_1 \sin t + c_2 \cos t.$$

Thus, all solutions are periodic with period 2π. The equation

$$x'' + x = \cos\beta t \tag{5.5}$$

is called the forced harmonic oscillator. It arises in a wide variety of applications in which an oscillating system is subjected to a periodic external force. For example, x might be the current in an electric circuit to which a periodic voltage is applied, or x might be the displacement of an automobile suspension system that is subjected to a periodic force (as could occur when driven over a washboard road).

Taking the Laplace transform of equation (5.5), with the initial conditions

$$x(0) = 0, \qquad x'(0) = 0$$

(which represent a system initially at rest), we have

$$s^2 X(s) + X(s) = \frac{s}{s^2 + \beta^2}$$

and hence

$$X(s) = \frac{s}{s^2 + 1} \frac{\beta}{s^2 + \beta^2}.$$

If $\beta \neq 1$, we use the partial fraction decomposition

$$X(s) = \frac{\beta}{1 - \beta^2} \frac{s}{s^2 + \beta^2} - \frac{\beta}{1 - \beta^2} \frac{s}{s^2 + 1}$$

to obtain the inverse transform

$$x(t) = \frac{1}{1 - \beta^2} \cos \beta t - \frac{\beta}{1 - \beta^2} \cos t. \qquad (5.6)$$

If $\beta = 1$ we use the convolution formula to invert

$$X(s) = \frac{s}{s^2 + 1} \frac{1}{s^2 + 1}$$

to obtain

$$x(t) = \int_0^t \cos(t - \tau) \sin \tau \, d\tau$$

$$= \int_0^t (\cos t \cos \tau + \sin t \sin \tau) \sin \tau \, d\tau$$

$$= \cos t \int_0^t \cos \tau \sin \tau \, d\tau + \sin t \int_0^t \sin^2 \tau \, d\tau$$

$$= (\cos t) \left(\frac{1}{2} - \frac{1}{2} \cos^2 t \right) + (\sin t) \left(\frac{1}{2} t - \frac{1}{2} \cos t \sin t \right)$$

or, upon simplification,

$$x(t) = \frac{1}{2} t \sin t. \qquad \blacksquare$$

Notice that when $\beta \neq 1$ the solution (5.6) is bounded, since it is the sum of two bounded cosine functions. The condition $\beta \neq 1$ means the forcing function $\cos \beta t$ does not have period 2π, the period of the unforced oscillator. When $\beta = 1$ the forcing function and the unforced oscillator both have period 2π. In this case the solution (5.6) is unbounded. This phenomenon is called resonance. This important effect can have important consequences in applications. An unbounded solution might mean, for example, that a mechanical system (such as a suspension system or a suspension bridge, etc.) will break down. Resonance effects are not always deleterious, however; for example, in tuning a radio or television the electronics are adjusted so that resonance occurs at the frequency of the desired station or channel.

If oscillations are undesirable, damping is usually introduced into the system. Damping forces generally introduce a first derivative term into the differential equation.

EXAMPLE 3   The equation

$$x'' + cx' + x = \cos t, \qquad c > 0$$

describes the case when a damping force (proportional to velocity) is added to the forced harmonic oscillator equation (5.5) at resonance. The positive coefficient c is the coefficient of friction. If the system is initially at rest, then

$$x(0) = 0, \qquad x'(0) = 0.$$

As an example, we use Laplace transforms to solve the case $c = 5/2$. Taking the Laplace transform of the equation

$$x'' + \frac{5}{2}x' + x = \cos t$$

and using the initial conditions, we obtain

$$s^2 X(s) + \frac{5}{2}s X(s) + X(s) = \frac{s}{s^2 + 1}$$

and hence

$$X(s) = \frac{1}{s^2 + \frac{5}{2}s + 1}\frac{s}{s^2 + 1}.$$

From the partial fraction decomposition

$$X(s) = \frac{4}{15}\frac{1}{s + 2} - \frac{4}{15}\frac{1}{s + \frac{1}{2}} + \frac{2}{5}\frac{1}{s^2 + 1}$$

we obtain the inverse

$$x(t) = \frac{4}{15}e^{-2t} - \frac{4}{15}e^{-\frac{1}{2}t} + \frac{2}{5}\sin t. \tag{5.7}$$

◼

The solution (5.7) is bounded. Indeed, the first two exponential terms tend to 0 as $t \to +\infty$ so that for large t the solution $x(t)$ is indistinguishable from $(2/5)\sin t$. The introduction of damping has eliminated the resonance. For more on damped oscillators and resonance, see Secs. 6.1.1 and 6.5.2 in Chapter 6.

9.5.2 Systems of Equations

The Laplace transform method is applicable to linear systems with constant coefficients. The homogeneous linear system

$$\begin{aligned} x' &= -2x + 2y \\ y' &= 2x - 5y \end{aligned} \tag{5.8}$$

is the system of the equations arising in the pesticide application discussed in Example 2 of the Introduction. Here x is the amount of pesticide present in the trees and y is the amount present in the soil. If no pesticide is initially present in the soil and an initial pesticide dosage $d > 0$ is applied to the trees, we have the initial conditions

$$x(0) = d, \quad y(0) = 0.$$

Taking the Laplace transform of both equations and using these initial conditions, we obtain the two equations

$$sX(s) - d = -2X(s) + 2Y(s)$$
$$sY(s) = 2X(s) - 5Y(s)$$

for the transforms

$$X(s) = \mathcal{L}(x), \quad Y(s) = \mathcal{L}(y)$$

of $x(t)$ and $y(t)$, respectively. These two linear, algebraic equations, which we can rewrite as

$$(s + 2) X(s) - 2Y(s) = d$$
$$-2X(s) + (s + 5) Y(s) = 0,$$

have solutions

$$X(s) = d \frac{s + 5}{(s + 6)(s + 1)}$$

$$Y(s) = d \frac{2}{(s + 6)(s + 1)}.$$

To invert these expressions we use the partial fraction decompositions

$$X(s) = d \left(\frac{1}{5} \frac{1}{s + 6} + \frac{4}{5} \frac{1}{s + 1} \right)$$

$$Y(s) = d \left(-\frac{2}{5} \frac{1}{s + 6} + \frac{2}{5} \frac{1}{s + 1} \right)$$

and obtain the solution pair

$$x(t) = d \left(\frac{1}{5} e^{-6t} + \frac{4}{5} e^{-t} \right)$$

$$y(t) = d \left(-\frac{2}{5} e^{-6t} + \frac{2}{5} e^{-t} \right).$$

EXAMPLE 4

Consider the initial value problem

$$x' = -2x + 2y + e^{-t}$$
$$y' = 2x - 5y$$
$$x(0) = 0, \quad y(0) = 0.$$

We can interpret this system in terms of the pesticide application problem in Example 2 of the Introduction. In this case there is initially no pesticide in the trees or the soil, but an application of pesticide is made to the trees at a rate that decreases (exponentially) over time. We can find formulas for the solution of this initial value problem by using Laplace transforms. Transforming both differential equations and using the initial conditions, we obtain

$$sX(s) = -2X(s) + 2Y(s) + \frac{1}{s + 1}$$
$$sY(s) = 2X(s) - 5Y(s)$$

or

$$(s + 2) X(s) - 2Y(s) = \frac{1}{s + 1}$$

$$2X(s) - (s + 5)Y(s) = 0.$$

These are two linear (algebraic) equations for the transforms X and Y. Their solutions are

$$X(s) = \frac{s + 5}{(s + 6)(s + 1)^2}$$

$$Y(s) = \frac{2}{(s + 6)(s + 1)^2}.$$

From the partial fraction decompositions

$$\frac{s + 5}{(s + 6)(s + 1)^2} = -\frac{1}{25}\frac{1}{s + 6} + \frac{1}{25}\frac{1}{s + 1} + \frac{4}{5}\frac{1}{(s + 1)^2}$$

$$\frac{2}{(s + 6)(s + 1)^2} = \frac{2}{25}\frac{1}{s + 6} - \frac{2}{25}\frac{1}{s + 1} + \frac{2}{5}\frac{1}{(s + 1)^2}$$

we calculate the inverse transforms

$$x(t) = -\frac{1}{25}e^{-6t} + \frac{1}{25}e^{-t} + \frac{4}{5}te^{-t}$$

$$y(t) = \frac{2}{25}e^{-6t} - \frac{2}{25}e^{-t} + \frac{2}{5}te^{-t}.$$

It is interesting to note that from these formulas we find that both x and y tend to 0 as $t \to +\infty$ in such a way that the fractions satisfy

$$\lim_{t \to \infty} \frac{x(t)}{x(t) + y(t)} = \frac{2}{3}$$

$$\lim_{t \to \infty} \frac{y(t)}{x(t) + y(t)} = \frac{1}{3}.$$

These are the same properties satisfied by solutions of the related homogeneous system (5.8). (See Example 2 in the Introduction.) ∎

Exercises

Use Laplace transforms to solve the following initial value problems. (a, b, p, q, β, x_0 are constants.)

1. $x' = x + \sin t, x(0) = 0$

2. $x' = x + \cos t, x(0) = 1$

3. $x' = 0.1x - e^{-0.3t}, x(0) = 1.25$

4. $x' = -0.2x + te^{-0.1t}, x(0) = -1.0$

5. $x' = -x + e^{-t} + e^{-2t}, x(0) = 0$

6. $x' = -2x + e^{-t} + e^{-2t}, x(0) = 0$

7. $x' = px + q, x(0) = x_0$

8. $x' = px + at, x(0) = x_0$

9. $x' = px + ae^{bt}, x(0) = x_0$

10. $x' = px + te^t, x(0) = x_0$

11. $x' = px + ate^{bt}, x(0) = x_0$

12. $x' = px + at, x(0) = x_0$

13. $x' = -x + u(t - 1), x(0) = 0$

14. $x' = -x + u(t - 2), x(0) = 3$

15. $x' = -x + u(t - 2\pi)\sin t, \, x(0) = 0$

16. $x' = -x + u(t - 2\pi)e^{-t}\sin t, \, x(0) = 0$

17. The initial value problem (5.1) describes the constant rate harvesting of an exponentially growing population. If the harvesting is ceased at $t = 2$, then the modified initial value problem is

$$x' = rx - H\left[u(t - 1) - u(t - 2)\right], \quad x(0) = x_0.$$

Use Laplace transforms to solve this problem. Compare your answer to the solution (5.2) of (5.1) by sketching a graph of both.

The inverse transforms below are used in Example 1. Calculate them.

18. $\mathcal{L}^{-1}\left(\dfrac{1}{s - \frac{1}{10}}\right)$

19. $\mathcal{L}^{-1}\left(\dfrac{1 - s}{s^2\left(s - \frac{1}{10}\right)}\right)$

20. $\mathcal{L}^{-1}\left(\dfrac{1}{s - \frac{1}{10}}\dfrac{1}{s^2}e^{-s}\right)$

21. Use the solution formula (5.4) to show the critical pesticide dosage in Example 1 is $H_0 = \frac{1}{2}\frac{e^{0.1}}{10 - 9e^{0.1}}$.

Assume x and x' are continuous functions that are exponentially bounded.

22. Use the derivative formula $\mathcal{L}(x') = s\mathcal{L}(x) - x(0)$ to derive a formula for $\mathcal{L}(x'')$. (*Hint:* A second derivative is a first derivative of the first derivative.)

23. Derive a formula for $\mathcal{L}(x'')$ directly from the definition of the Laplace transform.

Use Laplace transforms to solve the following second-order problems.

24. $x'' + 3x' + 2x = e^{-3t}, \, x(0) = 0, \, x'(0) = 0$

25. $x'' + x' - 2x = 1, \, x(0) = 1, \, x'(0) = 0$

26. $x'' + x = 1, \, x(0) = 0, \, x'(0) = 1$

27. $x'' + 4x = 1, \, x(0) = 0, \, x'(0) = 0$

Use Laplace transforms to solve the initial value problem $x(0) = 0, \, x'(0) = 0$ for the forced damped oscillator equations below.

28. $x'' + \frac{1}{2}x' + x = \cos t$

29. $x'' + x' + x = \cos t$

30. $x'' + 2x' + x = \cos t$

31. $x'' + cx' + x = \cos t, \, 0 < c < 2$

Use Laplace transforms to solve the following initial value problems.

32. $\begin{cases} x' = -2x + y + 2e^t \\ y' = x - 2y - 1 \\ x(0) = 0, \quad y(0) = 0 \end{cases}$

33. $\begin{cases} x' = \frac{1}{2}x - \frac{3}{2}y - t \\ y' = \frac{3}{2}x + \frac{1}{2}y \\ x(0) = 0, \quad y(0) = 0 \end{cases}$

34. $\begin{cases} x' = 3x - y \\ y' = x + y \\ x(0) = 1, \quad y(0) = -1 \end{cases}$

35. $\begin{cases} x' = 4x - 2y \\ y' = 7x - 5y \\ x(0) = 2, \quad y(0) = 1 \end{cases}$

9.6   Chapter Summary and Exercises

The Laplace transform $\mathcal{L}(x)(s) = X(s) \doteq \int_0^{+\infty} e^{-st}x(t)\,dt$ transforms a function x of t to a function X of s. This transform changes differentiation into multiplication in the sense that $\mathcal{L}(x') = s\mathcal{L}(x) - x(0)$. This property, together with the linearity of the transform, allows us to transform a linear differential equation for x into an algebraic equation for $X(s)$. Once $X(s)$ is found algebraically, we obtain the solution $x(t)$ of the differential equation by calculating the inverse Laplace transform of $X(s)$. To aid in transform and inverse transform calculations, it is helpful to use the fundamental properties of the Laplace transform displayed in Table 9.1 and to be familiar with the transforms of elementary functions (Table 9.2).

Exercises

Find the Laplace transforms of the following functions. (a, b, and c are constants.)

1. $te^{-t}\sin 2t$

2. $(5t - 2)e^{-3t}\cos 2\pi t$

3. $(3\sin 2\pi t + 2\sin 4\pi t)\,e^{-2t}$

4. $(\sin t - \cos t + \sin 2t - \cos 2t)\,e^{-t}$

5. $(at + b)e^{ct}$

6. $(at^2 + bt + c)\sin t$

7. $te^{-t}u(t - \pi)$

8. $t^2 u(t - a)$

9. $u(t - 1) - u(t - 2)$

10. $[1 - u(t - 1)]\,e^t$

Find the inverse Laplace transform $f(t)$ of the following functions. (a, b, and r are constants.)

11. $\dfrac{3 - 4s}{s^2 - 4s + 10}$

12. $\dfrac{3s + 5}{s^2 + 12s + 43}$

13. $\dfrac{1}{s^2 - 3s + 1}$

14. $\dfrac{5}{2s^2 - 3s - 4}$

15. $\dfrac{5s^2 - 24s - 4}{s^3 - 3s^2 - 4s}$

16. $\dfrac{s^2 + 3s - 6}{s^3 + 5s^2 + 6s}$

17. $\dfrac{a}{(s - r)(s^2 + 1)}$

18. $\dfrac{a}{(s - b)(s - r)}$

19. $\dfrac{a}{(s - r)(s - b)^2}$

20. $\dfrac{1}{s(s - r)}e^{-as}$, where $a > 0, r \neq 0$

21. $\dfrac{1}{s(s - r)}\left(e^{-as} - e^{-2as}\right)$, where $a > 0$ and $r \neq 0$

Use Laplace transforms to solve the initial value problem $x(0) = x_0$ for the following differential equations. (a, b, p, α, and β are constants).

22. $x' = x + ae^{bt}$

23. $x' = x + ate^{bt}$

24. $x' = px + \alpha\sin\beta t$

25. $x' = px + \alpha\cos\beta t$

26. $x' = px + e^{bt}u(t - a)$

27. $x' = px + q(t)$, where $q(t)$ is a step function that switches from 0 to 1 at $t = a > 0$, from 1 to -1 at $t = b > a$, and from -1 to 0 at $t = c > b$

Use Laplace transforms to solve the initial value problem $x(0) = x_0$, $x'(0) = x_1$ for the following second-order equations below.

28. $x'' + 3x' + 2x = 0$

29. $x'' + 3x' + 2x = \cos t$

30. $x'' + 4x' + 4x = e^{-t}$

31. $x'' + 4x' + 4x = 0$

Use Laplace transforms to solve the following initial value problems.

32. $\begin{cases} x' = 3x + 2y \\ y' = -4x - y + 2\sin t \\ x(0) = 0, \quad y(0) = 0 \end{cases}$

33. $\begin{cases} x' = 3x + 2y \\ y' = -4x - y + 2\sin t \\ x(0) = 1, \quad y(0) = 0 \end{cases}$

34. $\begin{cases} x' = x + 13y - 1 \\ y' = -2x - y + e^{-t} \\ x(0) = 0, \quad y(0) = 0 \end{cases}$

35. $\begin{cases} x' = x + 13y - 1 \\ y' = -2x - y + e^{-t} \\ x(0) = 0, \\ y(0) = 1 \end{cases}$

36. Consider the differential equation $x' = -x + u(t - a)$ where $a > 0$ is a positive constant.

  (a) Use a computer program to study the solution of the initial value problem $x(0) = 1$. Describe what happens to the solution as $t \to +\infty$ and how this depends on a.

  (b) Use Laplace transforms to solve the initial value problem $x(0) = 1$.

  (c) Use the formula obtained in (b) to corroborate your answer in (a).

Use the convolution formula to solve the initial value problem $x(0) = x_0$ for the following so-called integro-differential equations (in which the integral, as well as the derivative, of the unknown function x appears).

37. $x' = -x + \int_0^t x(\tau)\,d\tau$

38. $x' = -x + \int_0^t e^{-(t-\tau)}x(\tau)\,d\tau$

39. $x' = -x - \int_0^t e^{-(t-\tau)}x(\tau)\,d\tau$

40. $x' = -x + 2\int_0^t (t - \tau)x(\tau)\,d\tau$

41. Suppose $f(t)$ is a piecewise continuous and exponentially bounded function that is also periodic with period $T > 0$ [i.e., $f(t + T) = f(t)$ for all t]. Show that

$$\mathcal{L}(f) = \frac{1}{1 - e^{-sT}}\int_0^T e^{-st}f(t)\,dt.$$

(*Hint:* Write the integral over the infinite interval $0 \leq t < +\infty$ as an infinite sum of integrals over the intervals $nT \leq t < nT + T$, $n = 0, 1, 2, \ldots$ whose lengths are equal to the period. Then make a change of variables to change all of the integrals into integrals over the interval $0 \leq t < T$.)

Sketch a graph of the following functions for $t \geq 0$ and then use the formula in Exercise 41 to find their Laplace transforms.

42. $f(t) = \sin\beta t$

43. $f(t) = \cos\beta t$

44. $f(t)$ is the square wave function

$$f(t) = \sum_{n=0}^{+\infty} (-1)^n u(t-n)$$

45. $f(t)$ is a sawtooth function

$$f(t) = tu(t) - \sum_{n=1}^{+\infty} u(t-n)$$

9.7 APPLICATIONS

9.7.1 Bacterial Infection

In Sec. 1.5.1 of Chapter 1, we considered the problem of controlling the exponential growth of an infection by the bacterium *Staphylococcus aureus*. We considered an antibiotic treatment that removed bacteria at a rate h (cells/minute) and determined the effective dosage that eliminates the infection. From the known doubling time of *S. aureus* (approximately $\delta = 30$ minutes) we obtained the homogeneous linear differential equation $x' = 0.02310x$ for the number of cells $x(t)$ for an untreated infection. When the treatment is invoked, the equation becomes

$$x' = 0.02310x - h,$$

where h is the rate (per minute) at which the antibiotic kills *S. aureus* cells.

In Sec. 1.5.1 we considered two types of treatment: the case when h is a constant—as, for example, when the antidote is administered intravenously—and the case $h = \delta e^{-at}$ when the effectiveness of the dose decreases with time—as, for example, when the drug is administered orally by having the patient take a pill (or receive an injection). In both cases we found that there is a critical dosage level below which the antibiotic treatment fails to eliminate the infection.

However, when a patient is given antibiotics orally he or she is usually required to take several pills at specified intervals of time. This, of course, maintains antibiotics in the system for a longer period of time, and it also allows for a treatment consisting of several smaller doses, rather than one large dose that might have undesirable side effects. This section extends the model studied in Sec. 1.5.1 of Chapter 1 by allowing for the patient to receive more than one dose of antibiotic (by pill or injection) at several time intervals.

Consider a bacterial infection growing exponentially with rate r, which we treat with an oral (or injected) antidote at an initial time $t_0 = 0$ and at a later time $t = t_1 > 0$. Then the number of bacterial cells x satisfies the initial value problem

$$x' = rx - \delta e^{-at} - \delta e^{-a(t-t_1)} u(t - t_1) \qquad (7.1)$$
$$x(0) = x_0,$$

where x_0 is the size (number of cells) of the initial infection that occurred at time $t = 0$.

If the treatment involves more than one follow up dose of antibiotic, say n pills taken at equally spaced time intervals $t_1, 2t_1, 3t_1, \ldots, nt_1$, then we have the initial value problem

$$x' = rx - \delta e^{-at} - \sum_{i=1}^{n} \delta e^{-a(t-it_1)} u(t - it_1)$$
$$x(0) = x_0.$$

We can determine whether the treatment is successful [that is, whether $x(t) = 0$ at some time $t > 0$] from a formula for the solution. One way to find such a formula is by using Laplace transforms.

Consider the initial value problem (7.1) for a treatment using two pills. Taking the Laplace transform of the equation, we obtain (using the second translation formula) the equation

$$sX(s) - x_0 = rX(s) - \delta \frac{1}{s+a} - \delta \frac{1}{s+a} e^{-t_1 s}$$

for the transform $X(s)$ of the solution. Solving for $X(s)$, we find

$$X(s) = \left(x_0 - \frac{\delta}{a+r} \right) \frac{1}{s-r} + \frac{\delta}{a+r} \frac{1}{s+a} + \frac{\delta}{a+r} \frac{1}{s+a} e^{-t_1 s} - \frac{\delta}{a+r} \frac{1}{s-r} e^{-t_1 s}$$

and, by taking inverse Laplace transforms, the formula

$$x(t) = \left(x_0 - \frac{\delta}{a+r} \right) e^{rt} + \frac{\delta}{a+r} e^{-at} - \frac{\delta}{a+r} \left(e^{r(t-t_1)} - e^{-a(t-t_1)} \right) u(t-t_1) \quad (7.2)$$

for the solution of the initial value problem (7.1).

The infection is eliminated if $x(t) = 0$ at some time t. By using formula (7.2), we can determine the conditions under which a treatment succeeds; that is, for what dosage level δ and time interval t_1 the infection is eliminated at some time t. One expectation is that a two-pill treatment (i.e., a specified dose and a specified time interval) can succeed when a one pill treatment fails. Here is an example.

For the bacterium *S. aureus* we have $r = 0.02310$. In Sec. 1.5.1 of Chapter 1 we considered an antibiotic pill that kills cells at a rate of 10^{-6} (per gram per minute) but whose effectiveness decreases exponentially by 50% per hour. This implies $a = 0.01155$ and gives the initial value problem [see (5.5) in Chapter 1]

$$x' = 0.02310x - 0.01de^{-0.01155t}$$
$$x(0) = 1$$

for an infection of 1×10^6 cells treated with a pill containing a dosage of d grams of antibiotic. (Recall that x is measured in units of 10^6.) The solution formula for this problem is

$$x(t) = (1 - 0.2886d)\, e^{0.02310t} + 0.2886de^{-0.01155t}. \quad (7.3)$$

If a second pill is taken in two hours (120 minutes), we have from (7.1) the initial value problem

$$x' = 0.02310x - 0.01de^{-0.01155t} - 0.01de^{-0.01155(t-120)}u(t-120)$$
$$x(0) = 1$$

whose solution (7.2) is

$$x(t) = (1 - 0.2886d)\, e^{0.02310t} + 0.2886de^{-0.01155t} \quad (7.4)$$
$$- 0.2886d \left(e^{0.02310(t-120)} - e^{-0.01155(t-120)} \right) u(t-120).$$

One pill containing a dose of $d = 3.3$ grams of antibiotic is not successful in eliminating the infection. As shown in Fig. 9.5(a), where the solution (7.3) is plotted for this dosage, the infection initially decreases (for about an hour) but eventually regains its exponential growth.

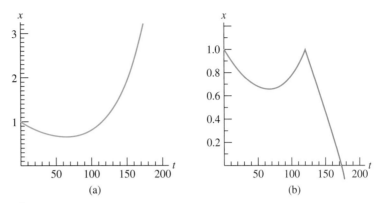

Figure 9.5 (a) The solution (7.3) with $d = 3.3$. (b) The solution (7.4) with $d = 3.3$.

On the other hand, from Fig. 9.5(b), where the solution (7.4) is plotted for this dosage, we see that a second pill, taken two hours after the first, will eliminate the infection in less than three hours. The effectiveness of a second pill taken two hours later depends, however, on the dosage d. For example, from Fig. 9.6 we find that the second pill of dosage $d = 3$ grams fails to control the infection. There is a critical dosage level (at approximately 3.26 grams) below which a second pill given in two hours will fail to work; see Fig 9.6.

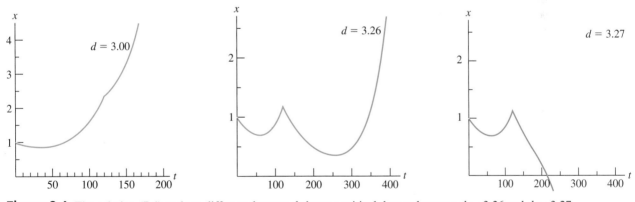

Figure 9.6 The solution (7.4) at three different dosages d shows a critical dosage between $d = 3.26$ and $d = 3.27$.

The timing of the second pill is also important. Figure 9.7 shows a dosage of $d = 3$ grams will in fact eliminate the infection if the second pill is given after one hour instead of after two hours.

9.7.2 Drug Kinetics

In Sec. 5.8.1 of Chapter 5 we considered the system

$$x' = -(r_1 + r_3)\,x + r_2 y$$
$$y' = r_3 x - r_2 y$$

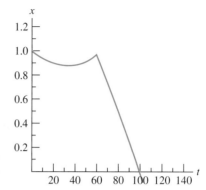

Figure 9.7 The solution (7.4) at dose $d = 3$ but with $t_1 = 120$ replaced by 60.

as a model for the kinetics of a drug in the bloodstream and body tissues of a patient. The component x is the amount of drug in the bloodstream at time t and y is the amount in the body tissues. See Fig. 9.8 for the compartmental diagram for this model.

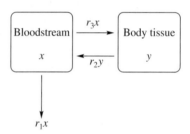

Figure 9.8

In Sec. 5.8.1 of Chapter 5 we studied the use of system (8.1) to stimulate a treatment of lidocaine, a drug used to treat irregular heart beats. For the rate coefficients in the model we used "normal" values estimated to be

$$r_1 = 2.40 \times 10^{-2} \text{ (per minute)}$$
$$r_2 = 3.80 \times 10^{-2} \text{ (per minute)}$$
$$r_3 = 6.60 \times 10^{-2} \text{ (per minute).}$$

The system

$$x' = -9.00 \times 10^{-2}x + 3.80 \times 10^{-2}y \qquad (7.5)$$
$$y' = 6.60 \times 10^{-2}x - 3.80 \times 10^{-2}y$$

determines the *amounts* of *lidocaine* (in mg) in the bloodstream and tissues.

An important point is that the *concentration* of lidocaine in the bloodstream must be at least 1.5 mg/liter in order to be effective. On the other hand, however, a concentration above 6 mg/liter has dangerous (possibly lethal) side effects. The effective and dangerous *amounts* of lidocaine in the bloodstream depend on the patient's blood volume. For example, the effective bloodstream amount for a patient with blood volume of 5 liters is

$$1.5 \text{ mg/liter} \times 5 \text{ liters} = 7.5 \text{ mg.}$$

The dangerous level for such a patient is

$$6 \text{ mg/liter} \times 5 \text{ liters} = 30 \text{ mg.}$$

In Sec. 5.8.1 of Chapter 5 and Sec. 6.5.1 of Chapter 6 we saw how the system (7.5) can be used to simulate two different types of treatment. In the first case, we considered a treatment consisting of a single injection of lidocaine and found that there is a range of injection dosages that result in lidocaine amounts in the bloodstream that lie between the effective and dangerous levels. In Sec. 6.5.1 we studied an intravenous treatment of lidocaine using the nonhomogeneous modification of (7.5)

$$x' = -9.00 \times 10^{-2}x + 3.80 \times 10^{-2}y + n$$
$$y' = 6.60 \times 10^{-2}x - 3.80 \times 10^{-2}y \tag{7.6}$$
$$x(0) = 0, \quad y(0) = 0$$

in which the drug was supplied to the bloodstream at a constant rate n. We calculated the intravenous rates n that keeps the bloodstream amount within the effective but safe range.

Another treatment is to have the patient take a pill containing a certain dosage of lidocaine. The pill supplies the drug to the bloodstream at a rate that decreases over time. Our goal is to modify the system (7.6) to describe this treatment and use the resulting model to determine those pill dosages that provide bloodstream amounts above the effective range but below the dangerous level. We also want to determine the time interval during which the drug levels lie in this range, so that we know approximately when to a second pill is needed.

We assume that the rate at which the pill supplies lidocaine to the bloodstream decreases exponentially over time. Under this assumption, n in the first equation of (7.6) is replaced by ne^{-at}, where $a > 0$ is the exponentially rate at which the pill's supply rate to the bloodstream decreases. The maximal (initial) delivery rate provided by the pill is denoted by n.

Suppose it is known that the delivery rate of the pill decreases by 75% every hour. Then $e^{-a60} = 0.25$ and $a = 2.310 \times 10^{-2}$. As a result, the model equations become

$$x' = -9.00 \times 10^{-2}x + 3.80 \times 10^{-2}y + ne^{-2.310\times10^{-2}t}$$
$$y' = 6.60 \times 10^{-2}x - 3.80 \times 10^{-2}y \tag{7.7}$$
$$x(0) = 0, \quad y(0) = 0.$$

In this model the initial condition arise from the assumption that there is no lidocaine in either the bloodstream or the body tissues at the time that the patient takes the pill.

Our goal is the determine those maximal supply rates n for which the amount of lidocaine in the bloodstream reaches the effective level for the patient, without exceeding the dangerous threshold amount, and to determine the time interval during which this occurs. We can obtain answers to these questions from solution formulas for the initial value problem (7.7).

One way to obtain solutions formulas is to use Laplace transforms. From (7.7) we have

$$sX = -\left(9.00 \times 10^{-2}\right)X + \left(3.80 \times 10^{-2}\right)Y + n\frac{1}{s + 2.310 \times 10^{-2}}$$
$$sY = \left(6.60 \times 10^{-2}\right)X - \left(3.80 \times 10^{-2}\right)Y.$$

If we solve these two (linear algebraic) for X and Y, we obtain

$$X(s) = n \frac{s + 0.038}{(s + 0.1204)\left(s + 2.310 \times 10^{-2}\right)\left(s + 7.573 \times 10^{-3}\right)}$$

$$Y(s) = 0.066n \frac{1}{(s + 0.1204)\left(s + 2.310 \times 10^{-2}\right)\left(s + 7.573 \times 10^{-3}\right)}.$$

Form the partial fraction decompositions

$$X(s) = -7.505n \frac{1}{s + 0.1204} - 9.854n \frac{1}{s + 2.310 \times 10^{-2}} + 17.36n \frac{1}{s + 7.573 \times 10^{-3}}$$

$$Y(s) = 6.009n \frac{1}{s + 0.1204} - 43.66n \frac{1}{s + 2.310 \times 10^{-2}} + 37.65n \frac{1}{s + 7.573 \times 10^{-3}}$$

we obtain the formulas

$$x(t) = -7.505ne^{-0.1204t} - 9.854ne^{-2.310 \times 10^{-2}t} + 17.36ne^{-7.573 \times 10^{-3}t} \tag{7.8}$$

$$y(t) = 6.009ne^{-0.1204t} - 43.66ne^{-2.310 \times 10^{-2}t} + 37.65ne^{-7.573 \times 10^{-3}t}$$

for the solution of the initial value problem (7.7).

To determine the time t_e at which the blood level amount attains the effective amount of 7.5 (mg) we must solve the equation $x(t_e) = 7.5$, that is, the equation

$$-7.505ne^{-0.1204t_e} - 9.854ne^{-2.310 \times 10^{-2}t_e} + 17.36ne^{-7.573 \times 10^{-3}t_e} = 7.5.$$

The solution t_e of this equation depends on the value of n. A graph of t_e as a function of n appears in Fig. 9.9. (One way to draw this graph is to solve the equation for n, plot n as a function of t_e, and reflect the graph through the 45-degree line $n = t_e$.)

The first observation we make from the graph in Fig. 9.9 is that the amount x of lidocaine in the bloodstream reaches the effective level of 7.5 mg only if n exceeds a threshold rate (of approximately 0.8474 mg/min).

Of course, as the pill is dissolved and its delivery rate decreases, the amount of lidocaine in the bloodstream eventually decreases and drops below the effective level of 7.5. mg. From Fig. 9.9 we see that the time interval during which the amount of lidocaine in the bloodstream remains above 7.5 increases as n increases. However, if n is too large (namely, larger than approximately 3.389 mg/min) then the lidocaine level in the bloodstream will exceed the dangerous threshold of 30 mg. This can be seen from the graph in Fig. 9.9 of the solution t_l of the equation $x(t_l) = 30$.

We conclude that the maximum delivery rate of a safe pill is $n = 3.389$ and that if such a pill is administered it will take approximately 2.5 minutes to supply an effective

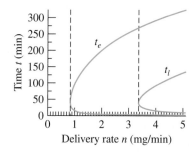

Figure 9.9 Times t_e at which $x(t)$ in (7.8) equals 7.5 are plotted against n. Also shown is a plot of the times t_l at which $x(t)$ equals 30. The vertical lines delineate the safe drug delivery range.

level of 7.5 mg of lidocaine. The pill will remain effective approximately 270 minutes, or six and one half hours. (Remember that these conclusions are for a patient with 5 liters of blood. The same threshold phenomena occur for other blood volumes, but there numerical values might change.)

Exercises

1. Use formula (7.2) to show that if a one pill treatment is successful, then a two-pill treatment (with the same dose δ) will succeed even faster.

2. Consider the harvesting problem (1.1)–(1.2) with harvest rate $H = 1$. Since $-q(t)$ is the rate at which the population is removed, the total amount harvested from the population over the harvesting interval $1 < t < 2$ is given by the integral $-\int_1^2 q(t)\,dt = H = 1$. In many applications it is of interest to consider an impulsive rate of removal (or addition) to a quantity. For example, we contemplate the possibility of harvesting the population instantaneously at a given time, say $t = 1$, in such a way that there is still a total amount of $-H$ removed. It is more realistic to imagine the harvesting occurring over a very small interval of time and then, if mathematically convenient, to pass the length of the interval to 0. One way to do this is to let the harvesting begin at the time just prior to 1, say $1 - \varepsilon/2$, and then cease at the time just after 1, say at $t = 1 + \varepsilon/2$, where $\varepsilon > 0$ is a small number. Such a switch is provided by

$$u\left(t - \left(1 - \frac{1}{2}\varepsilon\right)\right) - u\left(t - \left(1 + \frac{1}{2}\varepsilon\right)\right).$$

If we take the harvesting rate to be $q(t) = -\delta_e(t-1)$, where

$$\delta_\varepsilon(t) = \frac{1}{\varepsilon}\left[u\left(t - \frac{1}{2}\varepsilon\right) - u\left(t + \frac{1}{2}\varepsilon\right)\right],$$

then the total amount harvesting satisfies

$$-\int_{1-\varepsilon/2}^{1+\varepsilon/2} q(t)\,dt = \int_{1-\varepsilon/2}^{1+\varepsilon/2} \delta_e(t-1)\,dt = \frac{1}{\varepsilon}\varepsilon = 1.$$

To obtain an impulse harvesting rate, we will let $\varepsilon \to 0$.

(a) Show that

$$\lim_{\varepsilon\to 0}(-q(t)) = \begin{cases} 0 & \text{if } t \neq 1 \\ +\infty & \text{if } t = 1 \end{cases}.$$

(b) Compute the Laplace transform $\Delta_\varepsilon(s)$ of $\delta_\varepsilon(t)$.

(c) Calculate $\lim_{\varepsilon\to 0} \Delta_\varepsilon(s)$.

(d) Calculate $\lim_{\varepsilon\to 0} X_\varepsilon(s) = x_0(t)$, where $X_\varepsilon(s)$ is the Laplace transform of the solution $x_\varepsilon(t)$ of the initial value problem $x' = rx - \delta_\varepsilon(t-1)$, $x(0) = 0$.

(e) Find the inverse transform $x_0(t)$ of $X_0(s)$. Is $x_0(t)$ continuous?

The function $x_0(t)$ can be viewed as the solution of the impulse harvesting problem. But of what differential equation is it a solution? Since $x_0(t)$ is the limit as $\varepsilon \to 0$ of the solution $x_\varepsilon(t)$ of the initial value problem $x' = rx - \delta_\varepsilon(t-1)$, $x(0) = 0$, it seems natural to say that $x_0(t)$ is a solution of the initial value problem $x' = rx - \delta_0(t-1)$, $x(0) = 0$, where $\delta_0(t-1)$ is the limit of $\delta_\varepsilon(t-1)$. In (a) we see, however, that $\delta_\varepsilon(t-1)$ does not approach a function as a limit. Nonetheless, we refer to the limit

$$\delta_0(t-1) = \begin{cases} 0 & \text{if } t \neq 1 \\ +\infty & \text{if } t = 1 \end{cases}$$

as a *Dirac function* (or *delta function*) at $t = 1$. It is a shift of the Dirac function at $t = 0$, namely

$$\delta_0(t) = \begin{cases} 0 & \text{if } t \neq 0 \\ +\infty & \text{if } t = 0 \end{cases}.$$

(f) Define the Laplace transform of the Dirac function $\delta_0(t-a)$ to be

$$\mathcal{L}(\delta_0(t-a)) = \lim_{\varepsilon\to 0} \mathcal{L}(\delta_\varepsilon(t-a)).$$

Calculate $\mathcal{L}(\delta_0(t-a))$.

3. The homogeneous linear system

$$x' = -\alpha x - \beta y$$
$$y' = \gamma x - \delta y$$

arises as a model of the glucose/insulin regulation system in the bloodstream [see Example 1 in Sec. 5.2 of Chapter 5]. Here x and y are the excess concentrations of glucose and insulin, respectively, from their equilibrium levels. (Negative values are deficiencies below equilibrium.) If glucose is given to a patient intravenously at a rate $g(t) \geq 0$, then

$$x' = -\alpha x - \beta y + g(t) \qquad (7.9)$$
$$y' = \gamma x - \delta y.$$

In Chapter 6 a case when g is constant was considered. Specifically, the system

$$x' = -2.92x - 4.34y + 18$$
$$y' = 0.208x - 0.780y$$

with realistic coefficient values was solved and studied [see (1.3) in Sec. 6.1 of Chapter 6)]. Suppose the intravenous rate g is not constant but instead varies periodically with an average concentration of 18 as given by

$$g(t) = 18 + 10\cos t.$$

Assuming the glucose and insulin concentrations are initially at equilibrium, we have the initial value problem

$$x' = -2.92x - 4.34y + 18 + 10\cos t$$
$$y' = 0.208x - 0.780y$$
$$x(0) = 0, \quad y(0) = 0.$$

(a) Find the Laplace transforms $X(s)$, $Y(s)$ of the solution pair $x(t)$, $y(t)$.

(b) Using your answer in (a) and partial fraction decomposition calculate $x(t)$ and $y(t)$.

(c) From your answer in (b), determining the long-term behavior of $x(t)$ and $y(t)$. Relate the long-term behavior to the forcing function $18 + 10\cos t$.

4. Consider the initial value problem

$$x' = -2x + 2y + 1 - u\left(t - \frac{7}{365}\right)$$
$$y' = 2x - 5y$$
$$x(0) = 0, \quad y(0) = 0.$$

We can interpret this system in terms of the pesticide application problem in Example 2 of the Introduction. Initially there is no pesticide in the trees or the soil, but an application of pesticide is made to the trees at a rate of 1 unit per year beginning at time $t_0 = 0$ and lasting for one week. The unit of time is one year and this accounts for the step function appearing as a nonhomogeneous term in the first differential equation.

(a) Find the Laplace transforms $X(s)$, $Y(s)$ of the solution pair $x(t)$, $y(t)$.

(b) Using your answer in (a) and partial fraction decomposition, calculate $x(t)$ and $y(t)$.

(c) Calculate the long-term ratio of pesticide in the soil to that in the trees.

(d) Calculate the long-term fraction of pesticide that resides in the trees.

5. In the glucose and insulin regulation model in Exercise 3, suppose $g(t) = g_0 u(t - \delta)$ so that

$$x' = -2.92x - 4.34y + g_0 u(t - \delta)$$
$$y' = 0.208x - 0.780y$$
$$x(0) = 0, \quad y(0) = 0.$$

This models the situation when glucose and insulin levels are at equilibrium when at time $t_0 = 0$ glucose is ingested at a rate g_0 (grams/hour) for a time interval of length δ (hours).

(a) Using Laplace transforms find solution formulas for $x(t)$ and $y(t)$ when $g_0 = 10$ and $\delta = 1$.

(b) Show that both glucose and insulin levels asymptotically return to their equilibrium values.

(c) Show that the glucose level initially rises but then decreases and attains its equilibrium level in finite time. Show, however, that after this time the glucose level remains below its equilibrium level.

Answers to Selected Exercises

Introduction

Section 1

1. first order   **3.** first order   **5.** second order   **7.** first order
9. first order   **11.** first order   **13.** third order   **15.** e^{-3t} is a
solution   **17.** $-e^{-3t}$ is a solution   **19.** e^{2t} is not a solution
21. $-7e^{t^2}$ is a solution   **23.** $t^{-3/2}$ is a solution   **25.** $(t-1)^{-3/2}$ is a
solution   **27.** $t^{3/2}$ is not a solution   **29.** $(t-2)^{-2/3}$ is not a
solution   **31.** e^{-2t} is not a solution   **33.** e^{3t} is a solution
35. $5e^{2t}$ is a solution   **37.** $e^{2t} + e^{3t}$ is a solution   **39. (a)** e^{-5t} is a
solution **(b)** $3e^{-5t}$ is a solution **(c)** $5e^{-3t}$ is not a solution   **41. (a)** $\frac{1}{t}$ is
a solution **(b)** $\frac{2}{t}$ is not a solution **(c)** $\frac{1}{t-2}$ is a solution   **43. (a)** $\ln t$ is a
solution **(b)** 1 is a solution **(c)** t is not a solution   **44.** e^{4t} is a solution
46. ce^{4t} is a solution for any constant c **48.** $\frac{1}{2}e^{-2t}$ is a solution
50. e^{6t} is not a solution   **51.** e^{t} is a solution   **53.** $e^{t}e^{-2t} = e^{-t}$ is
not a solution   **55.** Yes   **57.** Is a solution pair.   **59.** Is not a
solution pair.   **61.** Is a solution pair.   **63.** Is a solution pair.
65. Is a solution pair for all c_1 and c_2.   **66.** e^{-3t} and $-e^{-3t}$ are
solutions for all t

68. $x' = y$, $y' = 3x - y$ **70.** $x' = y$, $y' = -4x^2 + 2xy + \frac{1}{3}$
72. $x' = y$, $y' = z$, $z' = -\frac{1}{2}x - 2y + 3z - \frac{3}{2}$ **74.** $x' = y$,
$y' = -4x - 2y + \cos t$ **76.** $x' = y$, $y' = -t^{-2}y^2 - t^{-2}\cos x$
78. $x' = y$, $y' = -2y - x + z$, $z' = w$, $w' = -w + 2x - z$
80. $x' = y$, $y' = -2x + \frac{1}{2}y + 4z - w$, $z' = w$,
$w' = x - 2y - 3z + w + \sin t$

82. linear   **84.** nonlinear (because of the x^2 term)   **86.** linear
88. nonlinear (because of the $\sin x$ term)   **90.** nonlinear (because of
the xx' term)   **92.** linear   **94.** nonlinear (because of the $(1-x)x$
term)   **96.** nonlinear (because of the e^{-x} term)   **98.** linear
100. nonlinear (because of the xy term)   **102.** linear
104. nonlinear [because for the equation to be linear $df(x)/dx$ would
have to be a constant and $d^2 f(x)/dx^2$ would equal 0]
106. nonlinear (because of the term $\sin x$)   **108.** nonlinear [because
of the term $\ln(ty)$]   **110.** linear   **111.** We can rewrite the equation
as the linear equation $x' = (\ln 2)x$.

Section 2

1. (a) $x' = rx$, $x(0) = d$ for a constant of proportionality $r > 0$
(b) $x' = rx - w$, $x(0) = d$ for a constant of proportionality $r > 0$.

3. (a) $c' = \frac{d}{v}c_{in} - \frac{d}{v}c$, $c(0) = 0$ **(b)** $c' = \frac{d}{v}c_{in} - \frac{d}{v}c - mec$, $c(0) = 0$,
for a constant of proportionality $m > 0$.   **4. (a)** $v' = 9.8 - k_0v^2$,

$v(0) = 0$ **(b)** Let $g = 9.8$ and write

$$v(t) = \sqrt{g}\,\frac{1 - \exp\left(-2t\sqrt{gk_0}\right)}{1 + \exp\left(-2t\sqrt{gk_0}\right)}.$$

Using the quotient rule from calculus, we calculate

$$v'(t) = g\,\frac{4\exp\left(-2t\sqrt{gk_0}\right)}{\left(1 + \exp\left(-2t\sqrt{gk_0}\right)\right)^2}.$$

Substituting v into the right-hand side of the differential equation, we

find $g - k_0(v(t))^2 = g - k_0 \left(\sqrt{\dfrac{g}{k_0}}\,\dfrac{1-\exp\left(-2t\sqrt{gk_0}\right)}{1+\exp\left(-2t\sqrt{gk_0}\right)}\right)^2$

$$= g\,\frac{4\exp\left(-2t\sqrt{gk_0}\right)}{\left(1 + \exp\left(-2t\sqrt{gk_0}\right)\right)^2}.$$

(c) $\lim_{t\to+\infty} v(t) = \lim_{t\to+\infty}\sqrt{\dfrac{9.8}{k_0}}\,\dfrac{1-\exp\left(-2t\sqrt{9.8k_0}\right)}{1+\exp\left(-2t\sqrt{9.8k_0}\right)} = \sqrt{\dfrac{9.8}{k_0}}$
The shuttlecock approaches a terminal velocity as it falls.
5. (a) $x' = $ inflow rate $-$ outflow rate $=$ birth rate $-$ death rate. So
$x' = bx - dx$ for positive constants b and d. **(b)** Let $d = cx$ for a
positive constant c. Then $x' = bx - cx^2$.   **7.** $v' = ae^{-bt}v$, $v(0) = v_0$
9. Let $x = x(t)$ be the temperature of the body. Then $x' = c(T_e - x)$,
where c is the constant of proportionality and T_e is the temperature of
the surrounding environment. The constant c must be positive so that
the temperature of the body increases when it is less than the external
temperature and decreases when it is greater.
10. $x' = -2x + 2y + p$, $y' = 2x - 5y$, $x(0) = 0$, $y(0) = 0$
12. $x' = (b_1 - c_1y)x$, $y' = (b_2 - c_2x)y$ where all (coefficients)
parameters are positive. The system is nonlinear.
13. $x' = 10 - \frac{1}{30}x$. This equation is linear.
15. $x' = bx - cx^2 - h$, $x(0) = x_0$ The differential equation is
nonlinear.   **17.** $p' = k\left(b\frac{1}{p} - ap\right)$, $p(0) = p_0$ The equation is
nonlinear.   **19. (a)** $x' = -\frac{r}{v}x$ **(b)** $x' = \frac{r}{v}s - \frac{r}{v}x$

Chapter 1

Section 1.1

1. $x = t + \frac{1}{3}t^3 + c$ **3.** $x = \frac{1}{2}e^{2t} + c$ **5.** $x = \frac{1}{3}t^3 + \frac{5}{3}$
7. $x = -te^{-t} - e^{-t} + 2$ **9.** Theorem 1 applies. There exists a
unique solution on an interval containing $t_0 = 0$.   **11.** Theorem 1
applies. There exists a unique solution on an interval containing
$t_0 = 0$.   **13.** Theorem 1 does not apply. No conclusion can be drawn
from the theorem.   **15.** Theorem 1 applies. There exists a unique
solution on an interval containing $t_0 = 0$.   **18.** When $a > 0$

Theorem 1 implies there is a unique solution on an interval containing $t_0 = 0$. For $a \le 0$ no conclusion can be drawn from the theorem.
20. Theorem 1 applies when $|a| > 2$ and there is a unique solution on an interval containing $t_0 = 1$. For $|a| \le 2$ no conclusion can be drawn from the theorem.   **22.** If $t_0 \ne 0$, $x_0 \ne 0$ then the initial value problem has a unique solution on an interval containing t_0. For $t_0 = 0$ and $x_0 = 0$ nothing can be concluded from Theorem 1.   **24.** If

$$x_0 \ne \frac{1}{2b}(2n + 1)\pi$$

for all $n = 0, \pm 1, \pm 2, \pm 3, \cdots$, then the initial problem has a unique solution on an interval containing t_0. For any other x_0 nothing can be concluded from the Theorem 1.   **27. (a)** For all t we have

$$x'(t) = 0 = \sqrt{1 - (\pm 1)^2} = f(t, x(t)).$$

(b) For $t > \pi/2$, we have $x'(t) = 0 = \sqrt{1 - 1^2} = f(t, x(t))$ and for $t < \pi/2$ we have $x'(t) = 0 = \sqrt{1 - (-1)^2} = f(t, x(t))$. For $-\pi/2 < t < \pi/2$, $x'(t) = \cos t = \sqrt{1 - \sin^2 t} = f(t, x(t))$. The function $x(t)$ is differentiable at the adjoining points $t = \pm\pi/2$ of these intervals, where $x'(t) = 0$ and at these points

$$x'\left(\pm\frac{\pi}{2}\right) = 0 = \sqrt{1 - (\pm 1)^2} = f(t, x(t)).$$

(c) $\partial f/\partial x = -x/\sqrt{1 - x^2}$ is not defined at $x_0 = 1$ and therefore Theorem 1 does not apply.

Section 1.2

1.

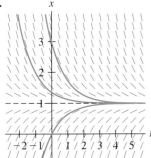

3.

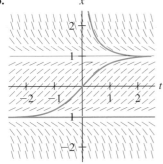

5.

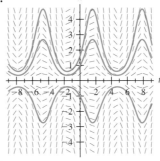

7.

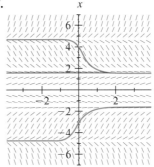

9.

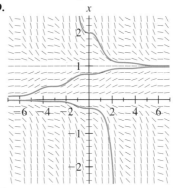

11.

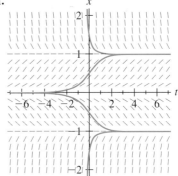

13. For $a > 1$ all solutions appear to decrease. For $a < 1$ there is a horizontal region (lying between two parallel, horizontal straight lines) in the t, x plane in which solutions increase and outside of which solutions decrease.

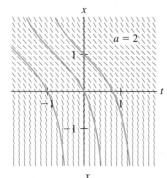

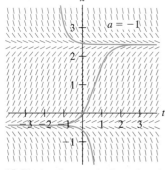

15. The isoclines are horizontal straight lines of the form $x = 1 - m$, where m, the associated slope, is any constant.

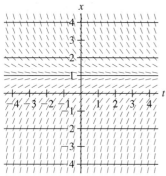

17. The isoclines are circles, centered at the origin, of the form

$$t^2 + x^2 = \frac{1}{m^2} - 1$$

where m, the associated slope, is any constant satisfying $|m| > 1$.

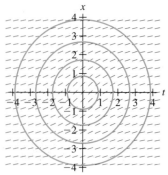

19. $x' = x - t$ **21.** $x' = \frac{1}{x-t}$ **23.** $x' = \left(2x^2 + 3t^2\right)^3$

Section 1.3

1.

Step Size s	$x(1) \approx$ Runge-Kutta	% Error
0.200	2.718251	1.134×10^{-3}
0.100	2.718280	6.262×10^{-5}
0.050	2.718282	0.0000
0.025	2.718282	0.0000

In reducing step size $s = 0.2$ by a factor of $1/2$, the percent error is reduced by a factor of approximately $(1/2)^4 = 6.25 \times 10^{-2}$. The Runge-Kutta method gives more accurate approximations at each step and converges faster than either the Euler algorithm or Heun's algorithm.

2. (a) $x(1) \approx 1.034967$ has one correct significant digit.

(b) $s = 0.001563$ and $s = 0.000391$.

(c)

Step Size s	$x(1) \approx$ Euler	Absolute Error
0.10000	1.034967	0.09893
0.05000	1.076373	0.05752
0.02500	1.102459	0.03143
0.01250	1.117387	0.01651
0.00625	1.125425	0.00847

The error goes down by approximately a fraction of $1/2$ at each step.
3. (a) $x(1) \approx 1.127950$ has two correct significant digits **(b)** $s = 0.05$ and $s = 0.025$.

(c)

Step Size s	$x(1) \approx$ Heun	Absolute Error
0.10000	1.127950	59.4×10^{-4}
0.05000	1.132298	15.9×10^{-4}
0.02500	1.133482	4.11×10^{-4}
0.01250	1.133789	1.04×10^{-4}
0.00625	1.133867	0.26×10^{-4}

The error goes down by approximately a fraction of $1/4 = (1/2)^2$ at each step.   **4. (a)** $x(1) \approx 1.1338883442$, which has four correct significant digits **(b)** $s = 0.1$ and $s = 0.1$.

(c)

Step Size s	$x(1) \approx$ Runge-Kutta	Absolute Error
0.10000	1.1338883442	5.075×10^{-6}
0.05000	1.1338931470	2.72×10^{-7}
0.02500	1.1338934041	1.5×10^{-8}
0.01250	1.1338934182	1.0×10^{-9}
0.00625	1.1338934190	0

The error goes down by approximately a fraction of $1/16 = (1/2)^4$ at each step.

10. (a) Predictor: $y_{i+1}^* = y_i + sf(t_i, y_i)$ Corrector:
$y_{i+1} = y_{i-1} + \frac{s}{3}\left[f(t_{i+1}, y_{i+1}^*) + 4f(t_i, y_i) + f(t_{i-1}, y_{i-1})\right]$

11. From the table of results below, the best Euler algorithm approximation is 0.04 and the best Heun's algorithm approximation is 0.041. The best approximation of all is the Runge-Kutta approximation 0.041791.

Step Size s	Euler $x(0.5) \approx$	Heun $x(0.5) \approx$	Runge-Kutta $x(0.5) \approx$
0.10000	0.030022	0.042621	0.041791
0.05000	0.035683	0.041999	0.041791
0.02500	0.038688	0.041843	0.041791
0.01250	0.040221	0.041804	0.041791
0.00625	0.041003	0.041794	0.041791

12. $s = 0.003125$ because there is virtually no change in the graph from $s = 0.00625$.

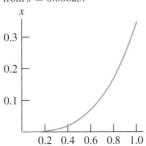

13. $s = 0.0125$ because there is virtually no change in the graph from $s = 0.025$.   **14.** $s = 0.025$ because there is virtually no change in the graph from $s = 0.05$.

Section 1.4

1. $x = \ln\left|\frac{t}{1-t}\right| + c$ **3.** $x = \tan^{-1} t + \frac{\pi}{4}$ **5.** Theorem 1 applies and the initial value problem has a unique solution on an interval containing $t_0 = 1$.   **7.** If $t_0 \neq x_0$ Theorem 1 applies and the initial value problem has a unique solution defined on an interval containing t_0. Nothing can be concluded from the theorem when $x_0 = t_0$.   **9.** If $x_0 \neq 0$ then Theorem 1 applies and the initial value problem has a unique solution on an interval containing t_0.   **11.** Theorem 1 does not apply and no conclusion can be drawn from this theorem.
13. Theorem 1 applies and there exists a unique solution on an interval containing $t_0 = 0$.   **14.** Theorem 1 applies to any initial value problem.

16.

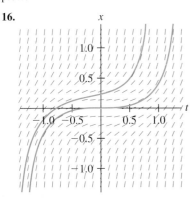

18.

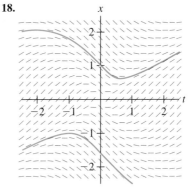

20. (a) and (3); (b) and (6); (c) and (1)   **21.** For the equation $x' = t^2 + 4x^2$ the isocline are eclipses centered at the origin for $m > 0$. For $m = 0$ the isocline is the origin. There are no isoclines for $m < 0$.

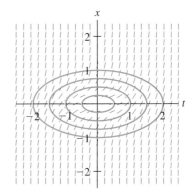

For the equation

$$x' = \frac{t^2 - x^2}{t^2 + x^2}$$

the isoclines are hyperbolae when $-1 < m < 1$ and ellipses when $m > 1$ or $m < -1$. When $m = -1$, there are two horizontal lines isoclines at $t = \pm\sqrt{2}/2$. There is no isocline for $m = 1$.

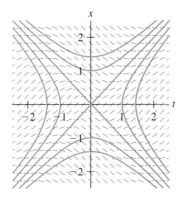

22.

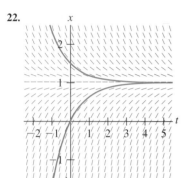

24.

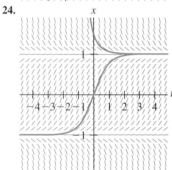

26. (a) none, because no digits have stabilized **(b)** 3, maybe 3.1, because these digits have stabilized for the last four step sizes **29.** The exact solution at $t = 0.6$ is 3.200540 (to seven significant digits). The Euler approximation is 3.026383, which has an absolute error of 0.174157 or a percent error of 5.44150%.   **30.** Stop at $s = 0.0015625$ since the graph changes very little from that at $s = 0.003125$. (Your answer might be different, depending on the computer you use.)   **33. (a)** The halving time is approximately 2 years, independent of the initial condition $x_0 > 0$.
(b) $\frac{1}{0.35} \ln 0.5 \approx -1.9804201$.   **35. (a)** $x' = ax^2 - dx$, $x(0) = x_0$

1.5
1. $h_{cr} = 3.4657 \times 10^6$ cells per minute.   **2.** h_{cr} is linearly related to a by approximately the formula $h_{cr} = 10^8 a + 3.4657 \times 10^6$.
6. $r = 0.5 \ln 2 = 0.3466$. $h_{cr} = 34.3$ (fish/year).
8. $h_{cr} = 866.5$ (fish/year)   **11.** Approximately $f = 13.52$.
14. Tommie Smith's predicted time is $10.31 + \frac{100}{10.46452} = 19.87$ sec and Jim Hines's predicted time is $10.56 + \frac{100}{10.88349} = 19.75$ sec.

Chapter 2

Section 2.1
1. linear, $p(t) = t^2$ and $q(t) = t$ **3.** nonlinear   **5.** nonlinear
7. nonlinear   **9.** linear $p(t) = \frac{1}{5}(t^2 + \sin t)$,
$q(t) = -\frac{1}{5}\left(\cos 3t + \frac{1}{t^2+1}\right)$ **11.** linear nonhomogeneous, $p(t) = t^2$
and $q(t) = -1$ **13.** linear homogeneous, $p(t) = t^2$ and $q(t) = 0$
15. linear nonhomogeneous, $p(t) = 3e^{-t}$ and $q(t) = -te^{-t}$
17. $0 < t < +\infty$ and $-\infty < t < 0$, respectively
19. $-\pi/2 < t < \pi/2$ and $\pi/2 < t < 3\pi/2$, respectively
21. $-\alpha < t < \alpha$ if $-\alpha < t_0 < \alpha$, $-\infty < t < -\alpha$ if $t_0 < -\alpha$,
$\alpha < t < +\infty$ if $t_0 > \alpha$ **23.** From computer simulations the solution might appear to have a vertical asymptote (near $t = 2$), but Corollary 1

implies that the solution exists on the entire real line $-\infty < t < +\infty$.
25. For equations in (0.1): **(a)** $p(t) = p$, $q(t) = 0$, homogeneous and autonomous **(b)** $p(t) = -c$, $q(t) = g$, nonhomogeneous and autonomous **(c)** $p(t) = ae^{-bt}$, $q(t) = 0$, homogeneous and nonautonomous **(d)** $p(t) = -a$, $q(t) = a(b_{av} + \alpha \sin(2\pi t/T))$, nonhomogeneous and nonautonomous **(e)** $p(t) = -r_{out}/(rt + V_0)$, $q(t) = c_{in} r_{in}$, nonhomogeneous and nonautonomous   **26.** $x = ce^{-3t}$
28. $x = c|t|$ **30.** $x = c \exp\left(-\frac{1}{3}e^{-3t}\right)$ **32.** $x = c\sqrt{1 + t^2}$
34. $x = c \exp(\sin t - t \cos t)$ **36.** $x = ce^{t/a}$
38. $\begin{cases} x = c \exp\left(\frac{1}{a}e^{at}\right) \text{ if } a \neq 0 \\ x = ce^t \text{ if } a = 0 \end{cases}$ **40.** $x = ce^{-2t} + 6$
42. $x = c \exp\left(\frac{1}{2}t^2\right) - 1$
44. $x = \begin{cases} ce^{at} + \frac{1}{a^2+b^2}(b \sin bt - a \cos bt) & \text{if } a^2 + b^2 \neq 0 \\ t + c & \text{if } a = b = 0 \end{cases}$
46. $x = ct^{-1} + \frac{3}{4}t^{1/3}$ **48.** $x = ce^{\frac{1}{2}t^2} - e^{\frac{1}{2}t^2} \int e^{-\frac{1}{2}u^2} du$ **50.** If $P_1(t)$ and $P_2(t)$ are any two integrals of $p(t)$, then $P_1(t) = P_2(t) + k$ for some constant k.

$$c_1 e^{P_1(t)} + e^{P_1(t)} \int e^{-P_1(t)} q(t)\, dt$$

$$= c_1 e^{P_2(t)+k} + e^{P_2(t)+k} \int e^{-P_2(t)-k} q(t)\, dt$$

$$= \left(c_1 e^k\right) e^{P_2(t)} + e^{P_2(t)} \int e^{-P_2(t)} q(t)\, dt.$$

Identify c_2 with $c_1 e^k$.   **51.** $x = -2e^{-\pi} e^{\pi t}$ **53.** $x = e^{3\pi/4} e^{\tan^{-1} t}$
55. $x = \exp\left(\frac{1-\cos at}{a}\right)$ **57.** $x = \frac{13}{3}e^{3t} + \frac{2}{3}$ **59.** $x = \frac{1}{3}t^4 - \frac{8}{3}t$
61. $x = \begin{cases} b\left(-1 + \exp\left(\frac{1}{a}\sin at\right)\right) & \text{if } a \neq 0 \\ be^t - b & \text{if } a = 0 \end{cases}$
63. $x = t^{-1}\left(\ln(1 + t^2) + \ln 4\right)$ **64.** $x = p_0 e^{(b-d)t}$ **65.** The doubling time $t - t_0 = \frac{1}{r} \ln 2$, is independent of x_0.   **68.** $x(t) = \left(x_0 - 2\pi T\alpha - \frac{h}{r}\right)e^{rt} + \frac{h}{r} + T\alpha\frac{2\pi}{4\pi^2+T^2r^2}\cos 2\frac{\pi}{T}t + T\alpha\frac{Tr}{4\pi^2+T^2r^2}\sin 2\frac{\pi}{T}t$
71. (a) The fourth-order Runge-Kutta method with step size $s = (0.1)/2^7 = 0.00078125$ gives $x(0.2) \approx 0.999999$. **(b)** The solution is very oscillatory. It is positive and bounded above. It appears to be nearly periodic, with a very small period. **(c)**
$x(t) = e^{\frac{5}{6\pi}(1-\cos 600\pi t)} \Rightarrow x(0.2) = 1$, **(d)** $x(t) = e^{\frac{5}{6\pi}(1-\cos 600\pi t)}$ is a positive periodic function of period $1/300$.   **73. (a)** The graph is below the t-axis and concave up, dropping to a minimum of approximately -0.55 at approximately $t = -0.575$ before increasing. As $t \to 0-$ the graph seems to approach the origin, that is, x seems to approach 0.

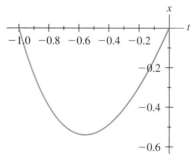

(b) From the fourth-order Runge-Kutta algorithm with step size 0.0001, we get

t	$x(t) \approx$
−0.1000	−0.161311
−0.0100	−0.017082
−0.0010	−0.001717
−0.0001	−0.000172

This table of values corroborates the assertion in (a) that x approaches 0 as t approaches 0. **(c)** $x(t) = et - te^{-t}$ and $\lim_{t \to 0^-} x(t) = 0$. Since $x'' = (2 - t) e^{-t} > 0$, the graph of $x(t)$ isconcave up. x has a minimum value of $x \approx -0.5419$ at $t \approx -0.5571$. **75. (a)** There appears to be a periodic solution with initial condition $x_0 = 0$. Its amplitude is approximately 3 and its period is approximately 1.25. All other solutions tend to this periodic solution as $t \to +\infty$.

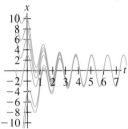

(b) $x(t) = (x_0 - 3)e^{-t} + 3 \cos 5t$ **(c)** The only solution in (b) that is periodic is obtained by choosing $x_0 = 3$. The resulting periodic solution $x(t) = 3 \cos 5t$ has period $2\pi/5$ and amplitude 3. Since e^{-t} tends to 0 as $t \to +\infty$, all solutions in (b) tend to $3 \cos 5t$.

Section 2.2

1. $x = ce^t + t^{100}e^t$ **3.** $x = ce^{-t} + 2e^{-3t} + e^t$ **5.** $x(t) = 10 - 5e^t$
7. $x(t) = -\frac{h}{r}e^{rt} + \frac{h}{r}$
11. $x' = k_1 x_1'(t) + k_2 x_2'(t) + k_3 x_3'(t)$
$= k_1 [p(t)x_1(t) + q_1(t)] + k_2 [p(t)x_2(t) + q_2(t)]$
$\quad + k_3 [p(t)x_3(t) + q_3(t)]$
$= p(t) [k_1 x_1(t) + k_2 x_2(t) + k_3 x_3(t)]$
$\quad + [k_1 q_1(t) + k_2 q_2(t) + k_3 q_3(t)]$
$= p(t)x + [k_1 q_1(t) + k_2 q_2(t) + k_3 q_3(t)]$

Section 2.3

1. applicable **3.** not applicable (p is not a constant)
5. applicable **7.** applicable **9.** applicable **11.** $x_p = ke^{0.2t}$
13. $x_p = kte^{3t}$ **15.** $x_p = k_1 e^{-t} \sin t + k_2 e^{-t} \cos t$
17. $x_p = k_1 t^4 \cos 2t + k_2 t^3 \cos 2t + k_3 t^2 \cos 2t$
$\quad + k_4 t \cos 2t + k_5 \cos 2t + k_6 t^4 \sin 2t + k_7 t^3 \sin 2t$
$\quad + k_8 t^2 \sin 2t + k_9 t \sin 2t + k_{10} \sin 2t$
19. If $a \neq p$, then $x_p = k_1 t^3 e^{at} + k_2 t^2 e^{at} + k_3 t e^{at} + k_4 e^{at}$ If $a = p$, then $x_p = k_1 t^4 e^{at} + k_2 t^3 e^{at} + k_3 t^2 e^{at} + k_4 t e^{at}$.
21. $x_p = k_1 \cos 2t + k_2 \sin 2t$
23. *Exercise 1* $x_p = -\frac{1}{3}t^2 e^{-t} - \frac{2}{9}e^{-t}t - \frac{2}{27}e^{-t}$
Exercise 3 Method not applicable
Exercise 5 $x_p = \frac{a}{\pi+2}e^{2t}$
Exercise 7 $x_p = -\frac{1}{2}e^{-2t}t^2$
Exercise 9 If $b \neq 1$, then $x_p = -\frac{3}{b-1}te^{bt} + \frac{3}{(b-1)^2}e^{bt}$. If $b = 1$ then
$x_p = -\frac{3}{2}t^2 e^t$.
24. *Exercise 11* $x_p = e^{0.2t}$
Exercise 13 $x_p = -15te^{3t}$
Exercise 15 $x_p = \frac{27}{32}e^{-t}\cos t + \frac{9}{32}e^{-t}\sin t$

Exercise 17 $x_p = \frac{984}{625}\cos 2t - \frac{912}{625}\sin 2t + \frac{168}{125}t\cos 2t + \frac{576}{125}t\sin 2t$
$\quad - \frac{132}{25}t^2 \cos 2t - \frac{24}{25}t^2 \sin 2t + \frac{12}{5}t^3 \cos 2t$
$\quad - \frac{16}{5}t^3 \sin 2t + t^4 \cos 2t + 2t^4 \sin 2t$
Exercise 19 If $p \neq a$, then $x_p = -\frac{2}{(p-a)^4}e^{at} - \frac{2}{(p-a)^3}te^{at}$
$- \frac{1}{(p-a)^2}t^2 e^{at} - \frac{1}{3}\frac{1}{p-a}t^3 e^{at}$. If $p = a$, then $x_p = \frac{1}{12}t^4 e^{at}$.
Exercise 21 $x_p = -\frac{2}{5}\cos 2t + \frac{4}{5}\sin 2t$
26. $x_p = e^t + \frac{3}{2}\cos t - \frac{3}{2}\sin t$ **28.** $x_p = 3te^t + 2\cos t - 2\sin t$

Section 2.4

1. $a \neq n\pi, n = 0, \pm 1, \pm 2, \pm 3, \ldots$ **3.** $a \neq 0$ **5.** $-\frac{7}{5}$ is a sink
7. 5 is a source **9.** $-2/(a-1)$ is a sink if $a < 1$ and a source if $a > 1$, and there is no equilibrium if $a = 1$. **11.** Since $p = 0$ the equation is nonhyperbolic. All solutions are unbounded if $a \neq n\pi$, $n = 0, \pm 1, \pm 2, \pm 3, \ldots$, and all solutions are equilibrium solutions if $a = n\pi$ for some integer $n = 0, \pm 1, \pm 2, \pm 3, \ldots$. **13.** All solutions are unbounded as $t \to -\infty$. **15.** All solutions are linearly unbounded as $t \to -\infty$. **18.** No, because the equilibrium remains a sink for all q when $p < 0$. There is also no bifurcation if $p > 0$ since the equilibrium remains a source for all q. **19.** There is one bifurcation point: $a \approx 0.56714$. For a less than this root the equilibrium $x = -\frac{1}{a-e^{-a}}$ is a sink; for a greater than this root it is a source.

Section 2.5

1. $x = c \exp\left(\frac{1}{2}at^2\right)$ **3.** $x = c \exp\left(\exp\left(-\frac{1}{2}t^2\right)\right)$
5. $x = c \exp\left(\frac{a}{r}e^{rt}\right)$ **7.** $x = ct^t e^{-t}$ **9.** $x = c\frac{1}{\cos t}$
11. $x = ce^t + \frac{1}{2}(\sin t - \cos t)$
13. $x = \begin{cases} -1 + ce^{\frac{1}{\beta}\sin \beta t} & \text{if } \beta \neq 0 \\ -1 + ce^t & \text{if } \beta = 0 \end{cases}$ **15.** $x = t^t + ct^t e^{-t}$
17. $x = \begin{cases} ce^t + \frac{1}{a-1}e^{at} & \text{if } a \neq 1 \\ ce^t + te^t & \text{if } a = 1 \end{cases}$
19. $x = ce^t + \left(-\frac{1}{101}\right)\sin 10t + \left(-\frac{10}{101}\right)\cos 10t$
21. $x = ce^{-2t} + \frac{1}{2}t^2 e^{-2t}$
23. $x = \left(k_1 t^3 + k_2 t^2 + k_3 t + k_4\right)e^{-t}\cos 2t$
$\quad + \left(k_5 t^3 + k_6 t^2 + k_7 t + k_8\right)e^{-t}\sin 2t$
25. $x = ce^t + \left(-\frac{2}{101}\right)\sin 10t + \left(-\frac{20}{101}\right)\cos 10t$
$\quad + \frac{3}{2}\sin t + \left(-\frac{3}{2}\right)\cos t$
27. $x = ce^{3t} + \left(-\frac{37}{27}\right) + \left(-\frac{19}{9}\right)t + \left(-\frac{5}{3}\right)t^2 + (-2)t^3$
29. $x = ce^{-t} + \frac{3}{2}\sin t + \left(-\frac{3}{2}\right)\cos t + \frac{2}{5}\sin 2t + \left(-\frac{4}{5}\right)\cos 2t$
31. $x = \pi e^{-25/2}\exp\left(\frac{1}{2}t^2\right)$
33. $x = \frac{10}{101}e^t + \left(-\frac{1}{101}\right)\sin 10t + \left(-\frac{10}{101}\right)\cos 10t$
35. $x = \left(x_0 + \frac{1}{16}\right)e^{-2t} + \frac{1}{4}te^{2t} + \left(-\frac{1}{16}\right)e^{2t}$
37. $x = x_0 \exp\left(\frac{a}{b^2}\left(1 - (bt + 1)e^{-bt}\right)\right)$ **39.** $x = x_0|1 + t|^a$
41. $x = x_0 \exp\left(a \tan^{-1} t\right)$ **44.** $q(t) = e^{-t}\sin t$ means immigration and emigration alternate periodically (with period 2π), but at an exponentially decreasing rate.
$x = \left(\frac{1}{(r+1)^2} + x_0\right)e^{rt} - \frac{1}{(r+1)^2}(\cos t + (r+1)\sin t)e^{-t}$ if $r \neq -1$ and $x = (x_0 + 1 - \cos t)e^{-t}$ if $r = -1$. **46.** $q(t) = 1 + 2\cos t$ means immigration/emigration oscillates periodically with mean 1 and amplitude 2. Note that it periodically becomes negative at which times emigration occurs.
$x = \left(x_0 + \frac{3r^2+1}{r(r^2+1)}\right)e^{rt} + \frac{2r}{r(r^2+1)}(\sin t - r\cos t) - \frac{1}{r}$
48. (a) $x \approx 2.99$ at $t = 2/3$. **(b)** The graph lies above the t-axis and appears periodic with a very shortperiod (i.e., high frequency).

(c) $x(t) = \exp\left(\frac{5}{2\pi}e^{\sin 40\pi t} - \frac{5}{2\pi}\right)$ and $x \approx 2.99257$ at $t = 2/3$. The numerical solutions found in (a) gave two decimals of accuracy. (d) $\sin 40\pi t$ is periodic with period equal to $1/20 = .05$ (or frequency 20) and hence so is the solution in (c). The solution x is always positive and hence the graph always lies above the t-axis.   **50.** The equilibrium 2 is a sink ($x \to 2$ as $t \to +\infty$).   **52.** The equilibrium 2 is a sink ($x \to 2$ as $t \to +\infty$).   **54.** The equilibrium $7/\pi$ is a sink ($x \to 7/\pi$ ast $\to +\infty$).   **56.** If $a < 1/2$, then the equilibrium $(1 - 2a)^{-1}$ is a sink. If $a > 1/2$, then the equilibrium $(1 - 2a)^{-1}$ is a source. If $a = 1/2$, then the equation is nonhyperbolic. $a = 1/2$ is a bifurcation point.   **58.** If $-1 < a < 1$, then the equilibrium $(1 + a)/(1 - a^2)$ is a sink. If $|a| > 1$, then the equilibrium $(1 + a)/(1 - a^2)$ is a source. If $a = +1$, then the equation is nonhyperbolic. If $a = -1$, then x remains constant. Bifurcation points are $a = \pm 1$.   **60. (a)** All solutions, except an equilibrium, are unbounded if a is less than about 0.7. For a greater than 0.7 all solutions tend to an equilibrium. The equilibrium depends on a. **(b)** For $a < 0.7$ there is a unique equilibrium and it is a source. For $a > 0.7$ there is a unique equilibrium and it is a sink. **(c)** Equation is nonhyperbolic at the root $a^* \approx 0.70347$ of $e^{-a} - a^2 = 0$. For other values of a the equation is hyperbolic and has a unique equilibrium $\left(e^{-a} - a^2\right)^{-1}$ which is a sink if $a > a^*$ and is a source if $a < a^*$. The root a^* is therefore a bifurcation point.   **61.** solution is $-\frac{1}{101}\sin 10t + \left(-\frac{10}{101}\right)\cos 10t$. All other solutions $x = ce^t - \frac{1}{101}\sin 10t + \left(-\frac{10}{101}\right)\cos 10t$ are exponentially unbounded.   **63.** A periodic solution is $\frac{1}{101}\sin 10t + \left(-\frac{10}{101}\right)\cos 10t + \frac{1}{2}\sin t + \frac{1}{2}\cos t$. All other solutions $x = ce^{-t} + \frac{1}{101}\sin 10t + \left(-\frac{10}{101}\right)\cos 10t + \frac{1}{2}\sin t + \frac{1}{2}\cos t$ tend to the periodic solution.   **65.** $x(t) = ce^{pt} - \frac{2p}{p^2+4\pi^2}\sin 2\pi t - \frac{4\pi}{p^2+4\pi^2}\cos 2\pi t$.   **68. (a)** The calculations

$$y'(t) = x'(t + T)$$
$$= px(t + T) + q(t + T)$$
$$= py(t) + q(t)$$

show that $y(t)$ is a solution. Moreover, $y(0) = x(T) = x(0)$ and the two solutions $x(t)$ and $y(t)$ satisfy the same initial value problem. Thus, they must be identical [i.e., $x(t) = x(t + T)$ for all t], which shows that $x(t)$ is periodic with period T. **(c)** The difference between the unique periodic solution $x_p(t)$ and any other solution $x(t)$, with initial condition x_0, is given by

$$x(t) - x_p(t) = \left(x_0 - \frac{e^{pT}}{1 - e^{pT}}\int_0^T e^{-ps}q(s)\,ds\right)e^{pt},$$

which is a multiple of the exponential e^{pt} This exponential tends to 0 as $p \to +\infty$ if $p < 0$ and is unbounded if $p > 0$.

Section 2.6

1. $t_{max} = k^{-1/2}$ is a decreasing function of k. The maximal intensity $x_0e^{-1/2}k^{1/2}$ is an increasing function of k.   **4. (a)** x increases to a maximum before decreasing to 0 as $t \to +\infty$. Maximum is higher and occurs later as r increases. **(b)** First-order, linear nonhomogeneous, nonautonomous. $x(t) = x_0\exp\left(rt - \frac{k}{p+1}t^{p+1}\right)$

(c) $x(t)$ has a maximum of $x_0\exp\left(r^{\frac{p+1}{p}}k^{-\frac{1}{p}}\frac{p}{p+1}\right)$ that occurs at

$t = \left(\frac{r}{k}\right)^{\frac{1}{p}}$. The maximum and the time to reach the maximum both are increasing functions of r.

$\lim_{t \to +\infty} x(t) = \lim_{t \to +\infty} x_0\exp\left(rt - \frac{k}{p+1}t^{p+1}\right) = 0$

(d) The rate kt^px equals $kt^p\exp\left(rt - \frac{k}{p+1}t^{p+1}\right)$. The maximum occurs at a point t_{max} that increases as r increases. **(e)** Set the derivative of the rate in (d) with respect to t equation to 0. t_{max} solves the equation $\frac{p}{t} + r = kt^p$. The graph of the left-hand side monotonically decreases to r as $t \to +\infty$ while the graph of the right-hand side starts at 0 and monotonically increases to $+\infty$ as $t \to +\infty$. It follows that there is a unique intersection of these two graphs at a point $t_{max} > 0$. As r increases the graph on the left rises an causes this intersection point to move to the right (i.e., causes t_{max} to increases).
7. $x(t) = 21.5e^{-\ln(43)t/7} + 78.5$. The temperature predictions made by this formula and their errors appear in the table below.

Time (minutes)	Data 0°F	Prediction 0°F	% Error
0.0	100.0	100	0
0.5	94.5	94.94	-4.60×10^{-1}
1.0	90.0	91.06	-1.18
1.5	87.0	88.10	-1.27
2.0	84.5	85.84	-1.59
2.5	83.5	84.11	-1.34
3.0	82.0	82.79	-9.63×10^{-1}
3.5	81.0	81.78	-9.61×10^{-1}
4.0	80.5	81.01	-6.29×10^{-1}
4.5	80.0	80.41	-5.20×10^{-1}
5.0	80.0	79.96	4.44×10^{-2}
5.5	79.5	79.62	-1.50×10^{-1}
6.0	79.5	79.36	-1.82×10^{-1}
6.5	79.0	79.15	-1.95×10^{-1}
7.0	79.0	79.0	0

8. $a = \frac{1}{\delta}\ln\left(\frac{x_1-x_0}{x_2-x_1}\right)$, $b = \frac{1}{2}\frac{x_1^2-x_0x_2}{x_1-\frac{x_0+x_2}{2}}$ provided $x_1 \neq \frac{x_0+x_2}{2}$

10. (a) $x(t) = \frac{845}{12}e^{-\frac{1}{3}(\ln 13)t} + \frac{955}{12}$

Time (minutes)	Data 0°F	Predicted 0°F	% error
0	150	150	0
1	109	109.5	-0.5
2	94	92.3	1.8
3	85	85	0
4	83	81.9	1.3
5	81	80.6	0.5
6	80	80	0

(b) Therefore, $x(t) = (150 - 79)e^{-0.7105t} + 79$.

Time (minutes)	Data 0°F	Predicted 0°F	% Error
0	150	150	0
1	109	113.9	-4.5
2	94	96.1	-2.3
3	85	87.4	-2.9
4	83	83.1	-0.2
5	81	81.0	$-.04$
6	80	80	0

(c) The fit in (a) is more accurate. Moreover, the fit in (b) is biased (toward overprediction).   **12. (a)** The time is independent of x_0 and b and is inversely related to a. **(b)** $t = \frac{1}{a}\ln\left(\frac{1}{1-\varphi}\right)$.

17. (a) $v(t) = \frac{F_0}{k_0}\left(1 - e^{-k_0 t}\right)$ **(b)** $x(t) = \frac{F_0}{k_0}\left(t + \frac{1}{k_0}\left(e^{-k_0 t} - 1\right)\right)$
(c) $k_0 \approx 0.784\,98$ and $F_0 \approx 7.3733$ **(d)** The graph of
$x(t) \approx 9.3930t + 11.966\left(e^{-0.78498t} - 1\right)$ appears in the accompanying figure. The solution appears to fit the data very well.

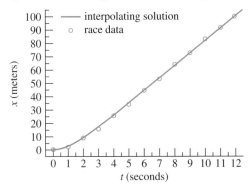

(e) The model fits the data during the later portion of the race quite well. It does less well at the beginning of the race, probably because the modeling assumption of a constant propulsive force exerted by the runner is less accurate during the high acceleration phase at the beginning of the race.

Time (s) t_i	Distance (meters) x_i	Predicted Distance $x(t_i)$	% Error $100\frac{x_i - x(t_i)}{x_i}$
0	0.0	0.0	0.0
1	2.0	2.9	− 44.3
2	8.5	9.3	− 9.5
3	15.5	17.3	− 11.9
4	25.5	26.1	− 2.4
5	34.0	35.2	− 3.6
6	44.5	44.5	0.0
7	53.0	53.8	− 1.6
8	64.0	63.2	1.2
9	72.5	72.6	− 0.1
10	83.0	82.0	1.2
11	91.5	91.4	0.2
11.92	100.0	100.0	0.0

21. (a) $t - t_0 = -\frac{H}{\ln 2}\ln\left(\frac{x(t)}{x_0}\right)$ **(b)** $t - t_0 = -\frac{H}{\ln 2}\ln\left(\frac{x'(t)}{x'(t_0)}\right)$
22. $-\frac{5568}{\ln 2}\ln\left(\frac{4.09}{6.68}\right) \approx 3,941$ years (1991 BC) **24.** The fraction is approximately 0.315 for all samples. **27.** The elapsed time is given by $t - t_0 = r^{-1}\ln f^{-1}$.
28. (a) $x' = rx$, $x(t_0) = x_0$ **(b)** $x = x_0 e^{r(t-t_0)}$ grows exponentially without bound as $t \to +\infty$. **(c)** $t - t_0 = \frac{1}{r}\ln 2$. This doubling time does not depend on x_0, only on r. **34. (a)**
$x(t) = \left(1 - 8.6561 \times 10^{-3}d\right)e^{2.3105 \times 10^{-2}t} + 8.6561 \times 10^{-3}d$ **(b)** $x(t)$ decreases if and only if $1 - 8.6561 \times 10^{-3}d < 0$, i.e.,
$d > d_{cr} = 115.53$. **(c)** $t_{cr} = \frac{1}{2.3105 \times 10^{-2}}\ln\left(\frac{8.6561 \times 10^{-3}d}{8.6561 \times 10^{-3}d - 1}\right)$ is a decreasing function of $d > d_{cr} \approx 115.53$

Chapter 3

Section 3.1.1

2. $x_e = 1, -3$. **4.** $x_e = 1$. **6.** $x_e \approx 0.70347$. **8.** There is one equilibrium (for each a). **10.** There is one equilibrium (for each

$a \neq 0$). **12.** Solutions are increasing if $x_0 < -1$ or $0 < x_0 < 1$ and decreasing if $-1 < x_0 < 0$ or $1 < x$. **14.** Solutions are increasing if $-\sqrt{g/c} < x_0 < \sqrt{g/c}$ and decreasing if $x_0 < -\sqrt{g/c}$ or $x_0 > \sqrt{g/c}$.
16. Solutions are increasing if $x_0 < 1/3$ or $x_0 > 1/2$ and decreasing if $1/3 < x_0 < 1/2$.

Section 3.1.2

1. $x_e = 1$ is a hyperbolic source. **3.** $x_e = 0$ is a (nonhyperbolic) shunt and $x_e = 1$ is a hyperbolic source. **5.** $x_e \approx 0.73909$ is a hyperbolic sink. **7.** $x_e = -1$ is a hyperbolic sink, 0 is a hyperbolic source, $x_e = 0.5$ is a shunt. **9.** $x_e = -1$ is a hyperbolic sink and 1 is a nonhyperbolic shunt. **11.** If $a < 0$, then $x_e = a, 0$ and 1 are hyperbolic (sink, source, and sink, respectively). If $0 < a < 1$ then $x_e = a$, 0 and 1 are hyperbolic (source, sink, sink, respectively). If $a > 1$, then $x_e = a, 0$ and 1 are hyperbolic (sink, sink, source respectively). If $a = 0$, then $x_e = 0$ is a shunt and $x_e = -1$ is a hyperbolic sink. If $a = 1$, then $x_e = 1$ is a shunt and $x_e = 0$ is a hyperbolic sink. **13.** The equilibria $x_e = n\pi$ are

hyperbolic sinks for n $= \cdots, -6, -4, -2$ and $n = 1, 3, 5 \cdots$
hyperbolic sources for n $= \cdots, -5, -3, -1$ and $n = 2, 4, 6 \cdots$
a nonhyperbolic shunt for $n = 0$.

The following are example answers only (based on using polynomials). There are infinitely many possible correct answers.
15. $x' = (x+3)(x-3)$ **17.** $x' = x^2(2-x)$ **19.** $x' = x^2(x-1)^2$
21. $x' = (x-a)(x-b)$ **23.** $x' = -(x-1)(x-2)(x-3)$
25. $x' = -(x-a)(x-b)(x-c)(x-d)^2$ **27.** The phase line portrait for $v \geq 0$ is $\longrightarrow \sqrt{9.8/k_0} \longleftarrow$ The equilibrium $\sqrt{9.8/k_0}$ is a hyperbolic sink.

Section 3.1.3

1. The linearizations are $u' = -u$ at $x_e = 0$ and $u' = 2u$ at $x_e = \pm 1$.
3. The linearizations are $u' = ru$ at $x_e = 0$ and $u' = -ru$ at $x_e = K$.
5. The linearization at $x_e = 0$ is $u' = 0$.
7. At $x_e = \left(1 + \sqrt{1-4h}\right)/2$ the linearization is $u' = -\sqrt{1-4h}u$. At $x_e = \left(1 - \sqrt{1-4h}\right)/2$ the linearization is $u' = \sqrt{1-4h}u$. **8.** The linearization at the positive equilibrium $v_e = \sqrt{9.8/k_0}$ is $u' = -2\sqrt{9.8k_0}u$. This linear equation has a sink $\left(-2\sqrt{9.8k_0} < 0\right)$. The linearization principle implies the positive equilibrium is also a sink.

Section 3.1.4

1. qualitatively equivalent $\longrightarrow \cdot \longleftarrow$ **3.** qualitatively equivalent $\longrightarrow \cdot \longleftarrow$ **5.** not qualitatively equivalent (source and sink)
7. qualitatively equivalent $\longleftarrow \cdot \longleftarrow \cdot \longrightarrow$ **9.** not qualitatively equivalent (first equation has two equilibria while the second has one)
11. qualitatively equivalent $\longleftarrow \cdot \longrightarrow \cdot \longleftarrow$ **13.** qualitatively equivalent $\longrightarrow \cdot \longleftarrow$ **15.** not qualitatively equivalent (first equation has one equilibrium while the second has three)
17. qualitatively equivalent $\longrightarrow$ **19.** A saddle-node bifurcation occurs at $p = 0$. **21.** A pitchfork bifurcation occurs at $p = -1$.
23. A transcritical bifurcation occurs at $p = 0$. **25.** A pitchfork bifurcation occurs at $p = 1$. **27.** There is a bifurcation at $p = 0$, but it is not of any of the types classified in the text. **29.** There are two bifurcations: a saddle-node bifurcation at $p = -2\sqrt{3}/9$ and a saddle-node bifurcation at $p = 2\sqrt{3}/9$. **31.** There is a saddle-node bifurcation at $p = 0$. **33.** A saddle-node bifurcation occurs at $p = 1$.

35. *Exercise* 19

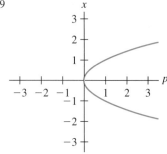

Exercise 21

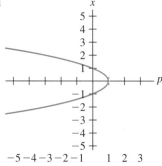

Exercise 23

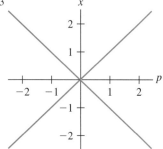

Exercise 25

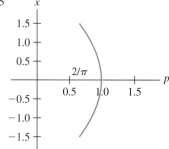

Exercise 27

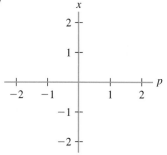

Exercise 29

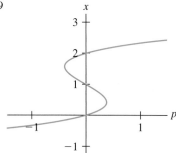

Exercise 31

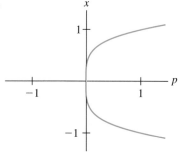

Exercise 33

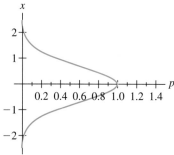

36. $\longleftarrow 0 \longrightarrow K \longleftarrow$ **38.** The equilibria are $x_e = 0$ and $x_e = \pm\sqrt{-p}$ for $p \leq 0$. The graphs of these three functions produce a pitchfork with nose at the origin and opening to the left. $x_e = 0$ loses stability as p is decreased through $p_0 = 0$ while the equilibria on the outer "forks" of the pitchfork are stable for all $p < 0$.

40. $x' = x(x^2 + p)$ **41. (a)** $h(x)$ is continuous because x^2 is continuous and because both one-sided limits as $x \to 0\pm$ are equal to

0. Similarly, its inverse

$$h(x) = \begin{cases} \sqrt{x} & \text{if } x > 0 \\ 0 & \text{if } x = 0 \\ \sqrt{-x} & \text{if } x < 0 \end{cases}.$$

is continuous.

Section 3.2

1. $x = \tan(t + c)$, c = arbitrary constant
3. $x^2 - 1 = ce^{2t}$, c = arbitrary constant, and the equilibria $x = \pm 1$
5. $-\ln(\cos x) = t + c$, c = arbitrary constant, and the equilibria
$x = \pi/2 \pm n\pi$, $n = 0, \pm 1, \pm 2, \ldots$ **7.** $x = \left(\frac{1}{3t-2}\right)^{1/3}$
9. $(x - 1)e^x = -t - 2e^{-1}$ **11.** $x = \left(1 - \frac{3}{4}e^{2t}\right)^{1/2}$
13. $x = -\left(1 - \frac{3}{4}e^{2t}\right)^{1/2}$

16. (a) $x = \begin{cases} b + \left(\frac{3}{at-c}\right)^3, & \text{where } c \text{ is an arbitrary constant} \\ b \end{cases}$

(b) $x(t) = b + \left(\frac{3(x_0-b)^{1/3}}{at(x_0-b)^{1/3}+3}\right)^3$ **(c)** $\lim_{t\to+\infty} x = b$.
18. $x = \frac{x_0}{x_0-(x_0-1)e^t}$ For $x_0 > 1$ this solution the maximal interval of
existence of this solution is $-\infty < t < \beta$, where $\beta = \ln\left(\frac{x_0}{x_0-1}\right)$.
20. $x = c \exp\left(\frac{1}{3}t^3\right)$, c = arbitrary constant

22. $\begin{cases} \frac{t}{1+ct}, & c = \text{arbitrary constant} \\ 0 \end{cases}$

24. $\begin{cases} \frac{1+c\exp(t^2)}{-1+c\exp(t^2)}, & c = \text{arbitrary constant} \\ \pm 1 \end{cases}$

26. $\begin{cases} \frac{2-c\exp(e^t)}{1-c\exp(e^t)}, & c = \text{arbitrary constant} \\ 1, 2 \end{cases}$

28. $\begin{cases} -a\frac{1+c\exp(2a\sin t)}{1-c\exp(2a\sin t)}, & c = \text{arbitrary constant} \\ \pm a \end{cases}$

30. $x = \exp\left(\frac{1}{3}t^3\right)$ **32.** $x^5 + 1 = \frac{5}{2}t^2 + 5t$
34. $x^{1+a} = (1+a)(t^2 - 1) + b^{1+a}$ if $a \neq -1$, $x = b\exp(t^2 - 1)$ if
$a = -1$ **36.** $x = x_0 e^{-at^{p+1}/(p+1)}$ and $\lim_{t\to+\infty} x(t) = 0$ **37.** Both
$x = x(t)$ and the equilibrium $x_e = x(t^*)$ satisfy the same initial value
problem, namely $x' = g(t)h(x)$, $x(t^*) = x_e$. Theorem 1 implies that
these solutions must be identical.

Section 3.3

1. $\begin{cases} \frac{2}{\sin t - \cos t + ce^{-t}}, & c = \text{arbitrary constant} \\ 0 \end{cases}$

3. $\begin{cases} \frac{1}{t}\frac{2}{2c+e^{-2t}}, & c = \text{arbitrary constant} \\ 0 \end{cases}$

5. $\begin{cases} \pm\left(1 + ce^{-2t}\right)^{-1/2}, & c = \text{arbitrary constant} \\ 0 \end{cases}$

7. $x = \frac{3t}{\sqrt{9-6t^3}}$ **9.** This is a Bernoulli equation. $x = \sqrt{\frac{3}{2e^{-t}+10e^{2t}}}$

13. $\begin{cases} (2ce^{-\cos t} - 3)(ce^{-\cos t} - 1)^{-1} \\ 2 \end{cases}$

15. $x = \frac{4ce^{-2t}-2t-1}{2t^2-t+4cte^{-2t}}$ **19. (a)** $\longleftarrow aK \longrightarrow K \longleftarrow$ **(b)** $x_0 = aK$ is
the threshold that must be exceeded in order to survive.

(c) $x = K\dfrac{cK(1-a)e^{r(1-a)t} + a}{cK(1-a)e^{r(1-a)t} + 1}$

(d) $x = K\dfrac{aK - x_0 - (aK - ax_0)e^{-r(1-a)t}}{aK - x_0 - (K - x_0)e^{-r(1-a)t}}$
(e) x vanishes at

$$t_1 = \frac{1}{r(1-a)}\ln\left(\frac{aK - ax_0}{aK - x_0}\right).$$

If $0 < x_0 < aK$ then $t_1 > 0$. **(f)** The denominator in x vanishes at

$$t_2 = \frac{1}{r(1-a)}\ln\left(\frac{K - x_0}{aK - x_0}\right) > t_1.$$

Thus, for $t > t_2$ the solution x is negative and approaches $-\infty$ as
$t \to t_2$. **(g)** Both numerator and denominator in x are negative and
hence x is defined for all $t > 0$.

$$\lim_{t\to+\infty} x = K\frac{aK - x_0 + 0}{aK - x_0 + 0} = K.$$

Section 3.4

1. $x_2(t) = 1 + (\sin 1)(t - \pi) + \left(\frac{1}{4}\sin 2\right)(t - \pi)^2$
3. $x_2(t) = x_0 + r\left(1 - \frac{x_0}{K}\right)x_0(t - t_0) + \frac{1}{2}x_0 r^2\left(1 - 2\frac{x_0}{K}\right)\left(1 - \frac{x_0}{K}\right)$
5. $x_2(t) = 1 + (\sin 1)(t - \pi) + \left(\frac{1}{4}\sin 2\right)(t - \pi)^2$ The accuracy of the
quadratic Taylor polynomial decreases as t becomes further away from
π.

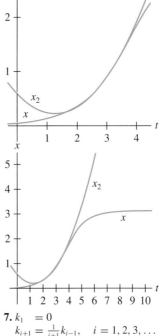

7. $k_1 = 0$
$k_{i+1} = \frac{1}{i+1}k_{i-1}$, $i = 1, 2, 3, \ldots$
$x(t) = 2\sum_{m=0}^{\infty}\frac{1}{2^m}\frac{1}{m!}t^{2m} = 2\exp\left(\frac{1}{2}t^2\right)$
8. $k_{i+1} = -2\frac{1}{i+1}k_i$, $i = 0, 1, 2, 3, \ldots$.
$x(t) = 1 - 2t + \frac{1}{2!}(2t) + \cdots = e^{-2t}$
11. $x = t + t^2 + \frac{2}{3}t^3 + \frac{5}{12}t^4 + \cdots$ **13.** $x = 1 + 2t + \frac{5}{2}t^2 + \frac{11}{6}t^3 + \cdots$
15. $x_1(t) = 2e^t + (-2 + 2e^t)\varepsilon$

17. $x_1(t) = e^{2t} + \left(\frac{1}{5}\cos t + \frac{2}{5}\sin t - \frac{1}{5}e^{2t}\right)\varepsilon$

19. $x_1(t) = 1 + \left(\frac{1}{2}e^{-t} - \frac{1}{2}\cos t + \frac{1}{2}\sin t\right)\varepsilon$.

22. $k_0' = -k_0$, $\qquad k_0(t_0) = x_0$
$k_1' = -k_1 - k_0\sin t$, $k_1(t_0) = 0$
$k_2' = -k_2 - k_1\sin t$, $k_2(t_0) = 0$
$x_2(t) = x_0 e^{-(t-t_0)} + x_0\left(\cos t - \cos t_0\right)e^{-(t-t_0)}\varepsilon$
$\qquad + \frac{1}{2}x_0\left(\cos^2 t - 2\cos t\cos t_0 + \cos^2 t_0\right)e^{-(t-t_0)}\varepsilon^2$

24. $k_0' = -k_0$, $\qquad k_0(0) = 1$
$k_1' = -k_1 + \sin t$, $k_1(0) = 0$
$k_i' = -k_i$, $\qquad k_i(0) = 0$ for $i \geq 2$
The initial value problem for $i \geq 2$ has the (unique) solution $k_i(t) = 0$.

28. (a) To first-order $x' = k_0' + k_1'\varepsilon + \cdots$ and

$$\left(1 - \frac{2}{9}x\right)x - (1 + \varepsilon\sin 2\pi t) = \left[\left(1 - \frac{2}{9}k_0\right)k_0 - 1\right]$$
$$+ \left[\left(1 - \frac{4}{9}k_0\right)k_1 - \sin 2\pi t\right]\varepsilon$$

Equate coefficients of ε^0 and ε. **(b)** The equilibria are $k_0 = 3$ and $3/2$.
(c) For $k_0 = 3$,

$$k_1 = -\frac{3}{1 + 36\pi^2}\sin 2\pi t + \frac{18\pi}{1 + 36\pi^2}\cos 2\pi t.$$

For $k_0 = 3/2$,

$$k_1 = \frac{3}{1 + 36\pi^2}\sin 2\pi t + \frac{18\pi}{1 + 36\pi^2}\cos 2\pi t.$$

(d) If $k_0 = 3$, then

$$x_1 = 3 + \left(-\frac{3}{1 + 36\pi^2}\sin 2\pi t + \frac{18\pi}{1 + 36\pi^2}\cos 2\pi t\right)\varepsilon.$$

If $k_0 = 3/2$, then

$$x_1 = 3/2 + \left(\frac{3}{1 + 36\pi^2}\sin 2\pi t + \frac{18\pi}{1 + 36\pi^2}\cos 2\pi t\right)\varepsilon.$$

30. $x_0 = 1$
$x_1 = 1 + pt$
$x_2 = 1 + pt + p^2t^2/2$

32. $x_0 = \frac{\pi}{2}$
$x_1 = \frac{\pi}{2} + (t - 1)$
$x_2 = \frac{\pi}{2} + \sin(t - 1)$

36. $x_0 = 1$
$x_1 = 2 - e^{-t}$
$x_2 = \frac{5}{2} - 2e^{-t} + \frac{1}{2}e^{-2t}$

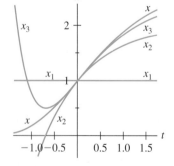

Section 3.5

1. Solutions increase for $x_0 > 0$ and decrease for $x_0 < 0$. **3.** The equilibria $x = \pi/2 + n\pi$, $n = 0, \pm 1, \pm 2, \ldots$ are nonhyperbolic shunts.

$$\cdots \longrightarrow -\frac{3}{2}\pi \longrightarrow -\frac{1}{2}\pi \longrightarrow \frac{1}{2}\pi \longrightarrow \frac{3}{2}\pi \longrightarrow \frac{5}{2}\pi \longrightarrow \cdots$$

5. If $a < 0$, there are no equilibria. If $a > 0$, there are two equilibria: $x_e = \pm\sqrt{a}$, both hyperbolic sinks. If $a = 0$, then the unique equilibrium $x_e = 0$ is a sink.

$a < 0$ implies $\quad \longrightarrow 0 \longleftarrow$

$a = 0$ implies $\quad \longrightarrow 0 \longleftarrow$

$a > 0$ implies $\quad \longrightarrow -\sqrt{a} \longleftarrow 0 \longrightarrow \sqrt{a} \longleftarrow$

Note: 0 is not an equilibrium when $a \neq 0$.
7. The linearization at the equilibrium 0 is $u' = 0$. The linearization at the equilibrium 1 is $u' = -u$. **9.** The linearization at the equilibrium $-(\ln b)/a$ is $u' = abu$.

12. (a) $\lim_{t\to+\infty} x'(t) = \lim_{t\to+\infty} f(x(t))$.
$\lim_{t\to+\infty} f(x(t)) = f\left(\lim_{t\to+\infty} x(t)\right) = f(x_L)$ and
$\lim_{t\to+\infty} x'(t) = f(x_L) = x_L'$. Since we are given that $x'(t) > 0$ for all t, the limit x_L' must be greater than or equal to zero.
(b) $x(t) - x(t_0) = \int_{t_0}^t x'(s)\,ds \geq (t - t_0)\frac{1}{2}x_L' > 0$. If $x_L' > 0$ then the right-hand side of this inequality tends to $+\infty$ as $t \to +\infty$, and so must the left-hand side. But this contradicts the given fact that $x(t)$ approaches a finite limit. **13.** $|x_0 - x_e| = |x_0| < \varepsilon$ implies $|x(t) - x_e| = |x_0|e^{pt} < |x_0| < \varepsilon$ for all $t > 0$. The definition of stability is fulfilled by taking $\delta = \varepsilon$. **14.** By Exercise 13, $x_e = 0$ is stable. Since

$$\lim_{t\to+\infty} |x(t) - x_e| = |x_0|e^{pt} = 0$$

for any x_0, the definition of asymptotic stability is satisfied for any $\delta_0 > 0$. **16.** By Exercise 15, $x_e = 0$ is stable. Since

$$\lim_{t\to+\infty} |x(t) - x_e| = |x_0|\left(1 + 2x_0^2 at\right)^{-1/2} = 0$$

for any x_0 the definition of asymptotic stability is satisfied for any $\delta_0 > 0$. **17. (a)** The equilibria $x_e = \pi/n$ where n is any positive or negative integer lie arbitrarily close to the equilibrium $x_e = 0$ which is therefore not isolated. **(b)** df/dx evaluated at 0 equals 0, which implies $x_e = 0$ is not hyperbolic.
18. (a) $\longrightarrow -\sqrt{2} \longleftarrow 0 \longrightarrow \sqrt{2} \longleftarrow$. **(b)** $x_e = 0$ is a source. $x_e = \pm\sqrt{2}$ are both sinks. **(c)** $x_e = 0$ is nonhyperbolic. Both $x_e = \pm\sqrt{2}$ are hyperbolic. **(d)** The linearization at $x = 0$ is $u' = 0$, which is nonhyperbolic. The linearizations at $x_e = \pm\sqrt{2}$ are both $u' = -8u/3$. **(e)** The linearization theorem does not apply for the nonhyperbolic equilibrium $x = 0$. It does apply for the equilibria $x_e = \pm\sqrt{2}$ and it asserts that both are sinks.
20. (a) $\longrightarrow$ for $p > 0$
$\longrightarrow 0 \longrightarrow$ for $p = 0$
$\longrightarrow -(-p)^{1/4} \longleftarrow (-p)^{1/4} \longrightarrow$ for $p < 0$.
(b) When $p = 0$, $x_e = 0$ is a shunt. When $p < 0$, $x = -(-p)^{1/4}$ is a sink and $x = (-p)^{1/4}$ is a source. **(c)** $x_e = 0$ is nonhyperbolic. $x_e = 1$ is hyperbolic. **(d)** The linearization at $x = 0$ is $u' = 0$. The linearization at $x_e = 1$ is $u' = 4u$. **(e)** When $p = 0$ the linearization theorem does not apply for the nonhyperbolic equilibrium $x = 0$. When $p < 0$, the theorem does apply for the equilibria $x_e = \pm(-p)^{1/4}$. It asserts that $x_e = -(-p)^{1/4}$ is a sink and $x_e = (-p)^{1/4}$ is a source.
The following are example answers only (based upon using polynomials). There are infinitely many possible correct answers.
22. $x' = x^2(x - 1)^2(2 - x)$ **24.** $x' = p(p - x)$ **26.** qualitatively equivalent $\longleftarrow \cdot \longrightarrow \cdot \longleftarrow$ **28.** not qualitatively equivalent (source and a sink)

30.
$$\begin{cases} \longrightarrow 0 \longleftarrow p \longrightarrow & \text{for } p > 0 \\ \longrightarrow 0 \longrightarrow & \text{for } p = 0 \\ \longrightarrow p \longleftarrow 0 \longrightarrow & \text{for } p < 0 \end{cases}$$
A transcritical bifurcation (with an exchange of stability) occurs at $p = 0$.

32.
$$\begin{cases} \longrightarrow & \text{for } p \geq 0 \\ \longleftarrow \bullet \longrightarrow \bullet \longleftarrow & \text{for } -1 < p < 0 \\ \longleftarrow \bullet \longleftarrow & \text{for } p = -1 \\ \longleftarrow & \text{for } p < -1 \end{cases}$$
A saddle-node bifurcation occurs at $p = -1$.

34.
$$\begin{cases} \longrightarrow & \text{for } p < 1 \\ \longrightarrow 1 \longrightarrow & \text{for } p = 1 \\ \longrightarrow (1 - \sqrt{p-1}) \longleftarrow (1 + \sqrt{p-1}) \longrightarrow & \text{for } p > 1 \end{cases}$$
A saddle node bifurcation occurs at $p = 1$.

36.
$$\begin{array}{ccc} \longleftarrow p^{-1} \longrightarrow 0 \longleftarrow & \longrightarrow 0 \longleftarrow & \longrightarrow 0 \longleftarrow p^{-1} \longrightarrow \\ p < 0 & p = 0 & p > 0 \end{array}$$
No interval containing $p_0 = 0$ has the same orbit structure throughout.

37. $x = \begin{cases} \dfrac{1}{1 - ce^t} & \text{for } c = \text{arbitrary constant} \\ 0 \end{cases}$

39. $x = 1 - c \exp\left(\frac{1}{2}t^2 - t\right)$ **41.** $x = -1/y$ where

$$y = \exp(at^2)\left(c - a^2 \int \exp(-at^2)\, dt\right)$$

43. $x^2 = ax_0^2\left((x_0^2 + a)e^{-2at} - x_0^2\right)$.   **45.** $x = ce^{\frac{2}{3}t^{5/2}}$

47. $x^2 = \dfrac{1}{1 + \frac{2}{5}\cos t + \frac{4}{5}\sin t + ce^{2t}}$ or

$$x = \pm\left(\dfrac{1}{1 + \frac{2}{5}\cos t + \frac{4}{5}\sin t + ce^{2t}}\right)^{1/2}$$

49. $x^2 = \dfrac{1}{ce^{-2t} - t + \frac{1}{2}}$

or $x = \pm\left(\dfrac{1}{ce^{-2t} - t + \frac{1}{2}}\right)^{1/2}$ **51.** $2\cos ax + at^2 = c$

53. $x = \left(2 - e^{-\frac{1}{2}t^2 + t + c}\right)\left(1 - e^{-\frac{1}{2}t^2 + t + c}\right)^{-1}$

54. $x = -\frac{1}{a}\ln\left(-\frac{1}{2}t^2 a + 1\right)$ and $x = -\frac{1}{a}\ln\left(-\frac{1}{2}t^2 a + e\right)$

56. $x = -\frac{1}{t}\frac{1}{1 + \ln t}$ and $x = -\frac{1}{t}\frac{1}{1 + \ln(-t)}$ **58.** $x = 1 - e^{t(t-2)/2}$

60. (a) $x_3(t) = 1 + t + t^2 + \frac{2}{3}t^3$ for $x_0 = 1$ **(b)** For $x_0 = 0$

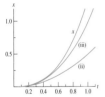

The third-order Taylor polynomial (iii) is a better approximation than the second-order Taylor polynomial (ii).

For $x_0 = 1$

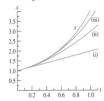

The sequence of Taylor polynomials shows increased accuracy.

62. $k_{i+1} = \frac{1}{i+1}k_{i-1}, \quad i = 2, 3, 4, \dots$
$x = x_0 + \frac{1}{2}(x_0 - 1)t^2 + \frac{1}{8}(x_0 - 1)t^4 + \cdots$
64. $k_{i+1} = \frac{1}{i+1}k_i, i = 0, 1, 2, 3, \dots$
$x = 1 + (t - 1) + \frac{1}{2}(t - 1)^2 + \frac{1}{6}(t - 1)^3 + \cdots$
66. $k_0' = a(\tau - k_0), \qquad k_0(0) = x_0$
$\quad\ k_1' = a(\tau \sin t - k_1), \quad k_1(0) = 0$
$\quad\ k_i' = -ak_i, \qquad\qquad k_i(0) = 0$ for $i \geq 2$
The solution of the equations for $i \geq 2$ are all $k_i = 0$ and hence
$x = k_0(t) + k_1(t)p$. **68.** $x_0 = \frac{K}{2}, x_1 = \frac{K}{2} + \frac{rK}{4}t,$
$x_2 = \frac{1}{2}K\left(1 + \frac{1}{2}rt + \frac{1}{8}r(r - 2)t^2 - \frac{1}{24}r^3t^3\right)$
70. $x_0 = 0, x_1 = \frac{1}{3}t^3, x_2 = \frac{1}{3}t^3 + \frac{1}{36}t^46$

Section 3.6
1. (a) $x(t) = 1.65e^{r(t-1900)}$ with $r \approx 8.4697 \times 10^{-3}$
(b)

Year	World (billions) Population (billions)	Interpolation (billions) at 1950 (billions)	% Error
1900	1.65	1.65	0
1910	1.75	1.80	−3
1920	1.86	1.95	−5
1930	2.07	2.13	−3
1940	2.30	2.32	−1
1950	2.52	2.52	0
1960	3.02	2.74	9
1970	3.70	2.99	19
1980	4.45	3.25	27
1990	5.30	3.54	33

3. (a) $x(t) \approx \dfrac{28.58}{8.82e^{-2.781\times10^{-2}(t-1950)} + 2.52}$ and $x(2100) \approx 10.76$ billion.
(b) $K = 11.34$ billion.
4. $x(t) = \dfrac{1.65}{(7.4939 - 3.4178\times10^{-3}t)^{1.6920}}$

Year	World Population (billions)	(6.9) Prediction (billions)	Error (%)
1900	1.65	1.65	0
1910	1.75	1.75	0
1920	1.86	1.86	0
1930	2.07	1.98	4
1940	2.30	2.12	8
1950	2.52	2.27	10
1960	3.02	2.43	19
1970	3.70	2.62	29
1980	4.45	2.83	36
1990	5.30	3.07	42

The errors get increasingly worse as the century progresses. The fit is considerably worse than that in Table 10.4. Doomsday occurs at $t = 2192.6$. **6.** The doubling time $t = 4\ln 6 \approx 7.2$ minutes. Under exponential growth the doubling time is $t = 4\ln 2 \approx 2.7$ minutes. **10. (a)** Define $y = x^{-1}$. Then y satisfies the linear, nonhomogeneous equation $y' = -ry + r/K(t)$. This linear equation a unique periodic solution $y_p(t)$ of period T; see Exercise 68, Chapter 2. From the variation of constants formula we see that $y_p(t)$ is positive and therefore $x_p(t) = 1/y_p(t)$ is the unique, positive periodic solution of the periodic logistic equation. **(b)** $x(t) = \frac{x_0}{(1-x_0 y_p(0))e^{-rt}+x_0 y_p(t)}$ and

$$\lim_{t\to+\infty}\left(x(t)-x_p(t)\right) = -\lim_{t\to+\infty} e^{-rt}\frac{1-x_0 y_p(0)}{(1-x_0 y_p(0))e^{-rt}+x_0 y_p(t)} = 0.$$

15. Since $F_f(v) < 0$ for $v > 0$ and $F_f(v) > 0$ for $v < 0$, we get the phase line portrait $\longrightarrow 0 \longleftarrow$. This means that no matter how the object is put in motion ($v(0) > 0$ or $v(0) < 0$), its velocity $v(t)$ tends to 0 as t increases and the object comes to rest. **17.** $k_0 \approx 1.128$ and $x(t) \approx 7.697e^{-1.129t} + 8.685t - 7.697$. Errors are relatively large and all positive (i.e., the model under predicts all data points).

Time t (seconds)	Distance Fallen x (meters)	Model Prediction (to nearest 100th)	Error %
0.000	0.00	0.00	0
0.347	0.61	0.52	15
0.470	1.00	0.92	9
0.519	1.22	1.09	10
0.582	1.52	1.35	11
0.650	1.83	1.64	10
0.674	2.00	1.75	12
0.717	2.13	1.96	8
0.766	2.44	2.20	10
0.823	2.74	2.49	9
0.870	3.00	2.74	9
1.031	4.00	3.66	8
1.193	5.00	4.67	7
1.354	6.00	5.73	4
1.501	7.00	6.75	4
1.727	8.50	8.40	1
1.873	9.50	9.50	0

22. (a) $w' = f(t) - p(t) - cw^q$, where $c > 0$ is a constant of proportionality. **(b)** If $p < f$, then the phase portrait is $\longrightarrow \left(\frac{f-p}{c}\right)^{1/q} \longleftarrow$ If $p > f$, then the phase portrait is $\longleftarrow$. **(c)** If $p < f$ then weight will monotonically (increase or decrease) to the equilibrium $w = \left(\frac{f-p}{c}\right)^{1/q}$. If $p > f$, then weight will drop to 0 in a finite amount of time. **(d)** Weight (in the long run) periodically fluctuates with period equal to 1. If p is too large, weight will equal zero in finite time. If p is less than a critical value, weight will always positive. **23. (a)** $c' = -\frac{mzc}{a+c}$, $c(0) = c_0$ **(b)** $\longleftarrow -a \longrightarrow 0 \longleftarrow$ Here $-a$ is not an equilibrium, but a singular point where the equation is not defined. **(c)** t_{half} linearly increases with c_0 **(d)** Implicit solution: $a \ln c + c + mzt = a \ln c_0 + c_0$ **(e)** $t_{\text{half}} = \frac{2a \ln 2 + c_0}{2mz}$ **24.** The two equilibria are given by the formula

$$c = \frac{1}{2d}\left(dc_{\text{in}} - ad - mz\right.$$
$$\left. \pm \sqrt{(dc_{\text{in}} - mz)^2 + 2d^2 c_{\text{in}}a + a^2 d^2 + 2admz}\right).$$

These equilibria are defined for all positive model parameters because the discriminant appearing under the radical is always positive. Only one of the equilibria is positive, namely the root using the $+$ sign.

Chapter 4

Section 4.1
1. Is a solution pair   **3.** Is not a solution pair   **5.** Is a solution pair   **7.** Is not a solution pair   **9.** Is a solution pair   **11.** Is a solution pair

14. For all t_0, and y_0 and for all $x_0 \neq -1$ there exists a unique solution pair on an interval containing t_0. For $x_0 = -1$, the fundamental theorem does not apply and no conclusions can be drawn.   **16.** For all initial conditions there exists a unique solution pair on an interval containing t_0.   **18.** For all initial conditions there exists a unique solution pair on an interval containing t_0.   **20.** For all initial conditions there exists a unique solution pair on an interval containing t_0.   **22.** For all t_0, and y_0 and all $t_0 \neq 0$ there exists a unique solution pair on an interval containing t_0. For $t_0 = 0$, the fundamental theorem does not apply and no conclusions can be drawn.   **24.** For all initial conditions there exists a unique solution pair on an interval containing t_0.

Section 4.2.1
1. From the given solution formulas, $x(2) = 0.1082695$ and $y(2) = 5.413166 \times 10^{-2}$ (rounded to 7 significant digits). The Euler algorithm approximations to the solution of (2.5) at $T = 2$ are given, together with their errors, in the following table.

s	$x(2) \approx$	Error	$y(2) \approx$	Error
0.4	-1.01344	-1.12171	2.18240	2.12827
0.2	0.0858994	-2.23705×10^{-2}	0.0429496	-1.11817×10^{-2}
0.1	0.0972613	-1.10085×10^{-2}	0.0486307	-5.50066×10^{-3}
0.05	0.102810	-5.45946×10^{-3}	0.0514046	-2.72666×10^{-3}
0.025	0.105551	-2.71846×10^{-3}	0.0527742	-1.35766×10^{-3}

The errors in the approximations for both $x(2)$ and $y(2)$ decrease by (at least) a factor of $1/2$. This is commensurate with the rate of convergence of the Euler algorithm since the step size is also decreased by a factor of $1/2$.

2. For Heun's algorithm, the errors are

s	$x(2) \approx$	Error	$y(2) \approx$	Error
0.4	1.53648	1.42821	-2.782171	-2.83630
0.2	0.110248	1.97855×10^{-3}	0.0544010	2.69344×10^{-4}
0.1	0.108662	3.92545×10^{-4}	0.0543216	1.90344×10^{-4}
0.05	0.108363	9.35446×10^{-5}	0.0541779	4.63444×10^{-5}
0.025	0.108293	2.35446×10^{-5}	0.0541430	1.13444×10^{-5}

The errors in the approximations for both $x(2)$ and $y(2)$ decrease by (at least) a factor of $1/4 = (1/2)^2$. This is commensurate with the rate of convergence of Heun's algorithm since the step size is also decreased by a factor of $1/2$.

6. (a) $x' = y$, $y' = -\alpha(x^2 - 1)y - x$ **(b)** The graph oscillates with a decreasing amplitude, appearing to approach the t-axis as t increases.

(c) The graph oscillates with a sustained amplitude (i.e., appears periodic).

(d) For $\alpha < 0$ the graphs oscillate but damp out as in (b). For $\alpha > 0$ the graphs oscillate with a sustained amplitude. The change occurs at $\alpha = 0$.

Section 4.2.2

1.

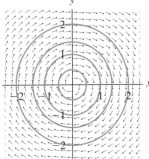

3.

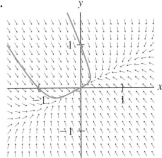

5.

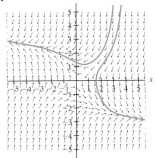

7. (a) and (6); (b) and (2); (c) and (1) **8. (a)** If (x_0, y_0) lies on both nullclines, then $f(x_0, y_0) = 0$ and $g(x_0, y_0) = 0$. **9.** The x-nullcline is the x-axis. The y-nullcline is the y-axis.

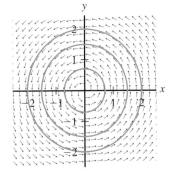

11. The x-nullcline is the straight line $y = x$. The y-nullcline is the straight line $y = -x$.

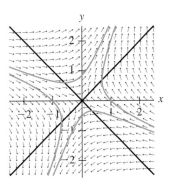

13. The x-nullcline is given by the y-axis and the horizontal straight line $y = 1$. The y-nullcline is the x-axis) and the vertical straight line $x = 1$.

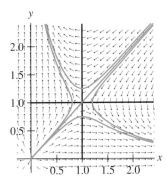

15. (a) The x-axis is a y-nullcline and the y-axis is an x-nullcline. The straight line given by the linear equation $r_1 - a_{11}x - a_{12}y = 0$ is an x-nullcline and the straight line given by the linear equation $r_2 - a_{21}x - a_{22}y = 0$ is a y-nullcline. **(b)** Either the two straight lines in (a) intersect or they don't. In the first case, there are two possibilities depending on which line lies above the other. In the second case, there are two possibilities depending on whether the x-nullcline or the y-nullcline has the smaller slope.

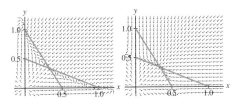

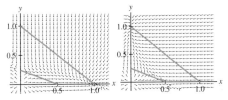

(d)

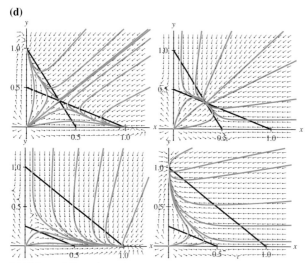

(e) If the nullclines do not intersect in the first quadrant (i.e., there is no equilibrium in this quadrant), all orbits tend to an equilibrium on either the x- or y- axis, depending on the configuration. This means that one species goes extinct. If the nullclines intersect, then in one case orbits tend to the equilibrium in the first quadrant, which means both species survive, and in the other case all orbits tend to an equilibrium on an axis, which means one species goes extinct. (Actually, in the latter case there are two exceptional orbits separating those orbits that go to the x-axis from those that go to the y-axis. These two orbits go to the positive equilibrium and are the only orbits that permit both species to survive.)   **17. (a)** The nullclines are the vertical line $y = 0$ and the curves given by the equation $-\alpha\left(x^2 - 1\right)y - x = 0$

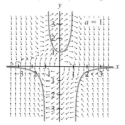

Section 4.3

1. First solution:

$$x_1' = 1 = \left(\frac{3}{2t}\right)t + \left(-\frac{1}{2}\right)1 = \left(\frac{3}{2t}\right)x_1 + \left(-\frac{1}{2}\right)y_1$$

and

$$y_1' = 0 = \left(-\frac{1}{2t^2}\right)t + \left(\frac{1}{2t}\right)1 = \left(-\frac{1}{2t^2}\right)x_1 + \left(\frac{1}{2t}\right)y_1.$$

Second solution:

$$x_2' = 2t = \left(\frac{3}{2t}\right)t^2 + \left(-\frac{1}{2}\right)(-t) = \left(\frac{3}{2t}\right)x_2 + \left(-\frac{1}{2}\right)y_2$$

and

$$y_2' = -1 = \left(-\frac{1}{2t^2}\right)t^2 + \left(\frac{1}{2t}\right)(-t) = \left(-\frac{1}{2t^2}\right)x_2 + \left(\frac{1}{2t}\right)y_2.$$

2. $x = tc_1 + t^2c_2$, $y = c_1 - tc_2$ **3.** $x = t + t^2$, $y = 1 - t$ **7.** We can construct the general solution from either the first and third or the second and third solution pairs. For example, the first and third yield

$$x = -\tfrac{1}{2}c_1e^{5t} - 3c_2e^{-3t}$$
$$y = -\tfrac{1}{2}c_1e^{5t} + c_2e^{-3t}.$$

9. We cannot construct the general solution from these solution pairs. **11.** We can use the first and third pairs to obtain the general solution

$$x = -c_1e^{-t} + c_2e^{t}$$
$$y = c_1e^{-t} - 2c_2e^{t}.$$

Other possibilities are to use the first and second or the second and third pairs.   **13.** We can use the first and third pairs to obtain the general solution

$$x = c_1e^{t}\cos 2t + c_2e^{t}\sin 2t$$
$$y = c_1e^{t}\sin 2t - c_2e^{t}\cos 2t.$$

Other possibilities are to use the first and second or the second and third pairs.   **15. (a)** For the first pair:

$$x' = 4e^{4t} = -2e^{4t} + 3\left(2e^{4t}\right) = -2x + 3y$$
$$y' = 8e^{4t} = 2e^{4t} + 3\left(2e^{4t}\right) = 2x + 3y$$

For the second pair:

$$x' = 9e^{-3t} = -2\left(-3e^{-3t}\right) + 3\left(e^{-3t}\right) = -2x + 3y$$
$$y' = -3e^{-3t} = 2\left(-3e^{-3t}\right) + 3\left(e^{-3t}\right) = 2x + 3y$$

(b) The pairs are not multiples of each other since they are exponentials with different exponents.
(c) $x = c_1e^{4t} - 3c_2e^{-3t}$ **(d)**
   $y = 2c_1e^{4t} + c_2e^{-3t}$
$x' = 8e^{4t} - 9e^{-3t} = -2\left(2e^{4t} + 3e^{-3t}\right) + 3\left(4e^{4t} - e^{-3t}\right) = -2x + 3y$
$y' = 16e^{4t} + 3e^{-3t} = 2\left(2e^{4t} + 3e^{-3t}\right) + 3\left(4e^{4t} - e^{-3t}\right) = 2x + 3y$
(e) Take $c = 2$ and $c_2 = -1$.
17. (a) $x(t) = \tfrac{3}{5}e^{5t} - \tfrac{8}{5}e^{-5t}$, $y(t) = \tfrac{4}{5}e^{-5t} + \tfrac{6}{5}e^{5t}$
18. (a) $x(t) = \cos\sqrt{3}t - \tfrac{2}{3}\sqrt{3}\sin\sqrt{3}t$, $y(t) = -\tfrac{1}{3}\sqrt{3}\sin\sqrt{3}t$
19. $x = c_1\cos t + c_2\sin t - \tfrac{1}{3}\sin 2t$, $y = -c_1\sin t + c_2\cos t - \tfrac{2}{3}\cos 2t$.
21. $x = c_1\cos t + c_2\sin t + 2$, $y = -c_1\sin t + c_2\cos t - 1$
24. The pairs are dependent means one is a constant multiple of the other. Suppose $x_2 = cx_1$ and $y_2 = cy_1$ for all $\alpha < t < \beta$ where c is a constant. Then

$$\det\begin{pmatrix} x_1 & x_2 \\ y_1 & y_2 \end{pmatrix} = x_1y_2 - y_1x_2 = x_1cy_1 - y_1cx_1 = 0$$

for all $\alpha < t < \beta$.

Section 4.4

1. $\tilde{x}(t) = \begin{pmatrix} \tfrac{6}{5}e^{-t} - \tfrac{1}{5}e^{-6t} \\ \tfrac{3}{5}e^{-t} + \tfrac{2}{5}e^{-6t} \end{pmatrix}$ **3.** $\tilde{x}(t) = \begin{pmatrix} 6e^{-t} + 4e^{-6t} \\ 3e^{-t} - 8e^{-6t} \end{pmatrix}$

5. $\tilde{x}(t) = \begin{pmatrix} -\cos t + \sin t \\ \sin t + \cos t \end{pmatrix}$ **7.** $\tilde{x}(t) = \begin{pmatrix} \cos t - \sin t \\ -\sin t - \cos t \end{pmatrix}$

9. *Exercise 7*: $\Phi(t)\tilde{c} = \begin{pmatrix} -\tfrac{1}{2}e^{5t} & -3e^{-3t} \\ -\tfrac{1}{2}e^{5t} & e^{-3t} \end{pmatrix}\begin{pmatrix} c_1 \\ c_2 \end{pmatrix}$

Exercise 9: We cannot construct a fundamental solution matrix from these solution pairs.

Exercise 11: $\Phi(t)\tilde{c} = \begin{pmatrix} -e^{-t} & e^{t} \\ e^{-t} & -2e^{t} \end{pmatrix}\begin{pmatrix} c_1 \\ c_2 \end{pmatrix}$

Exercise 13: $\Phi(t)\tilde{c} = \begin{pmatrix} e^t \cos 2t & e^t \sin 2t \\ e^t \sin 2t & -e^t \cos 2t \end{pmatrix} \begin{pmatrix} c_1 \\ c_2 \end{pmatrix}$

10. If $\tilde{x}(t)$ and $\tilde{y}(t)$ are solutions, then for any linear combination $k_1\tilde{x}(t) + k_2\tilde{y}(t)$

$$(k_1\tilde{x} + k_2\tilde{y})' = k_1\tilde{x}' + k_2\tilde{y}'$$
$$= k_1 A(t)\tilde{x} + k_2 A(t)\tilde{y}$$

is identical to

$$A(t)(k_1\tilde{x} + k_2\tilde{y}) = A(t)(k_1\tilde{x}) + A(t)(k_2\tilde{y})$$
$$= k_1 A(t)\tilde{x} + k_2 A(t)\tilde{y}.$$

Section 4.5

1. $x' = -e^{-t} = -2e^{-t} + e^{-t} = -2x + y$
$y' = -e^{-t} = e^{-t} - 2e^{-t} = x - 2y$

3. $x' = -e^{-t} - 3e^{-3t} = -2\left(e^{-t} + e^{-3t}\right) + \left(e^{-t} - e^{-3t}\right) = -2x + y$
$y' = -e^{-t} + 3e^{-3t} = \left(e^{-t} + e^{-3t}\right) - 2\left(e^{-t} - e^{-3t}\right) = x - 2y$

5. $x' = -6e^{-3t} = 4\left(2e^{-3t}\right) - 2\left(7e^{-3t}\right) = 4x - 2y$
$y' = -21e^{-3t} = 7\left(2e^{-3t}\right) - 5\left(7e^{-3t}\right) = 7x - 5y$

7. $x' = -6e^{-3t} - 2e^{2t} = 4\left(2e^{-3t} - e^{2t}\right) - 2\left(7e^{-3t} - e^{2t}\right) = 4x - 2y$
$y' = -21e^{-3t} - 2e^{2t} = 7\left(2e^{-3t} - e^{2t}\right) - 5\left(7e^{-3t} - e^{2t}\right) = 7x - 5y$

9. $x' = -2\sin t - 2\cos t = y$
$y' = -2\cos t + 2\sin t = -(3 + 2\cos t - 2\sin t) + 3 = -x + 3$
11. There exists a unique solution on an interval containing $t_0 = 0$.
13. We can conclude nothing from the theorem. **15.** For $x_0 \neq 1$ and $y_0 \neq 1$ there exists a unique solution on an interval containing t_0. For all other initial value problems the theorem does not apply and no conclusions. **17.** If $t_0 - x_0 - y_0 > 0$, $t_0 + x_0 + y_0 > 0$ then the initial value problem has a unique solution on some interval containing t_0. For all other initial value problems the theorem does not apply and no conclusion can be drawn. **19.** There exists a unique solution for any initial value problem and it exists for all t. **21.** All initial value problems have a unique solution defined on an interval containing t_0.
23. All initial value problems have unique solutions that exist for all t.
25. Any initial value problem for with $t_0 \neq 0$ has a unique solution. For $t_0 > 0$ the solution exists for $t > 0$ and for $t_0 < 0$ the solution exists for $t < 0$.
28. For Euler's algorithm:

s	$x(1)$	$t(1)$
0.10000	1.453298	−0.882508
0.05000	1.416965	−0.862285
0.02500	1.399212	−0.851938
0.01250	1.390452	−0.846718
0.00625	1.386102	−0.844098

The digits 1.3 for $x(1)$ and −0.8 for $y(1)$ appear to have stabilized.
29. For Heun's algorithm:

s	$x(1)$	$y(1)$
0.10000	1.381444	−0.842473
0.05000	1.381669	−0.841709
0.02500	1.381745	−0.841485
0.01250	1.381771	−0.841475
0.00625	1.381771	−0.841475

The digits 1.317 for $x(1)$ and −0.8414 for $y(1)$ appear to have stabilized.
30. For the Runge-Kutta algorithm:

s	$x(1)$	$y(1)$
0.10000	1.381773	−0.841470
0.05000	1.381773	−0.841471
0.02500	1.381773	−0.841471
0.01250	1.381773	−0.841471
0.00625	1.381773	−0.841471

The digits 1.381773 for $x(1)$ and −0.841471 for $y(1)$ appear to have stabilized.
31.

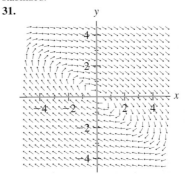

32. With Euler's algorithm and $s = 0.1$ the orbits spiral around the origin outwardly. For Heun's algorithm or the Runge-Kutta algorithm and $s = 0.1$ the orbits are closed loops (ellipses). If the step size s is decreased sufficiently Euler's algorithm will also give closed elliptical orbits. Conjecture: All orbits (except the equilibrium point at the origin) are closed elliptical loops around the origin which move clockwise. **39.** linear homogeneous **41.** nonlinear **43.** linear nonhomogeneous **45.** linear homogeneous

47. $A = \begin{pmatrix} 5 & -5 \\ -1 & -1 \end{pmatrix}$

$\tilde{h}(t) = \begin{pmatrix} 0 \\ -7 \end{pmatrix}$

$\begin{pmatrix} x \\ y \end{pmatrix}' = \begin{pmatrix} 5 & -5 \\ -1 & -1 \end{pmatrix} \begin{pmatrix} x \\ y \end{pmatrix} + \begin{pmatrix} 0 \\ -7 \end{pmatrix}$

49. $A = \begin{pmatrix} c-4 & 3c \\ -1 & d-1 \end{pmatrix}$

$\tilde{h}(t) = \begin{pmatrix} t^2 \\ -t \end{pmatrix}$

$\begin{pmatrix} x \\ y \end{pmatrix}' = \begin{pmatrix} c-4 & 3c \\ -1 & d-1 \end{pmatrix} \begin{pmatrix} x \\ y \end{pmatrix} + \begin{pmatrix} t^2 \\ -t \end{pmatrix}$

51. (a) $x(t) = \beta^{-1} \sin \beta t - \cos \beta t$, $y(t) = \cos \beta t + \beta \sin \beta t$
52. $x = c_1 e^{-3t} + c_2 e^t - 2$
54. $x = c_1 e^{-3t} + c_2 e^t + \left(-\frac{2}{5}\right) \sin t + \left(-\frac{1}{5}\right) \cos t$
56. $x = c_1 e^{-3t} + c_2 e^t - \frac{2}{9}k - \frac{1}{3}kt$

Section 4.6

1. Flight time: $t = 2v_0/g$. **2.** The distance traveled $x = 2u_0 v_0/g$.
3. Using $u_0 = s_0 \cos \theta$, $v_0 = s_0 \sin \theta$ the distance traveled $\left(s_0^2/g\right) \sin 2\theta$ is proportional to the speed square of the initial speed s_0. **4.** The distance $\left(s_0^2/g\right) \sin 2\theta$ is maximized when $\theta = \pi/4$ radians (45^0).
5. The maximum height is $y = v_0^2/2g$. **16.** $u' = -h - \frac{c}{m}u$,
$v' = -g - \frac{c}{m}v$
17. $u = \left(u_0 + \frac{mh}{c}\right)e^{-\frac{c}{m}t} - \frac{mh}{c}$, $v = \left(v_0 + \frac{gm}{c}\right)e^{-ct/m} - \frac{gm}{c}$
18. $x = \frac{m}{c}\left(u_0 + \frac{mh}{c}\right)\left(1 - e^{-ct/m}\right) - \frac{mh}{c}t$,

$y = \frac{m}{c}\left(v_0 + \frac{gm}{c}\right)\left(1 - e^{-ct/m}\right) - \frac{gm}{c}t$

19. The optimal angle is approximately the same (i.e., $\theta = 0.566$ radians). The maximal horizontal distance decreases from 507 ft to about 504.25 ft.

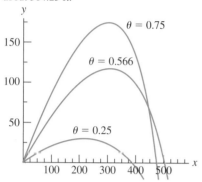

20. $\theta \approx 45^0$ at which the ball travels $x \approx 287$ feet.

Chapter 5
Section 5.1
1. First solution: $x_1' = 2e^{2t} = 4e^{2t} - 2e^{2t} = 4x_1 - 2y_1$ and
$y_1' = 2e^{2t} = 7e^{2t} - 5e^{2t} = 7x_1 - 5y_1$
Second solution: $x_2' = -6e^{-3t} = 4(2e^{-3t}) - 2(7e^{-3t}) = 4x_2 - 2y_2$ and
$y_2' = -21e^{-3t} = 7(2e^{-3t}) - 5(7e^{-3t}) = 7x_2 - 5y_2$
2. $x = c_1x_1 + c_2y_1 = c_1e^{2t} + 2c_2e^{-3t}$,
$y = c_1y_1 + c_2y_2 = c_1e^{2t} + 7c_2e^{-3t}$
3. $x_p' = e^t = e^t - e^t + e^t = x_p + y_p + e^t$ and
$y_p' = -e^t = e^{-t} - (-e^t) - 3e^t = x_p - y_p - 3e^t$
4. $x = c_1x_1 + c_2y_1 + x_p = c_1e^{2t} + 2c_2e^{-3t} + e^t$,
$y = c_1y_1 + c_2y_2 + y_p = c_1e^{2t} + 7c_2e^{-3t} - e^t$
5. $x = -\frac{4}{5}e^{2t} + \frac{4}{5}e^{-3t} + e^t$, $y = -\frac{4}{5}e^{2t} + \frac{14}{5}e^{-3t} - e^t$

Section 5.2
1. $x = c_1e^{2t} + 2c_2e^{-3t}$, $y = c_1e^{2t} + 7c_2e^{-3t}$
3. $x = c_1e^{t/2}\cos\frac{3}{2}t + c_2e^{t/2}\sin\frac{3}{2}t$, $y = c_1e^{t/2}\sin\frac{3}{2}t - c_2e^{t/2}\cos\frac{3}{2}t$
5. $x = 0.25c_1e^{-2.77t} + 0.77c_2e^{-0.39t}$, $y = 1.51c_1e^{-2.77t} + 0.64c_2e^{-0.39t}$
7. $x = c_1(1 + 2t)e^t + 2c_2te^t$, $y = -2c_1te^t + c_2(1 - 2t)e^t$
9. $x = c_1e^{3t} - c_2e^{-2t}$, $y = 4c_1e^{3t} + c_2e^{-2t}$
11. $x = 0.97c_1e^{-6.1t} + 0.044c_2e^{-1.5t}$, $y = 0.23c_1e^{-6.1t} + 1.04c_2e^{-1.5t}$
13. $x = c_1(\cos 5t - 5\sin 5t) + c_2(5\cos 5t + \sin 5t)$,
$y = -2c_1\cos 5t - 2c_2\sin 5t$ **15.** $x = c_1e^t + c_2te^t$,
$y = 2c_1e^t + c_2\left(\frac{4}{3} + 2t\right)e^t$
17. *Exercise 1* $x = \frac{9}{5}e^{2t} - \frac{4}{5}e^{-3t}$
$y = \frac{9}{5}e^{2t} - \frac{14}{5}e^{-3t}$
Exercise 3 $x = e^{t/2}\cos\frac{3}{2}t + e^{t/2}\sin\frac{3}{2}t$
$y = e^{t/2}\sin\frac{3}{2}t - e^{t/2}\cos\frac{3}{2}t$
Exercise 5 $x = -0.35e^{-2.77t} + 1.35e^{-0.39t}$
$y = -2.12e^{-2.77t} + 1.12e^{-0.39t}$
Exercise 7 $x = e^t$
$y = -e^t$
Exercise 9 $x = e^{-2t}$
$y = -e^{-2t}$
Exercise 11 $x = 1.06e^{-6.1t} - 0.060e^{-1.5t}$
$y = 0.25e^{-6.1t} - 1.25e^{-1.5t}$
Exercise 13 $x = \cos 5t - \frac{12}{5}\sin 5t$
$y = -\cos 5t - \frac{1}{5}\sin 5t$

Exercise 15 $x = e^t - \frac{9}{4}te^t$
$y = -e^t - \frac{9}{2}te^t$
18. *Exercise 1* $x = \frac{2}{5}e^{-3t} + \frac{8}{5}e^{2t}$
$y = \frac{2}{5}e^{-3t} + \frac{8}{5}e^{2t}$
Exercise 3 $x = 2e^{\frac{1}{2}t}\cos\frac{3}{2}t - 3e^{\frac{1}{2}t}\sin\frac{3}{2}t$
$y = 3e^{\frac{1}{2}t}\cos\frac{3}{2}t + 2e^{\frac{1}{2}t}\sin\frac{3}{2}t$
Exercise 5 $x = 1.75e^{-0.39t} + 0.25e^{-2.77t}$
$y = 1.46e^{-0.39t} + 1.54e^{-2.77t}$
Exercise 7 $x = 10te^t + 2e^t$
$y = -10te^t + 3e^t$
Exercise 9 $x = e^{-2t} + e^{3t}$
$y = -e^{-2t} + 4e^{3t}$
Exercise 11 $x = 0.11e^{-1.5t} + 1.89e^{-6.1t}$
$y = 2.55e^{-1.55t} + 0.45e^{-6.05t}$
Exercise 13 $x = 2\cos 5t + \frac{41}{5}\sin 5t$
$y = 3\cos 5t - \frac{7}{5}\sin 5t$
Exercise 15 $x = 2e^t - \frac{3}{4}te^t$
$y = 3e^t - \frac{3}{2}te^t$
21. $x' = $ inflow rate $-$ outflow rate $= 0 -$ liver absorption rate $-$ muscle absorption rate $= -\alpha x - \beta y$. $y' = $ inflow rate $-$ outflow rate $=$ pancreas production rate $-$ liver degradation rate $= \gamma x - \delta y$.

Section 5.3
1. $x = e^{-t/2}\left(c_1\cos\frac{\sqrt{3}}{2}t + c_2\sin\frac{\sqrt{3}}{2}t\right)$ **3.** $x = c_1e^{\sqrt{2}t/2} + c_2e^{-\sqrt{2}t/2}$
5. $x = c_1e^t + c_2e^{-4t}$ **7.** $x = c_1\cos\sqrt{5}t + c_2\sin\sqrt{5}t$
9. $x = c_1e^{-1.5t} + c_2e^{-2.3t}$ **14.** stable node and
$x = \frac{4}{3}x_0e^{-500t} - \frac{1}{3}x_0e^{-2000t}$ **16.** center and $x = x_0\cos 1000t$

Section 5.4
1. (a) roots $\lambda = 2, -3$ imply a saddle   **3. (a)** roots $\lambda = \frac{1}{2} \pm \frac{3}{2}i$ imply an unstable spiral   **5. (a)** roots $\lambda = -2.77, -0.39$ imply a stable node   **7. (a)** repeated root $\lambda = 1$ and $b = 2 \neq 0$ imply an unstable improper node   **9. (a)** roots $\lambda = 3, -2$ imply a saddle
11. (a) roots $\lambda = -6.0517, -1.5483$ imply a stable node

13. (a) roots $\lambda = \pm 5i$ imply a center   **15. (a)** repeated root $\lambda = 1$ and $b = 3/4 \neq 0$ imply an unstable improper node   **17.** stable spiral
19. saddle (unstable)   **21.** saddle (unstable)   **23.** center (neutrally stable)   **25.** stable node   **34.** $\delta < 0$ implies saddle (unstable); $\delta = 0$ implies stable improper node; $0 < \delta < 1$ implies stable node; $\delta = 1$ implies stable improper node; $1 < \delta$ implies stable spiral.
37. $x = c_1e^{-t}$, $y = c_2$.

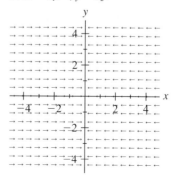

39. $x = c_1$, $y = c_1 t + c_2$.

Section 5.5

1. $\Phi(t) = \begin{pmatrix} e^{2t} & 2e^{-3t} \\ e^{2t} & 7e^{-3t} \end{pmatrix}$ and $\tilde{x} = \begin{pmatrix} c_1 e^{2t} + 2c_2 e^{-3t} \\ c_1 e^{2t} + 7c_2 e^{-3t} \end{pmatrix}$

3. $\Phi(t) = e^{t/2} \begin{pmatrix} \cos\frac{3}{2}t & \sin\frac{3}{2}t \\ \sin\frac{3}{2}t & -\cos\frac{3}{2}t \end{pmatrix}$ and

$\tilde{x} = e^{t/2} \begin{pmatrix} c_1 \cos\frac{3}{2}t + c_2 \sin\frac{3}{2}t \\ c_1 \sin\frac{3}{2}t - c_2 \cos\frac{3}{2}t \end{pmatrix}$

5. $\Phi(t) = \begin{pmatrix} 0.25e^{-2.77t} & 0.77e^{-0.39t} \\ 1.51e^{-2.77t} & 0.64e^{-0.39t} \end{pmatrix}$ and

$\tilde{x} = \begin{pmatrix} 0.25c_1 e^{-2.77t} + 0.77c_2 e^{-0.39t} \\ 1.51c_1 e^{-2.77t} + 0.64c_2 e^{-0.39t} \end{pmatrix}$

7. $\Phi(t) = e^t \begin{pmatrix} 1+2t & 2t \\ -2t & 1-2t \end{pmatrix}$ and $\tilde{x} = e^t \begin{pmatrix} c_1(1+2t) + 2c_2 t \\ -2c_1 t + c_2(1-2t) \end{pmatrix}$

9. $\Phi(t) = \begin{pmatrix} e^{3t} & -e^{-2t} \\ 4e^{3t} & e^{-2t} \end{pmatrix}$ and $\tilde{x} = \begin{pmatrix} c_1 e^{3t} - c_2 e^{-2t} \\ 4c_1 e^{3t} + c_2 e^{-2t} \end{pmatrix}$

11. $\Phi(t) \approx \begin{pmatrix} 0.97e^{-6.05t} & 0.05e^{-1.55t} \\ 0.23e^{-6.05t} & 1.04e^{-1.55t} \end{pmatrix}$ and

$\tilde{x} = \begin{pmatrix} 0.97c_1 e^{-6.05t} + 0.05c_2 e^{-1.55t} \\ 0.23c_1 e^{-6.05t} + 1.04c_2 e^{-1.55t} \end{pmatrix}$

13. $\Phi(t) = \begin{pmatrix} \cos 5t - 5\sin 5t & \sin 5t + 5\cos 5t \\ -2\cos 5t & -2\sin 5t \end{pmatrix}$ and

$\tilde{x} = \begin{pmatrix} (c_1 + 5c_2)\cos 5t + (-5c_1 + c_2)\sin 5t \\ (-2c_1)\cos 5t + (-2c_2)\sin 5t \end{pmatrix}$

15. $\Phi(t) = \begin{pmatrix} e^t & te^t \\ 2e^t & \left(\frac{4}{3} + 2t\right)e^t \end{pmatrix}$ and $\tilde{x} = \begin{pmatrix} c_1 e^t + c_2 te^t \\ 2c_1 e^t + c_2 \left(\frac{4}{3} + 2t\right)e^t \end{pmatrix}$

18. $e^{0.5125t} \begin{pmatrix} -4.5\cos 3.134t & -4.5\sin 3.1346t \\ -0.744\cos 3.134t & 3.134\cos 3.134t \\ -3.134\sin 3.134t & -0.744\sin 3.134t \end{pmatrix}$

21. $\Phi(t) = e^{-3t} \begin{pmatrix} -6+3t & -4 \\ 4 & 2+3t \end{pmatrix}$ and

$\tilde{x} = e^{-3t} \begin{pmatrix} -(6+t)c_1 - 4c_2 \\ 4c_1 + (2-t)c_2 \end{pmatrix}$

Section 5.6

1. $\Phi(t) = \begin{pmatrix} 3e^{3t} & e^t & -e^{5t} \\ e^{3t} & e^t & 0 \\ -3e^{3t} & -2e^t & e^{5t} \end{pmatrix}$

3. $\Phi(t) = \begin{pmatrix} -2e^{-t} & 7\cos t - \sin t & \cos t + 7\sin t \\ e^{-t} & -5\cos t & -5\sin t \\ 2e^{-t} & -5\cos t & -5\sin t \end{pmatrix}$

5. $\Phi(t) = \begin{pmatrix} e^t & e^{-t} & e^t\sin t & -e^t\cos t \\ e^t & e^{-t} & e^t\cos t + 2e^t\sin t & -2e^t\cos t + e^t\sin t \\ 0 & 0 & e^t\cos t & e^t\sin t \\ 0 & e^{-t} & e^t\cos t + 3e^t\sin t & -3e^t\cos t + e^t\sin t \end{pmatrix}$

7. For each initial value problem the solution is

$$\tilde{x}(t) = \Phi(t)\Phi(0)^{-1} \begin{pmatrix} 1 \\ 0 \\ -1 \end{pmatrix}.$$

Exercise 1 $\tilde{x}(t) = \begin{pmatrix} e^{5t} \\ 0 \\ -e^{5t} \end{pmatrix}$

Exercise 3 $\tilde{x}(t) = \begin{pmatrix} -\cos t + 3\sin t + 2e^{-t} \\ -2\sin t - e^{-t} + \cos t \\ -2\sin t - 2e^{-t} + \cos t \end{pmatrix}$

8. *Exercise 5* $\tilde{x}(t) = \begin{pmatrix} 6e^t - 8e^{-t} + 3e^t\cos t - 2e^t\sin t \\ 6e^t - 8e^{-t} + 4e^t\cos t - 7e^t\sin t \\ -3e^t\sin t - 2e^t\cos t \\ -8e^{-t} + 7e^t\cos t - 9e^t\sin t \end{pmatrix}$

9. $x = c_1 e^t + c_2 e^{-2t} + c_3 e^{-3t}$ **11.** $x = c_1 e^t + c_2 e^{2t} + c_3 e^{-3t} + c_4 e^{-t}$

13. *Exercise 9* $x = \frac{5}{12}e^t + \frac{4}{3}e^{-2t} - \frac{3}{4}e^{-3t}$

14. *Exercise 11* $x = \frac{17}{8}e^t - \frac{2}{5}e^{2t} + \frac{1}{40}e^{-3t} - \frac{3}{4}e^{-t}$

Section 5.7

1. $x = c_1 e^{-5t}$, $y = c_2 e^{-5t}$

3. $x = c_1 e^{-t}\cos\sqrt{3}t + c_2 e^{-t}\sin\sqrt{3}t$

$y = c_1 e^{-t}\left(-2\cos\sqrt{3}t + \sqrt{3}\sin\sqrt{3}t\right)$

$- c_2 e^{-t}\left(\sqrt{3}\cos\sqrt{3}t + 2\sin\sqrt{3}t\right)$

5. $x = c_1(-0.59\cos 5.98t - 0.41\sin 5.98t)$

$+ c_2(-0.41\cos 5.98t + 0.59\sin 5.98t)$

$y = c_2\cos 5.98t + c_1\sin 5.98t$

7. $x = c_1 e^{-0.39t}(-0.31\cos 1.53t - 0.08\sin 1.53t)$

$+ c_2 e^{-0.39t}(0.08\cos 1.53t - 0.31\sin 1.53t)$

$y = -c_1 e^{-0.39t}\sin 1.53t + c_2 e^{-0.39t}\cos 1.53t$

9. $x = c_1(1+2at)e^{at} - 2c_2 a^2 te^{at}$

$y = 2c_1 te^{at} + c_2(1-2at)e^{at}$

11. $x = 0.85c_1 e^{-0.71t}\cos 2.79t + 0.85c_2 e^{-0.71t}\sin 2.79t$

$y = c_1(-0.18e^{-0.71t}\cos 2.79t - 0.49e^{-0.71t}\sin 2.79t)$

$+c_2(0.49e^{-0.71t}\cos 2.79t - 0.18e^{-0.71t}\sin 2.79t)$

13. *Exercise 1* $x = e^{-5t}$

$y = -e^{-5t}$

Exercise 3 $x = e^{-t}\cos\sqrt{3}t - \frac{\sqrt{3}}{3}e^{-t}\sin\sqrt{3}t$

$y = -e^{-t}\cos\sqrt{3}t + \frac{5\sqrt{3}}{3}e^{-t}\sin\sqrt{3}t$

Exercise 5 $x = \cos 5.98t - 0.18\sin 5.98t$

$y = -\cos 5.98t - \sin 5.98t$

Exercise 7 $x = e^{-0.39t}\cos 1.53t + 0.59e^{-0.39t}\sin 1.53t$

$y = -e^{-0.39t}\cos 1.53t + 3.48e^{-0.39t}\sin 1.53t$

Exercise 9 $x = e^{at} + 2a(1+a)te^{at}$

$y = -e^{at} + 2(1+a)te^{at}$

Exercise 11 $x = e^{-1.47t}\cos 2.79t - 15.04e^{-1.47t}\sin 2.79t$

$y = -0.30e^{-1.47t}\sin 2.79t - e^{-1.47t}\cos 2.79t$

14. *Exercise 1* $x = 2e^{-5t}$
$\qquad y = 3e^{-5t}$

Exercise 3 $x = 2e^{-t}\cos\sqrt{3}t - \frac{7}{3}e^{-t}\sqrt{3}\sin\sqrt{3}t$
$\qquad y = \frac{20}{3}e^{-t}\sqrt{3}\sin\sqrt{3}t + 3e^{-t}\cos\sqrt{3}t$

Exercise 5 $x = -3.98\sin 5.98t + 2\cos 5.98t$
$\qquad y = 3\cos 5.98t + 5.44\sin 5.98t$

Exercise 7 $x = 2e^{-0.39t}\cos 1.53t - 0.47e^{-0.39t}\sin 1.53t$
$\qquad y = 3e^{-0.39t}\cos 1.53t + 5.76e^{-0.39t}\sin 1.53t$

Exercise 9 $x = 2(1 + 2at)e^{at} - 6a^2te^{at}$
$\qquad y = 4te^{at} + 3(1 - 2at)e^{at}$

Exercise 11 $x = 5.89e^{-0.71t}\sin 2.79t + 2e^{-0.71t}\cos 2.79t$
$\qquad y = -2.42e^{-0.71t}\sin 2.79t + 3e^{-0.71t}\cos 2.79t$

15. $x = c_1e^{\lambda_1 t} + c_2e^{\lambda_2 t}$ where

$$\lambda_1 = \left(-RC + \sqrt{R^2C^2 - 4LC}\right)/2LC$$

$$\lambda_2 = \left(-RC - \sqrt{R^2C^2 - 4LC}\right)/2LC.$$

17. *Exercise 15* stable node
18. $\quad x' = -0.5x + 0.01y$
$\qquad y' = 0.5x - 0.03y$
$\qquad x(0) = x_0, \; y(0) = 0.$

19. x and y approach 0 as $t \to +\infty$ for all $x_0 > 0$. The ratio y/x approaches the same limit (approximately 49) as $t \to +\infty$ for all $x_0 > 0$.

20. $x = 0.021x_0e^{\lambda_1 t} + 0.98x_0e^{\lambda_2 t}$
$\qquad y = 1.02x_0e^{\lambda_1 t} - 1.02x_0e^{\lambda_2 t}$
where $\lambda_1 = -0.02$, $\lambda_2 = -0.51$.

21. Since $0 > \lambda_1 > \lambda_2$, $\lim_{t\to+\infty} x = \lim_{t\to+\infty} y = 0$ and

$$\lim_{t\to+\infty}\frac{y}{x} = \frac{1.02}{0.021} \approx 48.57.$$

26. $\Phi(t) = \begin{pmatrix} 2e^{3t} + \frac{41}{36}te^{3t} - e^{-3t} & e^{3t} + \frac{5}{6}te^{3t} - e^{-3t} & 0 \\ -e^{3t} - \frac{5}{6}te^{3t} + e^{-3t} & \frac{31}{6}te^{3t} - \frac{1}{6}e^{-3t} & 0 \\ \frac{1}{6}e^{3t} + \frac{71}{36}te^{3t} - \frac{1}{6}e^{-3t} & \frac{35}{36}te^{3t} - \frac{1}{6}e^{-3t} & e^{3t} \end{pmatrix}$

28. $\Phi(t) = \begin{pmatrix} -e^{5t} - e^{t} + 3e^{3t} & -3e^{5t} + 3e^{3t} & -2e^{5t} - e^{t} + 3e^{3t} \\ -e^{t} + e^{3t} & e^{3t} & -e^{t} + e^{3t} \\ e^{5t} + 2e^{t} - 3e^{3t} & 3e^{5t} - 3e^{3t} & 2e^{5t} + 2e^{t} - 3e^{3t} \end{pmatrix}$

Section 5.8

2. The maximal safe dosage [for which $x(t)$ does not exceed 30] is approximately $y_0 = 90.27$. Using this dose, the effective level $x = 37.5$ is maintained for approximately $t_l = 216.1$ minutes or about 3.5 hours. **6.** Orbits satisfy the equation $\frac{1}{2}kx^2 + \frac{1}{2}my^2 = E$ for all t. This quadratic equation is that of an ellipse in the phase plane.

7. $x = -\frac{1}{28}\sqrt{30}\sin\frac{7}{15}\sqrt{30}t$, $a = \frac{1}{28}\sqrt{30}$, $p = \frac{\pi}{7}\sqrt{30}$

9. $x = -\frac{1}{2}\cos\frac{7}{15}\sqrt{30}t$, $a = \frac{1}{2}$, $p = \frac{\pi}{7}\sqrt{30}$

11. $x = -\frac{1}{56}\sqrt{30}\sin\frac{7}{15}\sqrt{30}t + \frac{3}{4}\cos\frac{7}{15}\sqrt{30}t$, $a = \frac{1}{56}\sqrt{1794}$,
$p = \frac{\pi}{7}\sqrt{30}$

13. $k = \frac{9}{200} = 0.045$ **15.** $k = \frac{8}{81}\pi^2 \approx 0.9748$

18. (a) $x = \frac{v_0 - x_0\lambda_2}{\lambda_1 - \lambda_2}e^{\lambda_1 t} + \frac{x_0\lambda_1 - v_0}{\lambda_1 - \lambda_2}e^{\lambda_2 t}$

Chapter 6

Section 6.1

1. $x_p = -1$, $y_p = 3$ **3.** $x_p = \frac{1}{2}e^{-t}$, $y_p(t) = 2e^{-t}$
5. $x_p = \cos 2t$, $y_p(t) = -2\sin 2t$
7. $x_p = 1 + 2t + t^2$, $y_p(t) = 2 + t^2$

9. *Exercise 1* $\quad x = -1 + \frac{1}{2}\left(1 - \sqrt{2}\right)e^{\sqrt{2}t} + \frac{1}{2}\left(\sqrt{2} + 1\right)e^{-\sqrt{2}t}$
$\qquad y = 3 + \left(\sqrt{2} - \frac{3}{2}\right)e^{\sqrt{2}t} - \left(\frac{3}{2} + \sqrt{2}\right)e^{-\sqrt{2}t}$

Exercise 3
$x = \frac{1}{2}e^{-t} - \frac{1}{4}\left(1 + 2\sqrt{2}\right)e^{(\sqrt{2}-1)t} + \frac{1}{4}\left(2\sqrt{2} - 1\right)e^{-(\sqrt{2}+1)t}$
$y = 2e^{-t} - \left(1 + \frac{1}{4}\sqrt{2}\right)e^{(\sqrt{2}-1)t} + \left(\frac{1}{4}\sqrt{2} - 1\right)e^{-(\sqrt{2}+1)t}$

Exercise 5 $\quad x = \cos 2t - \cos t$
$\qquad y = -2\sin 2t + \sin t$

Exercise 7 $\quad x = \frac{5}{4}e^{-2t} - \frac{9}{4}e^{2t} + 1 + t^2 + 2t$
$\qquad y = -\frac{5}{4}e^{-2t} - \frac{3}{4}e^{2t} + 2 + t^2$

10. $x = \frac{1}{4}\left(1 - \frac{1}{3}e^{-1}\right)e^{2(t-1)} + \frac{1}{4}\left(3 - e^{-1}\right)e^{-2(t-1)} + \frac{1}{3}e^{-t}$
$\qquad y = \frac{1}{4}\left(3 - e^{-1}\right)e^{2(t-1)} + \frac{1}{4}\left(e^{-1} - 3\right)e^{-2(t-1)}$

12. $x_p = \frac{2}{5}e^{-t}\cos t + \frac{1}{5}e^{-t}\sin t$ **14.** $x_p = \frac{1}{2}te^{t}$

16. *Exercise 12* $x = \frac{1}{5}\sin t - \frac{2}{5}\cos t + \frac{2}{5}e^{-t}\cos t + \frac{1}{5}e^{-t}\sin t$

Exercise 14 $x = \frac{1}{4}e^{-t} + \frac{1}{2}te^{t} - \frac{1}{4}e^{t}$

17. $y(t)$ is negative for $t > 0$ and tends to 0 as $t \to +\infty$. The root of $x(t)$ is approximately 1.1 and does not depend on a. $x(t)$ is positive for t less than the root, negative for t greater than the root, and tends to 0 as $t \to +\infty$.

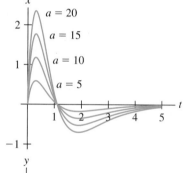

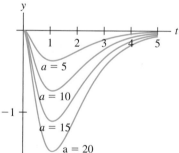

18. $\begin{cases} x = \frac{1}{2}a\left(-3e^{-3t} + 4e^{-2t} - e^{-t}\right) \\ y = -\frac{1}{2}a\left(e^{-3t} - 2e^{-2t} + e^{-t}\right) \end{cases}$

19. Both x and y tend to 0 as $t \to +\infty$. Rewrite the solution pair as $x(t) = \frac{1}{2}a\left(3e^{-t} - 1\right)\left(1 - e^{-t}\right)$, $y(t) = -\frac{1}{2}ae^{-t}\left(e^{-t} - 1\right)^2$

23. (a) $x_p = \frac{\delta}{\alpha\delta + \beta\gamma}g$, $y_p = \frac{\gamma}{\alpha\delta + \beta\gamma}g$

(b) $x = x_h + x_p = c_1(\lambda_1 + \delta)e^{\lambda_1 t} + c_2(\lambda_2 + \delta)e^{\lambda_2 t} + \frac{\delta}{\alpha\delta + \beta\gamma}g$
$\qquad y = y_h + y_p = c_1\gamma e^{\lambda_1 t} + c_2\gamma e^{\lambda_2 t} + \frac{\gamma}{\alpha\delta + \beta\gamma}g$

26. $x = 10^{-5}E_0 - \left(\frac{1}{75} \times 10^{-3}\right)E_0e^{-500t} + \left(\frac{1}{3} \times 10^{-5}\right)E_0e^{-2000t}$

28. $x = 10^{-5}E_0 - 10^{-5}E_0\cos 1000t$

30. $x = \frac{1}{2}e^{-t} + \frac{1}{2}te^{-t} - \frac{1}{2}\cos t$ **32.** $x = \frac{1}{2}t^2 e^{-t}$

Section 6.2

1. $x = c_1 e^t + 2c_2 e^{3t} + \frac{4}{3} - e^{-t}$, $y = c_1 e^t + 3c_2 e^{3t} + 1 - \frac{3}{4}e^{-t}$

2. $x = -\frac{1}{2}e^t + \frac{1}{6}e^{3t} + \frac{4}{3} - e^{-t}$, $y = -\frac{1}{2}e^t + \frac{1}{4}e^{3t} + 1 - \frac{3}{4}e^{-t}$

3. $x = \frac{7}{2}e^t - \frac{11}{6}e^{3t} + \frac{4}{3} - e^{-t}$, $y = \frac{7}{2}e^t - \frac{11}{4}e^{3t} + 1 - \frac{3}{4}e^{-t}$

7. $x = c_1 \sin t + c_2 \cos t + \frac{1}{2}t\sin t$ **8.** $x = \frac{1}{2}t\sin t$

9. $x = \cos t - \sin t + \frac{1}{2}t\sin t$

13. $x_p = \left(\dfrac{\alpha(k - \omega^2 m)}{c^2\omega^2 + (m\omega^2 - k)^2} \right) \sin \omega t + \left(-\dfrac{\alpha\omega c}{c^2\omega^2 + (m\omega^2 - k)^2} \right) \cos \omega t$

19. $q(t)$ decreases to a minimum at $t \approx 1.25$ and then increases and approaches 0.5 as $t \to +\infty$.

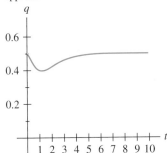

20. $q(t) = -e^{-t} + \frac{1}{2} + e^{-2t} + te^{-2t}$

21. $\lim_{t\to+\infty} q(t) = \lim_{t\to+\infty} \left(-e^{-t} + \frac{1}{2} + e^{-2t} + te^{-2t} \right) = \frac{1}{2}$. $q(t)$ has a minimum where $q'(t) = e^{-t} - e^{-2t} - 2te^{-2t} = 0$, i.e., at $t \approx 1.256431$.

Section 6.3

1. $\begin{pmatrix} x \\ y \end{pmatrix} = \begin{pmatrix} c_1 \cos t + c_2 \sin t - 1 \\ -c_1 \sin t + c_2 \cos t - 1 \end{pmatrix}$

3. $\begin{pmatrix} x \\ y \end{pmatrix} = \begin{pmatrix} \cos t + \sin t - 1 \\ -\sin t + \cos t - 1 \end{pmatrix}$

8. $x = c_1 \sin t + c_2 \cos t - (\cos t) \ln \left(\frac{1 + \sin t}{\cos t} \right)$

9. $x = \sin t - (\cos t) \ln \left(\frac{1 + \sin t}{\cos t} \right)$

15. $\widetilde{x}(t) = \begin{pmatrix} c_1 e^{-t} - c_2 e^t + 2c_3 e^{2t} - 3 \\ c_1 e^{-t} + 2c_2 e^t - 4 \\ c_1 e^{-t} + c_2 e^t + c_3 e^{2t} - 4 \end{pmatrix}$

Section 6.4

1. $x = c_1 e^{-\frac{1}{2}t} + c_2 e^{\frac{1}{2}t} - 7$
$y = c_1 e^{-\frac{1}{2}t} + 2c_2 e^{\frac{1}{2}t} - 10$

3. $x = c_1 (3\cos t - \sin t) + c_2 (\cos t + 3\sin t) - 1$
$y = -2c_1 \cos t - 2c_2 \sin t + 1$

5. $x = 4c_1 e^{-t} + 2c_2 e^t + 5r$
$y = -3c_1 e^{-t} - c_2 e^t - 3r$

7. *Exercise 1* $x = 7e^{-\frac{1}{2}t} + e^{\frac{1}{2}t} - 7$
$y = 7e^{-\frac{1}{2}t} + 2e^{\frac{1}{2}t} - 10$

Exercise 3 $x = 2\cos t - 4\sin t - 1$
$y = 2\sin t - 2\cos t + 1$

Exercise 5 $x = 2(1 - r)e^{-t} - (3r + 1)e^t + 5r$
$y = \frac{3}{2}(r - 1)e^{-t} + \frac{1}{2}(3r + 1)e^t - 3r$

11. When $b = c = 1$ the population with smaller initial size drops to 0 in finite time. The only possibility for coexistence is $x_0 = y_0$. When

$b = 4$, $c = 1$ population x disappears if $x_0 < 2y_0$ and population y disappears if $x_0 > 2y_0$. Both disappear as $t \to +\infty$ if $x_0 = 2y_0$.

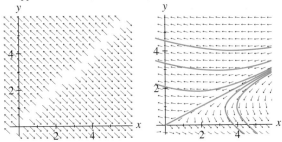

12. Let $\lambda_1 = 1 + \sqrt{bc}$ and $\lambda_2 = 1 - \sqrt{bc}$.

$$x = \frac{1}{2}\left(e^{\lambda_1 t} + e^{\lambda_2 t} \right) x_0 + \frac{1}{2}\sqrt{\frac{b}{c}} \left(-e^{\lambda_1 t} + e^{\lambda_2 t} \right) y_0$$
$$y = \frac{1}{2}\sqrt{\frac{c}{b}} \left(-e^{\lambda_1 t} + e^{\lambda_2 t} \right) x_0 + \frac{1}{2}\left(e^{\lambda_1 t} + e^{\lambda_2 t} \right) y_0$$

13. Rewrite the solution as

$$x = \frac{1}{2}\left[\left(x_0 - \sqrt{\frac{b}{c}}y_0 \right) + \left(x_0 + \sqrt{\frac{b}{c}}y_0 \right) e^{-2\sqrt{bc}t} \right] e^{(1+\sqrt{bc})t}$$
$$y = \frac{1}{2}\sqrt{\frac{c}{b}}\left[-\left(x_0 - \sqrt{\frac{b}{c}}y_0 \right) + \left(x_0 + \sqrt{\frac{b}{c}}y_0 \right) e^{-2\sqrt{bc}t} \right] e^{(1+\sqrt{bc})t}.$$

If $x_0 > \sqrt{\frac{b}{c}}y_0$, then $x > 0$ for all t, but y vanishes at

$$t = \frac{1}{2\sqrt{bc}} \ln \left(\frac{x_0 + \sqrt{\frac{b}{c}}y_0}{x_0 - \sqrt{\frac{b}{c}}y_0} \right).$$

If $x_0 < \sqrt{\frac{b}{c}}y_0$, then $y > 0$ for all t, but x vanishes at

$$t = \frac{1}{2\sqrt{bc}} \ln \left(\frac{x_0 + \sqrt{\frac{b}{c}}y_0}{-x_0 + \sqrt{\frac{b}{c}}y_0} \right).$$

17. $x = c_1 e^{-t}\cos t + c_2 e^{-t}\sin t + \frac{1}{5}\cos t + \frac{2}{5}\sin t$

19. $x = c_1 e^{-t}\cos t + c_2 e^{-t}\sin t + \frac{2}{5}\cos t + \frac{4}{5}\sin t + \frac{1}{2}te^{-t}\cos t$

21. If $a \neq 1$, or -5 then

$$x = c_1 e^{-t} + c_2 e^{-5t} + (a^2 + 6a + 5)^{-1}e^{at}.$$

If $a = 1$, then

$$x = c_1 e^{-5t} + c_2 e^{-t} - \frac{1}{16}e^{-t} + \frac{1}{4}te^{-t}.$$

If $a = -5$, then

$$x = c_1 e^{-5t} + c_2 e^{-t} - \frac{1}{4}te^{-5t} - \frac{1}{16}e^{-5t}.$$

23. $x = c_1 e^{-2t} + c_2 te^{-2t} - \frac{1}{4} + \frac{1}{4}t$

25. $x = c_1 e^{-2t} + c_2 te^{-2t} + \frac{3}{4}t - \frac{3}{4} - e^{-t} + t^2 e^{-2t}$

27. If $k \neq 1$, then

$$x = c_1 \cos kt + c_2 \sin kt + \frac{1}{k^2 - 1}\sin t.$$

If $k = 1$, then

$$x = c_1 \cos t + c_2 \sin t - \frac{1}{2}t\cos t.$$

28. x_1 vanishes in approximately $t_1 = 11$ minutes. x_2 vanishes in approximately $t_2 = 9$ minutes. x_3 vanishes in approximately $t_3 = 10.5$ minutes.

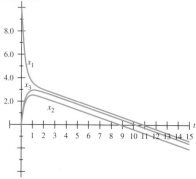

29. $x_1 = \frac{62}{9}e^{-3t} - \frac{1}{2}e^{-2t} + \frac{65}{18} - \frac{1}{3}t$

$x_2 = -\frac{31}{9}e^{-3t} + \frac{1}{2}e^{-2t} + \frac{53}{18} - \frac{1}{3}t$

$x_3 = -\frac{31}{9}e^{-3t} + \frac{31}{9} - \frac{1}{3}t$

30. Use a computer to find the roots t_i of each component: $t_1 \approx 10.8333$, $t_2 \approx 8.8333$, $t_3 \approx 10.3333$

Section 6.5

1. For 4 liters, the effective and lethal amounts (in the bloodstream) are 1.5 mg/liter×4 liter = 6 mg and 6 mg/liter×4 liter = 24 mg, respectively.

n Intravenous Delivery Rate (mg/min)	t_e Time (min) until Effective Amount in Bloodstream	t_l Time (min) until Lethal Amount in Bloodstream	x_e Maximal Amount (mg) in Bloodstream
0.1	∞	∞	4.17
0.2	147	∞	8.34
0.4	38.3	∞	16.7
0.6	18.2	401	25.0
0.8	11.4	147	33.4
1.0	8.25	92.3	41.7

For 6 liters, the effective and lethal amounts (in the bloodstream) are 1.5 mg/liter×6 liter = 9 mg and 6 mg/liter×6 liter = 36 mg, respectively.

n Intravenous Delivery Rate (mg/min)	t_e Time (min) until Effective Amount in Bloodstream	t_l Time (min) until Lethal Amount in Bloodstream	x_e Maximal Amount (mg) in Bloodstream
0.1	∞	∞	4.17
0.2	∞	∞	8.34
0.4	81.6	∞	16.7
0.6	38.3	∞	25.0
0.8	22.4	∞	33.4
1.0	15.2	242	41.7

For lower blood volumes the critical values n_1 and n_2 are lower. For higher blood volumes these critical values are higher. That is, these critical values seem to be increasing functions of the blood capacity b. The critical value n_1 is that value of n for which $\lim_{t \to +\infty} x(t) = 1.5b$ (i.e., $41.7n_1 = 1.5b$ or $n_1 = \frac{1.5}{41.7}b$). The critical value n_2 is that value of n for which $\lim_{t \to +\infty} x(t) = 6b$ (i.e., $41.7n_2 = 6b$ or $n_2 = \frac{6}{41.7}b$). These formulas show n_1 and n_2 are increasing functions of b.

10. $x = x_0 \cos \beta t + \left(\frac{v_0}{\beta} - \frac{\gamma_e \beta_e}{m(\beta^2 - \beta_e^2)\beta} \right) \sin \beta t + \frac{\gamma_e}{m(\beta^2 - \beta_e^2)} \sin \beta_e t$

12. When $\beta_e - \beta \approx 0$ we can view the solution formula (5.20) as the sinusoidal oscillation

$$\sin \left(\frac{\beta + \beta_e}{2} t \right) \approx \sin \beta t$$

multiplied by an amplitude

$$\frac{2\gamma_e}{m(\beta^2 - \beta_e^2)} \sin \left(\frac{\beta - \beta_e}{2} t \right)$$

which oscillates slowly [because its frequency $(\beta - \beta_e)/2$ is nearly equal to 0] and with large amplitude (because the denominator $\beta^2 - \beta_e^2$ is nearly equal to 0). **18.** The solution pair $x_1(t)$, $x_2(t)$ is eventually periodic with a period of approximately 6.25. **19.** The characteristic roots (eigenvalues) of the coefficient matrix

$$A = \begin{pmatrix} 0 & 1 & 0 & 0 \\ -2 & -1 & 1 & 0 \\ 0 & 0 & 0 & 1 \\ 1 & 0 & -1 & -1 \end{pmatrix}$$

of the equivalent first-order system all have negative real part equal to $-1/2$. Therefore, $\tilde{x}_h(t) \to 0$ as $t \to +\infty$.

20. $x_1(t) = -\frac{2}{5}\sin t - \frac{1}{5}\cos t$, $x_2(t) = -\frac{1}{5}\sin t - \frac{3}{5}\cos t$

Chapter 7

Section 7.1

1. $x_1 = 1 - t$, $y_1 = 1 + t$
$x_2 = 1 - t$, $y_2 = 1 + t - \frac{1}{2}t^2$
$x_3 = 1 - t - \frac{1}{6}t^3$, $y_3 = 1 + t - \frac{1}{2}t^2 + \frac{1}{3}t^3$
3. $x_1 = -2 - 2t$, $y_1 = 3t$
$x_2 = -2 - 2t + \frac{1}{2}t^2$, $y_2 = 3t + \frac{5}{2}t^2$
$x_3 = -2 - 2t + \frac{1}{2}t^2$, $y_3 = 3t + \frac{5}{2}t^2 + \frac{1}{2}t^3$
5. $x_1 = 1$, $x_2 = 1 - \frac{1}{2}t^2$, $x_3 = 1 - \frac{1}{2}t^2 + \frac{1}{6}t^3$
7. $x_1 = 1 - t$, $x_2 = 1 - t - \frac{1}{2}t^2$, $x_3 = 1 - t - \frac{1}{2}t^2 + \frac{1}{2}t^3$
9. $x_1 = 0$, $y_1 = 0$; $x_2 = 0$, $y_2 = \frac{1}{2}(t-1)^2$
$x_3 = -\frac{1}{3}(t-1)^3$, $y_3 = \frac{1}{2}(t-1)^2 - \frac{1}{3}(t-1)^3$
11. $x_1 = 1$, $x_2 = 1 - (t+1)^2$, $x_3 = 1 - (t+1)^2$
13. $x_3 = c_1\left(1 + t - \frac{1}{3}t^3\right) + c_2\left(-t - \frac{1}{2}t^2 + \frac{1}{6}t^3\right)$
$y_3 = c_1\left(t + \frac{1}{2}t^2 - \frac{1}{3}t^3\right) + c_2\left(1 - t^2 - \frac{1}{6}t^3\right)$
15. $x_3 = c_1\left(1 - \frac{1}{2}t^2 - \frac{1}{6}t^3\right) + c_2\left(t - \frac{1}{6}t^3\right)$
17. $x_0 = 1$, $y_0 = -1$
$x_1 = 1 + t + \frac{1}{2}t^2 + \frac{1}{3}t^3$, $y_1 = -1 + t - \frac{1}{2}t^2$
$x_2 = 1 + t + \frac{1}{2}t^3 + \frac{1}{4}t^4 + \frac{1}{10}t^5 + \frac{1}{18}t^6$, $y_2 = -1 + t + t^2 + \frac{1}{6}t^3 + \frac{1}{6}t^4$
19. $x_0 = 0$, $y_0 = 0$
$x_1 = -1 - \cos t$, $y_1 = \sin t$
$x_2 = \frac{1}{10}e^{2\pi} - \frac{1}{2}e^{2t} - \frac{2}{5}e^{2t}\cos t - \frac{1}{5}e^{2t}\sin t$,
$y_2 = -\pi - 1 + t + 2\sin t - \cos t$
21. $x_0 = 1$
$x_1 = 1 - t - \frac{1}{2}t^2 + \sin t$
$x_2 = \frac{25}{8} - 3t - \frac{3}{4}t^2 + \frac{1}{6}t^3 + \frac{1}{24}t^4$
$+\left(5 - t - \frac{1}{2}t^2\right)\sin t - 2(1+t)\cos t - \frac{1}{8}\cos 2t$
23. $x_0 = 0$, $y_0 = 1$
$x_1 = t$, $y_1 = 1 + \alpha t$
$x_2 = t + \frac{1}{2}\alpha t^2$, $y_3 = 1 + \alpha t + \frac{1}{2}\left(\alpha^2 - 1\right)t^2 - \frac{1}{3}\alpha t^3 - \frac{1}{4}\alpha^2 t^4$
25. $x_0 = 0$, $x_1 = t$, $x_2 = \frac{a+\pi}{\pi}t - \frac{a}{\pi^2}\sin \pi t$

Section 7.2

1. $x_1(t) = e^t + (e^t - t - 1)\varepsilon$
$y_1(t) = -te^t + (-te^t + 2e^t - t - 2)\varepsilon$
3. $x_1(t) = \frac{1}{3}e^t - \frac{1}{3}e^{-2t} + \left(\frac{1}{6} - \frac{1}{12}e^t - \frac{1}{6}e^{-2t} + \frac{1}{12}e^{-3t}\right)\varepsilon$
$y_1(t) = e^t + \left(\frac{1}{3} - \frac{1}{4}e^t - \frac{1}{12}e^{-3t}\right)\varepsilon$
5. $x_1(t) = e^{-2t} + \left(\frac{1}{10} + \frac{1}{2}e^{-2t} - \frac{3}{5}e^{-2t}\cos t - \frac{1}{5}e^{-2t}\sin t\right)\varepsilon$
$y_1(t) = -e^{-2t} + \left(\frac{1}{10} - \frac{1}{2}e^{-2t} + \frac{2}{5}e^{-2t}\cos t - \frac{1}{5}e^{-2t}\sin t\right)\varepsilon$
7. $x_1(t) = -\frac{1}{2} + \frac{1}{2}t^2 + \left(-4e^{-1} + \frac{5}{3}e^{-t} + te^{-t} + t^2 e^{-t} + \frac{1}{3}t^3 e^{-t}\right)\varepsilon$
$y_1(t) = \frac{1}{6}\left(2 - 3t + t^3\right)e^{-t}\varepsilon$
9. $x_1(t) = 2e^{-t} - e^{-2t} + \left(e^{-t} - 2te^{-2t} - e^{-3t}\right)\varepsilon$
11. $x_1(t) = \frac{3}{5}e^{-3t} + \frac{2}{5}e^{2t} + \left(-3e^{-3t} - 2e^{2t} + 6 - e^{5t}\right)\varepsilon$
13. $x_1(t) = \cos t + \frac{1}{4}\left(\sin t - t\cos t - t^2\sin t\right)\varepsilon$

Section 7.3

1. (a) $k_{i+2} = \frac{2}{i+2}k_{i+1} - \frac{2}{(i+2)(i+1)}k_i$, $i = 0, 1, 2, 3, \ldots$
(b) $x_6 = t + t^2 + \frac{1}{3}t^3 - \frac{1}{30}t^5 - \frac{1}{90}t^6$ **(c)** $x(t) = e^t\sin t$
3. (a) $k_{i+2} = \frac{3}{i+2}k_{i+1} - \frac{2}{(i+2)(i+1)}k_i$
(b) $x_6 = 1 - t^2 - t^3 - \frac{7}{12}t^4 - \frac{1}{4}t^5 - \frac{31}{360}t^6$ **(c)** $x(t) = 2e^t - e^{2t}$
5. (a) $k_{i+2} = \frac{2i-6}{(i+2)(i+1)}k_i$, $i = 0, 1, 2, 3, \ldots$ **(b)** $x_4(t) = t - \frac{2}{3}t^3$
7. (a) $k_0 = k_1 = 0$, $k_2 = \frac{1}{2}$,

$$k_{i+2} = -\frac{i}{i+2}k_{i+1} - \frac{1}{(i+2)(i+1)}k_{i-1} + \frac{1}{i!}\frac{1}{(i+2)(i+1)}$$

for $i = 1, 2, 3, \ldots$
(b) $x_4(t) = \frac{1}{2}t^2 + \frac{1}{24}t^4$
9. $k_2 = \frac{1}{2}k_1$, $k_{i+2} = \frac{1}{i+2}k_{i+1} - \frac{2}{(i+2)(i+1)}k_{i-1}$, for $i = 1, 2, 3, \ldots$
$x = k_0\left(1 - \frac{1}{3}t^3 - \frac{1}{12}t^4 \cdots\right) + k_1\left(t + \frac{1}{2}t^2 + \frac{1}{4}t^3 + \cdots\right)$
11. $k_2 = -\frac{1}{2}k_0$, $k_{i+2} = -\frac{i(i-1)+1}{(i+2)(i+1)}k_i$ for $i = 1, 2, 3, \ldots$
$x = k_0\left(1 - \frac{1}{2}t^2 + \frac{1}{8}t^4 + \cdots\right) + k_1\left(t - \frac{1}{6}t^3 + \frac{7}{120}t^5 + \cdots\right)$
14. The coefficients $p(t) = -\frac{t}{1-t^2}$ and $q(t) = \frac{\alpha^2}{1-t^2}$ are rational functions analytic at $t_0 = 0$ with radii of converenge equal to 1. Theorem 1 implies that all solutions are analytic at $t_0 = 0$ and their radii of convergence $R \geq 1$.

Section 7.4

1. Taylor: $x_4(t) = 1 - \frac{1}{2}t^2 - \frac{1}{3}t^3 + \frac{1}{12}t^4$
Picard: $x_0(t) = 1$, $x_1(t) = 1 - \frac{1}{2}t - \frac{1}{2}t^2 + \frac{1}{4}\sin 2t$, and
$x_3(t) = \frac{145}{128} - \frac{3}{4}t - \frac{9}{16}t^2 + \frac{1}{12}t^3 + \frac{1}{24}t^4 + \frac{1}{8}\sin 2t - \frac{1}{8}\cos 2t - \frac{1}{4}t\cos 2t - \frac{1}{8}t\sin 2t - \frac{1}{128}\cos 4t - \frac{1}{8}t^2\sin 2t$
3. Taylor: $x_3(t) = 1 - \frac{1}{6}t^3 + \frac{1}{24}t^4$, $y_3(t) = 0$
Picard: $x_0(t) = 1$, $y_0(t) = 1$
$x_1(t) = 1 - t + \sin t$, $y_1(t) = 1$
$x_2(t) = \frac{1}{4} - \frac{3}{2}t + t^2 - \frac{1}{3}t^3 + 3\sin t + \cos t - t\sin t - 2t\cos t + \frac{1}{4}\sin 2t - \frac{1}{4}\cos 2t$, $y_2(t) = 2 - \frac{1}{2}t^2 - \cos t$
5. $x_1(t) = \cos t + \left(-\frac{1}{2}\sin t + \frac{1}{4}t\cos t + \frac{1}{4}\cos^2 t\sin t\right)\varepsilon$
7. $x_1(t) = 1 + (-1 + \sin t + e^{-t})\varepsilon$, $y_2(t) = e^{-t} + (te^{-t})\varepsilon$
9. $x_1(t) = \frac{\sqrt{7}}{7}e^{t/2}\sin\left(\frac{\sqrt{7}}{2}t\right) + e^{t/2}\cos\left(\frac{\sqrt{7}}{2}t\right) +$
$\left(-\frac{4}{7}e^t - \frac{5\sqrt{7}}{14}e^{t/2}\sin\left(\frac{\sqrt{7}}{2}t\right) + \frac{1}{2}e^{t/2}\cos\left(\frac{\sqrt{7}}{2}t\right)\right.$
$\left. + \frac{1}{14}e^t\cos\left(\sqrt{7}t\right) + \frac{3\sqrt{7}}{14}e^t\sin\left(\sqrt{7}t\right)\right)\varepsilon$
$y_1(t) = \frac{4\sqrt{7}}{7}e^{t/2}\sin\left(\frac{\sqrt{7}}{2}t\right)$
$+\left(-\frac{8}{7}e^t + \frac{\sqrt{7}}{14}e^{t/2}\sin\left(\frac{\sqrt{7}}{2}t\right) + \frac{3}{2}e^{t/2}\cos\left(\frac{\sqrt{7}}{2}t\right)\right.$
$\left. - \frac{5}{14}e^t\cos\left(\sqrt{7}t\right) + \frac{\sqrt{7}}{14}e^t\sin\left(\sqrt{7}t\right)\right)\varepsilon$
10. $k_1 = m_0$, $m_1 = k_0$
$k_{i+1} = \frac{1}{i+1}\left(-k_{i-1} + m_i\right)$, $m_{i+1} = \frac{1}{i+1}\left(k_i - m_{i-1}\right)$ for $i = 1, 2, 3, \ldots$
$x = k_0\left(1 - \frac{1}{12}t^4 + \cdots\right) + m_0\left(t - \frac{1}{3}t^3 + \cdots\right)$
$y = k_0\left(t - \frac{1}{3}t^3 + \cdots\right) + m_0\left(1 - \frac{1}{12}t^4 + \cdots\right)$
12. The coefficients $p(t) = -2t$ and $q(t) = \alpha$ are polynomials and hence have radii of convergence $R = \infty$. Theorem 1 implies all solutions have radius of convergence $R = \infty$.

Section 7.5

1. $t \approx 0.1676$ year (or 61.17 days), which is slightly longer than the estimates obtained from the Taylor polynomial and Picard iterate approximations.   **4. (a)** $x_3 = t - \frac{3}{2}t^2 + \frac{1}{6}t^3$, $y_3 = t^2 - \frac{3}{8}t^3$

(b) $t \approx 0.1217$ (year). **(c)** $t = 0.25$ (year). **6. (a)** y first equals 0.01 at $t \approx 0.1162$ (year). **(b)** y attains its peak at $t \approx 1.20$ (year). **(c)** The answer in (a) is slightly sooner than that estimated by the Taylor polynomial and Picard iterate approximations. The answer in (b) is considerably later than that estimated by the Taylor polynomial and Picard iterate approximations. **9. (a)** $x_3(t) = t - \frac{1}{2}t^2 + \frac{\varepsilon}{3}t^3$ **(b)** The smallest root of $x_3'(t) = 1 - t + \varepsilon t^2$ is $t_m = \frac{1}{2\varepsilon}\left(1 - \sqrt{1 - 4\varepsilon}\right)$. **(c)** $x_3(t_m) = \frac{1}{2\varepsilon} - \frac{1}{12\varepsilon^2} + \frac{1}{12\varepsilon^2}\sqrt{(1 - 4\varepsilon)} - \frac{1}{3\varepsilon}\sqrt{(1 - 4\varepsilon)}$ **(d)** The approximate time is 25.5. The approximate height is 3191 meters.

Chapter 8

Section 8.2

1. (a) Since the differential equation $x'(t) = f(x(t), y(t))$ holds for all t, we can substitute $t + \tau$ for t and obtain $x'(t + \tau) = f(x(t + \tau), y(t + \tau))$. This shows that the pair $x(t + \tau)$, $y(t + \tau)$ satisfies the first equation of the system. A similar argument shows the pair also satisfies the second equation $y' = g(x, y)$. **(b)** The orbit is the set of points $\{(x(t), y(t))|$ for all $t\}$ in the (x, y)-plane. This set of points is identical to the set $\{(x(t + \tau), y(t + \tau))|$ for all $t\}$ for any τ. **3.** $(x_e, y_e) = (0, 0)$ and $(1, 1)$ **5.** $(x_e, y_e) = \left(\frac{3}{2}, \frac{1}{2}\sqrt{7}\right)$ and $\left(\frac{3}{2}, -\frac{1}{2}\sqrt{7}\right)$ **7.** $(x_e, y_e) = (2, 0)$ **9.** $x_e = n\pi$ for $n = 0, \pm1, \pm2, \pm3, \ldots$ **11.** $x_e = 0$ **13.** $(x_e, y_e) \approx (0.5671, 0.5671)$ **15.** $(x_e, y_e) = (0, 0)$, $(x_e, y_e) \approx (0.4452, -0.3339)$, and $(x_e, y_e) \approx (5.4848, -4.1136)$. **17.** $x_e \approx 0.3574$ and 2.1533 **19.** There are no equilibria if $r < 1$ and there are exactly two equilibria if $r > 1$. When $r = 1$, there is exactly one equilibrium. **21.** There are three equilibria if $r < 2$ and one equilibrium if $r > 2$. When $r = 2$ there are exactly two equilibria. **23.** If $r > e^{-1}$ there are no equilibria. If $r < e^{-1}$ there are two equilibria. If $r = e^{-1}$ there is exactly one equilibrium.

Section 8.3

1. $J(0, 0) = \begin{pmatrix} 1 & 0 \\ 1 & -1 \end{pmatrix}$, $J(1, 1) = \begin{pmatrix} 1 & -2 \\ 1 & -1 \end{pmatrix}$.

3. $J(0, 0) = \begin{pmatrix} 1 & 0 \\ 0 & 2 \end{pmatrix}$, $J\left(0, \frac{1}{2}\right) = \begin{pmatrix} \frac{1}{2} & 0 \\ -\frac{1}{2} & -2 \end{pmatrix}$,

$J(1, 0) = \begin{pmatrix} -1 & -1 \\ 0 & 1 \end{pmatrix}$, $J\left(\frac{2}{3}, \frac{1}{3}\right) = \begin{pmatrix} -\frac{2}{3} & -\frac{2}{3} \\ -\frac{1}{3} & -\frac{4}{3} \end{pmatrix}$.

5. $J\left(\frac{\sqrt{2}}{2}, \frac{\sqrt{2}}{2}\right) = \begin{pmatrix} -\sqrt{2} & -\sqrt{2} \\ 1 & -1 \end{pmatrix}$, $J\left(-\frac{\sqrt{2}}{2}, -\frac{\sqrt{2}}{2}\right) = \begin{pmatrix} \sqrt{2} & \sqrt{2} \\ 1 & -1 \end{pmatrix}$.

7. *Exercise 1* At $(x, y) = (0, 0)$ the linearization is
$$x' = x$$
$$y' = x - y.$$
At $(x, y) = (1, 1)$ the linearization is
$$x' = x - 2y$$
$$y' = x - y.$$
Exercise 3 At $(x, y) = (0, 0)$ the linearization is
$$x' = x$$
$$y' = 2y.$$
At $(x, y) = \left(0, \frac{1}{2}\right)$ the linearization is
$$x' = \frac{1}{2}x$$
$$y' = -\frac{1}{2}x - 2y.$$

At $(x, y) = (1, 0)$ the linearization is
$$x' = -x - y$$
$$y' = y.$$
At $(x, y) = \left(\frac{2}{3}, \frac{1}{3}\right)$ the linearization is
$$x' = -\frac{2}{3}x - \frac{2}{3}y$$
$$y' = -\frac{1}{3}x - \frac{4}{3}y.$$

Exercise 5 At $(x, y) = \left(\frac{\sqrt{2}}{2}, \frac{\sqrt{2}}{2}\right)$ the linearization is
$$x' = -\sqrt{2}x - \sqrt{2}y$$
$$y' = x - y.$$
At $(x, y) = \left(-\frac{\sqrt{2}}{2}, -\frac{\sqrt{2}}{2}\right)$ the linearization is
$$x' = \sqrt{2}x + \sqrt{2}y$$
$$y' = x - y.$$

8. *Exercise 1* $(0, 0)$ is unstable. Neither Theorem 1 nor Theorem 2 applies to the equilibrium $(1, 1)$.
Exercise 3 $(0, 0)$ is unstable, $\left(0, \frac{1}{2}\right)$ is unstable, $(1, 0)$ is unstable, and $\left(\frac{2}{3}, \frac{1}{3}\right)$ is stable.
Exercise 5 $\left(\frac{\sqrt{2}}{2}, \frac{\sqrt{2}}{2}\right)$ is stable. $\left(-\frac{\sqrt{2}}{2}, -\frac{\sqrt{2}}{2}\right)$ is unstable.
9. There are three equilibrium points: $(x_e, y_e) = (0, 0)$, $\left(\frac{3}{2}, 0\right)$ and $\left(\frac{1}{4}, \frac{5}{8}\right)$.
10. The linearization at $(0, 0)$ is
$$x' = \frac{3}{2}x$$
$$y' = -\frac{1}{4}y$$

The linearization at $\left(\frac{3}{2}, 0\right)$ is
$$x' = -\frac{3}{2}x - 3y$$
$$y' = \frac{5}{4}y$$

The linearization at $\left(\frac{1}{4}, \frac{5}{8}\right)$ is
$$x' = -\frac{1}{4}x - \frac{1}{2}y$$
$$y' = \frac{5}{8}x$$

11. $(0, 0)$ is unstable, $\left(\frac{3}{2}, 0\right)$ is unstable, and $\left(\frac{1}{4}, \frac{5}{8}\right)$ is stable

Section 8.4

1. $(x_e, y_e) = (0, 0)$ is a saddle and $(1, 1)$ is nonhyperbolic. **3.** $(x_e, y_e) = (0, 0)$ is an unstable node, $\left(0, \frac{1}{2}\right)$ is a saddle, $(1, 0)$ is a saddle, and $\left(\frac{2}{3}, \frac{1}{3}\right)$ is a stable node. **5.** $(x_e, y_e) = \left(\frac{\sqrt{2}}{2}, \frac{\sqrt{2}}{2}\right)$ is a stable spiral and $\left(-\frac{\sqrt{2}}{2}, -\frac{\sqrt{2}}{2}\right)$ is a saddle.   **13.** $(x_e, y_e) = (x_{in}, 0)$ and

$(1, x_{in} - 1)$ **14. (a)** $J(x_{in}, 0) = \begin{pmatrix} -d & -\frac{1}{\gamma} \frac{mx_{in}}{a+x_{in}} \\ 0 & \frac{mx_{in}}{a+x_{in}} - d \end{pmatrix}$

$J\left(\frac{ad}{m-d}, \left(x_{in} - \frac{ad}{m-d}\right)\gamma\right) = \begin{pmatrix} -d - ((d-m)x_{in} + da)\frac{d-m}{am} & -\frac{d}{\gamma} \\ \gamma((d-m)x_{in} + da)\frac{d-m}{am} & 0 \end{pmatrix}$.

(b) $(x_e, y_e) = (x_{in}, 0)$ is

$$\begin{cases} \text{a saddle if } \frac{mx_{in}}{a+x_{in}} > d \\ \text{a stable node if } \frac{mx_{in}}{a+x_{in}} < d \\ \text{nonhyperbolic if } \frac{mx_{in}}{a+x_{in}} = d \end{cases}$$

$(x_e, y_e) = \left(\frac{ad}{m-d}, \left(x_{in} - \frac{ad}{m-d}\right)\gamma\right)$ is a stable node.

Section 8.5
1. All orbits tend to $(0, 0)$ as $t \to +\infty$.

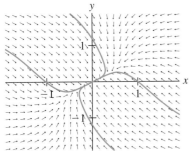

3. Each nonequilibrium orbit approaches a limit cycle.

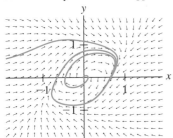

5. All nonequilibrium orbits tend to $(1, 0)$ as $t \to +\infty$.

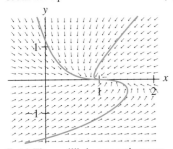

7. (a) The equilibrium equations are

$$x + y - x\left(x^2 + y^2\right) = 0$$
$$-x + y - y\left(x^2 + y^2\right) = 0.$$

If $y = 0$, the second equation implies $x = 0$. If $x = 0$ the first equation implies $y = 0$. If both x and y are nonzero, multiply the first equation

by y and the second by $-x$ and add the results to obtain $x^2 + y^2 = 0$, a contradiction. This shows there is no equilibrium other than $(0, 0)$.
(b) Orbits starting near the $(0, 0)$ spiral outward and orbits starting far from $(0, 0)$ spiral inward. All approach a limit cycle (which appears to be a circle of radius one centered at $(0, 0)$).

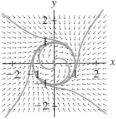

(c) The Jacobian at $(0, 0)$ is

$$J = \begin{pmatrix} 1 & 1 \\ -1 & 1 \end{pmatrix}$$

whose characteristic polynomial $\lambda^2 - 2\lambda + 2$ has complex conjugate roots $1 \pm i$ with positive real part 1. Thus, the phase portrait near $(0, 0)$ is an unstable spiral. **(d)** Differentiate $r^2 = x^2 + y^2$ with respect to t to obtain $2rr' = 2xx' + 2yy'$ and substitute for x' and y'. The phase line portrait of $r' = (1 - r^2)r$ is $0 \longrightarrow 1 \longleftarrow$. **(e)** Substitute x' and y' into the equation

$$\theta' = \frac{1}{1 + \left(\frac{y}{x}\right)^2} \frac{xy' - yx'}{x^2}.$$

(f) $\theta' = -1$ implies orbits spiral around the origin clockwise. The phase linear portrait for r implies $r \to 1$ as $t \to +\infty$ and hence all nonequilibrium orbits spiral into the circle $r = 1$ as $t \to +\infty$. Orbits starting on the unit circle remain there (i.e., the unit circle is a limit cycle).   **9. (a)** For any initial point $(x_0, 0)$ on the x-axis, the solution is $(x(t), 0)$ where $x(t)$ solves the equation $x' = (1 - x)x$. The phase line portrait of this equation is $\longleftarrow 0 \longrightarrow 1 \longleftarrow$, which therefore appears on the x-axis in the phase plane portrait. For any initial point $(0, y_0)$ on the y-axis, the solution is $(0, y(t))$ where $y(t)$ solves the equation $y' = (2 - y)y$. The phase line portrait of this equation is $\longleftarrow 0 \longrightarrow 2 \longleftarrow$, which therefore appears on the y-axis in the phase plane portrait. **(b)** Orbits cannot cross. Therefore, orbits cannot cross the x- or the y-axis and consequently must remain in the first quadrant. **(c)** A cycle must enclose an equilibrium and there are no equilibria in the first quadrant. **(d)** $(0, 0)$ is an unstable node and therefore cannot be in the set S of the bounded orbit. $(1, 0)$ is a saddle and therefore is approached only by the orbits on the x-axis. By the Poincaré-Bendixson theorem the only equilibrium in S is $(0, 2)$. This equilibrium is a stable node and thus the orbit approaches it as $t \to +\infty$.   **12.** $xf(x, y) + yg(x, y) = -x^2 - 2y^2 - 2x^2y^2 < 0$ for all $r > r_0 = 0$.
14. $xf(x, y) + yg(x, y) = -\left(x^2 + y^2 - 1\right)\left(x^2 + y^2\right) < 0$ for all $r > r_0 = 1$. **16.** Take D to be the first quadrant: $x > 0$, $y > 0$. In D, a calculation shows

$$\frac{\partial}{\partial x}(\mu f) + \frac{\partial}{\partial y}(\mu g) = -\frac{1}{y} - \frac{1}{x} < 0.$$

18. $\frac{\partial f}{\partial x} + \frac{\partial g}{\partial y} = 4 - y^2 - x^2 > 0$ if $x^2 + y^2 < 4$.   **20.** With $\mu = 1$ calculate $\frac{\partial}{\partial x}(\mu f) + \frac{\partial}{\partial y}(\mu g) = -3 - y^2 - x^2 < 0$ for all (x, y).
22. With $\mu = y$ calculate $\frac{\partial}{\partial x}(\mu f) + \frac{\partial}{\partial y}(\mu g) = 1 + x^2 + 6x^2y^2 > 0$ for all (x, y).   **24.** With $\mu = 1$ calculate $\frac{\partial}{\partial x}(f) + \frac{\partial}{\partial y}(g) = 2 + 4x^2 > 0$ for all (x, y).

Section 8.6

1. For $p < 0$, $(0, 0)$ is the only equilibrium and it is a stable node. For $p > 0$ there are three equilibria, $(x_e, y_e) = (0, 0)$, $(\pm\sqrt{p}, 0)$. This is characteristic of a pitchfork bifurcation. The equilibrium $(0, 0)$ is a saddle for $p > 0$ while both equilibria $(\pm\sqrt{p}, 0)$ are stable nodes. **3.** A saddle node bifurcation occurs at $p_0 = \frac{1}{2}$. **5.** A pitchfork bifurcation occurs at $p_0 = 1$. **7.** A transcritical bifurcation occurs at $p_0 = 0$. **9.** A saddle node bifurcation occurs at p_0 where p_0 is the value of p at which the straight line px and $\ln x$ are tangent. **14.** The Hopf bifurcation criteria hold at $p_0 = 0$. For $p < 0$, the origin is a stable spiral. For $p > 0$, the origin is an unstable spiral. A computer sketch of the phase portrait indicate a stable limit cycle exists for small values of $p > 0$. **16.** The Hopf bifurcation criteria hold at $p_0 = 4$. For $p < 4$, the origin is a stable spiral and computer sketches of the phase plane portrait show that there is also an unstable limit cycle encircling $(0, 0)$. There is also a stable limit cycle encircling the unstable limit cycle! For $p > 4$, the unstable cycle disappears (although the stable cycle remains), and the origin becomes unstable. **18.** The Hopf bifurcation criteria hold at $p_0 = 0$. For $p > 0$, the origin is a stable spiral. For $p < 0$ the origin is an unstable spiral. A computer sketch of the phase portrait indicate a stable limit cycle exists for small values of $p < 0$.

Section 8.7

1. $J(x, y, z) = \begin{pmatrix} 1 - 2x - y - z & -x & -x \\ y & -1 + x & 0 \\ z & 0 & -1 + x \end{pmatrix}$

The linearization at the origin is

$$x' = x$$
$$y' = -y$$
$$z' = -z.$$

The origin is unstable.

3. $J(x, y, z) =$
$\begin{pmatrix} -2x - y^2 - z & -2xy & -x \\ -y & 1 - x - 2y - z^2 & -2yz \\ -2zx & -z & -1 - x^2 - y - 2z \end{pmatrix}$

The linearization at the origin is

$$x' = 0$$
$$y' = y$$
$$z' = -z.$$

The linearization principle does not apply.

5. $J(x, y, z) = \begin{pmatrix} -2e^{y-2x} & e^{y-2x} & 0 \\ 0 & 0 & -1 \\ 0 & e^{y-2z} & -2e^{y-2z} \end{pmatrix}$

The linearization at the origin is

$$x' = -2x + y$$
$$y' = -z$$
$$z' = y - 2z$$

The origin is (locally asymptotically) stable.

7. There are three equilibria: $(x_e, y_e, z_e) = (0, 0, 0)$ is unstable, $(x_e, y_e, z_e) = (1, 2, 1)$ is (locally asymptotically) stable, and $(x_e, y_e, z_e) = (-5, 5, -5)$ is unstable. **11. (a)** There is a Hopf bifurcation to a limit cycle at $c \approx 6.1$. Another bifurcation occurs at $c \approx 1.4$ and orbits approach a double loop with twice the period. This

is called a period doubling bifurcation. Another period doubling bifurcation (to a quadruple loop limit cycle) occurs at $c \approx 0.85$ and another period doubling bifurcation (to an 8-fold loop) occurs at $c \approx 0.72$. **(b)** There is what appears to be a strange attractor at $c = 0.2$.

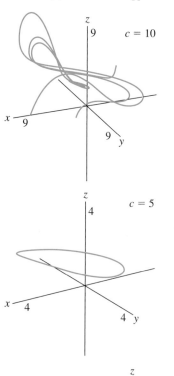

$c = 10$

$c = 5$

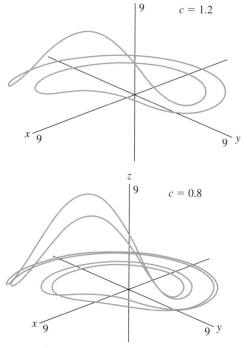

$c = 1.2$

$c = 0.8$

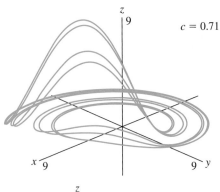

$c = 0.71$

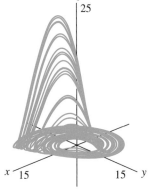

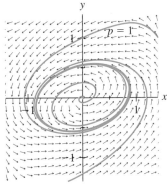

$p = 1$

17. (a) There are four equilibria : $(x, y) = (0, 0)$, $(1, 0)$, $(0, 1)$ and $\left(\frac{1}{3}, \frac{1}{3}\right)$. **(b)** $J(x, y) = \begin{pmatrix} 1 - 2x - 2y & -2x \\ -2y & 1 - 2x - 2y \end{pmatrix}$

(c) At $(0, 0)$:

$$x' = x$$
$$y' = y.$$

At $(1, 0)$:

$$x' = -x - 2y$$
$$y' = -y.$$

At $(0, 1)$:

$$x' = -x$$
$$y' = -2x - y.$$

At $\left(\frac{1}{3}, \frac{1}{3}\right)$:

$$x' = -\frac{1}{3}x - \frac{2}{3}y$$
$$y' = -\frac{2}{3}x - \frac{1}{3}y$$

19. (a) $(x, y) = (b/d, 0)$ satisfies the equilibrium equations

$$b - dx - c\frac{y}{x + y}x = 0$$

$$\left(c\frac{1}{x + y}x - (d + a)\right)y = 0.$$

Another solution is

$$(x, y) = \left(\frac{b}{c - a}, \frac{b}{c - a}\frac{c - d - a}{d + a}\right).$$

(b)

$$J(x, y) = \begin{pmatrix} -d + c\frac{y}{(x+y)^2}x - c\frac{y}{x+y} & -\frac{c}{x+y}x + c\frac{y}{(x+y)^2}x \\ -\frac{c}{(x+y)^2}x + \frac{c}{x+y} & -a - d + c\frac{x}{x+y} - cx\frac{y}{(x+y)^2} \end{pmatrix}$$

(c) $x' = -dx - cy$
$\quad y' = (c - a - d)\,y.$

(d) The equilibrium $\left(\frac{b}{d}, 0\right)$ is a stable node if $c < a + d$ and a saddle if $c > a + d$. It is nonhyperbolic if $c = a + d$. **21.** All orbits are cycles, but none are limit cycles (since no other orbits approach them as $t \to \pm\infty$). The equilibrium is therefore nonhyperbolic. As a result the linearization principle does not apply.

Section 8.8

1. The equilibrium $(x_e, y_e) = (0, 0)$ is an unstable node. The linearization principle does not apply to $(x_e, y_e) = (-1, 0)$. The equilibrium $(x_e, y_e) = (-1, -1)$ is a saddle. **3.** The equilibrium $(x_e, y_e) = (0, 0)$ is a saddle. The equilibrium $(x_e, y_e) = (1, -1)$ is a stable node. The equilibrium $(-1, -1)$ is an unstable node. **5.** The linearization principle does not apply to the equilibrium $(x_e, y_e) = (0, 0)$. **7.** There is a transcritical bifurcation at $p_0 = 0$. **9.** The is a saddle-node bifurcation at $p_0 = 0$ and two simultaneous transcritical bifurcations at $p_0 = 2$. **11.** There are no equilibrium bifurcations. **13.** There is a transcritical bifurcation at $p_0 = 2$. **15.** The Hopf bifurcation criteria hold at $p_0 = 0$. The figures below show that a Hopf bifurcation of a limit cycle occurs at $p_0 = 0$.

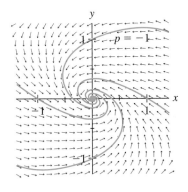

$p = -1$

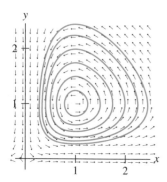

Section 8.9

2. For $S_0 = 0$ the solution pair is $S(t) = 0$, $I(t) = I(0)e^{-at}$, which, for $-\infty < t < \infty$ covers the positive I-axis.
4. (a) $\frac{dI}{dS} = \frac{rSI-aI}{-rSI} = \frac{a}{r}\frac{1}{S} - 1$ implies the result. **(d)** Since $(S(t), I(t)) \to (S_e, 0)$ as $t \to +\infty$ we get from the formula in (a) the equation

$$0 = S_{th} \ln S_e - S_e - S_{th} \ln S_0 + S_0 + I_0$$

for S_e. **(e)** $\frac{dS_e}{dS_0} = \frac{S_e}{S_0}\frac{S_{th}-S_0}{S_{th}-S_e} < 0$. **6.** A cycle must encircle an equilibrium. The only equilibria are points on the I-axis. A cycle encircling a point of the I-axis would intersect the I-axis. This cannot happen, since every point of the axis is an equilibrium (orbits cannot cross). **10. (a)** The solution pair satisfying the initial conditions $x(0) = x_0$, $y(0) = 0$ is $x(t) = (x_0 - b/d) e^{-dt} + b/d$, $y(t) = 0$.
(b) From the solution found in (a), we see that $x(t) \to b/d$ as $t \to +\infty$. **(c)** The orbit can leave the first quadrant only by crossing one of the positive axes. From the first equation in the system we find at points on the y-axis (where $x = 0$) that $x' = b > 0$. This means the direction field points *into* the quadrant at each point on the y-axis, and therefore the orbit cannot leave the quadrant by crossing the positive y-axis. Neither can the orbit cross the x-axis, since this axis consists of orbits. Thus, the orbit remains in the quadrant for all $t \ge 0$.
13. (a) By Exercises 10, 11, and 12 and by the Poincaré-Bendixson theorem, the set S of limit points of an orbit in the first quadrant contains an equilibrium. If $c < c_{th}$, the only equilibrium is $(x_e, y_e) = (b/d, 0)$. Since this equilibrium is stable when $c < c_{th}$, the orbit approaches the equilibrium as $t \to +\infty$. **(b)** If $c > c_{th}$, there exist two equilibria. The equilibrium $(x_e, y_e) = (b/d, 0)$ is a saddle and the only two orbits that approach it are those on the x-axis. Thus, no orbit in the quadrant can have this equilibrium as a limit point and the limit set S of all orbits must contain the only other equilibrium

$$(x_e, y_e) = \left(\frac{b}{c-a}, \frac{b}{c-a}\frac{c-a-d}{a+d} \right).$$

Since this equilibrium is stable when $c > c_{th}$, the orbit approaches the equilibrium as $t \to +\infty$. **14.** Since $y > 0$ for all $t > 0$ we have

$$y' = c\frac{x}{x+y}y - (d+a) y$$

$$\le cy - (d+a) y = (c - c_{th}) y < 0$$

16. $x' = y$, $y' = -\frac{g}{l} \sin x - \frac{c}{m} y$
18. The orbit for $y(0) = 7$ spirals to the equilibrium $(0, 0)$. The orbit for $y(0) = 8$ spirals to the equilibrium $(0, 2\pi)$. The motion differ in that the second orbit implies the pendulum swings over the top (2π radians) before oscillating and coming to rest.

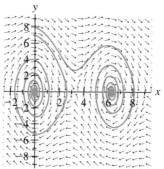

20. $\frac{\partial}{\partial x}f + \frac{\partial}{\partial y}g = \frac{-c}{m} < 0$ for all x and y. **22.** $E' = -l^2 cy^2 \le 0$.
24. The roots are

$$\lambda = \frac{1}{6K} \left(-4(7 - 3K) \pm 2\sqrt{4(7 - 3K)^2 - 54K\left(K - \frac{2}{3}\right)} \right).$$

The discriminant is positive and the roots are real if $0 < K < -\frac{11}{3} + \frac{1}{3}\sqrt{219}$. If $2/3 < K$, then

$$2\sqrt{4(7 - 3K)^2 - 54K\left(K - \frac{2}{3}\right)} < 2\sqrt{4(7 - 3K)^2} = 4|7 - 3K|$$

and the roots λ are both negative.
26. (a)

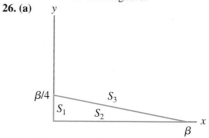

(c) The lines are translated, in parallel fashion, down toward the origin.
(d) Downward and into the triangle.
29. $(x_e, y_e) = (0, 0)$, $(K, 0)$ and $\left(\frac{1}{m-1}, \frac{r}{K}\frac{(m-1)K-1}{(m-1)^2} \right)$
31. $J(x, y) = \begin{pmatrix} r\left(1 - 2\frac{x}{K}\right) - m\frac{y}{(1+x)^2} & -m\frac{x}{1+x} \\ m\frac{y}{(1+x)^2} & -1 + m\frac{x}{1+x} \end{pmatrix}$
33. The equilibria

$$(x_e, y_e) = (K, 0) \quad \text{and} \quad \left(\frac{1}{m-1}, \frac{r}{K}\frac{(m-1)K-1}{(m-1)^2} \right)$$

undergo a transcritical bifurcation (with an exchange of stability) as K increases through the critical value $K_0 = \frac{1}{m-1}$. The second equilibrium loses stability at the critical value $K_1 = \frac{1+m}{m-1}$ at which the characteristic roots of the Jacobian are

$$\lambda = \pm \frac{1}{1+m}i\sqrt{(1+m)r(m-1)]}$$

This suggests that a Hopf bifurcation occurs at this critical value, a conjecture that can be corroborated by computer exploration.
35. (a) There are two such equilibria: $(x, y, z) = (0, 0, 0)$ and $(K, 0, 0)$. **36. (a)** $(x, y, z) = \left(\frac{2}{3}, \frac{5}{3K}\left(K - \frac{2}{3}\right), 0 \right)$. **(b)** The equilibrium is (locally) asymptotically stable for $\frac{2}{3} < K < \frac{130}{177}$.
39. (a) $x' = rx - mxy$, $y' = -dy + mxy$ **(b)** $(x_e, y_e) = (0, 0)$ and $(d/m, r/m)$. **(c)** The equilibrium $(x_e, y_e) = (0, 0)$ is a saddle. The

equilibrium $(x_e, y_e) = (d/m, r/m)$ is nonhyperbolic and the linearization principle does not apply. **(d)** All orbits in the first quadrant are closed loops, surrounding the equilibrium $(x, y) = (d/m, r/m)$.   **40. (b)** $(x, y) = (0, 0)$ and $(r_2/c_{21}, r_1/c_{12})$. **(c)** characteristic roots (eigenvalues) at $(0, 0)$ are $\lambda = r_1 > 0$ and $\lambda = r_2 > 0$. Thus, $(0, 0)$ is an unstable node. The characteristic roots (eigenvalues) at $(r_2/c_{21}, r_1/c_{12})$ are $\lambda = \pm\sqrt{r_1 r_2}$, one of which is positive and the other of which is negative. Thus, this equilibrium is a saddle.

Chapter 9

Section 9.1

1. $-2 + 5u(t - 2)$ **3.** $1 - u(t - 1) - u(t - 2)$
5. $1 + u(t - 1) + u(t - 2) + u(t - 3)$
7. $1 - tu(t) + [-1 + t]u(t - 1)$
9. $t^2 + \left[2 - t - t^2\right]u(t - 1) + [-3 + t]u(t - 3) + [\cos t + 1]u(t - 3\pi)$
11. *Exercise* 1 Not continuous.

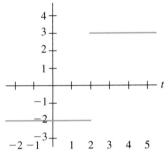

Exercise 3 Not continuous

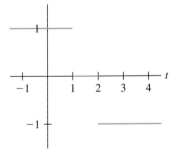

Exercise 5 Not continuous

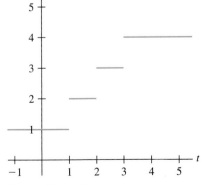

Exercise 7 Is continuous

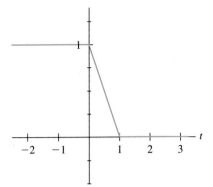

Exercise 9 Is continuous

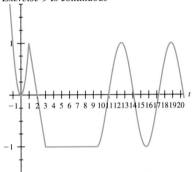

12. $2 - \frac{3}{2}(t - 1)u(t - 1) + \left[\frac{3}{2}(t - 1) - 3)\right]u(t - 3)$
14. $x = -1 + e^t + 2\left[1 - e^{t-1}\right]u(t - 1)$

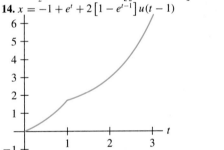

16. $x = \frac{1}{2}e^t + \frac{1}{2}e^{-t} + \left[-1 + \frac{1}{2}e^{t-1} + \frac{1}{2}e^{-t+1}\right]u(t - 1)$

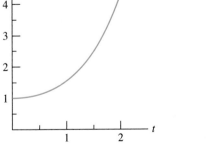

19. $x = \left(x_0 e^r - \frac{H}{r}\right)e^{-r}e^{rt} + \frac{H}{r}$ **20.** The union of all points of discontinuity from all the $q_i(t)$ is a finite set that divides the interval $a < t < b$ into a finite set of subintervals I_i on which every $q_i(t)$, and hence $s(t)$, is continuous. Since each $q_i(t)$ has one sided limits at each point in the union (even if it happens to be continuous there), so does the sum $s(t)$. This is because the one-sided limit of a sum is the sum of the one-sided limits. Thus, $s(t)$ is piecewise continuous on $a < t < b$.

23. We can construct a solution, subinterval by subinterval, in the piecewise fashion of Example 1. Or we can solve the initial value problem by the integrating factor method to arrive at the variation of constants formula. The integral of a piecewise continuous function is continuous, and hence the resulting formula defines a continuous solution, as required.   **26.** $f(t) = \sum_{n=0}^{+\infty} (-1)^n u(t-n)$.

Section 9.2

1. (a) A piecewise continuous function that tends to 0 as $t \to +\infty$ is necessarily bounded. Thus, there is a constant m such that $\left| e^{-\alpha t} x(t) \right| \le m$ and $|x(t)| \le m e^{\alpha t}$ for $t \ge 0$.
(b) $\lim_{t \to +\infty} \left| t e^{-t} \right| = \lim_{t \to +\infty} \left| \frac{t}{e^t} \right| = 0$
2. If

$$|f(t)| \le m_1 e^{\alpha_1 t} \qquad |g(t)| \le m_2 e^{\alpha_2 t}$$

then

$$|f(t) + g(t)| \le |f(t)| + |g(t)| \le m e^{\alpha t}$$

where $m = m_1 + m_2$ and α is the larger of α_1 and α_2.
5. $\left| e^{-st} \right| = \left| e^{-\alpha t} e^{-\beta t i} \right| = e^{-\alpha t} \left| e^{-\beta t i} \right| = e^{-\alpha t}$
7. $\mathcal{L}(e^{ct})(s) = \int_0^{+\infty} e^{-st} e^{ct}\, dt = \lim_{t \to +\infty} \frac{1}{c-s} e^{(\alpha-s)t} - \frac{1}{c-s} = 0 - \frac{1}{c-s}$ if $\text{Re}(c - s) < 0$ (i.e. if $\text{Re}\, s = \alpha > c$).   **9.** If $x(t)$ is exponentially bounded, then

$$0 \le \lim_{t \to +\infty} \left| e^{-st} f(t) \right| \le \lim_{t \to +\infty} m e^{(c-s)t} = 0$$

for $\text{Re}\, s = \alpha > c$. Thus, the limit (2.2) holds and the Laplace transform of x exists for $\alpha > 0$.   **12.** We have already seen that the formula holds for $n = 1$. Assume the formula holds for $n = k$, that is,

$$\mathcal{L}\left(x^{(k)}\right) = s^k \mathcal{L}(x) - s^{k-1} x(0) - s^{k-2} x'(0) - \cdots - x^{(k-1)}(0).$$

By the derivative formula we have

$$\mathcal{L}\left(x^{(k+1)}\right) = \mathcal{L}\left(\left(x^{(k)}\right)'\right) = s \mathcal{L}\left(x^{(k)}\right) - x^{(k)}(0)$$
$$= s \left(s^k \mathcal{L}(x) - s^{k-1} x(0) - s^{k-2} x'(0) - \cdots \right.$$
$$\left. - x^{(k-1)}(0)\right) - x^{(k)}(0),$$

which shows the formula holds for $n = k + 1$.

Section 9.3

1. $X = -\dfrac{5s - 8}{(s-1)(s+2)}$ **3.** $X = \dfrac{s^2 - s - 3}{(s-1)^2(s-2)}$

5. $X = \dfrac{-5s^2 - 2s - 5(s+1)e^{-3s}}{s(s+1)(s-10)}$ **7.** $X = \dfrac{s^2 + e^{-s}}{s(s^2+1)}$

9. $X = \dfrac{s^3 + 6s^2 - s - 4}{(s-1)(s+1)(s^2+4s+1)}$

Section 9.4

3. If the exponential e^{-st} is differentiated n times with respect to s we get

$$\frac{d^n}{ds^n} e^{-st} = (-1)^n t^n e^{-st}.$$

Then

$$\frac{d^n}{ds^n} X(s) = \frac{d^n}{ds^n} \int_0^{+\infty} e^{-st} x(t)\, dt = \int_0^{+\infty} \frac{d^n}{ds^n} e^{-st} x(t)\, dt$$
$$= (-1)^n \int_0^{+\infty} e^{-st} t^n x(t)\, dt = (-1)^n \mathcal{L}(t^n x(t))$$

Multiply both sides by $(-1)^n$ and use $(-1)^n (-1)^n = (-1)^{2n} = 1$.
4. $\int_0^\infty e^{-st} \left(\int_0^t x(u)\, du \right) dt = \dfrac{1}{-s} \int_0^{+\infty} \dfrac{d}{dt} (e^{-st}) \left(\int_0^t x(u) du \right) dt$

$= \dfrac{1}{-s} \left[e^{-st} \int_0^t x(u)\, du \Big|_{t=0}^{t=+\infty} - \int_0^{+\infty} e^{-st} x(t)\, dt \right]$

$= \dfrac{1}{-s} [0 - 0 - X(s)] = \dfrac{1}{s} X(s)$
6. We have already shown that the formula is valid for $n = 0$. Assume that the formula is valid for $n = m$, that is, assume

$$\mathcal{L}(t^m) = \frac{m!}{s^{m+1}}.$$

Then

$$\mathcal{L}(t^{m+1}) = \mathcal{L}(t \cdot t^m) = -\frac{d}{ds}\left(\frac{m!}{s^{m+1}}\right)$$
$$= -\left(-(m+1)\frac{m!}{s^{m+2}}\right) = \frac{(m+1)!}{s^{m+2}}$$

shows the formula is valid for $n = m + 1$.   **9.** Complete the square on the denominator to obtain

$$as^2 + bs + c = a\left[(s-\alpha)^2 + \beta^2\right]$$

where we have defined

$$\alpha = -\frac{b}{2a}, \qquad \beta = \sqrt{\frac{c}{a} - \frac{b^2}{4a^2}}.$$

Now we can write

$$X(s) = \frac{ms + k}{as^2 + bs + c}$$
$$= \frac{1}{a} \frac{ms + k}{(s-\alpha)^2 + \beta^2}$$
$$= \frac{m}{a} \frac{s - \alpha}{(s-\alpha)^2 + \beta^2} + \frac{k + m\alpha}{a\beta} \frac{\beta}{(s-\alpha)^2 + \beta^2}$$

so that by the first translation formula

$$\mathcal{L}^{-1}(X(s)) = \frac{m}{a} e^{\alpha t} \cos \beta t + \frac{k + m\alpha}{a\beta} e^{\alpha t} \sin \beta t$$

11. $2\dfrac{1}{s} + 3\dfrac{1}{s-2} = \dfrac{5s - 4}{s(s-2)}$ **13.** $2\dfrac{\pi}{s^2 + \pi^2} - 4\dfrac{s}{s^2 + 4\pi^2}$

15. $\dfrac{1}{s} - 2\dfrac{1}{s} e^{-s}$ **17.** $k_1 \dfrac{\beta}{s^2 + \beta^2} + k_2 \dfrac{s}{s^2 + \beta^2}$ where k_1, k_2 and β are

constants   **19.** $\dfrac{1}{(s+1)^2}$ **21.** $7\dfrac{1}{s+5} e^{-6(s+5)}$

23. $\dfrac{s - 2\pi}{(s - 2\pi)^2 + 4\pi^2}$

25. $\cos(1)\dfrac{s - \pi}{(s - \pi)^2 + 1} - \sin(1)\dfrac{1}{(s - \pi)^2 + 1}$

27. $\cos \varphi \dfrac{s - r}{(s - r)^2 + \beta^2} - \sin \varphi \dfrac{\beta}{(s - r)^2 + \beta^2}$ **29.** $e^{-2\pi s} \dfrac{1}{s^2 + 1}$

31. $e^4 e^{-2s} \dfrac{1}{s - 2}$ **33.** $e^{-s}\left[\dfrac{1}{s^2} + \dfrac{1}{s}\right]$ **35.** $e^{-cs}\left(\dfrac{1}{s^2} + (c - \beta)\dfrac{1}{s}\right)$

37. $\dfrac{1}{(s-1)^2}$ **39.** $\dfrac{2s}{(s^2+1)^2}$ **41.** $4\pi \dfrac{3s^2 - 4\pi^2}{(s^2 + 4\pi^2)^3}$ **43.** $\dfrac{6}{(s-c)^4}$

45. $2\dfrac{s+1}{(s^2+2s+2)^2}$ **47.** $16\dfrac{3s^2-6s-13}{(s^2-2s+17)^3}$

49. $\dfrac{1}{s}+\dfrac{2\pi}{s^2+4\pi^2}+\dfrac{s}{s^2+4\pi^2}$

51. $e^{-10}e^{-10s}\left[101\dfrac{1}{s+1}+20\dfrac{1}{(s+1)^2}+\dfrac{2}{(s+1)^3}\right]$

53. $e^{-\frac{\pi}{4}}\dfrac{s}{s^2+4}$ **55.** $\dfrac{s-\alpha}{(s-\alpha)^2+\beta^2}\cos\varphi-\dfrac{\beta}{(s-\alpha)^2+\beta^2}\sin\varphi$

57. $e^{2t}\frac{1}{2}t^2$ **59.** $e^{2t}\left(\frac{4}{3}\sin 3t+\cos 3t\right)$

61. $e^{-t}\left(a\cos t+(b-a)\sin t\right)$ **63.** $\left[-\frac{1}{10}+\frac{1}{10}e^{10(t-10)}\right]u(t-10)$

65. $\frac{1}{2}e^{4(t-0.5)}\left[3\cos 2(t-0.5)+\frac{11}{2}\sin 2(t-0.5)\right]u(t-0.5)$

67. $e^{a(t-3)}u(t-3)$ **69.** $e^{-t}-e^{-2t}$ **71.** $\dfrac{1}{2a}(e^{at}-e^{-at})$

73. $\frac{1}{2}\sin t+\frac{1}{2}t\cos t$ **75.** $t-\sin t$

77. $\dfrac{1}{a^2+\beta^2}(\beta e^{at}-\beta\cos\beta t-a\sin\beta t)$

79. $\left[\frac{1}{2}-\frac{1}{2}e^{-2(t-1)}\right]u(t-1)$ **81.** $(t-1)u(t-1)$ **83.** e^t-1

85. $\frac{3}{4}(1-\cos 2t)$ **87.** $\dfrac{1}{2a^2}(e^{at}+e^{-at}-2)$

89. $\frac{1}{2}+\frac{1}{2}e^t\sin t-\frac{1}{2}e^t\cos t$ **91.** $2e^{-3t}-5$ **93.** $ce^{pt}-\frac{p}{q}$

95. $\cos t-2\sin t$ **97.** $\sin t+\sin 2t+\sin 3t$ **99.** $e^{2t}+2e^t$

101. $2e^{-t}-3e^{-2t}+\sin t-\cos t$ **103.** $u(t-\alpha)-u(t-\beta)$

105. $(t-1)u(t-1)-(t-2)u(t-2)$

Section 9.5

1. $x=\frac{1}{2}e^t-\frac{1}{2}\cos t-\dfrac{1}{2}\sin t$ **3.** $x=2.5e^{-0.3t}-1.25e^{0.1t}$

5. $x=-e^{-2t}+te^{-t}+e^{-t}$ **7.** $x=x_0e^{pt}+\dfrac{q}{p}e^{pt}-\dfrac{q}{p}$

9. $x=\begin{cases}\left(\dfrac{a}{p-b}+x_0\right)e^{pt}-\dfrac{a}{p-b}e^{bt}&\text{if }p\neq b\\ x_0e^{pt}+ate^{pt}&\text{if }p=b\end{cases}$ **11.** $x=$

$\begin{cases}\left(x_0+a\dfrac{1}{(p-b)^2}\right)e^{pt}-\dfrac{1}{(p-b)^2}\left((p-b)t+1\right)ae^{bt}&\text{if }b\neq p\\ x_0e^{pt}+\dfrac{1}{2}at^2e^{pt}&\text{if }b=p\end{cases}$

13. $x=\left(1-e^{-(t-1)}\right)u(t-1)$

15. $x=\left(\frac{1}{2}\sin t-\frac{1}{2}\cos t+\frac{1}{2}e^{-(t-2\pi)}\right)u(t-2\pi)$ **18.** $e^{t/10}$

20. $\left(-100-10(t-1)+100e^{(t-1)/10}\right)u(t-1)$

22. $\mathcal{L}(x'')=\mathcal{L}((x')')$
$=s\mathcal{L}(x')-x'(0)$
$=s\left(s\mathcal{L}(x)-x(0)\right)-x'(0)$
$=s^2\mathcal{L}(x)-sx(0)-x'(0)$

24. $x=\frac{1}{2}e^{-t}+\frac{1}{2}e^{-3t}-e^{-2t}$ **26.** $x=\sin t+1-\cos t$

28. $x=-\dfrac{8\sqrt{15}}{15}e^{-\frac{1}{4}t}\sin\dfrac{\sqrt{15}}{4}t+2\sin t$ **30.** $x=-\frac{1}{2}te^{-t}+\frac{1}{2}\sin t$

32. $x=-\frac{1}{3}-\frac{5}{12}e^{-3t}+\frac{3}{4}e^t$ **34.** $x=e^{2t}+2te^{2t}$
$y=-\frac{2}{3}+\frac{5}{12}e^{-3t}+\frac{1}{4}e^t$ $y=-e^{2t}+2te^{2t}$

Section 9.6

1. $4\dfrac{s+1}{\left((s+1)^2+4\right)^2}$

3. $2\pi\dfrac{28+64\pi^2++28s+7s^2}{(s^2+4s+4+4\pi^2)(s^2+4s+4+16\pi^2)}$ **5.** $\dfrac{bs+a-bc}{(s-c)^2}$

7. $e^{-\pi(s+1)}\dfrac{\pi s+1+\pi}{(s+1)^2}$ **9.** $\dfrac{1}{s}e^{-s}-\dfrac{1}{s}e^{-2s}$

11. $f(t)=-\dfrac{5}{\sqrt{6}}e^{2t}\sin\sqrt{6}t-4e^{2t}\cos\sqrt{6}t$

13. $f(t)=\dfrac{\sqrt{5}}{5}\left(e^{\left(\frac{3}{2}+\frac{1}{2}\sqrt{5}\right)t}-e^{\left(\frac{3}{2}-\frac{1}{2}\sqrt{5}\right)t}\right)$

15. $f(t)=1-e^{4t}+5e^{-t}$

17. $f(t)=\dfrac{a}{1+r^2}e^{rt}-\dfrac{a}{1+r^2}\cos t-\dfrac{ar}{1+r^2}\sin t$

19. $f(t)=\dfrac{a}{2}t^2e^{rt}$

21. $f(t)=\dfrac{1}{r}\left(e^{r(t-a)}-1\right)u(t-a)-\dfrac{1}{r}\left(e^{r(t-2a)}-1\right)u(t-2a)$

22. $x=\begin{cases}\dfrac{a}{b-1}e^{bt}+\left(x_0-\dfrac{a}{b-1}\right)e^t&\text{if }b\neq 1\\ x_0e^t+ate^t&\text{if }b=1\end{cases}$

24. $x=$
$\begin{cases}\left(x_0+\dfrac{\alpha}{p^2+\beta^2}\beta\right)e^{pt}-\dfrac{\alpha}{p^2+\beta^2}(\beta\cos\beta t+p\sin\beta t)&\text{if }p^2+\beta^2\neq 0\\ x_0&\text{if }p^2+\beta^2=0\end{cases}$

26. $x=\begin{cases}x_0e^{pt}+\dfrac{1}{b-p}e^{ba}\left(e^{b(t-a)}-e^{p(t-a)}\right)u(t-a)&\text{if }b\neq p\\ x_0e^{pt}+(t-a)e^{p(t-a)}e^{pa}u(t-a)&\text{if }b=p\end{cases}$

28. $z=(2x_0+x_1)e^{-t}-(x_0+x_1)e^{-2t}$

30. $z=(x_0-1)e^{-2t}+(x_1+2x_0-1)te^{-2t}+e^{-t}$

32. $x=\frac{2}{5}\cos t+\frac{4}{5}\sin t-\frac{2}{5}e^t\cos 2t-\frac{1}{5}e^t\sin 2t$
$y=\frac{1}{5}e^t\cos 2t+\frac{3}{5}e^t\sin 2t-\frac{1}{5}\cos t-\frac{7}{5}\sin t$

34. $x=-\frac{1}{25}+\frac{1}{2}e^{-t}-\frac{23}{50}\cos 5t-\frac{1}{10}\sin 5t$
$y=\frac{2}{25}-\frac{1}{13}e^{-t}-\frac{1}{325}\cos 5t+\frac{12}{65}\sin 5t$

37. $x=x_0\dfrac{1}{r_1-r_2}(r_1e^{r_1t}-r_2e^{r_2t})$, where $r_1=-\frac{1}{2}+\frac{1}{2}\sqrt{5}$ and
$r_2=-\frac{1}{2}-\frac{1}{2}\sqrt{5}$ **39.** $x=x_0e^{-t}\cos t$

41. $\mathcal{L}(f)=\int_0^{+\infty}e^{-st}f(t)\,dt$
$=\sum_{i=0}^{+\infty}\int_{iT}^{iT+T}e^{-st}f(t)\,dt$
$=\sum_{i=0}^{+\infty}\int_0^T e^{-s(iT+\tau)}f(\tau+iT)\,d\tau$
$=\sum_{i=0}^{+\infty}\int_0^T e^{-s(iT+\tau)}f(\tau)\,d\tau$
$=\int_0^T e^{-s\tau}f(\tau)\,d\tau\sum_{i=0}^{+\infty}e^{-siT}$
$=\int_0^T e^{-s\tau}f(\tau)\,d\tau\dfrac{1}{1-e^{-sT}}$

42. $T=2\pi/\beta$ so

$$\mathcal{L}(f)=\dfrac{1}{1-e^{-2\pi s/\beta}}\int_0^{2\pi/\beta}e^{-st}\sin\beta t\,dt$$
$$=\dfrac{1}{1-e^{-2\pi s/\beta}}\beta\dfrac{1-e^{-2s\pi/\beta}}{s^2+\beta^2}$$
$$=\dfrac{\beta}{s^2+\beta^2}$$

44. $\mathcal{L}(f)=\dfrac{1}{s}\dfrac{e^{-s}-1}{e^{-2s}-1}$

Section 9.7

2. (a) If $t\neq 1$, then for ε sufficiently small t does not lie in the interval between $1-\dfrac{\varepsilon}{2}$ and $1+\dfrac{\varepsilon}{2}$ and therefore $-q(t)=0$ for $\varepsilon>0$ sufficiently small. It follows that

$$\lim_{\varepsilon\to 0}(-q(t))=0$$

for such a t. However, if $t = 1$, then $-q(t) = 1/\varepsilon$ and hence

$$\lim_{\varepsilon \to 0} (-q(t)) = +\infty.$$

(b) $\Delta_\varepsilon(s) = \dfrac{1}{\varepsilon s} \left(e^{-(1-\varepsilon/2)s} - e^{-(1+\varepsilon/2)s} \right)$ **(c)** $\lim_{\varepsilon \to 0} \Delta_\varepsilon(s) = e^{-s}$

3. (a) $X(s) = \dfrac{28.0s^3 + 18.0s + 21.8s^2 + 14.0}{s\,(s + 2.34)\,(s + 1.36)\,(s^2 + 1)}$

$Y(s) = \dfrac{5.82s^2 + 3.74}{s\,(s + 2.34)\,(s + 1.36)\,(s^2 + 1)}$

(b) $x = -17.0e^{-2.34t} + 10.6e^{-1.36t}$
$\quad + 4.41 + 0.382 \sin t + 2.93 \cos t$
$y = 2.39e^{-2.34t} - 3.81e^{-1.36t}$
$\quad + 1.18 + 0.417 \sin t + 0.246 \cos t$

(c) In the long run the glucose and insulin concentrations in the bloodstream are approximately

$$x \approx 4.41 + 0.382 \sin t + 2.93 \cos t$$
$$y \approx 1.18 + 0.417 \sin t + 0.246 \cos t$$

or

$$x \approx 4.41 + 2.95 \sin (t - 1.44)$$
$$y \approx 1.18 + 0.484 \sin (t - 0.533).$$

Thus, both concentrations oscillate sinusoidally, with the same period (namely 2π) as that of the input concentration, but with smaller amplitudes (2.95 and 0.484 grams, respectively) and with a time lag (1.44 and 0.533 hours, respectively).

5. (a) $x = 2.4526 - 6.7765e^{-2.3421t} + 4.3239e^{-1.3579t}$
$\quad + \left(-2.4526 + 70.498e^{-2.3421t} - 16.811e^{-1.3579t} \right) u(t - 1)$
$y = 0.65402 + 0.90231e^{-2.3421t} - 1.5563e^{-1.3579t}$
$\quad + \left(-0.65402 - 9.3869e^{-2.3421t} + 6.0509e^{-1.3579t} \right) u(t - 1)$

Index